Methods in Enzymology

Volume 207
ION CHANNELS

METHODS IN ENZYMOLOGY

EDITORS-IN-CHIEF

John N. Abelson   Melvin I. Simon

DIVISION OF BIOLOGY
CALIFORNIA INSTITUTE OF TECHNOLOGY
PASADENA, CALIFORNIA

FOUNDING EDITORS

Sidney P. Colowick and Nathan O. Kaplan

Methods in Enzymology

Volume 207

Ion Channels

EDITED BY

Bernardo Rudy

DEPARTMENT OF PHYSIOLOGY
NEW YORK UNIVERSITY MEDICAL CENTER
NEW YORK, NEW YORK

Linda E. Iverson

DIVISION OF NEUROSCIENCES
BECKMAN RESEARCH INSTITUTE OF THE CITY OF HOPE
DUARTE, CALIFORNIA

ACADEMIC PRESS, INC.

Harcourt Brace Jovanovich, Publishers

San Diego   New York   Boston
London   Sydney   Tokyo   Toronto

Academic Press, Inc.
1250 Sixth Avenue, San Diego, California 92101-4311

United Kingdom Edition published by
Academic Press Limited
24–28 Oval Road, London NW1 7DX

Library of Congress Catalog Number: 54-9110

International Standard Book Number: 0-12-182108-0

PRINTED IN THE UNITED STATES OF AMERICA
92 93 94 95 96 97   EB   9 8 7 6 5 4 3 2 1

Table of Contents

Section I. Modern Applications of Electrophysiological Techniques

Section II. Expression of Ion Channels

A. Expression of Ion Channels in *Xenopus* Oocytes

B. Expression of Ion Channels Using Other Systems

Section V. Recording of Ion Channels of Cellular Organelles and Microorganisms

Section VI. Data Storage and Analysis

Contributors to Volume 207

Article numbers are in parentheses following the names of contributors.
Affiliations listed are current.

JULIUS ADLER (47), *Departments of Biochemistry and Genetics, University of Wisconsin, Madison, Wisconsin 53706*

WILLIAM S. AGNEW (39), *Department of Cellular and Molecular Physiology, Yale University School of Medicine, New Haven, Connecticut 06510*

WOLFHARD ALMERS (9), *Department of Physiology and Biophysics, University of Washington, Seattle, Washington 98195*

OSVALDO ALVAREZ (56), *Departmento de Biologia, Facultado de Ciencias Universidad de Chile, Santiago, Chile*

CLAY M. ARMSTRONG (5,55), *Department of Physiology, University of Pennsylvania Medical School, Philadelphia, Pennsylvania 19104*

BRUCE P. BEAN (11), *Department of Neurobiology, Harvard Medical School, Boston, Massachusetts 02115*

TED BEGENISICH (4), *Department of Physiology, University of Rochester Medical Center, Rochester, New York 14642*

E. BLACHLY-DYSON (29), *Vollum Institute for Advanced Biomedical Research, Oregon Health Sciences University, Portland, Oregon 97201*

A. L. BLATZ (12), *Department of Physiology, University of Texas Southwestern Medical Center, Dallas, Texas 75235*

AMY L. BULLER (23), *Department of Pharmacology, University of Nebraska Medical Center, Omaha, Nebraska 68198*

MICHAEL CAHALAN (1), *Department of Physiology and Biophysics, College of Medicine, University of California, Irvine, Irvine, California 92717*

WILLIAM A. CATTERALL (35), *Department of Pharmacology, University of Washington Medical School, Seattle, Washington 98195*

AMITABH CHAK (36), *Center for Molecular Recognition, College of Physicians and Surgeons, Columbia University, New York, New York 10032*

GAVIN CHILCOTT (21), *Division of Biology, California Institute of Technology, Pasadena, California 91125*

TONI CLAUDIO (26), *Department of Cellular and Molecular Physiology, Yale University School of Medicine, New Haven, Connecticut 06510*

M. COLOMBINI (29), *Laboratories of Cell Biology, Department of Zoology, University of Maryland, College Park, Maryland 20742*

FRANCO CONTI (7,22), *Instituto di Cibernetica e Biofisica, CNR, I-16146 Genova, Italy*

ROBERTO CORONADO (49), *Department of Physiology, Services Memorial Institute, School of Medicine, University of Wisconsin, Madison, Wisconsin 53706*

GABRIEL COTA (55), *Department of Physiology, Biophysics, and Neurosciences, Cinvestav-IPN, Mexico DF 07000, Mexico*

MICHAEL R. CULBERTSON (47), *Laboratory of Molecular Biology and Department of Genetics, University of Wisconsin, Madison, Wisconsin 53706*

NATHAN DASCAL (21,25), *Department of Physiology and Pharmacology, Sackler School of Medicine, Tel Aviv University, Ramat Aviv 69978, Israel*

ANNE H. DELCOUR (47), *Department of Molecular and Cell Physiology, Beckman Center, Stanford University, Stanford, California 94305*

LISA EBIHARA (24), *Department of Pharmacology, Columbia University, New York, New York 10032*

FRANCES A. EDWARDS (13), *Department of Pharmacology, University of Sydney, N. S. W. 2006, Australia*

BARBARA E. EHRLICH (31), *Division of Cardiology, University of Connecticut, Farmington, Connecticut 06030*

R. S. EISENBERG (10,48), *Department of Physiology, Rush Medical College, Chicago, Illinois 60612*

GEORGE EISENMAN (56), *Department of Physiology, University of California School of Medicine, Los Angeles, California 90024*

VINCENT FLORIO (35), *Department of Pharmacology, University of Washington, Seattle, Washington 98195*

KIMBERLY FOLANDER (18), *Department of Pharmacology, Merck, Sharp and Dohme Research Laboratories, West Point, Pennsylvania 19486*

M. FORTE (29), *Vollum Institute for Advanced Biomedical Research, Oregon Health Sciences University, Portland, Oregon 97201*

GEORGES C. FRECH (40), *Department of Molecular Physiology, and Biophysics, Baylor College of Medicine, Houston, Texas 77030*

ROBERT J. FRENCH (50), *Department of Medical Physiology, University of Calgary, Calgary, Alberta T2N 4N1, Canada*

ANA MARIA GARCIA (33), *Eisai Research Institute, Andover, Massachusetts 01810*

WILLIAM F. GILLY (5), *Hopkins Marine Station, Stanford University, Pacific Grove, California 93950*

ALAN L. GOLDIN (15,16), *Department of Microbiology and Molecular Genetics, University of California, Irvine, Irvine, California 92717*

ANNE GROVE (34), *Departments of Biology and Physics, University of California, San Diego, La Jolla, California 92093*

MICHAEL C. GUSTIN (47), *Department of Biochemistry and Cell Biology, Rice University, Houston, Texas 77251*

RAINER HEDRICH (46), *Pflanzenphysiologisches Institut, Universität Göttingen, 34 Göttingen, Germany*

STEFAN H. HEINEMANN (7,22), *Max-Planck-Institut für Biophysikalische Chemie, am Fassberg, D-3400 Göttingen, Germany*

MARTIN D. HERMAN (43), *Division of Neurosurgery, Northwestern University Medical School, Chicago, Illinois 60611*

SUNIL R. HINGORANI (39), *Department of Cellular and Molecular Biology, Yale University School of Medicine, New Haven, Connecticut 06510*

RICHARD HORN (8), *Department of Physiology, Jefferson Medical College, Philadelphia, Pennsylvania 19107*

TAKEO IWAMOTO (34), *Department of Biochemistry, University of Southern California Medical School, and Children's Hospital, Los Angeles, California 90054*

MEYER B. JACKSON (51), *Department of Physiology, University of Wisconsin Medical School, Madison, Wisconsin 53706*

ROLF H. JOHO (40), *Department of Molecular Physiology and Biophysics, Baylor College of Medicine, Houston, Texas 77030*

PETER L. JØRGENSEN (38), *The August Krogh Institute, University of Copenhagen, DK-2100 Copenhagen OE, Denmark*

ALEXANDER KAMB (28), *Department of Biochemistry and Biophysics, University of California, San Francisco, San Francisco, California 94143*

ARTHUR KARLIN (36), *Center for Molecular Recognition, College of Physicians and Surgeons, Columbia University, New York, New York 10032*

ANDREAS KARSCHIN (27), *Division of Biology, California Institute of Technology, Pasadena, California 91125*

SEIKO KAWANO (49), *Department of Physiology, University of Wisconsin Medical School, Madison, Wisconsin 53706*

BERNHARD U. KELLER (46), *Max-Planck-Institut für Biophysikalische Chemie, Karl Friedrich Bonhoeffer Institut, D-3400 Göttingen, Germany*

JAN KITAJEWSKI (28), *Department of Microbiology and Immunology, University of California, San Francisco, San Francisco, California 94143*

DAN A. KLÆRKE (38), *Biomembrane Research Center, The August Krogh Institute, University of Copenhagen, DK-2100 Copenhagen, Denmark*

ARTHUR KONNERTH (13), *Max-Planck-Institut für Biophysikalische Chemie, D-3400 Göttingen, Germany*

JUAN I. KORENBROT (28), *Department of Physiology and Department of Biochemistry and Biophysics, University of California, San Francisco, San Francisco, California 94143*

STEPHEN J. KORN (8), *Department of Physiology and Neurobiology, University of Connecticut, Storrs, Connecticut 06268*

DOUGLAS S. KRAFTE (20), *Department of Cardiovascular Pharmacology, Sterling Winthrop Pharmaceuticals Research Division, Rensselaer, New York 12144*

CHING KUNG (47), *Laboratory of Molecular Biology, and Department of Genetics, University of Wisconsin, Madison, Wisconsin 53706*

PEDRO LABARCA (30,32), *Centro de Estudios Científicos de Santiago y Departmento de Biologia, Facultad de Ciencias, Universidad de Chile, Santiago 9, Chile*

RAMÓN LATORRE (30,32), *Centro de Estudios Científicos de Santiago y Departmento de Biologia, Facultad de Ciencias, Universidad de Chile, Santiago 9, Chile*

MICHEL LAZDUNSKI (37), *Institut de Pharmacologie Moléculaire et Cellulaire, Sophia Antipolis, 06560 Valbonne, France*

CHEOL J. LEE (49), *Department of Physiology, University of Wisconsin Medical School, Madison, Wisconsin 53706*

HENRY A. LESTER (20,21,27), *Division of Biology, California Institute of Technology, Pasadena, California 91125*

S. ROCK LEVINSON (45), *Department of Physiology, University of Colorado Health Science Center, Denver, Colorado 80262*

RICHARD A. LEVIS (2,3), *Department of Physiology, Rush Medical College, Chicago, Illinois 60612*

JEN-WEI LIN (42), *Department of Physiology and Biophysics, New York University Medical Center, New York, New York 10016*

ILANA LOTAN (41), *Department of Physiology and Pharmacology, Sackler School of Medicine, Tel Aviv University, Ramat Aviv, 69978 Israel*

K. L. MAGLEBY (12,53), *Department of Physiology and Biophysics, University of Miami School of Medicine, Miami, Florida 33101*

GAIL MANDEL (17), *Department of Neurobiology and Behavior, State University of New York at Stony Brook, Stony Brook, New York 11794*

BORIS MARTINAC (47), *Laboratory of Molecular Biology, University of Wisconsin, Madison, Wisconsin 53706*

PETER V. MINORSKY (47), *Laboratory of Molecular Biology, University of Wisconsin, Madison, Wisconsin 53706*

EDWARD MOCZYDLOWSKI (54), *Department of Pharmacology and Department of Cellular and Molecular Physiology, Yale University School of Medicine, New Haven, Connecticut 06510*

MAURICIO MONTAL (34), *Departments of Biology and Physics, University of California, San Diego, La Jolla, California 92093*

MYRTA S. MONTAL (34), *Departments of Biology and Physics, University of California, San Diego, La Jolla, California 92093*

TOSHIO NARAHASHI (43,44), *Department of Pharmacology, Northwestern University Medical School, Chicago, Illinois 60611*

DAVID NARANJO (32), *Centro de Estudios Científicos de Santiago y Departamento de Biologia, Facultad de Ciencias, Universidad de Chile, Santiago 9, Chile*

ERWIN NEHER (1,6), *Abteilung Membranbiophysik, Max-Planck-Institut für Biophysikalische Chemie, am Fassberg, D-3400 Göttingen, Germany*

YORAM ORON (25), *Department of Physiology and Pharmacology, Sackler School of Medicine, Tel Aviv University, Ramat Aviv 69978, Israel*

B. S. PALLOTTA (12), *Department of Pharmacology, University of North Carolina, School of Medicine, Chapel Hill, North Carolina 27599*

S. PENG (29), *Laboratory of Cell Biology, Department of Zoology, University of Maryland, College Park, Maryland 20742*

JAMES L. RAE (2,3), *Departments of Physiology and Biophysics and Ophthalmology, Mayo Clinic Foundation, Rochester, Minnesota 55905*

HUBERT REHM (37), *Pharmakologisches Institut, Universität Zürich, CH-8006 Zürich, Switzerland*

WILLIAM M. ROBERTS (9), *Department of Biology, Institute of Neuroscience, University of Oregon, Eugene, Oregon 97403*

YOSHIRO SAIMI (47), *Laboratory of Molecular Biology, University of Wisconsin, Madison, Wisconsin 53706*

SHLOMO SEIDMAN (14), *Department of Biological Chemistry, The Life Sciences Institute, The Hebrew University of Jerusalem, Jerusalem 91904, Israel*

F. J. SIGWORTH (52), *Department of Cellular and Molecular Physiology, Yale University School of Medicine, New Haven, Connecticut 06510*

TERRY P. SNUTCH (17), *Biotechnology Laboratory and Division of Neuroscience, University of British Columbia, Vancouver, British Columbia V6T 1W5, Canada*

HERMONA SOREQ (14), *Department of Biological Chemistry, The Life Sciences Institute, The Hebrew University of Jerusalem, Jerusalem 91904, Israel*

JÖRG STRIESSNIG (35), *Department of Pharmacology, University of Washington, Seattle, Washington 98195*

WALTER STÜHMER (19,22), *Abteilung Membranbiophysik, Max-Planck-Institut für Biophysikalische Chemie, D-3400 Göttingen, Germany*

KATUMI SUMIKAWA (16), *Department of Psychobiology, University of California, Irvine, Irvine, California 92717*

RICHARD SWANSON (18), *Department of Pharmacology, Merck, Sharp and Dohme Research Laboratories, West Point, Pennsylvania 19486*

J. M. TANG (10,48), *Department of Physiology, Rush Medical College, Chicago, Illinois 60612*

GARY THOMAS (27), *Department of Cell Biology and Anatomy, Vollum Institute for Advanced Biomedical Research, Oregon Health Sciences University, Portland, Oregon 97201*

BARBARA A. THORNE (27), *Department of Cell Biology and Anatomy, Vollum Institute for Advanced Biomedical Research, Oregon Health Sciences University, Portland, Oregon 97201*

WILLIAM B. THORNHILL (45), *Department of Physiology and Biophysics, Mount Sinai Medical Center, New York, New York 10029*

JOHN M. TOMICH (34), *Department of Biochemistry, University of Southern California Medical School and Division of Medical Genetics, Children's Hospital of Los Angeles, Los Angeles, California 90027*

CARMEN VALDIVIA (49), *Department of Physiology, University of Wisconsin Medical School, Madison, Wisconsin 53706*

HECTOR H. VALDIVIA (49), *Department of Physiology, University of Wisconsin Medical School, Madison, Wisconsin 53706*

ELEAZAR VEGA-SAENZ DE MIERA (42), *Department of Physiology and Biophysics, New York University Medical Center, New York, New York 10016*

ALFREDO VILLARROEL (56), *Department of Physiology, University of California School of Medicine, Los Angeles, California 90024*

J. WANG (10,48), *Department of Physiology, Rush Medical College, Chicago, Illinois 60612*

MICHAEL M. WHITE (23), *Department of Pharmacology, University of Pennsylvania School of Medicine, Philadelphia, Pennsylvania 19104*

WILLIAM F. WONDERLIN (50), *Department of Pharmacology and Toxicology, West Virginia University, Morgantown, West Virginia 26506*

J. ZHOU (52), *Department of Cellular and Molecular Physiology, Yale University School of Medicine, New Haven, Connecticut 06510*

Preface

The importance of ion channels in the generation and transmission of signals in the nervous system has been well known for over forty years, since the classical work of Hodgkin, Huxley, and Katz. The more recent introduction of new electrophysiological methods for the study of ion channels, in particular, the development of the patch clamp technique by Neher and Sakmann, has led to an explosion of research on ion channels in many different systems. It is now thought that ion channels are present in most, if not all, cell types, and are found in all organisms, both eukaryotes and prokaryotes. Furthermore, although it was initially thought that there may be relatively few different ion channel types, it is becoming increasingly clear that ion channels constitute an extremely large group of heterogeneous proteins having one feature in common, the ability to form a pore for the passive movement of ions across membranes. Electrophysiological techniques allowing the study of ion movements with both high sensitivity and temporal resolution are still the methods of choice for the functional study of ion channels. However, transport of ions through ion channels occurs with such high efficiency that the channels themselves need only be present in minute quantities, thus making the biochemical study of ion channels exceedingly difficult. The recent application of molecular biological techniques has facilitated the study of these relatively rare membrane proteins by allowing one to predict the amino acid sequence of the channel protein directly from the nucleotide sequence of the cloned ion channel gene and by allowing one to express both normal and mutant channel proteins in heterologous systems.

This book was conceived with the idea in mind that it would be useful not only to membrane biophysicists, but also to the many cell biologists, biochemists, molecular biologists, pharmacologists, geneticists, and microbiologists who are finding the study of ion channels important in their work. Given their fundamental significance in all organisms, it is essential that an accessible source be made available to scientists which describes not only techniques for recording and analysis of ion channels, but also potential sources of artifacts that may arise during electrophysiological experimentation. In addition, updates on classical techniques, as well as improved methods for expressing cloned ion channel genes and purifying and reconstituting ion channel proteins, should be helpful to all investigators.

Many excellent papers have been published describing the most important electrophysiological methods used today in the study of ion channels. Rather than attempt to duplicate these papers, we have chosen, instead, to focus on the more recent applications and modifications of these standard methods (Section I). Original papers describing these methods are referred to in chapters contained in this section and other sections of the volume. Many of the chapters in Section I address problems and sources of artifacts that one may encounter in the applications of electrophysiological techniques. We hope both newcomers to the field, as well as experienced membrane biophysicists, will find these chapters useful. A second fundamental aspect of electrophysiological experimentation emphasized in the volume is that of data analysis (Section VI). Here lies the power of electrophysiological study of ion channels. These methods of analysis allow the distinction between different channel types, the identification and elimination of potential common artifacts, and permit detailed analysis of ion channel function. Additional chapters throughout the volume also emphasize these two aspects of electrophysiological study of ion channels.

Methods to reconstitute ion channels in lipid bilayers are described in Section III. Although reconstitution methods are not as widely applied today as the patch-clamp technique, they continue to provide invaluable contributions to our present knowledge of ion channels. For example, the recent discovery that the ryanodine receptor is a calcium release channel was obtained from studies incorporating purified receptors into lipid bilayers and played a crucial role in our present understanding of excitation–contraction coupling. Reconstitution methods will undoubtedly continue to have wide application in the future discovery of new channels in membranes that remain inaccessible to other electrophysiological techniques and in functional studies of proteins isolated from native tissue or expressed from cloned ion channel genes. Furthermore, reconstitution methods are still the best means of examining the influence of lipid composition on ion channel function.

We did not include many standard molecular biological techniques because the application of these techniques to the study of ion channel genes is essentially no different than for any other gene, and many excellent books and papers describing these methods already exist and are referred to in chapters contained in Sections II and IV. Similarly, although we feel strongly that some of the most recent breakthroughs in the field are a direct result of years of research in *Drosophila* neurogenetics, we have chosen not to include chapters on these methods since their adequate description would require an entire volume. A few applications of molecular biological techniques particularly useful in the study of ion channel genes, for example, hybrid arrest and expression cloning, are described in Section IV. This section also contains chapters describing methods to purify some selected ion channel proteins. Although methods for purification of other channel proteins will, obviously, not be identical to those described here, these examples will be useful for the development of new purification protocols. Pharmacological tools have been, and will continue to be, extremely useful in the identification and examination of many distinct ion channel types. Two chapters in Section IV give overviews of drugs and toxins that interact with ion channels. In addition, an excellent chapter on the analysis of drug action at the single channel level can be found in Section VI. Section V deals with exciting applications of electrophysiological methods for studying ion channels in cellular organelles and single celled organisms including protozoa, bacteria, and yeast. Microbiologists should find these methods helpful in determining the functional role of ion channels in the physiology and pathology of microorganisms.

Section II deals with the expression of ion channels in heterologous systems. These methods serve as the bridge between the molecular biology and electrophysiology of ion channels. We have, therefore, elected to make this the most extensive and comprehensive section of the volume. Considerable focus is directed toward the description of methods for the expression of ion channels from RNA injected into *Xenopus* oocytes. The simplicity and utility of this system to express ion channels, demonstrated by the work of Miledi and colleagues, have resulted in it becoming the standard method for the heterologous expression of ion channel proteins. This section includes chapters describing sources and methods for handling frogs, preparation of oocytes, preparation of RNA from tissue, and *in vitro* synthesis of RNA from cloned cDNAs, as well as methods for recording from oocytes. The second part of this section includes chapters describing methods developed for the heterologous expression of ion channels in other cell types that have proved extremely useful. Using these heterologous expression systems, researchers are now able to address detailed questions concerning which channel structures are involved in determining or influencing specific channel properties such as kinetics, ion selectivity, voltage dependence, ligand binding sites, subunit composition, and modulation via second messenger systems, as well as the identification and isolation of new ion channel genes.

We are extremely grateful to our colleagues for their excellent contributions and the care and quality of their work. It is sometimes easy to forget how difficult the task of explaining subjects to those outside the field is. This is particularly true in the case of ion channels because many of the techniques used to analyze these molecules cannot be applied as recipes. Considerable effort was required on the part of the authors to describe complex methods and concepts to a broad audience. Without their dedication this volume would not have been possible. In an attempt to contain these chapters within a single *Methods in Enzymology* volume, many difficult and somewhat arbitrary decisions were necessary in the selection of topics. Inevitably, omissions have occurred due to oversight on our part, to several potential authors already being overcommitted, and to the rate of progress in this rapidly expanding area of research. We apologize to all those individuals who have made significant contributions to the field but whose work is not represented in this volume.

BERNARDO RUDY
LINDA IVERSON

METHODS IN ENZYMOLOGY

VOLUME XXVII. Enzyme Structure (Part D)
Edited by C. H. W. HIRS AND SERGE N. TIMASHEFF

VOLUME XXVIII. Complex Carbohydrates (Part B)
Edited by VICTOR GINSBURG

VOLUME XXIX. Nucleic Acids and Protein Synthesis (Part E)
Edited by LAWRENCE GROSSMAN AND KIVIE MOLDAVE

VOLUME XXX. Nucleic Acids and Protein Synthesis (Part F)
Edited by KIVIE MOLDAVE AND LAWRENCE GROSSMAN

VOLUME XXXI. Biomembranes (Part A)
Edited by SIDNEY FLEISCHER AND LESTER PACKER

VOLUME XXXII. Biomembranes (Part B)
Edited by SIDNEY FLEISCHER AND LESTER PACKER

VOLUME XXXIII. Cumulative Subject Index Volumes I–XXX
Edited by MARTHA G. DENNIS AND EDWARD A. DENNIS

VOLUME XXXIV. Affinity Techniques (Enzyme Purification: Part B)
Edited by WILLIAM B. JAKOBY AND MEIR WILCHEK

VOLUME XXXV. Lipids (Part B)
Edited by JOHN M. LOWENSTEIN

VOLUME XXXVI. Hormone Action (Part A: Steroid Hormones)
Edited by BERT W. O'MALLEY AND JOEL G. HARDMAN

VOLUME XXXVII. Hormone Action (Part B: Peptide Hormones)
Edited by BERT W. O'MALLEY AND JOEL G. HARDMAN

VOLUME XXXVIII. Hormone Action (Part C: Cyclic Nucleotides)
Edited by JOEL G. HARDMAN AND BERT W. O'MALLEY

VOLUME XXXIX. Hormone Action (Part D: Isolated Cells, Tissues, and Organ Systems)
Edited by JOEL G. HARDMAN AND BERT W. O'MALLEY

VOLUME 92. Immunochemical Techniques (Part E: Monoclonal Antibodies and General Immunoassay Methods)
Edited by JOHN J. LANGONE AND HELEN VAN VUNAKIS

VOLUME 93. Immunochemical Techniques (Part F: Conventional Antibodies, Fc Receptors, and Cytotoxicity)
Edited by JOHN J. LANGONE AND HELEN VAN VUNAKIS

VOLUME 94. Polyamines
Edited by HERBERT TABOR AND CELIA WHITE TABOR

VOLUME 95. Cumulative Subject Index Volumes 61–74, 76–80
Edited by EDWARD A. DENNIS AND MARTHA G. DENNIS

VOLUME 96. Biomembranes [Part J: Membrane Biogenesis: Assembly and Targeting (General Methods; Eukaryotes)]
Edited by SIDNEY FLEISCHER AND BECCA FLEISCHER

VOLUME 97. Biomembranes [Part K: Membrane Biogenesis: Assembly and Targeting (Prokaryotes, Mitochondria, and Chloroplasts)]
Edited by SIDNEY FLEISCHER AND BECCA FLEISCHER

VOLUME 98. Biomembranes [Part L: Membrane Biogenesis: (Processing and Recycling)]
Edited by SIDNEY FLEISCHER AND BECCA FLEISCHER

VOLUME 99. Hormone Action (Part F: Protein Kinases)
Edited by JACKIE D. CORBIN AND JOEL G. HARDMAN

VOLUME 100. Recombinant DNA (Part B)
Edited by RAY WU, LAWRENCE GROSSMAN, AND KIVIE MOLDAVE

VOLUME 101. Recombinant DNA (Part C)
Edited by RAY WU, LAWRENCE GROSSMAN, AND KIVIE MOLDAVE

VOLUME 102. Hormone Action (Part G: Calmodulin and Calcium-Binding Proteins)
Edited by ANTHONY R. MEANS AND BERT W. O'MALLEY

VOLUME 103. Hormone Action (Part H: Neuroendocrine Peptides)
Edited by P. MICHAEL CONN

VOLUME 104. Enzyme Purification and Related Techniques (Part C)
Edited by WILLIAM B. JAKOBY

Section I

Modern Applications of Electrophysiological Techniques

[1] Patch Clamp Techniques: An Overview

By Michael Cahalan and Erwin Neher

Historical Introduction

Patch-recording techniques originated from an effort to record currents through individual ion channels in biological membranes. By the early 1970s it had become clear that discrete molecular entities — integral membrane proteins — underlie the electrical signaling mechanisms of nerve and muscle. The selective actions of certain toxins, proteases, and protein-modifying agents had indicated that sodium and potassium channels constituted separate macromolecules.[1-3] Studies in artificial lipid membranes had shown that certain proteins isolated from bacteria[4] and some antibiotic polypeptides[5] were able to induce steplike changes in membrane conductance, which were attributed to the opening and closing of individual pore- or channellike structures. These discrete conductance changes were of the same order of magnitude as single-channel conductances inferred from analysis of current fluctuations in the neuromuscular junction and in nodes of Ranvier.[6-8]

Thus, it was tempting to look for similar discrete current changes in biological preparations. Bilayer studies had shown that electronic components available at that time were capable of handling such small signals. On the other hand, background noise in all standard voltage-clamp arrangements was higher by one to two orders of magnitude than the signals to be measured. Thus, it seemed straightforward to attempt to isolate a small area of membrane surface (a "patch") for localized electrical measurement by placing a measuring glass micropipette onto the surface of a voltage-clamped cell. Such arrangements had already been used for focal stimulation[9] or for measurement of local current density.[10-12] Simple considera-

[1] J. W. Moore and T. Narahashi, *Fed. Proc.* **26**, 1655 (1967).
[2] B. Hille, *J. Gen. Physiol.* **50**, 1287 (1967).
[3] C. M. Armstrong, F. Bezanilla, and E. Rojas, *J. Gen. Physiol.* **62**, 377 (1973).
[4] R. C. Bean, W. C. Shepherd, H. Chan, and J. Eichner, *J. Gen. Physiol.* **53**, 741 (1969).
[5] S. B. Hladky and D. A. Haydon, *Nature (London)* **225**, 451 (1970).
[6] B. Katz and R. Miledi, *J. Physiol. (London)* **224**, 665 (1972).
[7] C. R. Anderson and C. F. Stevens, *J. Physiol. (London)* **235**, 651 (1973).
[8] F. Conti, B. Hille, and W. Nonner, *J. Physiol. (London)* **353**, 199 (1984).
[9] A. F. Huxley and R. E. Taylor, *J. Physiol. (London)* **144**, 426 (1958).
[10] A. Strickholm, *J. Gen. Physiol.* **44**, 1073 (1961).
[11] K. Frank and L. Tauc, *in* "The Cellular Function of Membrane Transport" (J. Hoffman, ed.), p. 113. Prentice-Hall, Englewood Cliffs, New Jersey, 1963.
[12] E. Neher and H. D. Lux, *Pfluegers Arch.* **311**, 272 (1969).

METHODS IN ENZYMOLOGY, VOL. 207

tions of background noise led to the conclusion that such pipettes should allow resolution of picoampere-sized currents, such as acetylcholine-induced responses, whenever the "seal-resistance"—the resistance between the interior of the pipette and the bath—could be increased severalfold above the pipette internal resistance *and* be made to exceed a final value of 50–100 MΩ (for marginal resolution). Unfortunately, the initial experiments showed that it was very difficult to obtain satisfactory seals. In spite of systematic investigation using enzymes to clean cell surfaces, using a variety of pipette sizes and shapes, and using pipette glass with varying surface charge and hydropathy, it seemed impossible to improve the seal resistance beyond 100 MΩ. Nevertheless, optimizing enzymatic cleaning procedures and pipette geometries allowed single acetylcholine-induced currents to be resolved.[13]

The major drawbacks of measurements with low-resistance seals (between 1976 and 1980) included limited resolution, frequent occurrence of partial single-channel events owing to channels localized in the pipette rim area, and large leakage currents associated with excessive noise whenever small voltage differences occurred between pipette interior and bath. All these problems were eliminated or greatly improved when it was found[14,15] that the application of slight suction within a freshly prepared pipette readily induced the phenomenon now called gigaseal formation. This is a sudden transition of seal resistance from several tens of megohms to several gigaohms. The physical basis of the gigaseal is still not quite clear, but it certainly increased resolution by an order of magnitude, decreased the variability of channel step sizes, and allowed potentials to be applied across the seal for local voltage stimulation. Furthermore, it was recognized that the gigaseal provided mechanical stability, so that patches could be "excised" from the parent cell and studied in isolation.[16] Ironically, breaking patches paved the way to another present-day application of the patchclamp, tight-seal whole-cell recording. Breaking the patch (without the loss of the seal) provides electrical continuity between the patch pipette and the cell interior, thus leading to a configuration similar to conventional microelectrode impalement. It turned out that this form of impalement was much gentler than conventional microelectrode methods, and that it was tolerated by cells as small as a few micrometers in diameter. In small cells, the membrane impedance can be orders of magnitude higher than the

[13] E. Neher and B. Sakmann, *Nature (London)* **260**, 779 (1976).
[14] F. J. Sigworth and E. Neher, *Nature (London)* **287**, 447 (1980).
[15] O. P. Hamill, A. Marty, E. Neher, B. Sakmann, and F. J. Sigworth, *Pfluegers Arch.* **391**, 85 (1981).
[16] R. Horn and J. Patlak, *Proc. Natl. Acad. Sci. U.S.A.* **77**, 6930 (1980).

internal resistance of the measuring pipette. This is quite different from the case of conventional impalements, in which pipette resistance and cell impedance are usually of the same order of magnitude, requiring feedback circuitry for voltage clamping.

Procedures for obtaining the basic patch clamp configurations were well established by late 1980, as described by Hamill *et al.*[15] Some of the historical details of the development have been considered by Sigworth.[17] Experience from laboratories up to 1983 has been summarized in *Single-Channel Recording,* edited by Sakmann and Neher.[18]

Whereas the electrical aspects of patch clamp measurement consolidated within a year or two, it took somewhat longer to appreciate fully the biochemical implications of the technique. We now know that ion channels are gated and modulated not only by voltage and external ligands, but also by second messengers, regulatory proteins, and by phosphorylation. Many of these regulatory molecules are definitely lost when a patch is excised; even in the whole-cell recording mode, mobile intracellular ions may exchange with those in the pipette within seconds,[19,20] and second messengers may be lost within minutes. This feature may be utilized to investigate the influence of such regulators. For this purpose, techniques have been developed to perfuse the pipette interior (see [10] in this volume). On the other hand, "washout" of regulators may prevent prolonged study of physiological functions in small cells. For this reason, efforts have been made to find ways to permeabilize patches for small ions selectively, in order to provide electrical access to the cell without perturbing the cellular biochemistry. These techniques are known as slow whole-cell,[21] perforated-patch, or the nystatin method[22] (see also [8] in this volume).

Patch Configurations

Here, we provide an introduction to the variety of recording configurations that are possible following formation of a gigaseal. Four of them are essentially as described previously by Hamill *et al.*[15] and summarized in the flowchart of Fig. 1. In addition, perforated-patch recording and other minor variants are summarized briefly.

[17] F. J. Sigworth, *Fed. Proc.* **45**, 2673 (1986).
[18] B. Sakmann and E. Neher, eds., "Single-Channel Recording." Plenum, New York and London, 1983.
[19] A. Marty and E. Neher, *in* "Single-Channel Recording" (B. Sakmann and E. Neher, eds.). Plenum, New York and London, 1983.
[20] M. Pusch and E. Neher, *Pfluegers Arch.* **411**, 204 (1988).
[21] M. Lindau and J. M. Fernandez, *Nature (London)* **319**, 150 (1986).
[22] R. Horn and A. Marty, *J. Gen. Physiol.* **92**, 145 (1988).

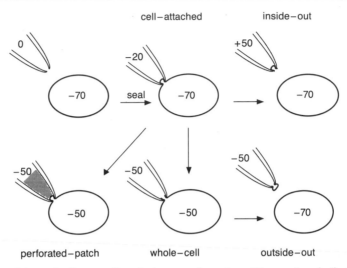

FIG. 1. Schematic diagram of patch clamp configurations. The numbers indicate potentials in millivolts and have been chosen so that a membrane potential of −50 mV is applied in all the configurations, assuming a cell resting potential of −70 mV. A pipette is moved toward the cell surface (upper left) to form a gigaseal. From the cell-attached configuration, one can attain (1) the *inside-out* configuration by withdrawing the pipette; (2) the *whole-cell* configuration by rupturing the patch using a pulse of suction or voltage; and (3) the *perforated patch* by including nystatin in the pipette filling solution. From the whole-cell configuration, one can attain the *outside-out* patch configuration by withdrawing the pipette (see text for details).

Seal Formation

As the pipette is advanced toward the cell, the pipette resistance is monitored by repetitively applying a small (usually < 10 mV) voltage step to the pipette. Typically, fire-polished glass pipettes with resistances of the order of $1-10$ MΩ are employed, corresponding to tip diameters of the order of 1 μm. Signals that are applied, and that can be observed during "sealing," are illustrated in Fig. 2. When the pipette touches the cell, the resistance increases. Application of suction to the interior of the pipette (typically $10-20$ cm H_2O) draws a small portion of cell membrane into the pipette, and, if all is "right," seal formation is visualized as a sudden increase in resistance such that virtually no current can pass between pipette and bath electrodes. When the resistance is measured carefully at high gain, seals of over 10^{10} Ω are common. Several factors promote seal formation, including Millipore-filtered solutions, positive pressure within the pipette during the approach to the cell, and a clean cell surface. Pipette glass and the composition of solutions also play a role in determining

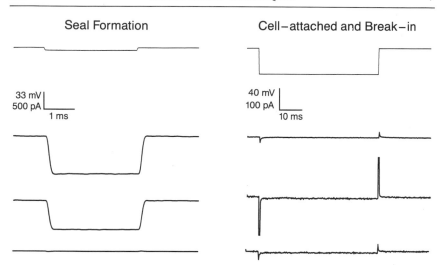

Fig. 2. Typical electrical signals observed during sealing and break-in. In the left-hand column (sealing), a 5-mV negative pulse is applied (top trace). The resulting currents are shown in the traces below. The top current trace was obtained before the pipette touched the cell. The amplitude of the current pulse indicates a pipette resistance of 4.4 MΩ (Ohm's law!). The center trace indicates a somewhat increased pipette resistance (smaller signal) after the pipette touched the cell. The bottom trace was recorded a few seconds after suction had been applied. The current signal virtually disappeared because a gigaseal had formed. The only signal visible at this resolution is a small transient of capacitive charging current due to the stray capacitance of the pipette. In the right-hand column, the voltage pulse was increased to 50 mV; current sensitivity was also increased. The capacitive transients on the current trace (see above) would be expected to be sizable under these conditions; however, they have been almost eliminated by fast capacitance neutralization. Subsequently, a pulse of suction was given, resulting in the third trace. Two changes can be seen when compared to the second trace: (1) an increase in noise and (2) large capacitive transients which have, in fact, been truncated. Both changes indicate patch rupture. The transients are due to the capacitance of the cell. The increased noise is due to the ionic conductances and the capacitance of the cell. In the bottom trace, the cell capacitance has been neutralized by electronic capacitance compensation. At this stage, a series of voltage pulses can be applied, as shown in Fig. 3.

success in forming seals. When conditions are favorable, seal formation occurs in nearly 100% of trials.

Cell-Attached Patch Recording

Currents through ion channels trapped within the pipette orifice can be recorded with subpicoampere resolution following seal formation. The overall sensitivity of the measurement depends on a variety of factors, including electronic noise, pipette capacitance, noise associated with the holder, and "pickup" of line frequency or computer-associated noise of higher frequency. Between 0 and 300 Hz, a background noise value of the order of 0.1 pA root mean square (rms) can be achieved using a 50-GΩ feedback resistor or capacitative feedback for current measurement, enabling detection of currents through single channels. If the cell is electrically active, action potentials will be detected as a capacitative current across the membrane patch. In this recording configuration, the cell remains intact, although seal formation involves some deformation as membrane is sucked into the pipette. The potential across the membrane is equal to the membrane potential of the cell minus the pipette potential. Applying a negative potential to the pipette depolarizes the patch of membrane. If channels within the patch are active, positive current flowing outward across the membrane goes into the pipette, and negative current flowing inward across the membrane goes out of the pipette. Because the seal will not permit leakage of bath constituents into the pipette, the extracellular surface of the patch is exposed only to the pipette solution; bath solution changes can be made without altering the solution exposed to the patch.

Inside-out Patch

An isolated patch of membrane can be torn off the cell simply by withdrawing the pipette after seal formation. Normally, the cell survives this insult. If the patch is excised from the cell-attached recording configuration, the resulting patch is "inside-out," with the cytoplasmic membrane surface facing the bath solution. Again, from the standpoint of the membrane, positive current flowing outward across the membrane goes into the pipette. However, in this case the membrane potential of the patch is the bath potential minus the pipette potential. Thus, to achieve a membrane potential of -50 mV, for example, a pipette potential of $+50$ mV must be applied. Single-channel resolution can be improved by lifting the patch near the surface of the bath solution in order to reduce the capacitance of the pipette. One problem which can occur is the formation of a "vesicle" if the patch of membrane seals over, enclosing a small volume of bathing solution. Sometimes this can be corrected by brief exposure of the pipette tip to air. Vesicle formation is inhibited by bath solutions containing low amounts of calcium. Excised patches do not consist solely of the membrane. Cytoskeletal elements and membranous organelles may also be present, as demonstrated vividly by a recent description of light-activated photoreceptor current in patches from retinal rods.[23] In this case the entire cytoplasmic transduction pathway including rhodopsin, transducin, phos-

---

[23] E. A. Ertel, *Proc. Natl. Acad. Sci. U.S.A.* **87**, 4226 (1990).

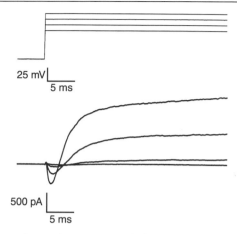

FIG. 3. Series of superimposed current records from whole-cell recording in response to depolarizing stimuli to −20, −10, 0, and +10 mV from a holding potential of −70 mV, using a bovine chromaffin cell at room temperature. The cell capacitance of 5 pF has been compensated (series resistance 7.2 MΩ).

phodiesterase, and cyclic GMP was present in the excised patch, in addition to the sodium channel present in the membrane.

Whole-Cell Recording

Following initial seal formation, it is also possible to rupture deliberately the patch of membrane trapped within the pipette by application of strong suction or a voltage pulse of several hundred millivolts. Patch rupture is detected as a sudden increase in capacitative current in response to a test potential step, as the membrane capacitance of the cell is "seen" through the series resistance of the pipette. This "break-in" results in electrical and diffusional continuity between the pipette and the cytoplasm, usually without altering the seal resistance between glass and membrane. If the access resistance between the pipette electrode and the cytoplasm is much lower than the membrane resistance of the cell, as is usually the case in small cells, the membrane potential of the cell is thereby voltage clamped to the pipette potential, with polarity opposite that of the inside-out patch; the membrane potential of the cell will be equal to the pipette potential minus the bath potential. The ensemble activity of ion channels can then be detected by monitoring the current.

In the case of whole-cell recording, positive, outward membrane current flows out of the pipette. In the example shown in Fig. 3, inward sodium currents and outward potassium currents in response to a family of voltage-clamp steps are illustrated. As pointed out above, whole-cell recording permits diffusional exchange of pipette contents with cytoplasmic constituents; ionic contents can be exchanged within seconds. From the point of view of being able to control both the electrical potential and cytoplasmic milieu, one major factor is maintaining good access between pipette and cytoplasm. If membrane currents are large (exceeding the nanoampere range), compensation for series resistance can help control the membrane potential at the desired level by electronic feedback. However, geometrical factors must also be considered in achieving adequate voltage control. These considerations are discussed in [5] in this volume.

Outside-out Patch

If the pipette is withdrawn from the cell during whole-cell recording, a tether of membrane is drawn away from the cell which normally breaks off and reseals on the pipette as an outside-out patch. Voltage and polarity conventions are identical to whole-cell recording, but the membrane surface area is greatly reduced, allowing subpicoampere resolution of current. In a variant of this technique, the nucleus of the cell is carried along during patch excision resulting in an outside-out "macropatch."[24]

Perforated-Patch Recording

An unavoidable consequence of whole-cell recording is the loss of cytoplasmic ions, nucleotides, and other diffusible constituents into the pipette. Several techniques have been invented to achieve electrical continuity between pipette and cytoplasm while minimizing dialysis. Pore-forming antibiotic molecules such as nystatin can be added to the pipette solution following seal formation, or backfilled at some distance from the tip to allow for normal seal formation. Nystatin will then spontaneously form conducting channels selective for monovalent ions, and thereby lower the electrical resistance of the patch to the point where the cell can be voltage clamped and current measured through the permeabilized patch of membrane beneath the pipette. This has the great advantage that second-messenger mechanisms within the cell can remain intact during the recording.

Electronic Enhancements

The design of modern, commercially available patch-clamp amplifiers incorporates low-noise current-measuring circuitry, voltage summing to deliver holding, offset, and applied command potentials to the pipette,

[24] P. Ascher, *Soc. Neurosci. Abstr.* **16,** 619 (1990).

capacity current subtraction incorporating at least two time constants, and electronic feedback compensation for series resistance. Low-pass filtering, internal pulse generation, voltage outputs to indicate gain and filter settings, and tone generators for auditory monitoring of seal resistances are additional features often provided. Some designs incorporate headstages with switchable feedback resistors to allow optimal signal-to-noise ratios for high-gain recording, while allowing large currents to be monitored without saturation at low gain. Other designs incorporate capacitative feedback to monitor current with great sensitivity. The voltage clamp can be interfaced to a digital computer enabling the programming of sophisticated stimulation and data acquisition protocols. In some designs, a computer can control the status of the amplifier directly, providing a more direct linkage between software and hardware.

Applications

Single-Channel Recording

Perhaps the most startling and inherently pleasing outcome of patch clamp methodology is simply the ability to monitor, in real time, the dynamics of a single ion channel within a membrane patch. Single-channel currents through most ion channel types have already been detected. Currents through single ion channels appear when conformational changes in the channel protein gate the flow of thousands of ions per millisecond. The detailed kinetic study of single-channel currents has provided unique insights into the mechanisms of ion-channel gating. A further example of the power of combining molecular and electrophysiological approaches is the investigation of the molecular genetics and function of ion channels. Patch clamp techniques are ideal for detailed analysis of channel kinetic states. The analysis of site-directed mutations with patch clamp techniques using *Xenopus* oocytes and transfected mammalian cells is a focus of much current research on the structure and function of ion channels. These topics are reviewed elsewhere in this volume.

Expanded Scope of Electrophysiology

Initially, patch clamp recordings were performed mainly on acutely isolated and enzymatically cleaned preparations. Soon it became clear, however, that the method was ideally suited for basically all cell culture preparations. At first, the major focus was to characterize ion channels at the single-channel level in order to determine unit conductance values and to investigate biophysical models for channel gating. However, the ability to investigate ion channels in a wide variety of cell types has vastly expanded the scope of modern electrophysiology. Quantitative voltage clamp measurements are no longer limited to a few preparations which, by virtue of size or special geometry, were especially favorable for voltage control by electronic feedback. Among the isolated or cultured cell preparations that have been investigated are a wide variety of neurons; cells of sensory transduction; glial cells; muscle cells including skeletal, cardiac, and smooth muscle types; endothelial and epithelial cells including those in the vasculature, the respiratory system, and the kidney; secretory cells including pancreatic acinar cells, lacrimal gland cells, and juxtaglomerular cells; hepatocytes; pancreatic β cells; keratinocytes, osteoblasts, and osteoclasts; and cells of hematopoietic origin including erythrocytes, lymphocytes, macrophages, mast cells, neutrophils, and blood platelets. Methods have also been adapted to extend measurements to oocytes,[25] plant protoplasts,[26] yeast,[27] and bacteria.[27] In addition to measurements on isolated cells, procedures have been described to obtain tight seals on defined cells in brain slices,[28] a technique which allows functional connections in the central nervous system (CNS) to be studied at unsurpassed resolution. Recently, explorations of ion channels in subcellular organelles have begun, including studies on mitochondria,[29] the nucleus,[30] and endoplasmic reticulum.[31] Techniques to record from liposomes or to form phospholipid bilayers on patch clamp pipettes have also been described.[32,33]

Discovery of Novel Ion Channels

The patch clamp has also expanded the catalog of recognized ion channel subtypes. It is now clear from patch clamp and from molecular biological approaches that voltage-dependent sodium, calcium, and potassium channels constitute a large superfamily composed of a wide variety of subtypes. A second major superfamily includes the transmitter-gated channel/receptors. In addition, it is now clear that distinct channel species

[25] C. Methfessel, V. Witzemann, T. Takahashi, M. Mishina, S. Numa, and B. Sakmann, *Pfluegers Arch.* **407**, 577 (1986).

[26] R. Hedrich and J. Schroeder, *Annu. Rev. Plant Physiol.* **40**, 539 (1989).

[27] Y. Saimi, B. Martinac, M. R. Culbertson, J. Adler, and C. Kung, *Cold Spring Harbor Symp. Quant. Biol.* **53**, 667 (1988).

[28] A. Konnerth, *Trends Neurosci.* **13**, 321 (1990).

[29] M. C. Sorgato, B. U. Keller, and W. Stuhmer, *Nature (London)* **330**, 498 (1987).

[30] M. Mazzanti, L. J. DeFelice, J. Cohen, and H. Malter, *Nature (London)* **343**, 764 (1990).

[31] A. Schmid, M. Dehlinger-Kremer, I. Schultz, and H. Gogelein, *Nature (London)* **346**, 374 (1990).

[32] D. W. Tank, C. Miller, and W. W. Webb, *Proc. Natl. Acad. Sci. U.S.A.* **79**, 7749 (1982).

[33] R. Coronado and R. Latorre, *Biophys. J.* **43**, 231 (1983).

that have not as yet been characterized at the molecular level are regulated by nucleotides, intracellular calcium and sodium ions, and by GTP-binding proteins. As a result, instead of the handful of channels recognized before the mid-1970s, hundreds of distinct channel subtypes that are regulated by a variety of mechanisms have now been described.

Capacitance Measurements

Current through a membrane consists of an ionic as well as a capacitative component. Because the fundamental structure of biological membranes consists of a lipid bilayer in which intrinsic membrane proteins are immersed, a value of approximately 1 μF/cm^2 is considered to represent the value of membrane capacitance. One picofarad of membrane capacitance represents approximately 100 μm^2 of membrane surface area. Because current measurement is extremely sensitive during patch recording, it is possible to measure the membrane area with great accuracy by monitoring membrane capacitance. This can be done by applying voltage steps and measuring the capacitative current response, in which case membrane capacitance is proportional to the integral of the charging transient. Alternatively, a lock-in amplifier which measures current in and out of phase with a sinusoidal applied potential provides resolution of capacitance to approximately 10 fF. The sensitivity of this measurement is such that the fusion of single secretory vesicles in mast cells has been measured as a step increase in membrane capacitance.[34,35] Thus, patch recording has enabled not only single channels to be resolved, but also single exocytotic events. In whole-cell recording, capacitance provides a real-time measure of secretion which has been applied to several cell types.

Cell-Signaling Mechanisms

Ion channels not only mediate electrical excitability in the nervous system and in the heart; they also appear to play important functional roles in the behavior of most cell types. One important factor in considering cell-signaling mechanisms is that the activity of ion channels can directly or indirectly affect the concentration of an important second messenger, namely, calcium ions. In most excitable cells, voltage-dependent calcium channels not only help to shape the action potential, but they gate the entry of calcium ions which in turn can activate kinases, contractile proteins, or ion channels. In a variety of electrically inexcitable cells, voltage-indepen-

[34] E. Neher and A. Marty, *Proc. Natl. Acad. Sci. U.S.A.* **79,** 6712 (1982).
[35] M. Lindau and E. Neher, *Pfluegers Arch.* **411,** 137 (1988).

dent calcium channels can be activated following the binding of a ligand to receptors on the cell surface. Furthermore, both receptor-linked GTP-binding proteins and second-messenger systems within the cytosol can affect the activity of ion channels. Thus, complex regulatory pathways linking surface receptors, the metabolism of the cell and ion channels can result in long-lasting changes in the behavior of the cell.

The patch clamp technique provides the experimental means for merging the tools of modern molecular and cellular biology with those of electrophysiology. Using the various recording configurations, it is possible to dissect the mechanisms of channel modulation. In cell-attached recording, modulation of channel activity in response to bath-applied agonist generally indicates a second-messenger mechanism. Candidate messengers can be tested directly on excised patches or in whole-cell recording. Perforated patch recording maintains the integrity of second-messenger systems while enabling the overall activity of ion channels in the cell to be evaluated following receptor stimulation. Current research on signaling pathways seeks to establish the functionally meaningful mechanisms through selective activation or inhibition of a portion of the pathway. The variety of patch clamp configurations, combined with single-channel resolution, provides a powerful experimental approach from the molecular level, in which channel genes are altered and expressed, to the cellular level, in which posttranslational signaling mechanisms are elucidated, to the systems level, in which cellular interactions in intact or slice preparations are revealed.

[2] Constructing A Patch Clamp Setup

By RICHARD A. LEVIS and JAMES L. RAE

Basic Components in Patch Clamp Setup

Different investigators have chosen to construct their patch clamp setups in very different ways, and it is clear that there is no one "best" way to configure the apparatus for patch clamping. There are, however, many features that good setups have in common, and there are some basic principles that one should consider when configuring and purchasing patch clamp hardware.[1,2] In this chapter we discuss these principles, describe

---

[1] J. L. Rae and R. A. Levis, *Mol. Physiol.* **6,** 115 (1984).

[2] J. L. Rae, R. A. Levis, and R. S. Eisenberg, *in* "Ion Channels" (T. Narahashi, ed.), p. 283. Plenum, New York and London, 1988.

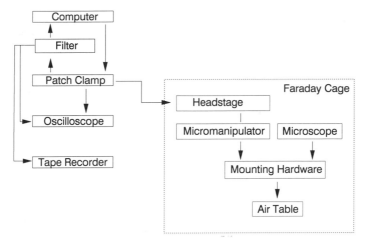

FIG. 1. Block diagram for a patch clamp setup.

ways in which investigators have implemented them, and suggest procedures that can result in very high quality patch clamp recordings.

Figure 1 shows a block diagram of a patch clamp setup that includes the important elements that any set should have. In the sections that follow, we take each of the elements from this diagram and discuss ways in which they might be implemented in a working patch clamp setup. We particularly emphasize the electronics and low noise techniques since it is through understanding these aspects of patch clamping that an investigator can most markedly improve the technical quality of the recordings.

The Microscope

Because the majority of patch clamp studies have been done on single cultured cells or on single freshly dissociated cells, inverted microscopes have been utilized most often. With these microscopes, visualization occurs from the side opposite to that from which the electrodes are positioned. With this arrangement, it is possible to use condensers whose working distance is sufficiently long that the patch clamp headstage, electrode holder, and electrode easily fit under the condenser with little worry about making mechanical contact with it.

With inverted microscopes, it is also possible to utilize a chamber whose bottom is optically ideal and in very close proximity to the objective. This allows one to employ high numerical aperture (NA) objectives with maximum resolving power. It is, however, not possible in general to make actual patch clamp measurements while utilizing the maximal resolving power (M_{res}) of the microscope. To do this would require the use of a condenser whose numerical aperture was as high as that of the objective since $M_{res} \cong 1.22\lambda_0/(NA_{obj} + NA_{cond})$ where λ_0 is the wavelength of light being used. High numerical aperture condensers do not have sufficiently long working distances that electrodes can fit under them. The long working distance condensers required have lower numerical aperture (0.6 or so) and thus lower resolution. Still, the total resolution of the microscope is quite good as long as the numerical aperture of the objective is large. Inverted microscopes also focus routinely by moving the nosepiece with its attached objective rather than by moving the stage. This offers the advantage that the chamber holding the cells can be rigidly attached to the stage for mechanical stability. In addition, it is often possible to mount the micromanipulators that hold the headstage and electrodes directly to the stage, an arrangement that is quite good mechanically. Other than these obvious advantages, there is little to recommend an inverted microscope over an upright microscope as both will accept the same wide range of accessories such as video cameras, photometers, and fluorescence attachments.

In some instances, preparations that have more than one cell layer must be utilized for patch clamping. With an inverted microscope, one is forced to look at the top layer of cells on which the patch clamping will be done through several layers of deep cells. This arrangement can cause substantial optical distortion at best and complete loss of visibility at worse. Under these circumstances, it is necessary to use an upright microscope with an objective that has a sufficiently long working distance that the patch electrodes and holder can be placed under the objective. Most microscope companies market a line of metallurgical objectives that have working distances in the 10–20 mm range while maintaining quite good numerical apertures. The Nikon (Garden City, NY) extralong working distance and super-long working distance objectives are most notable in this regard, although the 25× from Leitz (Rockleigh, NJ) is also an excellent choice. With these objectives, one can look directly at the surface of a multilayer preparation and have sufficient working distance to place an electrode on that cell layer under direct observation. In general, one must partially cover the chamber with a piece of microscope slide glass and place the electrode tip under it from the side. One then views the preparation through the microscope glass without the distortion caused by the meniscus of fluid that would otherwise exist near the edge of the electrode as shown in Fig. 2.

Most microscope companies are willing to make modifications to their microscopes in which a hinge is placed in the body of the microscope so that the binocular head and nosepiece can be folded back away from the

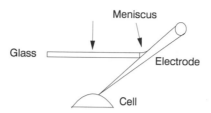

FIG. 2. Schematic drawing of the "glass bottom boat" approach for viewing cells from above with a noninverted compound microscope. The meniscus occurs at the edge of the glass and thus does not distort the field of vision.

chamber to give the investigator sufficient room for inserting the preparation in the chamber. If a long working distance, low power objective is used, microdissection can also be done as required. Figure 3 is a photograph of an inexpensive Nikon labophot modified in this way. This microscope utilizes an erect image binocular head which is imperative if one is to use it for microdissection.

Either inverted or upright microscopes have a variety of optical techniques available for visualizing the cells. Here again, inverted microscopes offer clear advantages optically both because of their ability to use short working distance, high numerical aperture objectives and because they can utilize the principle of transmitted light, Nomarski interference contrast. In this technique, when implemented with a high numerical aperture objective, an extremely thin optical slice is cut through the cell of interest with little interference from structures located either above or below the plane of focus. This is an exceptionally good way to visualize the *surface* of the cell about to be patch clamped, but it has perhaps an even greater virtue. Because of the very limited depth of field, it is possible to tell precisely when the electrode tip is located a very small distance above the cell surface. One focuses on the surface of the cell, defocuses to a position of a few micrometers above the cell, and then brings the electrode tip in focus by advancing the micromanipulator. This limits the distance over which high resolution but slow micromanipulator movements must be accomplished before the electrode makes mechanical contact with the cell surface. One is able to place the electrode tip onto the cell surface much more rapidly than with other optical principles. This technique is limited to use with inverted microscopes because of the Nomarski requirement for a relatively short objective working distance. To date, no one has implemented this technique with objectives that have the 10–20 mm working distances required for patch clamping with noninverted microscopes. Metallurgical microscopes with long working distance objectives can be used

FIG. 3. An upright compound microscope with a hinge in its stand to allow the image-erecting binocular head to be folded back. (Such microscopes were first designed in the laboratory of R. S. Eisenberg at UCLA.) The manipulators shown have x, y, horizontal rotational, vertical rotational, and tilt motions. The headstage is mounted on a final motor drive.

with incident Nomarski interference contrast where delivery of light and viewing both occur through the objective. This has not proved effective with cells because of reflections from the surface of the bathing solution.

Many of the long working distance metallurgical objectives used with upright microscopes can be modified to implement the technique of Hoffman modulation contrast.[3] This optical approach produces a Nomarski-like image that also has a quite limited depth of focus. The optical slices produced, while still quite thin, are thicker than those possible with Nomarski optics in the inverted microscope. Much of this is dependent on the fact that the metallurgical objectives, for the same magnification, have smaller numerical apertures and therefore greater depth of focus than objectives with shorter working distances.

For one building a patch clamp setup on a limited budget, it is possible to implement a Schlieren microscopy principle as described by Axelrod.[4] This in general requires simply putting a piece of tape over about one-third to one-fourth of the back aperture of the objective and then placing an opaque object such as a black piece of paper between the light source and the bottom of the condenser. When this subcondenser "stop" is aligned parallel to the tape in the objective so as to yield a small slit for light passage, it is possible to get a Nomarski-like image, again with limited depth of field. Although this technique does not offer all of the resolution of Nomarski or Hoffman modulation contrast, it is often sufficient for patch clamping.

Another useful accessory for either an inverted or an upright microscope is a video camera and display. Inexpensive charge-coupled-device (CCD) cameras are most attractive for this purpose because of their small size. The major advantage of video viewing is that it allows the objectives to be used at their full numerical aperture. In general, high numerical aperture objectives produce rather weak contrast unless the iris diaphragm of the condenser is stopped down to a point where only 75 to 80% of the back aperture of the objective is filled with light. This arrangement produces contrast but at the loss of resolution. With a video-based system, the required contrast can be generated from the video electronics. This allows the iris diaphragm to be opened to a level that the back aperture of the objective is fully filled with light to provide the highest resolution. Although the cells will look very washed out through the eyepieces, even their internal structures will be exceptionally visible on the video monitor. One does have to be certain, however, that the video camera and its cables are sufficiently shielded that they do not provide electrical interference to the patch clamp recordings.

Micromanipulators

There is tremendous versatility in the micromanipulators that are available for patch clamping. Virtually every company that produces mechanooptical equipment and every microscope manufacturer sells a wide

[3] R. Hoffman and L. Gross, *Nature (London)* **254**, 586 (1975).
[4] D. Axelrod, *Cell Biophys.* **3**, 167 (1981).

variety of micromanipulators (see, for example, Newport, Founatin Valley, CA; Klinger Scientific, Garden City, NY; Daedal, Harrison City, PA; Pacer Scientific, Los Angeles, CA; Aerotech, Pittsburgh, PA). Many of these are probably perfectly adequate for patch clamp recordings. The main requirements are that the micromanipulators have a geometry that allows them to be located close to the preparation and that they be capable of repeatable movements of the order of one-tenth the diameter of the cells being studied. Also of extreme importance is that they not drift once the electrode tip has been placed in its desired final position. Obviously, it is desirable that they have mass, stability, minimum backlash, and be capable of repeatable submicron movements, but many investigators have used with good successs quite inexpensive micromanipulators which do not possess all of these characteristics.

Several features of micromanipulators are useful for the implementation of an optimal patch clamp setup. One is that they be capable of horizontal *angular* movement. If the manipulator has a horizontal rotation stage, it is possible to move the headstage very rapidly to a position where it is 45° to 90° lateral to its patch clamping position. This makes it possible to change rapidly the electrode holder and electrode, a need which on some days occurs all too often. A second desirable motion is angular movement along the vertical plane. This movement, often performed by a device called a goniometer cradle, allows one to move the electrode from a position above the bath downward to very near the cell surface in a minimum of time. The best goniometer cradles have both coarse and fine controls over this movement, allowing one to place the electrode tip to within a few microns of the cell surface easily and rapidly (see Fig. 3).

Many investigators have found it useful to have the final movement of the electrode tip onto the cell surface be under motor control. Most manufacturers of manipulators provide either stepping motor drives or dc-driven motors that are capable of producing movements of less than 1 μm per second. These devices provide optimal control in this critical step of pressing the electrode against the cell membrane. In some cases, this final movement is not motor driven but is driven from a very high resolution hydraulic system which can also yield very fine movements. Some manufacturers provide final movement via piezoelectric translators. This provides what is probably the finest control presently available. In the best of systems, it is possible to get x, y, and z movements all driven essentially simultaneously by a joystick mechanism that controls three piezoelectric translators. Joystick arrangements are also possible with hydraulic systems and with three-dimensional (3-D) motor drives. Such joystick-based systems are desirable but not necessary for a quality patch clamp setup.

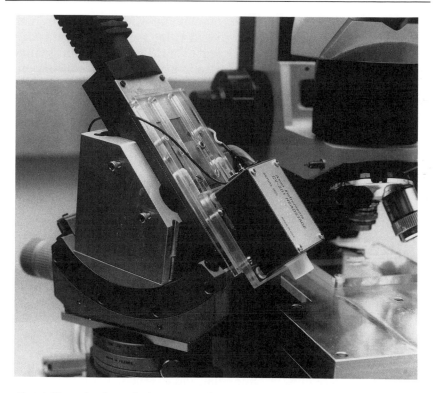

FIG. 4. Example of a method to mount a headstage to a final manipulator drive (see text).

Mounting the Headstage

In order for fine micromanipulators to do their jobs optimally, it is necessary that the patch clamp headstage be mounted on them in an optimal way and that the manipulators themselves be mounted on the rest of the setup in a way that their full potential is utilized. As a general principle, it is important that the headstage be rigidly attached to the micromanipulator. Some patch clamp companies provide their headstages with a long rod protruding from the back for mounting. With such a mount, it is possible for the headstage and the tip of the electrode to be suspended many inches from the center of mass of the micromanipulator. Although it is possible to do patch clamping under these circumstances, much better stability of the electrode tip is obtained if the body of the headstage is mounted to the manipulator via a rigid place. Figure 4 shows one such implementation. In this case the connection is made through a pair of Plexiglas plates, but an even better mount would be through a single plate of a light but strong metal. Also in the case shown, the input connector is located in the center of the end piece of the headstage. An asymmetric location would improve the ability to fit the electrode into tight geometry like that under a long working distance objective or an intermediate working distance, higher numerical aperture condenser.

Mounting Micromanipulators

Again, as a general principle, it is important that the preparation and the micromanipulator be mounted in a way that movement between the electrode tip and the cell surface be minimized. This is best facilitated by having the manipulator and the preparation mounted on the same structure so that any movement of the structure simultaneously moves both the electrode and the cell. Three quite different approaches have been utilized. In the first, the preparation is mounted in a chamber attached to the movable stage of a microscope that focuses by moving the nosepiece. There the micromanipulator (usually a 3-D joystick variety) is mounted directly to the nonmovable part of the microscope stage. Many investigators have used such a system successfully even though the preparation can, in principle, move independently of the micromanipulators if there is wobble in the mechanical stage. Also, the kind of manipulators that can be mounted directly to the stage must be quite small and in general less able to support the weight of the headstage than can a more substantial micromanipulator.

A second way to mount manipulators is that chosen in the "patch clamp towers" (Fig. 5) (List Electronics, Eberstadt, Germany). Here, the entire preparation, microscope and all, sits on a metal plate to which an optical rail is attached rigidly. Optical rails come from almost any of the optical–mechanical supply houses (see, however, Klinger Scientific) and are designed, through commercially available mounts, to allow the attachment of a wide variety of mechanical components. In the best systems, an optical rail exists on each side of the microscope, and the tops of the two rails are firmly connected so as to provide exceptional stability. When these systems are used for patch clamp setups, one of the mechanical components is a very fine $x-y$ stage suspended from the optical rail. This stage holds the chamber with the cells, takes the place of the microscope stage, and provides the source of $x-y$ movement of the preparation. Attached to the same optical rail is the micromanipulator which can be quite simple or very complex depending on budget and inclination. Horizontal and vertical rotation as well as 3-D motor drive capability can be implemented on

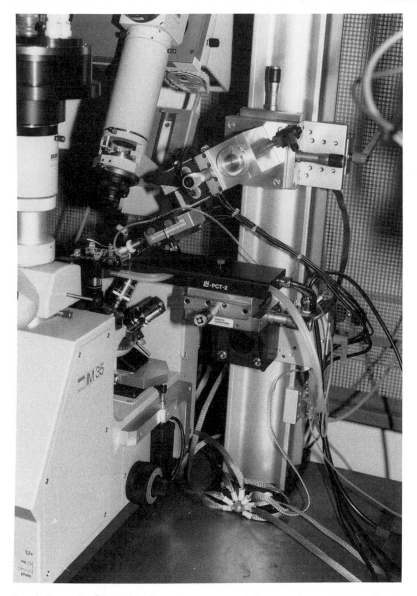

FIG. 5. Example of the patch clamp tower approach for mounting patch clamp hardware. Both the micromanipulator and specimen stage are connected to an optical rail. (Photo complements of Dr. J. Fernandez.)

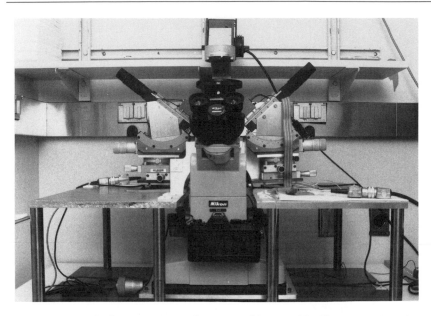

FIG. 6. Inverted microscope mounted on a movable $x-y$ table. The stage and manipulators are mounted to the metallic superstructure shown and do not move when the microscope is moved.

such systems. This is quite a favorable arrangement when both preparation and micromanipulators are mounted to the same optical rail to promote stability, but, in addition, the $x-y$ stages which move the preparation can be very much better than the mechanical stages supplied by microscope manufacturers. This results in a reduced tendency for the preparation and electrode to wobble with respect to each other.

At least potentially, the most rigid way of mounting the apparatus is to build a superstructure for holding the preparation and the manipulator very rigidly and then mount the microscope itself on a large toolmaker's $x-y$ stage. Under these circumstances, the cells are never moved. Rather, to find a new cell, the microscope is moved. One such implementation for an inverted microscope is shown in Fig. 6. When this superstructure is bolted to the tabletop, a very rigid structure is formed in which only the drift of the micromanipulator is important. In the other kind of mounts, drifts of the manipulators and the mechanical stages mounting the preparations are important.

It is not really known just how important the difference between these approaches is since all have been used successfully for patch and whole-cell recordings. They are presented as simple guidelines to aid readers in the implementation of their personal setups.

Vibration Isolation

Because of the sensitivity of patch clamp recordings to even the slightest movements, it is necessary to mount the microscope, micromanipulators, and headstage electronics on a vibration isolation table. Buildings differ in the natural vibrations that they contain, but virtually all buildings have vibrations that come from slamming of doors, rotating electrical equipment, and many other sources. If these vibrations are not damped out, they will produce artifacts in the current recordings. The easiest solution is to mount the apparatus on an air suspension table of a variety that can be obtained from standard optical companies like Newport, Barry (Watertown, MA), and Micro-G (Peabody, MA). Such tables damp out vibrations that occur at frequencies beyond a few hertz. Unless building vibrations are particularly bad, air tables near the bottom of the line of these manufacturers are quite adequate. Vibrational components below a few hertz can be damped out only with either extremely massive tables (2000 lb tops or so) or with tables using active feedback. One problem with any kind of vibration isolation is that the apparatus used will have one or more resonant frequencies. This means that the devices may quite adequately reduce some vibrational components but they may actually enhance others which coincide with their resonant frequencies. Whereas this is usually not a problem with air suspension tables with heavy tops, it is quite common with homemade isolation devices in which metal plates are placed on top of tennis balls, foam rubber, Styrofoam padding, or a combination of several such materials. Although these home-grown vibration remedies often can be made to work very effectively, one must be certain that the various components used actually damp vibrations and not enhance them.

Shielding and Grounding

It is very important that the headstage, electrode, and preparation be adequately shielded from stray interference. Sixty-hertz frequency signals coming from line outlets, lights, or other electronic equipment are the most common contaminant of current recording records, but interference from antennas, computer screens, and other high frequency sources also occurs. Many investigators have chosen to surround their apparatus and vibration isolation equipment completely with a Faraday cage. It is, of course, important that the Faraday cage not actually touch the top of the vibration isolating apparatus. This cage is simply a conductive enclosure which surrounds the preparation and is connected to ground. These enclosures interact with alternating electric fields to convert the signal to currents which are then shunted to ground through a low resistance connection.

The effectiveness of the shields depends on the frequency of the electric fields, being more effective for high frequency fluctuations than for low frequencies. Different materials have very different attenuation abilities for components of different frequencies. Copper and brass screening material which is commonly used for this purpose is less effective for 60-Hz signals than it is for high frequency signals. To attentuate 60 Hz substantially, the materials used for construction must provide electromagnetic shielding. Materials like mu metal are particularly good for this, although galvanized sheet metal and galvanized hardware cloth are also quite effective. In addition, there are new conductive plastic materials available that can be used. These plastic materials (like those used in shipping integrated circuits) can be obtained in large rolls and are particularly useful for constructing a drape at the front opening of the Faraday cage. This drape can be lifted out of the way while cells are being placed in the chamber, while the electrodes are being placed on the cells, etc., and then closed during recording. Many investigators have opted to make a Faraday cage of a good, solid electromagnetic shielding material on the top, back, and sides while utilizing a plastic drape in front. Another particularly useful feature of Faraday cages is that they can be built with small shelves on their internal walls which can be used for placing solution bottles, mounting pumps, or attaching other apparatus.

In many environments, Faraday cages are not required. Because patch clamp measurements are done in close proximity with either a microscope objective or a microscope condenser, excellent shielding can be achieved by simply grounding the microscope, the manipulator, and any conductive material in the region of the headstage, electrode, the cells. For this purpose, it is important that single-point grounding be used. The patch clamp headstage should have a high level signal ground point made available to the user. It is this point that must make low resistance contact with all of the conductive elements surrounding the cell chamber. If the setup is mounted on a metal frame and if the vibration isolation apparatus has a metallic top, these should also be connected to this ground through a resistance not to exceed 50Ω (ohms). It may be necessary to check all of the conductive elements around the setup with an ohmmeter to ensure that these low resistance contacts exist. Even if one uses a Faraday cage, this kind of single-point grounding should be used. In general, the Faraday cage is not connected to the high level signal ground but is connected to the instrument rack and/or to the ground terminal of a three-prong electrical plug.

Beyond these considerations, it is difficult to make general rules about grounding procedures. Because it is usually the case that one is trying to minimize 60-Hz interference, there are several approaches that one can use. To measure the impact of the procedures, connect the patch clamp current output to one input of the oscilloscope at sufficiently high gain that the 60-Hz signal largely fills the screen. The use of a 500-Hz bandwidth is good for this purpose. Couple via ac the input of the oscilloscope and synchronize the sweep to line so that the 60 Hz is synchronized with the screen sweep and appears not to shift in time. With proper grounding and shielding procedures, it should be possible to reduce the 60-Hz component so that it is somewhere between not detectable and 0.01 pA peak to peak. Quite often, it is adequate simply to plug the three-prong plug from each instrument into a grounded wall outlet, connect the structures on the vibration isolation setup to the single-point headstage ground, and connect the rack and Faraday cage to the ground pin of a three-prong outlet. If 60-Hz interference still remains, it is usually instructive to unplug one instrument at a time to locate one that is contributing 60 Hz. When such an instrument is found, it is common to reduce the interference by using a three-prong to two-prong adapter so as to isolate the ground of that particular instrument. If this approach does not work, it might be necessary to use three-prong to two-prong adapters on each of the power cords and then to test ways to ground each individual instrument to result in the lowest 60-Hz noise.

If problems persist, it can be useful to try to map the 60-Hz fields in the room where the apparatus is located. To do this, one can utilize a portable differential amplifier with a BNC cable extending from each of the inputs and with about 100 MΩ (megohms) of resistance connected between the two terminals at the opposite end of the cables. By connecting the amplifier output to the oscilloscope and by using the cables with the 100-MΩ resistor as an antenna, it is possible to probe various points in the room to identify sources of 60-Hz noise. When such sources are found, they can either be disconnected or somehow shielded so that their radiated 60 Hz is minimized. In most situations, one or more of the 60-Hz reducing procedures described here will be effective.

Major 60-Hz (or 120-Hz) interference may come from fluorescent room lights and from the microscope light itself. Often, the microscope light is driven by alternating current derived from the line and stepped down by a transformer in the microscope. Whereas in some cases it is possible to ground the microscope and the power cord to the light to adequately shield against this problem, more often one must rewire the microscope so that the light can be driven from a low ripple direct current power supply. Fluorescent lights in the ceiling over the preparation do not cause much interference in Faraday cage-based facilities but may if shielding comes from simply grounding the microscope and its surroundings. In this case, it is necessary to work with the room lights turned off or to have the room light switch rewired so that the lights specifically above the setup can be turned out. Some investigators have even installed dimmable incandescent lights near their setups since these do not produce the large fields that fluorescent lights do.

Patch Clamp Electronics

High quality patch voltage clamps are available from several manufacturers. Although most of these instruments have many features in common, some are exceptional in particular areas. Rather than describe the various features of patch clamps from specific manufacturers, we will here discuss some theoretical and practical aspects of patch clamp electronics in its present state of the art, leaving the choice of particular instruments to the reader. Our discussion will emphasize the patch clamp headstage, that is, the current to voltage converter which measures the patch and whole-cell currents, since in many respects this is the most critical portion of the overall electronics.

Headstage

There are two basic varieties of patch clamp headstages which we will denote as resistive feedback headstages and capacitive feedback headstages. Some manufacturers provide both resistive feedback and capacitive feedback in a single switchable headstage design; in this case the capacitive feedback mode is generally intended for ultra-low noise measurements of single channels, and the resistive feedback mode is intended for whole-cell voltage clamping.

Resistive Feedback Headstages. The resistive feedback circuit is well known, and its basic characteristics have been described in detail elsewhere.[1,5] Fundamentally, the circuit uses negative feedback from the output of the operational amplifier to its inverting input to maintain this input at a "virtual ground." The current source being measured, i, is attached to the inverting input. The JFET (junction field-effect transistor) input of the operational amplifier draws essentially no current (< 1 pA dc) at its gate (i.e., the inverting input) so that current entering this node is forced to flow through the feedback resistor, R_f. The amplifier output develops a voltage,

[5] F. J. Sigworth, *in* "Single-Channel Recording" (B. Sakmann and E. Neher, eds.), p. 3. Plenum, New York and London, 1983.

V_o, which is proportional to the input current, that is, $V_o = -iR_f$. When used in a patch voltage clamp, the patch pipette is attached to the inverting (−) input of the operational amplifier, and the circuit is modified so that command voltages, V_c, are applied to the noninverting (+) input; as a result of feedback the command voltage will also be imposed on the inverting input, that is, at the top of the patch pipette. It should be noted that most commercial patch clamps utilize a discrete JFET input stage in conjunction with a commercial operational amplifier to create the head-stage amplifier. This approach leads to lower noise than is possible by using any operational amplifier that is presently commercially available. Details of such designs have been presented elsewhere.[1,5]

Because the currents that are measured with a patch voltage clamp are so small, an extremely high valued feedback resistor is used both to provide adequate gain and to achieve a good signal-to-noise ratio (the input referred current noise of the feedback resistor is inversely proportional to the square root of its value). For single-channel measurements, the value of R_f is typically 50 GΩ; for whole-cell current measurements, 500 MΩ to 1 GΩ is typical in most commercial instruments. Most of the shortcomings of resistive feedback headstages arise from nonideal characteristics of these high valued resistors.

The power spectral density of the thermal current noise of an ideal feedback resistor, R_f, is given by $4kT/R_f$, where k is Boltzmann's constant and T is the absolute temperature. The root mean square (rms) noise in a given bandwidth, B, is given by $[(4kT/R_f)B]^{1/2}$. From these relationships, it is obvious that larger valued feedback resistors will produce less noise. However, all commercially available gigohm-range resistors with which we are familiar exhibit considerable amounts of noise in excess of the expected thermally induced noise fluctuations. Excess noise occurs at both low frequencies and high frequencies. Low frequency excess noise has the familiar $1/f$ spectral form; its amplitude is quite variable even among feedback resistors of the same type. $1/f$ noise is not a particularly important problem for patch clamp measurements, although it can place limits on the background noise when very small bandwidths are used. Excess high frequency noise in gigohm feedback resistors is a much more severe problem for patch clamp measurements. All commercially available gigohm-range feedback resistors have noise spectral densities which rise above the expected thermal noise level beginning at frequencies that range from a few hundred hertz to a few kilohertz. Even with the best resistors that we know of, this excess noise can account for about half the total noise of the headstage at a bandwidth of 10 kHz. For poorer resistors, high frequency excess noise can be the dominant noise source at all bandwidths above a few kilohertz.

Another drawback to using gigohm-range feedback resistors is the small bandwidth of the current signal directly at the headstage output. The frequency response of a resistive feedback headstage is limited by the stray capacitance, C_f, shunting the feedback resistor. For example, 0.1 pF of stray capacitance associated with a 50-GΩ resistor will lead to a time constant of 5 msec which corresponds to a bandwidth (-3 dB, one-pole RC filter) of only 32 Hz. Because this bandwidth is inadequante for almost all patch clamp measurements, a "boost" circuit is required to restore the high frequency components of the signal and, of course, of the background noise. It can easily be shown[1,6] that the effective band width can be increased by this circuit from B_{HS} (the headstage bandwidth, $B_{HS} = 1/(2\pi R_f C_f)$ to approximately $(B_{HS}B_u)^{1/2}$, where B_u is open loop unity gain bandwidth of the operational amplifier used in the boost circuit. For example, for $B_{HS} = 80$ Hz (time constant ~ 2 msec) and $B_u = 20$ MHz, a final "boosted" bandwidth as high as 40 kHz can be achieved. It should be noted, however, that the response will no longer be first order. Obviously, the final bandwidth can be larger for relatively small feedback resistors (e.g., 50–500 MΩ) than for very high valued resistors (e.g., 50 GΩ). Reducing the stray capacitance that shunts the feedback resistor could also increase the intrinsic headstage bandwidth (B_{HS}), although a low value of this capacitance can have adverse effects when the headstage has a large capacitive load at its input.[7]

In addition to the inherently limited bandwidth associated with gigohm-range resistors, it is also important to realize that the conductive path of most or all such resistors is nonuniform, so that the stray capacitance is distributed and the frequency response can be quite complex. This necessitates the use of two or more stages of boost circuitry to produce an adequately flat high frequency response. Such circuitry not only further restricts the final boosted bandwidth, but also provides more possible locations for drift with time and temperature which can slightly "detune" the boosted response. Finally, nonlinearities (see below) associated with high valued feedback resistors can somewhat change the shape of the boosted response as a function of signal level.

Three other limitations of high valued feedback resistors should also be noted: (1) relatively large voltage coefficients of resistance, (2) large temperature coefficients of resistance, and (3) rather poor stability with time (aging). The voltage coefficient of resistance of commercially available

[6] R. Levis, "Patch and Axial Wire Voltage Clamp Techniques and Impedance Measurements of Cardia Purkinje Fibers." University Microfilms International, Michigan and London, 1981.

[7] W. D. Niles, R. A. Levis, and F. S. Cohen, *Biophys. J.* **53**, 327 (1988).

gigohm resistors can be as high as 4–6% per volt which would mean that the resistance value would change by 40–60% as the headstage output swings from 0 to 10 V. Obviously, so great a nonlinearity is unacceptable for most patch clamp work. The chip resistors used by several commercial patch clamp manufacturers are substantially superior to this, but their voltage coefficients are not small enough to be ignored. In fact, the voltage coefficient of these resistors is itself nonlinear, varying from as much as 1% per volt for small fields (≤ 1 V) to less than 0.5% per volt for voltages in the vicinity of 10 V. Typical values are somewhat better, and lower value resistors (e.g., 500 MΩ) usually outperform higher value resistors (e.g., 50 GΩ). For the measurement of single-channel currents or of most whole-cell ionic currents, these nonlinearities are usually acceptable. However, when procedures like $P/4$ pulse protocols are used to eliminate linear capacity transients (e.g., for measurements of "gating" currents from whole cells[8]), such nonlinearities can become quite important, and it is necessary to use capacity and whole-cell compensation (see below) to eliminate as much of the linear transient as possible in order to minimize the output voltage excursions of the headstage. These nonlinearities can also lead to minor differences in the adequacy of the boost circuitry for signals of different amplitude.

The temperature coefficient of resistance (TCR) of all gigohm-range resistors is substantially higher than those of precision resistors with values less than about 100 MΩ. Such coefficients generally fall in the range of 0.1–0.5%/°; 0.2%/° is more or less typical. These values indicate that gain changes of as much 1% can occur with normal temperature variations in laboratory environments. Perhaps more importantly, it should be realized that during the first 30 min or so after the instrument is turned on, resistance values will slightly change as the headstage warms up. Some manufacturers have chosen to heat the feedback resistor (as well as nearby components) to a stable value above the normal range of room temperatures. This greatly reduces temperature-induced drift, but comes at the expense of greater gate current for the input JFET and increased low frequency noise.

Gigohm resistors are also generally less stable with time than lower valued precision resistors. Resistance changes of 2–5% per year are possible. Because of this it is a good idea to periodically (every 6 months or so) calibrate the gain of the patch clamp and, since resistance changes can effect the "tune" of the boosted response, check the performance of the boosted output at the same time.

[8] B. P. Bean and E. Rios, *J. Gen Physiol.* **94**, 65 (1989).

Despite the numerous shortcomings of high valued feedback resistors, resistive feedback headstages are quite adequate for most patch clamp measurements and, of course, have been used successfully for many years. In general, resistive feedback is still the only real option for whole-cell recording. However, recent advances in patch clamp technology have provided the user with a very effective alternative for single-channel measurements. This is the capacitive feedback (integrating) headstage. In the future, such headstage designs may also become more practical for whole-cell measurements.

Capacitive Feedback Headstages. Compared to highly imperfect gigohm-range feedback resistors, small capacitors are nearly ideal circuit elements. Small ceramic chip capacitors can have leakage resistances as high 10^{15} Ω. Such capacitors are almost perfectly linear over the typical range of voltages encountered in patch clamp instrumentation ($\pm 10 - 15$ V), and their frequency response is ideal to well beyond 1 MHz. Of even greater importance is the fact that capacitors introduce essentially no thermal or excess noise of the type described above for gigohm resistors. Because of this, the noise of a capacitive feedback headstage can be substantially lower than that which can be achieved with any resistive feedback element. Capacitive feedback headstages also offer wider bandwidth (with a perfectly flat transfer function up to the high frequency cutoff) and greatly extended dynamic range when compared to resistive feedback. The only major drawback of the new technology is the necessity to reset periodically the headstage output to zero; a lesser difficulty arises from dielectric absorption. These subjects are discussed below.

The current-to-voltage converter consists of an operational integrator followed by a differentiator. It is easily shown that the gain, R_g (ohms), of the integrator/differentiator combination is given by

$$R_g = R_d(C_d/C_{fi})$$

that is, $V_{od} = iR_d(C_d/C_{fi})$, where V_{od} is the output of the differentiator, C_{fi} is the feedback capacitor of the integrator, C_d is the input capacitor of the differentiator, and R_d is the feedback resistor of the differentiator. Typical values would be $C_{fi} = 1$ pF, $C_i = 10,000$ pF, and $R_d = 100$ kΩ, providing a current-to-voltage gain of 1 GΩ. As described below, the noise of the capacitive feedback headstage can be made to be essentially independent of its gain. This is very different from the situation with resistive feedback where gain and noise are intimately linked, that is, low noise requires a high valued feedback resistor.

The bandwidth of a capacitive feedback headstage can be very large relative to that which can be achieved using resistive feedback. The achievable bandwidth (or natural frequency, f_N) of the integrator/differentiator combination is well approximated in terms of the gain R_g ($R_g = C_d R_d / C_{fi}$), the feedback capacitance, C_{fi}, and the open loop unity gain frequency of the differentiator, f_{ud}, that is,

$$f_N = 1/[2\pi R_g (C_{fi}/f_{ud})]^{1/2}$$

For $R_g = 1$ GΩ, $C_{fi} = 1$ pF, and $f_{ud} = 80$ MHz, this indicates that a bandwidth of 110 kHz can be achieved; by reducing the gain to 100 MΩ the bandwidth can be extended to about 350 kHz. Clearly, wide bandwidth is associated with a fast differentiator amplifier, small values of C_{fi}, and low values of gain, R_g. Recall, however, that for the capacitive feedback headstage, low noise can be achieved even for low values of R_g.

The very wide bandwidth possible with capacitive feedback may well prove to be important in the future, but it should be noted that, at the present time, bandwidths of more than about 20 kHz are of questionable value to patch clamping or whole-cell measurements using the patch voltage clamp. This is because the noise of either resistive or capacitive feedback headstages increases steeply with increasing bandwidth, ultimately increasing as the bandwidth to the 3/2 power. The lowest noise we have achieved to date in an actual patch recording situation is about 0.25 pA rms (about 1.5 pA peak to peak) in a 10-kHz bandwidth. For a 20-kHz bandwidth, this should increase to about 0.7 pA rms (about 4.2 pA peak to peak), and for a 50-kHz bandwidth it should reach about 2.8 pA rms (about 17 pA peak to peak). Obviously even under ideal circumstances, such large bandwidths could only be used with extremely large single-channel currents. In the case of whole-cell clamping through a patch pipette, the actual bandwidth of current recording in the absence of series resistance compensation is limited to $1/(2\pi R_s C_m)$, where R_s is the series (pipette) resistance and C_m is the membrane capacitance. For $R_s = 10$ MΩ and $C_m = 50$ pF, this amounts to only about 320 Hz. A much larger recording bandwidth would only add noise to the measurement, not more information about the current signal. If series resistance compensation is used, then the actual bandwidth of current measurement will be $1/2\pi R_{sr} C_m$, where R_{sr} is the residual (uncompensated) series resistance (e.g., for $R_s = 10$ MΩ and 90% compensation, $R_{sr} = 1$ MΩ). In the previous example, 98.4% compensation would be required to justify a 20-kHz bandwidth. At this bandwidth (with $R_s = 10$ MΩ and $C_m = 50$ pF), the noise in the measurement would be more than 250 pA rms (1.5 nA peak to peak).

The noise of a capacitive feedback headstage can be substantially lower than that of a resistive feedback headstage. The reason for this is primarily due to the thermal and excess high frequency noise associated with gigohm-range resistors. At low to moderate frequencies, the floor of the resistive feedback headstage noise power spectral density is determined primarily by the thermal noise of the feedback resistor; with good input field-effect transistors (FETs) the input current noise associated with FET gate leakage current, i_g, is of lesser importance. In the capacitive feedback headstage, the low frequency noise floor is determined almost exclusively by the current noise of the FET gate current. For FETs that are also good selections for low noise at high frequencies, i_g can be as low as 0.1 pA or somewhat less, and the power spectral density of the input referred noise at low frequencies can be 10 times lower than that which is achieved with a 50-GΩ resistor. Thus, at low frequencies, the noise performance of the capacitive feedback headstage is definitely superior to that of a resistive feedback headstage. At higher frequencies, the noise of the capacitive feedback headstage is dominated by the input voltage noise of the FET input, which in conjunction with the input and stray capacitance at the input (plus the small feedback capacitor) produces current noise with a power spectral density that rises as frequency squared. This same noise source is present in resistive feedback headstages, but, as described above, excess high frequency noise from the feedback resistor typically produces as much or more noise at high frequencies.

Of course, real capacitors are not completely free of noise; some dielectric noise (see discussion below) will be present, but with high quality capacitors (dissipation factor less than 0.0001) dielectric noise of the feedback and compensation capacitors will be negligible by comparison with other noise sources. Noise associated with the differentiator can be made negligible by careful circuit design. In particular, noise associated with the differentiator is minimized by selecting a low-noise FET input operational amplifier (e.g., Burr-Brown, Tucson, AZ; OPA 101/102 or OPA 627) for this location and by keeping the ratio C_d/C_{fi} large.

On the basis of the above discussion it would be predicted that a capacitive feedback headstage could be built utilizing a differential U430 JFET (Siliconix, Santa Clara, CA) input stage with noise in the range of 0.12–0.15 pA rms in a 10-kHz bandwidth; this can be compared with 0.25–0.3 pA rms for high quality resistive feedback headstages. In actual practice, the noise of practical capacitive feedback headstages is found to be somewhat higher, typically in the range of 0.18–0.20 pA rms in a dc 10-kHz bandwidth (− 3 dB, 8-pole Bessel filter). There are several reasons for this. One additional noise source arises from the switch used to reset the headstage periodically. In addition, compensation signals used to balance currents injected by this switch during and just following reset and to eliminate the effects of dielectric absorption of the feedback capacitor inevitably add small amounts of noise. A certain amount of noise must also be expected to result from the packaging of critical components of the input stage and from the input connector. This noise results both from the addition of stray capacitance at the input and from dielectric noise (see discussion under electrode noise below) associated with the packaging and connector. Nevertheless, capacitive feedback can produce significantly lower noise patch clamp measurements than could previously be achieved using resistive feedback. We have achieved noise as low as 0.13 pA rms in a 5-kHz bandwidth in actual patch recording situations using capacitive feedback headstages, whereas with resistive feedback it is generally rare to go below 0.20 pA rms in the same bandwidth. Of course, to achieve these results it is essential to minimize all other sources of noise as described elsewhere in this chapter (see also [3] in this volume).

Further improvements in the noise of capacitive feedback headstages can be anticipated. It seems possible that headstage noise as low as 0.06 – 0.10 pA rms in a 10-kHz bandwidth can be achieved. Approaches to produce such noise reduction are not discussed here, but we believe that such low levels of headstage noise can be of practical significance to overall noise performance. The present and predicted limits of noise performance in patch voltage clamping are discussed below.

The dynamic range of a capacitive feedback headstage can also greatly exceed that of resistive feedback devices produced for low-noise measurements. The reason for this is basically the fact that with capacitive feedback gain and noise are not linked as they are in a resistive feedback headstage. Because the feedback capacitor lacks the thermal and excess noise associated with feedback resistors and because the noise contribution of the differentiator can be made negligible, it is possible to produce a capacitive feedback headstage with a gain of 100 μV/pA and input referred noise of less than 0.2 pA rms in a 10-kHz bandwidth. Even lower gains can be produced with the same input referred noise, although numerous precautions must be taken to ensure that noise of subsequent stages of the electronics (gain, filters, etc.) do not elevate the intrinsic headstage noise in such cases. To achieve noise in the range of 0.25 – 0.30 pA rms in a 10-kHz bandwidth with resistive feedback it is traditional to use a 50-GΩ resistor resulting in a minimum gain of 50 mV/pA. Thus, the dynamic range (defined, e.g., as the ratio of the full scale output to the minimum detectable signal) of the capacitive feedback headstage can be 500 – 1000 times larger than is possible with a 50-GΩ resistive feedback headstage. Of course for many measurements an output gain of 100 μV/pA may be undesirably small, particularly when the sensitivity of the final analog-to-digital converter is considered. In this situation it is only required that gain be provided following the integrator/differentiator.

At the present time, the very large dynamic range available with capacitive feedback has relatively limited utility for most biophysical measurements, but we feel that its existence will become more useful in the future. Additional benefits associated with capacitive feedback as compared to resistive feedback are (1) improved fidelity of current measurement, (2) improved linearity, (3) very low temperature coefficient of gain, and (4) the ability to pass very large transient currents.

As described in the previous section, resistive feedback headstages often require rather complicated high frequency "boost" circuitry to deal with the limited and somewhat complex frequency response of high valued feedback resistors. Even with such circuitry, aberrations of the step response with amplitudes as large as 2–3% of the total and durations as long as several milliseconds are sometimes unavoidable; the nature of these distortions can also vary somewhat as a function of signal level. The integrator/differentiator combination of capacitive feedback is completely free of such distortions.

In addition, the small feedback capacitor of the capacitive feedback headstage is almost perfectly linear over the range of voltages (± 10 V) encountered. The differentiator can also be easily made to be highly linear. This is a distinct improvement over the situation described above for most high valued feedback resistors. This capability for highly linear operation suggests to us that capacitive feedback may become useful in whole-cell recording as well as single-channel measurements. Improved linearity can be quite important in measurements such as whole-cell gating currents which rely on the subtraction of relatively large currents to reveal to smaller signal of interest. To be practical for most whole-cell measurements the size of the feedback capacitor should be increased (e.g., to 10 or 20 pF) to reduce the frequency of occurrence of resets.

The temperature coefficient *(TC)* of the gain of the capacitive feedback headstage can also be significantly less than that of resistive feedback. The *TC* of small ceramic chip capacitors suitable for use as the feedback element in the integrator is typically about 70–90 ppm/° (as compared to 2000 ppm/° for typical gigohm feedback resistors). By matching the *TC* of the input capacitor of the differentiator to that of the integrator (and using a low TCR resistors in the differentiator and subsequent electronics, which are readily available for resistors with values less than 1 MΩ) it should be possible to reduce the overall *TC* to about 10–20 ppm/°.

A capacitive feedback headstage can also readily supply large transient currents required to charge capacitive loads quickly without saturating the integrator. With a 1-pF feedback capacitor and a 10-V output voltage, 10 pC of charge (e.g., 1 nA for 10 μsec or 10 nA for 1 μsec) can be applied without saturation (or reset). With a 50-GΩ resistor and a 10-V output, the maximum current that can be passed is 200 pA. Of course the transient current passing ability of a resistive feedback headstage can be increased by placing a capacitor in parallel with the feedback resistor, but this is generally only practical for small (e.g., 500 MΩ) feedback resistors since with very large resistors the reduction of the intrinsic headstage bandwidth would be too great.

The only drawback to the capacitive feedback headstage is the necessity to reset periodically (i.e., discharge) the integrator feedback capacitor. The dc component of the current being measured, i, plus the gate leakage current, i_g, of the input FET will drive the output voltage of the integrator, V_{oi}, toward saturation at a rate given by

$$dV_{oi}/dt = (i + i_g)/C_{fi}$$

For $C_{fi} = 1$ pF and $i_g = 0.1$ pA, the output of the integrator will ramp upward at a rate of 100 mV/sec even in the absence of any measured current; saturation (about 10 V) would be reached in about 100 sec. Obviously, measured currents will drive the output toward saturation more quickly: a total average current of 10 pA (e.g., a 20-pA channel that is open half of the time or 100 mV across a 10-GΩ seal) would cause the output to reach 10 V in 1 sec, 100 pA of dc current would cut this time to 100 msec.

The solution to this problem is to discharge the integrator capacitor rapidly and thereby reset the output voltage of the integrator to zero. Of course during the period of time that the feedback capacitor is shorted, the current at the input of the headstage will not be measured. It is therefore important to keep the duration of reset as short as possible. It is not dificult to reset the integrator capacitor in 5–10 μsec. On the other hand, resetting the differentiator capacitor is somewhat more difficult. This is both because the differentiator input capacitor is much larger than the integrator feedback capacitor and because even if the differentiator capacitor itself is discharged in a few microseconds, transients associated with this reset as they appear at the differentiator output will have a duration that is in part determined by the bandwidth of the integrator/differentiator combination. For example, for a bandwidth at the differentiator output of 100 kHz something of the order of 20 μsec following the *end* of the actual reset (shorting) period is required for the transient to settle back into the noise. Because the measured output is replaced by the output of a modified sample and hold circuit as long as the actual output is perturbed by the reset transient, the need for a wide bandwidth at the differentiator output is dictated more by the requirement of keeping the total reset time (i.e., the time that the output does not accurately reflect the input current) as brief as possible than by the present usefulness of such wide bandwidths to biophysical measurements.

The total duration of the reset is typically some 30–50 μsec. During this period the output does not accurately reflect the input current. When the output is filtered with a cutoff frequency in the neighborhood of 5 kHz or less resets can be essentially undetectable, provided that dielectric absorption effects have been adequately canceled. Even so, data obtained during the brief reset period are not valid, and for this reason manufacturers of capacitive feedback headstages typically provide a "telegraph" signal indicating the occurrence of a reset. For pulsed data acquisition (e.g., as is typical for the measurement of voltage-gated ionic channels) resets can generally be avoided during the period in which data are recorded by forcing a reset just prior to the step voltage command.

Other Aspects of Patch Clamp Electronics

Aside from the type and quality of the headstage, there are a variety of other features that are important in patch clamp electronics. Many such features only provide added convenience, whereas others may be essential in various situations. We very briefly describe a few such features here.

Capacity Compensation. Stray capacitance at the headstage input plus the capacitance of the holder and pipette must be charged with the pipette potential is changed. As described in [3], this volume, the pipette capacity transient has a large fast component plus a smaller slower tail arising from the lossy dielectric characteristics of the glass. For resistive feedback headstages with R_f values of 50 GΩ even relatively small voltage steps can result in saturation of the headstage output during the charging of the fast capacity transient; for a 100-mV step, saturation will typically persist for 1 msec or more. Obviously single-channel data are lost during this period. The solution is to inject the charging current by a separate pathway, usually a 1-pF capacitor internally connected to the input. For capacitive feedback headstages, saturation is unlikely to occur, and so fast capacity compensation, although still desirable, is not as necessary. Essentially all commercial patch clamps provide fast capacity compensation. Some manufacturers also provide a slower component to help to cancel the tail of capacity current resulting from the pipette. Because the slow component arising from the pipette is not well described by a single exponential (see [3] in this volume), the cancellation provided is not perfect. For both slow and fast components of capacity compensation it is important for low-noise measurements that the compensation circuitry not add a significant amount of noise above that of the headstage alone.

Whole-Cell Compensation. It is often useful to cancel the current involved in charging the cell membrane capacitance during whole-cell recording. Several commercial patch clamps provide circuitry to accomplish this. It is important to realize, however, that by itself such compensation does nothing to speed the response time of the cell potential which is determined by the time constant $R_s C_m$. Some form of "supercharging" can speed up the membrane potential response but will not reduce membrane potential errors resulting from the flow of ionic current. Series resistance compensation will both speed up the membrane potential response and reduce errors arising from ionic currents. Whole-cell capacity compensation is usually designed so that the transient will continue to be appropriately canceled as series resistance compensation is advanced. The effects of series resistance on whole-cell voltage clamping are considered in greater detail below.

Compensation for whole-cell capacity transients can be particularly important when a "P/N" (e.g., the traditional P/4) procedure is used, for example, to study gating currents from whole cells. Nonlinearities of the feedback resistor (especially its voltage coefficient of resistance) can produce substantial artifacts in such situations unless the cell membrane capacity transient has been adequately reduced in size (i.e., canceled) so that the output excursions of the headstage remain relatively small.

Other features of patch clamp electronics, such as pipette offset, "tracking," current clamp, output gain stages, and filters, are not discussed here. The importance and use of such features are well explained in the manuals provided with commercial patch voltage clamps.

Noise Considerations for Patch Clamp Recording

Electrode Noise

In addition to the inevitable noise arising from the membrane–glass seal (see section on seal noise below), the pipette contributes noise to the measured current in patch voltage clamping by several mechanisms. In the first place, the holder and pipette are major sources of capacitance at the headstage input and will therefore react with the input voltage noise, e_n of the headstage to produce current noise with a power spectral density (PSD, A^2/Hz) that will rise as f^2 at frequencies above which e_n has become essentially constant (typically ≥ 1 kHz). This noise is perfectly correlated with the noise arising from the intrinsic capacitance, C_{in}, associated with the gate input of the JFET input stage of the headstage and its input voltage noise, and therefore the usual rules of root mean square noise addition do not apply. It should be noted that C_{in} consists of the JFET input capacitance ($C_{iss} = C_{gs} + C_{gd}$, where C_{gs} is the gate to source capacitance and C_{gd} is the gate–drain capacitance) plus $1-2$ pF of stray capacitance, plus the capacitance of the injection capacitor connected to the gate for compensation signals (typically 1 pF), and, for a capacitive feedback headstage, the feedback capacitor, C_{fi} (1–2 pF). If the holder plus pipette capacitance is denoted by C_{hp}, then the noise PSD associated with E_n will be given by $4\pi^2 e_n^2 (C_{in} + C_{hp})^2 f^2$. In general, e_n will be a function of frequency with $1/f$ noise dominating at low frequencies (< 100–1000 Hz for good JFETs) and a limiting high frequency value; it is often adequate to approximate e_n^2 by its high frequency limit when relatively wide bandwidth (say > 2–3 kHz) noise is considered.

The capacitance of the holder can range from about 1 to 5 pF depending on construction and the presence of absence of metallic shielding (which increases capacitance); the holders we use are unshielded and, by themselves, add only about 1–1.5 pF. The capacitance added by the pipette depends on a variety of factors including the depth of immersion of the pipette into the bath, the type of glass used, the ratio of outer to inner diameter, and the use of Sylgard coating. For uncoated pipettes, pipette capacitance ranges from about 0.5 to somewhat more than 2 pF per millimeter of immersion (see below). Sylgard coating can, however, significantly reduce these values. In general the total capacitance from the pipette alone will typically fall in the range of 1–5 pF. Thus, C_{hp} should range from about 2 to 10 pF. For well-designed differential headstages that are presently available e_n is generally in the range of 1.5–3 nV/Hz$^{1/2}$ ($f \geqslant 1$ kHz). C_{in} is typically about 15 pF. Assuming that e_n is 2 nV/Hz$^{1/2}$ and C_{in} is 15 pF, then in a 10-kHz bandwidth (-3 dB of an 8-pole Bessel filter) e_n–C_{in} noise will amount to about 0.15 pA rms. For C_{hp} equals 2 pF, the total noise arising from e_n would increase to 0.17 pA rms again for a 10-kHz bandwidth; for C_{hp} equals 10 pF this value would increase to 0.25 pA rms. Obviously, in terms of noise it is important to keep C_{hp} as low as possible.

It should also be obvious that if headstages become available with smaller values of e_n (without significantly increasing C_{in}), both the noise of the open-circuit headstage and the noise increment associated with C_{hp} will decrease; for example, for $e_n = 1$ nV/Hz$^{1/2}$ and the same C_{in} all of the root mean square values presented above would be cut in half. From the point of view of headstage design, the selection of the best JFET for the input stage should be based on the product of e_n and the total noise producing capacitance, C_T ($C_T = C_{in} + C_{hp}$), associated with it in an actual measurement situation. It can easily be seen that headstage amplifiers could be produced with essentially identical open-circuit noise which would behave differently when loaded with capacitance at the input. For example, a headstage with $e_n = 6$ nV/Hz$^{1/2}$ and $C_{in} = 5$ pF would produce the same

$e_n - C_{in}$ noise (and presumably the same total open-circuit headstage noise) as a headstage with $e_n = 1.2$ nV/Hz$^{1/2}$ and $C_{in} = 25$ pF; in a 10-kHz bandwidth (8-pole Bessel filter) e_n would be responsible for about 0.15 pA rms in each case. However, with $C_{hp} = 5$ pF, so that C_T is 10 pF for the first amplifier and 30 pF for the second amplifier, the "$e_n - C_T$" would be 0.30 pA rms in the same 10-kHz bandwidth for the amplifier with $e_n = 6$ nV/Hz$^{1/2}$ but only 0.18 pA rms for the amplifier with $e_n = 1.2$ nV/Hz$^{1/2}$.

A useful figure of merit for FETs is the ratio of their transconductance, g_m (mhos), to C_{iss}. The transconductance is approximately related to the input voltage noise (beyond the 1/f region) by $e_n^2 \cong 4kTv/g_m$, where v is theoretically equal to 0.67, but is usually higher. For the lowest noise this ratio should be as large as possible; for the best JFETs presently available, $g_m/C_{iss} \cong 1.0 - 1.5 \times 10^9$ sec^{-1}. Ratios about this good are, however, available from several different JFETs, some with relatively small g_m (and therefore relatively high e_n) and small C_{iss} and others with high g_m and C_{iss}. If g_m/C_{iss} were identical for all such FETs then the best selection would be a FET with C_{iss} approximately equal to the sum of the holder and pipette capacitance plus any compensation capacitors connected to the input, plus, for capacitive feedback headstages, the feedback capacitor, plus any stray capacitance at the input; this capacitance should total roughly 5 – 10 pF. However, unless the FET is cooled, the value of input gate current, i_g, must also be taken into account; some JFETs with very low e_n have rather high values of i_g and are thus impractical for patch clamp devices. In addition, some JFETs have higher values of e_n than would be predicted from their transconductance. It also seems likely that the choice of an optimum value of C_{iss} could be effected in a frequency dependent manner by the dielectric noise of the silicon itself. Of JFETs that are commercially available at the time of this writing, the U430 still is an excellent selection for the input stage of the patch clamp headstage. Nevertheless, improvements should be possible. In principle g_m/C_{iss} is linearly related to the carrier mobility of the material used to fabricate the FET; thus gallium arsenide FETs would be expected to have substantially lower e_n for a given value of C_{iss} than silicon FETs. This is in fact true at very high frequencies; however, at the present time all such FETs that we know of have very large amounts of 1/f noise which extends beyond the uppermost frequency used in patch clamping.

Short channel JFETs should also be attractive; g_m/C_{iss} theoretically varies as $1/L^2$, where L is the FET channel length. In most commercially available JFETs L is not less than roughly 10 μm, but it should be possible to significantly reduce this without too greatly increasing i_g. This suggests the possibility of producing silicon JFETs with g_m/C_{iss} ratios as high as

$3-5 \times 10^9$ sec^{-1}. Such a FET, with C_{iss} of $3-10$ pF, if it becomes available, would be ideal for the headstage amplifier and would decrease both its open-circuit noise and the noise increment associated with holder–pipette capacitance at the input.

It should also be noted that the noise associated with the command potential will react with C_{hp} (but not with C_{iss}) to produce noise in the measured current that once again has a PSD that rises as f^2. The command potential is normally attenuated (typically by 10:1 to 50:1) to reduce the noise arising from digital-to-analog converters or function generators. At the present time this is generally sufficient to reduce this potential source of noise to negligible levels. As other noise sources are reduced, however, more attention to such noise may become necessary. It should also be pointed out that the resistors in the attenuator are themselves a part of the command potential noise; therefore, low valued resistors are preferred.

In addition to the mechanisms just described, noise associated with the patch pipette arises from several other sources. Three potentially important sources of noise are considered here: (1) noise arising from the lossy dielectric characteristics of the glass, (2) noise arising from the thermal voltage noise of electrode resistance (distributed) in conjunction with the distributed capacitance of the pipette wall, and (3) noise arising from the patch capacitance in series with the resistance (lumped) of the pipette. Finally, the noise of the membrane–glass seal is briefly discussed.

Dielectric Noise. Thermal fluctuations in lossy dielectrics generate noise, the magnitude of which can be related to the real part of the admittance of the dielectric material (i.e., the loss conductance) by the fluctuation–dissipation theorem. More specifically for a dielectric with relatively low losses, the PSD, $i_d^2(f)$, of the noise current can be expressed in terms of the dissipation factor, D (also called the loss factor), and the capacitance, C_D, of the dielectric,[9] that is,

$$i_n^2(f) = 4kTDC_D(2\pi f) \qquad \text{A}^2/\text{Hz} \qquad (1)$$

Note that the PSD of the noise associated with the dielectric loss of the glass from which the pipette is fabricated is expected to rise linearly with frequency. In situations where the dielectric noise of the pipette dominates total noise, this expected spectral shape is usually quite well approximated (see [3] in this volume). The root mean square noise arising from the lossy dielectric for a given bandwidth, B, can then be computed as the square

[9] V. Radeka, *IEEE Trans. Nucl. Sci.* **NS-20,** 182 (1973).

root of the integral of Eq. (1) from dc to the cutoff frequency, that is,

$$\text{rms noise} = (4kTDC_D\pi B^2)^{1/2} \quad \text{A rms} \quad (2)$$

If it is assumed that the ratio of the inner and outer diameters of a pipette is approximately preserved during the pulling process, then for pipettes fabricated from glass with a given wall thickness the important parameters for determining dielectric noise are the dissipation factor, dielectric constant, and depth of immersion of the pipette tip into the bath. The importance of the dissipation factor is obvious from Eq. (1) and (2). The dielectric constant of the glass, its wall thickness, and the depth of immersion go together to determine C_D. A standard rule of thumb states that the capacitance associated with the pipette is about 1–2 pF per millimeter of immersion into the bath (in the absence of Sylgard coating), but this will, of course, depend on the dielectric constant of the glass and geometry and wall thickness near the tip.

We normally use glass with an outer diameter (OD) of 1.65 mm and an inner diameter (ID) of 1.15 mm (OD/ID $\cong$ 1.43). Assuming that these proportions are preserved as the glass is drawn out by pulling, pipettes fabricated from this glass should have a capacitance of about 0.15ε pF per millimeter of immersion, where ε is the dielectric constant of the glass (e.g., 3.8 for quartz, 4–5 for most borosilcates, 6–7 for aluminosilicates, about 7 for soda lime glass, and 7–10 for high lead glasses). For thin walled glass with an OD/ID ratio of 1.2 the capacitance should increase to about 0.30ε pF/mm of immersion, whereas for thick walled glass with an OD/ID of 2.0 the capacitance should drop to about 0.08ε pF/mm. Obviously these numbers are only approximate since the assumption that glass proportions remain constant during pulling is itself only approximately true; in particular we have observed that for some glasses (e.g., aluminosilicates) there is a pronounced thinning of the wall dimensions at the pipette tip. Moreover, as discussed below, coating the pipette with Dow Corning (Midland, MI) Sylgard 184 (which has a relatively low dissipation factor of about 0.006 and a dielectric constant of 2.9) will modify the capacitance (and dissipation factor) of the pipette.

From the above discussion and Eqs. (1) and (2) it is clear that if the OD/ID ratio and the depth of immersion are constant then the root mean square noise arising from the lossy dielectric will be proportional to $(D\varepsilon)^{1/2}$, that is, the lowest noise glass should minimize the product of the dissipation factor and the dielectric constant. For example, of the glasses we have been able to obtian and successfully pull, Corning 7760 has the lowest $D\varepsilon$ product ($D\varepsilon = 0.036$, $D = 0.008$, $\varepsilon = 4.5$). This is followed by 8161 ($D\varepsilon = 0.041$); 7040, 0120, EG-6, 1723, and 7052 are also reasonably low, with $D\varepsilon$

values of 0.048, 0.054, 0.056, 0.063, and 0.064, respectively. On the other hand, 7740 (Pyrex borosilicate) and 1720 (an aluminosilicate) are substantially higher with $D\varepsilon$ values of 0.133 and 0.194, respectively. Soda lime glasses have the highest $D\varepsilon$ product; 0.37 and 0.47 for R-6 and 0080, respectively.

Using Eq. (2) and assuming a "typical" OD/ID ratio of 1.43, it is instructive to compute the root mean square value of the dielectric noise for several types of glasses in a 10-kHz bandwidth (all values listed would be somewhat higher if the transfer function of an 8-pole Bessel filter were taken into account). For a 1 mm depth of immersion 7760 should produce about 0.16 pA rms in this bandwidth; 8161 would produce 0.18 pA rms; 7052 should produce about 0.21 pA rms. On the other hand, 7740 should produce about 0.31 pA rms, and the soda lime glasses R-6 and 0080 would produce 0.51 and 0.58 pA rms, respectively. Because with the assumption of a constant OD/ID ratio the value of C_D is a linear function of the depth of immersion, the root mean square noise in a given bandwidth should vary as the square root of immersion depth. Thus for a 2 mm depth of immersion all of these values would be increased by a factor of 1.4 (e.g., for 7760 the dielectric noise in a 10-kHz bandwidth would increase to about 0.23 pA rms). Clearly, with an excised patch it is advantageous to raise the pipette as close to the surface of the bath as possible, and with on-cell patches the bath should be as shallow as possible for the lowest noise. For example, if the depth of immersion was only 0.2 mm the noise contribution from dielectric loss in a 10-kHz bandwidth for a 7760 pipette would be 0.08 pA rms; for soda lime glass it would be about 0.25 pA rms.

Using thicker walled pipettes would reduce dielectric noise; for example, for an OD/ID ratio of 2.0 all of the above root mean square calculations would decrease to about 70% of the values listed. In addition, the above calculations do not include effects of Sylgard coating of the pipette. Even if pipettes were fabricated from materials with negligible dielectric loss, coating of the pipette with Sylgard (or some similar hydrophobic material) would be necessary for low noise recording since Sylgard can prevent the creep of a thin layer of solution up the outer wall of the pipette.[10] This can be the dominant source of noise in uncoated pipettes. Sylgard, however, confers additional advantages in terms of pipette noise. The Sylgard coating thickens the wall of the pipette and thus reduces its capacitance. Sylgard has a low dielectric constant of 2.9 and a dissipation factor of 0.0058, which is lower than that of most glasses used in the fabrication of patch pipettes. Thus, Sylgard coating can be expected to reduce the dielectric noise of the patch pipette. However, its effects will be

---

[10] Hamill et al., Pflugers Arch. **391**, 85 (1981).

difficult to quantify theoretically since the thickness of the applied coating is quite nonuniform; in particular it is difficult to produce very thick coatings in the final few hundred microns near the tip. It is expected and has been confirmed experimentally that the improvement associated with Sylgard coating will be greatest for relatively high noise glasses, especially soda lime glass. Smaller improvements should be expected for low loss glasses, but Sylgard coating will somewhat improve the noise of all glasses.

We have measured the noise of many types of glass and find that the agreement of the measured noise and the theoretical noise predicted for their dielectric loss is quite good. Plots of root mean square noise in a 10-kHz bandwidth for pipettes fabricated from 19 types of glass as a function of DC_D (more precisely $D\varepsilon/W$, where W is the OD/ID ratio, normalized to our standard ratio) are given in [3] in this volume. The relationship is monotonic and generally consistent with the theoretical predictions presented above. The depth of immersion was 1.5–2 mm, and all pipettes were covered with a moderate coat of Sylgard. In most cases, the measured value is somewhat less than predicted from Eq. (2): for low loss glasses the measured noise is only very slightly less than predicted; however, for high loss glasses (most notably the soda lime glasses) the departure is substantially higher. It seems almost certain that these departures arise from the Sylgard coating, which should have its largest effect on high-loss glasses. Nevertheless, considering the effects of Sylgard, a slightly variable depth of immersion, and somewhat variable tip geometry, we consider the agreement between theory and measurement to be excellent.

If ways become available to fabricate pipettes from them, several other glasses offer potentially lower dielectric noise, for example, 7070 ($D\varepsilon = 0.01$) and particularly quartz. At the time of this writing it appears that the ability to pull quartz capillaries into patch pipettes may soon become available. The dissipation factor of Corning 7940 (fused silica) has been variously reported to be as low as 3.8×10^{-5} and as high as 4×10^{-4}; its dielectric constant is 3.8. Thus, the $D\varepsilon$ product is in the range of 0.00014 to 0.0015. Arbitrarily selecting a value of D equal to 0.0002 ($D\varepsilon = 0.00076$), as a reasonable estimate, indicates that for the same OD/ID ratio the root mean square dielectric noise should be 7 times less for quartz pipettes than for pipettes fabricated from 7760 for the same depth of immersion, for example, only 0.023 pA rms for a 10-kHz bandwidth and a 1 mm depth of immersion. For a depth of immersion of only 0.2 mm this value would fall to 0.011 pA rms.

Noise Arising from Distributed Pipette Resistance and Capacitance.
The preceding discussion of dielectric noise might be taken to imply that pipettes fabricated from glass with very low dissipation factors might introduce only negligible amounts of noise into patch clamp measurements.

Unfortunately, however, the pipette has several other sources of noise that must be taken into account. One of these arises from the distributed resistance and capacitance of the pipette. In this section we consider the pipette capacitance to be lossless.

As described above, it is reasonable to assume that the capacitance of the pipette is more or less evenly distributed over the length of pipette immersed in the bath. On the other hand, the majority of the resistance of the pipette resides at or very near the tip. Nevertheless, a significant amount of resistance arises from the filling solution in the first few millimeters behind the tip. If the pipette is modeled as a shank region (with ID $\cong$ 1.2 mm) and a conical region approaching the tip with an angle of the cone of 11.4° (such that the ID increases to 100 μm at a distance 0.5 mm back from the tip, 200 μm at a distance of 1 mm from the tip, 400 μm at 2 mm, etc.) and a 1 μm diameter opening, then its total resistance should be about 3.2 MΩ when filled with a solution with a resistivity of 50 Ω cm. Of this about 2.4 MΩ will reside in the first 20 μm of the tip region; more than 3 MΩ will reside within the first 100 μm. However, significant resistances will be associated with regions further removed from the tip. For example, the region from 100 to 200 μm from the tip would have a resistance of 80 kΩ; the region from 200 to 300 μm from the tip would have a resistance of about 27 kΩ; and the region from 0.5 mm to 2 mm back from the tip would have a resistance of about 24 kΩ. Of course an Ag–AgCl wire extends into the solution in the pipette, and it is reasonable to assume that only a negligible amount of resistance is contributed by solution in the region into which the wire protrudes.

In any case, a feeling for the noise arising from this distributed capacitance and resistance can be gained by considering that roughly 0.5 pF of capacitance is associated with the final 0.5 mm of the electrode prior to the tip and perhaps 30 kΩ of resistance is associated with the pipette up to this point (i.e., the filling solution from the shank up to 0.5 mm from the tip). Clearly the 0.5 pF of capacitance is in series with this 30 kΩ. The 30-kΩ resistance has a thermal voltage noise PSD of about 22 nV/Hz$^{1/2}$ (about 10 times higher than the input voltage noise of a good headstage); in series with 0.5 pF this will produce current noise (PSD rises as f^2) which would have a magnitude of about 0.05 pA rms in a bandwidth of 10 kHz (8-pole Bessel filter) and 0.15 pA rms in a 20-kHz bandwidth.

Of course the actual situation is more complicated than this since both the pipette resistance and capacitance are distributed. Over the frequency range of interest to patch clamping the PSD (A^2/Hz) of this noise should rise approximately as f^2. If it is assumed that the resistance is negligible beyond 4 mm behind the tip (as would be expected if the Ag–AgCl — or particularly a platinized Ag–AgCl — wire protruded at least this close to the tip), then with the pipette geometry assumed above and assuming a uniformly distributed capacitance of 1 pF per millimeter of immersion, rough calculations indicate that the total noise arising from this mechanism would be about 0.13 pA rms for a bandwidth of 10 kHz (-3 dB, 8-pole Bessel filter) and an immersion of 2 mm. For an immersion depth of only 1 mm this value would fall to about 0.1 pA rms. For an electrode modeled as above but with a cone angle of 22.6° (with a tip opening of 1 μm the total resistance would be 1.6 MΩ) and with other parameters as assumed above, the noise from this mechanism should roughly fall by half, namely, for a 10-kHz bandwidth to about 0.07 pA rms for a 2 mm immersion depth and about 0.05 pA rms for 1 mm of immersion.

The above calculations are highly approximate. Noise arising from this mechanism would be expected to increase if the wire — which has the effect of shunting a portion of the electrode resistance — did not protrude as far toward the tip. This would be particularly true if there were any extended region behind the tip where the electrode taper became very shallow and resulted in significantly increased resistance distal to the tip. More generally, it seems clear that the geometry of the first few millimeters behind the tip will have important effects on this noise. It should also be noted that withdrawing the tip toward the surface would not reduce this noise as much as might be expected; this would decrease the capacitance of the pipette but not the resistance.

There should also be ways of reducing the noise arising from this mechanism. For even the best glasses presently used in patch clamping this might not seem particularly important since dielectric noise should be larger than the rough predictions listed above. However, as techniques become available to fabricate pipettes from lower loss glass, such reductions would be expected to take on greater significance. First, Sylgard coating can significantly reduce pipette capacitance. However, there are some limitations to this reduction in the first few hundred microns behind the tip since it is difficult to build up a thick coat of Sylgard in this region. Thicker walled glass would also reduce pipette capacitance. The use of glass with a low dielectric constant can also reduce the noise from this mechanism; for an OD/ID ratio of approximately 1.43, quartz ($\varepsilon = 3.8$) should have a capacitance of about 0.6 pF/mm of immersion and 7760 should have about 0.7 pF/mm. It should be noted that root mean square noise from this mechanism in a given bandwidth is expected to decrease linearly with decreasing capacitance per unit length. Finally, it should be possible to reduce this noise significantly without respect to the depth of immersion by using a fine wire (platinized Ag–AgCl would be preferred in this case) protruding as close to the tip as possible. For example, a 100 μm diameter wire slightly sharpened at the tip should be able to be advanced to within

0.5–1 mm of the tip. This would in effect short out the resistance of the pipette in the region into which the wire protrudes; the reduction of impedance would be frequency dependent but should be quite effective by 1 kHz. This should reduce noise from distributed pipette resistance and capacitance even if the pipette were immersed further than the wire extended toward the tip: the wall capacitance would remain (with its dielectric noise) in regions where the wire and bath overlap, but the resistance — and its thermal voltage noise — would be greatly reduced.

In summary, it should be possible to reduce noise associated with the distributed pipette resistance and capacitance to roughly 0.03 pA rms in a 10-kHz bandwidth by careful control of pipette geometry, use of low dielectric constant glass, careful coating with Sylgard, and using a fine wire protruding as close to the tip as possible. A minimum depth of immersion is also desirable. Although these precautions are probably unnecessary at the present time, they could become important as other sources of noise are reduced. In particular, for very low loss glasses (e.g., quartz) this mechanism would exceed dielectric noise.

Noise Arising from Lumped Pipette Resistance and Patch Capacitance. The entire resistance of the pipette is in series with the capacitance of the patch membrane. This series combination will lead to a current noise with a PSD denoted by $i_{pc}^2(f)$ that should be given by

$$i_{pc}^2(f) = 4\pi^2 e_p^2 C_p^2 f^2 \qquad (3)$$

where e_p^2 is the PSD of the voltage noise of the pipette resistance, R_p ($e_p^2 \cong 4kTR_p$), and C_p is the patch capacitance. Typical values of R_p range from about 1 to 10 MΩ. The value of C_p has been measured by Sakmann and Neher[11] to fall in the range of 0.01 to 0.25 pF. They found that despite a large amount of scatter, C_p was correlated with R_p. As expected C_p increased as R_p decreased; they estimated C_p to typically be 0.126 pF($1/R + 0.018$), where R is the electrode resistance in megohms. This would imply that "typically" for R_p values of 10, 5, 2, and 1 MΩ, C_p would be 0.015, 0.027, 0.065, and 0.128 pF, respectively; the root mean square noise in a 10-kHz bandwidth (8-pole Bessel filter) would be 0.03, 0.04, 0.06, and 0.08 pA rms, respectively, that is, typical noise form this mechanism would increase as R_p decreased. From the results of Sakmann and Neher it can also be predicted that even less noise is sometimes generated from R_p and C_p; in the most favorable situations in terms of noise ($R_p = 2.5–3$ MΩ, $C_p \cong 0.01$ pF) this noise is only about 0.01 pA rms in a 10-kHz bandwidth.

[11] B. Sakmann and E. Neher, eds., "Single-Channel Recording," p. 37. Plenum, New York and London, 1983.

However, in the least favorable situations (when a large bleb of membrane has been drawn into the pipette, e.g., $R_p \cong 2$ MΩ, $C_p \cong 0.25$ pF) the noise from this mechanism can be as high as 0.22 pA rms in a 10-kHz bandwidth. Clearly, the lowest amount of R_p–C_p noise will result from situations in which relatively little membrane in drawn into the pipette. Pipette geometries favoring this situation may not be optimal in terms of the distributed pipette resistance and capacitance noise considered in the previous section.

Seal Noise

The final mechanism of noise associated with the pipette which we will consider here is the noise associated with the membrane–glass seal. The seal resistance will be denoted by R_{sh}. If the only noise associated with the seal was the thermal current noise of R_{sh}, then for zero applied field across the seal the noise PSD arising from the seal would be given by $4kT/R_{sh}$. In a 10-kHz bandwidth this would amount to 0.4, 0.13, 0.04, and 0.028 pA rms for seal resistances of 1, 10, 100, and 200 GΩ, respectively. It should be noted that we have with some cell types frequently achieved seal resistances of 50–200 GΩ or more, whereas with other cell types typical seal resistance can be as low as 2–10 GΩ. Exceptional noise performance can only be obtained with very high resistance seals.

The simple assumption that seal noise can be approximated by the thermal current noise of R_{sh} may well be incorrect. More generally, the noise at equilibrium will be defined by $4kTRe\{Y_{sh}\}$, where $Re\{Y_{sh}\}$ is the real part of the seal admittance. Because the precise nature of the membrane–glass seal is unknown, we do not know how to estimate Y_{sh}. Clearly, the assumption that $Re\{Y_{sh}\} = 1/R_{sh}$ is a minimum estimate of noise. It is certainly possible that the PSD of the noise of the seal rises with increasing frequency owing to the capacitance of the glass and particularly of the membrane which makes up the wall of the seal.

Seal resistances of 100–200 GΩ are not uncommon. Using macroscopic calculations with reasonable estimates of the area involved in the seal thus leads to estimates of the separation between the membrane and the glass of the order of 0.1 Å. Such a result only indicates that macroscopic calculations of this sort are inappropriate. We therefore have no clear idea of how to model the seal electrically or precisely predict its noise. Moreover, it seems virtually impossible to dissect out the seal noise from patch clamp measurements empirically. Previous attempts to do this certainly overestimate this noise. For example, data from Sachs and Neher reported in Sigworth[5] indicate that the seal would produce a noise of about 0.13 pA rms in a 5-kHz bandwidth (assuming a very sharp filter cutoff;

with a Bessel filter this estimate should increase to at least 0.16 pA rms), whereas we have on many occasions achieved less noise than this for the entire system (headstage, holder, pipette, seal, etc.) in the same bandwidth. Besides, their measurement of seal noise would have included the R_s-C_p noise described above. We can only put an upper bound on seal noise by a root mean square subtraction of all noise sources we can account for from the total measured noise. For our lowest noise patches to date this upper bound is about 0.05–0.06 pA rms in a 5-kHz bandwidth.

Summary of Noise Sources and Limits of Noise Performance

Headstage. At the time of this writing, the best capacitive feedback headstages have noise of about 0.18 pA rms in a 10-kHz bandwidth (8-pole Bessel filter). As already noted, however, it seems possible that with present technology this could be reduced to as little as 0.06–0.10 pA rms. At present the input voltage noise, e_n, of the best available headstages is about 2 nV/Hz$^{1/2}$. Reduction of this value is also possible and will be important in reducing the noise contributed by the holder and pipette.

Holder. By itself, a well-designed holder adds relatively little noise to patch clamp measurements provided that it is periodically cleaned and maintained free of pipette filling solution. A holder which is not covered with metallic shielding will typically add only 1–2 pF of capacitance to the headstage input. If the holder is fabricated from a low-loss dielectric material this capacitance in conjunction with the input voltage noise of the headstage will be the dominant source of noise associated with the holder. It can be expected to increase the noise by about 10% over that of the headstage alone. For a good capacitive feedback headstage with 0.18 pA rms noise in a 10-kHz bandwidth, the noise should not increase to more than about 0.20 pA rms by the addition of the holder. For a headstage with lower e_n than is presently available both the open-circuit headstage noise and the noise increment associated with the holder will decrease.

Of course the holder will also produce some dielectric noise. However, for low-loss dielectric materials this is not expected to be too severe. In the future, as other sources of noise decline, the construction of the holder (and input connector) may need to be reevaluated to reduce dielectric noise further.

Pipette. The first mechanism to consider in terms of pipette noise is simply the lumped capacitance of the pipette in series with the input voltage noise of the headstage. As was the case with the holder, this noise is perfectly correlated with noise arising from e_n and the headstage input capacitance, and therefore the usual rules of root mean square addition of uncorrelated noise sources do not apply. The pipette can add 1–5 pF of capacitance to the headstage input, with the lowest values being associated with low dielectric constant glass, thick walled pipettes, heavy Sylgard coating, and a shallow immersion of the tip into the bath. With proper precautions this source of noise need not increase total noise by more than about 5–10% above that of the headstage–holder combination.

Three other sources of noise associated with the pipette were also described above and are briefly summarized here.

Dielectric Noise. At present, dielectric noise is probably the dominant source of noise for all but the very best glasses available. All else being equal, this noise will be the smallest for glasses with the smallest product of their dissipation factor, D, and dielectric constant, ε. This noise will also be minimized by using relatively thick walled pipettes, shallow depths of immersion, and, particularly for relatively lossy glasses, a heavy coating of Sylgard. Of commonly used glasses, Corning 7760 has the smallest value of $D\varepsilon$ of 0.036. With 7760 it can be expected that a noise component of about 0.2 pA rms in a 10-kHz bandwidth will result from a 1.5–2 mm immersion with "standard" wall thickness glass and a moderate coat of Sylgard. For thicker walled glass and shallower depths of immersion this value could fall to 0.06–0.08 pA rms.

It may soon become possible to fabricate patch pipettes from very low-loss (high melting temperature) glasses, most notably quartz. For quartz, the $D\varepsilon$ product is approximately 0.0008, and dielectric noise would be expected to fall to about 0.03 pA rms for a 1 mm depth of immersion.

Distributed Pipette Resistance and Capacitance. The capacitance of the pipette is more or less evenly distributed over the length which is immersed in the bath. The pipette resistance is located primarily at or very near the tip, but significant resistance still resides in the first few millimeters behind the tip. The thermal voltage noise of this distributed resistance in conjunction with the distributed capacitance of the pipette is a potentially large source of noise. Rough calculations indicate that for a 1–2 mm depth of immersion this noise source should produce 0.05–0.13 pA rms noise in a 10-kHz bandwidth for pipettes with relatively ideal geometry. For less ideal pipettes (particularly ones with relatively higher resistance more distal to the tip) the noise contribution could easily be larger. This noise should be minimized by using thick walled glass with a low dielectric constant, relatively heavy Sylgard coating, and careful control of pipette geometry. Extending a slender Ag–AgCl or platinized Ag–AgCl wire as close as possible to the tip, thereby shorting out most of the resistance up to the end of the wire, could also reduce this source of noise.

R_p–C_p *Noise.* The capacitance of the patch membrane is in series with the entire pipette resistance. The thermal voltage noise of the pipette resistance produces current noise with the patch capacitance. This noise will generally be quite small, but it can become significant when the area of the patch is large. Values can be as little as 0.01 pA rms in a 10-kHz bandwidth but can exceed 0.2 pA rms in the same bandwidth for large patches. A value of 0.05 pA rms in a 10-kHz bandwidth could be considered to be "typical" for a pipette of 3–5 MΩ.

Seal Noise

The noise associated with the seal is probably the most poorly understood of all noise sources in the patch voltage clamp technique. We have argued that previous attempts to quantify this noise almost certainly overestimate its value, but it is well known to anyone who has spent much time trying to achieve low noise measurements that there is a large degree of variability in total noise from seemingly identical recording situations, even when the dc seal resistance is very high. Because this variability is too large to be readily accounted for on the basis of the quantifiable sources of noise described above, it seems reasonable to blame it on the seal.

The minimum noise PSD associated with the seal in the presence of zero applied voltage is given by $4kT/R_{sh}$, where R_{sh} is the dc seal resistance; the root mean square value in a bandwidth, B, is given by $(4kTB/R_{sh})^{1/2}$. For a 100-GΩ seal this would amount to 0.04 pA rms in a 10-kHz bandwidth. It seems likely that the actual seal noise is larger than this, but we have no good theoretical basis on which to estimate total seal noise. From our lowest noise patches (with $R_{sh} \geq 50$ GΩ) we can estimate that an upper bound for seal noise is about 0.05–0.06 pA rms in a 5-kHz bandwidth and 0.10–0.12 pA rms in a 10-kHz bandwidth. Of course similar estimates for higher noise patches would yield higher values.

Limits of Noise Performance

With a headstage with a noise of 0.18 pA rms in a 10-kHz bandwidth we can predict that the best total noise that can be achieved at present is about 0.25 pA rms (10 kHz, 8-pole Bessel filter). This estimate arises from assuming that the noise associated with the holder and the lumped capacitance of the pipette increases the noise to 0.20 pA rms and assumes the following values for other (uncorrelated) noise sources (all in a 10-kHz bandwidth): dielectric noise; 0.08 pA rms; distributed resistance-capacitance pipette noise; 0.07 pA rms; $R_s - C_p$ noise; 0.02 pA rms; seal noise, 0.10 pA rms. The total is thus $(0.20^2 + 0.08^2 + 0.07^2 + 0.02^2 + 0.10^2)^{1/2}$, or approximately 0.25 pA rms. This is in good agreement with our best results to date.

It is worth noting that the headstage itself is the dominant source of noise in this situation; if headstage noise were reduced to zero, the total noise would be expected to fall to about 0.15 pA rms. With improved techniques it should be possible to reduce the dielectric noise of the pipette to 0.03 pA rms and the noise from the distributed resistance and capacitance of the pipette to a similar value, both in a 10-kHz bandwidth. It is also possible that seal noise might be as low as 0.03–0.05 pA rms in this bandwidth; using 0.04 pA rms as an optimistic (probably overly optimistic) estimate, it seems possible that under ideal circumstances all noise not directly associated with the headstage might produce as little as 0.06 pA rms in a 10-kHz bandwidth. With a headstage that contributes 0.20 pA rms, such improvements might hardly seem worthwhile since best case total noise would only fall from 0.25 to 0.21 pA rms. However, with a headstage with a total noise contribution of only 0.08 pA rms, total noise could fall to as little as 0.10 pA rms in a 10-kHz bandwidth. Thus, at least to our minds, further efforts to reduce both the noise of the electronics and noise associated with the pipette seem worthwhile and could be expected to reduce noise levels by a factor of 2 or somewhat more below the best that can be achieved at the present time.

Whole Cell Voltage Clamping

In the whole-cell variant of the patch voltage clamp technique direct access to the cell interior is provided by disrupting the patch membrane after the formation of a gigohm seal. Disruption is accomplished by either applying additional suction to the pipette or by applying a brief high voltage pulse (e.g., 1 V) to the pipette. If the procedure is successful the gigohm seal remains and the interior of the patch pipette directly communicates with the interior of the cell. An alternative approach is the so-called perforated patch technique in which the pipette contains amphotericin or nystatin; these channels incorporate into the patch membrane and, over a period of some 10 to 20 min, provide a low access resistance from the pipette to the cell interior. With either approach it is not uncommon to find that the final access resistance is 2 or even 3 times as large as the original resistance of the pipette. After a sufficient period of time the access resistance of the perforated patch technique can become quite stable. However, it is often found that following disruption of the patch the access resistance is not stable; if this is the case the resistance will typically increase with time.

Most of the general characteristics of the dynamic and noise performance of the whole-cell configuration of the patch voltage clamp are now well known and will not be elaborated here. Instead, we focus our attention on the effects of the access or series resistance arising from the pipette. The access resistance associated with the pipette is in series with the membrane capacitance of the cell being voltage clamped. This has very important effects on voltage errors associated with the flow of transmembrane current, on the actual bandwidth of the current measurement, and on the noise associated with the whole-cell voltage clamp.

Dynamic Effects of Series Resistance

It is well known that the pipette resistance, R_s, causes a voltage error in the presence of transmembrane ionic current, i_m, with a magnitude given by $i_m R_s$. With an access resistance of 10 MΩ, an ionic current of 2 nA will result in a 20-mV error in the absence of series resistance compensation. Such errors are compounded when dealing with voltage-activated channels. For example, inward currents through voltage-dependent sodium channels will lead to depolarizing error voltages which in turn will activate more channels. Thus, the whole-cell voltage clamp technique is often inadequate to study cells with large, voltage-dependent ionic currents. A typical mammalian ventricular myocyte, for example, can have sodium currents of 10 nA or more, and with an R_s of 10 MΩ it is necessary to compensate for 90–95% of this resistance to achieve marginally adequate voltage control.

It is also well known that uncompensated series resistance in conjunction with membrane capacitance will cause the true transmembrane potential, V_m, to respond to a step change of command potential, V_c, with a time course given by $V_m = V_c[1 - \exp(-t/\tau_s)]$, where $\tau_s = R_s C_m$. For $R_s = 10$ MΩ and $C_m = 50$ pF, $\tau_s = 500$ μsec, and it will require a V_m of about 2.3 msec to settle to within 1% of its final value following a step change in V_c. The capacity transient (prior to use of "whole-cell compensation" provided on most commercial patch clamps) has the shape of the derivative of the membrane potential, V_m.

Aside from this delay in establishing the desired transmembrane potential, it is often assumed that when studying relatively small ionic currents (say a few hundred pA) in the whole-cell configuration of the patch clamp series resistance presents no major limitations. Unfortunately, this ignores another important effect of series resistance, namely, the filtering effect of series resistance and membrane capacitance on the measured current. In the absence of series resistance compensation, the measured current is effectively filtered by a one-pole (RC) filter with a corner (-3 dB) frequency given by $1/(2\pi R_s C_m)$. Whether this filtering is important, of course, depends on the highest frequency components of interest in the signal to be measured.

In extreme situations this bandwidth restriction can be quite severe. For example, for a large cell with $C_m = 200$ pF, an uncompensated series resistance of 10 MΩ will reduce the effective bandwidth of current measurement to only about 80 Hz. For a more or less typical situation with $R_s = 10$ MΩ and $C_m = 50$ pF, the effective bandwidth is about 320 Hz, which is still too small for many types of measurements. In a more ideal situation with a small cell with $C_m = 10$ pF and a relatively low access resistance of 5 MΩ, the effective bandwidth is increased to about 3.2 kHz. In any case, setting the bandwidth of the external filter used with the patch clamp to much more than the bandwidth limitation imposed by R_s and C_m will not provide significant additional information about the current signal. It will, however, provide extra noise (see below).

Although open-loop techniques such as "supercharging"[12] can speed up the response time of the membrane potential, the only way to increase the bandwidth limitation is to employ series resistance compensation. Such compensation will reduce the effective resistance in series with the membrane from R_s to R_{sr}, where R_{sr} is the residual (uncompensated) series resistance. For example, with $R_s = 10$ MΩ, 80% compensation means that R_{sr} is 2 MΩ. In this situation the bandwidth limitation is increased to $1/(2\pi R_{sr} C_m)$. Thus, using "typical" parameters of $R_s = 10$ MΩ and $C_m = 50$ pF, 70% series resistance compensation will increase the effective bandwidth to about 1060 Hz; 80% compensation increases this bandwidth to about 1600 Hz; 90% compensation increases this further to about 3.2 kHz.

It must be noted, however, that achievement of series resistance compensation of 90% or better in whole-cell voltage clamping is often difficult and sometimes impossible. Even if such levels can be achieved, the response is often no longer first order so that the simple bandwidth calculations presented above will no longer be strictly accurate. Moreover, most commercial patch clamps provide a "lag" control to be used with series resistance compensation. Use of lag puts the signal fed back to compensate for series resistance through a filter (usually first order). In effect, this means that series resistance is only compensated up to some bandwidth determined by the lag circuit. For a "10-μsec lag" this bandwidth is 16 kHz. However, for a 100-μsec lag the bandwidth of series resistance compensation is reduced to only 1.6 kHz. The overall bandwidth of current measurement achieved by series resistance compensation is, of course, affected by the use of lag. Clearly, if wide bandwidths are desired the lag control should be set to the minimum level necessary to achieve stability.

It should also be noted that when series resistance compensation is used it is important to use "fast capacity compensation" to eliminate currents involved in charging stray and pipette capacitance at the headstage input.

[12] R. H. Chou and C. M. Armstrong, *Biophys. J.* **52,** 133 (1987).

Because this capacitance is not associated with any significant series resistance its effects must be eliminated if series resistance compensation is to be stable.

Noise Associated with Series Resistance in Whole-Cell Voltage Clamps

At moderate to high frequencies the noise performance of the whole-cell variant of the patch voltage clamp technique is generally dominated by current noise arising from the voltage noise of the series (pipette) resistance, R_s, in conjunction with the cell membrane capacitance, C_m. The power spectral density (PSD) of the voltage noise of the pipette excluding $1/f$ noise is given by

$$e_s^2 = 4kTR_s \quad \mathrm{V^2/Hz}$$

where k is Boltzmann's constant and T is the absolute temperature. For $R_s = 10$ MΩ, e_s is about 400 nV/Hz$^{1/2}$ which is more than 100 times the input voltage noise of a high quality headstage ($2-3$ nV/Hz$^{1/2}$ at frequencies above 1 kHz or so). In addition, $1/f$ noise will be associated with the pipette, particularly when current is flowing through it.[13] Even for an R_s of 1 MΩ the noise of the electrode will be at least 126 nV/Hz$^{1/2}$. Because this noise is directly in series with the headstage input voltage noise it should be obvious that the voltage noise of the electronics is irrelevant to noise performance in whole-cell voltage clamping. As will be seen below, the noise of the feedback resistor (typically only about 500 MΩ for whole-cell clamping) is also generally irrelevant to overall noise for all bandwidths above a few hundred hertz.

The power spectral density of the current noise arising from the electrode voltage noise (again ignoring any $1/f$ component) and the whole-cell membrane capacitance is given by

$$\mathrm{PSD} = (4\pi^2 f^2 e_s^2 C_m^2)/(1 + 4\pi^2 f^2 \tau_{sr}^2) \quad \mathrm{A^2/Hz} \qquad (4)$$

where f is the frequency in hertz, $\tau_{sr} = R_{sr}C_m$, and R_{sr} is the residual (uncompensated) series resistance. If it were possible to compensate for 100% of series resistance then R_{sr} would be zero and Eq. (4) would simplify to $4\pi^2 f^2 e_s^2 C_m^2$. Because it is generally not possible to achieve series resistance compensation levels much above 90% without introducing excessive ringing into the capacitive current, the full form of Eq. (4) should generally be used. Note that the form of the power spectral density rises (as f^2) with increasing frequency and then plateaus at frequencies above $1/(2\pi\tau_{sr})$. The

---

[13] L. J. DeFelice and D. P. Firth, *IEEE Trans. Biomed. Eng.* **18**, 339 (1971).

high frequency asymptote is given by $e_s^2 C_m^2/\tau_{sr}^2$. As an example, consider a cell with $C_m = 50$ pF voltage clamped through a series resistance of 10 MΩ (recall that this is typical for pipettes with an original resistance in the range of 3–5 MΩ). Without any series resistance compensation the actual bandwidth of current measurement is limited to about 320 Hz $[1/(2\pi R_s C_m)]$. By a frequency of 100 Hz the current noise power spectral density has risen to 1.6×10^{-28} A²/Hz which is equivalent to the thermal current noise of a 100-MΩ resistor; indeed at any frequency above about 45 Hz the noise PSD from this mechanism will exceed that of a 500-MΩ resistor, which is the typical value used in commercial patch clamps for whole-cell measurements. At frequencies above 320 Hz the noise PSD approaches a steady level of 1.6×10^{-27} A²/Hz which is equivalent to the noise of a 10-MΩ resistor. An external filter must be used to roll off this noise; a cutoff frequency much larger than 320 Hz is not justified on the basis of the effective *signal* bandwidth limitation introduced by the series resistance.

When series resistance compensation is employed the value of τ_{sr} is decreased and the noise power spectral density continues to rise to higher frequencies. For example, with the same parameters just considered 90% series resistance compensation will produce an effective signal bandwidth of 3200 Hz, and the noise will continue to rise up to this frequency; by 1000 Hz the noise PSD reaches 1.6×10^{-26} A²/Hz (equivalent to the current noise of a 1-MΩ resistor), and the high frequency plateau will reach 1.6×10^{-25} A²/Hz (equivalent to the current noise PSD of a 100-kΩ resistor). With about 97% compensation (if this could be achieved) the effective signal bandwidth would increase to 10 kHz, and the high frequency asymptote of the noise would reach a level equivalent to the PSD of a 10-kΩ resistor (1.6×10^{-24} A²/Hz). Assuming that sufficient series resistance compensation is used to extend the effective signal bandwidth to somewhat more than the -3 dB bandwidth of an external 8-pole Bessel filter the noise expected from this mechanism alone will be somewhat more than 3 pA rms (about 18 pA peak to peak) for a filter cutoff frequency of 1 kHz. This would increase to 9 and about 36 pA rms (about 220 pA peak to peak) for filter bandwidths of 2 and 5 kHz, respectively. These values are equivalent to the root mean square current noise arising from resistances of about 1.5 MΩ, 400 kΩ, and 60 kΩ for the bandwidths 1, 2, and 5 kHz, respectively. It is obvious in this situation that the electronics will contribute only a very tiny fraction of the overall noise at any of these filter settings even if the feedback resistor were only 50 MΩ.

For any particular cell, that is, for any particular value of membrane capacitance, the only way to reduce the noise arising from the access resistance is to reduce the access resistance itself. In the specific example considered above the value of R_s was 10 MΩ. For a value of R_s of 2.5 MΩ with the same value of C_m (50 pF) the root mean square noise values listed above would be reduced by a factor of 2. Pipettes can be fabricated from several types of glass with tip geometries intended to minimize R_s (see [3], this volume). Unfortunately, however, in our experience, even with seemingly ideal pipettes it is relatively rare to establish stable whole-cell recordings with access resistances much below 2–3 MΩ. Thus when series resistance compensation is used to extend the signal bandwidth above $1/2\pi R_s C_m$ large levels of noise must be anticipated. It is important to note that series resistance compensation restores the high frequency components of both the signal and the associated noise. In a well-designed system the increase in noise is only the restoration of the noise that would have been present as a result of the series combination of the electrode voltage noise and cell capacitance in the absence of the filtering effects of R_s already described.

The specific example used above (i.e., R_s = 10 MΩ, C_m = 50 pF) may seem overly pessimistic. However, worse situations, in terms of both noise and the need for series resistance compensation, have been reported in the literature. For example, mammalian ventricular myocytes often have capacitances as large as 200–300 pF and have been clamped with access resistances of 10 MΩ or more. Nevertheless, it is worth briefly considering a more ideal situation, namely, a small cell with C_m = 5 pF and a relatively low value of R_s of 5 MΩ (e.g., chromaffin cells, see Marty and Neher[14]). The bandwidth limitation arising from C_m and completely uncompensated R_s is 6.4 kHz in this case, which is large enough for most whole-cell measurements. The power spectral density of the noise arising from R_s and C_m is, of course, much less than in the previous example. However, it will still exceed the PSD of the thermal current noise of a 500-MΩ resistor (typical for whole-cell voltage clamps) at all frequencies above about 640 Hz. At a frequency of 1 kHz the "R_s–C_m" noise is equivalent to the current noise PSD of a 200-MΩ resistor, and by about 5 kHz it has risen to a level equivalent to an 8-MΩ resistor. Thus, even in this situation it is obvious that except at rather narrow bandwidths (less than about 1 kHz) the noise of the feedback resistor will not dominate total noise. For a bandwidth of current measurement of 5 kHz, the root mean square noise arising from R_s and C_m (5 MΩ and 5 pF) will be at least 6 times greater than that of a 500-MΩ resistor in the same bandwidth, so that the total

[14] A. Marty and E. Neher, *in* "Single-Channel Recording" (B. Sakmann and E. Neher, eds.), p. 107. Plenum, New York and London, 1983.

noise would be reduced by only about 1% by completely eliminating the noise of the feedback resistor.

Filter

Filter characteristics are important in determining the amount of noise present in a given measurement and the resolution of the signal. Obviously, the bandwidth of a filter is adjusted to reduce noise to tolerable levels so that the desired signal can be adequately observed. Filtering prior to digitization is also required to prevent aliasing (see discussion in next section). When the desired signal is large relative to the background noise, the selection of filter bandwidth (and digitization rate) can be determined simply on the basis of the time resolution required in the measurement; wider bandwidths, of course, allow the observation of more rapid events. However, when the signal-to-noise ratio is relatively small, compromises between the amount of noise that is allowed and the time resolution achieved are often necessary. This second situation is usually the case for single-channel recordings.

The ideal filter would be one which possesses both a sharp cutoff in the frequency domain as well as a rapid smooth settling step response in the time domain. Unfortunately such a filter is theoretically impossible. There are well-defined limits on the degree to which a signal can be simultaneously "concentrated" in both the time domain and the frequency domain. Filters with narrow smooth impulse responses and therefore rapid rise times with minimal overshoot have rather gradual roll-offs in the frequency domain, whereas filters with a sharp cutoff in the frequency domain will be characterized (for the same -3 dB bandwidth) by more spread out impulse responses and a step response with a slower rise time and rather severe overshoot and ringing.

Of commonly used filter types, the Gaussian and Bessel filters provide the best resolution with minimum overshoot and ringing of the step response. The impulse response of a Gaussian filter has the shape of the Gaussian distribution. It has an essentially ideal step response with no overshoot; its 10–90% rise time is approximately $0.34/f_c$, where f_c (Hz) is the -3 dB bandwidth, and it settles to within 1% of its final value in about $0.8/f_c$ sec. Unfortunately, in the frequency domain the roll-off of a Gaussian filter is rather gradual. If the transfer function of a filter is denoted by $H(f)$, then for a Gaussian filter at $f = f_c$, $H(f) = 0.707$ (-3 dB); at $f = 2f_c$, $H(f) = 0.25$ (-12 dB); at $f = 3f_c$, $H(f) = 0.044$ (-27 dB); and at $f = 4f_c$, $H(f) = 0.004$ (-48 dB). An eighth order Bessel filter closely approximates the response of a Gaussian filter in both the time and the frequency domains; in fact as the order of the Bessel filter becomes large the two filters become essentially identical.

From the point of view of noise reduction at high frequencies, it would be desirable to have a filter with a transfer function that rolls off much more rapidly after it reaches f_c. A wide variety of analog filters with such characteristics are available (e.g., Chebyshev and elliptical filters). In fact, analog filters which have rolled off to $H(f)$) $= 0.01$ (-40 dB) by $f = 1.06 f_c$ are available, and digital filters can achieve even sharper cutoffs. Unfortunately such sharp cutoff filters have very undesirable characteristics in the time domain. In particular, for the same f_c their rise time will be longer than that of a Gaussian or Bessel filter, and their step response will have a large overshoot and prolonged ringing. For example, the step response of an eighth order Chebyshev filter (0.5 dB ripple) will have a 10–90% rise time of about $0.55/f_c$ (i.e., about 1.6 times that of a Gaussian or Bessel filter), a peak overshoot of about 23%, and would require more than $8/f_c$ sec to settle to within 1% of its final value. An eighth order Butterworth filter has frequency domain performance that can be thought of as lying between that of the Gaussian and Bessel types and the extremely sharp cutoff filters such as Elliptical and Chebyshev. Nevertheless, its time domain performance is rather poor, and these filters should also generally be avoided for most patch clamp studies.

To achieve their excellent performance in the frequency domain, sharp cutoff filters have sacrificed time domain performance. In fact, if we operationally define the time domain resolution of a filter as $1/T_r$, where T_r is the 10–90% rise time, it would be found that in order to achieve the same time resolution with a sharp cutoff filter as that achieved with a Gaussian or Bessel filter it is necessary to use a higher value of f_c for the sharp cutoff filter. In this case the sharp cutoff filter would generally pass as much or more noise than the Gaussian or Bessel filter if they have been set to achieve essentially the same time resolution. Thus, the presumed noise advantage of sharp cutoff filters is an illusion if the objective is to achieve the minimum noise for a given time resolution. Although some rather exotic filter types can provide the same rise time with minimal overshoot as the Gaussian or Bessel filter and reduce typical patch clamp noise by a small amount, the improvement is only a few percent with realistic noise power spectral densities.

Thus the Gaussian filter or a high order Bessel filter is the best choice for most patch clamp work. Although it is quite simple to produce a digital Gaussian filter, this type is more difficult to produce with analog electronics. Thus Bessel filters (fourth or preferably eighth order) are typically used with patch voltage clamps. There are many commercial sources of highly adjustable Bessel filters which are quite inexpensive. Several commercial patch clamps also provide Bessel filters with a few values of f_c as an integral part of the instrument.

Aliasing

The sampling theorem states that a signal can be completely determined by a set of regularly spaced samples at intervals of $T = 1/f_s$ (where f_s is the sampling frequency) only if it contains no components with a frequency greater than or equal to $f_s/2$. Here we will denote $f_s/2$ by f_n; f_n is often called the Nyquist or folding frequency. Another way of stating the sampling theorem is that, for data sampled at evenly spaced intervals, frequency is only defined over the range from 0 to f_n. Noting that at least two points per cycle are required to define a sine wave uniquely, it should be obvious that components with frequencies higher than f_n cannot be described. Of course, there is nothing to stop an experimenter from digitizing a signal (plus noise) with frequency components that extend far beyond f_n. If this is done, it is reasonable to wonder what will happen to those frequency components above f_n in the final digitized data. The answer is that higher frequency components "fold back" to produce "aliases" in the range of frequencies from 0 to f_n.

The Nyquist frequency, f_n, is also referred to as the folding frequency because the frequency axis of the signal plus noise power spectral density will fold around f_n in a manner similar to folding a carpenter's scale. Frequency components lying above f_n are shifted to frequencies below f_n. If the frequency of a signal or noise component above f_n is denoted by f_x, then the frequency of its alias, f_a ($0 \leq f_a \leq f_n$) is given by

$$f_a = |f_x - kf_s|$$

Where f_s is the sampling frequency, k is a positive integer which takes on whatever value required so that f_a will fall into the frequency range from 0 to f_n (recall $f_n = f_s/2$), and the vertical bars indicate absolute value. For example, with $f_s = 10$ kHz ($f_n = 5$ kHz), a frequency component at 19 kHz will alias to a component at 1 kHz ($f_a = |19$ kHz $- 2 \times 10$ kHz$| = 1$ kHz). Similarly, frequency components at 9, 11, 19, 21, 29, 31, 39 kHz, etc., will all produce aliases at 1 kHz in the sampled data. Antialiasing filters are used when digitizing data to eliminate such aliases by attenuating the amplitude of all frequency components of the signal (plus noise) to negligible levels at frequencies above f_n.

We now present two examples of aliasing to show the kinds of problems it can create. First, consider noise with a white (i.e., constant) power spectral density of 10^{-14} V^2/Hz (100 nV/Hz$^{1/2}$) extending from dc to a sharp cutoff at 1 MHz. The total noise in a 1-MHz bandwidth is 100 μV rms or about 0.6 mV peak to peak. If this noise were sampled at a rate of 1 point per 100 μsec (f_s = 10 kHz) without the use of an antialiasing filter, a little reflection should indicate that the root mean square (or peak-to-peak) value of the sampled points will be the same as it was in the original data, namely, 100 μV rms. However, the sampled data cannot describe frequency components that are greater than 5 kHz, that is, greater than f_n. In fact, if a smooth curve were fitted through the sampled points it would be found that the noise process appeared to be band limited from dc to 5 kHz but that its amplitude was the same as in the original data. The PSD of the sampled data will have increased to 2×10^{-12} V^2/Hz (1.4 μV/Hz$^{1/2}$), which is 200 times greater than that of the original data, because the act of sampling at 10 kHz had folded over the original PSD 200 times; all frequency components above f_n (5 kHz) have been aliased into the range from dc to 5 kHz.

It is important to note that the effects of aliasing cannot be undone by any subsequent digital operations on the digitized and aliased waveform. For example, subsequent digital filtering of the sampled waveform with a cutoff frequency of 1 kHz will only reduce the noise-to-amplitude ratio to about 45 μV rms. On the other hand, passing the original data through an analog filter with an f_c value 1 kHz would have reduced its amplitude to 3.16 μV rms. There are two possible solutions to this problem: either sample the data at a much higher rate (here at least 2 MHz), or, if the original 10-kHz sample rate is desired, use an analog antialiasing filter prior to sampling which will reduce the amplitude of all frequency components above 5 kHz to an acceptably small level.

In the patch voltage clamp at frequencies above a few kilohertz the PSD of the noise is not flat, but instead rises with increasing frequency, eventually approximately as f^2. In this situation the consequences of aliasing can be even worse than those considered in the previous example. Consider a voltage noise source of 10 nV/Hz$^{1/2}$ in series with a capacitance of 10 pF. In a 100-kHz bandwidth this would result in a current noise of 11.4 pA rms; assume that 100 kHz is the highest frequency of this current noise process. Once again assume that this noise is sampled at a rate of 10 kHz with no antialiasing filter. This will mean that all of the noise above 5 kHz will be aliased into the frequency range from dc to 5 kHz, that is, the sampled data will still have an amplitude of 11.4 pA rms (and an altered PSD), even though the original noise process would only have had an amplitude of about 0.13 pA rms in a bandwidth from dc to 5 kHz. In addition, subsequent digital filtering of the sampled noise with a cutoff frequency of 1 kHz would only reduce its amplitude to about 5 pA rms. However, analog filtering of the original noise with a filter cutoff frequency of 1 kHz would have reduced the noise to about 0.011 pA rms, that is, more than 400 times less than achieved by digital filtering of the aliased digitized waveform. Once again, much faster sampling or the use of an appropriate antialiasing filter is the solution to this problem.

In the examples presented above it has been assumed that the filter used had a very sharp cutoff beyond its -3 dB bandwidth. As already described, such filters have very undesirable time domain characteristics. When using Gaussian or Bessel filters, which have much more gradual roll-offs beyond f_c, as antialiasing filters it is not appropriate to set f_c equal to f_n. The requirement to avoid significant aliasing and preserve the shape of the original PSD in the frequency range from 0 to f_n is that all frequency components higher than f_n must be adequately attenuated. When using a 4- or 8-pole Bessel filter for antialiasing, the selection of the cutoff frequency, f_c, relative to f_n should take into account the spectral characteristics of the noise as well as just how much aliasing can be tolerated. We advise that at most $f_c < 0.5f_n$ $(0.25f_s)$; we typically use $f_c \simeq (0.2 - 0.4)f_n$.

Cascaded Filters

Care must be taken when two "filters" are used in series. For two Bessel filters, the composite frequency f_c is approximated by

$$1/f_c^2 = 1/f_1^2 + 1/f_2^2$$

This relationship is a reasonable approximation for other filters whose roll-offs are not very steep. Therefore, a resistive headstage patch clamp with a 20-kHz inherent bandwidth filtered by a 10-kHz Bessel filter will result in a bandwidth of about 8.9 kHz. That same patch clamp when recorded by a tape recorder through the 10-kHz Bessel filter and then replayed through a 5-kHz Bessel filter would have a final bandwidth of about 4.4 kHz. It is important not to overlook this result of cascading filters. This problem is a little less severe with integrating patch clamps owing to their high inherent bandwidth, which can easily be 50–100 kHz. A 100-kHz patch clamp filtered through a 10-kHz Bessel filter will have a final bandwidth of about 9.95 kHz, very close to that of the filter itself.

Tape Recorder

For the measurement of steady-state single-channel currents, it is often convenient to use a tape recorder. Up until the last few years, this required instrumentation-type tape recorders that could be very expensive and have quite limited bandwidth. If they had bandwidth, they would go through a great deal of tape to achieve that bandwidth since the tape had to be moved by the recording heads at high speed.

Bezanilla[15] described a technique for using a digital audio processor (DAP) and a standard high-fidelity video tape recorder (home electronics) to do a high quality analog recording. These devices have become so popular that they can now be obtained commercially from a number of different manufacturers. They, in general, sample at 44 kHz, can with some kinds of electronic filtration support up to 20 kHz of continuous bandwidth, and contain from 2 to 8 channels. For the Bessel or Gaussian filters used for patch and whole-cell clamping (see section on filters above); one must sample at 4–5 times the corner frequency in order to eliminate aliasing and to produce data optimal for single-channel analysis. Therefore, the data to the tape recorder must be filtered through a Bessel filter of not more than 9–10 kHz. If measurements are to be made in the frequency domain, sharp roll-off elliptical filters of up to 20 kHz can be supported by the 44-kHz sampling rate of the recorder. These recorders are generally 14- to 16-bit digital devices that use standard video cassettes for their recording, each cassette holding up to 1.5 gigabytes of 16-bit data. For channel current recordings that last up to several minutes, this may be the instrumentation of choice. This technology is now being challenged by computer-based systems with "tape recorder" software (see, e.g., Axotape from Axon Instruments, Foster City, CA).

Computer

Many kinds of patch clamp recording are simply not possible to do adequately without computer control. Many, if not most, channels show transient behavior on switching from one voltage to another. These transients are often finished within 1 to 100 msec after a voltage step, and so this behavior is lost on systems that are tape recorder based and look at only steady-state behavior. To measure this transient behavior, it is necessary that the voltage steps be applied to the patch clamp from digital-to-analog converter hardware located in a computer and that essentially simultaneously the currents coming from the patch clamp be digitized by an analog-to-digital converter also residing in the computer. Many computer systems with real-time interface hardware aimed specifically at this task are now commercially available. The majority of the systems utilize either PC/AT or Macintosh compatible hardware.

One of the newest and most capable systems runs on an Atari microcomputer. Many of these systems use computer graphics hardware to allow

15 F. Bezanilla, *Biophys. J.* **47**, 437 (1985).

whole-cell currents and single-channel currents to be visualized in essentially real time on the screen of the computer terminal. Some of these systems are beginning to approach the resolution of analog and digital oscilloscopes. The majority of the systems have several channels of analog-to-digital converters and several channels of digital-to-analog converters used to sense the configurations of the patch clamps and to deliver control signals to the electronics, respectively. Some patch clamp manufacturers have implemented "sender" outputs so that the position of key switches like gain switches, filter bandwidth switches, and configuration switches can be read by the computer and stored with the data files being collected. Needless to say, it has become impossible to do state-of-the-art whole-cell recording and single-channel recordings without a system allowing on-line computer support.

Several of the companies that provide hardware support also provide quite complete software packages for the analysis of both whole-cell currents and single-channel currents collected through the use of their interface and computer systems. Consequently, a user getting into patch clamping today has far less to do to be up to speed technically with the field than was necessary a few years ago. These systems all contain software so that raw data files can be reduced, placed in attractive plot format, and plotted on a laser printer to often produce figures that are of publication quality. Anyone getting into patch clamping in the serious way will want to consider the purchase of one or more of these computer hardware and software systems.

Oscilloscope

In addition to the computer display, it is often useful to have a standard oscilloscope connected to the patch clamp output. These instruments provide much greater resolution than computer displays and allow easy change of gain and sweep speeds. One difficulty with standard oscilloscopes is that the screen persistence is short, and so channel currents appear briefly and are not retained. Within the last few years it has become possible to purchase quite inexpensive, digital oscilloscopes that utilize analog-to-digital converters and digital-to-analog converters and internal memory to produce a storage capability not possible with technology that simply alters the screen persistence. Virtually all major oscilloscope manufacturers now produce inexpensive digital oscilloscopes. These are particularly valuable for patch clamp setups because they can at one moment be operated in analog mode and used for all of the electronic purposes that oscilloscopes are good for and then with the push of a single front panel control provide the digitally derived persistence necessary for optimal viewing of single-channel currents.

Summary

It should be obvious that there are many ways to construct clamp setups that are either equivalent or sufficient for the experiments planned. The hardware and electronics can be obtained from several manufacturers, as can analysis software. What we have presented here are guidelines primarily meant to point a new experimenter in the right direction and, we hope, to guide more experienced investigators toward techniques that can improve the resolution of their measurements.

[3] Glass Technology for Patch Clamp Electrodes

By JAMES L. RAE and RICHARD A. LEVIS

Introduction

In the simplest sense, a patch clamp electrode is just a fluid bridge of proper geometry to connect a reference electrode to the surface or interior of a cell. The glass envelope which accomplishes this is a passive component of the overall circuit which records currents and applies voltages, yet the properties of the glass electrode can be an important determinant of the quality of the recordings.

Several properties of glasses are important when trying to construct effective electrodes for patch clamping. Thermal properties dictate how easily desired tip shapes can be produced and determine the extent to which the tips can be heat polished. Optical properties determine if the tip can be heat polished to a visually distinct end point. Electrical properties determine the noise the glass produces in a recording situation and determine the size and number of components in the capacity transient following a change of potential across the pipette wall. Noise and capacitance properties are correlated. Good electrical glasses minimize both. Finally, glasses are complex substances composed of many compounds (see Table II). Glass composition may influence how easily a glass seals to membranes but may also yield compounds that can leach into the pipette filling solution to inhibit, activate, or block channel currents.

In this chapter, we expand the present literature concerning patch clamp electrode technology[1-6] by discussing practical issues about glasses that affect the quality of patch clamp recordings. Those glass properties that optimize single-channel recordings may not optimize whole-cell recordings or may be irrelevant to them and vice versa. It is a reasonable assertion that regardless of the type of patch clamping being done, the use of glasses with good electrical properties is desirable.

Overview of Patch Electrode Fabrication

There are a wide variety of glasses available for patch clamping. Garner Glass (Claremont, CA) has been particularly instrumental in supplying specialty glasses for this purpose. Glass tubing of whatever variety selected for the fabrication of patch electrodes should have walls of substantial thickness (0.2 to 0.3 mm). Thick walls result in decreased electrical noise and increased bluntness at the tip, which prevents penetration of the cell during seal formation. Most investigators use glass tubing with a 1.5 to 2.0 mm outside diameter and a 1.15 to 1.2 mm inside diameter. With an inside diameter this large, it is possible to utilize commercially available 1 mm Ag/AgCl pellets (In-Vivo Metrics, Healdsburg, CA; E. W. Wright, New Haven, CT) which will easily fit into the back of the electrode. For smaller inside diameters, one is constrained to use smaller chlorided wire as the internal reference electrode. These electrodes can be constructed easily from small diameter silver wire which has been rendered free of oxide on its surface by use of fine sandpaper. By immersing this cleaned tip into Clorox bleach for about 20 min, a substantial coating of AgCl can be formed to produce a very good Ag/AgCl internal reference electrode. In either case, any nonchlorided silver wire associated with the electrode should be isolated from the filling solution. This is usually done by surrounding it with a small Teflon tube filled with either Sylgard or epoxy which is subsequently cured to encapsulate the wire.

It is probably a good general idea to clean the glass tubing before using

[1] O. P. Hammill, A. Marty, E. Neher, B. Sakmann, and F. J. Sigworth, *Pflugers Arch.* **391**, 85 (1981).
[2] B. Sakmann and E. Neher, *in* "Single-Channel Recording" (B. Sakmann and E. Neher, eds.), p. 37. Plenum, New York and London, 1983.
[3] D. P. Corey and C. F. Stevens, *in* "Single-Channel Recording" (B. Sakmann and E. Neher, eds.), p. 53. Plenum, New York and London, 1983.
[4] J. L. Rae and R. A. Levis, *Biophys. J.* **45**, 144 (1984).
[5] J. L. Rae and R. A. Levis, *Mol. Physiol.* **6**, 115 (1984).
[6] J. L. Rae, R. A. Levis, and R. S. Eisenberg, *in* "Ion Channels" (T. Narahashi, ed.), p. 283. Plenum, New York and London, 1988.

it to make patch electrodes. In our experience, this is often unnecessary, but at other times we have found it imperative to clean the glass for the best noise performance. Sonicating the glass in 100% ethanol in an ultrasonic cleaner is often effective for this purpose. Another approach suggested in the glass literature[7] is to etch the glass for 10 min with a 1% sodium hydroxide solution at 95°. This is followed by a 2-min cleaning with 5% hydrochloric acid at 50° and then meticulous washing with distilled water. We recommend that all of this be done in an ultrasonic cleaner to assure agitation and movement of the solution inside the glass tubing. Following any cleaning procedure, place the glass in an oven at around 200° for 10 to 30 min to achieve complete drying. Heat treatment of this sort has also proved necessary if low noise recordings are to occur in environments where the humidity is exceptionally high. This is a good idea in high humidity even for glass tubing that either has not been previously cleaned or has been meticulously cleaned at some time in the past.

Patch electrodes require much blunter tips than standard intracellular microelectrodes, and it is usually not possible to pull them adequately on single-stage electrode pullers. Many laboratories have modified standard vertical electrode pullers so that they pull in multiple stages. This modification involves placing stops in the pulling apparatus. These stops can be simple metal or wooden blocks which stop the puller movement after it has experienced a displacement of a few millimeters. One of the puller clamps is then loosened and the glass is moved so that the now hourglasslike tapered region is repositioned near the filament. The glass is then repulled. This stopping and repositioning can occur several times before the pull is allowed to separate the two pieces of electrode glass. With this approach, it is possible to produce quite different tip tapers. Pulling electrodes, however, has become very much easier with the advent of microprocessor-driven microelectrode pullers like the model P-87/PC from Sutter Instruments (Novato, CA). Similar pullers are also made by a number of other manufacturers. With these pullers, it is possible to implement very complicated multistage pulls of electrode glass and to store all of the parameters required in memory. These pullers allow multiple programs to be stored, and consequently it is possible with the push of a button or two to set up the puller to pull optimally almost any kind of glass. The Sutter puller is particularly versatile because it contains a solenoid valve that allows gating of a burst of gas to cool the filament rapidly. This feature gives one a great

[7] L. D. Pye, H. J. Stevens, and W. C. LaCourse, "Introduction to Glass Science," p. 513. Plenum, New York and London, 1972.

deal of control over the final taper of the tips, which is very important for patch clamp recordings.

For the lowest noise recordings, electrodes must be coated with a hydrophobic material to within 100 μm or less of their tip. This prevents bathing solution from creeping up the wall of the electrode and thus limits what would be substantial noise source. A commonly used compound is Sylgard 184 from Dow Corning (Midland, MI). This compound also has exceptional electrical properties (see Table I) and so improves the electrical properties of the glass when it is painted on the glass surface. Sylgard, meticulously mixed until it is frothy with bubbles, can be stored at simple freezer temperatures in small capped centrifuge tubes. The thorough mixing is very important because pockets of the compound not adequately exposed to polymerizer can flow to the electrode tip (even against gravity) and render the tips difficult to seal to cells. When handled in this way, the Sylgard can be stored for several weeks. A tube of this freezer-stored Sylgard, when brought to room temperature for use in painting electrodes, will last for several hours before it begins to polymerize.

The Sylgard is painted on the electrode tip using a small utensil like a piece of capillary tubing pulled to a reasonably fine tip in a flame. This painting can be done using magnifications available with standard dissecting microscopes. It is useful, but not required, to modify the dissecting microscope to work in dark field. This can be done fairly inexpensively by purchasing a fiber optic ring illuminator that can be connected to a standard fiber optic light source. At a location of 3 to 4 inches above the ring illuminator placed on the stage of the microscope, dark-field illumination is achieved, and the walls of the electrode glass show up as bright lines of light. The location of the Sylgard being painted and the tip of the electrode can be very easily discerned with this dark-field illumination. It is important that the Sylgard be directed away from the tip by gravity at all times during the painting procedure. Otherwise, the Sylgard will flow into the tip and make fire polishing and/or sealing impossible. The Sylgard can be cured by holding the tip for 5 to 10 sec in the hot air stream coming from a standard heat gun of the variety used in electronics. Again, the Sylgard must be gravitationally directed away from the tip during the curing process.

Finally, to promote gigohm seals and to reduce the possibility of tip penetration into the cell during seal formation, electrode tips should be fire polished. In some cells, fire polishing has proved unnecessary, but we have found, as a general rule, that sealing of difficult to seal cells is promoted by fire polishing the electrode tip. Fire polishing can be done either using an upright or an inverted microscope. In fact, many investigators have chosen to Sylgard coat their pipettes and fire polish them using an inverted microscope with a $40\times$ or so long working distance objective. Another very useful approach is to utilize a standard upright microscope converted to the 210 mm tube length that is standard for metallurgical microscopes and objectives. Several microscope companies, but particularly Nikon (Garden City, NY), make extralong working distance and superlong working distance high-magnification metallurgical objectives. Most noteworthy is the $100\times$ ELWD or $100\times$ SLWD objectives that have 1 and 2 mm working distances, respectively. With these objectives and $15\times$ eyepieces and with the electrode mounted on a slide held in the mechanical stage of the microscope, it is possible to move the electrode tip into the optical field and visualize directly the electrode tip at $1500\times$ magnification (Fig. 1).

At such high magnifications, it is possible to fire polish the tip to a very distinct optical end point under direct visualization. This approach ensures very repeatable electrodes from one electrode to the next. The fire polishing itself is accomplished by connecting to a micromanipulator a rod of inert material to which has been fastened a short loop of platinum iridium wire. The ends of this wire must be soldered to two other pieces of wire that can be connected to a voltage or current source to allow current to be passed through the platinum wire. The platinum loop is generally bent into a very fine hairpin so that it can be brought to within a few microns of the electrode tip under direct observation. Because of early reports that platinum can be sputtered from the wire onto the electrode tip and prevent sealing, the platinum wire is generally coated with a glass like Pyrex (Corning 7740) or Corning 7052 to prevent such sputtering. This is done by overheating the platinum wire and pushing against it a piece of electrode glass that has been pulled into an electrode tip. At high temperatures, the glass melts and flows over the platinum wire and ends up thoroughly coating it and forming a distinct bead of glass. With an arrangement like this, it is possible to fire polish electrode tips very precisely (see Fig. 2).

If the Sylgard has been coated too near the tip, fire polishing causes the tip to droop downward at the juncture where the Sylgard coating ends. If one desires to paint Sylgard extremely close to the tip, it may be necessary to do most of the fire polishing before Sylgard coating and then to fire polish lightly again after Sylgard coating.

Electrode Properties for Single-Channel versus Whole-Cell Recording

Electrodes for patch and whole-cell recordings have some properties in common but other properties which can be very different. First, the noise of the electrode is very much more important in single-channel recordings than in whole-cell recordings. In whole-cell recordings, the dominant noise

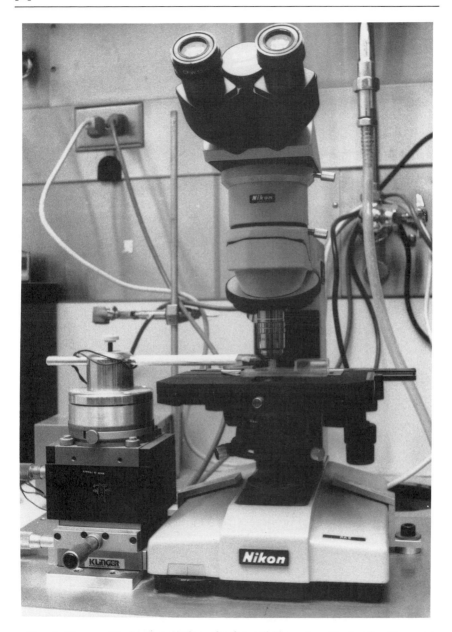

FIG. 1. One possible implementation of a fire polishing setup. The heating filament is attached to the micromanipulator. The electrode rests in a groove cut in a thick acrylic plastic microscope slide and is moved by the mechanical stage of the microscope. The objective is 100× metallurgical and the eyepieces 15×.

source at moderate to high frequencies comes from the resistance of the electrode in series with the capacitance of the entire cell, and so electrode noise is relatively less important. On the other hand, the resistance of a whole-cell electrode should be as low as possible (a few megohms at most) to minimize both dynamic errors associated with series resistance and noise. This is not a requirement for single-channel recording, nor does higher electrode resistance result in much additional noise there until the electrodes become some tens of megohms in resistance. In either single-channel recordings or whole-cell recordings, it is necessary that capacity currents which flow during voltage steps be sufficiently small and simple in time course that they can be corrected by simple circuitry in the patch clamp. In addition, both kinds of electrodes must be made of glasses which do not leach compounds from their walls that can alter the currents being measured from the particular channels of interest. The particular requirements for these two kinds of recording require also that the electrodes be pulled somewhat differently to optimize their use in the particular recording configuration being used.

Types of Glasses and Their Properties

There are several ways in which patch clamp glasses can be classified. One is on the basis of the temperature at which they soften. Another is based on their electrical properties. A third and perhaps more common way is on the basis of their major chemical constituents. Many of these properties are itemized in specification sheets from the manufacturer, and so it is often possible to choose glasses which should be effective for patch clamping just from examining their specifications.

In Table I, we list the properties of a number of glasses that have been used for patch clamping. We also list the properties of quartz and Sylgard because of their relevance to the issues discussed in this chapter. The glasses are listed in increasing order of loss factor times dielectric constant (see section on electrical properties below). We also somewhat arbitrarily classify them into four categories based on their primary chemical content: soda lime, high lead, borosilicate, or aluminosilicate.

Several important points can be noted from Table I. The first is that the electrical properties and the thermal properties of the glasses bear no obligatory relationship. It has been a misconception among some biophysicists that "soft" glasses, those that soften at relatively low temperatures, are poor glasses electrically, whereas "hard" glasses, those that soften at relatively high temperatures, are good glasses electrically. A comparison of 8161, a very soft glass, and 7760, a medium hard glass, quickly dispels that notion. Both glasses have very low loss factors yet soften at substan-

TABLE I
ELECTRICAL AND THERMAL PROPERTIES OF GLASSES

Glass	Loss factor (LF)	Log vol. r.	Dielectric constant (DC)	LF × DC	Softening temperature (°C)	Description
7940	0.0038	11.8	3.8	0.01444	1580	Quartz (fused silica)
1724	0.0066[a]	13.8	6.6	0.04365	926	Aluminosilicate
7070	0.25	11.2	4.1	1.025	—	Low loss borosilicate
Sylgard	0.58	13.0	2.9	1.682	—	#184 Coating compound
7059	0.584	13.1	5.8	3.387	844	Barium borosilicate
7760	0.79	9.4	4.5	3.555	780	Borosilicate
8161	0.50	12.0	8.3	4.15	604	High lead
7040	1.00	9.6	4.8	4.8	700	Kovar seal borosilicate
0120	0.80	10.1	6.7	5.36	630	High lead
EG-6	0.80	9.6	7.0	5.6	625	High lead
7720	1.30	8.8	4.7	6.11	755	Tungsten seal borosilicate
1723	1.00	13.5	6.3	6.3	910	Aluminosilicate
7052	1.30	9.2	4.9	6.37	710	Kovar seal borosilicate
EN-1	1.30	9.0	5.1	6.63	716	Kovar seal borosilicate
KG-12	1.00	9.9	6.7	6.7	632	High lead
0010	1.07	8.9	6.7	7.169	625	High lead
3320	1.50	8.6	4.9	7.35	780	Tungsten seal borosilicate
7050	1.60	8.8	4.9	7.84	705	Series seal borosilicate
7056	1.50	10.2	5.7	8.55	720	Kovar seal borosilicate
EG-16	0.90	11.3	9.6	8.64	580	High lead
KG-33	2.20	7.9	4.6	10.12	827	Kimax borosilicate
7740	2.60	8.1	5.1	13.26	820	Pyrex borosilicate
1720	2.70	11.4	7.2	19.44	915	Aluminosilicate
N-51A	3.70	7.2	5.9	21.83	785	Borosilicate
R-6	5.10	6.6	7.3	37.23	700	Soda lime
0080	6.50	6.4	7.2	46.8	695	Soda lime

[a] We question the loss factor given for 1724. It seems to be too low.

tially different temperatures. An even more dramatic comparison is that of KG-12, a high lead glass, with 1723, an aluminosilicate glass. They have the same low loss factor and yet soften at temperatures which differ by almost 300°. The high lead glasses which soften at the lowest temperatures of any glasses included in Table I, have, as a group, the lowest loss factors.

A second significant point is that Sylgard, a coating compound commonly used to paint patch electrodes, has better electrical properties than most glasses shown in Table I. It is therefore not surprising that placing a heavy Sylgard coating on pipettes fabricated from many glasses improves their electrical properties. The wall of the electrode ends up having properties intermediate between those of its glass and those of Sylgard. It is also

TABLE II

COMPOSITION OF GLASSES

Glass	SiO$_2$	B$_2$O$_3$	Al$_2$O$_3$	Fe$_2$O$_3$	PbO	BaO	CaO	MgO	Na$_2$O	K$_2$O	Li$_2$O	As$_2$O$_3$	Sb$_2$O$_3$	S(
1724	NA[a]	NA	NA	NA	NA	NA	NA	NA	NA	NA	NA	NA	NA	N.
7070	70.7	24.6	1.9	—	—	0.2	0.8	0.8	—	—	0.56	—	—	—
8161	38.7	—	0.2	—	51.4	2.0	0.3	0.04	0.2	6.6	—	0.04	0.38	—
7059	50.3	13.9	10.4	—	—	25	—	—	0.08	—	—	—	—	—
7760	78.4	14.5	1.7	—	—	—	0.1	0.1	2.7	1.5	—	0.18	—	—
EG-6	54.1	—	1.0	3.9	27.1	—	0.1	0.1	3.4	9.2	—	0.2	—	—
0120	55.8	—	—	0.03	29.5	—	0.25	—	3.6	8.9	—	0.4	—	—
EG-16	34.8	—	0.3	—	58.8	—	0.05	0.05	0.1	5.5	—	0.2	0.3	—
7040	66.1	23.8	2.9	—	—	—	0.1	0.1	4.1	2.7	—	0.1	—	—
KG-12	56.5	—	1.5	—	28.95	—	0.1	0.1	3.7	8.6	—	0.4	0.25	—
1723	57.0	4.0	16.0	—	—	6.0	10.0	7.0	—	—	—	—	—	—
0010	61.1	—	—	—	22.5	—	0.3	0.1	7.2	7.3	—	—	—	—
7052	65.0	18.3	7.4	—	—	2.7	0.2	0.1	2.4	2.9	0.6	—	—	—
EN-1	65.0	18.0	7.6	—	0.01	2.7	0.1	0.1	2.3	3.2	0.6	—	—	—
7720	71.4	15.2	2.0	—	6.1	0.3	0.2	0.1	3.7	0.3	—	—	0.5	—
7056	69.0	17.3	3.9	—	—	—	0.12	—	0.91	7.5	0.68	0.48	—	—
3320	75.3	14.3	—	—	—	—	0.1	0.1	4.0	—	—	—	0.8	—
7050	67.6	23.0	3.2	—	—	0.1	0.1	0.1	5.1	0.2	—	—	—	—
KG-33	80.4	12.9	2.6	—	0.005	—	0.05	—	4.0	0.05	—	—	—	—
7740	80.4	13.0	2.1	—	—	—	0.1	0.1	4.1	—	—	—	—	—
1720	62.0	5.3	17.0	—	—	—	8.0	7.0	1.0	—	—	—	—	—
N51-A	72.3	9.9	7.3	—	0.02	—	0.9	0.05	6.5	0.7	—	0.02	—	—
R-6	67.7	1.5	2.8	—	—	2.0	5.7	3.9	15.6	0.6	—	—	—	0.2
0080	73.0	0.04	—	—	—	0.1	4.8	3.2	16.8	0.4	—	—	—	0.2

[a] NA, not available.

expected and can be shown experimentally that poor electrical glasses are helped more by Sylgard coating than are good glasses. A third notable point is that fused silica (quartz) has substantially better electrical properties than any glass that has been used to date and so offers a potential way to further reduce patch clamp noise when a reliable method is available to fabricate patch clamp electrodes from it. Corning glasses 1724 and 7059 would also appear to be very good glasses electrically, but we are unaware of reports of their use to date for low noise patch clamp recordings. Corning 7070, low loss electrical, appears to be an excellent glass electrically, but to date no way has been found to pull electrodes from it. The glass changes its properties when it is heated.

Table II shows the chemical constituents of many of the same glasses shown in Table I. We do not presently have a way to predict which of these glasses will be useful for patch clamping simply based on these constituents, but Table II may be useful in deciding which glasses have a high probability of containing leachable components that might affect channel currents.

A number of interesting observations can be made about Table II. A small number of the glasses contain antimony compounds. Corning 3320 is most notable in this regard, but substantial content is also found in 8161, a highly utilized patch clamp glass. The majority of the high lead glasses and 7760, a borosilicate glass, contain arsenic compounds. These particular glasses are noteworthy because of their low noise properties. Yet other glasses contain barium compounds. This is true of 8161, 1723, and 7052, three glasses that have found considerable use for patch clamping. Corning 7059, a barium borosilicate glass of extremely good electrical properties, contains very much more barium than other kinds of glass. Notice also that high lead glasses do not contain the boron compounds found at high concentrations in most other glasses. These glasses make up for this lack by having exceptionally high contents of lead compounds. Corning 8161 and EG-16 are most notable in this regard, each having PbO_2 amounting to more than 50% of the total composition. The high lead glasses not only have elevated lead compounds but also reduced levels of SiO_2.

The kind of information shown in Table II is not made available by companies that manufacture these glasses but rather comes from direct assay of the glasses. The variability in the composition of the glasses from one batch to another is unknown, and in several cases the percentages shown here do not add up to 100%. Therefore, Table II cannot be considered to be a highly accurate assessment of the composition of these glasses but is presented only as a guideline. It is the best information that we could obtain, but we are sure that there are many other trace compounds that exist in these glasses that do not show up in Table II. It seems unlikely that one will ever know all of these trace compounds since, in general, the material composition is considered proprietary information by the glass industry.

Thermal Properties

It is clear from experience that glasses which soften at lower temperatures offer several advantages in fabricating patch clamp electrodes. This is particularly true of the high lead glasses like 8161, EG-6, 0120, EG-16, KG-12, and 0010. Soda lime glasses such as R-6 and 0080 also offer many of these thermal advantages, but we do not recommend them because of their poor electrical properties. First, low softening temperature glasses are easy on microelectrode pullers. Because of the low filament current required to pull these glasses, filaments rarely change their properties with

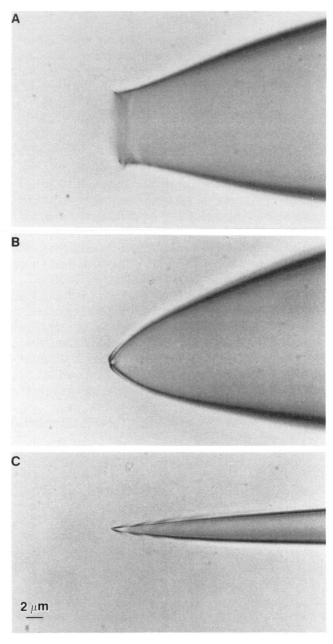

FIG. 2. Versatility of Corning 8161 (or other high lead glasses) in fabricating patch pipette tips. (A) Tip just after pulling. (B) Tip from (A) after fire polishing. (C) An 8161 tip pulled and fire polished for small tip diameter.

extended use and do not require replacement even after a year or two of continued operation. Second, they allow the fabrication of extremely blunt tips much more readily than glasses with higher softening temperatures. We illustrate this point in Fig. 2 wherein we show high magnification photographs of electrode tips pulled from Corning 8161 before and after fire polishing. This is one of the high lead glasses which as a group soften in the 580°–632° range. This is more than 200° lower than the softening temperature of 7740 (Pyrex), the glass most commonly used for intracellular microelectrodes. Corning 8161 also offers a fire polishing ability not provided by higher softening temperature glasses.

Figure 2A shows a very blunt tip immediately after pulling; Fig. 2B is the same tip after fire polishing. Such blunt tips, which are formed exceedingly easily with low softening temperature glasses, offer several advantages. They provide the lowest access resistance for whole-cell recordings. Also, their blunt taper makes them less likely to penetrate when they are pressed against the cell during seal formation. The cells can be indented to a larger extent than with sharper electrodes, and this often helps in seal formation. These high lead glasses are so amenable to fire polishing that it is possible to pull electrodes at such low temperature that the resulting tips are broken and jagged with diameters in excess of 50 μm and yet are easily fire polished into usable patch electrodes. The resulting tips are exceedingly blunt but have proved sealable to cells even when their final resistance is less than 0.5 MΩ.

Blunt tips are very important for perforated patch recordings (see [8], this volume). Such tips draw in large omega-shaped pieces of membrane when suction is applied. This large membrane area maximizes the number of parallel amphotericin or nystatin channels than can be inserted and thus minimizes the final access resistance achievable.

On the other hand, these same glasses can be pulled at slightly higher temperature to yield tips that are very sharp with resistances exceeding 20 MΩ (Fig. 2C). Such electrodes can be useful, for example, for trying to minimize the number of channels in a membrane patch by reducing the size of the patch. Therefore, these low softening temperature glasses are extremely versatile with respect to achievable tip geometries.

Borosilicate glasses soften at temperatures in the 700°–850° range. Those at the low end of the softening range (see Table I) are quite easily pulled and fire polished although they are clearly not in the same class with the high lead glasses in this regard. Fire polishing of these glasses is much more dependent on the shape of the tip after pulling than with the high lead glasses. In general, most pullers fabricate sharper tipped pipettes from hard glasses than from soft. In fact, there were several reports in the early patch clamp literature that whole-cell electrodes could not be made from these glasses. With the advent of multistage, computerized electrode pullers, that restriction no longer holds. One can routinely make both patch and whole-cell pipettes from almost any glass. Corning 7070, 7760, 7040, 7052, EN-1, 7720, 7056, 3320, 7050, KG-33 (Kimax), N51-A, and 7740 (Pyrex) are examples of glasses in this borosilicate, intermediate softening temperature category. This class of glasses contains those most often used for patch clamp recordings.

Aluminosilicate glasses, which are very hard, high softening temperature glasses, were found in early work to produce low noise single-channel recordings and so were highly recommended. That low noise came at a high price, however. Glasses in this class soften at temperatures above 900° and so pulling them is quite hard on puller coils and filaments. The coils change their properties with time, and so they must be replaced or readjusted frequently. In addition, these glasses have had the undesired property of being very thin at the tip after pulling. This, along with their high softening temperature, has made them much more difficult to fire polish than softer glasses. The thin wall at the tip may in part offset the inherently good electrical and noise properties of this glass.

Noise Properties

To date, there has been no convincing theoretical analysis of a patch electrode as a noise source in current recordings. At present, we know of no way with simple equations to predict the noise that will come from a particular glass when it is used for patch clamping (but see [2] in this volume for a useful approximation). It is clear both theoretically and in practice that the glass is not a major noise contributor in whole-cell recordings. There, at moderate to high frequencies, the major noise comes from the series resistance associated with the electrode and the whole-cell capacitance (see [2], this volume). Therefore, our discussion of noise here is aimed primarily at single-channel recordings.

Figure 3 shows the power spectral density of noise measured from a particularly low noise resistive patch clamp headstage and the noise that comes from several patch clamp electrodes pulled from different glasses, Sylgard-coated to within 100 μm of the tip and sealed to Sylgard in the bottom of a chamber containing about a 2 mm depth of solution. Several important points are evident in Fig. 3. First, the noise of the headstage approaches a limiting value at low frequency but then rises with increasing frequency. Although the three electrode glasses in Fig. 3 show similar behavior, all have higher noise levels at low frequencies than does the headstage alone, and the noise increments even more steeply, in comparison to the headstage, as frequency increases. An extremely good electrical glass like Corning 1723 rises less steeply with frequency than does Corning

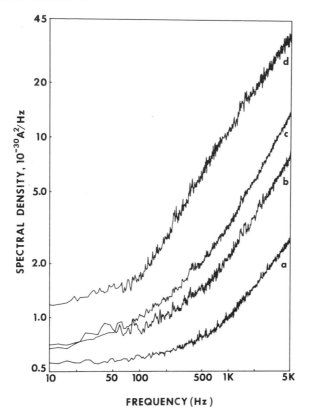

FIG. 3. Power spectral density of headstage (a) and headstage and electrode (b–d) noise. For (b–d), electrodes are coated with Sylgard 184 coating and sealed to Sylgard at the bottom of a 2 mm deep chamber filled with normal saline. b, Corning 1723 (aluminosilicate), c, Corning 7052 (borosilicate), d, Kimble R-6 (soda lime). [From J. L. Rae and R. A. Levis, *Biophys. J.* **45**, 144 (1984).]

7052, and both of these glasses are very much better in this regard than Kimble R-6 glass. Therefore, substantially greater noise comes from the electrode plus the headstage than comes from the headstage alone. This is true of all frequencies shown but particularly at high frequencies.

It is expected and borne out by experiments that the noise coming from a patch electrode is related to the product of the loss factor and the dielectric constant of the glass. The loss factor is a parameter used by manufacturers to describe the dielectric properties of a glass. It is technically defined as the tangent of the loss angle at a frequency of 1 MHz.[8]

[8] R. H. Doremus, "Glass Science," p. 190. Wiley, New York, 1973.

Simply stated, if the glass wall could be modeled as a perfect capacitor, a sinusoidal voltage applied across it would produce a sinusoidal current through it that was 90° out of phase with the voltage. If on the other hand the glass wall were a lossy capacitor, the phase angle would differ from 90°. The difference between the actual angle and the 90° angle for a perfect capacitor is what is defined as the loss angle. The loss angle should and does depend on frequency. The 1-MHz frequency is often used by manufacturers in their specification sheets. It is less in the few kilohertz range important for patch clamping, but unfortunately the data describing the loss angle as a function of frequency are not available for most glasses. One usually has only the value at 1 MHz to use in predicting which glasses might be most useful for low noise single-channel recordings.

Figure 4 demonstrates the dependence of noise in a patch electrode on the loss factor of the glass and its dielectric constant. In these experiments, patch electrodes were fabricated from various glasses, coated with Sylgard to within $100 \, \mu$m of the tip, filled with solution, and sealed to Sylgard lining the bottom of a fluid-filled chamber about 2 mm in depth. The root mean square (rms) noise was measured with the pipette in air just above the bath and then again with the electrode sealed to Sylgard. Many commercial patch clamp amplifiers contain root mean square meters to allow experimenters to assess noise at desired points in their experiments. The 10-kHz noise on Fig. 4 was measured through an 8-pole Bessel filter and comes from the root mean square subtraction of the noise in air from the noise sealed to Sylgard:

$$rms_{total} = (rms^2_{Sylgard} - rms^2_{air})^{1/2}$$

The noise is plotted against loss factor times capacitance rather than loss factor times dielectric constant so that each glass can be normalized for its wall thickness (see [2], this volume). The major finding is that the glasses with the lowest loss factor–dielectric constant product show the lowest noise. Realize, however, that the noise shown here is not exclusively glass noise. It also includes the thermal noise of the Sylgard seal, but this is expected to be quite small because of the exceedingly high resistance of the Sylgard–glass seal.

Several other important points are apparent. First, the noise decreases as the wall thickness increases (compare 7052 with 7052TW, thin walled). Therefore, there is a noise advantage that comes from the use of thick walled pipettes. Second, the shape of the curve at low loss factor–dielectric constant is undetermined since no data are available for glasses with loss factor–dielectric constant products lower than 7760. The data show little tendency to flatten so it seems likely that significant improvement in this component of patch clamp noise could occur with the use of, for example,

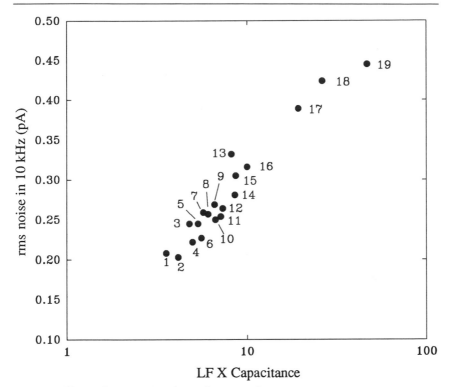

FIG. 4. Glass noise versus loss factor times capacitance. Each electrode was coated with Sylgard 184 coating and sealed to Sylgard at the bottom of a 2 mm deep chamber filled with normal saline. The root mean square noise shown is the root mean square difference between the noise when the electrode is sealed to Sylgard and the noise with the tip in air just above the surface of the bath. The capacitance was obtained by multiplying the dielectric constant by 0.225/wall thickness. [Adapted from J. L. Rae, R. A. Levis, and R. S. Eisenberg, *in* "Ion Channels" (T. Narahashi, ed.), Vol. 1, p. 288. Plenum, New York, 1988.] The glasses shown are as follows: (1) 7760, (2) 8161, (3) 7040, (4) 1723, (5) 0120, (6) EG-6, (7) 7052, (8) 7720, (9) EN-1, (10) KG-12, (11) 0010, (12) 3320, (13) 7052TW, (14) 7056, (15) EG-16, (16) 7740, (17) 1720, (18) R-6, (19) 0080.

quartz (fused silica) electrodes. Corning 7070 (low loss electrical) or Corning 1724 (aluminosilicate) also offer the potential for improvements if one were to work out proper techniques to fabricate patch electrodes from them.

It is clear from Fig. 4 that the noise does not scale with volume resistivity. One notable example is Corning 1720, an aluminosilicate glass with very high volume resistivity but quite bad noise properties. It is also clear that noise is not a direct function of dielectric constant alone. Corning 8161 has a very high dielectric constant and yet produces very low noise. Plots of noise against either volume resistivity or dielectric constant would not produce a monotonic relationship as found for noise against loss factor times dielectric constant.

In summary, the high lead glasses (particularly 8161) which have low loss factors at present offer the best noise performance for single-channel patch clamping. However, Corning 7760 and 1723 have produced quite comparable results. Unfortunately, 1723 is no longer available, but perhaps 1724 will prove to be effective in the future. Corning 7760, which in our opinion is the best all-around electrode glass available, is also scarce and expensive at the present time and can only find extensive use in patch clamping if future demand warrants the effort for commercial concerns to make it available.

Capacity Compensation

Another important electrical property associated with the dielectric characteristics of the pipette wall is the shape of the capacity transient resulting from the application of a step voltage command to a sealed pipette. Obviously a large portion of this transient arises from stray capacitance at the headstage input. However, a significant portion comes from the pipette itself. This is particularly true of relatively slow components of the transient.

If the glass of the pipette wall were an ideal capacitor, the capacity transient associated with a step of voltage applied to the pipette would be a simple rapid spike with a shape that would essentially reflect the time derivative of the command voltage waveform (as modified, of course, by the bandwidth of the headstage electronics). It is well known, however, that the capacity transient associated with tight-seal patch clamping has slower components that can be of significant amplitude and with durations of many milliseconds. When studying voltage-gated ionic channels, it is traditional to observe single-channel currents immediately following step changes in the command potential. The presence of the capacity transient in the current measured at the headstage output can obscure or distort single-channel currents for some time following a voltage step, especially when there is a relatively large, slow tail of capacity current. Commercial patch clamps provide capacity compensation with one or two time constants to eliminate a major portion of the capacity current from the headstage output. Provided that the residual transient in the current output is of sufficiently small amplitude and is essentially constant, it is possible in the fortunate situation that a few blank records (i.e., records which contain no channel openings) are obtained to average several blanks and then subtract this average from each record containing channel openings. This procedure is frequently quite satisfactory for removing capacity current from channel records. However, if the channels do not cooperate by periodically not opening in response to a step change in command voltage, the problem can be more difficult to deal with.

The slow component of the capacity transient of all glass types we have tested is *not* well described by a single exponential, and therefore electronic compensation even from commercial patch voltage clamps which provide a slow component of capacity compensation (and some manufacturers do not) is not really adequate to eliminate the transient. The best solution is to choose a glass for pipette fabrication that shows a minimum amount of slow component in its response to step changes in potential. Because this slow component arises from the lossy dielectric characteristics of the glass wall of the pipette, it is to be expected that glasses with low loss factors (e.g., Corning 8161, 7760) will display smaller slow components than glasses with relatively high loss factors (e.g., soda lime). We have verified this expectation experimentally. Fortunately such low loss factor glasses also display less noise and are therefore the natural selection for the fabrication of pipettes for high quality patch clamp measurements.

To study the capacity transients arising from patch pipettes, we pulled pipettes from several different types of glass. Pipettes were normally coated with Sylgard 184 to approximately 100 μm from the tip. The pipettes were then sealed to Sylgard at the bottom of a chamber that was always filled to the same depth (about 2 mm) with Ringer's solution. Typical pipette resistance prior to sealing was about 3 MΩ. Because we were interested in slow components of the capacity transient, the fast component was electronically canceled as completely as possible, and the residual transient was studied at times greater than 50–100 μsec following the start of a step voltage command. Command steps had an amplitude of 200 mV to maximize the measured signal; pulse durations varying from 2.5 to 500 msec were used. We studied pipettes fabricated from Corning 7052, 7040, 7760, 8161, and 0010 and Fisher Blue Dot (soda lime).

Although there was some variability among pipettes fabricated from the same glass, we found, as expected, that glasses with low loss factors showed significantly less slow component in their capacity transients than glasses with relatively high loss factors. Not only were the slow components of low loss factor glasses such as 8161 and 7760 smaller in amplitude than those of glasses like Fisher Blue Dot, but the duration of the slow component was also very much less. However, in no case was it found that the slow component of any glass was well described by a single exponential decay. In fact, even for the best glasses it was found that the decay more closely approximated a logarithmic function of time than an exponential, as might be anticipated for a lossy dielectric. Figure 5 shows some typical results.

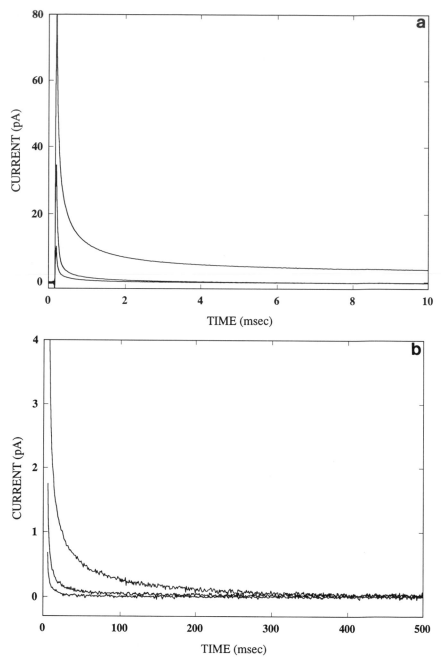

Fɪɢ. 5. Slow capacity transients from representative patch electrodes constructed from three different glasses. In each plot, 8161 is the lowest trace, 7052 is the middle trace, and Fisher Blue Dot is the uppermost trace. Electrodes were coated with Sylgard 184 coating and sealed to Sylgard at the bottom of a 2 mm deep chamber filled with normal saline. The fast capacity transient was negated using circuitry inherent to the patch clamp, and so the traces

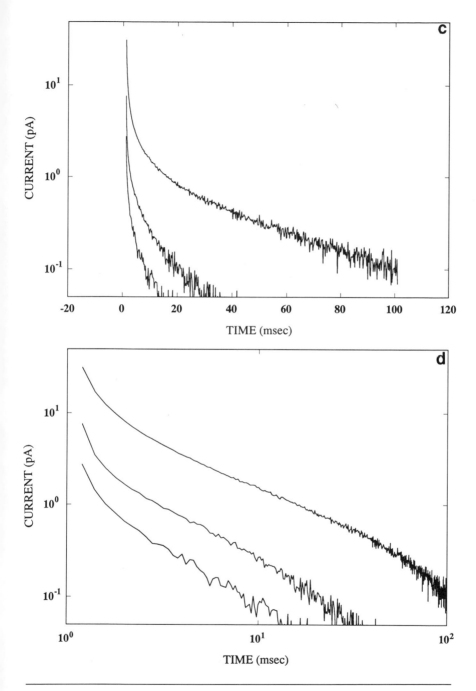

shown are residual transients. (a) Fast time base recording; (b) slow time base recording, linear scale; (c) slow time base recording, linear–log scale; (d) slow time base recording, log–log scale.

Figure 5a shows, on a fast time scale, that different glasses have very different amplitudes and durations of their slow components. Figure 5b–d illustrates the nonexponential nature of the slow component of the capacity transient for several representative glasses. It is obvious that a single exponential component of electronic capacity compensation is not adequate for even the best glasses. In fact for all glasses that we studied, a minimum of three exponential components was required to achieve reasonable (although not perfect) fits to the data. Clearly with existing commercial patch clamps which have at most one exponential component for electronic compensation of these slow transients, the magnitude and duration of the residual (uncompensated) transient will be a function of the quality of the glass.

Of the glasses tested, 8161 showed the least slow component of its capacity transient. Corning 7760 typically had a slightly larger slow component, 7052 and 7040 displayed somewhat larger and longer lasting slow components, and Fisher Blue Dot (soda lime) glass showed by far the largest and longest lasting slow component in its capacity transient. It is instructive to compare the results of a good 8161 pipette and the worst of the four soda lime pipettes that we tested. For a 200-mV pulse, the slow component for 8161 had an amplitude of about 9 pA 100 μsec after the start of the step; it declined to less than 1 pA after about 1.4 msec and fell to 0.5, 0.2, and less than 0.1 pA after about 3, 8, and 17 msec, respectively. For the same size command step, the slow component of the soda lime pipette transient amplitude was about 50 pA 1 msec after the start of the pulse, was still somewhat larger than 10 pA in amplitude after 10 msec, and was about 4 pA after 50 msec. Even after 200 msec, the transient was still about 1.2 pA above its final value. The amount of capacitance, C_{slow}, associated with the slow component for each glass can be estimated by integrating the slow component of the transient to determine the amount of charge and then using the relationship $C = Q/V$ to determine the capacitance. For the pipettes described above this capacitance is only about 0.05 pF for 8161 but is about 1.2 pF for the soda lime pipette.

It should also be noted that in the absence of Sylgard coating, the slow component of the capacity transient of all glasses studied increased. The increase was smaller for low loss glasses than for soda lime glass. However, in the case of low loss factor glasses, applying an exceptionally heavy coat of Sylgard and painting very close to the tip did not yield significant improvement relative to results obtained with "normal" Sylgard coating. On the other hand, applying a heavy Sylgard coating does reduce the magnitude of the slow component of the capacity transient for high loss factor glasses such as Fisher Blue Dot, and such a coating is recommended if one must use glasses of this variety.

Finally, we observed (but did not systematically study) that for soda lime glass the magnitude of the slow component was not constant for repeated pulses delivered at rates of once every 1 to 3. Over a series of such pulses, the magnitude of the slow component was often seen to decline with repeated pulses, reaching a steady level only after some 20–30 pulses had been delivered. As expected, this effect was decreased if the repetition frequency was decreased. No such phenomena were observed for low loss factor glasses. This effect could prove to be troublesome for some types of patch clamp measurements. We did not examine the linearity of the slow component of the capacity transient over a significant range of voltages.

From these results, it is reasonable to conclude that the use of Sylgard-coated, low loss factor glasses is desirable for high quality patch clamp recording. This is particularly true for pulsed single-channel recordings.

Sealability

It is not clear what happens physically when a gigohm seal forms between a membrane and a glass. Any fluid pathway that remains between the two is highly restricted and greatly curtails the movement of even small ions like Na^+ and K^+. It largely excludes the movement of larger molecules. When patch clamping was a new technique, there was hope that there existed some optimal glass that would promote this sealing to cells. We initially reported that Corning 7052 was a glass with exceptional sealing properties[4] and continue to believe that it is a solid choice for use in patch and whole-cell clamping when all of its properties are considered. Subsequent work, however, has shown that essentially any glass is capable of making a seal with cell membranes, and there is not much solid evidence that one glass seals better than another. Such evidence would require that tips of electrodes fabricated from glasses being tested be of similar shape and wall thickness and fire polished to the same end point, etc., to rule out factors other than glass composition in the promotion of seals. In short, we have not done careful experiments aimed at quantifying sealability nor are we aware of any such experiments in the literature. There is much anecdotal information, but to our knowledge there are no controlled experiments published on this aspect of patch clamping. Even if such experiments were to be done, there is no reason to believe that what is found for one cell type would necessarily be applicable to others.

One of us (J.L.R.) has experience in sealing about 40 different cell types. During the course of that experience, there were many specific examples where one glass seemed to seal a particular cell better than other types. One notable example was isolated bile duct epithelium wherein we failed to obtain a single seal with 7052 in over 20 attempts but never failed to obtain seals with 8161 in 12 attempts. Since our early publication of 7052 results, some investigators have reported that after switching to 7052 their sealing frequency remarkably improved. Others have found no effect, and yet others have obtained worse results. It seems clear that there is no final answer available concerning sealability of glasses. It is a complex issue that depends on tip geometry, details of suction application, etc., but probably more on the quality of the cells or what might be coating them than on the glass. We have found that seals form readily with many different kinds of cells we have used when electrodes are constructed from 7040, 7050, 8161, 7760, or 7052. More than one seal with the same pipette has been achieved with 7040, 7050, and 7052 but not with most other glasses. It has always been true that we could optimize sealability by changing the tip geometry and extent of fire polishing of any particular glass. It is possible that any glass might be made quite sealable by such an effort. Therefore, we are cautious about conclusions concerning sealability. Clearly, there has not been enough detailed work on each glass for one to make general rules about sealability.

Leachable Component

Glasses are complicated substances made of many different compounds. Table II contains a listing of the major constituents found in many glasses that have been used for patch clamping. Although glasses have major constituents that lead to their classification as soda lime, aluminosilicate, borosilicate, etc., they have many trace compounds as well. The location of these compounds in the glass is itself a complicated phenomenon. It is difficult to predict which of the particular compounds may be at the surface of the glass. It is clear, however, that glasses can have components at their surface which can leach into an aqueous environment in which they are in contact.[9] There are also many reports in the literature of atomic absorption measurements of cell constituents that have been contaminated by Na^+ or Ca^+ that leached into the solutions from glass containers.

Leachable components could be particularly problematic in patch clamp and whole-cell recordings owing to the close proximity of the channels to the glass. Several glass constituents such as Ba^{2+}, Rb^+, and As^{2+} are known blockers of ionic channels which could alter the recording of chan-

---

[9] R. H. Doremus, "Glass Science," p. 229. Wiley, New York, 1973.

nel currents were they to reach a sufficiently high concentration in the solution in the immediate proximity of the channels. Glasses contain many such compounds that might have these kinds of effects.

The literature now contains several reports that these undesirable effects do occur in patch and whole-cell recordings. Cota and Armstrong[10] reported that Corning 8161 blocked K^+ currents in single pituitary cells. Furman and Tanaka,[11] in the most complete study to date, reported that several glass types caused blockade of or otherwise altered currents in photoreceptor cells when used to construct electrodes for excised patch recordings. In that particular study, they found that Corning 0010, a high lead glass, was best for their currents but that several other high lead glasses caused blockade. The use of Corning 7052 resulted in larger inward currents at negative voltages than obtained with glasses thought to be inert to these particular currents. Rojas and Zuazaga[12] reported substantial kinetic differences in nicotinic acetylcholine receptors when a "hard" pipette glass was used for recording rather than a "soft" glass. In our own unpublished experiments, we found that 8161 produced a flickery blockade of lens inward rectifier single-channel currents at large negative voltages. Others have found that 8161 activates chloride channels in tracheal epithelium. Yet another investigator has found that solutions perfused onto neurons through a 7052 pipette prevent the activation of some receptor types by their agonists. Although many of these findings have not been formally published, there is enough published information to merit that one seriously consider the possibility that the glass used for the electrode may modify the currents in some way. It therefore seems imperative that one record currents with several different kinds of pipette glass to investigate this possibility.

Low Noise Recording Techniques

Modern patch clamps, particularly those implemented with integrating technology (see [2] in this volume) are capable of very low noise, particularly below 10 kHz of bandwidth. To utilize this performance, the user must pay close attention to other sources of noise. The total root mean square noise of a patch clamp recording is the square root of the sum of the

[10] G. Cota and C. Armstrong, *Biophys. J.* **53,** 107 (1988).
[11] R. E. Furman and J. C. Tanaka, *Biophys. J.* **53,** 287 (1988).
[12] L. Rojas and C. Zuazaga, *Neurosci. Lett.* **88,** 39 (1988).

individual squared root mean square noise sources. This means that any particular noise source that is large will dominate the total noise. Therefore, all potential contributory noise sources must be minimized. Specifically, the headstage, the electrode glass, the holder, and the seal contribute significantly even under circumstances where extraneous noise pickup from the environment is negligible. It is, of course, necessary that the entire preparation be properly shielded and hum from power supply mains, etc., be made negligible. Here, we suggest some approaches to low noise recording of single channels. Whereas these same approaches are a good idea for whole-cell recording, they are less important there since in whole cell recording the dominant noise source, at bandwidths above a few hundred hertz, comes from the access resistance in series with the whole-cell capacitance.

The noise from electrode glass itself arises from the lossy characteristics of its walls. Therefore, it is expected that glasses with the lowest inherent loss factors will have the lowest noise, and it is expected that the thicker the wall the lower the noise will be.

Even if one uses electrically superior glasses, low noise will not result unless the outer surface of the glass is coated with a hydrophobic substance like Dow Corning Sylgard 184 coating to prevent bathing solution from creeping up the outer wall of the electrode glass. A thin film of solution produces a distributed resistance which interacts with the glass capacitance to produce a noise source which rises with frequency. It becomes the dominant noise source and so must be eliminated. The Sylgard also decreases the capacitance of the electrode wall and so reduces the lossiness of the wall as well. It has been shown experimentally that Sylgard coating will improve the noise of any glass but will not turn a poor electrical glass into a good one. Low loss glasses coated with Sylgard give significantly less noise than poor glasses coated with Sylgard. The Sylgard should be painted as close to the tip as is practically possible, but the majority of the noise improvement is achieved if one paints to within $50-100$ μm from the tip.

Holders must be made of low noise materials. Polycarbonate has been found experimentally to produce the lowest noise in limited tests of several likely materials, but it was only slightly better than polyethylene, polypropylene, and Teflon. When constructed from one of these materials, holders contribute only a small fraction of the total noise. We cannot, however, exclude the possibility that the holder material has some further effect on the noise associated with the holder–electrode combination. Holders should avoid metal and shielding which are noise sources. Holders do become a significant noise source if they get fluid in them. Therefore, great care must be taken in filling electrodes with solution. They should be filled only far enough from the tip so that the end of the internal reference electrode is immersed. Any solution that gets near the back of the electrode should be dried with dry air or nitrogen to keep it from getting into the holder. Holders that become contaminated with solution should be disassembled and sonicated in ethanol or pure deionized water and allowed to dry thoroughly before being used again. It is also a good idea to clean the holders periodically this way even if no solution has been observed in them.

The noise of the holder and electrode can be checked before each attempt at a seal. When the holder and filled electrode has been inserted in the headstage connector and the electrode tip is positioned just above the bathing solution, the root mean square current noise seen on the meter of most commercially available patch clamp amplifiers should not be much above 0.1 pA in a 5-kHz bandwidth when using an integrating patch clamp and 0.2 pA for a standard resistive feedback headstage.

The seal will usually be the dominant noise source if it is only a few gigohms (at least up to bandwidths of several kilohertz). Seal resistances in excess of 20 GΩ must be obtained if exceptionally low noise single-channel recordings are to be routinely possible. The quality of the seal can also be tested each time by looking again at the root mean square noise meter. The noise also depends on the depth of the electrode tip below the surface of the bathing solution since the effective electrode capacitance increases as the depth of immersion increases. The voltage noise of the headstage interacts with the electrode capacitance to produce a noise source which rises with frequency. With integrator technology and with excised membrane patches lifted to just under the surface of the bathing solution, it has been possible for the authors to produce background noise as low as 0.13 pA rms in a 5-kHz band in a membrane patch with channels from several preparations using, for example, 7052 or 7760 glasses. A background noise of 0.15–0.17 pA rms was routinely possible.

One last potential noise source to consider is the noise in the signal generator which provides the command. In most patch clamps, this noise is reduced by heavily attenuating the external command, but it is possible, particularly if the command signal comes from a digital–analog (DA) converter, for this noise source to be significant.

Summary

Based on all of the properties of glass described here, it is obvious that no one glass can be recommended for all purposes and for all cells. Borosilicate glasses like 7760, 7052, and 7040 are good general purpose glasses for both single-channel and whole-cell recordings. They are good initial choices but, of course, must be checked for each cell type for problems associated with leaching of blockers, etc., from the glass. Corning 8161 is the best glass studied to date with respect to electrical and thermal properties but must be checked carefully for leachable components. If perforated-patch whole-cell recordings are to be used, 8161, KG-12, or some other high lead, low melting point glass are probably the best choices.

[4] Ion Channel Selectivity, Permeation, and Block

By TED BEGENISICH

Introduction

One of the first steps in an effort to understand how cells perform their observed function is to determine what ion channels are in the membrane of the cell of interest and the properties of those channels. The channels can be classified by the ion that is most permeant and by what other ions can also permeate the channel. Further information can be obtained from a proper analysis of blockage of the channel by impermeant ions. The procedures described here assume the ion permeation pathway is a water-filled pore and that ions diffuse through the pore without associated large movements of the protein channel as might occur in "carrier"-mediated transport. As we have learned more of the details of these two general types of mechanisms, the distinction between them has diminished. However, single-channel currents that represent more that 10^5 ions/sec (0.02 pA) clearly fall into the pore category.[1,2]

Most of the permeation properties of pores can be determined through the use of macroscopic currents obtained with the whole-cell variant of the patch clamp technique. In most cases, use of these macroscopic currents is preferred but there are situations where single-channel currents more easily reveal specific permeation properties.[3] Unless specifically stated, the pro-

[1] B. Hille, "Ionic Channels of Excitable Membranes." Sinauer Associates, Sunderland, Massachusetts, 1984.
[2] B. P. Bean, this volume [11].
[3] E. Moczydlowski, this volume [54].

cedures to follow apply to either macroscopic or single-channel currents. The references in this chapter were chosen to provide specific examples and not to critically review the field.

Experimental Procedures

General Considerations

The theoretical background for studies of ion channel permeation has followed two general courses. One of these uses the electrodiffusion approach developed by Nernst[4,5] and Planck,[6,7] and the other employs absolute reaction-rate theory.[8] Much of the characterization of ion channel permeation can be done independently of the particular theoretical formalism, especially if many of the terms are considered empirical values.

Ion Channel Selectivity

One of the first properties to determine for a channel is its selectivity: which ions are permeant. If a pore is permeable to a single ion (X) of valence z, the net current through the pore is zero at a potential (V_X) given by the Nernst (or equilibrium) potential for that ion [Eq. (1)]:

$$V_X = (RT/zF) \ln(a_o/a_i) \tag{1}$$

where R, T, and F have their usual thermodynamic meanings; RT/F at 20° is approximately 25 mV. This voltage is an equilibrium property and does not depend on the theoretical framework from which it is derived. It is a function only of the ion concentrations and is independent of any impermeant ions and independent of the presence of any fixed charges on the membrane.

The terms a_o and a_i represent the external and internal activities of ion X related to the concentration by activity coefficients, γ_o and γ_i:

$$V_X = (RT/F) \ln(\gamma_o[X]_o/\gamma_i[X]_i) \tag{2}$$

[4] W. Nernst, Z. Phys. Chem. **2**, 614 (1888).
[5] W. Nernst, Z. Phys. Chem. **4**, 129 (1889).
[6] M. Planck, Ann. Phys. Chem. Neue Folge **49**, 161 (1890).
[7] M. Planck, Ann. Phys. Chem. Neue Folge **40**, 561 (1890).
[8] H. Eyring, R. Lumry, and J. W. Woodbury, Rec. Chem. Prog. **100**, 100 (1949).

These activity coefficients depend on the ionic strength of the solution. In a dilute salt solution, a monovalent ion has an activity coefficient of approximately 0.85, and so the ion activity is not much different from the ion concentration. At higher salt concentrations, the coefficients become rather small and very small for divalent ions.[9] However, if the internal and external solutions are similar, the activity coefficients are often ignored:

$$V_X = (RT/F) \ln([X]_o/[X]_i) \tag{3}$$

not because they are near unity but rather because they are approximately equal. If the solutions are not of similar concentration or if only one contains a moderate concentration of divalent ions, then these coefficients should not be ignored. With this caution, the remaining discussion will use concentrations rather than activities.

Equation (3) suggests a test that a pore is permeable to a specific ion, say K^+: measure the channel zero-current potential (or reversal potential) with known internal and external concentrations of the test cation and see if this potential is equal to the Nernst potential for the ion. To allow a valid comparison, the measured potential must be corrected for the appropriate liquid junction potential.[10] In addition, the solutions should not contain other ions that may be permeant. Candidates for possibly impermeant cations include Tris, tetramethylammonium, and N-methyl-D-glucamine; large anions may be impermeant to anion channels and include isethionate and glucuronate. It may be difficult to find impermeant ions for some very nonselective channels (e.g., see Adams et al.[11]).

The use of Eq. (3) requires knowing the intracellular concentration of the ion of interest. Unfortunately, in many experimental situations, this concentration is not accurately known. This is sometimes true even for the usual whole-cell variant of the patch clamp technique since it relies on diffusion of the pipette contents into the cell interior.[12] So, rather than compare the computed Nernst potential to a single measurement, the reversal potential is measured at several external concentrations of the test ion and (often) plotted as a function of the logarithm of the external ion concentration. Equation (3) predicts a linear relationship in such a plot with a $58/z$ mV/decade slope (at 20°), and such a finding demonstrates

[9] R. A. Robinson and R. H. Stokes, "Electrolyte Solutions." Butterworth, London, 1965.
[10] E. Neher, this volume [6].
[11] D. J. Adams, T. M. Dwyer, and B. Hille, J. Gen. Physiol. 75, 494 (1980).
[12] R. S. Eisenberg, this volume [10].

that the pore is permeant to the tested ion. Furthermore, with such data the internal concentration of the ion can be computed.

Seldom is a perfectly linear relationship found when using physiological solutions (but see Lucero and Pappone[13] for a recent example). Rather, the results of many such experiments are in agreement with Eq. (3) only at high concentrations of the test ion. At low concentrations the slope is often much less than predicted (see Sah *et al.*[14] for a recent example). Such a result suggests that the pore may not be perfectly selective for the tested ion.

Indeed, few channels are permeant to only a single type of ion, and the expected dependency of pore zero-current potential (V_{rev}) on ion concentration is, in general, both model dependent and more complicated. However, if two ions (X and Y) of the same valence (z) are considered, a relatively simple equation can be derived (see Hille[1] for some of the details of these computations) from either of the two limiting theoretical formalisms:

$$V_{rev} = (RT/zF) \ln \frac{P_X [X]_o + P_Y [Y]_o}{P_X [X]_i + P_Y [Y]_i} \qquad (4)$$

where P_X and P_Y are the permeabilities of ions X and Y. The reversal potential of Eq. (4) actually depends only on the relative permeabilities:

$$V_{rev} = (RT/zF) \ln \frac{[X]_o + (P_Y/P_X)[Y]_o}{[X]_i + (P_Y/P_X)[Y]_i} \qquad (5)$$

At high external concentrations of ion X, Eq. (5) predicts a linear relationship between the measured reversal potential and the logarithm of $[X]_o$. Deviations from linearity are expected at lower concentrations owing to the second term in the numerator of Eq. (5). The relative X to Y permeability can be obtained by fitting Eq. (5) to the data.

Biionic Conditions. While Eq. (5) suggests experiments that can determine the relative ion permeability of a pore, a simpler method can be used if the solution on the cytoplasmic face of the membrane can be accurately controlled (e.g., perfused giant axons, excised patches, whole cell measurements with perfused pipettes[12]). This experimental control allows the use of a simplified form of Eq. (4) if ions of type X are the only permeant ones

[13] M. T. Lucero and P. A. Pappone, *J. Gen. Physiol.* **94**, 451 (1989).
[14] P. Sah, A. J. Gibb, and P. W. Gage, *J. Gen. Physiol.* **92**, 264 (1988).

in the external solution and Y ions are the only permeant species in the solution on the cytoplasmic side of the membrane:

$$V_{rev} = (RT/zF) \ln (P_X[X]_o/P_Y[Y]_i) \qquad (6)$$

or

$$P_X/P_Y = ([Y]_i/[X]_o) \exp[(zV_{rev}F)/RT] \qquad (7)$$

When experimental conditions allow such a situation, the relative permeabilities are easily computed from the measured reversal potential and the known ion concentrations. Chandler and Meves[15] used this technique to show that the squid axon Na^+ channel is permeable to several other cations including Li^+, K^+, Rb^+, and Cs^+.

Internal Ions Not Known but Constant. Hille[16] showed how relative permeability can be determined even if the internal medium cannot be controlled but is constant. In this technique, the channel reversal potential is measured twice: once with only ion X in the external solution ($V_{rev,X}$) and again with only ion Y in the external solution ($V_{rev,Y}$). Under these conditions, the relative permeability can be computed [using a form of Eq. (4)] from the difference between these two measurements:

$$P_X/P_Y = ([X]_o/[Y]_o) \exp[zF(V_{rev,X} - V_{rev,Y})/RT] \qquad (8)$$

Using this technique, Hille[17] showed that the selectivity of the Na^+ channel in myelinated nerve is similar to that of the squid axon Na^+ channel and extended the list of permeant ions to include several organic cations.[16]

It is important to note that the Eqs. (4)–(8) are valid only for ions of the same valence and for a pore that obeys independence.[18] Independence means that the movement of each type of ion is independent of the presence of the other. Very few ion channel pores allow independent ion movement. Rather, ions transiently bind during passage to sites within the pore. Ions compete for these sites, and so the ion movements do not obey independence. Consequently, these equations are not, in general, expected to be valid but still have considerable utility.

[15] W. K. Chandler and H. Meves, *J. Physiol. (London)* **180**, 788 (1965).
[16] B. Hille, *J. Gen. Physiol.* **58**, 599 (1971).
[17] B. Hille, *J. Gen. Physiol.* **59**, 647 (1972).
[18] B. Hille, *in* "Membranes—A Series of Advances, Volume 4: Lipid Bilayers and Biological Membranes: Dynamic Properties" (G. Eisenman, ed.), p. 255. Dekker, New York, 1975.

Classification of Ion Pores

Pores that do not allow independent ion flow can be classified by the maximum number of ions that can simultaneously occupy the pore. In one-ion pores, the presence of a single ion prevents entry (perhaps through electrostatic repulsion) of any other ion. Multiion pores allow simultaneous occupancy by more than one ion.

One-Ion Pores. Hille[18] has shown that Eq. (7) applies to one-ion pores even if independence is not obeyed. The permeabilities in Eq. (7) may not be constant for such a pore but, rather, may be functions of membrane potential but not functions of ion concentration. However, in many situations the voltage dependence may be small or nonexistent. Consequently, Eq. (7) is very useful for determining relative selectivities even for one-ion pores.

The nonindependent nature of such pores makes the current through such pores very sensitive functions of the concentrations of the ions. Consequently, ion selectivity cannot be judged by the relative current carried by different types of ions but can be determined from measurements of the V_{rev}. Ion currents are also affected by many pore-blocking ions (see below). If these ions are impermeant, however, they will not contribute to the measured reversal potential, and Eq. (7) may still be used.

Multi-ion Pores. Not surprisingly, the permeation properties of multi-ion pores are more complex than those of one-ion pores. Nevertheless, Hille and Schwarz[19] have shown that Eq. (7) may apply even to these more complicated situations but that the permeability ratio may be a function ion concentration. Consequently, a finding of concentration-dependent permeabilities in Eq. (7) indicates that the pore under study has multi-ion characteristics. It may be difficult to separate the effects of voltage and concentration since changing ion concentrations will necessarily shift the reversal potential to a new value. However, it is sometimes possible to separate these two effects: squid axon Na^+ channel selectivities are concentration but not (significantly) voltage dependent.[20]

Pore-Blocking Studies

Valuable information on ion channels can be obtained from studies of current inhibition by impermeant ions (see Moczydlowski[3]). For example, Armstrong[21] used tetraethylammonium (TEA) ions and similar analogs to

[19] B. Hille and W. Schwartz, *J. Gen. Physiol.* **72,** 409 (1978).
[20] T. Begenisich and M. Cahalan, *J. Physiol. (London)* **407,** 217 (1980).
[21] C. M. Armstrong, *J. Gen. Physiol.* **58,** 414 (1971).

probe the structure of the cytoplasmic portion of the squid axon, delayed rectifier K$^+$ channel. In one of the best studies of this type, Miller[22] used monovalent and bis-quaternary ammonium compounds to determine the spatial dependence of the electric potential in part of the pore of a K$^+$ channel from sarcoplasmic reticulum.

One experiment in these types of studies is to determine the voltage dependence of current inhibition produced by a charged ion. As described by Woodhull[23] (and see Moczydlowski[3] for more details) the dissociation constant for the blocking ion, of valence z, binding to a site within the membrane electric field is given as

$$K_d(V_m) = K_d(0) \exp(-\delta z F V_m / RT) \tag{9}$$

where $K_d(0)$ is the zero-voltage dissociation constant and δ is the location of the binding site as a fraction (measured from the inside) of the membrane voltage. Then, the fraction of current (channels) blocked, $f(V_m)$, will not only be a function of the concentration of blocking ions, [B], but also a function of voltage:

$$f(V_m) = \frac{[B]}{[B] + K_d(V_m)} \tag{10}$$

A determination of the fraction of current blocked as a function of membrane potential will, through the application of Eqs. (9) and (10), yield a value for the electrical distance to the blocking site.

There are, however, two potential problems with such an analysis: (1) a voltage dependence of block is neither a necessary nor sufficient condition to identify the block as within the pore; and (2) multiple ion occupancy of the pore makes it impossible to determine the fraction of the electric potential present at the blocking site.[19]

A better test of the presence of the blocking site within the pore is to demonstrate that increases in permeant ion concentration on the side opposite to that with the blocking ion interacts with the blocking ion. Armstrong[21] showed that increased external K$^+$ increased the rate of recovery from block of squid axon K$^+$ channels by internal TEA analogs. Similar results have been obtained for relief of internal Cs$^+$ block of squid axon K$^+$ channels by external K^{+}[24] and Begenisich and Danko[25] showed that increased intracellular Na$^+$ reduces block of squid axon Na$^+$ channels by external H$^+$. In perhaps the most elegant study of this type, MacKinnon

[22] C Miller, *J. Gen. Physiol.* **79**, 869 (1982).
[23] A. Woodhull, *J. Gen. Physiol.* **61**, 687 (1974).
[24] W. J. Adelman, Jr., and R. J. French, *J. Physiol. (London)* **276**, 13 (1978).
[25] T. Begenisich and M. Danko, *J. Gen. Physiol.* **82**, 599 (1984).

and Miller[26] studied block of single high-conductance, Ca^{2+}-activated K^+ channels by the scorpion toxin charybdotoxin. External application of this toxin inhibits current through these channels. McKinnon and Miller found that internal application of the permeant ions Rb^+ and K^+ increased the toxin dissociation rate but that impermeant ions like Cs^+, Na^+, and Li^+ were without effect. These authors have termed this phenomenon "trans-enhanced dissociation" and, rightly, argue that, while not "air-tight proof," such findings are simply and naturally explained by physical occlusion of the pore by the blocking ion.

A good example of the difficulty of obtaining the fractional field location of a blocking site can be found in the work of Adelman and French.[24] A literal application of the Woodhull-type analysis reveals that the blocking site reached by an external Cs^+ is more than 100% across the membrane voltage drop. This apparently nonsensical result can be a consequence of a multi-ion pore,[27] and, indeed, the data of Adelman and French can be reproduced by a specific multi-ion pore model.[28] In such a pore, any permeant ions must vacate the blocking ion binding site before block can occur. Consequently, the voltage dependence of block will reflect not only the voltage-dependent binding of the blocking ion with this site but also the voltage-dependent evacuation of the pore by permeant ions.[19]

In these modeling exercises, specific locations for permeant and blocking ion binding sites are specified, but, because of the large number of adjustable, offsetting parameters, these locations cannot be unambiguously determined. Consequently, any effort to determine the position of a blocking ion binding site must be preceded by experiments demonstrating the lack of multi-ion effects in the pore under investigation. A description of these types of experiments is beyond the scope of this chapter but can be found in Hille and Schwartz[19,27] and Begenisich.[29,30]

Conclusion

As described here, ion channel pore permeation can be a complicated process, and a detailed analysis may be model dependent. However, a determination of pore selectivity is relatively straightforward, and suitable

[26] R. MacKinnon and C. Miller, *J. Gen. Physiol.* **91,** 445 (1988).

[27] B. Hille and W. Schwartz, *Brain Res. Bull.* **4,** 159 (1979).

[28] T. Begenisich and C. Smith, *in* "Current Topics in Membranes and Transport," Vol. 22. Academic Press, New York, 1984.

[29] T. Begenisich *in* "Membranes and Transport" (A. N. Martonosi, ed.), Vol. 2, p. 274. Plenum, New York, 1982.

[30] T. Begenisich, *Annu. Rev. Biophys. Chem.* **16,** 247 (1987).

experimental techniques are readily available. Information on pore structure can be obtained from an analysis of block produced by ions, but such an analysis is complicated and model dependent unless it can be shown that the pore does not exhibit multi-ion properties.

Acknowledgments

This work was supported by a grant (NS 14138) from the U.S. Public Health Service.

[5] Access Resistance and Space Clamp Problems Associated with Whole-Cell Patch Clamping

By CLAY M. ARMSTRONG and WILLIAM F. GILLY

Introduction

The exhilaration of forming a seal and recording from a living cell with the whole-cell patch clamp technique often makes one forget two important sources of error. Series resistance and space clamp artifacts can render worthless the handsomest results once they are subjected to critical review. Stated briefly, without proper compensation for series resistance and demonstration of a good space clamp, all measurements of membrane current are qualitative. In many cases, difficulties associated with these two problems cannot be completely eliminated, and separating justified from unjustified conclusions necessarily becomes a matter of judgment, experience, and cautiously working through every possibility for artifactual explanations of the phenomenon in question.

This chapter discusses these impediments that have harried investigators since the first development of the voltage clamp. A careful analysis of series resistance errors was performed by Cole and Moore[1] over thirty years ago, and series resistance and space clamp problems have been clearly recognized since that time. Examples exist, however, that clearly show the dangers of ignoring these problems and forging ahead. Investigations of ionic currents in cardiac muscle, for example, were handicapped for many years by lack of an adequate space clamp.[2] Consequently, the fundamental principles outlined here are not original, but we hope to present them in a new way. We write this chapter to provide a helpful guide for separating fact from fiction in evaluating voltage-clamp data, and as a reminder not to

[1] K. S. Cole and J. W. Moore, *J. Gen. Physiol.* **44**, 123 (1960).
[2] E. A. Johnson and M. Lieberman, *Ann. Rev. Physiol.* **33**, 479 (1971).

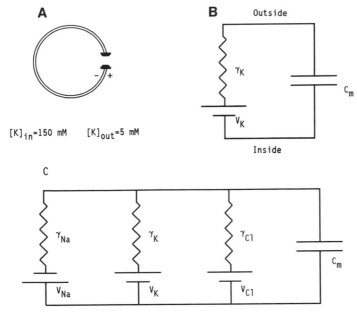

FIG. 1. A model cell and its electrical equivalent circuits. (A) A single K^+-selective channel is inserted into the membrane of a small, spherical cell. The K^+ concentrations inside and outside at the moment of channel insertion are indicated. As potassium ions leave the cell, an excess of positive charge (+) is deposited on the external surface of the membrane, and a deficit of charge is left behind on the inner surface (−). This corresponds to the development of an internal negative voltage. (B) The equivalent circuit of the cell in (A). (C) Additional channels for passing Na^+ and Cl^- are incorporated into the model cell of (B).

forget, willfully or otherwise, these problems. Eventually the tax collector will come!

Equivalent Circuit of a Cell

The normal function of many ionic channels is to change the membrane voltage (V_m) by catalyzing electric current flow. Figure 1A shows a small cell or vesicle, with a single potassium-selective ionic channel in its membrane and the indicated internal and external K^+ concentration. The channel, let us suppose, was just inserted, and V_m was zero at the moment of insertion. Potassium ions will flow out through the channel, because their concentration is higher inside, thus producing an outward current. As potassium ions flow out of the vesicle, a small excess of positive charge (simply called charge hereafter) develops externally, and an equal-sized deficit of charge (or, equivalently, an excess of negative charge) develops internally. A pair of these charges is shown in Fig. 1A. This separation of charge across the membrane defines or "creates" a transmembrane voltage, V_m. By convention, $V_m = V_{in} - V_{out}$.

Potassium ions will continue to flow out in our example until the developing negative internal voltage, which attracts K ions, just balances their tendency to diffuse out. This balance occurs when V_m is equal to the electrochemical equilibrium potential for K ions, V_K. V_K is defined by the Nernst equation [Eq. (1)], which expresses the balance of the chemical concentration "force" on the right and the electrical "force" on the left:

$$V_K F = RT \log([K^+]_o/[K^+]_i) \tag{1}$$

where F is Faraday's constant, R is the gas constant, T is absolute temperature, and $[K^+]_o$ and $[K^+]_i$ are the external and internal K ion concentrations, respectively. The quantity RT/F is equal to approximately 60 mV at room temperature. The current through the K^+ channel is given (with sufficient accuracy for our purposes) by the following equation:

$$I_K = \gamma_K(V_m - V_K) \tag{2}$$

where γ_K is the conductance (reciprocal of resistance) of the open K^+ channel.

Both the internal and external solutions are excellent conductors, and they are separated from each other by a good but not perfect insulator, the cell membrane. To a physicist, a capacitor consists of two conducting plates separated by an insulator and acts to separate charge. In our case of the cell, the two solutions and the membrane constitute the "membrane capacitance," with the membrane serving as the insulating dielectric. Positive charge on the outside is attracted to, and gets as close as possible to, the negative charge on the inside of the membrane. The result is that opposite charges are lined up at the outer and inner surfaces of the membrane, unable to cross because the lipid bilayer is an insulator.

The charge (q) that accumulates in the external (and internal) fluid at the membrane interfaces, as described above, is the actual source of V_m, and the two quantities are related at all instants of time:

$$q = C_m V_m \tag{3}$$

where C_m is the membrane capacitance of the cell. This important equation requires that whenever a net current flows across the membrane through the channel, V_m must change. This can be seen mathematically by differentiating Eq. (3) with respect to time:

$$dq/dt = C_m(dV_m/dt) = I_C \qquad (4)$$

and I_C is the current flowing onto (and off of) the capacitor plates. In our special case I_C is equal in magnitude to I_K.

Equations (2)–(4) are always valid, and they give a complete picture of the charging of the membrane, which will stop when V_m reaches V_K. The point to be emphasized is that all of the current flowing through the channel is accumulating on the outer plate of the membrane capacitor, and there is an equal and opposite charge left behind on the inner plate.

Figure 1B gives the "equivalent circuit" for the vesicle in Fig. 1A. There is a tendency for a current of K ions to flow out of the channel, and this tendency is expressed in electrical terms by a battery, the voltage of which is given by the Nernst equation [Eq. (1)]. This battery forces current outward whenever V_m is more positive than V_K, according to Eq. (2). The current flow is limited by the small size of the channel, and this is expressed by the finite conductance of the channel, γ_K.

The outer end of the channel deposits charge in the external medium, which is the outer plate of the membrane capacitor. This is expressed in the circuit by the straight line, representing a perfect conductor, drawn between the outer end of the "channel" to the outer plate of the capacitor. Likewise, the inner end of the "channel" is connected to the inner plate of the capacitor by another perfect conductor. The conductors represent the external and internal fluid, through which current can flow very easily but not really with perfect ease. The straight lines are thus approximations, acceptable unless the current path is long or restricted (see below). The voltage is everywhere the same in a perfect conductor.

If two other pores, one that carries Na^+ and one that carries Cl^-, are inserted into the membrane, the equivalent circuit would be modified simply by adding the representation of each pore, that is, a battery and conductor, in parallel with the "K^+ channel," to yield the circuit shown in Fig. 1C. The net membrane current, however, is now the sum of the currents flowing through the three sets of channels.

Finally, let us suppose that we have many Na^+ and K^+ channels, and that the fraction of open channels varies with voltage and/or time. This means the total Na conductance (g_{Na}) and total K conductance (g_K) are each variable, and this is expressed by drawing arrows through the resistor symbols, as has been done in Fig. 2. It is quite instructive to build such a circuit with flashlight batteries and potentiometers and to see how V_m varies as the Na and K conductances are adjusted. You will find that the

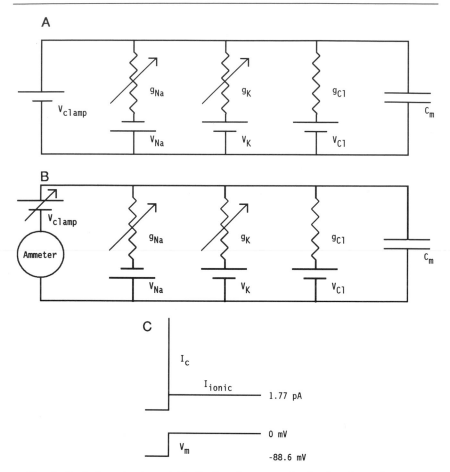

FIG. 2. Equivalent circuits and idealized performance of a simple voltage clamp. (A) A clamp battery sets V_{clamp} and thereby fixes V_m to be the identical value, regardless of the values of the ionic conductances. (B) An ammeter added in series to the clamp battery allows measurement of the current necessary to hold V_m to be equal to V_{clamp} without interfering with its flow. (C) V_m and I_{clamp} traces for a voltage step applied to a membrane with a single open K^+ channel in it. An initial surge of current charges the membrane capacitance (I_c), and a maintained current indicates flow of ions through the channel and the clamp battery.

battery connected to the highest conductance (lowest resistance) is most influential in determining V_m, according to the following equation:

$$V_m = (g_{Na}V_{Na} + g_K V_K + g_{Cl}V_{Cl})/g_{total} \tag{5}$$

where $g_{total} = g_{Na} + g_K + g_{Cl}$.

Simple Voltage Clamp

It was noted in conjunction with Fig. 1A that all of the current flowing through the channel accumulates as charge on the membrane capacitance. Suppose we wanted to measure the current, rather than letting it accumulate. We could then imagine a machine that would draw away all the outflowing charges and count them. Such a machine is the voltage clamp. A consequence of the drawing away action is that none of the charges are left to accumulate, so V_m does not change with time; it is clamped.

One way to make such a machine (though it would not be very convenient) is to connect a battery to the membrane as shown in Fig. 2A. If the voltage clamp battery is connected to the inner and outer solutions by perfect conductors as shown, V_m is simply equal to the clamp-battery voltage, V_{clamp}. Any current flowing through the ion channels is drawn away by the battery, rather than accumulating on the capacitor plates. Seen another way, by Eq. (4) if I_C is zero, V_m cannot be changing; it is clamped. All of the channel current is drawn away through the battery, which overrides any attempt to change V_m by pushing in or pulling away whatever current is necessary. Further, for this ideal clamp $V_m = V_{clamp}$, always.

The arrangement in Fig. 2A clamps the voltage, but it does not measure current. An ammeter can be added (see Fig. 2B) to measure the current flowing through the battery without interfering with its flow. Finally, because we may want to change the clamp voltage, an arrow is drawn through the clamp battery to indicate that it is variable.

It is worth taking a moment to examine the behavior of an ideal clamp. Suppose that we change V_{clamp} nearly instantaneously (a typical experiment), and that the membrane contains only one K^+ channel which is already open and has a conductance of 20 pS. Two types of current would flow. First, there would be an essentially instantaneous surge as the clamp supplied current to charge C_m to the newly imposed voltage. Equation (4) states that the faster the change in voltage is, the larger the surge of I_C will be. This spike of capacity current would be followed by the second type of current, a steady ("dc") flow of ions through the channel, with a magnitude given by Eq. (2). If, for example, we stepped from $V_m = V_K = -88.6$ mV to 0 mV, $V_m (= V_{clamp})$ and I_m (the current through the ammeter) would look like the records shown in Fig. 2C. Except for the instant during the voltage transition, the capacitive current is zero, and I_m is equal to the current through the channel, as given by Eq. (2). This is zero before the step, and 1.77 pA afterward.

To summarize, an ideal voltage clamp has two functions: first, it imposes a voltage on the membrane such that $V_m = V_{clamp}$, and, second, it measures ionic current flow. (We shall see in the next section that V_m and

V_{clamp} are not exactly equal in a real clamp). When applied to a relatively complex circuit like that in Fig. 2B, the voltage clamp performs the same two functions, but it measures the sum of the current flowing through all of the open channels. Separating out the current for each channel type from all of the others requires special tricks, like applying toxins that block one channel type without affecting the others.

Simple Patch Clamp and Origin of Series Resistance

Figure 3A shows the arrangement of a patch clamp experiment, in the whole-cell configuration. The electrode forms a tight seal with the cell membrane, so the cytoplasm is electrically continuous with the saline in the pipette and completely isolated from the external medium. The pipette tip is small, and for this reason there is considerable resistance to current flow from pipette to cytoplasm (R_{access}). A silver/silver chloride electrode in the pipette solution connects the pipette to the negative input of an operational amplifier. Externally, a silver/silver chloride electrode grounds the bath through a low resistance saline column (usually an agar bridge).

The operational amplifier serves a dual function: it is the variable clamp battery described above, and it also serves as an ammeter to measure current. Consider first its role as a variable battery. For our purposes it is enough to say that an operational amplifier keeps the voltage of its negative input exactly equal to the voltage of the positive input. It is not necessary to worry about how the electronics of the amplifier does this. Thus, if a pulse generator is attached to the positive input of the amplifier, the negative input, connected to the pipette, follows exactly the imposed voltage and serves as the variable clamp battery. Such operational amplifiers have very high frequency response, and the clamp voltage can therefore be switched very quickly.

The operational amplifier also serves to measure current, as follows. In Fig. 3A, outward ionic current (by convention this is a positive current; inward current is negative) flows through the open channels in the membrane, and this current also flows out of the pipette, which is connected to the negative input of the amplifier. For an ideal operational amplifier, current flowing into (or out of) the inputs (I_{input}) is zero, and all of the current flowing into the cell must therefore pass through the feedback resistor, R_f, which is connected back to the output of the amplifier. I_{clamp} is therefore defined as $(V_{out} - V_{clamp})/R_f$, and the output voltage of the amplifier is thus the sum of V_{clamp} (set by the pulse generator) and the voltage drop caused by the current flowing through R_f, given by the product $I_{clamp}R_f$. I_{clamp}, the quantity of interest, is measured by electronically subtracting V_{clamp} from this sum in a subsequent stage (not illustrated).

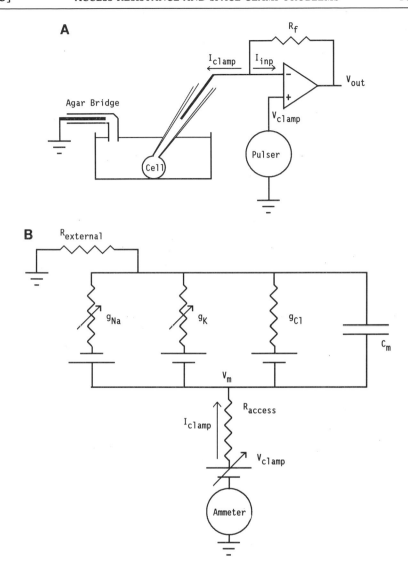

FIG. 3. Schematic diagrams for a whole-cell patch clamp. (A) Current flowing out of the output of the operational amplifier passes through R_f and into the cell interior via the pipette. This corresponds to outward cellular current. The voltage drop across R_f ($I_{clamp}R_f$) is thus a measure of membrane current. See text for additional details. (B) Equivalent circuit of the experimental arrangement in (A). Resistance in the pipette tip (R_{access}) and in the agar bridge ($R_{external}$) is indicated. These elements both contribute to the effective series resistance that exists between the voltage clamp battery and the membrane. Note that V_m will not equal V_{clamp} unless the series resistance or I_{clamp} is zero.

The equivalent circuit of this patch clamp is shown in Fig. 3B. The membrane contains many channels and is represented by a circuit like that in Fig. 1C. Its external side is attached to ground (0 mV) through $R_{external}$, which is mainly the resistance of the agar bridge. In practice, it is essential that the resistance of the agar bridge be measured, for it adds to the pipette resistance (R_{access}) in determining the total effective series resistance (R_{series}). In the following discussion pertaining to whole-cell patch clamping, it will be assumed that $R_{external}$ is small enough to ignore and that R_{access} and R_{series} are synonymous. The internal surface of the cell membrane is attached to the clamp battery through R_{access} and to the external solution through the seal resistance (not illustrated), which we assume to be infinite. I_{clamp} in the diagram is all of the current flowing through the open channels in the cell membrane, plus the current flowing to and from the "plates" of the membrane capacitance.

In this circuit, the pipette interior also electrically contacts the external solution through the capacitance of the glass wall. This capacitance (not illustrated) is generally quite small compared to the capacitance of the cell in most whole-cell experiments, and it can usually be ignored. With some commercial patch clamps, however, the series resistance compensation (see below) and capacity compensation circuitry are interactive, and this makes it difficult to compensate the series resistance properly if the pipette capacity cannot be properly compensated (e.g., if it is too large). It is a good practice, especially in the latter case, to minimize the pipette capacitance by applying Sylgard coating to the outside of the glass.

Errors Introduced by Series Resistance

In relation to the circuit in Fig. 3B, it is important to remember that unless either I_{clamp} or R_{access} (and $R_{external}$ as well) is zero, V_m, the controlled parameter of scientific interest, is *not* equal to V_{clamp}, the parameter under our direct control. The "series resistance error" is defined as the difference, $V_{clamp} - V_m$. The error is negative when I_{clamp} is outward and positive when current is inward. Because I_{clamp} will rarely be zero during an experiment, a series resistance error is basically always present. Whether the problem is significant is the question that must be reckoned with.

It is worth dwelling for a time on these errors, which can seriously distort observations on the properties of voltage-dependent channels. Let us begin by simulating an ideal clamp experiment (R_{series} is very small) on a membrane containing only K^+ channels, and then compare the results to those obtained after including a sizable R_{series}. The simulation uses the Hodgkin and Huxley model for K^+ channel properties and assumes a channel density found in the squid giant axon. In this preparation, the effective series resistance lies primarily in $R_{external}$, rather than in R_{access} as it does in whole-cell patch clamp experiments. Neither of these specific features of the model affects the conclusions to be drawn here.

Figure 4 shows I_{clamp} for a voltage clamp step from -70 to $+20$ mV. In Fig. 4A, R_{access} is small (0.2 Ω), and V_{clamp} and V_m are indistinguishable at all times. I_{clamp} shows a large capacity transient coincident with the voltage step, and current then drops to near zero until, after a lag, some K$^+$ channels begin to open. I_K is directly proportional to the number of open channels, and I_K continues to rise to a maintained value.

The effects of a 50-fold higher R_{access} are shown in Fig. 4B. The first thing to note is that V_m and V_{clamp} are no longer equal. V_m rises more slowly than does V_{clamp} (same time course as in Fig. 4A), because the membrane capacitance is being charged through R_{access}. The capacity current transient thus has a finite time course: it is a decaying exponential, with a time constant equal to the product $C_m R_{access}$. After the capacity current transient, I_{clamp} falls to near zero, and the series resistance error momentarily vanishes. As K$^+$ channels open and outward current develops, however, V_m becomes smaller than V_{clamp}. This has several effects on the observed I_{clamp}. First, because V_m (the quantity experienced by the membrane channels) is not as large as V_{clamp}, fewer channels open then when R_{series} is negligible (Fig. 4A), and they open more slowly. Second, the current through each channel is smaller than expected, because the driving force $V_m - V_K$ is reduced by the series resistance error (negative in sign for outward current). Both of these factors act to make the observed total I_K artifactually small.

Figure 4C compares I_K versus V_{clamp} curves with (filled symbols) and without (open symbols) a significant series resistance error (values used in Fig. 4A,B). Deviation between the two data sets becomes more serious as the current gets larger, as expected. Nonetheless, there is a good qualitative similarity between the two curves, and one might (unwisely) conclude that the series resistance error is not so bad.

How does this amount of series resistance error compare to that typically encountered during whole-cell recordings, for example, on one of the squid giant fiber lobe neurons that gives rise to the giant axon? Such a cell might have an input capacity of 25 pF, and if the I_K density were equal to that in the axon, the corresponding amplitude of I_K at $+20$ mV in the cell body would be 200 nA. The low and high R_{access} values would be 8 kΩ and 400 kΩ, respectively. These values are considerably lower than those that can be realistically achieved, even with pipettes of large diameter. In reality, I_K density in these cell bodies is lower than that in the axon, however, and the amount of series resistance corresponding to the same relative amount of error is therefore proportionately larger. For example, if I_K at $+20$ mV in the whole cell patch clamp experiment is 20 nA (rather

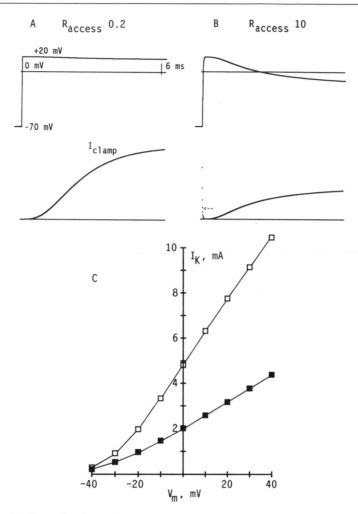

FIG. 4. Effects of series resistance error on measurement of I_K in a voltage-clamped membrane. (A) With a low value for series resistance (R_{access}), the calculated V_m trace is indistinguishable from the applied V_{clamp} (upper traces) for a step from -70 to $+20$ mV. (B) With an increased value for R_{access}, V_m deviates markedly from V_{clamp} [illustrated in (A)], and I_{clamp} is attenuated and kinetically distorted. (C) I_K-V_{clamp} relations calculated for the conditions given in (A) and (B) are given by open and filled symbols, respectively. V_{clamp} is plotted as V_m. See text for additional details. This model calculates current amplitudes for 1 cm² of squid giant axon membrane. If a giant fiber lobe neuron from the squid had the same K⁺ ·channel density as the giant axon, I_K ($+20$ mV) would equal 200 nA in a cell of 25 pF input capacitance, and the pipette resistance equivalent to the values of R_{access} used in (A) and (B) would be 8 and 400 kΩ, respectively.

than 200 nA), then the traces in Fig. 4A,B would correspond to R_{access} values of 80 kΩ and 4 MΩ, respectively. These values are more realistic. A clogged pipette can easily be higher than 4 MΩ in resistance, and the illustrated problems are not implausibly exaggerated.

Series resistance errors are even more serious when dealing with voltage-dependent inward currents. Figure 5 shows similar calculations for a membrane with only Na^+ channels of the Hodgkin and Huxley type. A voltage step from −70 to −45 mV is applied to the membrane, and results are compared for R_{access} values of 0.2 Ω (Fig. 5A) and 20 Ω (Fig. 5B). After the capacity transient, which again is much slower with the higher R_{access}, Na^+ channels open with a lag. As inward current develops, V_m becomes larger (more positive) than V_{clamp} owing to the series resistance error. Because of the effectively larger voltage command felt by the membrane, more Na^+ channels open than in the absence of significant R_{access}, and they open more rapidly. With a 20-Ω series resistance, the error is very large, and the membrane voltage undergoes what is almost an action potential. In no way can the I_m waveform in Fig. 5B be imagined to reflect the properties of the membrane at −45 mV.

Figure 5C,D shows calculations for a larger pulse to 0 mV and also illustrates the normal (Fig. 5C) and distorted (Fig. 5D) "tail" currents that flow when V_{clamp} is returned to −70 mV at a time when many Na^+ channels are conducting. With low R_{access} (Fig. 5C), the current magnitude jumps to a larger value coincident with the V_{clamp} step, because of the increased driving force. V_m is an almost perfect step, but, because of the small series resistance error, has a rounded corner. After the step there is a rapidly decaying current tail as the Na^+ channels close. With the larger R_{access} (Fig. 5D), V_m returns to −70 mV with a slow and complex time course. The tail current under these circumstances is obviously distorted and bears little resemblance to that recorded with a low R_{access}. Clearly the picture of the properties of the sodium channels based on these data would again be seriously flawed by the series resistance error.

Peak I_{Na} versus V_{clamp} relations for the same values of R_{access} are plotted in Fig. 6. The effect of high R_{access} is to steepen the downward slope of the I_{Na} versus V curve. This makes the channels appear to be more voltage sensitive than they really are. Note that positive to −25 mV, I_{Na} calculated for the larger R_{access} is smaller than I_{Na} corresponding to the case for the negligibly small R_{access}. This is because the series resistance error pushes the actual V_m positive from V_{clamp} and toward V_{Na}, thus reducing the driving force.

An easy way to view the clamp failure just described is given in Fig. 7. When the clamp is perfect ($R_{access} = 0$; Fig. 7A), the voltage clamp successfully draws away all of the current flowing inward through the Na^+ chan-

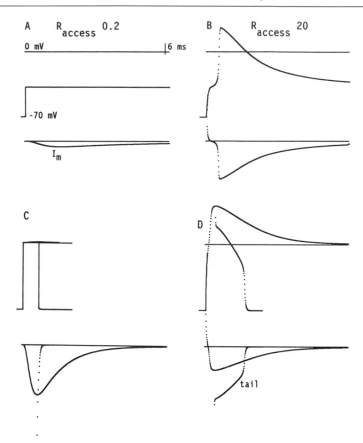

FIG. 5. Calculations of the effects of series resistance error when membrane current is inward. (A) A step in V_{clamp} from -70 to -40 mV (upper trace) elicits a well-controlled inward current (lower trace) when R_{access} is low. V_{m} is indistinguishable from V_{clamp} at all times. (B) When R_{access} is increased 100-fold, V_{m} escapes control of the clamp and follows a trajectory similar to that of an action potential. The current waveform is greatly distorted and bears no relationship to the well-clamped trace in (A). (C) Simulated records for a larger pulse to 0 mV with low R_{access}. Calculations for both a long (6 msec) and short (to time of peak inward current) pulse are superimposed. V_{m} is almost perfectly controlled, and a faithful representation of I_{Na} is obtained. (D) Calculations analogous to (C) after increasing R_{access} reveal the total lack of control for V_{m} (upper traces) and the resulting distortion of the recorded I_{Na}. This is especially evident in the tail current.

nels: none is left to flow onto the capacitor plates. When present (Fig. 7B), R_{access} prevents the clamp from successfully drawing away all of the inflowing current. Some of it flows onto the capacitor plates, causing the change in V_{m} that we refer to as the series resistance error. The amount of current diverted to the capacitor plates depends on the magnitude of R_{access}. The direction of the voltage error depends on the direction of I_{clamp}. With

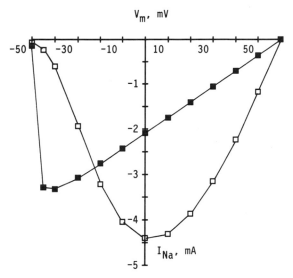

FIG. 6. Calculated I_{Na}–V_{clamp} relations from the models used in conjunction with Fig. 5 to show the distortion caused by series resistance errors. Open symbols are for the low value of R_{access}, 0.2 Ω; filled symbols are for the higher value, 20 Ω. V_{clamp} is plotted as V_m.

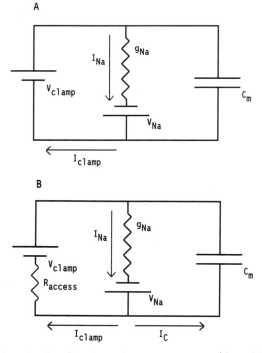

FIG. 7. Equivalent circuits of a voltage clamp arrangement without (A) and with (B) the R_{access} element included. See text for details.

inward current, as illustrated, V_m will be more positive than V_{clamp}; with outward current, V_{clamp} will be more negative.

Measurement of Series Resistance

Total series resistance ($R_{access} + R_{external}$) can be measured in two ways. In the first method, a brief and rapidly rising current pulse is applied to the membrane. On application of the current step, the measured membrane potential (V) will "jump" by an amount given by the product of the current step amplitude (I) and the total value of the series resistance. Membrane capacitance (C_m) can be obtained in such an experiment from the measured rate of voltage change (dV/dt) immediately after the jump. If the current pulse is short compared to the membrane time constant ($R_m C_m$), and if the amplitude of the current pulse is small (or negative going) so that no voltage-dependent ion channels open, C_m can be approximated as

$$C_m = I/(dV/dt) \qquad (6)$$

Although C_m is relatively easy to obtain with this method, reliable measurement of series resistance is more difficult, primarily because of limitations in applying a sharply rising current pulse.

The second and generally more useful method is to apply a voltage clamp step to the membrane and measure the time constant (τ) of the capacity current transient, which is given by the product of C_m and the parallel combination of the membrane resistance (R_m) and the series resistance (R_{series}):

$$\tau = C_m(R_m R_{series})/(R_m + R_{series}) \qquad (7)$$

When the membrane resistance has its resting value, R_m is typically hundreds of megohms and much larger than R_{series}. In this case, the time constant is closely approximated by

$$\tau = C_m R_{series} \qquad (8)$$

C_m can be determined by integrating the capacity current transient and applying Eq. (3). R_{series} is obtained by dividing the measured time constant (τ) by C_m.

In some cases, C_m can be read from the calibrated capacity compensation controls on commercially available patch clamps, provided the compensation has been properly adjusted. Often this is not possible, however, for example, if a cell is too large. Most commercial patch clamps also have a calibrated readout for R_{access}, but this reading normally indicates how much series resistance has been compensated, not how much remains uncompensated and left to alter the experimental results in the ways described above. Thus, direct measurement of series resistance is necessary. It is also critical to realize that the relevant value of R_{access} is the one obtained after breaking into the cell, rather than the pipette resistance measured with the electrode in free solution. R_{access} is usually at least twice the pipette resistance, presumably because of cellular material that partially occludes the pipette tip. Moreover, the effective R_{access} typically changes continuously, and unpredictably, during an experiment, and its value should therefore be measured periodically.

Minimizing Access Resistance in Experiments

Low values of R_{access} can be obtained in only one way, by using low resistance electrodes of the appropriate shape for whole-cell clamping. Desirable pipette shapes and the effects of different glass compositions have been discussed elsewhere.[3,4] In our laboratories we find that the most critical factor in the reliable pulling of low resistance pipettes of the proper shape lies in the application of heat during the second stage of pulling. Heating only a short length of glass is key, and to this end we use a narrow heating coil (3 mm internal diameter), with only two turns, and control the heating current with fanatical care. It is also essential that the contacts with the heater coil be clean and tight: a fraction of an ohm can lead to poor repeatability.

Compensation for Series Resistance

Once R_{access} is known, the series resistance error can be calculated from the product $R_{access}I_{clamp}$. This quantity can be electronically computed by analog circuitry and fed back to the command signal to compensate for the error. Unfortunately, the feedback in this case is positive, and the result is instability that makes the clamp "ring" or oscillate when a substantial fraction of the error has been compensated. A compensation of 100% is possible only with special tricks, such as slowing the onset of the compensating signal, but this degrades frequency response of the system. Thus, even when R_{access} has been optimally compensated in an experiment, the residual, effective series resistance must be directly measured.

---

[3] B. Sakmann and E. Neher, "Single-Channel Recording" (B. Sakmann and E. Neher, eds.), p. 37. Plenum, New York, 1983.

[4] A. Marty and E. Neher, "Single-Channel Recording" (B. Sakmann and E. Neher, eds.), p. 107. Plenum, New York, 1983.

A

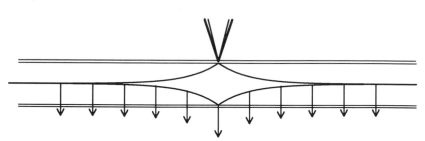

B

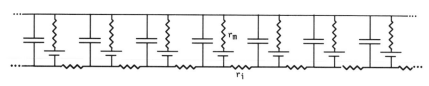

C

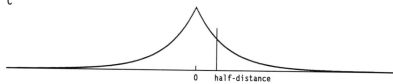

D

E

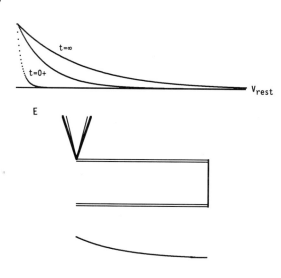

Measurements with Distributed Systems

Thus far we have only considered the whole-cell method applied to a small spherical cell, in which the voltage in the interior (cytoplasm) is uniform. The behavior of the whole cell in this case is simply like that of a single patch of membrane. If the cell is not spherical, the voltage of the interior will not be the same everywhere, and interpretation of the data obtained from such a system involves careful consideration of the relevant "cable properties."[5]

The idealized case is a uniform, cablelike cell that is cylindrical and infinitely long with a surface that has uniform properties and no voltage-dependent channels. If the membrane potential of such a cell cable is controlled at one spot and driven away from the resting potential (e.g., by current injection from a pipette in contact with the cytoplasm), the steady-state voltage as one progresses away from the control point decays back to the resting level as an exponential function of distance (Fig. 8A).

Figure 8B shows the equivalent circuit for this cable. The illustrated symbols represent the membrane conductance per unit length ($1/r_m$, units of Ω^{-1}/cm), the internal resistance to flow through the cytoplasm (r_i, units of Ω/cm) and the membrane capacitance per unit length. The dotted lines indicate that repeating elements of the circuit go on indefinitely. The spatial deviation of voltage from the resting potential owing to the injection of current at $x = 0$ in this structure can be characterized by an "electrical" half-distance; that is, the voltage displacement falls from its value at $x = 0$ (the control point; Fig. 8C) to one-half this value in one half-distance, to one-quarter in two half-distances, etc. This "electrical" half-distance depends on r_m, r_i, and the resistance to flow through the

---

[5] J. J. B. Jack, D. Noble, and R. W. Tsien, "Electric Current Flow in Excitable Cells," Oxford University Press, London (1983).

FIG. 8. Cable properties of a cell and how they affect the voltage distribution during a voltage clamp experiment. (A) The diagram shows the site of current injection in an infinitely long cable. (B) The equivalent circuit of an infinite cable with uniform electrical properties. See text for details. (C) The voltage imposed on the cable at $x = 0$ decays (in the steady state) exponentially with distance from the site of current injection. This decay can be characterized by an "electrical half-distance" as described in the text. (D) The voltage distribution described in (C) takes time to be established following a step voltage change at $x = 0$. Immediately after the voltage step ($t = 0+$), only the area immediately around the site of current injection experiences a significant voltage displacement. This disturbance spreads as a wave down the cable until the steady-state, exponential profile of voltage decay is attained ($t = \infty$). (E) The steady state voltage distribution is shown for a truncated cable cell.

extracellular fluid. For a cell immersed in large fluid bath, the last resistance is usually negligible, and, if so, the formula for half-distance is

$$d_{1/2} = \ln(1/2)(r_m/r_i)^{1/2} = 0.693(r_m/r_i)^{1/2} \qquad (9)$$

The quantity $(r_m/r_i)^{1/2}$ is often called the length constant and is symbolized by λ. It is the value of distance where the voltage at the control point falls to $1/e$ of its original value.

Because voltage varies from point to point along a cable cell like this one, interpretation of current–voltage data derived with the whole-cell patch clamp method for voltage-dependent ion conductances would clearly be difficult. In practice, only indirect calculations can be made,[6] and even these must be regarded in practice as qualitative.

Spatial nonuniformity of imposed voltage changes is not the only problem inherent with nonspherical cells. The voltage distribution just given applies only to the steady state. If the voltage at the point $x = 0$ (Fig. 8C) is suddenly changed from the resting level as in a voltage clamp step, the voltage distribution changes with time as the imposed disturbance spreads out from the point of application, as indicated in Fig. 8D. If one wishes to change the membrane voltage suddenly, the time dependence of the voltage distribution compounds the difficulties in interpretation of the data, because the voltage-dependent conductances under study are also time-dependent.

Many cells explored with whole-cell recording methods, such as large myotubes or neurons with dendritic arborizations (or axons), bear similarity to an experimental preparation called a "truncated cable" of finite length. In such a cell cable, the ends are sealed by surface membrane. The steady-state voltage distribution in this case (see Fig. 8E) is mathematically more complicated than that for the infinite cable,[5] but it resembles the latter in form. Again, the voltage falls smoothly from the point of current injection, but the deviation in voltage at the end of the truncated cable is not nearly so bad. If the length of the truncated cell cable is short compared to the half-distance (computed as above for an infinite cable with the same properties), the change of the membrane potential with distance may be small enough to ignore. This requires quantitative knowledge of r_m and r_i, however, and these cannot be predicted with certainty in most normal situations.

Even with an apparently very short, truncated cable, however, there are two points of caution. The first is that the half-distance drops when a resting membrane becomes active. Opening the sodium channels in an

[6] K. S. Cole, "Membranes, Ions and Impulses," p. 152. Univ. California Press, Berkeley, California, 1972.

axon, for example, cuts the half-distance by a factor of 10 to 100. The resting half-distance thus has almost no value as a predictor of voltage uniformity in an active cell. The second caveat is that the steady-state voltage distribution is not established instantaneously, as was shown in Fig. 8D for an infinite cable. Both of the points are evident in calculations based on the two-patch model described below.

Space Clamp

Even Hodgkin and Huxley did not attempt to interpret data from a distributed system like an axon described by the circuit in Fig. 8B. Instead, they followed the lead of Marmont[7] and pushed a wire axially through the interior of the axons they examined. The wire, in addition to passing the current supplied by the voltage clamp, also shorted out the resistance to current flow through the cytoplasm (r_i in Fig. 8B). In this way, the entire axon can be regarded as a single large patch of membrane. The relevant equivalent circuit is thereby greatly simplified and becomes the one shown in Fig. 2. In terms of Eq. (9), r_i drops to zero, and the half-distance becomes infinite.

Unfortunately, the axial wire technique is not universally applicable. Even when there is no overt attempt to alter properties of the cell under study by inserting a wire, however, it is customary (and indeed necessary) to ask whether the preparation is well "space clamped." By this we mean does it have a long "electrical distance" (electrical half-distance compared to physical length) and an approximately uniform voltage distribution under all conditions of study. In general, the only secure method of testing for voltage uniformity under real experimental conditions is direct measurement, using a second electrode inserted at various points along the cell cable. In a whole-cell patch clamp experiment, an intracellular microelectrode or a second whole-cell pipette operating in the current clamp mode can be utilized to carry out such checks.[8,9] In too many cases, such as in brain slice preparations, even this may not be practicable, and one is left guessing and hoping—a dangerous situation.

Two-Patch Model

Even the relatively simple model cell described by equations pertinent for the truncated cable (which are not trivial to solve mathematically)

[7] G. Marmont, *J. Cell. Comp. Physiol.* **34**, 351 (1949).
[8] W. F. Gilly and T. Brismar, *J. Neurosci.* **9**, 1362 (1989).
[9] G. Cota and C. M. Armstrong, *J. Gen. Physiol.* **94**, 213 (1989).

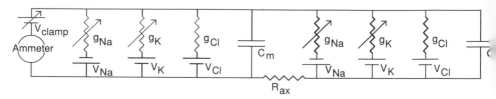

Fig. 9. Equivalent circuit for the two-patch model. The left-hand half of the circuit represents a cell in perfect contact with the voltage clamp (circuit of Fig. 2). The interior of this cell is connected to another part, for example, a dendrite or axon, by R_{ax} which represents resistance to current flow through the cytoplasm (analogous to r_i in Fig. 8B).

becomes intractably complex when voltage- and time-dependent conductances are incorporated into the membrane.[5] A less rigorous, but more illustrative, approach to the complexities of an extended cell, such as one bearing a dendrite, is to use the "two-patch" model. That is, we consider one patch, as previously, with a close connection to the voltage clamp (separated at most by the access resistance of the pipette) and a second patch that simulates a part of the cell that is further away. The equivalent circuit of the two-patch model is given in Fig. 9, which is thus similar to 2 but has an added patch. The two patches are separated by a resistance R_{ax}, analogous to r_i in Fig. 8B, representing the resistance to flow through the cytoplasm that joins the two patches.

This circuit is already moderately complicated, but it pales in comparison to a more realistic model of a morphologically elaborate dendrite, which might have a thousand or more patches. If one has any trouble with the circuit in Fig. 9, please hesitate before attempting to interpret voltage clamp data from an extended structure like a neuron in a brain slice preparation. We think it fair to say that few, if any, people can interpret such data without a lot of guessing. If, on top of the spatial distribution problem, the properties of the various parts of the structure vary from region to region, as most careful investigations have revealed, the problem becomes very difficult indeed.

Space Clamp "Errors"

Errors arising from space clamp problems are harder to define than those associated with series resistance. Perhaps a reasonable operational definition, or question, would be the following: How do the properties inferred from voltage clamp data of the complex preparation compare to those from a well-clamped single patch? This question assumes that the properties of the structure are uniform, which may not be so. There is no general answer, and we content ourselves here with providing an indication of the severity of this problem by showing computed solutions from two identical patches electrically separated by an internal resistance, R_{ax} (circuit of Fig. 9).

Figure 10 shows calculated traces for the two-patch model with two values of R_{ax}, 25 Ω (Fig. 10A–C) versus 0.5 Ω (Fig. 10D). Figure 10A is for a hyperpolarizing voltage step. The voltages of the two patches, V_1 and V_2, are plotted separately, as though a separate electrode were monitoring voltage of the second patch. The voltage uniformity for this step is excellent: V_2 has a rounded corner, but after 50 μsec it coincides with V_1. One would say that the preparation has a short electrical distance and is "well space clamped."

Figure 10B shows the same "preparation" during a clamp step to −40 mV, and the result is much less pleasing. V_2 again follows V_1 at first

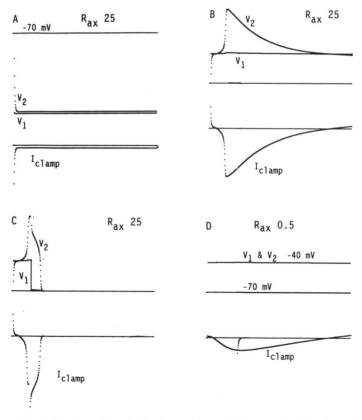

FIG. 10. Calculated results with the two-patch model. V_1 (Patch 1) is clamped by the pipette, but V_2 (Patch 2) is not. (A) A hyperpolarization does not reveal any space-clamp problem despite the large R_{ax}. (B) A depolarization excites Na conductance, and Patch 2 cannot be properly controlled because of the large R_{ax}. (C) A brief depolarizing pulse excites an action potential in Patch 2. (D) When R_{ax} is small, both patches behave as one well-controlled element. See text for details; time base matches that in Fig. 5.

(note the rounding of V_2) but then rockets off toward V_{Na}. This happens because the Na^+ channels of the second patch escape control of the voltage clamp and open regeneratively as V_2 goes positive, exactly as occurs during the rising phase of an action potential. V_1 is well controlled. For a maintained step, V_2 eventually approaches V_1 as the Na^+ channels inactivate, and the traces cross near the end of the sweep as K^+ channels open, producing an error in the opposite direction. The time course of I_{clamp} has a strong resemblance to that of V_2. If the step is terminated near the peak of the inward current (Fig. 9C), V_1 immediately returns to the resting level, but V_2 follows only after a considerable delay and shows a complex time course, again reflected in I_{clamp}.

Just how "bad" the results in Fig. 10B,C are is indicated by the analogous traces in Fig. 10D computed for the identical voltage step to -40 mV after effectively shorting out R_{ax} by decreasing its value 50-fold. This would correspond to the insertion of an axial wire in a squid axon. The two voltages are indistinguishable at all times. The current is much smaller now, because V_2 does not escape control, and thus only a small fraction of the Na^+ channels in the second patch are opened. When the pulse is terminated near the time of peak inward current, both voltages are again well behaved, and I_{clamp} decays quickly to the resting level with a simple, exponential time course.

To summarize, Fig. 10 illustrates several significant points. (1) Spatial uniformity of voltage under resting conditions does not imply spatial uniformity in the active cell. (2) Steady-state uniformity of voltage does not imply uniformity after a quick change of voltage. (3) Measured current in the absence of voltage uniformity bears no resemblance to the current from a well space-clamped cell.

Conclusions

We would like to end with a message of hope regarding the significance of current measurements in extended structures, but we cannot. Interpretation of voltage clamp data where adequate "space clamping" is impossible is extremely difficult unless the membrane is passive and uniform in its properties and the geometry is simple. If voltage- and time-dependent channels are present, there is no general method, and the burden of proof that results are interpretable must be on the investigator.

[6] Correction for Liquid Junction Potentials in Patch Clamp Experiments

By ERWIN NEHER

Introduction

It is standard practice in patch clamp experiments to indicate membrane potentials with respect to the zero-current potential, as measured when the patch pipette is positioned near the cell but not yet sealed. This procedure can be considered as valid in the case of a cell-attached or inside-out patch measurement provided that the pipette contains a solution identical to that of the bath. Whenever these two solutions are different, however, there will be a liquid junction potential at the tip of the pipette during the determination of the zero-current potential; this complicates matters.[1-3]

Liquid junction potentials are due to different mobilities of ions at interfaces between different solutions.[4] Intuitively they can be understood as a result of some inbalance of charge, which results when the more mobile ion diffuses more rapidly across the concentration gradient at the interface. Liquid junction potentials are typically 2–12 mV for the solutions in general use in electrophysiological experiments.

In conventional microelectrode recordings the problem of liquid junction potentials is minimized by using concentrated KCl as the electrode filling solution. KCl is chosen because K^+ and Cl^- have almost equal ionic mobilities, and, as a consequence, liquid junction potentials are in the range of 1 mV or smaller. In patch clamping, on the other hand, the choice of solutions is dictated by the physiological requirements of the experiment. Although there are good reasons for tolerating liquid junction problems, the necessity remains to compensate for their effects. Unfortunately, authors of patch clamp papers only rarely describe in detail how they handle these problems.

Electrochemical Account of Patch Clamp Measurements

Strictly speaking the patch clamp measurement consists of two parts: (1) a reference measurement under current clamp conditions which results

[1] E. M. Fenwick, A. Marty, and E. Neher, *J. Physiol. (London)* **331,** 577 (1982).

[2] P. H. Barry, this series, Vol. 171, p. 678.

[3] P. H. Barry and J. W. Lynch, *J. Membr. Biol.* **121,** 101 (1991).

[4] J. O'M. Bockris and A. K. N. Reddy, "Modern Electrochemistry," Vol. 1. Plenum, New York, 1970.

in the zero-current potential and (2) a test measurement under voltage-clamp conditions, in which membrane current is measured as a function of applied voltage (after seal formation). Figure 1 (left-hand side) depicts schematically the reference measurement. It can be seen that the potential, V, which can be measured under zero-current conditions consists of three terms: V_{EI1}, the electrode potential of the silver chloride electrode in the pipette; $-V_{LJ}$, the liquid junction potential; and $-V_{EI2}$, the electrode potential of the bath electrode. V_{LJ} appears with negative polarity here, because the convention of Barry[2] is followed, which defines V_{LJ} as the potential of the bath solution with respect to the pipette solution. V_{EI2} also has been denoted with negative polarity, since the orientation of the reference electrode is opposite to that of the pipette electrode.

Figure 1 (right-hand side) depicts the situation after formation of an inside-out patch. Any concentration gradients which gave rise to the V_{LJ} on the left-hand side now occur across the patch and give rise to the membrane currents of interest. It is immediately apparent that a potential of absolute magnitude V_{LJ} is being applied to the patch if V is left constant during patch formation. Thus, it is incorrect to accept the zero-current potential as the origin of the voltage axis. Concerning the polarity, considering, as in the above case (and throughout this chapter), V_{LJ} as the potential of the bath with respect to the inside of the pipette, the potential applied would be V_{LJ}, since, by convention, the potential of the inside-out patch is the negative of the pipette potential. Thus, the membrane potential V_M should be calculated from the reading provided by the patch-clamp

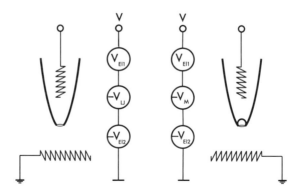

FIG. 1. Schematic representation of the reference measurement (left-hand side) and an inside-out patch measurement (test measurement, right-hand side). On the far left a patch pipette and the electrodes are depicted symbolically. Adjacent is a representation of the relevant electrochemical electromotive forces V_{EI1}, $-V_{LJ}$, and $-V_{EI2}$ (see text). For zero current the condition $V_{EI1} - V_{LJ} - V_{EI2} = V$ must be fulfilled. In the test measurement (right-hand side), $-V_{LJ}$ is replaced by the patch potential $-V_M$.

amplifier V according to

$$V_M = -V + V_{LJ} \tag{1}$$

for inside-out. The above treatment assumes that all the conditions stay constant in between the reference measurement and the test measurement and, in particular, that the electrode potentials are stable.

The arguments above also apply to the cell-attached configuration, except that the potential of the cell, V_c, has to be taken into account. With the usual conventions,[5] Eq. (1) then takes the form

$$V_M = V_c - V + V_{LJ} \tag{2}$$

for cell attached. In the outside-out configuration the absolute magnitude of the membrane potential is the same as in the inside-out case, but polarity is inverted:

$$V_M = V - V_{LJ} \tag{3}$$

for outside-out patch and small whole-cell.

The whole-cell configuration is equivalent to the outside-out case except for an additional problem. For the outside-out patch it has been assumed that there is no liquid junction potential between some residual cytoplasm attached to the membrane and the pipette solution, because concentration gradients at this junction are expected to equilibrate rapidly. In whole-cell measurements this may also be the case if small cells are employed. Small mobile ions equilibrate across the pipette orifice by diffusion within a few seconds in cells of $10-20$ μm diameter.[6-8] However, for larger cells this may not be the case. The situation can be even more complex since the cytoplasm may contain immobile anions, giving rise to Donnan potentials. These complexities have been discussed in detail.[3-7]

Patch Clamp Amplifiers and Liquid Junction Potentials

Most patch clamp amplifiers provide separate settings for handling holding potential (V_{Hold}) and an offset. The sum of both is applied to the

---

[5] By convention, the liquid junction potential V_{LJ} is given as the potential of the bath with respect to that inside the pipette. Membrane potential V_M is given as that of the cytoplasmic side with respect to the extracellular side. The voltage reading V of the amplifier is set to zero in the zero-current condition by means of an offset. This offset stays fixed during the rest of the measurement.

[6] O. P. Hamill, A. Marty, E. Neher, B. Sakmann, and F. J. Sigworth, *Pfluegers Arch.* **391,** 85 (1981).

[7] A. Marty and E. Neher, *in* "Single-Channel Recording" (B. Sakmann and E. Neher, eds.), p. 107. Plenum, New York, 1983.

[8] M. Pusch and E. Neher, *Pfluegers Arch.* **411,** 204 (1988).

headstage, but only V_{Hold} is displayed on a meter and is available at an output. During the reference measurement V_{Hold} is set to zero, and the offset is adjusted for zero current. Then the test measurement is done at an appropriate V_{Hold} setting. Finally the correction as indicated in Eqs. (1)–(3) is applied. For convenience some amplifiers provide a switch-selectable operating mode (setup mode, search mode, or something similar) in which V_{Hold} is disenabled to avoid the necessity of repeatedly resetting V_{Hold} during cycles of reference and test measurements.

Alternatively, the following procedure can be adopted in order to avoid the *a posteriori* correction. The offset is adjusted for zero current during the reference while V_{Hold} is active and set to $-V_{LJ}$. Thereby the setting of V_{Hold} anticipates the potential that will be seen by the patch after seal formation (compare Fig. 1). This procedure exactly implies the corrections of Eqs. (1)–(3) if subsequently V_{Hold} readings are interpreted as $-V_M$ for inside-out and cell-attached configurations or as $+V_M$ for the other configurations.

Solution Changes: Puffer Pipettes, "Sewer Pipes," and U Tools

So far, it was assumed that the bath solution does not change in between reference and test measurements. Very often, however, it is desirable to change ionic concentrations in the bathing medium. In such cases it has to be strictly avoided that the reference electrode "sees" the solution change, particularly if it involves a change in chloride concentration, since V_{El2} (Fig. 1) depends strongly on chloride. Thus, the use of a salt bridge is indicated (but note the caveat on salt bridges below!). It may be simpler to change only the solution in the immediate vicinity of the cell by local perfusion using a puffer pipette, "sewer pipe" arrangement, or U tool. This also prevents concentration changes at the reference electrode. However, under these conditions, concentration gradients exist within the experimental chamber, giving rise to liquid junction potentials wherever they occur.[3]

The situation is depicted in Figure 2 for a whole-cell measurement. It turns out that

$$V_M = V - \Delta V_{El} - V_{2,1} \qquad (4)$$

where $V_{2,1}$ is the liquid junction potential of solution 2 (close to the cell) with respect to the solution further away, the ΔV_{El} is $V_{El1} - V_{El2}$. If we call V_{ref} the potential applied to the pipette during the reference measurement (performed with the original solution) we see from Fig. 1 that

$$V_{ref} = -V_{LJ} + \Delta V_{El} \qquad (5)$$

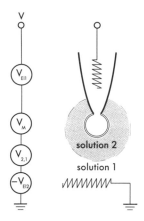

Fig. 2. Local perfusion. The shaded area (right) represents the milieu in the vicinity of a cell in which the bath solution (solution 1) has been replaced by a test solution (solution 2) through local perfusion. The left-hand side schematically gives the electrical potential differences involved.

From Eqs. (4) and (5) we obtain (after elimination of ΔV_{El})

$$V_M = (V - V_{ref}) - V_{LJ} - V_{2,1}$$

During the reference measurement the amplifier offset had been adjusted, such that V_{ref} appears to be zero. Therefore

$$V_M = V - V_{LJ} - V_{2,1} \tag{6}$$

for whole-cell and outside-out patch recording, in analogy to Eq. (3). Thus the *a posteriori* correction is the sum of the two liquid junction potentials.

Let us assume for simplicity that the pipette solution contains predominantly potassium glutamate, that solution 2 is sodium glutamate, whereas solution 1 is Ringer's solution. In this instance, V_{LJ} and $V_{1,2}$ are similar in absolute magnitude, since in both cases the anion (glutamate) is much less mobile than the cation (K^+ or Na^+). However, with the conventions specified the polarities are different, thus the *a posteriori* correction for the part of the experiment performed in sodium glutamate is very small. The correction would be approximately -10 mV for measurements in Ringer's solution and -3 mV during local sodium glutamate perfusion. Disregarding these differences would lead to serious errors in the determination of permeability ratios. If, for instance, the shift of the reversal potential for a 10-fold change in $[Cl^-]$ is measured as 47 mV instead of 55 mV, one would estimate a P_{Cl}/P_{Glu} of 16 instead of 70 (based on the Goldman–Hodgkin–Katz equation).

For the inside-out configuration Eq. (1) would change by analogy to Eq. (6) to

$$V_M = -V + V_{LJ} + V_{2,1} \tag{7}$$

for inside-out patch. Similarly Eq. (2) would change to

$$V_M = V_c - V + V_{LJ} + V_{2,1} \tag{8}$$

for cell-attached, where V_c, the intracellular potential, has to be known for the condition of the local ionic milieu.

Using Salt Bridges

Careful investigators might prefer to avoid the problems associated with local solution changes for the reasons described above and use salt bridges. However, even with salt bridges a situation very similar to that described above may arise. The optimal salt bridge involves a compromise: on the one hand, it should contain 3 M KCl and should be "freely flowing" for a low liquid junction potential[9]; on the other hand, it should not contaminate the bath with KCl. The compromise very often is a relatively thin agar-filled piece of tubing, which at the beginning of the experiment contains KCl throughout. As the experiment proceeds, KCl diffuses out of the agar, and the interface between Ringer-like agar and KCl–agar recedes toward the interior of the bridge. If then a rapid solution change in the bath is performed, one is left essentially with a double-bridge arrangement as depicted in Fig. 3. If the electrode potentials and $V_{1,3}$ are corrected for during the reference measurement, exactly the same situation as discussed in the last section arises.

This means that a bridge, unless used with great care, is not much better than local perfusion. This point was already stressed in 1970 by Barry and Diamond[9] in the context of measurements of epithelial potentials and more recently by Barry and Lynch.[3] These authors also pointed out that electrochemically well-defined junctions in a physiological context can be obtained in two special cases. One is the biionic case, in which salts at equal concentrations are present at both sides of the junction, and either the cation or the anion is the same on both sides; the other one is the dilution case, in which the salts on both sides are identical but may have different concentrations. It can be shown that for these two cases the total junction potential does not depend on the particular concentration profile and thus does not change appreciably with time.[9]

[9] P. H. Barry and J. M. Diamond, *J. Membr. Biol.* **3**, 93 (1970).

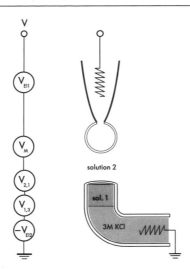

Fig. 3. A nonoptimal salt bridge. The right-hand side schematically shows a salt bridge which is initially filled with KCl–agar. When the experiment proceeds for an extended time using predominantly Ringer's solution in the bath, the terminal portion of the agar bridge will slowly equilibrate with Ringer's solution (solution 1). If then, for a short time, the bath solution is switched to solution 2, there will be effectively a double-bridge arrangement with an interface solution 2–solution 1 between the salt bridge and the bath, and an interface solution 1–KCl in the interior of the salt bridge. The associated liquid junction potentials are indicated schematically at left.

Measurement of Liquid Junction Potentials

Accurate measurement of liquid junction potentials requires a salt bridge with an abrupt boundary between the 3 M KCl solution and the test solution (see above). Such a device is readily made from a wide-bore patch pipette ($\sim 2 \mu$m diameter) filled with 3 M KCl. This pipette, with a well-chlorided silver wire inside, is used as the bath ground. Another patch pipette is filled with the solution to be tested, and mounted onto the patch clamp headstage.

Three conditions have to be fulfilled for an accurate measurement: (1) The salt bridge should be the only ground connection of the bath, that is, any liquid leaks to grounded parts of the setup have to be avoided. This point is easily checked by temporarily removing the salt bridge from the bath. The patch-clamp amplifier should then read zero current, usually with large noise and movement artefacts superimposed. (2) The electrode potentials should be stable, that is, no drift in zero-current potential should occur when the bath solution is stationary (<1 mV in 10 min). If this is not the case, then the silver wires need to be rechlorided. (3) Both patch pipettes should be filled sufficiently to guarantee a small steady outflow of

TABLE I

Liquid Junction Potentials V_{LJ} of Standard Saline with Respect to Exemplary Solutions[a]

Solution	Composition[b]	V_{LJ} (mV)
Low sodium Ringer's	32 NaCl, 108 Tris-Cl, 2.8 KCl, 2 MgCl$_2$, 1 CaCl$_2$, 10 NaOH–HEPES	−3
Sulfate Ringer's	70 Na$_2$SO$_4$, 70 sorbitol, 2.8 KCl, 2 MgCl$_2$, 1 CaCl$_2$, 10 NaOH–HEPES	+6
Glutamate internal solution	145 Potassium glutamate, 8 NaCl, 1 MgCl$_2$, 0.5 ATP, 10 NaOH–HEPES	+10
Chloride internal solution	145 KCl, 8 NaCl, 1 MgCl$_2$, 0.5 ATP, 10 NaOH–HEPES	+3
Cesium glutamate internal solution	145 Cesium glutamate, 8 NaCl, 1 MgCl$_2$, 0.5 ATP, 10 NaOH–HEPES	+11
Cesium citrate internal solution	60 Cesium citrate, 10 CsCl, 8 NaCl, 1 MgCl$_2$, 0.5 MgATP, 20 CsOH–HEPES	+12

[a] Composition of saline (numbers in mM): 140 NaCl, 2.8 KCl, 2 MgCl$_2$, 1 CaCl$_2$, 10 NaOH–HEPES.

[b] Values given as millimolar; for convention on polarity see the example on p. 127.

solution at the tip, otherwise the "double-liquid junction effect" previously discussed will result.

To start the measurement, the bath is first filled with the test solution (which also is in the pipette). The instrument is switched to current clamp mode, and the voltage reading on the amplifier is set to zero by adjusting the offset. The bath fluid is then exchanged for Ringer's solution (or other reference solution). The voltage reading should rapidly approach a stable value, which is the desired liquid junction potential with the polarity opposite to that required for use in Eqs. (1)–(7) (because the convention considers bath potential with respect to pipette). Following this the bath solution should be switched back to the test solution to check for reversibility.

Values of several test solutions against Ringer's solution are given in Table I. They range from −5 to +12 mV. The predominant salt of the test solution sets the value; minor additions of other salts and buffers usually have negligible effects. More values for liquid junction potentials are given in Ref. 3, together with equations and values to calculate V_{LJ} theoretically.

Summary

This chapter describes corrections that have to be applied to measured membrane potentials in patch clamp experiments. Some of them [Eqs. (1)–(3)] are required regardless of the nature of the reference electrode (in the Ringer's solution bath) whenever the pipette-filling solution is different from the bath solution. They represent the liquid junction potentials that are present at the pipette tip before patch formation. In addition, corrections have to be applied when the bath solution is being changed during a measurement (i.e., after seal formation). In that case the following rules apply. (1) The new solution should never get into contact with the bare silver/silver chloride wire of the reference electrode. This requirement is best met by using a salt bridge. (2) The "best" salt bridge is a 3 M KCl bridge with an abrupt KCl–bath fluid boundary at its tip (see above). This bridge does not require any additional potential corrections, but it may lead to KCl poisoning of the bath or become contaminated by solutions used previously. (3) Local solution changes (microperfusion by puffer pipette, U tool or sewer pipe arrangements) as well as recessed KCl bridges require additional corrections, which (together with the simple liquid junction potential correction) are approximately given by Eqs. (6)–(8).

It should be stressed that all equations given here represent approximate corrections, since liquid junction potentials are thermodynamically ill-defined. This is particularly relevant for Eqs. (6) and (7) where the sum of two liquid junction potentials appears.

[7] Nonstationary Noise Analysis and Application to Patch Clamp Recordings

By Stefan H. Heinemann and Franco Conti

Introduction

Noise analysis was the first type of measurement that yielded reliable quantitative estimates of single-channel parameters[1-3] and thereby provided evidence for the very existence of channel proteins embedded in the lipid matrix of excitable membranes. It is still a powerful tool for investigation of ion transport mediated by channels.

The more direct way of measuring single-channel properties is the recording of unitary events using the patch clamp technique.[4,5] Single-channel recordings provide the richest information on the kinetics of the

[1] B. Katz and R. Miledi, Nature (London) 226, 692 (1970).
[2] F. Conti, L. J. DeFelice, and E. Wanke, J. Physiol. (London) 248, 45 (1975).
[3] E. Neher and C. F. Stevens, Annu. Rev. Biophys. Bioeng. 6, 345 (1977).
[4] E. Neher and B. Sakmann, Nature (London) 260, 799 (1976).
[5] O. P. Hamill, A. Marty, E. Neher, B. Sakmann, and F. J. Sigworth, Pfluegers Arch. 391, 85 (1981).

conformational transitions which underlie the operation of individual channel proteins, but they are extremely difficult to perform when these kinetics are fast for the channel under investigation, as in the case of the voltage-activated sodium channel. Furthermore, they can only be analyzed easily for membrane patches containing few channels, possibly only one channel, and analysis requires a long time. For this reason most studies of single sodium channel events have been limited to fixed temperature conditions and to the narrow voltage range in which the frequency of channel openings is not too high.[6-8] Also, only single-channel events larger than the background noise can be evaluated directly, whereas macroscopic fluctuations can reveal much smaller unitary signals, as in the case of gating noise experiments where charges translocated in an elementary conformational transition of voltage-gated ion channels are measured.[9,10]

Analysis of nonstationary fluctuations of macroscopic currents are a classic way of estimating single-channel conductance.[2,11-13] Whole-cell recordings from small cultured cells provide an ideal preparation for this type of study, owing to the very high ratio between channel noise and background noise which is obtained with this technique.[14] The greatest advantages of this method as compared to single-channel recordings are (1) it yields fairly accurate estimates of the single-channel conductance for most preparations; (2) enough useful recordings can be obtained from the same preparation for several different experimental conditions; and (3) the analysis does not involve any subjective selection of events, it requires much shorter time, and it is easily automated by computer programs.

In patch clamp recordings, transient artifacts due to instabilities of the seal as well as rundown and drift phenomena, especially while measuring under extreme environmental conditions, are not uncommon. Therefore it is advisable to introduce further improvements to the method of nonstationary noise analysis as used earlier. Among these are (1) a better strategy for analysis of recordings to obtain variance estimates, (2) a more rigorous fitting procedure of variance versus current plots to account for weights of the data, and (3) automatic, objective tests for discarding "bad" records. As an application of the analysis method we shall report some studies of nonstationary fluctuations of sodium currents in bovine adrenal chromaf-

[6] F. J. Sigworth and E. Neher, *Nature (London)* **287**, 447 (1980).
[7] E. M. Fenwick, A. Marty, and E. Neher, *J. Physiol. (London)* **331**, 577 (1982).
[8] R. Horn and C. A. Vandenberg, *J. Gen. Physiol.* **84**, 505 (1984).
[9] F. Conti and W. Stühmer, *Eur. Biophys. J.* **17**, 53 (1989).
[10] S. H. Heinemann, F. Conti, and W. Stühmer, this volume [22].
[11] F. Conti and E. Wanke, *Q. Rev. Biophys.* **8**, 451 (1975).
[12] F. J. Sigworth, *Nature (London)* **270**, 265 (1977).
[13] F. J. Sigworth, *J. Physiol. (London)* **307**, 97 (1980).
[14] A. Marty and E. Neher, *in* "Single-Channel Recording" (B. Sakmann and E. Neher, eds.), p. 107. Plenum, New York, 1983.

fin cells for estimating the temperature and pressure dependence of the conductance of voltage-activated sodium channels.

Nonstationary Noise Analysis

The purpose of noise analysis is to relate macroscopic observables, such as the total ionic current, to microscopic parameters like the single-channel current i, the number of functional channels in the membrane N, and the probability that the channels are open under a given condition, p_{open}. For a homogeneous population of statistically independent channels, the mean, $I(t)$, and the curent variance, $\sigma_I(t)^2$, are given by[15,16]

$$I(t) = Nip_{open}(t) \tag{1}$$

$$\sigma_I(t)^2 = Ni^2 p_{open}(t)[1 - p_{open}(t)] \tag{2}$$

Provided that I and σ^2 are estimated for various open probabilities, i and N can be determined by data fitting procedures according to

$$\sigma_I(I)^2 = Ii - I^2/N \tag{3}$$

as first introduced by Sigworth.[12,13,17] It is customary to convert i estimates to a single-channel conductance γ, obtained by knowledge of the specific reversal potential, E_{rev}, and the potential set by the voltage clamp, E_{com},

$$\gamma = i/(E_{com} - E_{rev}) \tag{4}$$

Note that the latter equation implies a linear relationship between i and the voltage only if γ is assumed to be constant.

Effect of Series Resistance

In practice the single-channel current in Eq. (3) may vary because the variable mean current $I(t)$ gives rise to a variable voltage drop across the series resistance R_s which is determined by the electrical access from the pipette electrode to the cell membrane. For small values of IR_s we have

$$i = i^* - \gamma^* IR_s \tag{5}$$

where the asterisk denotes quantities ideally measurable for a membrane potential equal to E_{com}. Thus, it follows

$$\sigma_I^2 = i^*I - \gamma^* I^2 R_s - I^2/N$$

$$= i^*I - \frac{1 + \Gamma_{max} R_s}{N} I^2 \tag{6}$$

[15] G. Ehrenstein, H. Lecar, and R. Nossal, *J. Gen. Physiol.* **55**, 119 (1970).
[16] T. Begenisich and C. F. Stevens, *Biophys. J.* **15**, 843 (1975).
[17] F. J. Sigworth, *in* "Membranes, Channels and Noise" (R. S. Eisenberg, M. Frank, and C. F. Stevens, eds.), p. 24. Plenum, New York, 1984.

where $\Gamma_{max} = \gamma^*N$ is the maximal macroscopic conductance. This result implies that a finite value of R_s does not affect the estimates of i^* derived from nonstationary noise analysis, provided a uniform voltage clamp exists all over the cell membrane. For small cells without big membrane invaginations this is a fair approximation, because R_s is mainly determined by the pipette tip opening.[14]

Equation (6) can be written formally as Eq. (3):

$$\sigma_I^2 = i^*I - I^2/N^* \tag{7}$$

showing that R_s causes a wrong estimate of the number of channels by the factor $(1 + N\gamma R_s)^{-1}$. In most cases this correction factor is expected to be close to unity, except for whole-cell measurements from large cells where N exceeds 1000. For example, $\gamma^* = 20$ pS, $R_s = 5$ MΩ, and $N = 1000$ would yield an error of 10% in the estimate of N.

A correct fitting procedure according to Eq. (7) requires proper account to be taken of the errors involved in the determination of σ^2 and I, errors which vary very strongly as a function of I. Another important weighting factor in fitting σ^2 versus I data is due to the strong correlation expected to exist between σ^2 measurements separated in time by intervals of the order of the time constants which characterize the time course of the voltage clamp responses. The theoretical considerations underlying the fitting procedure that we have used to cope with these problems are detailed in a later section.

Bandwidth Limitations

Because the bandwidth of the recordings is limited, the measured variance will be smaller than the true variance. To estimate the effect of bandwidth consider the case of a channel whose gating is described by a single relaxation process with the time constant τ. The variance of the channel noise is then the total integral of a Lorentzian spectrum

$$\sigma^2 = \int_0^\infty \frac{S_0}{1 + (f/f_N)^2} \, df$$

$$= \frac{\pi}{2} S_0 f_N \tag{8}$$

where f is the frequency and $f_N = 1/(2\pi\tau)$.

Further assuming a filter with an infinitely sharp cutoff at f_c the measured variance is

$$\sigma_{(m)}^2 = S_0 f_N \tan^{-1}(f_c/f_N) \tag{9}$$

yielding a relative deviation from σ^2 of

$$\frac{\sigma^2 - \sigma_{(m)}^2}{\sigma^2} = 1 - \frac{2}{\pi} \tan^{-1}(f_c/f_N) \tag{10}$$

Thus, with $f_N = 1$ kHz ($\tau \cong 160$ μsec) and a recording bandwidth of 10 kHz, the measured variance is only 6.3% smaller than the theoretical variance. The deviation is 9.1% for a one-pole filter.

Stimulation Protocol

Following the succession of operations actually performed in the off-line analysis of the data we shall describe the protocol of analysis in two steps: (1) determination of mean currents and variances and (2) fitting of these data according to Eq. (7). Before, however, we describe the stimulation protocol tailored to noise measurements.

The protocol of voltage stimulation for noise measurements was designed to optimize the subsequent off-line analysis of the data. It consisted of repeated alternating sequences of test stimulations — from a stationary holding potential, E_H, of -90 mV to test potentials between -10 and $+30$ mV — and control stimulations having the same temporal pattern but voltage amplitudes of only ± 30 mV relative to a control holding potential of -100 mV. Each test sequence consisted of 20 identical stimulations, and each control sequence consisted of 4 stimulations. Successive stimulations were repeated after a pause period of about 1 sec. For a typical noise measurement at any test potential these alternating sequences were applied successively at least 7 times for collecting a total of at least 140 test records, yielding a theoretical 10% accuracy of the variance measurements and, therefore, of the single-channel conductance estimates (see below). Thus, the measurement lasted about 3 min. In between successive noise measurements at different membrane potentials and/or under different temperature or pressure conditions, standard protocols of stimulation were applied to collect information about the macroscopic current–voltage relationship and the voltage dependence of channel inactivation. These protocols allowed us to characterize the size of the overall drifts which had occurred in the preparation during the noise measurements, owing to rundown or, in the case of whole-cell recordings, to diffusion of intracellular substances initiated by the pipette perfusion.[7]

Analysis of Noise Records

Very generally, fluctuation analysis yields estimates of the parameters which characterize elementary events with an accuracy increasing linearly with the square root of the number of independent samples and therefore with the square root of the time during which fluctuations are measured. When applied to biological preparations this basic principle finds itself in conflict with the intrinsic property of these systems of undergoing steady, irreversible modifications. Therefore, it is mandatory to organize the analysis of the data so that there is a minimum contamination of the estimated properties of reversible fluctuations from systematic irreversible changes of the system.

Difference Records

Ideally one could calculate the ensemble variance as the mean-square deviation of each record from the mean. However, particularly in patch clamp recordings, slow drifts in the electrode potentials, changes in cell capacitance, washout phenomena, and irreversible shifts in the voltage dependence of ion channels are known to occur as function of time.[14] Such linear drifts are largely eliminated if the variance is calculated from the ensemble average of the squared differences of successive records, $\delta\xi(t)$[18]:

$$\sigma_I^2 = \langle \delta\xi(t)^2 \rangle /2 \qquad (11)$$

For the estimate of the variance according to Eq. (11) all couples of successive records [i.e., (1,2), (2,3), . . .] can be used rather than only the nonoverlapping ones, because it can be shown that by this procedure the accuracy of the variance estimates is improved by a factor of $2/(3^{1/2})$ (see Appendix I). An example of analysis of an experiment with large leak currents and a changing, uncompensated capacitance is shown in Fig. 1.

Discarding "Bad" Records

All test and control records were analyzed for background noise by calculating the variance of the baseline current σ_B^2 sampled during the first few milliseconds of the records at the holding potential. A first upper limit for the maximum acceptable variance in any record is initially set by the analyzer in order to discard records that for some reason (transient deterioration of the pipette seal, extraneous electrical artifacts, etc.) have an obvious abnormal background noise. For the remaining records the mean, $\langle \sigma_B^2 \rangle$, and the root mean square (rms) deviation of σ_B^2 from the mean, $\Delta\sigma_B^2$, was calculated and used for a second selection, based this time on the objective principle that any record with $\sigma_B^2 > \langle \sigma_B^2 \rangle + 4\Delta\sigma_B^2$ is most likely affected by some artifactual extra noise. In the majority of our measurements both of these tests were passed by all records, but a few percent of noise sequences contained one to four records to be discarded according to these tests. In all these cases σ_B^2 was at least 50% higher than its maximum allowed value, clearly legitimating the selection performed by the algorithm.

Analysis of Leak Records

First a further check for the absence of occasional artifacts was performed by discarding any record which differed at any point from the next

[18] F. Conti, B. Neumke, W. Nonner, and R. Stämpfli, *J. Physiol. (London)* **308**, 217 (1980).

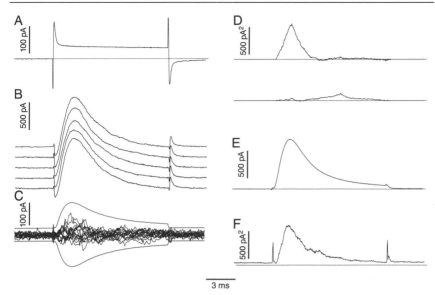

Fig. 1. Steps in nonstationary noise analysis of records in response to test pulses to -10 mV at 14°. (A) Averaged leak record ($N = 72$) yielding a leak reversal potential of 18 mV. The maximal accepted baseline noise was 5.96 pA (rms); the estimated baseline variance was 20.5 pA². Four leak records did not pass the criterion and were discarded. (B) Consecutive individual current traces during test pulses without leak correction. (C) Difference records with a boundary of 7 times the expected standard deviation [see Eq. (17)] determining the criterion for discarding records with excessively large noise. (D) Two covariance functions centered at the time to the peak current, t_p, and at $3t_p$, respectively, showing a correlation time of 1.1 msec. (E) Leak-corrected mean current ($N = 219$); one record had to be discarded. (F) Ensemble variance record.

one by more than expected on the basis of the previous measurement of $\langle \sigma_B^2 \rangle$ (apart from a constant difference arising from a baseline shift). The analysis of the remaining records was accomplished in two steps. For each record, the difference between the mean baseline current, I_i^H, and the mean current, I_i^C, during the second half of the main control segment at voltage E_C is computed in order to obtain an estimate of the leak reversal potential, E_L:

$$E_L = \frac{1}{N_L} \sum_{i=1}^{N_L} \frac{I_i^C E_H - I_i^H E_C}{I_i^C - I_i^H} \tag{12}$$

Then, the ensemble mean of all leak records $\xi_i^L(t)$ was calculated

$$\langle \xi_i^L(t) \rangle = \frac{1}{N_L} \sum_{i=1}^{N_L} \xi_i^L(t) \tag{13}$$

as illustrated in Fig. 1A. The purpose of calculating E_L is to correct possible artifacts in the analysis of test records occurring when successive records have differences in the mean baseline current which are statistically unexpected, indicating changes in leak or seal conductance, γ_L, between the records. It would be incorrect to cope with these changes by simply assuming that the difference record has a constant bias, because the variation of leak current is expected to be given by

$$\Delta I_L = \Delta \gamma_L (E - E_L) \qquad (14)$$

so that it will be different for the holding potential and the test potential.

Analysis of Test Records

The first step in analyzing test records is simply to obtain an ensemble average from all of them, which is then corrected for linear leakage and capacitive components by subtracting the ensemble average of the leak records appropriately scaled. The mean record obtained in this way was then used as a first estimate of the time course of the mean currents to perform a further test of the quality of each individual record and discard those which contained deviations from adjacent records much larger than statistically expected. This test was based on the expectation that at any time t,

$$\text{Prob}\{|\delta\xi(t)| > 5[2(\sigma_I^2 + \langle \sigma_B^2 \rangle)]^{1/2}\} < 10^{-6} \qquad (15)$$

so that any couple of records for which $|\delta\xi(t)| > 7(\sigma_I^2 + \langle \sigma_B^2 \rangle)^{1/2}$ can be safely discarded as being affected by some artifact. At this state, before any actual measurement of σ_I, the latter quantity could only be estimated theoretically from its upper limit:

$$\sigma_I^2 = iI(1 - p_{\text{open}}) < iI \qquad (16)$$

For estimating i we also had to rely on already analyzed data yielding a reasonable estimate for the single-channel conductance. Therefore, the test for accepting any couple of records was

$$|\delta\xi(t)| < 7[\gamma(E_{\text{com}} - E_{\text{rev}})I + \langle \sigma_B^2 \rangle]^{1/2} \qquad (17)$$

In Fig. 1C consecutive difference records are shown together with the envelope determined by the criterion of Eq. (17). In 70% of the measurements this test was passed by all test records. In the worst experiment 5 records out of 120 were discarded. For the final analysis only the selected records were used to evaluate the mean current (Fig. 1E). Then, for each pair of successive selected records a "pure fluctuation record" was constructed by taking the difference of the original records. If the difference in the time averaged baseline current of the two records exceeded its maximum expected random fluctuation $[\pm 4(\langle \sigma_B^2 \rangle T_0 B)^{1/2}$, where B is the bandwidth of the recordings and T_0 the duration of the recorded baseline trace], the difference record was corrected on the basis of the assumption that the baseline change was due to a change in leak conductance. The fluctuation records were then used to compute the ensemble variance according to Eq. (11) (see Fig. 1F) and autocovariance functions centered at different times, t_i,

$$\phi(t_i, t) = \langle \delta\xi(t_i)\delta\xi(t) \rangle \tag{18}$$

Figure 1D shows two covariances centered at the time to the peak current, t_p, and at $3t_p$, respectively. The covariances can be used for consistency checks that the measured noise has correlation features expected for the fluctuations under investigation. This is of particular importance, for example, for gating noise experiments, where one wants to verify that no ionic currents contaminate the gating currents.[9] The covariances are also used as a criterion for grouping the variance versus current data in bins as discussed in the next section.

Analysis of Variance versus Current Plots

For the correct attribution of weights to the estimates of I and σ_I^2 two important considerations must be made. (1) Each estimate of I and σ_I^2 is affected by an error proportional to σ_I and to σ_I^2, respectively, which depends on I. Therefore, estimates which were obtained near the peak of the mean current, I_p, must be weighted less than those obtained for small values of I if the maximal p_{open} is smaller than 0.5. (2) Fluctuations in the current measured at any time are strongly correlated with those measured within time intervals of the order of the intrinsic correlation times of channel fluctuations, which also characterize the macroscopic current kinetics.

For a constant sampling interval, a plot of all $\{I, \sigma_I^2\}$ estimates would show a much higher density of points for I values collected during the slow phase of the response, attributing excessive independent weight to data that are not independent. To account at least partially for correlations the mean current and variance data are grouped into a smaller number of bins corresponding to the division of the duration of the main pulse segment into successive intervals of variable length such that (1) I does not vary by more than $I_p/20$ within each interval and (2) the interval length does not exceed the estimated correlation time, t_c. Mean variances within any bin were plotted against the corresponding mean current. The standard error of any such individual data point was then estimated taking into account correlations of data only within each bin (see Appendix II). This assured that the data points were now approximately statistically independent such that least-squares fitting methods could be used. The data, weighted according to the inverse of the sum of the squares of the standard errors of the

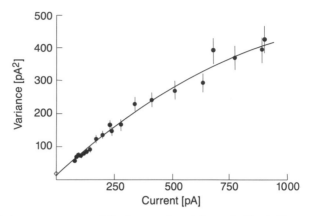

FIG. 2. Variance versus current for the experiment shown in Fig. 1. The result of the fit was $\gamma = 10.2$ pS, $p_{max} = 0.32$, $N = 4630$ channels. The curve represents a nonconstrained fit. The diamond shows the estimated background variance during the baseline before the pulse.

variance and the mean, were finally fitted by the relationship of Eq. (7) employing an effective variance least-squares method[19] which accounts for both errors in I and σ_I^2 and allowing a constant background variance, expected to be close to $\langle \sigma_B^2 \rangle$.

A plot of variance versus current corresponding to the raw data shown in Fig. 1 is given in Fig. 2, yielding an estimate of the single-channel current of 1.18 pA. The data fit according to Eq. (7) yield an estimate of the current flowing through an open channel at the test voltage. This value was converted to an estimate of the apparent single-channel conductance by dividing it by the driving force applied to the channels, using estimates of the reversal potential, E_{rev}, obtained from macroscopic current–voltage measurements. If patches had a too small number of channels, current –voltage plots were not reliable, and E_{rev} was assumed to have the standard value of $+55$ mV, the mean of measurements from large patches and whole cells. Reliable estimates of the number of channels could only be obtained for voltage steps which elicited large currents, that is, such that the open probability near the peak was larger than 0.2. This is seen in Fig. 3, which shows the analysis of current fluctuations at low potential and high temperature. An estimated maximal open probability of 0.1 does not allow an unconstrained fit according to Eq. (7). For this purpose we introduced two possible constraints; first, we fixed optionally the zero-current point to the measured baseline noise variance. The other constraint was a fixed number of channels, which can be estimated roughly. This does

[19] J. Orear, *Am. J. Phys.* **50**, 912 (1982).

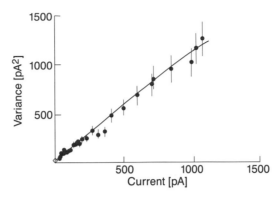

FIG. 3. Variance versus current from a measurement at 27° with test pulses to −30 mV. The correlation time was 0.56 msec. The result of the fit was $\gamma = 13.8$ pS, $p_{max} = 0.10$. Because the open probability of the channels was so low, no unconstrained fit could be performed. The background noise level was fixed to the value estimated from the baseline (diamond), and the number of channels was set to 9000.

not appreciably affect the estimate of i^*, but it obviously excludes the possibility of estimating N independently.

It should be stressed that, no matter how laborious the whole procedure might appear, the actual computer time required by our PDP 11/73 computer to perform the whole analysis, starting from 200 records of raw sampled data of 256 points each and ending with an estimate for γ and for N^* was 1 min for step 1 and 10 sec for step 2. With computers used in the laboratory nowadays the process could be faster by another order of magnitude such that the time for analysis poses no real limit.

Application to Patch Clamp Data

Recordings of single-channel events undoubtedly contain more information than noise measurements about the gating kinetics and the conductance of specific ion channels.[20] However, in spite of the perfected recording technique currently available, there is still a broad variety of cases where application of single-channel analysis is not feasible. Among the possible reasons are a too small signal-to-noise ratio making it impossible to identify single channel openings and the amount of time needed to analyze such recordings, particularly if ion channel screening is desired.

To illustrate the application of nonstationary noise analysis, we investigated the effect of temperature and hydrostatic pressure on the single-

[20] F. J. Sigworth and J. Zhou, this volume [52].

channel conductance of voltage-activated sodium channels in bovine adrenal chromaffin cells. The experimental conditions were such that single-channel analysis would have failed because of prohibitively large background noise (high-pressure recordings) or because of too short durations of the single-channel open state (high temperature).

Materials and Methods

Chromaffin cells from the medulla of bovine adrenal glands were prepared and cultured as described by Fenwick et al.[7] All measurements were performed in either whole-cell configuration or on excised outside-out patches, in both instances using the methods described by Hamill et al.[5] Standard patch pipette holders were used in all experiments, except for those on the effect of pressure, for which we used a specially designed holder and cable connection to the headstage of the recording amplifier (EPC-7, List Electronics, Darmstadt, Germany). The methods for pressurization of cells and membrane patches under patch clamp control are described by Heinemann et al.[21]

Pipettes with large tip openings and resistances in the range of 1 to 3 $M\Omega$ were selected in order to reduce the access resistance, R_s, for whole-cell measurements. However, we did not make any further attempt to reduce the effective value of R_s by electronic compensation, since (as discussed above) the most important artifact generated by R_s in whole-cell recordings of voltage clamp currents turns out to have no consequences for the estimates of single-channel conductances from nonstationary noise analysis.

Current records were sampled and stored on-line on a Winchester disk with the aid of a PDP 11/73 computer, which also generated the voltage stimulation protocol. Before being sampled (at a rate of 25 kHz) the current records were passed through an 8-pole Bessel filter with a -3 dB cutoff frequency at 10 kHz.

Solutions. In most of the experiments described here the extracellular solution had the following composition (in mM): 140 NaCl, 2.8 KCl, 2 $MgCl_2$, 1 $CaCl_2$, 10 HEPES–NaOH, pH 7.2. In some control experiments with whole cells Co^{2+} was used to replace Ca^{2+} in order to ensure that our records of sodium currents were not contaminated by calcium currents. We found no obvious difference in the results obtained with the Ca^{2+}-free solutions, most likely because the contribution of the sodium channels to the total currents is in any case overwhelming. Furthermore, it is known that calcium channels subside after 10 min of cell dialysis.[7] Co^{2+} was used

---

[21] S. H. Heinemann, W. Stühmer, and F. Conti, *Proc. Natl. Acad. Sci. U.S.A.* **84,** 3229 (1987).

routinely in measurements from excised outside-out patches, where a significant contribution from few calcium channels might have occasionally and unpredictably biased our data. The pipette solution which was invariably used had the following composition (in mM): 70 CsCl, 70 CsF, 1 $MgCl_2$, 0.5 $CaCl_2$, 10 HEPES–NaOH, 11 EGTA–NaOH, pH 7.2.

Effect of Temperature and Pressure on Sodium Channels

The dependence of the ionic permeabilities of excitable membranes on intensive thermodynamic parameters like temperature and pressure, besides having important physiological consequences on the adaptation of biological systems to different environments, yields information about the molecular mechanisms underlying the operation of ion channels. Changes with temperature and pressure in the time course of the relaxation of ionic currents following a step in membrane potential in a classic voltage clamp experiment can be directly ascribed to variations of the rate constants of the conformational transitions of ion-channel proteins, and, within any assumed kinetic scheme, can be interpreted unambiguously in terms of activation enthalpies and activation volumes of these transitions.[22] However, variations in the absolute values of the ionic currents yield only information about the product of two quantities which can both be strongly dependent on the external parameters. One is the total number of activatable ion channels; the other is the conductance or permeability of a single open channel. Most literature on temperature effects has assumed that N is constant, apart from irreversible rundowns of the biological preparation. However, slow inactivation phenomena, by which the number of normally functioning ion channels strongly depends on the holding membrane potential, are known.[23,24] It has also been clearly demonstrated that transitions of functional ion channels to conformational states from which they cannot be easily resumed for normal activity are brought about by decreasing temperature[25] or increasing pressure.[26] The only possibility to quantify these phenomena reliably is to measure the single-channel conductance directly.

The effect of temperature on the conductance of single sodium channels was estimated by averaging over 14 experiments at temperatures between 9° and 28°. Measurements at various temperatures on the same

[22] F. Conti, *Neurol. Neurobiol.* **20**, 25 (1986).
[23] J. M. Fox, *Biochim. Biophys. Acta* **426**, 245 (1976).
[24] B. Rudy, *J. Physiol. (London)* **283**, 1 (1978).
[25] D. R. Matteson and C. M. Armstrong, *J. Gen. Physiol.* **79**, 739 (1982).
[26] F. Conti, R. Fioravanti, J. R. Segal, and W. Stühmer, *J. Membr. Biol.* **69**, 23 (1982).

TABLE I
PRESSURE EFFECT ON SODIUM CHANNELS[a]

Potential (mV)	Pressure (MPa)	Temperature (°C)	N_{rec}	N_{ch}	γ (pS)	p_{open}	τ (msec)
−10	0.1	23.2	89	3850	12.9	0.16	0.64
−10	45.5	23.8	70	2170	11.9	0.13	1.12
−10	0.1	23.5	88	1792	11.7	0.13	0.64
−10	0.1	21.0	128	7370	11.5	0.29	0.48
−10	43.3	21.6	89	3000	12.6	0.26	0.64
−10	0.1	21.3	126	2600	11.5	0.33	0.24
−10	0.1	21.6	114	5550	11.4	0.21	0.64
−10	41.7	22.0	114	2350	13.9	0.18	1.04
−10	0.1	21.7	114	2060	14.3	0.23	0.56
10	0.1	16.1	67	1670	10.5	0.27	0.64
10	40.0	16.3	72	910	11.6	0.22	1.28
10	0.1	16.0	72	920	11.5	0.30	0.64

[a] The effect of hydrostatic pressure on the single-channel conductance γ, the maximal open probability p_{open}, the correlation time τ, and the estimated number of functional channels under investigation N_{ch}. N_{rec} denotes the number of raw current traces which passed all tests for unexpectedly large noise and that were used for the final computation of the mean currents and ensemble variances.

cell very rarely succeeded because of the long time required to obtain a new equilibrium. The determined temperature dependence of γ_{Na} of 1.3 (±0.1)/10° at −10 mV test potential compares well with values reported for other ion channels measured by single-channel analysis.[21]

The determination of the effect of hydrostatic pressure on the sodium channel is experimentally more involved, because various experimental manipulations are needed to pressurize the patch clamp holder with the pipette and the attached cell.[21] However, because changes in hydrostatic pressure could be obtained much faster than changes in temperature, experiments could be performed on single cells, which is of particular importance to show the reversibility of the effects. In Table I four such experiments with pressure applications of approximately 40 MPa (1 MPa ≅ 10 atm) are listed. It is seen that within the experimental error the single-channel conductance is not affected by this pressure (11.9 ± 1.2 pS at ambient pressure and 12.5 ± 1.0 pS at 40.0–45.5 MPa) as reported for the alamethicin channel[27] and the acetylcholine receptor channel.[21] The correlation times increase on pressure application in a reversible manner as already shown by the measurement of mean current kinet-

[27] L. J. Bruner and J. E. Hall, *Biophys. J.* **44**, 39 (1983).

ics.[26,28,29] The increase of the time constant by the factor of 1.8 ± 0.1 between 0.1 and 43 MPa corresponds to an apparent activation volume of 60 Å³.

The consistent irreversible decrease in the number of functional channels amounting to approximately 50% of the initial number on pressurization to 40 MPa is remarkable. This phenomenon was anticipated by the earlier measurements of peak currents. However, it could not be demonstrated conclusively, because no reliable information about the single-channel conductance was available.

Conclusions

The method of nonstationary noise analysis was automated employing statistical considerations regarding the "quality" of individual records. The procedure presented allows a rapid analysis of noise records obtained under nonideal experimental conditions based on objective selection criteria.

Although single-channel analysis has surpassed noise analysis in many instances, there are still regimes where it cannot be successfully applied for various reasons. In this sense nonstationary noise analysis will retain its value for electrophysiological research in particular, since ever fainter electrical signals are being investigated in biological membranes.

To demonstrate the methods we measured the temperature and pressure dependence of the sodium channel conductance. In both respects the sodium channel showed features similar to other ion channels (e.g., the acetylcholine receptor channel), namely, having a weak temperature dependence of γ_{Na} of $1.3/10°$ and no appreciable pressure dependence in the range explored. Both findings are in accord with the physical picture of a rather free ion diffusion through the channel pore which, unlike the channel gating mechanism, does not involve protein rearrangements associated with measurable activation volumes.

Appendix I: Errors of Variance Estimates

In this appendix we calculate the error involved in the estimate of the variance using difference records. Let us first write the measured signal of record i as sum of its expectation, assumed to be independent of i, and a fluctuating term:

$$\xi_i(t) = x_0(t) + \delta x_i(t) \tag{A.I.1}$$

[28] J. V. Henderson and D. L. Gilbert, *Nature (London)* **258**, 351 (1975).
[29] S. H. Heinemann, F. Conti, W. Stühmer, and E. Neher, *J. Gen. Physiol.* **90**, 765 (1987).

Arising from the superposition of a large number of independent elementary fluctuations, $\delta x_i(t)$ can be safely assumed to be a Gaussian stochastic variable with zero mean. Using angular brackets to denote expectation values and omitting the explicit writing of the time dependence, unless necessary, we have $\langle \xi_i \rangle = x_0$; $\langle \delta x_i \rangle = 0$; $\text{Var}(\xi_i) = \langle \delta x_i^2 \rangle = \sigma^2$; $\langle \delta x_i^4 \rangle = 3\sigma^4$. It follows for the variance of the squared fluctuations

$$\text{Var}(\delta x_i^2) = \langle \delta x_i^4 \rangle - \langle \delta x_i^2 \rangle^2 = 2\sigma^4 \tag{A.I.2}$$

Now we consider three possible strategies for estimating σ^2.

Case I: Independent Records

Using upper bars to denote ensemble averages over N records we have

$$\bar{\xi} = \frac{1}{N} \sum_{i=1}^{N} \xi_i = x_0 + \overline{\delta x} \tag{A.I.3}$$

If all δx_i and δx_j are statistically independent the following equations hold: $\langle \delta x_i \delta x_j \rangle = 0$, $\langle \overline{\delta x} \delta x_j \rangle = \sigma^2/N$, $\langle \delta x_i^2 \rangle = \sigma^2$, $\langle \overline{\delta x}^2 \rangle = \sigma^2/N$, $\langle \delta x_i^2 \delta x_j^2 \rangle = \sigma^4$, $\langle \delta x_i^4 \rangle = 3\sigma^4$, $\langle \overline{\delta x}^4 \rangle = 3\sigma^4/N$. Now we introduce V_ξ, the experimental estimate of σ^2:

$$V_\xi = \frac{1}{N-1} \sum_{i=1}^{N} (\xi_i - \bar{\xi})^2$$

$$= \frac{1}{N-1} \left[\left(\sum_{i=1}^{N} \delta x_i^2 \right) - N\overline{\delta x}^2 \right] \tag{A.I.4}$$

$$\langle V_\xi \rangle = \frac{N\sigma^2 - N(\sigma^2/N)}{N-1} = \sigma^2 \tag{A.I.5}$$

$$\text{Var}(V_\xi) = \langle V_\xi^2 \rangle - \langle V_\xi \rangle^2$$

$$= \frac{\langle (\Sigma \, \delta x_i^2)^2 \rangle + N^2 \langle \overline{\delta x}^4 \rangle - 2N \langle \overline{\delta x}^2 \Sigma \, \delta x_i^2 \rangle}{(N-1)^2} - \sigma^4 \tag{A.I.6}$$

Thus, as expected, the standard error squared of the "variance estimate from N records,"

$$\text{Var}(V_\xi) = \frac{2\sigma^4}{N-1} \tag{A.I.7}$$

is $1/(N-1)$ times the variance of the "squared fluctuation of only one record."

Case II: Nonoverlapping Difference Records

Now we define the difference between two consecutive records:

$$\eta_i = \xi_i - \xi_{i+1} = \delta x_i - \delta x_{i+1} \tag{A.I.8}$$

It can be shown that $\langle \eta_i \rangle = 0$, $\langle \eta_i^2 \rangle = 2\sigma^2$, $\langle \eta_i^2 \eta_{i+1}^2 \rangle = 6\sigma^4$, $\langle \eta_i^4 \rangle = 12\sigma^4$, and $\text{Var}(\eta_i^2) = \langle \eta_i^4 \rangle - 4\sigma^4 = 8\sigma^4$. We estimate σ^2 experimentally by taking the average

$$V_\eta = \frac{1}{N} \sum_{i=0}^{N/2-1} \eta_{2i+1}^2 \tag{A.I.9}$$

Because no overlapping traces are used for averaging, η_{2i+1} is independent of η_{2j+1}. It follows that $\langle V_\eta \rangle = \sigma^2$ and

$$\text{Var}(V_\eta) = \frac{1}{N^2} \sum_{i=0}^{N/2-1} \text{Var}(\eta_{2i+1}^2) = \frac{4\sigma^4}{N} \tag{A.I.10}$$

Therefore, in this case the error in the experimental estimate of σ^2 is the same that would occur in Case I if we processed $N/2 + 1$ independent records without drifts.

Case III: Overlapping Difference Records

In the case of overlapping difference records, which is what we did in our analysis, we take all η_i values for estimating σ^2; that is, we use the information of all pairs of consecutive records. Let us call this estimate V_η':

$$V_\eta' = \frac{1}{2(N-1)} \sum_{i=1}^{N-1} \eta_i^2 \tag{A.I.11}$$

Again we obtain $\langle V_\eta' \rangle = \sigma^2$. The variance, however, is

$$\text{Var}(V_\eta') = \frac{\left\langle \left(\sum\limits_{i=1}^{N-1} \eta_i^2 \right)^2 \right\rangle}{4(N-1)^2} - \sigma^4 \tag{A.I.12}$$

After some algebra one obtains

$$\text{Var}(V_\eta') = \frac{3N-4}{(N-1)^2} \sigma^4$$

$$\cong \frac{3\sigma^4}{N} \tag{A.I.13}$$

for large values of N. Thus, in this case the error in the estimate of σ^2 is equivalent to that obtainable from the processing of approximately $2/3\ N$ independent records without drift. Compared with using nonoverlapping difference records only, the precision for the estimate of σ^2 is increased by a factor of $2/(3^{1/2})$.

Appendix II: Variance of Time Averages

Let $I(t_1)$, $I(t_2)$, . . . , $I(t_n)$ be n measurements of the stochastic variable $I(t)$ at different times. Letting c_{ij} be the cross-correlation between $I(t_i)$ and

$I(t_j)$, $c_{ij} = \langle \delta I(t_i)\delta I(t_j)\rangle$ where δI indicates the deviation from the expectation value. If we take the time average of $I(t)$ in the interval (t_1, t_n), we get a new random variable $\bar{I}^t = \frac{1}{n}\sum_{i=1}^{n} I(t_i)$. The variance of $\bar{I}^t$ is

$$\mathrm{Var}(\bar{I}^t) = \left\langle \left[\frac{1}{n}\sum_{i=1}^{n} \delta I(t_i) \right]^2 \right\rangle$$

$$= \frac{1}{n^2}\sum_{i,j}^{n} c_{ij} \tag{A.II.1}$$

Assume that for the set of times under consideration $(t_1, t_2, \ldots, t_n)$ which constitute a bin the covariance can be written as:

$$c_{ij} = \sigma^2 \exp\left\{ -\frac{|t_i - t_j|}{t_c} \right\} \tag{A.II.2}$$

where t_c is the correlation time and c_{ii} $(=\sigma^2)$ is assumed to be fairly independent of i within the bin of the data. Then,

$$\mathrm{Var}(\bar{I}^t) = \frac{\sigma^2}{n^2}\sum_{i,j}^{n} \exp\left\{ -\frac{|t_i - t_j|}{t_c} \right\} \tag{A.II.3}$$

Equation (A.II.3), showing incidentally that $\mathrm{Var}(\bar{I}^t) > \sigma^2/n$, can be applied directly to estimate the error of the time-averaged current within each bin, given the knowledge of the correlation time t_c obtained from autocovariance measurements.

To calculate the error of the time average of the estimated variance within a bin we proceed in a similar manner, but we need first to evaluate the autocovariance, ϕ_{ij}, of the estimated variances:

$$\phi_{ij} = \langle \delta I^2(t_i)\delta I^2(t_j)\rangle - \langle \delta I^2(t_i)\rangle\langle \delta I^2(t_j)\rangle \tag{A.II.4}$$

If, as assumed throughout these appendices, $\delta I(t)$ is a Gaussian stochastic variable, it can be shown that

$$\phi_{ij} = 2c_{ij}^2 = 2\langle \delta I(t_i)\delta I(t_j)\rangle^2 \tag{A.II.5}$$

Equation (A.II.5) shows that the correlation between squared fluctuations dies out twice as steeply as that of simple fluctuations. Thus, by the same argument leading to Eq. (A.II.3),

$$\mathrm{Var}(\overline{\delta I^{2t}}) = \frac{2\sigma^4}{n^2}\sum_{i,j}^{n} \exp\left\{ -\frac{2|t_i - t_j|}{t_c} \right\} \tag{A.II.6}$$

[8] Prevention of Rundown in Electrophysiological Recording

By RICHARD HORN and STEPHEN J. KORN

Introduction

The invention of the gigaseal patch clamp technique[1] has led to stupendous strides in the study of electrophysiological phenomena. Of the several variants of this methodology, whole-cell recording is the most popular, largely because of the ease and speed of obtaining and analyzing macroscopic currents. Nonetheless, standard whole-cell recording suffers from several problems related to the need to break the membrane at the interface between the cytoplasm and the interior of the patch pipette. The exchange of material between the interior of the cell and the pipette solution results in the dilution, or "washout," of cytoplasmic constituents that may affect the functional properties of ion channels. Common examples of the deleterious consequences of the diluted cytoplasm include the decline (i.e., rundown) of calcium (Ca^{2+}) channel function and the disappearance of receptor-activated second-messenger responses shortly after breaking into a cell. Other, perhaps more subtle, changes accompanying whole-cell recording include alterations in the activity of cytoplasmic buffers and enzymes. We shall discuss and evaluate several alternative methods that reduce or avoid the problems presented by whole-cell recording. We do not, however, consider methods used for single-channel recording, which include the cytoplasm-preserving cell-attached patch configuration.

Microelectrodes

Patch clamp recording has largely supplanted the use of microelectrodes, especially in experiments with small mammalian cells that are difficult to penetrate benignly with a microelectrode. Microelectrode recording nonetheless retains several advantages, including (1) washout and dialysis of cytoplasm are minimized, (2) it is possible to impale cells that are inaccessible to a patch electrode, for example, cells deeply embedded in tissue; and (3) for large currents in very large cells, such as *Xenopus* oocytes, a two-microelectrode voltage clamp avoids the series resistance

---

[1] O. P. Hamill, A. Marty, E. Neher, B. Sakmann, and F. J. Sigworth, *Pluegers Arch.* **391,** 85 (1981).

METHODS IN ENZYMOLOGY, VOL. 207

errors arising from passing current and recording voltage through the same electrode, as done in whole-cell recording.

Disadvantages in the use of microelectrodes include (1) the electrical resistance between the cytoplasm and interior of the recording pipette is usually higher in microelectrode recording than in whole-cell recording. A high electrode resistance reduces the quality of the recording, especially in voltage clamp experiments; (2) microelectrodes are frequently more noxious than a patch pipette and may damage a cell[2]; and (3) there is some exchange between the pipette and cytoplasm. For example, microelectrodes filled with CsCl, instead of KCl, are effective in reducing potassium currents in impaled cells. Also, because large molecules (e.g., messenger RNA) may be pressure injected through microelectrodes, other substances may exchange at unknown rates between the cytoplasm and the electrode. Technical details for microelectrode recording have been described elsewhere.[3]

Big Patches: Loose and Tight

Cell-attached patch recording usually implies that cytoplasm is maintained on the intracellular surface of the plasma membrane. With this configuration one usually observes single-channel currents. However, if channel density is very high (e.g., see Ref.[4]) or if very large pipettes are used, macroscopic currents may be observed in cell-attached patches. Presumably such currents are modulable by second messengers and do not wash out.

One of the technical difficulties with large pipettes is that the probability of obtaining a gigohm seal is inversely related to pipette size. The current jargon thus differentiates macropatch recording (with a seal) and loose patch recording (without a seal). The seal improves the signal-to-noise ratio and increases the dynamic range of the recording amplifier. In certain experiments, however, loose patch recording may be preferable, for example, when multiple recordings are obtained from different locations on the same cell. Because the macroscopic currents through a large patch may be substantial, compared with the input impedance of the cell under study, it may be necessary to voltage clamp the cell with separate elec-

---

[2] E. M. Fenwick, A. Marty, and E. Neher, *J. Physiol. (London)* **331**, 557 (1982).
[3] T. G. Smith, H. Lecar, S. J. Redman, and P. W. Gage, eds., "Voltage and Patch Clamping with Microelectrodes." American Physiological Society, Bethesda, Maryland, 1985.
[4] J. Patlak and M. Ortiz, *J. Gen. Physiol.* **86**, 89 (1985).

trodes, in order to maintain the membrane potential during current flow through the patch. (For further details see Refs. 5 and 6).

Whole-Cell Recording with Supplemented Pipette Solutions

The most direct method to prevent washout in whole-cell recording is to supplement the pipette solution with ingredients that will maintain the normal function of the ionic currents under study. This approach has been used successfully with Ca^{2+} currents, which may be maintained with a soup containing substances that inhibit proteases and support phosphorylation by cyclic AMP (e.g., ATP, Mg^{2+}, catalytic subunit of cAMP-dependent protein kinase, leupeptin, and Ca^{2+} buffers[7]). This method may not, however, prevent the loss of hormone or transmitter responses, either because the essential ingredients are unknown or because they are added in the wrong concentrations. Furthermore, a second-messenger response to an extracellular stimulus involves transient changes in the concentration of intracellular substances; such transients are expected to be suppressed in whole-cell recording because the pipette solution tends to clamp cytoplasmic concentrations. The extent of suppression will depend on the time course of the transient and the rates of diffusion of relevant substances between the cell and the pipette. In one report a cytoplasmic extract, added to the pipette solution in whole-cell recording, inhibited the rundown of a hormone response in pituitary cells.[8]

Whole-Cell Recording with Small Pipettes

The speed of washout depends on the resistance of a whole-cell pipette and the size of the cell.[9] Therefore small pipettes have been favored in some experiments.[10] The obvious penalty in such an approach is the concomitant increase in series resistance between the pipette and the cell

[5] W. M. Roberts and W. Almers, this volume [9].

[6] W. Stühmer, this volume [19].

[7] J. Chad, D. Kalman, and D. Armstrong, in "Cell Calcium and the Control of Membrane Transport" (D. C. Eaton and L. J. Mandel, eds.), Vol. 42, p. 167. Society of General Physiologists Series, Rockefeller Univ. Press, New York, 1987.

[8] B. Dufy, A. MacDermott, and J. L. Barker, Biochem. Biophys. Res. Commun. 137(1), 388 (1986).

[9] A. Marty and J. Zimmerberg, Cell. Signalling 1, 259 (1989).

[10] L. O. Trussell and M. B. Jackson, J. Neurosci. 7, 3306 (1987).

interior. Another problem with this approach is that the exchange is slowed, but not prevented. This means that the concentrations of some cytoplasmic substances are changing slowly during the course of an experiment, and that the rate of exchange will depend on the molecular weight of each constituent.[9,11] A final problem is that the series resistance often increases spontaneously when using small pipettes (i.e., they tend to "seal over").

Permeabilized Patch Recording

The prevention of rundown in whole-cell recording clearly requires a diminution of the exchange of material between the recording pipette and the cytoplasm. The methods in this section rely on the plasma membrane itself as a dialysis membrane. Ordinarily the membrane under a patch electrode during cell-attached patch recording is far too resistive either to measure the membrane potential of a cell or to control the cell's voltage while recording current. Nonetheless we have occasionally measured a fraction of the membrane potential through a cell-attached patch during spontaneous action potentials (Fig. 1). Of course this method can only yield qualitative results.

Peering through a patch into the cell was finally accomplished with the inclusion in the pipette solution of ionophores that permeabilized the patch.[12] Lindau and Fernandez[13] used ATP, which induces a conductance on the external surface of mast cells. When placed into the patch pipette in the cell-attached configuration, the series resistance is decreased from unmeasurably high to $100-5000$ MΩ. This is, however, approximately $100-200$ times greater than observed in whole-cell recording, resulting in slow capacitative currents in response to voltage steps, hence the name slow whole-cell recording.

The large series resistance and the specificity of ATP for a selected range of cell types limits the use of slow whole-cell recording. Both of these problems were obviated by the introduction of the polyene antibiotic nystatin as the ionophore.[14] This method is called perforated patch recording. Recently another polyene antibiotic, amphotericin B, has been introduced for perforated patch recordings.[15] The theory and methods for per-

[11] M. Pusch and E. Neher, *Pfluegers Arch.* **411**, 204 (1988).
[12] B. D. Gomperts and J. M. Fernandez, *Trends Biochem. Sci.* **10**, 414 (1985).
[13] M. Lindau and J. M. Fernandez, *Nature (London)* **319**, 150 (1986).
[14] R. Horn and A. Marty, *J. Gen. Physiol.* **92**, 145 (1988).
[15] J. Rae, K. Cooper, P. Gates, and M. Watsky, *J. Neurosci. Methods* **37**, 15 (1991).

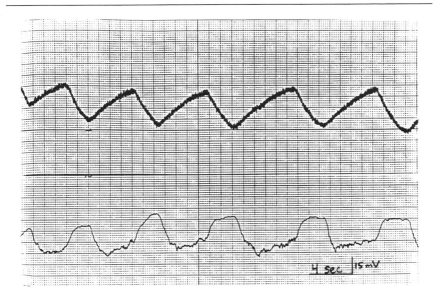

Fig. 1. Picoleak recording of fluorescence and membrane potential from an AtT-20 pituitary cell [see Ref. 17 and S. J. Korn, A. Bolden, and R. Horn, *J. Physiol. (London)* **439,** 423 (1991) for technical details]. The cell was loaded with the membrane-permeant form of Fura-2. The Fluorescence (top) shows a continuous measurement at an excitation wavelength of 380 nm; at this wavelength a decrease (downward trend) in fluorescence indicates an increase in intracellular Ca^{2+} concentration. The voltage (bottom), which was recorded from a cell-attached patch using a pipette filled with the normal Ringer's bath solution, shows a series of spontaneous long duration action potentials accompanied by a prolonged Ca^{2+} influx. The membrane potential is an unknown fraction of that inside the cell, and it depends on the resistance of the seal and the membrane patch under the patch pipette.

forated patch recording have been described extensively.[14-17] We present only a brief overview here.

In lipid bilayers and cell membranes nystatin and amphotericin B form channels that are selective to monovalent or uncharged molecules smaller than about 0.8 nm. Because these antibiotics cannot pass through a bilayer membrane, they do not enter the cell in perforated patch experiments. This is consistent with the longevity and high resistance of cells examined with this method. Recent experiments[18] also suggest that nystatin is incapable of diffusing laterally through the plasma membrane from the perforated

[16] S. J. Korn and R. Horn, *J. Gen. Physiol.* **94,** 789 (1989).
[17] S. J. Korn, A. Marty, J. A. Connor, and R. Horn, *in* "Methods in Neurosciences" (P. M. Conn. ed.), Vol. 4, p. 364. Academic Press, San Diego, California, 1991.
[18] R. Horn, *Biophys. J.* **60,** 329 (1991).

patch, under the glass–membrane seal, and into the surrounding plasma membrane. The series resistance achieved in perforated patch recordings is not much larger than observed for whole-cell recordings under similar conditions, and the polyene antibiotics are apparently promiscuous in their ability to form channels in a wide variety of membranes.

A detailed description of the methods used for perforated patch recording may be found in Refs. 15 and 17. A few new technical details[15] are worth mentioning: (1) the stock solution of antibiotic in dimethyl sulfoxide (DMSO) may lose its potency in the freezer and should be made up fresh every time the pipette solution is prepared, and (2) blunt-shaped pipettes are preferable to thin pipettes, presumably because of the membrane areas of patches. An abbreviated protocol follows.

1. Dissolve either nystatin or amphotericin B in DMSO (60 mg/ml). Add 20 μl of this stock to 5 ml of pipette solution and sonicate briefly ($\leq$ 20 sec). This solution retains potency for around 2 hr.

2. Fill the electrode tip (e.g., by dipping it for a few seconds) with antibiotic-free solution. Increase the amount of this solution in the electrode tip only if seals are a problem. Backfill the rest of the pipette with the solution containing nystatin or amphotericin B.

3. Get the patch pipette sealed onto a cell as quickly as possible, to prevent the movement of the antibiotic to the pipette tip, where it may inhibit seals.

4. Set a holding potential and observe capacitative transients in response to 10-mV pulses. Usually a slow transient appears within a few minutes. The time course of this transient becomes increasingly rapid as the patch becomes perforated with antibiotic. The series resistance reaches steady state within about 30 min. The final value of series resistance depends on the size and shape of the pipette and on the amount of antitiotic-free solution in the tip. Values below 3 MΩ have been reported for both nystatin and amphotericin B. It is not yet clear if either of these antibiotics has an advantage over the other, although it is possible that some membranes may be able to differentiate them.

Perforated patch recording preserves many of the components of the cytoplasm, including buffers, enzymes, and multivalent ions, while permitting the exchange of small monovalent ions, such as K^+, Na^+, Cs^+, and Cl^-. Initial reports indicate that the rundown of both ionic currents and second-messenger responses is reduced or eliminated. It is even possible to excise vesicles and study the modulation of single channels in the nonperforated membrane facing the bath.[15,19]

[19] E. S. Levitan and R. K. Kramer, *Nature (London)* **348,** 545 (1990).

Improvements in perforated patch recording are possible. For example, it would be useful to have a battery of ionophores with different selectivities, so that it would be possible to choose the ingredients filtered by the pores. Many ionophores are presently available, of course, but few have all the advantages of nystatin and amphotericin B, namely, fixed pore size, inability to pass through the patch into the cell, high enough conductance (and/or density of channels incorporated into the membrane) to allow reasonable voltage clamp of whole-cell currents, and weak dependence of the conductance on transmembrane voltage.

Conclusion

We have evaluated several methods that have been used to prevent rundown in electrophysiological experiments. These techniques preserve, to varying extents, the integrity of the cytoplasm. Each has its own virtues and pitfalls, and it is unlikely that one method will suffice for all purposes. In our laboratory, for example, we use most of the above methods, depending on the particular experiment. Usually the choice of technique involves a compromise between the ability to determine the composition of the intracellular solution and the preservation of the cytoplasm.

Acknowledgments

We thank Dr. James Rae for sending us a preprint of his paper.[15]

[20] S. J. Korn, A. Bolden, and R. Horn, *J. Physiol. (London)* **439**, 423 (1991).

[9] Patch Voltage Clamping with Low-Resistance Seals: Loose Patch Clamp

By WILLIAM M. ROBERTS and WOLFHARD ALMERS

Introduction

Patch voltage clamp methods employ smooth-tipped glass micropipettes to record electric currents across small regions of cell membrane. Unlike sharp pipettes, used to penetrate the lipid bilayer of the plasma membrane for intracellular recording, the smooth rim of a patch pipette nestles against the extracellular surface of the membrane, forming a seal that impedes the flow of electric current between the lumen of the pipette and the rest of the extracellular medium. The presence of a seal distinguishes a patch recording from an extracellular field-potential recording and makes it possible to collect current and control the membrane potential across a circumscribed patch of cell membrane.

Patch voltage clamp methods can be divided into two broad classes based on how tightly the rim of the recording pipette adheres to the cell. A "tight seal" forms under conditions that promote a strong attraction between atoms on the glass and lipid surfaces. Once initial contact is made, this attraction draws the fluid lipid bilayer into contact with the glass and forces nearly all of the water out of the space between. Tight seals have high electrical resistances (> 1 GΩ), extremely low electrical noise (< 1 pA), and cannot be broken without tearing the cell membrane. A "loose seal" forms when the attraction between glass and lipid is weak or obstructed by material on one or both surfaces. Loose seals usually inflict less damage on the cell membrane than tight seals, but at the cost of increased electrical noise. This chapter is concerned with methods that fall into the latter class, which have been synonymously termed "loose-seal" and "loose-patch" voltage clamp techniques.

Nearly all loose-seal methods involve recording from intact cell membranes, similar to the cell-attached configuration of tight-seal techniques.[1] Because loose seals are mechanically weak, it is not feasible to excise patches of loosely sealed membrane, as can be done with tight seals. However, a similar configuration has been achieved by sealing against the surface of a cut-open axon.[2-6] When studying very large cells, it is sometimes possible to rupture the patch of membrane beneath the tip of the pipette, to obtain a loose-seal analog of the tight-seal whole-cell configuration.[7]

Loose-seal methods are among the oldest recording techniques in electrophysiology. One of their first important uses was to record the extracellular potentials produced by synaptic currents.[8] Later, loose-seal voltage clamp recordings were made on crayfish muscle,[9] frog muscle,[10] snail

[1] O. P. Hamill, A. Marty, E. Neher, B. Sakmann, and F. J. Sigworth, *Pfluegers Arch.* **391**, 85 (1981).
[2] H. Fishman, *J. Membr. Biol.* **24**, 265 (1975).
[3] I. Llano and F. Bezanilla, *J. Gen. Physiol.* **83**, 133 (1984).
[4] J. M. Bekkers, N. G. Greef, and R. D. Keynes. *J. Physiol. (London)* **377**, 463 (1986).
[5] C. L. Schauf, *Can. J. Physiol. Pharmacol.* **65**, 568 (1987).
[6] C. L. Schauf, *Can. J. Physiol. Pharmacol.* **65**, 1220 (1987).
[7] P. G. Kostyuk and O. A. Krishtal, *J. Physiol. (London)* **270**, 545 (1977).
[8] Del Castillo and Katz, *J. Physiol. (London)* **132**, 630 (1956).
[9] A. Strickholm, *J. Gen. Physiol.* **44**, 1073 (1961).
[10] E. Neher, B. Sakmann, and J. H. Steinbach, *Pfluegers Arch.* **375**, 219 (1978).

neurons,[11-14] and squid axons.[2,15] In 1981, the momentous discovery by Hamill *et al.*,[1] namely, that freshly heat-polished glass micropipettes seal tightly onto clean cell membranes, breathed new life into the middle-aged field of ion channel biophysics and sent physiologists scurrying to investigate the electrical properties of small cells using the whole-cell configuration. While the excitement about tight-seal methods diverted attention away from other voltage clamp techniques, the decade of the 1980s also brought improvements and new uses for loose-seal methods that are the main subject of this review.

Loose-seal methods are now used in situations in which the advantages of low noise and mechanical stability provided by tight seals are outweighed by other factors. Some cell membranes, especially those covered by a basal lamina or other protective layers, simply will not seal tightly to glass. The mechanical stability of tight seals can also be a hindrance when the objective is to map the spatial distribution of ionic currents on the surface of a cell, because the pipette cannot be repositioned once a tight seal is made.[16] Even if no attempt is made to break the tight seal, it probably causes local disruptions. By contrast, loose-seal recording causes minimal perturbation. Seals can be made and broken many times, and the same pipette can be used to record from many locations on the same cell. Loose-seal recording is the method of choice to explore the distribution ion channels over cell membranes,[17-23] to make stable voltage clamp recordings lasting hours, and to study properties of ion channels that depend on the microscopic structure of the cell membrane and its spatial relationship to the cytoskeleton.[24,25]

[11] K. Frank and L. Tauc, *in* "The Cellular Function of Membrane Transport," p. 113. Prentice-Hall, Englewood Cliffs, New Jersey, 1964.

[12] E. Neher and H. D. Lux, *Pfluegers Arch.* **311,** 272 (1969).

[13] E. Neher, *Pfluegers Arch.* **322,** 35 (1971).

[14] R. Eckert and H. D. Lux, *J. Physiol. (London)* **254,** 129 (1976).

[15] I. Llano and F. Bezanilla, *Proc. Natl. Acad. Sci. U.S.A.* **77,** 7484 (1980).

[16] D. S. Corey and C. F. Stevens, *in* "Single-Channel Recording" (B. Sakmann and E. Neher, eds.), p. 107. Plenum, New York, 1983.

[17] W. Almers, P. R. Stanfield, and W. Stühmer, *J. Physiol. (London)* **336,** 261 (1983).

[18] W. J. Betz, J. H. Caldwell, and S. C. Kinnamon, *J. Physiol. (London)* **352,** 189 (1984).

[19] K. G. Beam, J. H. Caldwell, and D. T. Campbell, *Nature (London)* **313,** 558 (1985).

[20] J. H. Caldwell, D. T. Campbell, and K. G. Beam, *J. Gen. Physiol.* **87,** 907 (1986).

[21] W. M. Roberts, *J. Physiol. (London)* **388,** 213 (1987).

[22] J. H. Caldwell and R. L. Milton, *J. Physiol. (London)* **401,** 145 (1988).

[23] R. L. Milton and J. H. Caldwell, *J. Neurosci.* **10,** 885 (1990).

[24] W. M. Roberts, R. A. Jacobs, and A. J. Hudspeth, *J. Neurosci.* **10,** 3664 (1990).

[25] W. M. Roberts and A. J. Hudspeth, *Soc. Neurosci.* **13** *(Abstr.),* 177 (1987).

In this chapter, we discuss several loose-seal methods, concentrating on the possible difficulties associated with each and how they can be avoided. We focus on applications in which the potential across the membrane is controlled (voltage clamp). The basic loose-patch clamp,[26] which employs a single loosely sealed extracellular pipette and no intracellular electrodes, is appropriate for studying rapidly activating currents, such as voltage-gated Na^+ and K^+ currents, in muscles and other large cells with stable resting potentials. This technique is not well suited to studying currents that are much smaller or slower, such as the voltage-gated Ca^{2+} current in skeletal muscles, because of artifacts associated with the low resistances of the loose seals. However, more elaborate variations on the basic method can be used to eliminate these artifacts, allowing the measurement of all the known types of membrane currents, including the resting "leak current" that is difficult to measure by any other means. Many of these variations involve the use of an intracellular voltage clamp in conjunction with a loose-seal pipette.

Comparison between Loose and Tight Seals

A "loose seal" is distinguished from a "tight seal" by the larger separation between the rim of the pipette and the cell surface. In a loose seal, the distance is usually 20–100 nm,[27] too far for atoms on the surface of the glass to interact directly with lipids in the plasma membrane, or even with large membrane glycoproteins. The separation is probably maintained by the basal lamina on intact muscle cells; it is not known that prevents closer contact on some enzymatically cleaned cell surfaces. Loose seals are mechanically weak and, when proper care is taken, do not injure the cell membrane. In contrast, a tight seal forms a strong bond, with a separation that is probably less than 1 nm.[16] A tight seal thus provides nearly perfect electrical isolation of the patch of membrane that it encircles, but in so doing it disrupts the local membrane structure. Damage is apparent in light microscopic observations that show a bleb being pulled several micrometers into the tip of the pipette as the seal forms.[28,29] The cytoskeleton within the bleb may be disrupted, as evidenced by cytoplasmic particles undergoing Brownian motion.[24] Within blebs, membrane proteins may be released from restraints on their lateral diffusion.[30]

[26] W. Stühmer, W. M. Roberts, and W. Almers, in "Single-Channel Recording" (B. Sakmann and E. Neher, eds.), p. 123. Plenum, New York, 1983.

[27] W. Stühmer and W. Almers, Proc. Natl. Acad. Sci. U.S.A. 79, 946 (1982).

[28] B. Sakmann and E. Neher, in "Single-Channel Recording" (B. Sakmann and E. Neher, eds.), p. 37. Plenum, New York, 1983.

[29] M. Sokabe and F. Sachs, J. Cell Biol. 111, 599 (1990).

[30] D. W. Tank, E. S. Wu, and W. W. Webb, J. Cell Biol. 92, 207 (1982).

Loose seals can be made and broken without damaging the membrane. Cells with basal laminae, such as skeletal muscle, can withstand remarkably rough treatment, including strong suction (1 meter H_2O) applied to the pipette[31] to stabilize the seal. However, we recommend gentler suction (<50 mm H_2O); during long-term measurements it is prudent to forgo suction.[32] Greater care must be exercised when working with cells that lack an intact basal lamina because their plasma membranes are easily damaged.[23,24] When recording from such delicate cells, it is usually necessary to watch for blebs through a compound microscope. If necessary, the cell membrane can be protected by a collagen plug in the tip of the pipette (see below).

Techniques

Pipettes: Pulling, Polishing, and Plugging

Loose-seal pipettes are pulled in essentially the same way as tight-seal pipettes used for whole-cell recordings.[16] Pipettes with tip diameters in the $1-20$ μm range that have low internal resistances and seat well against the membrane must be pulled in two stages. The first, at high heat, is stopped when the glass diameter is reduced to approximately 250 μm. The glass is then raised to recenter it in the heating filament and pulled at low heat until it separates. The tubing should separate cleanly to produce two symmetrical pipettes that taper steeply to flat, smooth, crackless tips.

A vertical pipette puller, such as the List model L/M-3P-A (List-Electronics, Darmstadt, Germany) works well. The Kopf model 700 (David Kopf Instruments, Tujunga, CA) can also be used with a few modifications. It is not necessary that the puller have a well-regulated power supply, as long as there is some means to monitor accurately and adjust the heater current; a digital voltmeter connected across the heating coil is sufficient to measure the heat. The other requirements are a weak pull (we use ~ 160 g weight), a windscreen, and mechanical stops that reproducibly determine the distance of the first pull and the raise before the second pull. Large diameter (~ 5 mm) heating coils generally work better than smaller ones. The heat of the second pull is varied to give the desired tip diameter; higher temperatures produce smaller tips.

After pulling, pipette tips must be heat polished to smooth and thicken their rims. We mount the pipette and a 350-μm-diameter platinum wire

[31] W. Almers, P. R. Stanfield, and W. Stühmer, *J. Physiol. (London)* **339**, 253 (1983).
[32] R. E. Weiss, W. M. Roberts, W. Stühmer, and W. Almers, *J. Gen. Physiol.* **87**, 955 (1986).

heating filament in micromanipulators that position them in the field of view of a 40× objective lens in a horizontally mounted microscope. The duration and temperature of polishing are determined, by visual inspection, to produce the desired shape of the tip and thickness of the rim. To obtain the lowest possible internal resistance, is often best to start with a tip that has about twice the desired diameter. Extensive heat polishing then reduces the diameter of the orifice and thickens the rim. However, in some applications a thinner rim is preferred (see section on space-clamp errors below).

To avoid damaging delicate cells, a collagen plug can be deposited inside the pipette by drawing a few microliters of a solution of rat tail collagen (1 mg/ml) dissolved in acetic acid into its tip.[24] As the fluid evaporates, the collagen becomes concentrated in the decreasing volume until it precipitates at the tip to form a meshwork that supports the cell membrane and prevents it from being pulled into the pipette. Such collagen plugs do not appreciably increase the internal electrical resistance of the pipette. The treatment also prevents the formation of tight seals, which can be difficult to avoid on enzymatically dissociated cells. Using collagen-plugged pipettes, we studied patches on dissociated hair cells without disrupting the spatial relationships between ion channels that communicate through local diffusion of an intracellular messenger.[24]

Forming a Seal

The recording pipette is held in a micromanipulator and positioned under visual guidance, using either a dissecting microscope or a compound microscope. For best results, the pipette should be oriented and advanced along an axis approximately perpendicular to the cell surface. If necessary, the pipette can be bent near its tip[21] to allow clearance around the microscope objective or condenser. The formation of a seal is monitored electrically by measuring the resistance between the top of the pipette (i.e., the big end) and the bath. Before touching the cell, nearly all of this resistance is the internal resistance of the pipette (R_p), which is concentrated inside the pipette near its tip. The increase seen when the pipette touches the cell surface is the seal resistance (R_s). After the initial increase at first contact, the pipette is advanced slowly until R_s has grown to some criterion value that depends on the particular application; usually R_s should be at least twice R_p (see section on accuracy of series resistance compensation below). This may be difficult to achieve unless the pipette pushes the cell against a firm substrate. Suction through the pipette is often used to increase and stabilize R_s.

The surface of the cell should be as clean as possible but may be covered by an intact basal lamina. Recordings have even been made from neurons ensheathed by glia,[26] but the results must be interpreted with caution. When intracellular microelectrodes were used to apply voltage steps in the cell body of a crayfish stretch receptor neuron, a loosely sealed pipette on the cell body sometimes recorded currents that probably originated about 100 μm away in the dendrite and were channeled between the layers of glia to the recording site.[33] Although this type of artifact probably was not present in published recordings from other neurons[11-14,34-37] because their glial sheaths were disrupted during the dissection or subsequent enzyme treatment, fingers of glial remnants must have remained on some deeply folded cells. More serious consideration of the effects of glia or glial remnants seems warranted, particularly when loose-seal recordings are used to explore variations in channel densities over distances of a few micrometers.

Estimating Patch Area

Patch currents are usually expressed in units of current density, calculated by dividing the patch current by the patch area. A lower limit of the patch area can be obtained from the tip area of the pipette measured in a light micrograph. On a cell with a smooth surface, this estimate is probably about 20% too low, since it does not take into account the area of contact between the membrane and the outside of the pipette. On skeletal muscles, the somata of large neurons, and other crenulated cells, such estimates may be too low by factors of 2–15. The only reliable measure of membrane area in such cells is the patch capacitance, which can be determined by applying voltage steps through the loosely sealed pipette[21,23] or, more easily, by applying steps through a separate intracellular voltage clamp.[24,35,36]

Loose-Seal Voltage Clamp of Large Cells

The basic loose-seal method, which uses a single-barreled extracellular pipette and no intracellular electrodes, is readily applied to large or electrically coupled cells that are difficult to study using whole-cell or other intracellular voltage clamp methods. Rather than attempting the difficult task of changing the intracellular voltage, the loose-seal voltage clamp changes the extracellular voltage locally. The basic loose-seal method has

[33] W. M. Roberts, unpublished observations (1982).
[34] M. Westerfield and H. D. Lux, *J. Neurobiol.* **12,** 507 (1982).
[35] S. Thompson and J. Coombs, *J. Neurosci.* **8,** 1929 (1988).
[36] J. W. Johnson and S. Thompson, *Biophys. J.* **55,** 299 (1989).
[37] W. M. Roberts and W. Almers, *Pfluegers Arch.* **402,** 190 (1984).

been used to study muscles and several other cell types. It is probably the simplest voltage clamp method in existence. Making a seal and recording currents need take no longer than 2–3 min. With attention to mechanical stability, recordings can be made from the same patch for hours.[27,31]

Compensating for Electrical Properties of Pipette and Seal

The recording pipette is connected to a voltage clamp amplifier that compensates for R_p, R_s, and the pipette capacitance (C) to the bath. There is usually no reason to use an ultra-low-noise current monitor because the low resistance of the seal makes the recordings inherently noisy. Most commercially available tight-seal voltage clamps cannot be used for such applications because they do not fully compensate for the pipette resistance and make no provision for subtracting the large leakage current through the seal.

Two loose-patch voltage clamps are now commercially available. The LPC3 (Medical Systems, Inc., Greenvale, NY) was designed specifically for this purpose. The EPC-9 (List-Electronics) is also suitable for loose-seal applications if used with the low-resistance (5 mΩ) headstage.

In the discussion that follows, we consider a typical recording from a skeletal muscle cell, in which a 100-mV depolarizing voltage step from the resting potential is applied through a 0.5-MΩ pipette with a 1-MΩ seal to cause a 10-nA peak Na$^+$ current through a 100-μm^2 patch. Superimposed on the Na$^+$ current is a 10-fold larger leakage current across the seal (100 mV/1 MΩ = 100 nA) that must be subtracted to reveal the Na$^+$ current. Thus, even allowing a 10% error in the measured Na$^+$ current, leak subtraction must be accurate to 1%. To measure smaller membrane currents, the accuracy of leak subtraction often must be better than 0.1%. Because errors inherent in leak subtraction and series resistance compensation limit the accuracy of loose-seal measurements, we shall consider them in detail below.

Accuracy of Series Resistance Compensation

It is easy to control the voltage (V_t) inside the tip of the pipette to within a few percent; the major source of error in V_t is the uncertainty in R_p. V_t is determined from the voltage (V) at, and current (I) entering, the top of the pipette, using Ohm's law:

$$V_t = V - I_t R_p \tag{1}$$

where

$$I_t = I - C(dV/dt) \tag{2}$$

V and I can be measured at the top of the pipette with essentially no error. The voltage clamp that we developed[21,26] (Figs. 1 and 2) determines V_t by analog circuitry and injects current into the pipette as required to force V_t to equal the command voltage.

While V_t is changing, the current (I_t) that flows out the pipette tip is smaller than I because some escapes into the bath across the capacitance of the pipette wall. The capacitance of the pipette is distributed uniformly along the submerged length of its shaft, while its resistance is concentrated near the tip. Therefore, essentially all of the capacitance is located above the resistance, as shown schematically in Fig. 1A (see below). We usually coat pipettes at the waterline with a hydrophobic material, such as Sylgard coating (Dow Chemical Co., Midland, MI), to stop a film of water from creeping up the shaft. Without this treatment, the portion of the capacitance in series with the resistance of the thin fluid film cannot easily be compensated, and it fluctuates as air currents evaporate the fluid. An error in C has a significant effect on V_t only during the brief times when V is changing rapidly in response to a step change in the command voltage. At other times, the only important source of error in determining V_t is the uncertainty in R_p.

To estimate R_p, one measures the current during small voltage steps applied to the pipette before it contacts the cell. Although most authors assume that R_p equals V/I, this equation overestimates R_p by approximately 10%, since it includes the convergence resistance of the bath outside the pipette.[9,21] We use the equation $R_p = 0.9V/I$ which, given the variation in the geometry of pipette tips, is accurate to about 5% of R_p.

The value of R_p is entered as a dial setting on the front panel of our voltage clamp amplifier, although compensation is not activated until the pipette touches the cell surface. Because R_p cannot be measured reliably after a seal is formed, the value determined before sealing is assumed to hold for the duration of the recording. This assumption is justified as long as the temperature remains constant and the solutions in the pipette and bath are identical. Increasing the ambient temperature increases R_p by approximately 2%/°. Using different solutions in the bath and pipette can lead to serious errors, which will be considered later. After the seal is established and series resistance compensation is activated, the 5% error in R_p causes an error in V_t that is smaller by the ratio of R_p/R_s. The voltage clamp can thus maintain excellent control over V_t as long as R_s meets the minimum criterion of $R_p > 2R_s$ and the ambient temperature does not change by more than a few degrees.

"Space-Clamp" Errors

A more serious voltage error occurs in the annulus of membrane beneath the rim of the pipette, where the extracellular voltage is intermediate between V_t and 0. Some of the current from this region enters the pipette, causing an artifact similar to space-clamp errors in other voltage clamps. The size of this region increases if the pipette has a thick rim, or if the pipette is advanced until it causes a large indentation of the cell surface. When we recorded Na^+ currents from skeletal muscle, rim currents caused little error in the peak Na^+ current but did interfere with measurements of reversal potentials.[26,37]

The voltage gradient under the rim can be eliminated by using concentric-barreled pipettes.[37,38] Separate series resistance compensation for the two barrels ensures that no voltage difference develops across the rim of the inner barrel, which then registers the membrane current through the central portion of a larger isopotential patch. The concentric pipette thus eliminates space-clamp errors, as well as some of the artifacts associated with leak subtraction described below. The inner barrel should protrude slightly to improve its seal with the membrane. This was difficult using the original method for making concentric pipettes. It may help to use glasses of different melting points for the two barrels and triangular tubing for the inner barrel.[39]

Uncertainty in Intracellular Potential

The ultimate goal of the voltage clamp is to control the membrane potential (V_m) across the patch: $V_m = V_i - V_t$, where V_i is the intracellular potential. Without an intracellular recording electrode, the error in V_m is thus at least as large as the uncertainty in the resting potential. In addition, V_i changes when currents flow through the patch. Fortunately, this change occurs exponentially with the time constant of the cell membrane, which is often much longer than the duration of the current being studied. In the example presented above, a 10-nA Na^+ current that reaches its peak in less than 1 msec will depolarize a typical muscle by less than 2 mV, but a longer-lasting current flowing into a cell with a higher input impedance will grossly alter V_i. The voltage change caused by a rectangular current pulse of duration t into a cell with input resistance R_{in} and time constant τ_m is approximately

$$\Delta V = I_m R_{in}, \qquad \text{for } t > \tau_m \tag{3}$$

$$\Delta V = I_m R_{in} t/\tau_m, \qquad \text{for } t < \tau_m \tag{4}$$

[38] W. Almers, W. M. Roberts, and R. L. Ruff, J. Physiol. (London) 347, 75 (1984).
[39] H. Yamaguchi, Cell Calcium 7, 203 (1986).

A *steady* 10-nA current into a 10-MΩ cell will thus depolarize it by 100 mV, whereas a *transient* 10-nA Na$^+$ current that lasts less than 1 msec will cause only a 3-mV error if the time constant of the cell is 30 msec.

Voltage errors associated with the input resistance usually restrict the use of the basic loose-seal method to large cells. The input resistance of some small cells can be lowered by placing them in a high-potassium medium that opens a large K$^+$ conductance. Otherwise, to be used for high-impedance cells, loose-seal methods must incorporate an intracellular electrode (see below).

Leak Subtraction

Usually, less than 10% of I_t crosses the cell membrane. The rest, which leaks through the loose seal, is of no interest to the physiologist and must be subtracted from I_t to reveal the membrane current (I_m). This procedure, called "leak subtraction," must be done precisely because I_m is such a small fraction of I_t. The first step is to determine the seal resistance (R_s), so that Ohm's law can be applied:

$$I_m = I_t - V_t/R_s \tag{5}$$

The procedure for calculating V_t and I_t were discussed above. Calculating R_s is more difficult because it is essentially impossible to distinguish current leaking through the seal from current through the resting membrane conductance. The best that can be accomplished is to find a "linear range" of membrane potentials in which the membrane conductance does not change with voltage or time, assume that $I_m = 0$ within this range, and calculate R_s from Eq. (5): $R_s = V_t/I_t$. In practice, R_s is measured by applying small voltage steps ("control pulses") in the linear range, while adjusting the leak-subtraction dial on the amplifier until the steady-state value of I_m does not change when the voltage changes. After making this adjustment, the I_m measured during "test pulses" that activate voltage-gated currents contains no information about the resting membrane conductance.

The inability to measure the conductance of the resting cell is a limitation of many voltage clamp methods. For example, an intracellular microelectrode produces a leak at the site of impalement that may be much larger than the resting membrane conductance. Despite their obvious physiological importance, ion channels responsible for this conductance have therefore received little attention. By using an intracellular voltage clamp in conjunction with the extracellular patch clamp, it is possible to measure the membrane conductance of a resting cell[24] and explore its distribution over the cell surface.

Capacitance Compensation

After adjusting the leak-subtraction dial, capacitance compensation is adjusted to minimize the transient currents at the beginning and end of the control pulse. Ideally, separate adjustments should be provided for the capacitances of the pipette and patch which, by virtue of being located at opposite ends of the resistance of the pipette, require slightly different compensation. However, we use a single adjustment and rely on digital signal processing to subtract any uncompensated capacitive transients.

The patch capacitance can be determined by subtracting the capacitance measured with the pipette sealed against a thick substrate, such as Sylgard or Parafilm, from the capacitance when it is sealed against the cell.[9,21,23] The two measurements should be made close together in the bath to keep the submerged length of pipette constant.

Fine-Tuning Leak Subtraction

To compensate effectively for fluctuations in the seal resistance, the time interval between control and test pulses must be made as short as possible, usually below 100 msec. The manual leak-subtraction adjustment described above thus provides only a first approximation that must be refined by an automated process, in which control pulses are delivered immediately before or after the test pulses. This correction can be done in real time by the voltage clamp amplifier,[40] but it is usually done by a computer operating on digitized responses to the control and test pulses.

In the most commonly used digital leak-subtraction procedure, currents during control pulses are multiplied by a scale factor and subtracted, point by point, from currents during the larger test pulses. This method assumes that the former flow entirely through incompletely nulled resistive and capacitive pathways across the membrane and seal that remain constant for the duration of the control and test pulses. To simplify the subtraction procedure, control pulses and test pulses usually have the same waveform and differ only in amplitude. Often, a so-called P/N protocol is used, in which one or more control pulses are followed by a test pulse that is N times as large. Control pulses may be offset by a constant voltage to keep them in the linear range of the cell; their polarity may be the same as (N positive), or opposite to (N negative) that of the test pulse. The control responses are then averaged, multiplied by N, and subtracted from the test response. The P/N procedure works well on skeletal muscle cells, which have a large range of membrane potentials in which few voltage-gated channels open. Before attempting to apply the basic loose-

[40] H. Antoni, D. Böcker, and R. Eickhorn, *J. Physiol. (London)* **406,** 199 (1988).

seal method to other cells, it is important to consider whether they have such a linear range that can be used for leak subtraction.

Although the computer program that implements the P/N procedure is usually designed to operate on-line, allowing the experimenter to view the leak-subtracted data while the experiment is in progress, there is a risk associated with making this complicated process so "transparent" that its pitfalls are forgotten. The most common problem occurs when the control pulses trespass into a range of membrane potentials where ion channels begin to open or close. Even small voltage-gated currents that contaminate the control response can cause large artifacts after being scaled up by a factor of N. If on-line leak subtraction is employed, control responses must be stored along with the processed data for later verification.

Electrophoresis of Membrane Proteins

Extracellularly applied electric fields can move ion channels from one end of a cell to the other by electrophoresis within the membrane.[41,42] When a steady holding potential of a few millivolts or more is applied to V_t, the extracellular voltage gradient across the seal is, in principle, large enough to redistribute membrane proteins that are mobile and have a net charge on their extracellular domains. A 25-mV holding potential should cause an e-fold change in the equilibrium concentration of proteins in the patch that carry one net charge. Although this type of focal electrophoresis has yet to be demonstrated, it should be considered whenever changes in the holding potential cause slow changes in the size of the patch current.[31]

Current Noise Eliminated by Signal Averaging

The random movements of ions across the loose seal adds noise to I_m. For a 1-MΩ seal, the root mean square (rms) current noise is approximately 13 pA if I_m is filtered with a bandwidth of 10 kHz.[43] This corresponds to a peak-to-peak noise of ~ 80 pA. However, P/N leak subtraction can greatly increase this noise. To reduce the noise contributed by the control pulses to the same or less than that of the test pulse, at least N^2 control responses should be averaged for each test response. From the standpoint of noise reduction, it is best to make control pulses as large as possible within the linear range, rather than using a P/N protocol, in which they are proportional to the test pulses.

[41] M.-M. Poo, *Annu. Rev. Biophys. Bioeng.* **10**, 245 (1981).
[42] J. Stollberg and S. Fraser, *J. Cell Biol.* **107**, 1397 (1988).
[43] F. J. Sigworth, *in* "Single-Channel Recording" (B. Sakmann and E. Neher, eds.), p. 3. Plenum, New York, 1983.

Errors in I_m Not Eliminated by Signal Averaging

Overestimating the pipette resistance by including the convergence resistance of the bath causes I_m to be overestimated by a factor of $\Delta R_p/R_s$, where ΔR_p is the error in R_p. Thus, if $R_s > 2R_p$, a 10% error in R_p causes at most 5% error in I_m.

More serious errors arise if the seal resistance varies in a systematic way during the test pulse. In theory, the passage of current across the seal should not alter R_s as long as the pipette is filled with the bath saline, the distance between the pipette rim and the cell surface does not change, and the current is not so large that it heats the medium. However, a careful experimental verification that the seal behaves like a constant resistance should be done before attempting to study currents that are much smaller or slower than Na^+ currents in skeletal muscle.

It is probably not feasible to use media with significantly different ionic compositions in the bath and pipette unless the ratio R_p/R_s is 10 or higher, because any difference in the electrical conductivities of the bath and pipette salines can cause the seal resistance to vary depending on the size and direction of current across it. If a 100-mV step is applied across a 1-MΩ seal, the ensuing 100-nA current transfers about 1 pmol/sec of ions into or out of the small space[27] between the pipette rim and the cell surface, which probably has a volume below 10 μm^3. This flux is therefore large enough to change the ionic concentrations in this region at a rate of over 100 M/sec, totally exchanging the ions within a few milliseconds. If the pipette contained 120 mM KCl and the bath 120 km$M$ NaCl (which differ in conductivity by 20%), the seal resistance could easily change by about 10% during the first few milliseconds after a voltage step from 0 to 100 mV. After leak subtraction, this change would appear as a time-varying membrane current with a maximum amplitude of approximately 10 nA, an error that is one or two orders of magnitude larger than acceptable! The importance of this error diminishes as the seal resistance increases because leak subtraction then becomes less important; with an R_s/R_p ratio of 10, Garcia *et al.*[44] were able to use significantly different pipette and bath solutions, which allowed them pharmacologically to isolate Na^+, K^+, and Ca^{2+} currents in loose-patch recordings. Apart from these electrical artifacts, mixing of the bath and pipette solutions in the pipette tip can make it difficult to know the precise composition of the extracellular solution at the patch, particularly if suction is used to increase the seal resistance.

Another artifact, which can be eliminated, is caused by the polarization of the liquid/metal junctions in the bath and the pipette holder. Polarization of a few millivolts can occur during prolonged passage of a 100-nA

---

[44] U. Garcia, S. Grumbacher-Reinert, R. Bookman, and H. Reuter, *J. Exp. Biol.* **150**, 1 (1990).

current, causing an error in I_m of a few nanoamperes. To eliminate this error, separate voltage-sensing and current-passing junctions should be placed in the bath. A simple circuit, which effectively voltage clamps the bath to ground, is shown in Fig. 2. For the same reason, the pipette holder should use separate voltage-sensing and current-passing junctions.

Amplifier Design

Our voltage clamp utilizes an "image circuit" to perform an analog computation equivalent to Eqs. (1), (2), and (5). Figure 1A is a simplified schematic diagram, drawn to emphasize the principle of operation. The left-hand side of Fig. 1A depicts items associated with the experimental preparation: the membrane patch, the pipette and seal, and the first-stage amplifier. Each object in the left-hand side has a counterpart located inside the voltage clamp, and shown in the right-hand side of Fig. 1A. The three variable elements (R_p', C', and R_s') comprise an electrical model of the

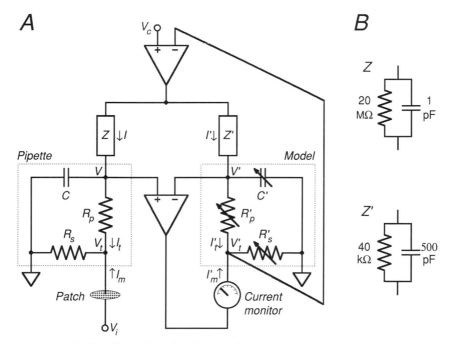

FIG. 1. (A) Simplified schematic diagram of the loose-seal voltage clamp. See text for definitions of symbols. (B) The impedances Z and Z' each consist of a resistor and capacitor in parallel; the capacitor improves the high frequency response but is not necessary. Even if no capacitor is used in Z, a trimming capacitor should be included in Z' to compensate for the stray capacitance across Z. C' can be a variable air-gap capacitor or, for convenience, can be simulated using a fixed capacitor and variable-gain amplifier, as shown in Fig. 2.

pipette and seal. They are adjusted, as described in preceding sections, by turning calibrated dials on the front panel. When properly adjusted, each element on the left-hand side is proportional to its image on the right; the proportionality constant is the ratio, Z/Z', of the two fixed elements, which is 500 (Fig. 1B). The model is placed in a feedback loop that holds the voltage at the top of the pipette equal to the voltage at the corresponding point in the model ($V = V'$). Currents I and I' are injected into the pipette and model through impedances Z and Z', which have equal voltages across them. I' is therefore proportional to I, and currents through the resistive and capacitive pathways in the pipette and seal are similarly proportional to the currents through corresponding elements in the model. Thus, $V_t = V_t'$ and $I_m I_m' Z'/Z$. Feedback forces V'_t to equal the command voltage. Figure 2 shows a more complete circuit, including the bath ground.

We favor this simple design because it uses a minimum of components and is easy to modify to suit different applications. The values shown allow compensation for R_p and R_s up to 10 MΩ. The capacitor in parallel with the resistor in Z allows the amplifier to pass large transient currents without increasing its current noise. This design is similar to integrating patch clamp amplifiers operating in "mixed RC" mode. The resistive part of Z ideally should be about 3 times the seal resistance. Smaller values significantly increase the current noise above the limit imposed by the seal resistance; larger values needlessly decrease the maximum steady current that can be passed. Other components can be added to the model as needed to make it correspond more closely to the experimental conditions. For example, to compensate for the membrane capacitance, a variable capacitance to ground can be added at the corresponding place in the model.

Simultaneous Intracellular and Loose-Seal Voltage Clamps

The two most significant sources of errors in loose-seal methods, imperfect leak subtraction and the uncertainty in V_i, can be eliminated by using an intracellular voltage clamp in addition to the loose-seal voltage clamp. This arrangement is extremely versatile because it allows separate control over the voltages at the intracellular and extracellular faces of the membrane. It can be applied to small cells that have high input impedances, or to large cells that cannot be adequately space clamped. The method can resolve currents of a few picoamperes through patches as small as a few square micrometers.

Usually, one voltage clamp applies the stimulus, while the other records the membrane current at a constant voltage. By thus eliminating the need

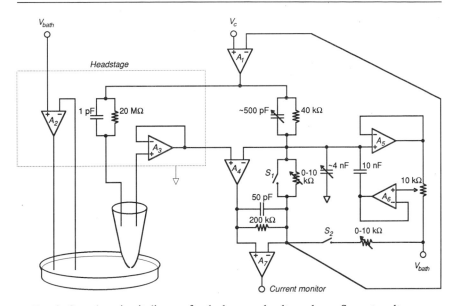

FIG. 2. Complete circuit diagram for the loose-seal voltage clamp. Separate voltage commands are provided for the pipette (V_c) and bath (V_{bath}). In applications in which the bath is held at ground potential, V_{bath} should be connected directly to the ground reference for V_c. The headstage is a grounded box mounted on a micromanipulator; the pipette holder plugs into a connector on the headstage. When closed, switch S_1 disables compensation for the pipette resistance, and S_2 enables leak subtraction. The approximately 4-nF variable capacitor compensates for the input capacitance of amplifier A_3 and other stray capacitances to ground; it is constructed by placing a fixed capacitor in parallel with a trimming capacitor. To compensate for the capacitance between the pipette and bath (20 pF maximum), a variable capacitance is simulated by connecting a fixed 10-nF capacitor to the output of a variable-gain amplifier (A_6). The current monitor uses a precision instrumentation amplifier (A_7: AD521, Analog Devices, Norwood, MA) to measure the voltage across a 200-kΩ resistor in parallel with 50 pF ($RC = 10$ μsec). The gain of the current monitor is selected from a range of values by a switch on the front panel that selects the gain-setting resistor for A_7 (not shown). A small (~ 50 Ω) resistor should be placed in series with the output of A_7 to isolate it from capacitive loads. Amplifier A_1 is a low-noise precision operational amplifier (OP-27, Precision Monolithics, Santa Clara, CA); A_2, A_3, A_4, A_5, and A_6 are field-effect transistor (FET)-input operational amplifiers (LFT356, National Semiconductor, Santa Clara, CA). Offset adjustments (not shown) are provided for amplifiers A_2, A_4, and A_7. The three variable resistances are 10-turn potentiometers. It is important that the approximately 500-pF capacitor in Z' be carefully trimmed to match the capacitance in Z. This is easiest to do after removing the pipette holder from the headstage and connecting together the two leads that would normally go to the pipette. With S_1 and S_2 open, $V_{bath} = 0$, and the simulated capacitance set to zero, large voltage steps are applied to V_c while viewing the current monitor output on an oscilloscope. The two variable capacitors are then adjusted to null both the fast and slow components of the transient current.

for leak subtraction, it is possible to record faithfully patch currents smaller than 10 pA and to determine the patch capacitance to an accuracy better than 10 fF.[24] If noise minimization is of paramount importance, the best arrangement is to use the loose-seal pipette to apply voltage steps and a whole-cell tight-seal voltage clamp to record I_m. However, space-clamp errors from the membrane beneath the rim will still be present.

Space-clamp errors are minimized by applying the voltage steps through the intracellular voltage clamp and recording currents through the loose-seal pipette. Signal averaging can then reduce the current noise associated with the seal resistance, leaving only small artifacts caused by capacitive coupling between the electrodes[36] and by the extracellular field potential that is generated across the resistance of the bath electrolyte by the current from the cell.[24] Errors associated with capacitive coupling can be observed by moving the patch pipette over 10 cell diameters away from the cell and recording current through the patch pipette (without series-resistance compensation) while stimulating a large current from the cell. At this distance the field potential should be small, and the pipette should record no significant current. If the bath is grounded as shown in Fig. 2, capacitive coupling between the intracellular voltage clamp and the patch pipette is the most likely source of a significant current recorded this far from the cell. If such coupling is found, it can be reduced by using low resistance, shielded electrodes.[36] After verifying that capacitive coupling does not cause a significant artifact, the effect of the field potential can be observed by moving the pipette toward the cell. As it advances within about 1 cell diameter, the patch pipette should record an increasingly large current that is driven into the pipette by the field potential.

Once a seal is made, the field potential artifact diminishes in proportion to $1/R_s$, and it is usually a small enough fraction of the patch current that it can be ignored. However, if ionic currents are distributed nonuniformly over the cell surface, the artifact may be relatively large in patches that contain few or no ion channels. In such a situation, it is important to make a rough estimate of the size of the field potential artifact.[24] It may also be possible to measure the field potential artifact by selectively blocking either the patch current or the current through the rest of the cell. For example, if a blocker of the ionic current is available and effective at low concentrations, it can be added to the pipette solution to block the patch current without altering the field potential artifact, making it possible to measure the size of the artifact as a function of seal resistance. Other possible approaches involve applying combinations of holding potentials and command potentials simultaneously to the intracellular and loose-seal pipette voltage clamps to inhibit selectively either the patch current or the current through the rest of the cell.

Applications

From the time of the earliest electrical recordings, it has been recognized that local variations in the electrical properties of a neuron are essential to its function. More recently, it has become clear that ion channels can be modulated in response to changes in the physiological state of a neuron. Many of these modulatory pathways involve diffusion of chemical messengers, which are certain to be localized to particular regions of a large neuron. To study these important phenomena, it is necessary to use methods that can reveal properties of ion channels with good spatial resolution, under highly physiological conditions, and over long periods of time. Loose-seal techniques are important tools for these measurements because they can achieve spatial resolution of a few micrometers and do little damage to the cell membrane.

Before any patch voltage clamp method, be it tight or loose, can be applied to a cell, it is necessary to gain access to the plasma membrane at the location of interest. For this reason, muscle cells that are surrounded by only thin basal laminae have been widely used for patch recordings. Other convenient preparations include cell bodies of neurons located on the surface of invertebrate ganglia, axons teased out of peripheral nerves, cells grown in culture or dissociated from tissues, and brain slices. In many of these preparations, loose-seal methods are easier to apply than tight-seal methods because they tolerate more debris on the cellular surface, but even loose seal methods cannot be used to voltage clamp through thick glial sheaths. Thus, a patch pipette pushed against a squid giant axon could not control the membrane potential of the axon or reliably assess local variations in current density unless the tip diameter of the pipette were much larger than the thickness of the sheath; even then the pipette might not yield reliable results because current can probably flow parallel to the axonal surface more easily than across the layers of glial membranes. In this case, it is irrelevant that the seal resistance seems adequate if there is a low-resistance pathway within the sheath that bypasses the seal. To overcome this problem, loose-seal patch pipettes have been applied to the *inside* surface of cut-open squid axons,[2-6] where they can seal directly against the cytoplasmic face of the plasma membrane. Because the presence of a glial sheath usually cannot be assessed in the light microscope, electron micrographs should be obtained before applying loose-seal methods to preparations in which glia or glial remnants may cover the cell surface. The inaccessibility of the surface membrane of most central nervous system (CNS) neurons makes the study of their electrical properties a challenging problem.

The second important requirement for patch voltage clamping is a means of controlling, monitoring, or inferring the intracellular potential behind the patch. Muscle cells are ideal subjects for all types of cell-attached patch recording because they have low input impedances and stable resting potentials that usually remain constant without intracellular control. Accordingly, the basic loose-seal method, which assumes a constant intracellular potential, has been applied most frequently to muscle cells, including vertebrate skeletal muscles,[17-23,26,27,31,32,37,45,46] invertebrate muscles,[47,48] and cardiac muscles.[40] The first voltage clamp recordings of ionic currents in human skeletal muscles[38,49] were also made using a loose-patch technique. While making loose-seal voltage clamp recordings, the pipette can also be used as a light guide that allows determination of the lateral mobility of voltage-gated ion channels by measuring the recovery of patch currents after local photoablation.[27,32] Loose-patch studies of muscle cells have generally used patch pipettes $2-20$ μm in diameter and have shown that voltage-gated Na^+ and K^+ channels are unevenly distributed,[17] with especially high concentrations of Na^+ channels at the motor end plate.[19-21] No correlations have been found between the variations in Na^+ and K^+ channel densities, and K^+ channels are never found at densities that approach the Na^+ channel density at end plates. The basic loose-patch clamp has also been used to study large axons[50-52] and glial cells.[53]

Recently, the same method has been applied to neurons. Retzius cells[44,54] were removed from the CNS of leeches by a procedure that preserved primary and some secondary neurites, then placed into a culture medium that promoted regrowth of neurites. On each cell, a single pipette $(3-10$ μm in diameter) was used to map the spatial distribution of ion channels at several locations, including cell bodies, axons, and growth cones. Voltage-gated Na^+ channels were initially found to be concentrated in distal axons, and later at the tips of regenerating axon stumps; K^+ channels were distributed more uniformly. Currents from the cell body of a different leech neuron (the P cell) have also been measured *in situ*, using concentric-barreled loose-patch pipettes.[37] To ensure that the intracellular potential in Retzius neurons was not disturbed by the largest patch

[45] L. Simoncini and W. Stühmer, *J. Physiol. (London)* **383**, 327 (1987).
[46] R. L. Ruff, L. Simoncini, and W. Stühmer, *J. Physiol. (London)* **383**, 339 (1987).
[47] L. Schwartz and W. Stühmer, *Science* **225**, 523 (1984).
[48] R. S. Zucker and L. Landó, *Science* **231**, 574 (1986).
[49] F. Zite-Ferenczy, K. Mathias, S. R. Taylor, and R. Rüdel, *in* "Progress in Zoology" (H. C. Lüttgau, ed.), Vol. 33. Gustav Fischer, Stuttgart, 1986.
[50] P. Shrager, *J. Physiol. (London)* **392**, 587 (1987).
[51] P. Shrager, *J. Physiol. (London)* **404**, 695 (1988).
[52] P. Shrager, *Brain Res.* **483**, 149 (1989).
[53] H. Marrero, M. L. Astion, J. A. Coles, and R. K. Orkand, *Nature (London)* **339**, 378 (1989).
[54] J. G. Nicholls and U. Garcia, *Q. J. Exp. Physiol.* **74**, 965 (1989).

currents (4 nA amplitude, 20 msec duration), an intracellular electrode was placed close to the patch pipette in a few cells.[44] This type of control experiment is important, even when recording from large muscle cells,[38] because it establishes the largest size patch that can be studied without intracellular voltage control. The success of this application of the basic loose-seal method to Retzius neurons offers hope for qualitative analysis of the ion channels in the regions of a neuron membrane that determine its integrative properties.

With few exceptions, the basic loose-seal method is not suited for use with smaller cells because patch currents of only a few picoamperes can significantly perturb the intracellular potential in these high-impedance cells. If one tried to use patch pipettes small enough to avoid disturbing the intracellular potential, leak-subtraction artifacts would then obscure the small patch currents. One class of exceptions includes cardiac muscles[40] and other cells with extensive electrical coupling that lowers their input impedances. It is also possible to lower the resistance of a cell artificially, for example, by placing it in a high-K^+ saline to depolarize it and open voltage-gated K^+ channels. This approach has been used for tight-seal cell-attached patch recordings from small cells, and it could be used with loose seals as well. An effective, albeit extreme, means of lowering the input resistance of a cell is to cut it open. This approach has been shown to work for squid axons.[3-6,15]

When intracellular voltage control is necessary, the loose-seal method can be used in conjunction with conventional one- or two-electrode voltage clamps, or with a whole-cell tight-seal voltage clamp. Even when not absolutely necessary, an intracellular voltage clamp is sometimes used because it greatly increases the accuracy and flexibility of the loose-seal method. Such hybrid methods have been used to record currents from voltage-clamped patches on large neurons,[11-14,34-37] skeletal muscles,[10,31] electric organs,[55] mechanosensory hair cells,[24,25] and taste receptors.[56] In one study, pipettes filled with sperm cells were used to make loose seals on egg cells, and the capacitance and conductance changes that occurred at fertilization were recorded.[57] These techniques can measure currents as small as a few picoamperes. They have shown that spatial heterogeneities in channel density and other channel properties can exist over distances of a few micrometers or less, and that the spatial distributions of different types of ion channels can be independent or tightly correlated. In one study, a combination of loose-seal patch recordings, tight-seal whole-cell recordings, and freeze–fracture electron micrographs indicated that the

[55] Y. Dunant and D. Muller, *J. Physiol. (London)* **379**, 461 (1986).
[56] S. C. Kinnamon, V. E. Dionne, and K. G. Beam, *Proc. Natl. Acad. Sci. U.S.A.* **85**, 7023 (1988).
[57] D. H. McCulloch and E. W. Chambers, *J. Gen. Physiol.* **88**, 33a (1986).

array of membrane particles seen at presynaptic active zones on hair cells are clusters of ion channels that contain a carefully regulated mixture of voltage-gated Ca^{2+} channels and Ca^{2+}-activated K^+ channels.[24] Future uses of the loose-seal technique will probably address a wider range of problems, such as local mechanisms of channel modulation that require preservation of the microscopic structure of the membrane or its spatial relationship to the cytoskeleton.

Acknowledgments

The writing of this chapter and some of the work described herein were supported by National Institutes of Health Grant NS27142 and a Sloan Foundation Fellowship to W.M.R.

[10] Perfusing Patch Pipettes

By J. M . TANG, J. WANG, and R. S. EISENBERG

Introduction

The routine recording of currents through single ionic channels has transformed membrane biology into a molecular science. The patch clamp technique measures current flow through individual protein molecules, thus specifying the type of channel with little ambiguity. The currents through individual channels are not hidden in the sum of thousands of unitary currents through many types of channels, as they are in most macroscopic recordings. Experimental energy can be spent studying molecules rather than identifying them.

The patch clamp technique depends on the electrical and physical isolation of one compartment, the pipette lumen, from another, the surrounding bath. This isolation occurs because the membrane binds tightly to the glass of the pipette, restricting the shunt pathway (between pipette lumen and bath) to only about 100 pS (picosiemens). Current through an ionic channel in the isolated patch of membrane winds up in the lumen of the pipette, where it can be collected, recorded, and amplified by suitable electronics.

The isolation of the lumen of the pipette is the essential feature of the patch clamp technique; failures in isolation introduce artifacts, always excess electrical noise and sometimes distortion in the time course of currents. The necessary isolation of the lumen is a serious constraint on the patch clamp method: it severely limits the types of materials that can touch the fluid in the lumen; it determines the electronics that can be used to collect and amplify the currents through the membrane patch; and it makes changing the solution in the pipette quite difficult.

Nonetheless, changing solutions around channels is of great experimental importance. Only in that way (1) can the concentration and type of permeating ion be varied and (2) can modulators or drugs be applied to both sides of the membrane. Nearly every investigation of channels is improved if solutions can be changed on both sides of the channel; but changing solution must be convenient if it is to be widely used: patch clamp experiments fail often enough even without adding complex apparatus or procedures. It is necessary to find a convenient way to change pipette solutions without compromising its necessary isolation, and without much adding noise or capacitive distortion.

Our system for perfusing the patch pipette only slightly modifies the "standard" patch clamp apparatus. We construct a perfusion capillary (with as little fuss as possible), insert it into the patch pipette (with as high a success rate as possible), and change the perfusion solution (without adding noise or complexity).

Patch Pipette Holder

The patch pipette holder is modified only by drilling a small hole to allow entry of polyethylene (PE) tubing (Fig. 1A).[1] A hole is drilled (with a #73 drill bit, nominal diameter 0.024 inches or 0.61 mm) in the patch pipette holder, at approximately a 45° angle, between the BNC pin and the suction line outlet. (We use the polycarbonate holder EPC-PHP from Medical Systems Corporation, Greenvale, NY: see Hamill *et al.*[2]) Through this hole we thread about 10 cm of polyethylene tubing (type PE-10, outside diameter 0.61 mm, inside diameter 0.28 mm; Clay Adams, Parsippany, NJ). The gap between the hole in the pipette holder and the PE tubing is sealed with grease. Using a grease seal (instead of a O ring) simplifies construction and allows easy positioning of the perfusion capillary: the PE tubing can slide back and forth in the grease seal.

Tubing for Perfusion

One end of the PE tubing is placed in a small reservoir, open to the air, containing the initial perfusion fluid. We found it quite convenient to use capsules designed for embedding specimens for electron microscopy (e.g., Beem capsule, size 00; Ted Pella, Tustin, CA). These capsules can be filled

[1] J. M. Tang, J. Wang, F. N. Quandt, and R. S. Eisenberg, *Pfluegers Arch.* **416**, 347 (1990).
[2] O. P. Hamill, A. Marty, E. Neher, B. Sakmann, and F. J. Sigworth, *Pfluegers Arch.* **391**, 85 (1981).

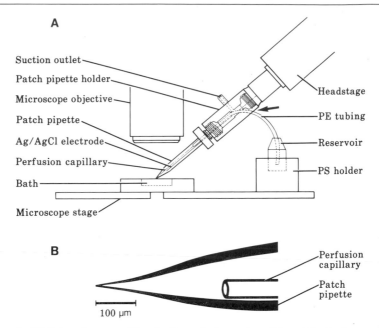

FIG. 1. (A) Setup diagram of the pipette perfusion system. The patch pipette was made from Kimax 51 glass capillary tubing and was connected to the modified patch pipette holder (see text). The perfusion capillary was constructed as described in the text and inserted into the patch pipette so its tip filled approximately one-half the cross section of the patch pipette; here it was approximately 300 μm from the tip of the patch pipette (see B). The Ag/AgCl wire electrode was 2 to 3 mm shorter (further from the tip of the patch pipette) than the perfusion capillary so it did not interfere with perfusion of the pipette tip. The other end of the perfusion capillary was connected to polyethylene (PE) tubing that had been threaded through a hole drilled in the pipette holder (see arrow). Grease was used to fill gaps between the PE tubing and the pipette holder. The other end of the PE tubing was placed in one of two solution reservoirs, Beem capsules in a holder made of expanded polystyrene (PS). The suction outlet of the pipette holder was connected with silicone tubing to a pneumatic transducer that monitored and controlled pressure/suction. (B) Tracing from a video screen image of the tip of a patch pipette containing a perfusion capillary. The tip of the perfusion capillary was about 300 μm from the tip of the patch pipette. (Redrawn from Tang *et al.*[1])

with solution and capped, preventing evaporation. Just before the experiment, the capsule is inverted, and the pointed end is cut, giving enough room for the *PE* tubing. Other perfusion solutions are kept in nearby ($\sim$ 1 cm) capsules. The far end of the *PE* tubing slips over the back end of the perfusion capillary. The front end of the perfusion capillary must reside in the lumen of the patch pipette, violating its isolation as little as possible.

The properties of the perfusion capillary are critical. It is made from quartz tubing (synthetic fused silica) coated with polyimide, kindly shown to us by Gabor Szabo.[3] The enormous volume resistivity (10^{16} Ω-cm) and surface resistance (10^{15} Ω) of the polyimide presumably allow the continued electrical isolation of the pipette lumen, even once it is compromised by the perfusion capillary. Other properties of the quartz tubing are helpful: it is surprisingly flexible and durable. The same perfusion capillary can usually be used for several days, in a number of experiments with many patch pipettes, taking only a reasonable amount of care. We purchase tubing (product number TSP 100/245; outside diameter 245 μm, inside diameter 100 μm) from Polymicro Technologies (Phoenix, AZ) that fits snugly into polyethylene tubing, size PE-10.

Making Perfusion Capillary

The quartz tubing is made into a perfusion capillary by drawing it out after it is softened by heat. The tubing is held vertically, and two alligator clips are attached as weights. The tubing can be softened rapidly (in a few seconds) in a hot microflame made from hydrogen and oxygen (using, e.g., the MicroWelder A$^+$ of Johnson Matthey, Wayne, PA), or it can be softened slowly (in a few minutes) using a flame made from "natural gas" (i.e., gaseous alkanes). Either flame burns the plastic coating incompletely, and the resulting debris need to be removed from the capillary. The drawn out tubing is put under a stereomicroscope, and the debris are scraped away with a microknife. This procedure is easier than it sounds: scraping a quartz tube is much easier than dissecting a muscle. The drawn out tubing is cut so its inner diameter is some 35 μm. The cut tubing, now called a perfusion capillary, is further cleaned by sonication.

The diameter of the perfusion capillary is important in our system, which does not include a valve or constrictor on the perfusion line. The capillary diameter (inner diameter 35 μm) is small enough to allow suction to create a gigaseal between membrane and patch pipette but large enough to allow good perfusion.

Procedures

Simple procedures can be used to mount the perfusion capillary, position it within the patch pipette, and control perfusion. The cleaned perfusion capillary is simply inserted into the end of the PE tubing, after that tubing is threaded through the hole in the pipette holder. The perfusion capillary and PE tubing is filled with the first solution to be perfused (typically the same solution that is in the pipette initially) using a syringe.

[3] J. V. LaPointe, and G. Szabo, *Pfluegers Arch.* **410**, 212 (1987).

The far end of the PE tubing is placed in the Beem capsule and the nearby Beem capsule(s) is filled with the second (or later) solutions to be used.

The silver/silver chloride wire that collects current is inserted into the patch pipette along with the perfusion capillary. The patch pipette is held down by the O ring of the patch pipette holder, as usual. The perfusion capillary is positioned by sliding it through its grease seal (by hand) while watching its location in the microscope of the patch clamp setup.

Bullet-shaped patch pipettes are used to allow the perfusion capillary to approach the tip as closely as possible, while obstructing the lumen of the patch pipette as little as possible. As LaPointe and Szabo[3] point out, a good location is where the perfusion capillary outer diameter is half the inner diameter of the patch pipette (Fig. 1B). Typically, the perfusion capillary was some 300 μm from the tip of the patch pipette in our experiments.[1] Fairly long patch pipettes are used (4–5 cm) because we do not want solution to actually enter the pipette holder or suction line. In our perfusion system, pipette solution is displaced up the pipette and not actually drawn down the suction line. We are afraid that fluid in the pipette holder or suction line would add noise and severely compromise the electrical isolation of the pipette.

The detailed description of our procedures should not mislead the reader. Changing pipettes could easily be done in 30 sec and was almost always done without damaging the perfusion capillary.

Perfusion was initiated with suction of 10–20 mm Hg and maintained with 2–4 mm Hg suction throughout perfusion. Control of this suction was critical for the success of our procedure. When suction was applied by mouth or hand (i.e., with a syringe), perfusion often damaged the gigaseal and preparation. When suction was applied with a pneumatic transducer (we used Model DPM-1 from Bio-Tek Instruments, Burlington, VT), we almost never lost a seal or preparation.

Perfusion solutions were changed simply by lifting the far end of the PE tubing out of one Beem capsule and placing it in another: in particular, suction was turned off, the gain of the amplifier was decreased to its minimum, and after the tubing was transferred, the gain was returned to its original value. Suction was started when we wanted to start perfusion. Bubbles did not form when we changed perfusion solutions, presumably because of the parameters of our system, chiefly the diameter and length of the perfusion line.

Properties and Evaluation of Technique

The best way to evaluate most experimental techniques is to use them in real experiments, and that is what we have done.[1] In several hundred experiments, we had no difficulty making or positioning perfusion capillaries, and we hardly ever lost a gigaseal or damaged a preparation because of perfusion.

The rate of perfusion depends on many variables, only one of which (position of perfusion capillary) was carefully controlled. Nonetheless, measurements of the rate of perfusion show that it is complete in 1 min, judging by the change in reversal potential with time of a channel in the patch.[1] The perfusion time might be decreased by using beveled perfusion capillaries that could be placed closer to the membrane patch, or by applying pressure to the perfusion tube in a closed version of the system. In the latter case, it would be important to maintain the transmembrane pressure as small as possible while increasing the pressure difference between suction and perfusion line.

The noise level of our apparatus is comparable to that of a typical patch clamp system[1]; below some 2 kHz it is indistinguishable, and above 2 kHz it is somewhat elevated. That increase in noise seems entirely due to the increase in capacitance to ground, which lessens the isolation of the lumen of the patch pipette. Noise is reduced in our setup by placing the Beem capsules of solution in an expanded polystyrene block a few centimeters thick. It could be reduced further by taking obvious precautions, for example, (1) replacing nearby metal with plastic, (2) using as small a Beem capsule as possible (using a size 3 capsule) placed as far from grounded metal as possible, or (3) by taking the perfusion line (i.e., the PE tubing) out of the Beem capsule after perfusion and moving it away from grounded metal.

The perfusion technique described here has worked well and conveniently in our laboratory and several others, and it should prove generally useful and convenient.

[11] Whole-Cell Recording of Calcium Channel Currents

By BRUCE P. BEAN

Introduction

The techniques most useful for whole-cell recording of calcium channels can be summarized easily: Use barium as the charge carrier. Use cesium as the internal cation. Include ATP in the internal solution. Study Fenwick *et al.*[1]

[1] E. M. Fenwick, A. Marty, and E. Neher, *J. Physiol. (London)* **331**, 599 (1982).

METHODS IN ENZYMOLOGY, VOL. 207

Preparation of Cells

Freshly isolated cells are generally preferable to cultured cells for recording whole-cell Ca^{2+} currents. Many important discoveries have been made using cultured cells, including neurite-forming neurons, but there is frequently a risk of insidious artifacts from poor space clamping. Measurements of current kinetics and reversal potentials are especially risky in cultured cells that may be imperfectly space clamped.

For overall ease of use, bullfrog sympathetic ganglion and dorsal root ganglion neurons are the best vertebrate cell types for recording Ca^{2+} currents. They are easy to isolate,[2,3] hardy, and have Ca^{2+} currents that are unusually resistant to rundown. Ventricular cells from guinea pig, rabbit, rat, or bullfrog hearts also have large, robust Ca^{2+} currents. Faster settling of capacity transients can be obtained with atrial cells, which are smaller and lack a T-tubule system. Cardiac cells are best prepared by Langendorff perfusion with enzymes; the procedure we use[4] was adapted from those of Isenberg and Klockner[5] and Mitra and Morad.[6]

Enzyme treatment during dissociation might alter the properties of the Ca^{2+} channels. So far, at least, there is no evidence for such alteration. In cardiac muscle, where L-type Ca^{2+} channels were well studied in intact tissue before the advent of single-cell techniques, channel properties such as ionic selectivity, kinetics, drug sensitivity, and modulation by transmitters do not seem to be altered by enzymatic dissociation. On the other hand, similar enzymatic dissociation does affect other ion channels (e.g., destroying NMDA receptor channels). When possible, it would be prudent to compare properties of Ca^{2+} channels in dissociated cells with those that have "recovered" in culture for several days or can be studied without any enzyme treatment (e.g., in non-enzyme-treated neuronal slices).

Internal Solutions

For the best isolation of Ca^{2+} channel currents, the major internal cation should be impermeant in most K^+ channels. Cesium, tetraethylammonium (TEA), Tris, and N-methyl-D-glucamine (NMDG) all work well; a caution for the use of NMDG has been noted.[7] Cesium is not completely impermeant in K^+ channels and may leave a small outward K^+ channel current at very positive potentials. However, Cs^+ has the advantage over TEA and NMDG of carrying current quite well through the Ca^{2+} channel,

[2] S. W. Jones, *J. Physiol. (London)* **389,** 605 (1987).
[3] B. P. Bean, *Nature (London)* **340,** 153 (1989).
[4] B. P. Bean and E. Rios, *J. Gen. Physiol.* **94,** 65 (1989).
[5] G. Isenberg and U. Klockner, *Pfluegers Arch.* **395,** 6 (1982).
[6] R. Mitra and M. Morad, *Am. J. Physiol.* **249,** H1056 (1985).
[7] C. O. Malecot, P. Feindt, and W. Trautwein, *Pfluegers Arch.* **411,** 235 (1988).

allowing the definition of a true reversal potential (near +60 mV with 5 mM Ba^{2+} versus 145 mM Cs$^+$) and recording of outward Ca^{2+} channel currents. Such currents are carried through Ca^{2+} channels, because they can be inhibited by Cd^{2+} (Fig. 1) or by nitrendipine (for L-type channels) or ω-conotoxin (for N-type channels). Because of their lower mobilities, TEA, Tris, and NMDG solutions produce series resistances larger (by about 30–50%) than equivalent solutions with Cs$^+$. With any internal solution, there may be small contaminating non-Ca^{2+} channel currents at large depolarizations; for the most accurate measurement of the reversal potential and of outward Ca^{2+} channel current, it is necessary to subtract the currents resistant to Cd^{2+}, nitrendipine, or ω-conotoxin.

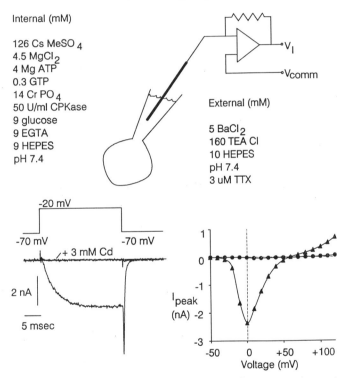

FIG. 1. Internal and external solutions for recording Ca^{2+} channel currents in a freshly isolated rat dorsal root ganglion neuron. Triangles, current measured at the end of a test pulse in control (5 mM Ba^{2+}) solution, corrected for leak current recorded for a -70 to -80 mV hyperpolarization; circles current after addition of 3 mM Cd^{2+} to the external solution. Internal solution (in mM): 126 cesium methanesulfonate, 4.5 MgCl$_2$, 4 adenosine triphosphate (Mg salt), 0.3 guanosine triphosphate (tris salt), 14 creatine phosphate (tris salt), 50 units/ml creatine phosphokinase, 9 glucose, 9 ethylene glycol-bis(aminoethylether) N,N,N′, N′-tetraacetic acid, 10 HEPES buffer, pH 7.4 with CSOH. External solution (in mM): 5 BaCl$_2$, 160 tetraethylammonium chloride, 10 HEPES buffer, pH 7.4 with tetraethylammonium hydroxide, 3 μM tetrodotoxin.

Buffering Internal Calcium

Calcium currents (and the cells themselves) last longer if the internal Ca^{2+} levels are buffered to low values. We generally use 9–10 mM EGTA in all internal solutions, with no added Ca^{2+}. Lower buffer concentrations are feasible; in heart muscle cells, 100 μM EGTA works well and preserves cell contraction for many minutes. For rapid buffering, BAPTA [1,2-bis(amino-ethane N,N,N,N'-tetraacetic acid] is preferable to EGTA. However, the series resistance often increases slowly with time when BAPTA is used in place of EGTA.

Rundown

Calcium currents tend to disappear after internal dialysis is begun.[8,9] The "rundown" of Ca^{2+} currents can be very rapid (tens of seconds) in small, well-dialyzed neurons,[10] but it is slower in large neurons[11] or small neurons studied with smaller-tipped patch pipettes. It is extremely rapid in inside-out patches, where Ca^{2+} channel activity disappears within seconds.[1,12,13] In the early experiments from the laboratory of Kostyuk, addition of cAMP together with MgATP to the internal solution was found to slow and even reverse the rundown of Ca^{2+} channels.[10] Various combinations of ATP, cAMP, and the catalytic subunit of kinase A have since been found to reduce rundown in many cell types.[11,14–19]

Including ATP in the internal solution at 2–15 mM is the single most important factor in preventing rapid rundown of Ca^{2+} currents; it is now standard in almost all whole-cell Ca^{2+} current recording. Including cAMP or kinase A is more problematical, since (at least for L-type channels) these may be better regarded as modulating the channels[20–22] rather than preserving them. An additional ATP-regenerating system of creatine phos-

[8] P. G. Kostyuk, O. A. Krishtal, and V. I. Pidoplichko, *J. Neurosci. Methods* **4**, 201 (1981).
[9] P. G. Kostyuk, *Neuroscience* **13**, 983 (1984).
[10] S. A. Fedulova, P. G. Kostyuk, and N. S. Veselovsky, *Brain Res.* **214**, 210 (1981).
[11] L. Byerly and S. Hagiwara, *J. Physiol. (London)* **322**, 503 (1982).
[12] A. Cavalie, R. Ochi, D. Pelzer, and W. Trautwein, *Pfluegers Arch.* **398**, 284 (1983).
[13] D. Armstrong and R. Eckert, *Proc. Natl. Acad. Sci. U.S.A.* **84**, 2518 (1987).
[14] R. Eckert and J. E. Chad, *Prog. Biophys. Mol. Biol.* **44**, 215 (1984).
[15] L. Byerly and B. Yazejian, *J. Physiol. (London)* **370**, 631 (1986).
[16] J. Chad, D. Kalman and D. Armstrong, *Soc. Gen. Physiol. Ser.* **42**, 167 (1987).
[17] B. Belles, C. O. Malecot, J. Hescheler, and W. Trautwein, *Pfluegers Arch.* **411**, 353 (1988).
[18] V. J. Schouten and M. Morad, *Pfluegers Arch.* **415**, 1 (1989).
[19] Y. M. Shuba, B. Hesslinger, W. Trautwein, T. F. McDonald, and D. Pelzer, *J. Physiol. (London)* **424**, 205 (1990).
[20] G. Brum, W. Osterrieder, and W. Trautwein, *Pfluegers Arch.* **401**, 111 (1984).
[21] A. B. Cachelin, J. E. De Peyer, S. Kokubun, and H. Reuter, *Nature (London)* **304**, 462 (1983).
[22] W. Trautwein and J. Hescheler, *Annu. Rev. Physiol.* **52**, 257 (1990).

phate and creatine phosphokinase may slow rundown even more than ATP alone.[23]

Rundown (as distinct from reversible slow inactivation) may be faster at depolarized membrane potentials.[18] In some cells, large (>20 mM) concentrations of HEPES buffer may be necessary to prevent rundown owing to internal acidification.[24,25] The protease inhibitor leupeptin (0.1 mM) dramatically reduces rundown in snail neurons.[14]

GTP should be included in the pipette if neurotransmitter modulation is being studied. Without added GTP, GTP-dependent transmitter effects disappear within 5–15 min in well-dialyzed small cells,[26] but they may persist longer in large or less well dialyzed cells. We generally include 0.3 mM GTP, which may be higher than necessary.

Preparation and Storage of Internal Solutions

If stored for more than a day, ATP- and GTP-containing solutions must be kept frozen to prevent hydrolysis. We make 0.5-ml aliquots of concentrated nucleotide-regenerating stock solution consisting of 140 mM creatine phosphate (Tris salt), 40 mM MgATP, 3 mM GTP (Tris salt), and 500 U/ml creatine phosphokinase, with the pH adjusted to 7.4 with Tris base. The aliquots are kept at $-70°$ until use. They are then diluted (most commonly with 140 mM cesium methane sulfonate, 5 mM MgCl$_2$, 10 mM EGTA, 10 mM HEPES, pH adjusted to 7.4 with CsOH) to make 5 ml of internal solution, which is stored on ice during the experiment. The creatine phosphokinase seems not to dissolve completely, and formation of gigohm seals is virtually impossible if pipettes are dipped directly into the resulting scummy solution. We therefore dip the electrodes into enzyme-free solution before backfilling the pipette with the enzyme-containing solution. All internal solutions are filtered through 0.2-μm filters immediately before use.

External Solution

Main External Cation

Tetraethylammonium seems to be the best choice of external cation in most cases. Besides being impermeant in both Na$^+$ and K$^+$ channels, TEA

[23] P. Forscher and G. S. Oxford, *J. Gen. Physiol.* **85,** 743 (1985).
[24] W. J. Moody and L. Byerly, *Can. J. Physiol. Pharmacol.* **65,** 994 (1987).
[25] L. Byerly and W. J. Moody, *J. Physiol. (London)* **376,** 477 (1986).
[26] P. J. Pfaffinger, J. M. Martin, D. D. Hunter, N. M. Nathanson, and B. Hille, *Nature (London)* **317,** 536 (1985).

blocks many K^+ channels. It is as good as any other cation for the formation and maintenance of gigohm seals. In principle, TEA might partially block Ca^{2+} channels. This does not happen in cardiac L-type currents, where current carried by 5 mM Ba^{2+} was superimposable in two solutions, one with 140 mM NaCl and the other with 140 mM TEA-Cl, both with 4 mM $MgCl_2$, 10 mM HEPES, and buffered to pH 7.4.

Different batches of TEA-Cl make different colored solutions, varying from clear to amber. This is unsettling, but direct comparisons (using the same cell) have always failed to show differences in Ca^{2+} currents. Filtering the TEA solutions does not seem to affect their properties for electrophysiological recordings.

The pH of the external solution should be adjusted using TEA-OH, not NaOH or KOH, since even small amounts of Na^+ or K^+ can give confounding currents through Na^+ or K^+ channels. Currents through K^+ channels can be especially insidious. If the cell is dialyzed with TEA or Cs, only $1-2$ mM K^+ in the external solution can result in sizable inward current carried through K^+ channels. I have often puzzled over odd, slowly activating and deactivating "Ca currents" recorded from cells dialyzed with Cs or TEA and bathed in physiological saline containing $2-4$ mM K^+. Agar bridges containing 3 M KCl can soon liberate millimolar concentrations of potassium if the chamber is small and stagnant. To avoid such problems, bridges can be equilibrated with isotonic TEA-Cl or perfusion.

Divalent Cation

Barium is usually preferable to Ca^{2+}. Background and leak currents tend to be smaller with Ba^{2+} than with Ca^{2+}. Also, unlike Ca^{2+}, Ba^{2+} is nearly completely impermeant in Na^+ channels. With 10 mM Ca^{2+}, small Na^+ channel currents can superficially resemble T-type currents [and may be incompletely blocked by tetrodotoxin (TTX) if cells have TTX-resistant Na^+ channels].

Ca^{2+} channel current kinetics can be greatly different in the presence of Ba^{2+} compared to Ca^{2+}, especially for Ca^{2+} channels demonstrating Ca^{2+}-dependent inactivation. Barium should be avoided by those aiming at physiological conditions. True purists may also wish to avoid voltage clamping the cell and instead spend the afternoon contemplating action potentials. The slowing of inactivation kinetics by Ba^{2+} of some channels (L-type and probably N-type channels) but not others (T-type channels) can be an advantage in distinguishing different components of current.[27]

[27] B. P. Bean, *J. Gen. Physiol.* **86**, 1 (1985).

Contamination of External Solutions by Heavy Metals

We add 0.1 mM EGTA to all external solutions as a precaution to chelate any small amounts of contaminating blockers (perhaps lead) that may be present. The risk of such contaminants was suggested by noticing a 10–15% increase in Ba^{2+} current when tubing was rinsed with an EGTA solution instead of the usual distilled water. Calcium channels may be a highly sensitive bioassay for heavy metal contamination. The use of a small amount of added EGTA has the additional benefit of producing much faster recovery when reversing the blocking effects of La^{2+}- or Cd^{2+}-containing solutions (e.g., when recording gating currents).

Changing External Solutions

For most purposes, the best method of making solution changes is to expose the cell to the outflow of solutions from multiple microcapillaries.[28] We use an array of microcapillaries with an internal diameter of about 145 μm (1 μl Drummond microcaps, 64 mm special length, available from VWR) glued together by epoxy. Each capillary is fed by a reservoir 50 cm above the bath. For freshly dissociated cells, the microcapillary array is bent (using the flattened flame from a Bunsen burner) so that the ends of the tubes are parallel to the bottom of the chamber, a millimeter or so above the bottom. After establishing a whole-cell recording, the cell is lifted and placed in front of one of the capillaries. Rapid solution changes can be made by moving the cell from the opening of one capillary to the next. If the cell is moved abruptly, the solution bathing the cell changes with a time constant of about 10 msec if small, spherical cells are used.[29] The speed of solution change seems to be limited by the shape of the cell, since the changes in ionic current are slower for larger cells.

For cultured neurons that cannot be lifted, the microcapillary array is not bent and is aimed at the bottom of the chamber at an angle of about 45°. The solutions from adjacent pipes remain unmixed as long as the array is moved close enough to the cell on the bottom of the chamber, within a distance equal to the internal diameter of a tube.

Speed of Clamp

The speed and accuracy of voltage control depend critically on two factors: resistance between different regions inside the cell and the series resistance of the pipette tip. Both should be as low as possible. For isolated cells, the resistance between different regions inside the cell will be signifi-

---

[28] D. D. Friel and B. P. Bean, *J. Gen. Physiol.* **91**, 1 (1988).
[29] B. P. Bean, *J. Neurosci.* **10**, 1 (1990).

cant only for neurons with large dendritic trees or a substantial length of axon. It is much more likely to be significant for cultured cells that have formed processes. The existence of such internal resistances can be detected by slow components of the capacity transient. It is very difficult to make quantitative calculations of just how seriously the presence of some slow-charging cell regions might distort the Ca^{2+} current that is measured. Even currents from cells that have neurites but do not show obvious loss of control should be interpreted with great caution, especially for measurements of activation kinetics, tail current kinetics, and reversal potential.

The effects of pipette series resistance are easier to calculate. For a spherical or near-spherical freshly isolated cell, the only limitation in the speed and accuracy of the voltage clamp is likely to arise from the pipette series resistance. If the series resistance for the pipette is much less than the parallel combination of the seal resistance and the membrane resistance (as is almost always true), and if the cell is isopotential, then the capacity transient for a step change in voltage will decay with a single exponential time constant given by R_sC_{cell}, where R_s is the series resistance and C_{cell} is the capacitance of the cell. C_{cell} can be quickly calculated by on-line integration of the capacity transient (divided by the size of the voltage step). R_sC_{cell} is then calculated by fitting an exponential to the decay of the capacity transient, and R_s is calculated by dividing the time constant by C_{cell}. The effective series resistance can be reduced if a series resistance compensating circuit is included in the patch clamp amplifier. For LIST (Medical Systems, Inc., Greenvale, NY) and Axopatch (Axon Instruments, Inc., Fullerton, CA) amplifiers, it is critical to reduce the capacitance of the pipette by coating with Sylgard resin (Dow Corning, Midland, MI) in order to be able to use the maximal amount of series resistance compensation. The pipette capacitance can also be reduced by keeping the solution level as low as possible. With these precautions, 80–90% of the series resistance can usually be compensated for.

For maximum speed of clamping, which is especially critical for resolving fast components of tail current, only spherical, freshly isolated cells should be used. Pipettes should be as large as possible, and the maximum possible series resistance compensation should be used. The current signal should be filtered as little as possible. It is easily possible to obtain conditions so that the capacity transient decays with a time constant of 20–30 μsec, allowing resolution of tail components with time constants of 100 μsec or so. Even faster settling of the voltage can be achieved if "supercharging"[30] is implemented (either by software or hardware) by transiently increasing the command voltage beyond the desired final voltage. With

---

[30] C. M. Armstrong and R. H. Chow, *Biophys. J.* **52**, 133 (1987).

standard resistive patch clamps, the amount of charging current that can be passed is limited by the feedback resistor (since no more than 12 or 15 V can be imposed across it), which should therefore be kept low for maximum speed. New patch clamp circuits using a feedback capacitor do not have this limitation, and somewhat faster clamps and better resolution of tail currents can be achieved with these amplifiers.

Series resistance errors (the maximal current times series resistance remaining after compensation) of less than 5 mV are acceptable. Simulations show that even quantitative measurements of kinetics, drug block, etc., are insignificantly affected by such errors.

Distinguishing Components of Calcium Current

Whole-cell recording without single-channel recording is not well suited for distinguishing components of current carried by different types of Ca^{2+} channels. In neurons, for example, high-threshold N-type and L-type channels can be distinguished fairly clearly by their different single-channel conductances, but, at the whole-cell level, overlap considerably in their kinetics and voltage dependence. In several types of neurons now known to possess multiple types of high-threshold channels, initial kinetic analyses could be performed that seemed consistent with a single type of channel. Distinctions between different channel types can best be made by a combination of single-channel and whole-cell recording, with channel pharmacology forming a crucial link between the two.

A good strategy is to evoke families of currents from several different holding potentials and to look for distinct kinetic components that are inactivated by holding potential. The strategy is most effective in the case of T-type current, which in many cell types completely disappears at only mildly depolarized holding potentials and can be recognized by its rapid inactivation and lower threshold.

Apparent separation of different current components by voltage dependence and kinetics can be misleading if one component of current happens to have complicated voltage-dependent gating behavior. One cautionary case is that of N-type current, which can inactivate rapidly but only partially.[31,32] Another is frog skeletal muscle, where distinct rapidly and slowly activating components of current[33] could be reasonably interpreted as reflecting different channel types yet seem to arise from a single population of L-type channels.[34]

[31] T. Aosaki and H. Kasai, *Pfluegers Arch.* **414**, 150 (1989).
[32] M. R. Plummer, D. E. Logothetis, and P. Hess, *Neuron* **2**, 1453 (1989).
[33] G. Cota and E. Stefani, *J. Physiol. (London)* **370**, 151 (1986).
[34] D. Feldmayer, W. Melzer, B. Pohl, and P. Zollner, *J. Physiol. (London)* **425**, 347 (1990).

In cells containing T-type currents and L-type currents, the two components can be distinguished nicely by the much slower tail currents of the T-type channels.[35] Except for T-type current, however, tail current kinetics may not be generally useful for separating different components of Ca^{2+} current. L-type and N-type channels in neurons have tail current kinetics that differ by less than a factor of two (L. J. Regan, D. W. Y. Sah, and B. P. Bean, unpublished results, 1990); multiexponential fitting cannot distinguish components that are so similar.

There are small differences in ionic selectivity of different channel types. T-type channels pass Ba^{2+} and Ca^{2+} equally well, whereas L-type channels carry more current with Ba^{2+}. Whether differences in selectivity are large enough to be useful in distinguishing other Ca^{2+} channel types remains to be seen.

Inorganic Blockers

Cobalt and manganese are rather weak blockers and even at several millimolar do not completely block currents carried by $5-10$ mM Ba^{2+} or Ca^{2+}. Cadmium and lanthanum are much more useful blockers. Cadmium at $100-500$ μM usually blocks Ca^{2+} channel current nearly completely. However, Cd^{2+} is itself very slightly permeant in the Ca^{2+} channel, and tiny currents remain even with Cd^{2+} as the only divalent cation present, both in neurons[36] and in cardiac cells (B. P. Bean, unpublished results). Block by low Cd^{2+} concentrations is partially relieved by both hyperpolarization and large depolarization,[37,38] so if Cd^{2+} is being used to isolate Ca^{2+} channel currents it should be used at a high concentration (>0.5 mM). At millimolar concentrations, Cd^{2+} effectively blocks both inward and outward currents through Ca^{2+} channels (Fig. 1).

Lanthanum is not even sparingly permeant in the Ca^{2+} channel, so block is complete,[39] and La^{2+} is slightly more potent than Cd^{2+}.[37] However, block by La^{2+} is generally less reversible; also, in some cells, exposure to La^{2+} for more than 1 or 2 min seems to result in irreversible loss of Ca^{2+} currents. Block by both Cd^{2+} and La^{2+} reverses faster if the cell is washed by a solution containing EGTA or EDTA.[37] Both Cd^{2+} and La^{2+}, especially La^{2+}, tend to stick to tubing and glassware, which should be rinsed with EGTA- or EDTA-containing solutions.

[35] C. M. Armstrong and D. R. Matteson, *Science* **227**, 65 (1985).

[36] W. R. Taylor, *J. Physiol. (London)* **407**, 433 (1988).

[37] S. W. Jones and T. N. Marks, *J. Gen. Physiol.* **94**, 151 (1989).

[38] D. Swandulla and C. M. Armstrong, *Proc. Natl. Acad. Sci. U.S.A.* **86**, 1736 (1989).

[39] R. D. Nathan, K. Kanai, R. B. Clark, and W. Giles, *J. Gen. Physiol.* **91**, 549 (1988).

L-type channels, and probably N-type channels, are somewhat more sensitive to block by Cd^{2+} than are T-type channels.[40] T-type channels are somewhat more sensitive to block by Ni^{2+} than are L-type channels.[41] However, the differences in sensitivity are too small to allow complete block of one component while sparing another. Gadolinium (Gd^{2+}) was proposed to block N-type channels selectively in neurons,[42] but more recent work shows potent block of L-type channels as well[42a]; a complicated interaction with bicarbonate buffer gives the appearance of selectively blocking different components of current.

Organic Blockers

Dihydropyridine drugs like nifedipine, nitrendipine, and nimodipine are potent blockers of L-type Ca^{2+} channels. Dihydropyridine block is very potent at depolarized holding potentials, with a K_d or 1 nM or less.[43] At negative holding potentials, block is much weaker, and it takes $2-10$ μM nitrendipine, nimodipine, or nifedipine to block L-type current completely. At 10 μM, the drugs are not completely selective for L-type channels but also significantly depress T-type channels[44] and even Na^+ [45] and K^+ [46] channels. Similar considerations probably apply to diltiazem and verapamil.

Dihydropyridine "agonists" like BAY K 8644 produce dramatic enhancement of L-type current for small depolarizations and induce slow components of tail current, corresponding to the lengthening of L-type channel openings seen at the single-channel level. The stimulatory effects of BAY K 8644 seem highly selective for L-type channels.[47,48] BAY K 8644 is a mixture of optical enantiomers. It may be best to use pure isomers, for example, $(+)$-202-791. However, in direct comparisons we found that the mixed isomer form of BAY K 8644 was actually somewhat more effective than $(+)$-202-791 in producing enhancement of L-type current (D. W. Y. Sah and B. P. Bean, unpublished results). Concentrations of $1-3$ μM of either BAY K 8644 or $(+)$-202-791 produce saturating effects.

[40] A. P. Fox, M. C. Nowycky, and R. W. Tsien, *J. Physiol. (London)* **394**, 149 (1987).
[41] N. Hagiwara, H. Irisawa, and M. Kameyama, *J. Physiol. (London)* **395**, 233 (1988).
[42] R. J. Docherty, *J. Physiol. (London)* **398**, 33 (1988).
[42a] L. M. Boland, T. A. Brown, and R. Dingledine, manuscript in preparation.
[43] B. P. Bean, *Proc. Natl. Acad. Sci. U.S.A.* **81**, 6388 (1984).
[44] C. J. Cohen and R. T. McCarthy, *J. Physiol. (London)* **387**, 195 (1987).
[45] A. Yatani and A. M. Brown, *Circ. Res.* **56**, 868 (1985).
[46] J. R. Hume, *J. Pharmacol. Exp. Ther.* **234**, 134 (1985).
[47] M. C. Nowycky, A. P. Fox, and R. W. Tsien, *Proc. Natl. Acad. Sci. U.S.A.* **82**, 2178 (1985).
[48] A. P. Fox, M. C. Nowycky, and R. W. Tsien, *J. Physiol. (London)* **394**, 173 (1987).

Conotoxin is a potent blocker of N-type channels in neurons,[49] and recent work suggests that it is highly selective for N-type over L-type channels[31,32] as well as T-type channels.[49] Amiloride is a somewhat selective blocker of T-type channels in some cells,[50] but its selectivity is far from complete; it is not very potent.

Fluctuation Analysis

As shown by Fenwick *et al.,*[1] whole-cell recording using patch pipettes is ideally suited for the technique of nonstationary fluctuation analysis of voltage-dependent currents, originally developed by Sigworth[51] for studying Na^+ channels in the node of Ranvier. Background noise is low, and the number of channels in a cell (typically 1000–10,000) is low enough that the fluctuations from channel gating are relatively large. Fluctuation analysis can give information about single-channel current size and channel kinetics at voltages or divalent ion concentrations where single-channel currents are too small to measure directly.

Unlike the case of the voltage-clamped node of Ranvier,[51] the background variance in whole-cell recording does not change significantly with membrane conductance or membrane potential; the variance at holding potentials of −80 to −60 mV is virtually identical to the variance at the Ca^{2+} channel reversal potential (without Ca^{2+} current blocked) or at 0 mV or so after the Ca^{2+} current has been blocked by La^{2+} or nitrendipine. Thus, background variance measured at the holding potential can be used to correct that measured during test pulses in order to isolate variance arising from Ca^{2+} channels.

The major technical difficulty comes from filtering. Excessive low-pass filtering of the current signal reduces fluctuations (and the measured variance) without affecting mean current. This results in an artifactually low value of the calculated single-channel current. Even if the external filter is set at a high bandwidth (say, 5–10 kHz), the filtering by the R–C combination of cell capacitance and series resistance can be much more severe. The amount of filtering from this source should be estimated, either from the time constant of the capacity transient or (if a LIST or Axopatch amplifier is used, where the capacity transient is nulled) by calculation from the original uncompensated capacity transient and the amount of series resistance compensation. The extent of filtering from the cell–series resistance combination can be checked by examining the effect of addi-

[49] E. W. McCleskey, A. P. Fox, D. H. Feldman, L. J. Cruz, B. M. Olivera, R. W. Tsien, and D. Yoshikami, *Proc. Natl. Acad. Sci. U.S.A.* **84**, 4327 (1987).
[50] C. M. Tang, F. Presser, and M. Morad, *Science* **240**, 213 (1988).
[51] F. J. Sigworth, *J. Physiol. (London)* **307**, 97 (1980).

tional filtering by an external filter. For example, if additional filtering with a bandwidth of 1 kHz has little effect on the variance of the current signal, it was already filtered at least this severely by the cell–series resistance combination.

Gating Current

The whole-cell method is very well suited for measuring gating current signals. Because of the high resistance of the electrode-to-cell connection, the background noise relative to the signal is lower than for a squid axon clamp, a frog node clamp, or a petroleum-jelly gap clamp of skeletal muscle. Asymmetric charge movement can be measured in single sweeps, without signal averaging, as first shown by Beam and Knudson[52] in cultured skeletal muscle. Both Na^+ channel gating current[4,53] and Ca^{2+} channel gating current [4,54] signals can be measured in cardiac myocytes.

To record gating currents, ionic currents through channels must be eliminated. This seems best accomplished by using TEA as the predominant cation both inside and out and blocking Ca^{2+} channel ionic currents by a combination of Cd^{2+} and La^{2+}.[4] Removing divalents altogether makes the leak resistance too low. Cadmium alone is not ideal, because it is very slightly permeant in the Ca^{2+} channel.

Cardiac myocytes are especially favorable for recording Ca^{2+} channel gating current; voltage-activated potassium channels are sparse and open slowly, and the signal from Na^+ channels can be largely eliminated by depolarized holding potentials. Isolation of Ca^{2+} channel gating current in neurons is more difficult.

Acknowledgments

I thank my former and present colleagues Linda Boland, Michael Brainard, Paul Ceelen, David Friel, Jim Huettner, Isabelle Mintz, Laurie Regan, and Dinah Sah.

[52] K. G. Beam and C. M. Knudson, *J. Gen. Physiol.* **91,** 799 (1988).
[53] D. A. Hanck, M. F. Sheets, and H. A. Fozzard, *J. Gen. Physiol.* **95,** 439 (1990).
[54] R. W. Hadley and W. J. Lederer, *J. Physiol. (London)* **415,** 601 (1989).

[12] Recording from Calcium-Activated Potassium Channels

By B. S. PALLOTTA, A. L. BLATZ, and K. L. MAGLEBY

Introduction

Calcium-activated potassium channels represent a ubiquitous class of ion channels that are activated by nanomolar to micromolar concentrations of Ca^{2+} at the intracellular membrane surface (Ca_i^{2+}).[1-5] When Ca^{2+}-activated K^+ channels open, K^+ flows out of cells, down its electrochemical gradient, making the cell more negative inside. With widespread distributions in both excitable and nonexcitable cells, these channels appear to participate in a variety of physiological functions that require either a hyperpolarization or an efflux of K^+ in response to increased Ca_i^{2+}. These functions include action potential repolarization,[6-10] afterhyperpolarizations,[6,7,11-13] insulin release,[14] pacemaker activity,[1,15] contribution to resting potential generation,[16-18] and K^+ efflux in secretory[2,14] and nonexcitable[19] cells.

Within the general class of Ca^{2+}-activated K^+ channels there appear to be many different types, as indicated by conductances which range from 4

[1] R. W. Meech, *Annu. Rev. Biophys. Bioeng.* **7**, 1 (1978).
[2] O. H. Petersen and Y. Maruyama, *Nature (London)* **307**, 693 (1984).
[3] A. L. Blatz and K. L. Magleby, *Trends Neurosci.* **10**, 463 (1987).
[4] B. Rudy, *Neuroscience* **25**, 729 (1988).
[5] R. Latorre, A. Oberhauser, P. Labarca, and O. Alvarez, *Annu. Rev. Physiol.* **51**, 385 (1989).
[6] P. Pennefather, B. Lancaster, P. R. Adams, and R. A. Nicoll, *Proc. Natl. Acad. Sci. U.S.A.* **82**, 3040 (1985).
[7] P. R. Adams, S. W. Jones, P. Pennefather, D. A. Brown, C. Koch, and B. Lancaster, *J. Exp. Biol.* **124**, 259 (1986).
[8] C. D. Benham, T. B. Bolton, R. J. Lang, and T. Takewaki, *J. Physiol. (London)* **371**, 45 (1986).
[9] J. J. Singer and J. V. Walsh, Jr., *Pfluegers Arch. Eur. J. Physiol.* **408**, 98 (1987).
[10] J. M. Velasco and O. H. Petersen, *Biochim. Biophys. Acta* **896**, 305 (1987).
[11] E. F. Barrett and J. N. Barrett, *J. Physiol. (London)* **255**, 737 (1976).
[12] G. Romey and M. Lazdunski, *Biochem. Biophys. Res. Commun.* **118**, 669 (1984).
[13] G. Romey, M. Hugues, H. Schmid-Antomarchi, and M. J. Lazdunski, *J. Physiol. (Paris)* **79**, 259 (1984).
[14] O. H. Petersen, I. Findlay, K. Suzuki, and M. J. Dunne, *J. Exp. Biol.* **124**, 33 (1986).
[15] A. Hermann and C. J. Erxleben, *J. Gen. Physiol.* **90**, 27 (1987).
[16] A. L. F. Gorman and A. Hermann, *J. Gen. Physiol.* **333**, 681 (1982).
[17] J. W. Johnson and S. H. Thompson, *Soc. Neurosci. Abstr.* **9**, 1187 (1983).
[18] J. F. Ashmore and R. W. Meech, *Nature (London)* **322**, 368 (1986).
[19] E. K. Gallin, *Biophys. J.* **46**, 821 (1984).

to 350 pS.[3-5] For convenience of discussion Ca^{2+}-activated K^+ channels can be arbitrarily divided into at least three subclasses, each of which may be further subdivided. One subclass consists of the large conductance Ca^{2+}-activated K^+ channels, also called BK channels or maxi K^+ channels.[20-22a] BK channels are highly selective for K^+ over Na^+, have single-channel conductances of about 180–350 pS, are activated by depolarization as well as intracellular Ca^{2+}, and are typically blocked by extracellular TEA (tetraethylammonium ion) and not blocked by extracellular apamin.[12,23,24] A second subclass includes those Ca^{2+}-activated K^+ channels of intermediate conductances (typically 20–180 pS).[3] A third subclass includes those Ca^{2+}-activated K^+ channels with small conductances (10–14 pS). These small conductance (or SK) channels are highly sensitive to intracellular Ca^{2+}, are blocked by extracellular apamin, and appear to have minimal voltage sensitivity.[12,24,25] There also appear to be even smaller conductance (about 4 pS) Ca^{2+}-activated K^+ channels.[24]

The purpose of this chapter is to describe some of the techniques used to record from and identify Ca^{2+}-activated K^+ channels. The emphasis will be on techniques that have successfully been used to study BK and SK channels.

Identification of Calcium-Activated Potassium Channels

The unambiguous identification of a Ca^{2+}-activated K^+ channel requires a demonstration that the channel is both activated by intracellular Ca^{2+} in a reversible manner and selective for K^+ ions. A demonstration of reversible activation by Ca^{2+} is shown in Fig. 1 for SK and BK channels in the same excised patch of membrane from cultured rat skeletal muscle.[24] Currents are recorded with the patch clamp technique.[26] At the downward arrow in Fig. 1a, Ca^{2+} at the intracellular membrane surface, Ca_i^{2+}, is increased from 0.1 to 1 μM. After a delay, mainly due to the perfusion time, small conductance Ca^{2+}-activated K^+ channels (SK) become active, as indicated by the small downward steps in the current record. After a further delay large conductance Ca^{2+}-activated K^+ channels (BK) also

[20] A. Marty, *Nature (London)* **291,** 497 (1981).
[21] B. S. Pallotta, K. L. Magleby, and J. N. Barrett, *Nature (London)* **293,** 471 (1981).
[22] R. Latorre and C. Miller, *J. Membr. Biol.* **71,** 11 (1983).
[22a] R. Latorre, C. Vergara, and C. Hidalgo, *Proc. Natl. Acad. Sci. U.S.A.* **79,** 805 (1982).
[23] A. L. Blatz and K. L. Magleby, *J. Gen. Physiol.* **84,** 1 (1984).
[24] A. L. Blatz and K. L. Magleby, *Nature (London)* **323,** 718 (1986).
[25] D. G. Rang and A. K. Ritchie, *Pfluegers Arch. Eur. J. Physiol.* **410,** 614 (1987).
[26] O. P. Hamill, A. Marty, E. Neher, B. Sakmann, and F. J. Sigworth, *Pfluegers Arch. Eur. J. Physiol.* **391,** 85 (1981).

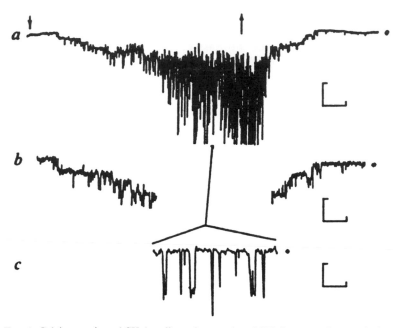

Fig. 1. Calcium-activated SK (small conductance) and BK (large conductance) channels in the same excised patch of membrane. (a) An excised inside-out patch of membrane from primary cultures of rat skeletal muscle (myotubes) was placed in a microchamber so that the inner membrane surface would be perfused. At the upward arrow the Ca^{2+} in the solution flowing to the microchamber was raised from 0.1 to 1 μM. At the downward arrow the solution was switched back to 0.1 μM Ca^{2+}. The delay in response is consistent with the time required for the new solution to reach the microchamber. SK channels open first (small downward current steps) followed by the opening of BK channels (large downward current steps). (b) Current trace in (a) at higher gain and faster sweep speed to show SK channels. (c) Indicated part of current trace in (a) at lower gain to show BK channels. The dots to the right of the records indicate the current levels when all channels are closed. Vertical bar: a, 2 pA; b, 1 pA; c, 5 pA. Horizontal bar: a, 5 sec; b, 2.5 sec; c, 50 msec. [Reprinted by permission from A. L. Blatz and K. L. Magleby, *Nature (London)* **323,** 718 (1986), Copyright © 1986 Macmillan Magazines Ltd.]

become active, as indicated by the large downward steps. Figure 1b shows the SK channels at increased gain, and Fig. 1c shows the BK channels at decreased gain. Notice that the current through a BK channel is about 20 times that through an SK channel.

At the upward arrow in Fig. 1a the Ca^{2+} is switched back from 1 to 0.1 μM, and, after a perfusion delay, first the BK channels and then the SK channels are no longer active. The observation that BK channels become active later as Ca^{2+} is increased and become inactive sooner as Ca^{2+} is decreased suggests that they require a higher Ca^{2+} concentration for activation than SK channels. This is the case, since mixing in the perfusing system causes the Ca^{2+} concentration to rise and fall with a delay. Methods for changing solutions will be considered in a later section.

Once it is determined that a channel is Ca^{2+} activated, the next step in identifying a Ca^{2+}-activated K^+ channel is to examine whether it is selective for K^+ over other ions. Figure 2 presents plots of the amplitudes of single-channel currents versus the membrane potential for an excised membrane patch with 100 mM KCl outside and the indicated concentrations of KCl inside. In symmetrical KCl the currents reverse at 0 mV as expected. With 10 mM KCl inside, the currents reverse at about +40 mV. This observed value is almost identical to the value of +40.9 mV that would be calculated with the Nernst equation assuming that the channel is permeable to K^+ and not Cl^- [40.9 mV = 25.4 ln(100/20)]. A positive

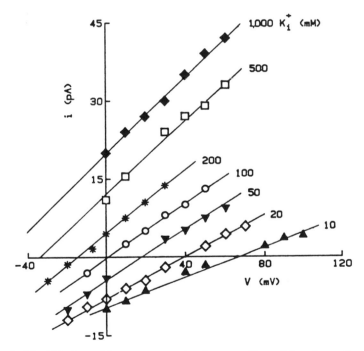

FIG. 2. The BK channel from rat myotubes is permeable to K^+ and not Cl^-, and its conductance increases with the concentration of K_i^+. The symbols plot the current through a single BK channel as a function of membrane potential. The K_o^+ concentration was 100 mM and the K_i^+ level is indicated for each type of symbol. The shift in reversal potential is consistent with K^+ rather than Cl^- permeability (see text). Single-channel conductance, as defined by Eq. (1), increases with K_i^+ [Reproduced from A. L. Blatz and K. L. Magleby, *J. Gen. Physiol.* **84,** 1 (1984), by copyright permission of the Rockefeller University Press.]

reversal is consistent with K^+ permeability, since a positive potential at the inner cell membrane is necessary to counter the inward concentration gradient of K^+. If the channel were permeable to Cl^- instead of K^+, then a negative reversal would have been observed, since a negative potential would be needed to counter the concentration gradient of Cl^-. All the observed reversal potentials in Fig. 2 are consistent with the channel being essentially impermeable to Cl^-.

Having established that the channel is not an anion channel, the next step is to determine whether it is permeable to cations besides K^+. Figure 3 shows how this can be done. Figure 3a presents single-channel currents with 140 mM KCl on both sides of the membrane. The currents are outward at positive potentials, inward at negative potentials, and reverse at about 0 mV. Figure 3b presents single-channel currents with the 140 mM KCl at the inner membrane surface replaced with 140 mM NaCl. With Na^+ inside, the currents are inward at +30 mV (Fig. 3b) instead of outward, as was the case with K^+ inside (Fig. 3a). Thus, even though there is

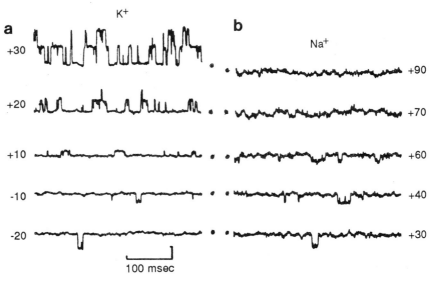

Fig. 3. The BK channel from rat myotubes is impermeable to Na^+. (a) Single-channel currents with 140 mM KCl on both sides of the membrane of an excised patch. (b) Single-channel current with the 140 mM K^+ at the inside membrane surface replaced with 140 mM Na^+. The membrane potentials for each current trace are indicated. At positive membrane potentials K^+ can carry outward current [upward current steps in (a)]. In contrast, Na^+ cannot carry outward current, even when the driving force on Na^+ is much higher [absence of upward current steps in (b)]. The dots indicate the current levels when all channels are closed. Vertical bar: 5 pA for K^+ and 2.5 pA for Na^+. [Reproduced from A. L. Blatz and K. L. Magleby, *J. Gen. Physiol.* **84**, 1 (1984), by copyright permission of the Rockefeller University Press.]

an outward voltage gradient (+30 mV) that would act to drive Na^+ out and oppose the influx of K^+, the inward currents indicate that more K^+ flows into the channel than Na^+ out of the channel. Thus, the channel is much more permeable to K^+ than Na^+. Even with the large outward driving force of +90 mV, there are still no detectable outward currents. Thus, it appears that the channel is essentially impermeable to Na^+, since there is no indication that Na^+ can carry current through the channel.

Experiments like those summarized in Figs. 1–3 can then be used to test for Ca^{2+} activation and K^+ permeability, in order to identify Ca^{2+}-activated K^+ channels.

Classifying Calcium-Activated Potassium Channels on Basis of Conductance

The primary classification of Ca^{2+}-activated K^+ channels has been on the basis of conductance. That BK channels have a much greater conductance than SK channels is clearly evident from Fig. 1. The conductance of a channel, g, is given by the single-channel current amplitude, i, divided by the driving force, which is given by the membrane potential minus the reversal potential, $(V - V_{reversal})$:

$$g = i/(V - V_{reversal}) \qquad (1)$$

From Fig. 2 it can be seen that the conductance for the BK channel increases with the concentration of K^+. With 100 mM K_o^+ and 10 mM K_i^+ the conductance is 150 pS. Increasing the K_i^+ to 100 or 500 mM raises the conductance to 270 and 360 pS, respectively. Thus, when stating the conductance of a channel it is always necessary to state the concentrations of K^+. Furthermore, since common ions such as Na^+ and Mg^{2+} reduce the observed conductance of BK channels,[27–29] it is necessary to consider all ions when making comparisons between conductance of channels. Finally, the conductance of BK channels, as well as most channels, increases with increasing temperature,[30] so it is also necessary to take temperature into consideration.

Techniques for Changing Calcium Concentrations

An important feature of different Ca^{2+}-activated K^+ channels, in addition to their conductance, is their Ca^{2+} sensitivity. This sensitivity is usually expressed as the Ca^{2+} concentration required for half-maximal activity

[27] A. Marty, Pfluegers Arch. Eur. J. Physiol. **396**, 179 (1983).
[28] G. Yellen, J. Gen. Physiol. **84**, 187 (1984).
[29] W. B. Ferguson and K. L. Magleby, Biophys. J. **55**, 546a (1989).
[30] J. N. Barrett, K. L. Magleby, and B. S. Pallotta, J. Physiol. (London) **331**, 211 (1982).

at a given voltage, and it is obtained from a plot of channel open probability, P_{open}, versus Ca_i^{2+}. The plot is constructed from epochs of channel activity obtained in solutions with different free calcium ion concentrations. Because Ca_i^{2+}-activated K^+ channels can switch between different gating modes which typically have different values of P_{open},[31] it is important to ensure that suitably long and representative periods of channel activity are obtained at each Ca_i^{2+}.

Although the ability to change solutions rapidly is not essential for these types of measurements, rapid changes allow more confidence that the final Ca^{2+} concentration has been reached and allow more Ca^{2+} concentrations to be investigated in a shorter period of time. Another consideration is the uniformity of the Ca^{2+} concentration in the recording chamber. Although new solutions might rapidly enter the recording chamber, chamber geometry and the whims of fluid flow might result in incomplete washout of the old solution that would lead to incorrect results.

A variety of devices have been used to ensure rapid and uniform solution changes at or near an excised membrane patch. Excised patches of membrane have been placed into a microliter glass chamber with a small volume (about 0.1 μl),[30,32] so that complete solution changes can be made relatively quickly, without disturbing the membrane seal, as shown in Fig. 4A. The miniature chamber is made from a patch pipette whose narrowed end is sealed shut by passage through a small flame. One side of the tip is then slowly brought toward the edge of the flame as positive pressure is applied to the inside of the pipette by mouth via a length of tubing. As the glass melts a small bubble forms at the tip and then bursts, forming a nearly round chamber approximately 0.5 mm in diameter with a 0.3 mm opening. The patch pipette with the attached excised membrane patch is inserted through this opening, which allows the fluid to escape. During the experiment the solution continuously flows through the microchamber, with changes in Ca^{2+} made by switching to a new solution. Typical solution changes can be made in a few seconds. With this system it is frequently possible to record from a membrane patch for 1–6 hr during dozens of solution changes.

To obtain faster solution changes, Yellen[33] used several small-bore glass tubes through which different solutions flowed continuously (Fig. 4B). With this approach, excised patches can be rapidly moved from the mouth of one tube to another for rapid and uniform solution change. Stepper motors attached to a micromanipulator can make these movements in a few milliseconds.

[31] O. B. McManus and K. L. Magleby, *J. Physiol. (London)* **402,** 79 (1988).
[32] P. T. A. Gray, S. Bevan, and J. M. Ritchie, *Proc. R. Soc. London B* **221,** 395 (1984).
[33] G. Yellen, *Nature (London)* **296,** 357 (1982).

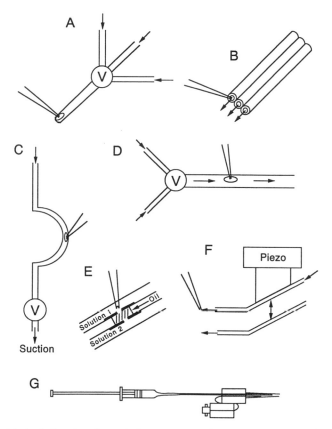

FIG. 4. Methods for changing solutions. (A) Drawing based on the description of the microchamber method of J. L. Barrett, K. L. Magleby, and B. S. Pallotta, *J. Physiol. (London)* **331**, 211 (1982). (B) Redrawn after the multiple outlet method of G. Yellen, *Nature (London)* **296**, 357 (1982). (C) Redrawn after the U-tube method of E. M. Fenwick, A. Marty, and E. Neher, *J. Physiol. (London)* **331**, 577 (1982). (D) Redrawn after the rapid flow method of R. S. Brett, J. P. Dilger, P. R. Adams, and B. Lancaster, *Biophys. J.* **50**, 987 (1986). (E) Redrawn after the oil-gate method of D. Qin and A. Noma, *Am. J. Physiol.* **255**, H980 (1988). (F) Redrawn after the liquid filament method of C. Franke, H. Hatt, and J. Dudel, *Neurosci. Lett.* **77**, 199 (1987). (G) Injection syringe method described in the text.

Several concentration-jump devices have also been developed to change solutions rapidly. A U tube[34,35] in which the solution of choice suddenly exits from a micron-sized hole can be readily fabricated from polyethylene (PE) 10 tubing (Fig. 4C). If an excised patch is placed within 150 μm of the hole, solution changes can occur with time constants of the

[34] O. A. Krishtal and V. I. Pidoplichko. *Neuroscience* **5**, 2325 (1980).
[35] E. M. Fenwick, A. Marty, and E. Neher, *J. Physiol. (London)* **331**, 577 (1982).

order of 10–20 msec[35] (B. Pallotta, unpublished observations, 1989). In a different type of solution changer[36] the excised patch on the tip of a patch pipette is placed either within a small outflow tube or through a small hole into the center of the tube (Fig. 4D). This system allows complete solution changes in 5–15 msec. Rather than using pressure, as in Fig. 4D, it is possible to use suction to control the flow of the solution.[37,38] Another method for rapid solution changes is to move the patch pipette between two solutions through an oil gate,[39] as in Fig. 4E.

Exchange times of 1 msec or less are apparently possible by directing a laminar stream (or liquid filament) of solution onto an excised outside-out patch using the system of Franke et al.,[40] as in Fig. 4F. Unfortunately for the study of Ca^{2+}-activated K^+ channels, such rapid exchanges might not be possible for inside-out patches where the membrane patch is typically withdrawn a short distance into the pipette tip, leading to diffusion delay. Diffusion and hindered diffusion (owing to cytoplasm) in inside-out patches can alter the interpretation of results.[39,41]

Methods like those described in this section, then, provide means to rapidly change Ca^{2+}_i in order to identify Ca^{2+}-activated K^+ channels and study their kinetics.

Calcium Buffering

Calcium-activated channels typically respond to micromolar concentrations of Ca^{2+}. Achieving such low concentrations of free Ca^{2+} in saline solutions without calcium buffering is not possible because of the typical Ca^{2+} contamination of glass-distilled water and analytical grade salts, which can be as high as 20 μM.[42] In principle, the needed low concentrations of free Ca^{2+} can be achieved under equilibrium conditions with a calcium chelator such as EGTA [ethylene glycol bis(β-aminoethyl ether)-N,N,N',N'-tetraacetic acid]. In practice, however, several factors contribute to an appreciable uncertainty in the actual free Ca^{2+} concentration.

Miller and Smith[42] considered three principal sources of error in preparing EGTA-buffered solutions. Probably the most significant is the assumption of an incorrect value for EGTA purity. Commercially available EGTA can be several percentage points less pure than is stated on the label,[42,43] resulting in underestimation of the free Ca^{2+} level by perhaps

[36] R. S. Brett, J. P. Dilger, P. R. Adams, and B. Lancaster, *Biophys. J.* **50**, 987 (1986).
[37] M. Kakei and F. M. Ashcroft, *Pfluegers Arch. Eur. J. Physiol.* **109**, 337 (1987).
[38] Y. Ikemoto, O. Kyoichi, A. Yoshida, and N. Akaike, *Biophys. J.* **56**, 207 (1989).
[39] D. Qin and A. Noma, *Am. J. Physiol.* **255**, H980 (1988).
[40] C. H. Franke, H. Hatt, and J. Dudel, *Neurosci. Lett.* **77**, 199 (1987).
[41] A. L. Zimmerman, J. W. Karpen, and D. A. Baylor, *Biophys. J.* **54**, 351 (1988).
[42] D. J. Miller and G. L. Smith, *Am. J. Physiol.* **246**, C160 (1984).
[43] D. M. Bers, *Am. J. Physiol.* **242**, C404 (1982).

severalfold. A second source of error is inaccurate pH determination, since H^+ competes with Ca^{2+} for the EGTA. Many pH electrodes in normal use have mean errors of about 0.2 pH units owing to liquid junction potentials at the porous ceramic plug.[44] The error typically arises because the ionic strengths of the buffers used to calibrate the electrodes are different from the solutions whose pH is being adjusted. Electrodes with a free-flowing liquid junction are less susceptible to this error, and a simple test for defective electrodes based on diluting a buffer is available.[44]

A third source of error is the use of stability constants that are inappropriate for the ionic strength of the solution[45] or are simply in error. There is still little agreement on appropriate stability constants in the physiological literature. Keeping these limitations in mind, free Ca^{2+} can be estimated by calculation with buffering programs.[46,47] The typical buffering program calculates a list of free Ca^{2+} for a range of added amounts of Ca^{2+} or EGTA (and other ingredients). The investigator then reads down the list to find how much Ca^{2+} and/or EGTA to use for the desired free Ca^{2+}. An alternative approach for those who have not yet written a buffering program is to use commercially available personal computer software (e.g., EQCAL from BioSoft, Ferguson, MO) that calculates (estimates) free Ca^{2+} from total EGTA, pH, ATP, and Mg^{2+} concentrations. When buffering programs are employed, it is always necessary to state the values of the stability constants that were used (see Refs. 46–48 for estimates of stability constants).

One way around some of the errors considered above is to measure the free Ca^{2+}. Bers[43] described a simple method that uses a calcium electrode and does not require accurate knowledge of ionic strength, stability constants, EGTA purity, or pH. An overall stability constant is determined which can then be used to project to lower Ca^{2+}. More recent techniques involve the use of BAPTA [bis-(o-aminophenoxy)-ethane-N,N,N',N'-tetraacetic acid] for Ca^{2+} buffering[45,49] and various Ca^{2+} indicators, such as Fura-2[50,51] and antipyrylazo III.[52] Newer techniques should, of course, be used with caution, since potential artifacts often take time to surface.

In the absence of direct measurements of free Ca^{2+}, and perhaps even when direct measurements are made, the purity of the calcium buffer should be determined and stated in the publication together with the

[44] J. A. Illingworth, *Biochem. J.* **195**, 259 (1981).
[45] S. M. Harrison and D. M. Bers, *Biochim. Biophys. Acta* **925**, 133 (1987).
[46] A. Fabiato and F. Fabiato, *J. Physiol. (Paris)* **75**, 463 (1979).
[47] D. Chang, P. S. Hsieh, and D. C. Dawson, *Comput. Biol. Med.* **18**, 351 (1988).
[48] S. K. B. Donaldson and W. G. L. Kerrick, *J. Gen. Physiol.* **66**, 427 (1975).
[49] R. Pethig, M. Kuhn, R. Payne, E. Adler, T. H. Chen, and L. F. Jaffe, *Cell Calcium* **10**, 491 (1989).
[50] G. Grynkiewicz, M. Poenie, and R. Y. Tsien, *J. Biol. Chem.* **260**, 3440 (1985).
[51] M. Konishi, A. Olson, S. Hollingworth, and S. M. Baylor, *Biophys. J.* **54**, 1089 (1988).
[52] S. Hollingworth, R. W. Aldrich, and S. M. Baylor, *Biophys. J.* **51**, 383 (1987).

amount of added and contaminating Ca^{2+} in each solution (typically determined by flame spectrophotometry). From this information, laboratories can better compare results, even though there might be disagreement about the estimated free Ca^{2+} concentration.

Because of the voltage sensitivity of BK channels, the amount of Ca^{2+} required to activate them is considerably greater at negative membrane potentials,[30] allowing the use of sufficiently high Ca^{2+} at negative voltages so that Ca^{2+} buffering is not necessary. Studies at negative membrane potentials may be useful for kinetic studies, since the problems associated with using Ca^{2+} buffers could then be avoided.

Step Changes in Concentration for Buffered and Unbuffered Ions

When the solutions are heavily buffered for Ca^{2+}, erroneous results can arise if it is assumed that the time course of the change in free Ca^{2+} during a solution change follows the more easily measured time course of a change in an unbuffered ion, such as K^+. (The time course of the change in the concentration of K^+ can be made by changing the concentration of K^+ and following the time course of current through a BK channel or by monitoring the tip potential of an electrode in the changing solution.) When raising the free Ca^{2+} concentration, the unbound calcium buffer in the leading solution will bind Ca^{2+} from the new solution and slow the rise of free Ca^{2+} when compared to an unbuffered ion. When decreasing the free Ca^{2+}, the excess buffer in the new solution will bind Ca^{2+} from the previous solution and increase the rate of decay of free Ca^{2+}, when compared to an unbuffered ion. Thus, caution must be used in interpreting results from step changes in buffered solutions.

Separation of Small Conductance from Large Conductance Channels

Tetraethylammonium ion is commonly used to block K^+ currents in a variety of preparations,[53] and patch clamp experiments on tissue-cultured rat skeletal muscle revealed that the large conductance Ca^{2+}-activated K^+ channel, the BK channel, was blocked completely by 1 mM externally applied TEA.[12,23] Yet, even with the BK channels blocked with as much as 5 mM TEA, the Ca^{2+}-dependent afterhyperpolarization following action potentials was still seen.[12,54] Thus, the afterhypolarization could not be due to BK channels, even though it had all the properties of a Ca^{2+}-activated K^+ current.

Romey and Lazdunski[12] did find that the afterhyperpolarization was blocked by external application of apamin. (Apamin is a basic polypeptide

[53] P. R. Stanfield, *Rev. Physiol. Biochem. Pharmacol.* **97**, 1 (1983).
[54] J. N. Barrett, E. F. Barrett, and L. B. Dribin, *Dev. Biol.* **82**, 258 (1981).

consisting of 18 amino acids with two disulfide bridges, and it is one of the neurotoxic components of honey bee venom.[55]) Thus, the channel that gives rise to the afterhyperpolarization should be blocked by apamin but not TEA. We used this pharmacological profile to identify a small conductance Ca^{2+}-activated K^+ channel (the SK channel in Fig. 1) as a potential candidate for the channel giving rise to the afterhyperpolarization in cultured rat skeletal muscle[24]; the SK channel was blocked by extracellular apamin and not by TEA.

Excised inside-out patches of membrane were used for the identification of the SK channel. In these patches, the membrane exposed to the solution in the patch pipette is the normal extracellular membrane. The experimental approach was to block the BK channels in the membrane patch by adding 5 mM TEA to the solution in the patch pipette. The SK channels were then identified as Ca^{2+}-activated K^+ channels by changing the intracellular concentrations of Ca^{2+} and K^+. A pressure injection of apamin through the recording pipette then convincingly demonstrated block of the SK channels by apamin.[24] Because the SK channels were blocked by apamin and not by the TEA, they have the necessary properties to produce the afterhyperpolarization. The following section describes in detail how apamin was applied to the normal extracellular membrane surface that was exposed to the solution in the patch pipette.

Changing Composition of Solution in Patch Pipettes

The patch clamp technique, when compared to recording from channels in lipid bilayers,[56] has the major drawback of providing only limited access to the membrane surface exposed to the solution inside the pipette. This can present a problem if the agent of interest, such as a neurotransmitter or a blocking drug, acts on the channel from the extracellular membrane surface, which for inside-out membrane patches is located facing the pipette solution. One way to overcome this problem is to use outside-out membrane patches and apply the drug of interest via the bath solution. Unfortunately, for many cell membranes, outside-out patches are much more fragile than inside-out patches and do not last through enough solution exchanges to complete the required experiments. Another option is to use inside-out patches and employ pipette perfusion techniques[57,58] to perfuse the membrane surface in the patch pipette.

A third option, useful only for application of relatively potent neuro-

[55] E. Habermann, *Pharmacol. Theor.* **25**, 255 (1984).
[56] C. Miller, ed., "Ion Channel Reconstitution." Plenum, New York, 1986.
[57] J. M. Tang, J. Wang, F. N. Quandt, and R. S. Eisenberg, *Pluegers Arch. Eur. J. Physiol.* **416**, 347 (1990).
[58] J.-Y. Lapointe and G. Szabo, *Pfluegers Arch. Eur. J. Physiol.* **410**, 212 (1987).

transmitters or blockers, is simply to introduce a concentrated solution of the drug into the patch pipette with a syringe, as close to the membrane as possible. It is this last technique that we used to apply apamin to SK channels, as described in the previous section. The following sections explain the fabrication of the injection syringe and then the injection process.

Although we did not originate the idea of pulling a plastic syringe, and do not know who did, it seems worthwhile presenting this technique for those who are not yet familiar with it. Although any plastic 1-cm^3 disposable syringe will probably work for the injection syringe, we have had great success with those made by Sherwood Medical Industries (Deland, FL 32720; Monoject, Cat. No. 501S-TB). A two-stage heat-pulling process is used to form the required thin tube which will be inserted into the patch pipette. To fashion the injection syringe, the plunger is pulled almost totally out of the syringe and is used as a handle for one end of the syringe while the needle end is used as the other handle. A standard laboratory burner is used to soften the syringe barrel so that it can be pulled. The only critical part of this process is to allow the plastic to melt sufficiently to become transparent over about 1 cm of the length of the syringe. Often the plastic actually catches fire before this occurs, but the fire can simply be blown out before the plastic becomes scorched. When the syringe plastic is sufficiently melted, the needle end is let go and allowed to fall by gravity to the floor while holding the syringe vertically with the plunger handle pointing upward. The syringe should be held vertically for several minutes to allow the plastic to cool, which is indicated by the plastic turning from clear to translucent.

At this point, the thinned portion of the stretched syringe can be cut to separate it from the needle portion, which is discarded. The pulled tubing made from the syringe must be sufficiently thin so that it easily fits within the bore of a patch pipette. The thinned portion of the syringe is now pulled again, but with a smaller burner, such as one made from a 16-gauge hypodermic needle. Trial and error, in terms of how much to heat the plastic and how fast and how far to pull, is the only method that works at this stage, and several attempts are usually made before the tubing pulls so that the thinnest portion of the second pull will fit almost to the tip of a patch pipette, with the length of the secondary thinned part less than 1–2 cm. If the secondary thinned length of tubing is longer than this, it will be difficult to insert it into the patch pipette. Care must be taken to ensure that the end of the injection tube is cut at a steep angle cleanly with a new razor blade so that bits of plastic do not occlude the tip.

Once the injection syringe is pulled to the desired outside diameter, it is filled with about 50–100 μl of the appropriately concentrated drug or blocker solution, taking care to ensure that an air gap does not exist between the plunger and the solution column. An easily available electrode holder (Model PC-S3, E. W. Wright, 760 Durham Road, Guilford, CT 06437) is used to connect the patch clamp electrode to the input state of the current–voltage converter. The holder is modified by cutting it into two pieces, one containing the BNC connecter and silver wire, and the other consisting of the front end of the holder with the suction port and the screw clamp for the patch pipette. The front end of the holder is glued piggyback style onto the BNC end so that the back of the patch electrode can be accessed when the holder is attached to the headstage. The silver wire is threaded from the BNC connecter into the back of the top holder and extends about 1 cm into the patch pipette so that it can contact the solution in the patch pipette (Fig. 4G). To maintain the airtight seal required to apply the negative suction to obtain the seal with the membrane, a wax plug is inserted into the back of the top end of the holder.

A gigohm seal is then obtained in the conventional manner between the electrode tip and the membrane of the myotube, and the patch is excised into the inside-out configuration by passing it through the air–bath interface and back into the bath solution. The tip of the patch pipette is then placed into the microchamber described in a previous section so that the solution can be changed at the inner membrane surface. The wax plug is then removed, and the tip of the pulled injection syringe is inserted into the back of the patch pipette and advanced until it bumps against the tip of the pipette. (The tip is cut at an angle so that it does not make a tight seal with the inside tip of the pipette.) At this point, the solution contained in the syringe is slowly injected into the patch pipette. Once the drug is injected, the injection syringe tube can be removed from the recording pipette. Thus, little if any additional noise is added by this technique because there are no permanent attachments to the headstage and because the volume added to the recording pipette solution is small. Within about 30 sec after the injection, the apamin typically started to block channels, and within several minutes the blocking effect of apamin was typically complete.[24] Although the above technique is rather primitive compared to those described by Lapointe and Szabo[58] and Tang et al.,[57] it is easily applied and can be sufficient for those instances where it is not necessary to know the precise concentration of the injected drug or ion.

Summary

The techniques described in this chapter together with many of the techniques described elsewhere in this volume should be useful toward identifying and characterizing Ca^{2+}-activated K^+ channels.

[13] Patch-Clamping Cells in Sliced Tissue Preparations

By F. A. EDWARDS and A. KONNERTH

Introduction

The introduction of patch clamp techniques[1] allowed an enormous increase in the resolution of membrane current recording in small cells and thus opened the possibility of studying phenomena not previously measurable. The technique was, however, limited by the apparent need for direct contact of a glass pipette with a smooth cell membrane. Thus, use of patch clamp techniques was restricted to isolated or cultured cell preparations. Although much has been achieved with such preparations, the study of synaptic currents in central neurons was impossible, except using synapses formed between cultured cells in which cell types and the stages of cell maturity are undefined. In addition, in studies of single-channel currents, questions can always be raised as to the effects of enzymes on the preparation during isolation procedures. Many of these problems were overcome by the application of patch clamp techniques to brain slices.[2] Most recently these techniques were extended to nonneural preparations.[3,4]

Slice Techniques

Introduction of Brain Slices

Brain slices were originally used early in the twentieth century mostly for biochemical measurements. Although they were first used for electrophysiological recordings in the late 1950s by McIlwain (for review, see Alger *et al.*,[5] brain slices did not become widely established as a tool for electrophysiology until the 1970s (for review, see Andersen and Langmoen[6]). Studies using brain slices dominated central nervous system syn-

---

[1] O. P. Hamill, A. Marty, E. Neher, B. Sakmann, and F. J. Sigworth, *Pfluegers Arch.* **391,** 85 (1981).

[2] F. A. Edwards, A. Konnerth, B. Sakmann, and T. Takahashi, *Pfluegers Arch.* **414,** 600 (1989).

[3] N. A. Burnashev, F. A. Edwards, and A. N. Verkhratsky, *Pfluegers Arch.* **417,** 123 (1990).

[4] M. B. Jackson, A. Konnerth, and G. Augustine, *Proc. Natl. Acad. Sci. U.S.A.* **88,** 380 (1991).

[5] B. E. Alger, S. S. Dhanjal, R. Dingledine, J. Garthwaite, G. Henderson, G. L. King, P. Lipton, A. North, P. A. Schwartzkroin, T. A. Sears, M. Segal, T. S. Whittingham, and J. Williams, *in* "Brain Slices" (R. Dingledine, ed.), p. 381. Plenum, New York, 1984.

[6] P. Anderson and I. A. Langmoen, *Q. Rev. Biophys.* **13,** 1 (1980).

aptic research in the 1980s. Techniques have of course gradually improved with advanced technology, and the stability of intracellular recordings is such that several laboratories have achieved long-lasting simultaneous impalements of two synaptically connected neurons.[7,8] This was a remarkable achievement allowing analysis of a single synaptic connection.

Such recording and analysis were, however, limited by difficulties in finding and identifying cells and by the relatively low signal-to-noise ratio of central synaptic currents recorded intracellularly. Various approaches have been used to overcome these problems. A technique was described by Barnes and Werblin[9] for making whole-cell patch clamp recordings from amacrine cells in salamander retina. Patch recordings from mammalian hippocampus were also described using a slice preparation which, after enzyme treatment, was split along the cell line allowing access to the somas of CA1 and CA3 pyramidal cells.[10] Recently we have described a technique for making patch clamp recordings in slices without the use of enzymes, which is applicable to a wide range of tissues.[2] The use of high-resolution optics allowed visualization of the cells to be recorded.[11-13] It was thus possible to remove tissue covering visually identified cells and record currents in synaptically connected neurons *in situ* with patch clamp techniques. This introduced a 10-fold improvement in resolution compared to that which was previously possible using single-electrode voltage clamp techniques.

Survival of Neurons in Slices: Technical Tips

It is perhaps surprising that brain tissue can be removed from the living animal and be maintained *in vitro* over many hours. Of course it is necessary to remember that changes often occur; however, the tissue does appear to stay "alive," and comparison with *in vivo* recordings suggests that resting membrane potentials and other physiological parameters such as the time course and size of action potentials are similar.[14] The basic methods of slicing brain tissue have been repeatedly described (e.g., for review of methods, see Alger *et al.*[5]). However, except for the few studies in which cells could be visualized, only ease and stability of recording and the

[7] R. Miles and R. K. S. Wong, *J. Physiol. (London)* **356,** 97 (1984).

[8] R. J. Sayer, S. J. Redman, and P. Anderson, *J. Neurosci.* **9,** 840 (1989).

[9] S. Barnes and F. Werblin, *Proc. Natl. Acad. Sci. U.S.A.* **83,** 1509 (1986).

[10] R. Gray and D. Johnston, *J. Neurophysiol.* **54,** 134 (1985).

[11] C. Yamamoto, *Experientia* **31,** 309 (1975).

[12] T. Takahashi, *Proc. R. Soc. London* **202,** 417 (1978).

[13] R. Llinas and M. Sugimori, *J. Physiol. (London)* **305,** 171 (1980).

[14] P. A. Schwartzkroin, *Brain Res.* **85,** 423 (1975).

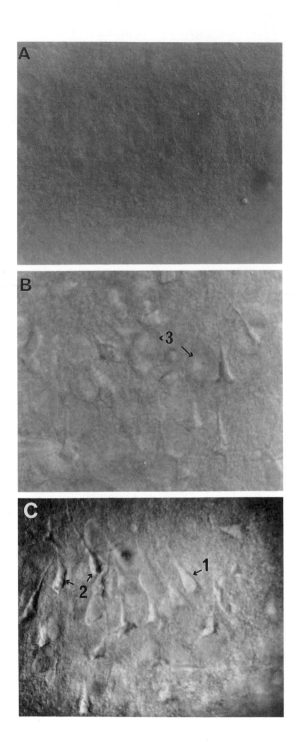

physiological parameters of successfully impaled cells could be used to assess the quality of slices. As all these parameters are affected not only by the quality of the slice but also by the quality of electrodes and all other parts of the recording system, a bank of unsubstantiated superstition has grown up over which parameters are important for survival of tissue. With the possibility of visualizing the tissue and individual cells, a much more direct assessment of the number of healthy cells compared to dead or dying tissue is available (Fig. 1).

We include here only a relatively brief description of the methods we use as these have been described previously in great detail[2,15] and our method is certainly not the only way of preparing slices. We do, however, also include a few hints where particular parts of the preparation seem to be critical and where we have improved our methods since the publication of the earlier papers.

Preparation of Tissue

As mentioned, the preparation of the slices is fairly standard:

Animal decapitated
Brain hemisected and removed
Brain placed in ice-cold saline
(<1 min from decapitation to this stage)
Pause ($2-10$ min) while tissue cools
Tissue trimmed to correct angle
Trimmed tissue glued to slicer stage
[place on glue (cyanoacrylate); cover with solution as quickly as possible]
Cut slices with vibrating slicer ($60-400$ μm thick) (~ 10 min)
Move each slice to holding chamber immediately after slicing (using cut and polished Pasteur pipette)
Incubate at $32-37°$ for at least 30 min before recording

[15] A. Konnerth, *Trends Neurosci.* **13**, 321 (1990).

FIG. 1. Choosing healthy cells in a slice. (A) A very poor slice of the CA1 region of a rat hippocampus. The pyramidal cell line runs across the upper middle of the photo but is barely visible. Such a slice should be immediately discarded. (B) A usable but still not ideal slice of rat visual cortex. An electrode tip is visible at lower left, attached to a healthy cell. Several round swollen cells are visible on the surface (3). (C) A very good slice of rat hippocampus showing the cortical end of the CA1 region. A few "shiny" cells which would probably not allow seal formation are visible at left (2). In the center are healthy cells, patching of which would be very likely to be successful. The cell of choice is indicated (1). (Width of photo ~ 250 μm.)

One of the most critical points about preparing the tissue for slicing seems to be the time taken from killing the animal to having the brain submerged in ice-cold physiological saline. This procedure should not take more than 1 minute. It may be partly for this reason that it is easier to prepare tissue from younger animals (0–3 weeks) as the skull is softer and fast removal of the brain is easier. Moreover, the fact that the young brain is itself smaller implies that the tissue will become cold throughout in a shorter time. For laboratories wishing to establish the technique of making patch clamp recordings from slices, it is thus a good idea to begin with young rats and if necessary progress to older tissue when the method is well established. Once the brain is in ice-cold solution the rest of the procedure can continue at a more relaxed pace.

Most vibrating slicers seem to be suitable for cutting brain slices although they generally must be used at the maximum frequency setting. Some commonly used slicers have an insufficient maximum width of vibration. In this case the forward movement of the blade during slicing must be extremely slow to avoid "pushing" the tissue. Although, if the tissue is kept very cold throughout slicing, time is not very critical, it is preferable to avoid taking more than about 20 min for the whole procedure. An insufficient width of vibration prolongs slicing (e.g., 30–40 min to produce 8–10 slices), and it is possible that this may have an influence, especially if the temperature of the preparation increases during the procedure.

Another feature which seems to make a substantial difference to the quality of slices is the use of very sharp blades. In general steel blades (e.g., slicer blades from Campden Instruments, Loughborough, UK) are sharper than stainless steel blades.

Lastly, the survival of a neuron seems to be very dependent on the survival of its dendritic tree. In fact, it is rare on successfully forming the whole-cell configuration and filling cells with Lucifer yellow to find a cell which does not have an extensive dendritic tree. This presumably implies that cells with substantially truncated dendritic trees die during incubation or tend not to form seals. Further, during removal of tissue covering a cell ("cleaning"), damage to dendrities even 30 or 40 μm from the soma will cause immediate death (seen as fading away of the cell soma). Fortunately, for most structures an optimal slicing angle can be found resulting in minimal dendritic damage (e.g., transverse slices of hippocampus,[6] parasagittal visual cortex slices[16,17]). In addition to the angle of slicing, the thickness of the slice will also have a strong influence. In the original

[16] P. Stern, F. A. Edwards, and B. Sakmann, *Eur. Neurosci. Assoc. Abstr. Turin,* 62 (1989).
[17] P. Stern, F. A. Edwards, and B. Sakmann, *J. Physiol. (London)* **449**, 247 (1992).

methods paper,[2] 120-μm slices were mostly used, though it was mentioned that slices up to 300 μm thick were possible when using a video camera. We now use slices 200–400 μm in thickness and find that the thicker the slice, the better the survival of the tissue. With optimal slices and good optics, neurons and overlying tissue can still be clearly seen even without use of a camera (Fig. 1C).

Incubation of Slices

There are undoubtedly many different types of chambers suitable for incubating slices. In the original methods paper,[2] we described an easily disposable incubation chamber in great detail. Although this chamber usually worked well, it was somewhat unstable, and slices sometimes ended up on top of each other in the corners of the chamber or stuck to the bottom, resulting in a considerable deterioration. It was also mentioned that even under the best conditions, slices deteriorated to some extent after 3–4 hr, although they could still be used for 10–12 hr. We illustrate here (Fig. 2a) another incubation chamber developed by Alasdair Gibb (University College London, personal communication). The Gibb chamber is stable, and hippocampal slices consistently survive for the whole of an experimental day with minimal deterioration. Survival of less robust tissue such as neocortical slices is also much improved by use of this method.

Making Patch Clamp Recordings in Slices

Preparing Tissue for Patch Clamp Recordings

To make high-quality patch clamp recordings, it is necessary to have access to a clean cell membrane. As previously described,[2] this can be achieved without use of any enzyme treatment (Fig. 2b). Ideally, a pipette is pulled from soft glass to a diameter of about 5 μm, filled with normal physiological saline, and placed in a normal patch pipette holder. By placing the cleaning pipette on the surface of the tissue near a selected neuron and applying positive pressure, the covering tissue can be broken and disturbed. The debris is then removed by gentle suction. This procedure results in exposure of the soma membrane while the dendrites remain buried in the slice. In healthy slices cleaning of a cell takes about 15–20 sec. If it is not convenient to pull cleaning pipettes, a patch pipette may be simply broken to a diameter of about 5 μm.

A recent paper reported a version of the technique which requires no cleaning.[18] Under a dissecting microscope, positive pressure is applied to

[18] M. G. Blanton, J. J. L. Turco, and A. R. Kriegstein, *J. Neurosci. Methods* **30**, 203 (1989).

a   **b**

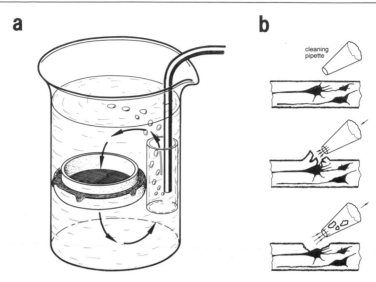

FIG. 2. Incubation and "cleaning" of slices. (a) The Gibb chamber: an improved slice incubation chamber. The inner holding chamber itself is as previously described.[2] In brief, the top and bottom are broken out of a 35-mm petri dish and lid, forming two rings which fit into each other such that a piece of cotton or other suitable fabric can be clamped between them. This small holding chamber can then be wedged halfway down a 100-ml beaker with a stiff plastic tube (e.g., a cutoff disposable syringe) such that the tube extends well below and a little bit above the chamber. When a bubbler is placed about two-thirds of the way down inside the tube, the bubbles rise up the tube to the surface, drawing solution after them from the bottom of the beaker. This creates a circulation through the beaker such that freshly oxygenated solution flows constantly down over the slices, through the cotton, and back into the lower half of the beaker. Under these conditions slices always lie stably on the bottom of the chamber, never sticking, and remain in uniformly good condition throughout the experiment (>8 hr). (b) Schematic diagram of the exposure of a neuron. [From F. A. Edwards, A. Konnerth, B. Sakmann, and T. Takahashi, *Pfluegers Arch.* **414**, 600 (1989).] The cleaning pipette is brought close to the surface of the slice above a neuron. Tissue which covers the cell is broken up by a stream of physiological saline applied through the cleaning pipette. Resulting debris is removed by gentle suction through the cleaning pipette.

the patch electrode as it is lowered through the tissue until an increase in resistance is evident. Letting go the pressure results in sudden formation of a high-resistance seal. Although this goes against all expectations, it does work. In our hands, however, there is a clear tendency toward an increased series resistance value compared to recordings using the original technique. Further, it is much more difficult to identify a particular cell, and thus the direct control of cell type and quality is generally lost. Nevertheless, the recording quality is still considerably better than intracellular recording, and where the superior optics necessary for cleaning cells is not available this is a good compromise. Moreover, for particular applications where high-quality recording is not the priority, the advantages of speed and ability to patch deep in the slice with minimal disturbance of neurites may make this the method of choice. For example, when stimulating in the whole-cell mode, it is necessary to monitor the occurrence of action potentials but not to record accurately in the voltage clamp mode. In such a case easy access to deep cells without cleaning is a distinct advantage.

Choosing a Cell for Patching

In good slices cells are clearly visible both on the surface and deep within the slice (Fig. 1B,C). Even without filling cells with fluorescent dyes, cell bodies and often substantial portions of the dendritic tree can easily be seen. It is thus possible to choose a cell type by its morphology. For example, in a hippocampal slice, cells can be distinguished which have a pyramidal structure, while others clearly show transverse dendrites as one might expect from interneurons (Fig. 1C).

Another consideration is how to choose a "patchable" cell which is likely to result in a stable recording. Naturally with experience this becomes very clear. The general rule is that healthy cells are clearly visible presenting an apparently smooth surface. Certain features present a clearly unfavorable prognosis for obtaining good recordings. On one hand, cells which have an uneven, "spotty" appearance or stand out from the slice with very high contrast (extremely "shiny" cells) tend not to form seals. At the other end of the spectrum, the surface of the slice sometimes features rather transparent, swollen cells with the nucleus clearly visible (Fig. 1B). Although it is frequently very easy to form high-resistance seals on such cells and even to see single-channel currents in the cell-attached configuration, these cells are generally lost immediately on trying to break through into whole-cell configuration.

Patch Pipettes

Techniques for pulling pipettes and forming high-resistance (> 10 GΩ) seals on cleaned neurons in slices are standard,[1] and all configurations are possible. We find that seals are most easily formed with thin walled (0.3 mm wall thickness, 2 mm diameter) borosilicate glass electrodes having resistances of $2-7$ MΩ (when filled with 140 mM KCl). In whole-cell recordings the upper end of this range ($5-7$ MΩ) may cause series resistance problems depending on the type of current measured. Electrodes of around 4 MΩ easily form seals and are electrically suitable for small neurons (e.g., hippocampal granule cells). For single-channel recordings it may be convenient to use somewhat higher resistance thick walled glass pipettes (0.5 mm thickness, 2 mm diameter), but it begins to be difficult to break through to the whole-cell configuration with electrodes of more than about 12 GΩ.

Applications of Patch Clamp Recording in Slices

Application of Patch Clamp Techniques to Slices of Different Tissues

The technique of making patch clamp recordings in slices is applicable to a wide variety of cells. In the original methods paper[2] we had already tried many brain areas in the rat (e.g., spinal cord, hippocampus, retina, olfactory bulb, frontal and visual cortex, cerebellum, striatum, and brain stem). In addition, recordings were made on slices from cat and mouse hippocampus and visual cortex. Since that time the technique has been applied to other structures, for example, the pituitary.[4]

The successful application of this technique to rat cardiac cells[3] opens the possibility of its general use in various peripheral tissues. Although the fact that heart muscle is a syncytium makes effective clamping of whole-cell currents impossible, the cells were shown to have resting potentials ranging from -30 to -70 mV, and in cell-attached patches K^+ and Na^+ single-channel currents were recorded with high resolution (Fig. 3). Note that in the case of this muscle preparation it was not possible to remove covering tissue. However, visualization of the cells made it possible to select healthy myocytes on the surface of the slice.

Improved Resolution of Synaptic Currents

As mentioned above, one of the greatest advantages of applying patch clamp techniques to brain slices is the improved resolution of central synaptic currents. Previous patch clamp studies of synaptic currents in central neurons were restricted to synapses formed in culture. The developmental stage of such synapses or how closely they are related in structure and function to synapses formed *in vivo* is unclear. Thus, most of the information gathered about central synaptic currents comes from intracellular recordings using the single-electrode voltage clamp (SEVC; for review, see Edwards and Stern[19]).

The possibility of applying patch clamp techniques to synaptic currents has made it possible to apply direct quantal analysis to the central nervous system (CNS).[20] We chose the inhibitory synapses on hippocampal granule

[19] F. A. Edwards and P. Stern, *in* "Excitatory Amino Acids and Synaptic Function" (H. Wheal and A. Thomson, eds.). Academic Press, London, 1991.
[20] F. A. Edwards, A. Konnerth, and B. Sakmann, *J. Physiol. (London)* **430,** 213 (1990).

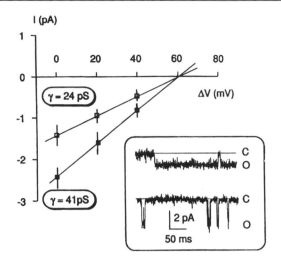

FIG. 3. Single-channel currents recorded in cell-attached patches from slices of newborn rat heart. $I-V$ curve for single-channel currents through two types of inward rectifying potassium channels, recorded in the cell-attached mode. The slice was bathed in physiological saline, and the patch pipette was filled with high potassium solution. Solid lines are the slopes of the linear regression. Slope conductance values are shown near each line. Potentials are given with respect to the resting potential. (Inset) Records of single-channel currents through two types of inward rectifying potassium channels at resting potential. Current records were filtered at 1 kHz (-3 dB) and sampled at 3 kHz. [From N. A. Burnashev, F. A. Edwards, and A. N. Verkhratsky, *Pfluegers Arch.* **417**, 123 (1990).]

cells for analysis. These synapses are particularly suitable for this purpose as the cells are small and have high input resistances (>1 GΩ) and the synapses are situated on or near the cell soma.[21] The currents can be measured with a considerably higher time and amplitude resolution than was possible with the single-electrode voltage clamp (Fig. 4), revealing very fast rise times and slower two-exponential decays. On constructing amplitude distributions of stimulated currents, multiple equidistant peaks were evident (Fig. 5). The quantal size was very small, falling into one group around 20 pA and another group around 10 pA (at -50 mV in symmetrical Cl$^-$). The data resulting from the possibility of recording these currents with such high resolution suggested that the size of quantal currents must be determined postsynaptically. Although such a detailed study of amplitude distributions has not yet been reported for excitatory currents, their similarly small size and fast rise times[16,17,22] suggest that a similar model

[21] M. Frotscher, *in* "Neurotransmission in the Hippocampus" (M. Frotscher, P. Kugler, U. Misgeld, and K. Zilles, eds.). Springer-Verlag, Berlin, Heidelberg, and New York, 1988.

[22] A. Konnerth, B. U. Keller, and A. Lev-Tov, *Pfluegers Arch.* **417**, 285 (1990).

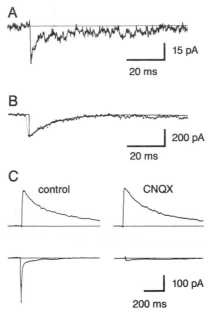

FIG. 4. Synaptic currents. Comparison of single-electrode voltage clamp recording and patch clamp recording of the same type of current under similar conditions. (A) Whole-cell patch clamp recording of a γ-aminobutyric acid (GABA)-mediated inhibitory postsynaptic current (IPSC) from a 6-week-old rat hippocampal CA1 cell at -50 mV membrane holding potential. The recording pipette contained the following (in mM): KCl 140, MgCl$_2$ 1, CaCl$_2$ 1, K$^+$ATP 2, EGTA 10, HEPES 10 (pH 7.3). The trace was filtered at 2 kHz (-3 dB). (B) Single-electrode voltage clamp recording of a GABA-mediated IPSC from a 3-month-old rat CA1 pyramidal cell at -60 mV membrane holding potential. The recording pipette contained 3 M KCl and had a resistance of 50 MΩ. The trace was filtered at 1 kHz (-3 dB, not limiting to rise time). Note the difference in scale bars for traces (A) and (B). Despite the microelectrode recording being more heavily filtered and the resolution being one-tenth that of the patch clamp recording, the noise bands of the two traces are similar. Two exponentials must be included to fit trace (A), one of which is completely lost in trace (B). (C) Pharmacological separation of N-methyl-D-aspartate (NMDA)- and non-NMDA-mediated components of stimulated excitatory postsynaptic currents (EPSCs) recorded from a 3-week-old rat hippocampal granule cell. The left-hand side shows traces of currents in control solution, and the right-hand side shows traces of currents recorded in 5 μM cyano-nitroquinoxaline-dione (CNQX) to reveal the pure NMDA-mediated component. The upper two traces were recorded at $+40$ mV membrane potential, where NMDA and non-NMDA components would be expected to be evident in control solution. The lower traces were recorded at $+70$ mV, where NMDA-mediated currents would be expected to be largely blocked by Mg^{2+}.

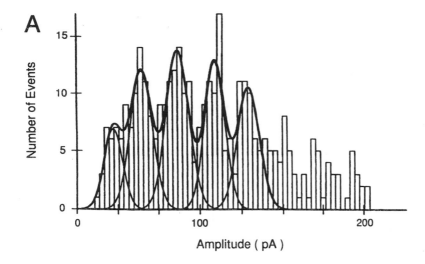

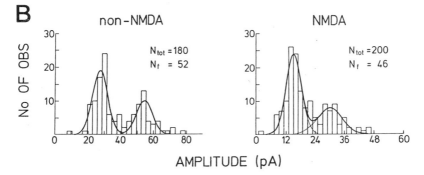

FIG. 5. Distributions of synaptic current amplitudes in CNS neurons show quantal distributions with very small quantal sizes. (A) Amplitude distribution of GABA-mediated IPSCs recorded in the granule cell region of rat hippocampus slices at −50 mV. The solid lines superimposed on the histogram are the best least-squares fit of the data using a modified Simplex routine. Means ± S.D. for peaks 2–5 were (in pA) 39.2 ± 6.6, 61.5 ± 6.9, 83.8 ± 5.9, and 104.0 ± 6.5. The mean ± S.D. of the peak separation (fitted mean of each peak divided by its peak number) was 20.7 ± 2.4 pA. [Reprinted from F. A. Edwards, A. Konnerth, and B. Sakmann, *J. Physiol. (London)* **430,** 213 (1990).] (B) Amplitude distributions of the NMDA- and non-NMDA-mediated components of EPSCs recorded in a motoneuron in the spinal cord of a newborn rat. The histograms were fitted with summated Gaussian functions showing bimodal distributions. The means of the distributions for the non-NMDA component were (in pA) 27.2 and 54.5 and those for the NMDA-receptor-mediated component, 15.1 and 30. [Reprinted from A. Konnerth, B. U. Keller, and A. Lev-Tov, *Pfluegers Arch.* **417,** 285 (1990).]

may also be applicable. Extensive intracellular work has led to similar conclusions (for review, see Redman[23]).

Measurement of Single-Channel Currents in Slices

Ligand-Gated Channels. In parallel with recording of synaptic currents, use of patch clamp recording in slices allows recording of the underlying single-channel currents in the same cells under the same conditions. In addition, the avoidance of enzyme treatment may be particularly important where the extracellular domain of the protein is important for function. One example in which results in slices appear to be somewhat different from results in dissociated or cultured cells is the diversity of subconductance states seen in single-channel recordings of γ-aminobutyric acid ($GABA_A$) -mediated currents. In earlier reports of different cultured or enzyme-dissociated preparations from rodents, varied results have been noted.[24-26] The consistent finding was that $GABA_A$ -mediated channels showed multiple subconductance states. However, different preparations varied greatly as to which states occurred and at what relative frequency. In contrast, using patch clamp recordings in rat hippocampal slices; $GABA_A$ -mediated currents in granule cells show considerably less variation.[20] Two types of channels are evident, each showing a single conductance state (14 and 23 pS, Fig. 6). These conductances could then be directly compared with conductances of quantal synaptic events, allowing an estimate that less than 30 and perhaps as few as 10 channels open during a quantal event.

Voltage-Gated Channels. Although it is perhaps less likely that voltage-gated channels would be affected by enzyme dissociation, the possibility of positively identifying cells makes slices again the preparation of choice. In a series of recent papers, Takahashi has illustrated this with a thorough characterization of both voltage-gated and transmitter-gated channels in identified motoneurons in neonatal rat spinal cord slices.[27-29] He describes tetrodotoxin (TTX)-sensitive sodium channels and separates, among others, three potassium channels which probably contribute to action potential repolarization. In addition he describes the properties of currents

[23] S. Redman, *Physiol. Rev.* **70,** 165 (1990).
[24] J. Bormann, O. P. Hamill, and B. Sakmann, *J. Physiol. (London)* **385,** 243 (1987).
[25] R. L. MacDonald, C. J. Rogers, and R. E. T. Twyman, *J. Physiol. (London)* **410,** 479 (1989).
[26] S. M. Smith, R. Zorec, and R. N. McBurney, *J. Membr. Biol.* **108,** 45 (1989).
[27] T. Takahashi, *J. Physiol. (London)* **423,** 27 (1990).
[28] T. Takahashi, *J. Physiol. (London)* **423,** 47 (1990).
[29] T. Takahashi and A. J. Berger, *J. Physiol. (London)* **423,** 63 (1990).

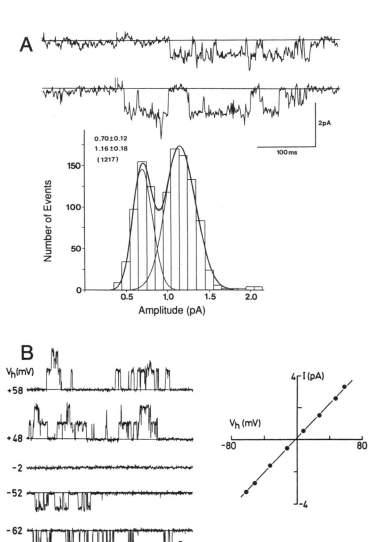

FIG. 6. Single-channel currents measured in outside-out patches from hippocampal granule cells in slice preparations. (A) GABA-mediated currents measured at −50 mV from the same cell type, under the same conditions as the currents for the IPSC amplitude distribution shown in Fig. 5A. The amplitude distribution of single-channel events shows two conductance states (14 and 23 pS). The proportion of the two different types of openings is very different between patches, with three of nine patches showing only 14-pS openings and the other six patches ranging from 10 to 40% of openings having the lower conductance state. Thus, two different channel types are presumably present. [Reprinted from F. A. Edwards, A. Konnerth, and B. Sakmann, *J. Physiol. (London)* **430**, 213 (1990).] (B) NMDA-mediated currents. The conductance calculated from the $I-V$ curve is 52 pS. {Reprinted from B. Keller, A. Konnerth, and Y. Yaari, *J. Physiol. (London)* **435**, 275 (1990).]

contributing to inward rectification in spinal motonneurons and compares these with 5-hydroxytryptamine (5-HT)-mediated currents in the same cell type.

Optical Measurements of Internal Ca²⁺ Concentration Combined with Patch Clamp Recordings

Another application of patch clamp recordings in slices is the possibility of combining high-resolution electrophysiological recording with the recent advances in optical techniques allowing relatively fast and well spatially resolved measurement of internal neuronal Ca^{2+} levels. The influx of Ca^{2+} could then be pictured simultaneously using Ca^{2+} imaging.[30] It would thus be possible to determine in response to presynaptic stimulation the location of synapses from the stimulated neuron or pathway and the extent in both space and time of spread of Ca^{2+} in the cell. In principle, in combination with laser confocal microscopy, this could be extended to see spines on neurons and resolve questions as to the exact location of synaptic N-methyl-D-aspartate (NMDA) receptors. Thus, assessment could be made of the extent to which it is possible to resolve currents which may be isolated from the main neuronal compartments by physical features such as the apparently narrow neck of the synaptic spines.

[30] A. Konnerth, J. Dreessen, and G. J. Augustine, *Pfluegers Arch.* **415** (Suppl. 1), 312, R84 (1990).

Section II

Expression of Ion Channels

A. Expression of Ion Channels in *Xenopus* Oocytes
Articles 14 through 25

B. Expression of Ion Channels Using Other Systems
Articles 26 through 29

[14] *Xenopus* Oocyte Microinjection: From Gene to Protein

By Hermona Soreq and Shlomo Seidman

I. Introduction

The biosynthesis of biologically active proteins in microinjected oocytes has been extensively reviewed,[1] especially with respect to the implications of the system for molecular neurobiology.[2–5] The recognized capacity of the oocytes to translate efficiently and faithfully foreign genetic information combined with its ability to assemble oligomeric receptor/channel complexes and insert them into the plasma membrane to generate elicitable electrophysiological responses makes them a powerful tool for the molecular neurobiologist. *Xenopus* oocyte microinjection has already led to the heterologous expression of numerous ion channels, pumps, and receptors, and the cloning and characterization of the genes encoding them (reviewed by Lester[3]). Site-directed mutagenesis, chimeric protein engineering, and *in ovo* reconstitution of multisubunit molecules are subsequently providing the means with which to dissect the functional roles of various intra- and intermolecular components in specifying the biological properties of these conglomerates. Forthcoming advances in molecular neurobiology will need to take further advantage of this remarkable system to pursue questions regarding the regulation of gene expression, posttranslational processing, and biological functions of important nervous system proteins.

The purpose of this chapter is to present an overview of the types of experimental protocols which have been exploited in the oocyte expression system and, in particular, to highlight some of the novel approaches to *Xenopus laevis* oocyte microinjection which are pushing forward the frontiers of molecular neurobiology. With an eye toward sparing the reader a mountain of overlapping references, we have attempted to limit the cited papers to those most recent and those exemplifying multiple aspects of the issues at hand, expecially methodologies. Therefore, the reader will find papers which are cited several times in different contexts. The general

[1] H. Soreq, *Crit. Rev. Biochem.* **18,** 199 (1985).
[2] N. Dascal, *Crit. Rev. Biochem.* **22,** 317 (1987).
[3] H. A. Lester, *Science* **241,** 1057 (1988).
[4] R. Miledi, I. Parker, and K. Sumikawa, *in* "Fidia Research Foundation Neuroscience Award Lectures," p. 57. Raven, New York, 1989.
[5] S. Seidman and H. Soreq, *Int. Rev. Neurobiol.* **32,** 107 (1991).

organization represents an attempt to present an introduction to the versatile applications of the *Xenopus* oocyte as an experimental system, followed by specific examples of experimental strategies and approaches to the question of detection. Each section is intended to stand independently, forming the basis of an easily accessible reference section.

II. Applications

A. mRNA Bioassay

As a preliminary step in molecular cloning studies, the microinjection of total poly(A)$^+$ mRNA from tissue homogenates into *Xenopus* oocytes serves to provide evidence for the presence of translatable mRNA encoding the desired polypeptide while establishing the ability of the oocyte to express the biologically active protein. The early expression of acetylcholinesterase from total unfractionated rat brain poly(A)$^+$ RNA[6] formed, for example, the foundations for the subsequent cloning and expression of both the human acetylcholinesterase and butyrylcholinesterase genes. In the case of neurotransmitter receptors or ion channels, oocyte microinjection may additionally demonstrate the correct assembly and plasma membrane insertion of the functional receptor or channel. Reports of the successful expression of the fMet-Leu-Phe receptor[7] and a human dopamine transporter[8] from tissue-extracted RNAs offer recent examples of preliminary studies which it is hoped will lead to the molecular cloning of these pursued genes. The fidelity of the oocyte in mRNA translation may also permit the relative "quantitation" of specific mRNAs in various tissues or physiological states. The tissue-specific expression of the labile enzyme testosterone 5α-reductase[9] and the ligand-induced downregulation of thyrotropin-releasing hormone[10] were both shown to be regulated at the level of available mRNA using oocyte expression as a quantitative mRNA bioassay. It is important to note, however, that some channel or receptor activities are curiously omitted from oocytes injected with tissue mRNAs, suggesting yet undefined limitations to the potential of the oocyte to process certain mRNAs or their polypeptide derivatives (discussed in detail by Snutch[11]).

The "transplantation" of active heterologous channel or receptor com-

[6] H. Soreq, R. Parvari, and I. Silman, *Proc. Natl. Acad. Sci. U.S.A.* **79**, 830 (1982).

[7] W. D. Coats, Jr., and J. Navarro, *J. Biol. Chem.* **265**, 5964 (1990).

[8] M. J. Bannon, C.-H. Xue, K. Shibata, L. J. Dragovic, and G. Dapatos, *J. Neurochem.* **54**, 706 (1990).

[9] Y. Farkash, H. Soreq, and J. Orly, *Proc. Natl. Acad. Sci. U.S.A.* **85**, 5824 (1988).

[10] Y. Oron, R. E. Straub, P. Traktman, and M. C. Gershengorn, *Science* **238**, 1406 (1987).

[11] T. P. Snutch, *Trends Neurosci.* **11**, 250 (1988).

plexes into *Xenopus* oocytes has been shown in many cases to generate novel membrane conductances closely resembling those in the native tissues. Receptor subtypes such as those of the muscarinic acetylcholine receptor,[12] the β-adrenergic receptor,[13,14] and the neuronal nicotinic acetylcholine receptor[15] are thus distinguishable in microinjected oocytes based on their unique electrophysiological characteristics. Exploiting this attribute of the oocyte expression system, Stuhmer *et al.*[16] correlated the functional diversity of voltage-gated potassium channels in mammalian brain with the existence of a polymorphic family of K⁺ channel-forming proteins. In a further application of this feature to the exploration of allelic polymorphism, brain RNA from mouse strains differing in their genetically determined sensitivities to ethanol was shown to elicit γ-aminobutyric acid (GABA$_A$) receptors in microinjected *Xenopus* oocytes exhibiting diametric responses to externally applied ethanol.[17] The production of heterologous proteins demonstrating biological activities characteristic of native molecules thus offers the exceptional opportunity to evaluate the existence of multiple tissue-specific or stage-specific mRNAs encoding polymorphic protein variants. In this vein, the employment of inhibitory "antisense" oligonucleotides to block specifically the translation of adult spinal cord glycine receptor mRNAs demonstrated developmentally regulated heterogeneity of strychnine-sensitive glycine receptors in the rat central nervous system[18] (see Section III,F).

    Related enzymes and their encoding RNAs may also be accurately differentiated in microinjected oocytes. Thus, synthetic mRNAs encoding the homologous acetylcholine hydrolyzing enzymes acetylcholinesterase (ACHE) and butyrylcholinesterase (BCHE) induce, in oocytes, biochemically distinct catalytic activities characteristic of the native human enzymes.[19,20] Using specific substrates and selective inhibitors, these two

[12] K. Fakuda, T. Kubo, I. Akiba, A. Maeda, M. Mishina, and S. Numa, *Nature (London)* **327,** 623 (1987).

[13] T. Frielle, S. Collins, K. W. Daniel, M. G. Caron, and R. J. Lefkowitz, *Proc. Natl. Acad. Sci. U.S.A.* **84,** 7920 (1987).

[14] B. K. Kobilka, H. Matsui, S. Kobilka, Y. L. Yang-Feng, U. Caron, and R. J. Lefkowitz, *Science* **238,** 650 (1987).

[15] E. S. Deneris, J. Connolly, J. Boulter, E. Wada, K. Wada, L. W. Swanson, J. Patrick, and S. Heinemann, *Neuron* **1,** 45 (1988).

[16] W. Stuhmer, J. P. Ruppersberg, K. H. Schroter, B. Sakmann, M. Stocker, K. P. Giese, A. Pershke, A. Baumann, and O. Pongs, *EMBO J.* **8,** 3235 (1989).

[17] K. A. Wafford, D. M. Burnett, T. V. Dunwiddie, and R. A. Harris, *Science* **249,** 291 (1990).

[18] H. Akagi, D. E. Patton, and R. Miledi, *Proc. Natl. Acad. Sci. U.S.A.* **86,** 8103 (1989).

[19] H. Soreq, S. Seidman, P. A. Dreyfus, D. Zevin-Sonkin, and H. Zakut, *J. Biol. Chem.* **264,** 10608 (1989).

[20] H. Soreq, R. Ben-Aziz, C. Prody, S. Seidman, A. Gnatt, L. Neville, Lieman-Hurwitz, E. Lev-Lehman, D. Ginzberg, Y. Lapidot-Lifson, and H. Zakut, *Proc. Natl. Acad. Sci. U.S.A.* **87,** 9688 (1990).

recombinant cholinesterases are distinguishable even in homogenates from oocytes coinjected with both messages (H. Soreq and S. Seidman, unpublished observations, 1990). Indeed, these expression studies confirmed accumulating evidence indicating the existence of independent messenger RNAs encoding these two enzyme species (reviewed in Seidman and Soreq[5]). Significantly, the ability to quantitate accurately discrete biological activities contributed by polymorphic variants of a gene family in coinjected oocytes opens the possibility of studying the role of mRNA competition in the regulation of gene expression as well.

The microinjection of size-fractionated mRNA may provide preliminary data indicating the existence of multiple mRNAs encoding polymorphic proteins. Using this approach, the heterogeneity of cholinesterase mRNAs was predicted several years prior to their cloning.[21] Similarly, the contributions of multiple heterologous subunits to the generation of native biological activities has been deduced following the microinjection of fractionated mRNAs. In this way, evidence implicating the involvement of multiple gene products in controlling the inactivation process of rat and rabbit Na^+ channels[22] and in determining the kinetic and pharmacological properties of rat brain A-type K^+ channels[23] was derived. In the case of an unusual, $3'$-extended mRNA species encoding butyrylcholinesterase in brain tumors, size-fractionated poly(A)$^+$ RNA demonstrated the translatability of the aberrant high molecular weight transcript, and it provided evidence for mRNA instability as the source of its underrepresentation in cDNA libraries prepared from these tissues.[24] In other cases, RNA size fractionation allowed the identification of the size class of mRNA encoding the sought after protein and an estimation of its expected molecular weight. Thus, the 5-hydroxytryptamine (5-HT_{1C}) receptor in rat brain was found to be encoded by a 5- to 6-kilobase (kb) mRNA,[25] whereas the mRNA for testosterone 5α-reductase was shown to be approximately 1.6–2.0 kb in size.[9]

The use of the oocyte for specific mRNA bioassays has proved to be an invaluable asset to molecular cloning efforts. The possession of a tool for the relatively rapid screening of RNA preparations for the mRNA of

[21] H. Soreq, D. Zevin-Sonkin, and N. Razon, *EMBO J.* **3,** 1371 (1984).

[22] D. S. Krafte, T. P. Snutch, J. P. Leonard, N. Davidson, and H. A. Lester, *J. Neurosci.* **8,** 2859 (1988).

[23] B. Rudy, J. H. Hoger, H. A. Lester, and N. Davisdon, *Cell (Cambridge, Mass.)* **1,** 649 (1988).

[24] A. Gnatt, C. A. Prody, R. Zamir, J. Lieman-Hurwitz, H. Zakut, and H. Soreq, *Cancer Res.* **50,** 1983 (1990).

[25] H. Lubbert, T. P. Snutch, N. Dascal, H. A. Lester, and N. Davidson, *J. Neurosci.* **7,** 1159 (1987).

interest facilitates the selection of an appropriate tissue source for the preparation of a cDNA library. For example, the construction of a cDNA library from adult brain basal ganglia in the search for the human acetylcholinesterase gene followed the demonstration of translatable ACHE mRNA in this tissue using oocyte microinjection.[20] Similarly, oocyte expression is routinely employed to monitor the enrichment of specific mRNAs, including those encoding proteins with detectable biological effects yet undefined biochemical properties. For instance, an approximately 2-kb mRNA species in liver homogenates was shown to encode an unidentified factor capable of enhancing transcription mediated by the phosphoenolpyruvate carboxykinase (PEPCK) promoter in microinjected oocytes.[26] Increasingly popular is the use of oocyte expression to screen cDNA expression libraries via pooled synthetic RNA, a strategy successfully exploited in the cloning of the rat substance P receptor.[27]

Finally, oocyte microinjection may establish the irrefutable identity of a cloned DNA sequence by demonstrating its ability to direct the production of the biologically active protein which it is purported to encode. A recent cloning endeavor in which oocyte expression provided the biological confirmation of a sequence tentatively identified by strong circumstantial evidence is that of the thyrotropin receptor.[28] It is clear that *Xenopus* oocyte microinjection is currently providing the basis for the molecular cloning of nonabundant genes for which no comparable screening paradigm is available.

B. Structure–Function Studies of Cloned Genes

Once a gene is cloned, engineered modifications in the coding sequence may allow detailed analyses of the contributions of specific polypeptide regions or even single amino acids to the biological activity of the encoded gene products. In an early study, controlled deletions introduced into various regions of the cDNA encoding the α subunit of the nicotinic acetylcholine receptor (αnAChR) were monitored for their effects on the induced acetylcholine response in microinjected oocytes.[29] Site-directed mutagenesis of specific amino acids has subsequently been shown, via

[26] N. Benvenisty, T. Shoshani, Y. Farkash, H. Soreq, and L. Reshef, *Mol. Cell Biol.* **9**, 244 (1989).

[27] Y. Yokota, Y. Sasai, K. Tanaka, T. Fujiwara, K. Tsuchida, R. Shigemoto, A. Kakizuka, H. Ohkubo, and S. Nakanishi, *J. Biol. Chem.* **264**, 17649 (1989).

[28] M. Parmentier, F. Libert, C. Maenhaut, A. Lefort, C. Gerard, J. Perret, J. Van Sende, J. E. Dumont, and G. Vassart, *Science* **246**, 1620 (1989).

[29] M. Mishina, T. Tobimatsu, T. Imoto, K.-I. Fujita, D. Fakuda, M. Durasaki, T. Takahashi, Y. Morimoto, T. Hirose, S. Inayama, T. Takahashi, M. Kuno, and S. Numa, *Nature* (*London*) **313**, 364 (1985).

oocyte microinjection, to confer tetrodotoxin and saxitoxin insensitiviy on the rat sodium channel II,[30] to disrupt the catalytic phosphorylation and ATP-binding sites of the subunit of *Torpedo* Na^+,K^+-ATPase,[31] and to delete the covalent attachment site of the irreversible band 3 protein inhibitor H_2DIDS.[32] A combination of single amino acid substitutions, deletion mutagenesis, and patch clamp recording facilitated a detailed structural analysis of the components involved in activation and inactivation of the rat brain sodium channel II.[33]

In the case of human butyrylcholinesterase, cloned allelic variants were exploited to derive detailed biochemical characterizations of naturally occurring and genetically engineered mutant BCHEs.[34,35] These studies revealed that only when the previously identified Asp-70 → Gly substitution in human BCHE[36] is linked to the novel Ser-425 → Pro replacement[24] are the biochemical characteristics associated with the defective "atypical" BCHE phenotype displayed by recombinant enzyme produced *in ovo*. Perhaps a more conceptually sophisticated analysis is the construction of hybrid molecules such as the chimeric $\delta-\gamma$ *Torpedo*-mouse nAChR cDNA built in an attempt to pinpoint regions within these two relatively homologous polypeptides which stipulate their mutually exclusive functional properties.[37]

The coinjection of cloned DNAs or cRNAs encoding the various subunits of heteropolymeric complexes facilitates the elucidation of the role of various oligomeric subcomponents in conferring species-specific, tissue-specific, or developmentally regulated properties of multisubunit receptors and channels. Mishina *et al.*[38] demonstrated that although mRNAs encoding all four subunits of the *Torpedo* acetylcholine receptor were required to induce a significant *de novo* electrophysiological response to acetylcholine

[30] M. Noda, H. Suzuki, S. Numa, and W. Stuhmer, *FEBS Lett.* **259**, 213 (1989).
[31] M. Ohtsubo, S. Noguchi, K. Takeda, M. Morohashi, and M. Kawamura, *Biochim. Biophys. Acta* **1021**, 157 (1990).
[32] D. Bartel, H. Hans, and H. Passow, *Biochim. Biophys. Acta* **985**, 355 (1989).
[33] W. Stuhmer, F. Conti, H. Suzuki, X. Wang, M. Noda, N. Yahagi, H. Kubo, and S. Numa, *Nature* (*London*) **339**, 597 (1989).
[34] L. F. Neville, A. Gnatt, R. Padan, S. Seidman, and H. Soreq, *J. Biol. Chem.* **265**, 20735 (1990).
[35] L. F. Neville, A. Gnatt, Y. Loewenstein, and H. Soreq, *J. Neurosci. Res.* **27**, 452 (1990).
[36] M. C. McGuire, C. P. Nogueira, C. F. Bartels, H. Lightstone, A. Hajra, A.F.L. Van der Spek, O. Lockridge, and B. N. LaDu, *Proc. Natl. Acad. Sci. U.S.A.* **86**, 953 (1989).
[37] F. Mixter-Mayne, K. Yoshii, L. Yu, H. A. Lester, and H. Davidson, *Mol. Brain Res.* **2**, 191 (1987).
[38] M. Mishina, T. Kurosaki, T. Tobimatsu, Y. Morimoto, M. Noda, T. Yamamoto, M. Terao, J. Lindstrom, T. Takahashi, M. Kuno, and S. Numa, *Nature* (*London*) **307**, 604 (1984).

in microinjected oocytes, the α subunit alone was sufficient to induce α-bungarotoxin binding. Others have employed cloned acetylcholine receptors to construct hybrid mouse–*Torpedo*[39] calf–*Torpedo*,[40] or cat–*Torpedo*[41] receptors (see also Section III,C,2) by coinjection protocols. Furthermore, substitution of the calf-specific, ϵ subunit-encoding RNA for that of synthetic RNA encoding the homologous adult bovine δ subunit was shown to result in oocyte-expressed receptors characteristic of fetal as opposed to adult acetylcholine receptor.[42] Among other things, these studies provided convincing evidence that the δ subunit is the primary determinant in specifying species-specific dose responses and desensitization characteristics of the nicotinic acetylcholine receptor complex in mammalian and nonmammalian nervous tissue. These pioneering studies with the nAChR further demonstrated that the degree of interspecies homology within this gene family permits the formation of functional hybrid receptors provided that all four receptor subunits are represented. In an advanced development of our understanding of the nAChR, heteropolymer reconstitution demonstrated the interchangeability of neuronal nAChR $\beta2$ subunit with the $\beta1$ subunit of the acetylcholine receptor found at the neuromuscular junction[15] (see also Section III,C,1).

Note. The tendency of research employing *Xenopus* oocyte microinjection to develop from cloning and structure–function studies to studies of biosynthetic and gene expression pathways is a natural one indeed. Thus, although the primary applications of this system to the study of ion channels may currently lie in the realm of the former category, a move in the latter direction is certainly to be expected. The observation that genes encoding ion channels and neurotransmitter receptors are often expressed in developmentally regulated and/or tissue-specific manners together with the conspicuous requirement for correct assembly, transport, and membrane insertion makes questions relating to the undoubtedly complex biosynthetic pathways of these molecules particulary interesting. The following sections therefore address the potential of oocyte microinjection in the study of protein biogenesis, both at the levels of posttranscriptional and posttranslational processing and at the level of gene expression.

[39] M. White, K. Mixter-Mayne, H. A. Lester, and N. Davidson, *Proc. Natl. Acad. Sci. U.S.A.* **82,** 4852 (1985).

[40] B. Sakmann, C. Methfessel, M. Mishina, T. Takahashi, T. Takai, M. Kurasaki, K. Fukuda, and S. Numa, *Nature (London)* **318,** 538 (1985).

[41] K. Sumikawa and R. Miledi, *Proc. Natl. Acad. Sci. U.S.A.* **86,** 367 (1989).

[42] M. Mishina, T. Takai, K. Imoto, M. Noda, T. Takahashi, S. Numa, C. Methfessel, and B. Sakmann, *Nature (London)* **321,** 406 (1986).

C. Studying Biosynthetic Pathways

The biosynthetic pathways leading to the expression of heterologous proteins in oocytes have been studied with regard to various posttranslational processing events. Using tunicamycin as a block to glycosylation (see also Section III, E), the effect of N-linked glycosylation on the expression of various receptors and channels from rat brain or chick optic lobe poly(A)+ RNA was studied.[43] The results of that study indicated an unequal dependence on glycosylation among different receptors. Sumikawa and Miledi[44] followed up this work by demonstrating that although N-glycosylation is not necessary for the assembly of heterologous *Torpedo* nAChR subunits, it appears to be a prerequisite for their efficient implantation into the plasma membrane. Using a similar strategy, Takeda *et al.*[45] demonstrated the dispensible role of glycosylation in the expression of functional *Torpedo* Na+,K+-ATPase in microinjected oocytes.

Other factors involved in the posttranslational processing of nascent polypeptides, such as protein sorting and transport and oligomeric assembly, have also been assessed in oocytes. Ackermann and Geering[46] demonstrated the stabilization of the Na+,K+-ATPase α subunit by the β subunit and the mutual dependence of the two subunits for transport out of the endoplasmic reticulum. Employing an immunocytochemical staining approach (see Section IV,C), Dreyfus *et al.*[47] observed polarized compartmentalization of newly synthesized oocyte-produced BCHE and the intracellular accumulation of nascent BCHE around cytoplasmic vesicles in tunicamycin-treated oocytes. Schmale *et al.*[48] employed biochemical analyses to examine the impact of altered protein structures on the intracellular trafficking of mutant vasopressin precursor. Coinjection experiments (see Section III,C,2) demonstrated the existence of tissue-specific accessory factors involved in specifying the polymorphic assembly patterns characteristic of human cholinesterases.[19] At the level of protein–membrane interactions is the study, in oocytes, of the aggregation of membrane-associated cholinesterases[47] and the establishment of cell–cell gap junction channels.[49,50]

[43] K. Sumikawa, I. Parker, and R. Miledi, *Mol. Brain Res.* **4,** 191 (1988).

[44] K. Sumikawa and R. Miledi, *Mol. Brain Res.* **5,** 183 (1989).

[45] K. Takeda, S. Noguchi, A. Sugino, and M. Kawamura, *FEBS Lett.* **238,** 201 (1988).

[46] U. Ackermann and K. Geering, *FEBS Lett.* **269,** 105 (1990).

[47] P. A. Dreyfus, S. Seidman, M. Pincon-Raymond, M. Murawsky, F. Rieger, E. Schejter, H. Zakut, and H. Soreq, *Cell. Mol. Neurobiol.* **9,** 323 (1989).

[48] H. Schmale, B. Borowiak, H. Holtgreve-Grez, and D. Richter, *Eur. J. Biochem.* **182,** 621 (1989).

[49] R. Werner, E. Levine, C. Rabadan-Deihl, and G. Dahl, *Proc. Natl. Acad. Sci. U.S.A.* **86,** 5380 (1989).

[50] K. I. Swenson, J. R. Jordan, E. C. Beyer, and D. L. Paul, *Cell (Cambridge, Mass.)* **57,** 145 (1989)

Other investigators have employed oocytes to study specific biosynthetic components. For example, Thompson *et al.*[51] coinjected a monoclonal antibody raised against the hexapeptide repeat of RNA polymerase II with DNA encoding various human genes and demonstrated specific inhibition of Pol II-directed transcription (see also Section III,F). Others are finding the oocyte a convenient system in which to probe the nucleic acid–protein interactions in the riboprotein complexes comprising the spliceosome[52,53] (see also Section III,D).

D. Studying Regulation of Gene Expression

The ability of *Xenopus* oocytes to express foreign DNA as well as RNA has prompted their use in the analysis of mammalian gene expression (see also Section III,B). Cloned promoters linked to a reporter gene such as that encoding bacterial chloramphenicol acetyltransferase (CAT) may be dissected with respect to both cis- and trans-acting factors. In this way, cis-acting sequences of the sea urchin histone H_2A modulator element were shown to interact with oocyte transcription factors when injected into oocyte germinal vesicles.[54]

Nuclear injection of promoter-containing plasmids followed by RNA extraction and analysis offers another approach to the study of transcription mechanisms in oocytes. Using this approach, Middleton and Morgan[55] have identified sequences required for transcription initiation of the *Xenopus* XαT14 α-tubulin gene. Cytosine methylation within the recognition sites for the transcription factors CTF and Sp1 was similarly shown, in this fashion, to inhibit *in vivo* transcription from the herpes simplex virus thymidine kinase (HSV*tk*) promoter.[56] The overall integrity of the oocyte transcription/translation pathway permitted the *in ovo* reconstitution of cell-specific regulatory patterns involved in the expression of the rat PEPCK gene[26] (see Section III,C,3).

Other regulatory processes such as transcription termination[57–59] and

[51] N. E. Thompson, T. H. Steinberg, D. B. Aronson, and R. R. Burgess, *J. Biol. Chem.* **264**, 11511 (1989).
[52] J. Hamm, N. A. Dathan, D. Scherly, and L. W. Mattaj, *EMBO J.* **9**, 1237 (1990).
[53] J. Hamm, N. A. Dathan, and I. W. Mattaj, *Cell (Cambridge, Mass.)* **59**, 159 (1989).
[54] F. Palla, C. Casano, I. Albanese, L. Anello, F. Gianguzza, M. G. DiBernardo, C. Bonura, and G. Spinelli, *Proc. Natl. Acad. Sci. U.S.A.* **86**, 6033 (1989).
[55] K. M. Middleton and G. T. Morgan, *Nucleic Acids Res.* **17**, 5041 (1989).
[56] J. Ben-Hattar, P. Beard, and J. Jiricny, *Nucleic Acids Res.* **17**, 10179 (1989).
[57] D. L. Bentley and M. Groudine, *Cell (Cambridge, Mass.)* **53**, 245 (1988).
[58] K. M. Middleton and G. T. Morgan, *Mol. Cell. Biol.* **10**, 727 (1990).
[59] C. A. Spencer, R. C. LeStrange, U. Novak, W. S. Hayward, and M. Groudine, *Gen. Dev.* **4**, 75 (1990).

TABLE I
OOCYTE MICROINJECTION IN GENE EXPRESSION

Process	Site	Experimental strategy	Example[a]
1. Structural changes in DNA topology	Nucleus	Nuclear injection of labeled DNA in linear or super-coiled states and electro-phoretic autoradiography followup with time	Ballas et al. (1)
2. Trans-element regulation of specific transcription	Nucleus	Nuclear coinjection of examined promoter linked to a reporter gene and cytoplasmic mRNA from tested tissues	Benvenisty et al. (2)
3. Cis-element effects	Nucleus	Nuclear injection of examined promoter–enhancer sequences followed by RNA extraction and analysis	Middleton and Morgan (3)
4. Transcriptional fidelity	Nucleus	Nuclear injection of examined DNA sequence followed by RNA extraction, gel electrophoresis, and autoradiography of RNA product following RNA blot hybridization	Bentley et al. (4)
5. Transcriptional shutoff	Nucleus	Injection of E1a or heat shock, then followup of general transcription levels by autoradiography of electrophoresed RNAs	Bienz and Gurdon (5)
6. RNA stability	Nucleus	Microinjection of labeled synthetic RNA followed by RNA extraction, gel electrophoresis, and autoradiography	Galili et al. (6)
7. Splicing	Nucleus	(a) Antisense oligonucleotide-mediated degradation of U snRNA, followed by microinjection of cloned replacements (b) Microinjection of labeled synthetic precursor mRNA followed by gel analysis	Hamm et al.(7a), Inoue et al. (7b)
8. Nuclear cytoplasmic transport of processed transcripts	Nuclear envelope	Cytoplasmic injection of labeled snRNA followed by extraction and analysis of RNA from isolated oocyte nuclei	Fischer and Luhrmann (8)

TABLE I *(continued)*

Process	Site	Experimental strategy	Example[a]
9. Translation	Cytoplasmic polysomes	Cytoplasmic mRNA injections and followup of protein product formation by various assays	Reviewed by Soreq *(9)*
10. Termination of translation	Cytoplasmic polysomes	Test of nonsense suppression and readthrough mechanisms by gel electrophoresis	Bienz *et al.* *(10)*
11. Posttranslational processing	Golgi apparatus and rough endoplasmic reticulum	Microinjection of synthetic mRNA, biosynthetic labeling, immunoprecipitation, and gel analysis of protein products	Cerotti and Colman *(11)*
12. Protein secretion	Cytoplasmic membrane	Synthetic mRNA microinjection following mutagenesis of signal sequences and subcellular compartmentalization of products	Colman and Morser *(12)*
13. Membrane and/or extracellular positioning of protein products	Surface membrane and extracellular material	Immunohistochemistry of oocyte sections at light and electron microscopy levels	Dreyfus *et al.* *(13)*
14. Subunit assembly of produced proteins	Intracellular	Coinjection of synthetic and poly(A)$^+$ mRNA followed by enzyme activity and sucrose gradient centrifugation analyses	Soreq *et al.* *(14)*
15. Structure–function characteristics of product proteins	Depending on protein	Microinjection of mutagenized synthetic RNAs and analysis of activity on ligand binding measurements	Neville *et al.* *(15)*

[a] Key to references:
(1) N. Ballas, S. Broido, H. Soreq, and A. Loyter, *Nucleic Acids Res.* **17**, 7891 (1989); *(2)* N. Benvenisty, T. Shoshani, Y. Farkash, H. Soreq, and L. Reshef, *Mol. Cell. Biol.* **9**, 244 (1989); *(3)* K. M. Middleton and G. T. Morgan, *Nucleic Acids Res.* **17**, 5041 (1989); *(4)* D. L. Bentley, W. L. Brown, and M Groudine, *Gen. Dev.* **3**, 1179 (1989); *(5)* M. Bienz and G. Gurdon, *Cell (Cambridge, Mass.)* **27**, 811 (1982); *(6)* G. Galili, E. E. Kawata, L. D. Smith, and B. A. Larkins, *J. Biol. Chem.* **263**, 5764 (1988); *(7a)* J. Hamm, A. Dathan, D. Scherly, and L. W. Mattaj, *EMBO J.* **9**, 1237 (1990); *(7b)* K. Inoue, M. Ohno, H. Sakamoto, and Y. Shimura, *Gen. Dev.* **3**, 1472 (1989); *(8)* U. Fischer and R. Luhrmann, *Science* **249**, 786 (1990); *(9)* H. Soreq, *Crit. Rev. Biochem.,* **18**, 199 (1985); *(10)* M. Bienz, E. Kubli, J. Kohli, S. deHenan, G. Huez, G. Marbaix, and H. Grosjean, *Nucleic Acids Res.* **9**, 3835 (1981); *(11)* A. Ceriotti and A. Colman, *EMBO J.* **7**, 633 (1988); *(12)* A. Colman and J. Morser, *Cell (Cambridge, Mass.)* **17**, 517 (1979); *(13)* P. Dreyfus, S. Seidman, M. Pincon-Raymond, M. Murwasky, F. Rieger, E. Schejter, H. Zakut, and H. Soreq, *Cell. Mol. Neurobiol.* **9**, 323 (1989); *(14)* H. Soreq, S. Seidman, P. Dreyfus, D. Zevin-Sonkin, and H. Zakut, *J. Biol. Chem.* **264**, 10608 (1989); *(15)* L. F. Neville, A. Gnatt, R. Padan, S. Seidman, and H. Soreq, *J. Biol. Chem.* **265**, 20735 (1990).

RNA capping,[60,61] polyadenylation,[62,63] and transport[63] are also being successfully approached using microinjected oocytes as an assay system (see Section III,D). After a lapse, some research groups are returning to oocytes as an *in vivo* splicing complementation assay, particularly in the study of U snRNAs and their role in spliceosome assembly and function[52,53,64] (see also Section III, D). One might observe that oocyte microinjection has been used in the study of essentially every step in the pathway for gene expression for which an experimental paradigm is available (Table I). It will be most interesting to follow the development of research directed toward the study of ion channel genes at these various levels.

III. Experimental Strategies

A. RNA Injections

1. Total Unfractionated Poly(A)+ RNA. The injection of tissue-extracted mRNA may serve (1) to achieve the expression of specific yet unisolated genes or (2) to supply unspecified or unknown accessory protein factors to a defined expression system by coinjection with a cloned cDNA or synthetic mRNA. Coupled to a sensitive bioassay, the oocyte can give rise to a detectable product whose specific mRNA comprises as little as 0.001% of total cellular message. Indeed, the quantitation of nonabundant mRNA species via the microinjection of total cellular RNA may be quite precise, as demonstrated by the example of the human cholinesterases following molecular cloning of their corresponding cDNAs.[6,19,20] RNA may be extracted from cultured cells[65] and fresh[9] or frozen[66] tissues. We have successfully prepared oocyte-translatable poly(A)+ RNA from postmortem human tissue maintained up to 24 hr at 4° after autopsy.[67] Frozen tissues or homogenates are stored at −70° until used. Careful dissection of brain subregions permits the isolation of region-specific RNA populations as was performed in the study by Blakely *et al.*[68] of rat brain neurotransmitter receptors expressed in oocytes.

[60] K. Inoue, M. Ohno, H. Sakamoto, and Y. Shimura, *Gen. Dev.* **3**, 1472 (1989).

[61] U. Fischer and R. Luhrmann, *Science* **249**, 786 (1990).

[62] U. Z. Littauer and H. Soreq, *Prog. Nucleic Acid Res. Mol. Biol.* **27**, 53 (1982).

[63] G. Galili, E. E. Kawata, L. D. Smith, and B. A. Larkins, *J. Biol. Chem.* **263**, 5770 (1988).

[64] Z. Q. Pan and C. Prives, *Gen. Dev.* **3**, 1187 (1989).

[65] F. Vilijn and N. Carrasco, *J. Biol. Chem.* **264**, 11901 (1989).

[66] M. J. Bannon, C.-H. Xue, K. Shibata, L. J. Dragovic, and G. Kapatos, *J. Neurochem.* **54**, 706 (1990).

[67] K. M. Dziegielewska, N. R. Saunders, E. J. Schejter, H. Zakut, D. Zevin-Sonkin, R. Zisling, and H. Soreq, *Dev. Biol.* **115**, 93 (1986).

[68] R. D. Blakely, M. B. Robinson, and S. G. Amara, *Proc. Natl. Acad. Sci. U.S.A.* **85**, 9846 (1988).

Two principal methods of RNA preparation predominate among laboratories using oocyte microinjection: (1) guanidine thiocyanate extraction followed by CsCl gradient centrifugation and (2) phenol–sodium dodecyl sulfate (SDS)–chloroform extraction followed by selective precipitation of RNA.[69] Sumikawa *et al.*[70] report better success with the latter method for the expression of neurotransmitter receptors and channels. To enrich messenger RNA, unwanted poly(A)$^-$ RNA, consisting mostly of ribosomal and transfer RNAs, are discarded. For this purpose, the total RNA preparation is passed over an oligo(dT) column to which poly(A)$^+$ RNA is bound in the presence of high salt and subsequently eluted with low-salt buffer.

For microinjection, poly(A)$^+$ mRNA is generally redissolved in double-distilled water (0.5–2.0 μg/μl) and injected (50 nl/oocyte). Active cholinesterase has been detected in microinjected oocytes as early as 30 min postinjection, continuing for at least 4 days and with maximum activity observed within 18 hr[47] (H. Soreq and S. Seidman, unpublished observations, 1990). For membrane-associated neurotransmitters and channels, biologically active complexes may only be observed after several days to a week.[70] For detailed discussions of the use of tissue RNA as a source of ion channels, see the reviews by Snutch[71] and Goldin and Sumikawa.[71a]

2. Size-fractionated RNA. The microinjection of size-fractionated poly(A)$^+$ RNA serves primarily as an assay to monitor the enrichment of specific mRNA species and estimate their size. However, it may also provide information regarding the expected molecular weight of the encoded polypeptide, the existence of multiple mRNAs encoding polypeptides with similar activities, or the requirement for multiple nonhomologous components in the reconstitution of native activities (see Section II, A). In the event that heterologous protein factors are involved in the modulation of fundamental biological activities, the possibility that subtle properties may be lost in the course of expressing proteins from fractionated RNA need be taken into consideration.[25] Injection of recombined fractions can be expected to clarify discrepancies and provide a basis for identifying the complementary factor(s) in such cases.

Size fractionation of poly(A)$^+$ RNA is most commonly achieved via sucrose density gradient centrifugation[21,24,72] but may also be accomplished by agarose gel electrophoresis. The latter method has been praised for its

[69] M. Gilman, *in* "Current Protocols in Molecular Biology" (F. M. Ausubel, R. Brent, R. E. Kingston, D. D. Moore, J. G. Seidman, J. A. Smith, and D. Struhl, eds.), Chap. 4. Wiley, New York, 1987.

[70] Sumikawa, I. Parker, and R. Miledi, *in* "Methods in Neuroscience," Vol. 1, p. 30. Academic Press, New York, 1989.

[71] T. P. Snutch and G. Mandel, this volume [17].

[71a] A. L. Goldin and K. Sumikawa, this volume [16].

[72] P. Sehgal, H. Soreq, and I. Tamm, *Proc. Natl. Acad. Sci. U.S.A.* **75,** 5030 (1978).

high resolution of relatively high molecular weight RNA species.[11] Briefly, Lubbert et al[25] reported expression of rat brain 5-HT$_{1C}$ receptors from 5- to 6-kb mRNA fractionated in a 1% ultralow-melting agarose gel. Prior to loading, 70 μg poly(A)$^+$ RNA was heated to 65° for 1 min in 80% formamide, 0.1% SDS, 1 mM EDTA to dissociate aggregates. Following electrophoresis the gel was sliced, heated briefly to 65°, diluted into binding buffer, and passed over oligo(dT)-cellulose. RNA was eluted with salt and precipitated with ethanol. One-quarter of each fraction was subsequently used for microinjection.

3. Synthetic mRNA. A cloned DNA may be subcloned into a transcription vector from which the in vitro production of synthetic mRNA may be directed by the RNA polymerase II binding site of phages such as SP6, T3, or T7. In addition to facilitating much higher levels of expression than those achievable with tissue-extracted RNAs, the use of single-message synthetic RNA permits a detailed analysis of a specific RNA transcript and its protein derivative(s). Furthermore, the successful preparation of synthetic mRNA opens the door to studies directed toward assessing the interactions between discrete messenger RNAs or their unique polypeptide products and other molecular or cellular components (see Sections II,B, II,C, and III,C).

Naturally, an essential requirement of any in vitro transcription system is the efficient initiation and completion of full-length transcripts capable of being effectively translated in microinjected oocytes. In a good transcription reaction 1 μg DNA may yield 5–10 μg RNA. For the in vitro transcription of GC-rich genes such as that encoding human acetylcholinesterase, we have found the use of single-strand-binding protein helpful when employed in conjunction with additional modifications to the transcription protocol.[73] The m^{7}G(5′)ppp(5′)G capping of synthetic mRNA appears nonessential but highly recommended.[74] Expression of capped RNA may be up to 10-fold more efficient than that of uncapped RNA (H. Soreq, R. Ben-Aziz, and Y. Loewenstein, unpublished observation, 1991), which is probably attributable to the protection afforded capped RNAs against nucleolytic degradation.[60] We find capping performed concomitantly to transcription significantly more efficient than post facto capping. Poly(A) tailing, either by the inclusion of poly(T) in the transcription vector or by "tailing" RNA in vitro with poly(A) polymerase, is generally advised in light of the enhanced stability attributed to most species of poly(A)$^+$ over poly(A)$^-$ RNA.[62,63]

For expression studies, 50 nl/oocyte of 0.1–0.5 μg RNA/μl (i.e., 5–25 ng/oocyte) has proved optimum in our laboratory, although protocols

[73] R. Ben-Aziz and H. Soreq, Nucleic Acids Res. 18, 3418 (1990).
[74] R. E. McCaman, L. Carbini, V. Maines, and P. M. Salvaterra, Mol. Brain Res. 3, 107 (1988).

employing injections of 0.5–50 ng synthetic RNA/oocyte have been reported.[15,30,74,75] Significantly, we have observed that excess quantities of injected synthetic RNA may be inhibitory (H. Soreq, R. Ben-Aziz, and S. Seidman, unpublished observation, 1991). Such inhibition may reflect a tip in the balance between the intrinsic translatability of the specific sequences involved and the potential toxicity of high concentrations of injected nucleic acids. Becuase this balance depends on the specific nucleotide composition and primary sequence of the injected RNA, microinjection protocols should be independently optimized for each particular message. For detailed discussions of *in vitro* synthesis of RNA for expression in *Xenopus* oocytes, see Goldin and Sumikawa[71a] and Swanson and Folander.[76]

The expression of a protein from a homogeneous RNA preparation is obviously expected to produce a higher quantity of detectable product per nanogram injected RNA than that obtainable with tissue poly(A)$^+$ RNA. The question is, however, what level of enhancement may be achieved and at what point does the system become saturated either for specific polysomes[1] or for some other limiting endogenous component such as second-messenger elements, as demonstrated for the expression of rat brain 5-HT$_{1C}$ receptors.[25] The suggestion has also been made that restricted diffusion of injected RNA may impede efforts to saturate completely the biosynthetic capacity of the oocyte.[47] Microinjection of synthetic mRNAs encoding human cholinesterases results in levels of catalytically active enzyme approximately 3–4 orders of magnitude above those achieved with tissue-extracted poly(A)$^+$ RNA.[19,20] However, assuming that cholinesterase mRNAs comprise only 0.001% of total cellular mRNA,[20,77] it is evident that the enrichment attained with synthetic mRNA does not give rise to a directly proportional increment in measureable cholinesterase activities.

B. DNA Expression Vectors

Microinjection of DNA expression plasmids containing coding sequences inserted downstream from an efficient eukaryotic promoter may bypass the requirement for *in vitro* transcription of RNA. Although most of the injected DNA is probably degraded shortly after its introduction into the oocyte and thus never transcribed, the mRNA which is successfully produced *in ovo* should be better protected than its microinjected counterpart against nucleolytic degradation. This factor, in combination with the

[75] K. Yoshi, L. Yu, K. Mixter-Mayne, N. Davidson, and H. A. Lester, *J. Gen. Physiol.* **90,** 553 (1987).

[76] R. Swanson and K. Folander, this volume [18].

[77] C. A. Prody, A. Gnatt, D. Zevin-Sonkin, O. Goldberg, and H. Soreq, *Proc. Natl. Acad. Sci. U.S.A.* **84,** 3555 (1987).

prolonged transcription expected of the surviving vector DNAs, facilitates production of the encoded polypeptide. In the following, the requirements and potential of DNA vectors for oocyte microinjection will be briefly discussed.

1. Structural Requirements. DNA vectors designed for oocyte microinjection need be both efficient and versatile. Thus, the following considerations may be relevant: (1) Nontranslated sequences 5' of the coding sequence are not essential and may, in fact, hamper expression. The exception to this rule is, of course, the translation initiation signal, which should conform to the consensus sequence derived from eukaryotic mRNAs known to be efficiently translated *in vivo*.[77a] (2) Similarly, nontranslated 3' sequences are essentially superfluous and should be removed. However, at least one and preferably several translational *stop* codons should be present downstream of the coding sequence to ensure correct termination of translation (reviewed by Soreq[1]). Furthermore, an efficient polyadenylation signal should be included downstream of the *stop* codon(s). Where the cloned DNA lacks an intrinsic polyadenylation signal an effective sequence such as that of the simian virus (SV40) early gene[77b] may be engineered into the vector. (3) Naturally, efficient expression of injected DNA sequences will be dependent on the presence of a potent eukaryotic promoter upstream of the coding sequence. Examples of such promoter elements include that of the cytomegalovirus (CMV) IE gene[77c] and that of the SV40 early gene.[77d] (4) Finally, unique restriction sites should be included in the expression vector upstream and downstream of the coding sequence. Such foresight will provide for the future insertion and shuffling of mutant sequences into the same vector, thus maximizing its versatility.

2. Comparison to Synthetic mRNA. An expression vector including a human acetylcholinesterase coding sequence linked to the CMV IE gene enhancer–promoter element directed the production of catalytically active acetylcholinesterase both in human embryonic kidney cells[78] and in microinjected *Xenopus laevis* oocytes (H. Soreq, R. Ben-Aziz, and S. Seidman, unpublished observation, 1990). Enzymatic ACHE activity attained in oocytes injected with 1–2 ng ACHE DNA reached up to 4–5 times that of oocytes injected with 5 ng synthetic ACHE mRNA. In contrast to the

[77a] M. Kozak, *Nucleic Acids Res.* **12,** 857 (1984).
[77b] C. Benoist and P. Chambon, *Nature (London)* **290,** 304 (1981).
[77c] M. K. Foeking and H. Hofstetter, *Gene* **45,** 101 (1986).
[77d] D. Bertrand, M. Ballivet, and D. Rungger, *Proc. Natl. Acad. Sci. U.S.A.* **87,** 1993 (1990).
[78] B. Velan, C. Kronman, H. Grosfeld, M. Leitner, Y. Gozes, Y. Flashner, T. Sery, S. Cohen, R. Ben-Aziz, S. Seidman, A. Shafferman, and H. Soreq, *Cell. Mol. Neurobiol.* **11,** 143 (1991).

expression of synthetic ACHE mRNA, where maximum activity is observed following overnight incubation, maximum activity was observed in oocytes injected with ACHE DNA only after a period of 3 days. Such a delay may be accounted for by considering a requisite association of the injected DNA with the oocyte nuclear elements coupled with the inherent rate limitations to RNA transcription and extranuclear transport. Alternatively, or additionally, it is possible that the high G–C content of the ACHE coding sequence[20,79] imposes restrictions on its *in vivo* transcription. This latter possibility is strengthened by comparison with the neuronal nAChR, which has also been expressed using DNA expression vectors.[77d] In these experiments plasmids carrying cDNAs encoding the $\alpha 4$ and non-α subunits of nAChR were placed under the control of the SV40 early gene promoter. Oocytes coinjected with 1–2 ng of each plasmid demonstrated acetylcholine-induced currents within 24 hr.

2. Applications. A variety of eukaryotic promoters have been successfully expressed in *Xenopus* oocytes and employed in the study of transcription regulatory processes. Expression of cloned promoter sequences has been quantitatively monitored directly in terms of *de novo* RNA synthesis, or indirectly, in terms of ultimate protein synthesis. For example, Ballas *et al.*[80] demonstrated that 50 ng of supercoiled plasmid DNA containing plant regulatory elements linked to the bacterial chloramphenicol acetyltransferase gene supported the production of catalytically active CAT in microinjected oocytes. One of the plant promoter elements investigated was equally effective as the SV40 early gene promoter in directing oocyte-mediated transcription, although the converse was not observed, namely, cultured plant protoplasts did not effectively recognize the SV40 promoter element. A functional polyadenylation signal, either plant or animal, was also shown in this study to be essential for efficient CAT biosynthesis. When linearized plasmid DNA was injected into oocytes, active CAT was essentially undetectable. DNA blot analysis of labeled plasmid sequences recovered 20 hr postinjection indicated that linear plasmid DNA was either degraded or aggregated into high molecular weight components. In contrast, circular supercoiled DNA, although partially transformed to relaxed or linear forms, remained relatively stable. These studies demonstrated a clear sequence-independent correlation between DNA topology and transcription in oocytes, and a curious nonreciprocal recognition of plant promoter elements by an animal cell. Furthermore, they lay the

[79] R. Ben-Aziz, A. Gnatt, C. Prody, E. Lev-Lehman, L. Neville, S. Seidman, D. Ginzberg, and H. Soreq, *in* "Proceedings of the Third International Meeting on Cholinesterases" (J. Massoulie, ed.), p. 172. American Chemical Society Books, Washington, D.C., 1991.
[80] N. Ballas, S. Broido, H. Soreq, and A. Loyter, *Nucleic Acids Res.* **17**, 7891 (1989).

groundwork for an interesting study of plant channels and receptors in oocytes, a possibility not yet exploited.

Site-directed modification of known promoter elements followed by microinjection into *Xenopus* oocytes is helping elucidate the specific and general mechanisms by which gene expression is regulated in eukaryotic cells. Middleton and Morgan[55] have studied the expression of a *Xenopus laevis* α-tubulin gene (XαT14) by introducing controlled deletions in the 5′ leader sequence of the cloned DNA and injecting the resultant plasmids into oocytes. Nuclear injections of as little as 15 pg DNA/oocyte gave rise to levels of RNA which were readily analyzed by RNase protection and primer extension. Ben-Hattar *et al.*[56] used oocyte microinjection and RNA primer extension to assess the effects on transcription of cytosine methylation within the HSV *tk* promoter. Palla *et al.*[54] analyzed the role of the sea urchin histone H_2A modulator in activating the *tk* basal promoter. Plasmid DNA (240 ng) containing the *tk* promoter linked to the bacterial CAT gene and the H_2A enhancer element was coinjected with various amounts of plasmid carrying only the H_2A element. This "homologous competition" assay demonstrated the ability of excess H_2A sequences to compete for endogenous trans-acting oocyte factors and eliminate the transcription-stimulating effect of H_2A.

C. Coinjections

The successful isolation and expression of a cloned gene permits the subsequent coinjection of RNAs encoding specific or nonspecific heterologous factors capable of interacting with its gene product either to confer biological activity or to generate polymorphic catalytic or structural characteristics. Coinjection experiments may therefore be conducted with the various cloned subunits of an oligomeric molecule or with a cloned gene in conjunction with tissue-derived mRNA.

1. Heteropolymer Reconstitution and Subunit Mixing Experiments. The first successful reconstitution of a heteropolymeric receptor was that achieved by Mishina *et al.*[38] employing pure synthetic mRNAs encoding all four subunits of the *Torpedo* nicotinic acetylcholine receptor. Since that time, a variety of reconstitution experiments have been performed with the cloned subunits of the *Torpedo* nAChR, including subunit substitutions[42] and genetic engineering of individual subunits.[29,37] The outcome of such studies has been the characterization of individual receptor components and the demonstration of functional homology among parallel receptor components from different tissues or species (see also Section II,B).

A particularly good example of the implementation of heteropolymer reconstitution is its role in the stepwise buildup of a model for the rat neuronal nicotinic acetylcholine receptor. Following the isolation of several cDNA clones representing transcripts expressed in different regions of the brain, and encoding polypeptides sharing sequence similarities and structural features of the nAChR family, oocyte microinjection was employed to demonstrate their ability to direct the synthesis of AChR subunits which interact to generate functional receptors with ligand-binding characteristics of neuronal AChRs.[15,81,82] $\alpha2$, $\alpha3$, and $\alpha4$ were all shown to form functional heteropolymers with $\beta2$ when coinjected in a pairwise fashion. Furthermore, $\beta2$ was shown to be interchangeable with the muscle $\beta1$ subunit in *Xenopus* oocytes injected with synthetic mRNAs encoding all four subunits comprising the mouse muscle nAChR. Other combinations of $\beta2$ and various muscle receptor subunits did not give rise to functional receptors, confirming the identification of $\beta2$ as a nonagonist-binding, β-like member of the nAChR supergene family. Together, these studies identified a gene family of neuronal nAChRs capable of generating at least three different ligand-binding subunits which can recombine with a single $\beta2$ subunit to form three potentially variant receptors differentially distributed in the mammalian brain. It is important to note here that several reconstitution studies have indicated that intersubunit interactions may impart stabilizing effects on otherwise unstable component polypeptides.[46,83]

2. Synthetic mRNA plus Tissue-Extracted Poly(A)+ RNA. Tissue-specific processing of human butyrylcholinesterase was demonstrated in microinjected oocytes by coinjecting SP6-derived synthetic BCHE mRNA with poly(A)+ RNA extracted from fetal human brain, muscle, and liver.[19,47] Whereas BCHE mRNA injected alone generated exclusively dimeric BCHE molecules, coinjection of tissue RNA (25 ng/oocyte) was shown to facilitate higher level oligomeric assembly in a manner consistent with the cholinesterase polymorphism observed in the donor tissue. Furthermore, immunocytochemical studies (see Section IV,C) demonstrated cell-surface association patterns of recombinant BCHE which resembled those observed in either brain or muscle depending on the source of the coinjected RNA. Because cholinesterases constitute a minor fraction of total protein even in highly enriched tissue sources, cholinesterase mRNA contributed by tissue RNAs was considered negligible, and, indeed, induced catalytic activity in coinjected oocytes remained essentially equiva-

[81] J. Boulter, J. Connolly, E. Deneris, D. Goldman, S. Heinemann, and J. Patrick, *Proc. Natl. Acad. Sci. U.S.A.* **84**, 7763 (1987).
[82] K. Wada, M. Ballivet, J. Boulter, J. Connolly, W. Etsuko, E. Wada, E. S. Deneris, L. W. Swanson, S. Heinemann, and J. Patrick, *Science* **240**, 330 (1988).
[83] S. Noguchi, M. Mishina, M. Kawamura, and S. Numa, *FEBS Lett.* **225**, 27 (1987).

lent to that observed in oocytes injected with BCHE mRNA alone. These results were therefore taken to indicate the existence of tissue-specific accessory factors involved in posttranslational processing events of human cholinesterases.

Sumikawa and Miledi[44] took an alternate approach to this type of coinjection experiment. Rather than supplying noncharacterized gene products to an excess of synthetic mRNA product, they supplied an excess of specific *Torpedo* AChR subunits to cat muscle AChR induced by the microinjection of cat muscle mRNA. This experimental strategy allowed them to generate hybrid cat–*Torpedo* receptors despite the fact that the genes encoding the cat AChR have not yet been isolated. When combined with 7 ng synthetic mRNA encoding the β or δ subunit of *Torpedo* AChR, 50 ng poly(A)$^+$ RNA from denervated cat muscle produced functional hybrid receptors with the desensitization properties characteristic of the native cat AChR. However, oocytes coinjected with *Torpedo* γ subunits displayed a rapid desensitization uncharacteristic of cat and resembling but not identical to that observed in oocytes injected with mRNAs from all four *Torpedo* AChR subunits. The difference between pure *Torpedo* AChR and cat–*Torpedo* γ AChR was attributed to the presence of pure cat receptors even in coinjected oocytes. Thus, as in other subunit mixing experiments, this study led to the association of specific electrophysiological properties with a particular subunit of a heteropolymeric receptor/channel complex.

3. DNA Vectors plus Tissue-Extracted Poly(A)$^+$ RNA. To study the liver-specific expression of the rat phosphoenolpyruvate carboxykinase gene, the PEPCK promoter was conjugated to the bacterial CAT reporter gene and injected into *Xenopus* oocytes alone (50 ng supercoiled plasmid DNA/oocyte) or together with 50 ng poly(A)$^+$ RNA from PEPCK-expressing or -nonexpressing tissues.[26] Coinjection of liver RNA stimulated PEPCK-directed expression of CAT in the oocytes, but not expression directed by promoter–enhancer elements from other cloned genes expressed in different tissues. Only liver and kidney RNAs displayed this PEPCK enhancement effect, and controlled deletions in the PEPCK promoter demonstrated the sequence dependence of this phenomenon. Furthermore, coinjection of sucrose gradient-fractionated liver RNA implicated a 1600 to 2000-nucleotide RNA species in encoding the putative liver-specific trans-acting factor. These experiments underscore the potential of the oocyte system in the faithful reconstitution of tissue-specific patterns of gene expression, providing a relatively simple and convenient assay for both cis- and trans-activating factors involved in modulating eukaryotic transcription mechanisms.

D. RNA Processing Assays

Following the nuclear microinjection of DNA expression vectors or synthetic precursor mRNAs, RNA processing events may be assessed by the extraction and analysis of the resultant RNA products (see Section IV,B). Bentley and Groudine[57] utilized *Xenopus* oocytes to identify sequence requirements involved in the early transcription termination signal of the c-*myc* gene using two approaches. The first was to inject plasmids containing controlled deletions in the c-*myc* gene and assay the transcription products by either RNA blot or RNase protection analyses of total RNA recovered from 10–20 oocytes after overnight incubation. The second was to perform nuclear runoff transcription assays of manually dissected germinal vesicles from 40 injected oocytes to assess the relative density of RNA polymerases on the microinjected sequences 5' and 3' of the termination sites. Coinjection of α-amanitin ($\sim$ 1 ng/oocyte) subsequently demonstrated the dominant role of RNA polymerase II in transcribing this gene.[84] With the advent of polymerase chain reaction (PCR) technology, detection of mRNA transcripts may now be performed with far greater sensitivity.[84a]

Inoue *et al.*[60] used similar RNA analyses to investigate the role of 5' capping of mRNA on hnRNA splicing. To stabilize rapidly degraded noncapped control RNAs a 5' ApppG cap was employed. Precursor mRNAs endowed with the ApppG cap were as stable in the oocytes as m^7 GpppG-capped RNAs, but they behaved like uncapped pre-mRNA in both *in vitro* and *in vivo* splicing reactions. A 70-fold molar excess of m^7 G-capped competitor RNA but not that carrying the ApppG cap effectively inhibited the splicing reaction in oocyte nuclei, reinforcing the idea that the cap structure itself plays a positive role in facilitating splicing *in ovo*.

RNA splicing complementation assays are being employed to study the components of U snRNAs involved in the formation of functional snRNPs.[52,53] Nuclear injection of 300 μM "antisense" oligodeoxynucleotide complementary to accessible regions of U snRNAs results in RNase H-mediated degradation of the targeted RNA and consequent inhibition of splicing performed on synthetic template pre-mRNA. Coinjection of DNA encoding the corresponding wild-type *Xenopus* U snRNA restores splicing activity and provides a standard against which controlled mutations in

[84] D. L. Bentley, W. L. Brown, and M. Groudine, *Gen. Dev.* **3,** 1179 (1989).

[84a] Y. Lapidot-Lifson, D. Patinkin, C. A. Prody, G. Ehrlich, S. Seidman, R. Ben-Aziz, F. Benseler, F. Eckstein, H. Zakut, and H. Soreq, *Proc. Natl. Acad. Sci. U.S.A.* **89,** 579 (1992).

these genes may be evaluated. The requirement for two independent nuclear injections makes this experiment a bit tricky, but the approach has been successfully exploited to delineate domains of U1 and U2 snRNAs required for snRNP assembly and function.

E. Tunicamycin as Block to Glycosylation

Tunicamycin is an inhibitor of N-linked glycosylation.[84b] Because tunicamycin has been shown not to interfere with protein synthesis in oocytes,[44] this agent can be employed to study the role of glycosylation in the biosynthesis, transport, and biological function of expressed heterologous proteins. Several approaches have been taken to tunicamycin treatment. Dreyfus et al.[47] microinjected oocytes with 10 μM tunicamycin 24 hr prior to RNA microinjection and showed the intracellular accumulation of nascent BCHE around cytoplasmic vesicles (see Fig. 3). Takeda et al.[45] incubated oocytes for 3 days in 2 μg/ml tunicamycin prior to concomitant injection of 40 μg tunicamycin and 1 μg/μl RNA followed by continuous incubation in externally applied drug.

Work by Sumikawa et al.[43] implied, however, that the external application of 2 μg/ml tunicamycin 24 hr prior to and continuously following RNA injection imposes a block to glycosylation comparable to that achieved by its intracellular introduction by microinjection. In a study of the role of N-glycosylation on AChR subunit assembly and plasma membrane insertion,[44] it was noted that tunicamycin failed to block glycosylation of the α subunit completely. Concanavalin A-Sepharose was therefore employed to eliminate AChR complexes containing residual glycosylated α subunits. It was further noted that tunicamycin-treated oocytes appeared less healthy than their nontreated counterparts, an observation with which we wholeheartedly concur. One approach which has been taken to overcome the technical difficulties in working with tunicamycin is the site-directed mutagenesis of potential N-linked glycosylation sites such as was employed with the acetylcholine receptor[42] and vasopressin proteins.[48]

F. Use of Antisense Oligonucleotides

Antisense oligodeoxynucleotides are finding rapidly expanding applications in the targeted in vivo inhibition of specific gene products.[85,86] However, antisense oligonucleotides of varying lengths or directed toward

[84b] A. Elbein, Trends Biochem. Sci. **August,** 219 (1981).
[85] E. L. Wickstrom, T. A. Bacon, A. Gonzalez, D. L. Freeman, G. H. Lyman, and E. Wickstrom, Proc. Natl. Acad. Sci. U.S.A. **85,** 1028 (1988).
[86] H. Zheng, B. M. Sahai, P. Kilgannon, A. Fotedar, and D. R. Green, Proc. Natl. Acad. Sci. U.S.A. **86,** 3758 (1989).

various sequence domains on a given mRNA exhibit unequal efficacies in exerting specific translational arrest. The discrepancies observed in the inhibition exerted by different oligonucleotides are probably due to uncharacterized steric restraints, and they may be difficult to predict. *Xenopus* oocytes may, therefore, provide a suitable assay for the preliminary evaluation of the potency of a given oligonucleotide to suppress expression of the gene product against which it is designed.

As unmodified oligodeoxynucleotides are susceptible to endogenous cellular nucleases, chemically modified phosphorothioate[87] or structurally modified α-DNA[88] analogs are being introduced. Injection of femtomolar to nanomolar quantities of a 17 to 30-mer antisense oligonucleotide (10^3 to 10^4-fold excess) has been shown sufficient to inhibit translation of synthetic[88] or native[89] mRNAs in oocytes. Inhibition has been observed in pre-mRNA, co-mRNA, and post-mRNA oligo-injected oocytes, and varying results have been reported vis à vis the region on the message most vulnerable to antisense attack.[89-91] In addition to the prognostic value of testing oligos on the microscale of an oocyte before attempting the laborious task of tissue culture experimentation, the microinjection of antisense oligodeoxynucleotides is also serving the purpose of discriminating between polymorphic mRNAs in heterogeneous RNA populations.[89] Dahl *et al.*[92] employed "hybrid arrest" to verify the identity of a cloned gap junction protein by demonstrating the ability of its antisense RNA to block expression of a junctional conductance induced by liver poly(A)$^+$ RNA in microinjected oocytes. For a detailed discussion of the use of hybrid arrest to assess the functional roles of cloned cDNAs, see Lotan.[92a]

G. Use of Monoclonal Antibodies to Block Specific Biosynthetic Components

Fischer and Luhrmann[61] coinjected m_3G-capped, ^{32}P-labeled, synthetic U1 snRNA [25 fmol at 2×10^6 counts/min (cpm)/pmol] together with varying concentrations of a monoclonal anti-m_3G IgG antibody and demonstrated suppression of transport of the injected RNA to the nucleus. In a similar approach, Thompson *et al.*[51] assessed the role of the C'-terminal hexapeptide repeat which is associated with eukaryotic RNA polymerase

[87] F. Eckstein, *Annu. Rev. Biochem.* **54**, 367 (1985).
[88] C. Cazenave, C. A. Stein, N. Loreau, N. T. Thuong, L. M. Neckers, C. Subasinghe, C. Helene, J. S. Cohen, and J.-J. Toulme, *Nucleic Acids Res.* **17**, 4255 (1989).
[89] H. Akagi, D. E. Patton, and R. Miledi, *Proc. Natl. Acad. Sci. U.S.A.* **86**, 8103 (1989).
[90] D. A. Melton, *Proc. Natl. Acad. Sci. U.S.A.* **82**, 144 (1985).
[91] E. S. Kawasaki, *Nucleic Acids Res.* **13**, 4991 (1985).
[92] G. Dahl, T. Miller, D. Paul, R. Voellmy, and R. Werner, *Science* **236**, 1290 (1987).
[92a] I. Lotan, this volume [41].

II. Here, 12.5–60 ng monoclonal antibody was microinjected together with template DNA (1 ng supercoiled plasmid DNA/oocyte) and 0.25 μCi [α-^{32}P]GTP. Plasmid DNA (0.25 ng) containing the Pol III-specific 5 S maxigene served as a control for successful nuclear injections. RNA from individual oocytes was extracted and analyzed by gel electrophoresis.

IV. Detection Methods

A. Whole-Cell Assays

The expression of biologically active proteins whose functional detection essentially depends on the maintenance of an intact living cell, or whose stability in tissue homogenates is highly restricted, requires the development of sensitive assays which do not disrupt the integrity of the injected oocyte. In the case of neurotransmitter receptors and ion channels this prerequisite has been met through the decades of cellular electrophysiology which have preceded the current era in molecular neurobiology. In other instances, a novel approach must be introduced. A recent example is that of the oocyte-mediated biosynthesis of catalytically active rat testosterone 5α-reductase. This enzyme is known to be highly unstable in tissue homogenates, and it was therefore assayed using a whole-cell assay approach.[9]

As may be expected with single living cells, even individual oocytes simultaneously obtained from the same frog vary considerably in the quantity of protein they will synthesize from a given aliquot of a particular microinjected RNA. This difficulty has been encountered by numerous investigators and has prompted the development of various approaches to the question of "quality control" in microinjection experiments. The need to establish quality control standards may be particularly apparent where mRNA is expressed for receptors coupled to second messengers, the presence of which may themselves be expected to affect the metabolism of the oocyte and overall protein synthesis. One example of such an assay is the concomitant microinjection of an mRNA for a secreted mutant form of human placental alkaline phosphatase, an enzyme readily measured in the oocyte incubation medium by a standard colorimetric assay.[93] However, this type of internal standard may only be useful where the experimental RNA is of the same type of preparation (i.e., both are synthetic RNAs). In other cases, alternative standards may be required. Finding an effective standard may be especially important where assay protocols call for the analysis of individual oocytes.

1. Electrophysiology. Because many of the other chapters in this volume concern themselves specifically with the question of the electrophysi-

---

[93] S. S. Tate, R. Urade, R. Micanovic, L. Gerber, and S. Udenfriend, *FASEB J.* **4**, 227 (1990).

ological detection of expressed ion channels (see especially Stühmer[93a]), we shall not venture far into this subject. Suffice it to say that microinjected oocytes lend themselves to a variety of electrophysiological recording techniques, and may in many cases be more convenient to work with than native cells or tissues. Intracellular, extracellular, and single-channel recordings have been performed. In a novel technique pioneered by Dahl and collaborators[94] the formation of gap junctions between paired oocytes has been observed following the microinjection of cloned gap junction proteins.[49,50] Oocytes microinjected with RNAs encoding either connexin 32 or connexin 43 were immobilized and juxtaposed vegetal pole to vegetal pole. Induced junctional conductances were then measured by dual-voltage clamp in pairwise combinations of RNA- and water-injected oocytes (see Ebihara[95] for review of the expression of gap junction channels in *Xenopus* oocytes).

2. Transport Assays. The expression of anion or neurotransmitter transport proteins is typically monitored via influx assays based on uptake, by microinjected oocytes, of radiolabeled analogs. The induction of a variety of neurotransmitter transport activities has been demonstrated with ^{3}H analogs of glutamate, GABA, glycine, dopamine, norepinephrine, and 5-HT.[68] Individual oocytes incubated in the presence of 1 μM labeled substrate for 60 min at room temperature were rinsed, solubilized in 500 μl SDS, and subjected to liquid scintillation counting. To examine the Na^+ dependence of transporter activities, choline chloride was substituted for NaCl in the incubation buffer. Iodide transport was similarly monitored using ^{125}I.[65] Here, oocytes injected with poly(A)$^+$ RNA from FRTL-5 cells were incubated in 100 μl buffer containing 50 μM Na^{125}I for 45 min, after which the reaction was terminated with ice-cold quenching buffer (100 mM choline chloride, 10 mM HEPES, pH 7.5, 1 mM methimazole) followed by rapid filtration through nitrocellulose filters.

Anion efflux in oocytes injected with RNA encoding mouse erythroid band 3 protein was measured in an opposite fashion. ^{36}Cl$^-$ (50–70 nl, 0.113 mCi/ml) was injected into oocytes 16 hr following microinjection of RNA. Individual oocytes were then placed on the window of a Geiger–Müller tube and perfused with buffer while remaining radioactivity was continuously monitored.[32,96] In a different approach, Ca^{2+} efflux was followed by incubating injected oocytes with 20 μCi/ml ^{45}Ca^{2+} in a well of a 96-well microtiter plate, replacing the incubation medium at regular intervals, and determining secreted radioactivity by scintillation counting.[97]

[93a] W. Stühmer, this volume [19].

[94] G. Dahl, R. Azarnia, and R. Werner, *Nature (London)* **289,** 683 (1981).

[95] L. Ebihara, this volume [24].

[96] D. Bartel, S. Lepke, G. Layh-Schmitt, B. Legrum, and H. Passow, *EMBO J.* **8,** 3601 (1989).

[97] L. C. Mahan, R. M. Burch, F. J. J. R. Monsma, and D. R. Sibley, *Proc. Natl. Acad. Sci. U.S.A.* **87,** 2196 (1990).

3. Monitoring Intracellular Calcium Mobilization. The mobilization of intracellular stores of Ca^{2+} via pathways mediated by inositol trisphosphate (IP_3) has been monitored by the microinjection of 2 $\mu g/\mu l$ aequorin, a calcium-specific photoprotein which emits a chemiluminescent signal in the presence of calcium.[98] Here, Ca^{2+}-depleted modified Barth's medium was prepared by substitution of $MgCl_2$ for $CaCl_2$ and supplementation with 200 μM EGTA. Emitted light was monitored in a liquid scintillation counter with the coincidence gate switched off. Others have used a photomultiplier diode to quantitate Ca^{2+}-stimulated luminescence.[99] The use of albino *Xenopus* oocytes addresses the problem of quenching by the endogenous black pigment vesicles[98,100] of the oocyte.

B. Biochemistry

Classic biochemical analyses of microinjected *Xenopus* oocytes are conveniently performed where induced polypeptides lend themselves to catalytic activity assays in whole-cell extracts or subcellular fractions. In some instances, it may be important to determine the levels of nonbiologically active molecules resulting from premature termination products, defective posttranslational processing, site-directed mutagenesis, or incomplete oligomeric assembly.

Xenopus oocytes are easily homogenized, 10 oocytes/150–300 μl homogenization buffer, with a glass–Teflon homogenizer. Where nondenaturing solubilization of membrane proteins is desired, 1% Triton X-100 is included. A short (15–20 min) centrifugation in a precooled microcentrifuge is usually sufficient to pellet insoluble cellular materials (including black pigment-carrying structures), and separate the lipid-containing overlay from the clear supernate. The clear supernatant is gently removed with a drawn-out Pasteur pipette after as short a delay as possible, then used immediately or stored at $-20°$. For studies of oocyte-produced human cholinesterases, groups of as few as 10 oocytes have proved amenable to detailed catalytic activity analyses and inhibitor studies.[19,20,34,35]

Subcellular fractionations have also been described. For example, plasma membranes have been prepared as follows: After making an incision in the oocyte membrane with the tip of a syringe needle, occytes were gently squeezed to discharge cytoplasmic materials. Membranes were then suspended in 50 mM imidazole-HCl buffer (pH 7.5) containing 250 mM sucrose, 1 mM EDTA, and 1 mM phenylmethylsulfonyl fluoride (PMSF), and homogenized. The homogenate was centrifuged over a 50% sucrose

[98] K. Sandberg, A. J. Markwick, D. P. Trih, and K. J. Catt, *FEBS Lett.* **241**, 177 (1988).

[99] I. Parker and R. Miledi, *Proc. R. Soc. London B* **228**, 307 (1986).

[100] K. Sandberg, M. Bor, H. Ji, A. Markwick, M. A. Millan, and K. J. Catt, *Science* **249**, 298 (1990).

cushion and the supernatant recentrifuged at 160,000 *g* for 30 min.[45,83] Others have prepared animal pole/vegetal pole specific homogenates by dividing frozen oocytes along the equatorial midline with a hand-held razor.[101]

1. Enzyme Assays. With a modified version of the thiocholine ester hydrolysis method of detection[102] (Fig. 1A) to assay recombinant human cholinesterases produced in microinjected oocytes, biochemical characterization of these enzymes and a number of their naturally occurring and engineered mutants has proceeded rapidly.[19,34,35] Samples of total oocyte homogenates (10–20 μl) are assayed in individual wells of a 96-well microtiter plate where the kinetics of substrate hydrolysis may readily be monitored by an automated microtiter plate reader. With this technique up to 96 kinetic assays can be performed and managed within a matter of hours. We have recently adapted this protocol to the measurement of cholinesterase activities in single microinjected oocytes (H. Soreq and S. Seidman, unpublished, 1990). A further modified protocol has been employed for the analysis of oligomeric assembly to oocyte-produced cholinesterases by linear sucrose density gradient centrifugation[19] (Fig. 1B).

ATPase activities assayed on the microsomal (plasma membrane) fraction of injected oocytes has been performed by determining release of P_i after incubating 20–80 μg protein with 50 mM imidazole-HCl buffer (pH 7.5), 140 mM NaCl, 14 mM KCl, 5 mM MgCl$_2$, 1 mM ATP in the presence and absence of 1 mM ouabain, a specific inhibitor of Na$^+$, K$^+$-ATPase activity.[45] Induced choline acetyltransferase activity in microinjected oocytes was assayed by measuring the conversion of [^{14}C]acetyl-CoA to [^{14}C]acetylcholine.[74] Assays were performed on detergent extracts (2% Triton X-100) equivalent to as little as one-tenth of an oocyte. To monitor the intracellular accumulation of acetylcholine in this study, oocyte extracts were prepared by homogenization of one to several oocytes in ice-cold 0.15 M formic acid in acetone, microfuge centrifugation, and aliquoting of portions representing one-tenth to one-hundredth of an oocyte to individual microtubes. Assays were performed according to a radioenzymatic procedure developed for submicromolar measurements.[103]

2. Immunoprecipitation. Immunoprecipitation of oocyte-synthesized proteins is a common preparatory step for gel electrophoretic analysis of *de novo* protein synthesis. Specific precipitation of desired polypeptides permits the relative quantitation of particular nonbiologically active products, be they alternatively processed components of a heterogeneous population[45] or the independent components of a coinjection experiment.[44] Pre-

---

[101] A. Cerioti and A. Colman, *EMBO J.* **7**, 633 (1988).
[102] G. L. Ellman, K. D. Courtney, V. Andres, Jr., and R. M. Featherstone, *Biochem. Pharmacol.* **7**, 88 (1961).
[103] R. E. McCaman and J. Stetzler, *J. Neurochem.* **28**, 699 (1977).

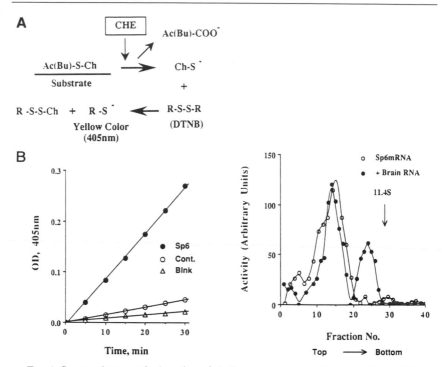

FIG. 1. Spectrophotometric detection of cholinesterases produced in microinjected *Xenopus* oocytes. (A) Schematic presentation of the thiocholine ester hydrolysis method of G. L. Ellman, K. D. Courtney, V. Andres, Jr., and R. M. Featherstone [*Biochem. Pharmacol.* **7**, 88 (1961)] (top) and graphic presentation of results acquired from a representative microinjection experiment (bottom). Oocytes were injected with either 5 ng synthetic butyrylcholinesterase mRNA (Sp6) or Barth's medium (Cont) and incubated overnight at 18° in a humidified incubator. Samples (10 μl) of total oocyte homogenates [10 mM Tris-HCl, 1M NaCl, 1 mM EGTA, 1% Triton X-100 (pH 7.4), 10 oocytes/300 μl] were preincubated for 30–45 min in individual wells of a 96-well microtiter plate with Ellman's reagent [100 mM phosphate buffer (pH 7.0), 0.5 mM 5,5'-dithiobis-2-nitrobenzoic acid (DTNB)]. The reaction was initiated by addition of 10 μl butyrylthiocholine (20X stock). Hydrolysis of substrate releases free acid and thiocholine. Thiocholine reacts 1:1 with DTNB to generate the yellow anion 5-thio-2-nitrobenzoic acid, whose high extinction coefficient (ϵ_{405} 13,600 M^{-1} cm^{-1}) renders this assay highly sensitive (path length ~0.5 cm). The kinetics of substrate hydrolysis were monitored by the V_{max} automated microtiter plate reader (Molecular Devices Corp., Menlo Park, CA) linked to an IBM-compatible computer equipped with specially adapted software for the calculation of rate constants. This assay accurately detects cholinesterase (CHE) activities down to nanomoles substrate hydrolyzed/minute/oocyte in 30 min–1 hr. (B) Sucrose gradient (5–20%) analysis of oocyte homogenates (10 oocytes in 200 μl; 12 ml gradient) following injection of synthetic BCHE mRNA with or without poly(A)$^+$ RNA from human fetal brain. Spectrophotometric detection of fractionated enzyme was as described above, with some minor modifications.

cipitation of newly synthesized proteins is accomplished by "targeting" the protein of interest with an appropriate antiserum or monoclonal antibody and subsequently sedimenting the antibody–antigen complex with second antibody or protein A coupled to an inert solid support such as Sepharose beads. The immunoreaction is performed in IP buffer [such as 0.15% Tween 80 (w/v), 20 mM Tris-HCl (pH 7.5), 0.5% Triton X-100, 0.5% sodium deoxycholate, 0.1% L-amino acid (according to the labeled amino acid used)] for 2 hr at 37° followed by overnight incubation at 4°.[104] Binding of coupled second antibody or protein A is accomplished by a 30-min incubation at 37° followed by 1 hr of incubation at 4°. The total suspension is then filtered over glass wool by centrifugation in a microfuge tube and washed 5–8 times with washing buffer [10 mM phosphate (pH 6.8), 1% Triton X-100, 1% sodium deoxycholate, 0.2% L-amino acid, 0.1% bovine serum albumin (BSA)]. Immunoreactive material is subsequently eluted with SDS-PAGE sample buffer and boiling.

3. Gel Electrophoresis. Detection of oocyte-produced polypeptides by denaturing polyacrylamide gel electrophoresis (SDS-PAGE) may be crucial where nonbiologically active molecules are under study. Owing to the minute quantities of protein synthesized by the oocytes in a microinjection experiment, radioactive labeling and autoradiography may be employed to facilitate its visualization. Where appropriate antisera are available, Western blotting may not only increase sensitivity, but it permits positive identification of the polypeptide of interest and its by-products.

Biosynthetic labeling of *de novo* proteins induced by the microinjection of foreign RNA or DNA may be accomplished by coinjection of a radiolabeled amino acid such as [^{35}S]methionine or [^{3}H]leucine (2–25 μCi/oocyte) or by incubation of injected oocytes in medium containing the label (1–10 μCi/well for [^{35}S]Met). The former approach may be favored where high specific activity labeling of the foreign protein product is desired or where the injected mRNA is expected to have a short half-life. Labeled oocytes are then homogenized and applied directly to a 10% polyacrylamide gel or first subjected to purification by immunoprecipitation[46] (see also Section IV,B,2) or affinity binding.[44] Employment of a fluorographic reagent such as Amplify (Amersham, UK) and an intensifying screen for autoradiography should permit detection of even low abundance proteins after a 6-hr to overnight exposure at −70° (see Fig. 2).

Following SDS-PAGE, oocyte proteins may be transferred to a nitrocellulose filter and subjected to immunodetection with an appropriate antiserum (Western blotting). For example, we have employed commer-

[104] M. Burmeister and H. Soreq, *in* "Molecular Biology Approach to the Neurosciences" (H. Soreq, ed.) p. 201. Wiley, New York, 1984.

cially available polyclonal rabbit antiserum cholinesterase antibodies (Dako, Denmark) to detect inactive mutant BCHEs produced in *Xenopus* oocytes after site-directed mutagenesis of specific amino acids.[104a] Samples representing as little as one-third of an oocyte permitted detection of an approximately 70K molecular weight band in mRNA-injected oocytes which were not present in mock-injected oocytes. With horseradish peroxidase (HRP)-linked second antibodies and enhanced chemical luminescence (ECL, Amersham) detection, the entire procedure can be completed in one day.

Nondenaturing gel electrophoresis has also been used to detect the products of oocyte microinjection. A single band representing oocyte-produced human butyrylcholinesterase was detected by nondenaturing gel electrophoresis followed by activity staining based on the procedure for cytochemical staining of cholinesterases developed by Karnovsky and Roots[105] (see Fig. 2). In an earlier study, crossed immunoelectrophoresis was employed to analyze cholinesterase polymorphism in *Xenopus* oocytes injected with mRNA from human fetal brain, muscle, and liver[106] (Fig. 2).

C. Immunocytochemistry

Immunocytochemical staining has proved instructive in the study of posttranslational sorting and ultimate subcellular localization of *de novo* heterologous proteins. Immunofluorescent probes detectable by fluorescent light microscopy were employed in a study of the subcellular transport of recombinant BCHE in microinjected oocytes[47] (Fig. 3). This approach permitted monitoring the kinetics of appearance of recombinant BCHE at the periphery of injected oocytes and the identification of two distinct types of BCHE aggregates ("patches" and "clusters") at the oocyte extracellular surface. A higher resolution was obtained by substituting protein A conjugated to 5-nm colloidal gold beads for the fluorescent second antibody and viewing the sections by electron microscopy (Fig. 3). The latter approach revealed that BCHE deposits were primarily associated with the exterior surface of the oocyte extracellular material, not the plasma membrane.

Quantification of the total number of patches and clusters from a complete series of sections from two representative oocytes indicated polarized animal–vegetal compartmentalization of synthetic BCHE. Noteworthy is the observation that the density of BCHE molecules in clusters or patches was estimated in this study to be approximately 5000 molecules/

[104a] L. F. Neville, A. Gnatt, Y. Loewenstein, S. Seidman, G. Ehrlich, and H. Soreq. *EMBO J.* in press (1992).

[105] M. J. Karnovsky and L. Roots, *J. Histochem. Cytochem.* **12,** 219 (1964).

[106] H. Soreq, K. M. Dziegielewska, D. Zevin-Sonkin, and H. Zakut, *Cell. Mol. Neurobiol.* **6,** 227 (1986).

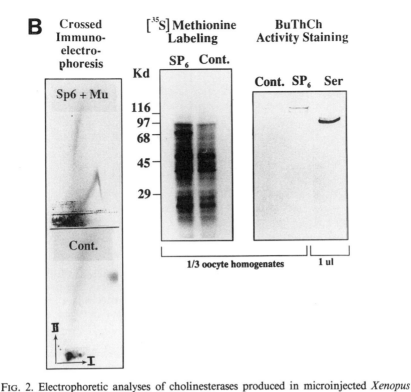

FIG. 2. Electrophoretic analyses of cholinesterases produced in microinjected *Xenopus* oocytes. Left: Crossed immunoelectrophoresis of oocytes injected with Barth's medium (Cont, bottom) or coinjected with synthetic BCHE mRNA and poly(A)$^+$ RNA from human fetal muscle (SP6 + Mu, top). Oocyte proteins (labeled with [^{35}S] methionine) were applied to the first-dimension gel together with human fetal plasma, containing both ACHE and BCHE activities, as an unlabeled carrier. Electrophoresis was in 1% (w/v) agarose in Tris–barbitone buffer (pH 8.6). Second-dimension gels were impregnated with 2.5 μl/cm^2 commercial polyclonal antibodies. First-dimension electrophoresis was at 10 V/cm for 1 hr; the second dimension was at 2 V/cm overnight at room temperature. Middle: SDS–polyacrylamide gel electrophoresis analysis (10% polyacrylamide) of oocytes injected with BCHE mRNA and incubated overnight in Barth's medium containing 0.1 μCi/μl [^{35}S] methionine. Total oocyte homogenates equivalent to one-third of an oocyte were applied to each lane in 1.5× sample buffer; electrophoresis was run at 120 V for 4 hr. The gel was fixed with 40% methanol, 10% acetic acid overnight then treated for 30 min with Amplify (Amersham) fluorographic reagent and dried. Exposure was overnight at −70°. Right: Nondenaturing gel electrophoresis and butyrylthiocholine activity staining of oocyte-produced BCHE. Tris–Triton X-100 extracts of injected oocytes (20 μl; two-thirds of an oocyte) were loaded in sample buffer (62.5 mM Tris-HCl, pH 6.8, 3% glycerol, 0.4% bromphenol blue) to individual lanes of a continuous 7% polyacrylamide gel made up in Tris–glycine running buffer (0.1 M Trizma base, 38 mM glycine, 1.0% Triton X-100, pH 8.9) following a 1-hr prerun. The gel was electrophoresed for 4 hr in the cold, rinsed several times in double-distilled water, and incubated in staining buffer for 1 hr at room temperature, then overnight at 4°. One microliter human serum (Ser) served as the standard. Improved results have been achieved using a 6% resolving gel and a short 3% stacking gel [B. Velan, C. Kronman, H. Grosfeld, M. Leitner, Y. Gozes, Y. Flashner, T. Sery, S. Cohen, R. Ben-Aziz, S. Serdman, A. Shafferman, and H. Soreq, *Cell. Mol. Neurobiol.* **11**, 143 (1991)].

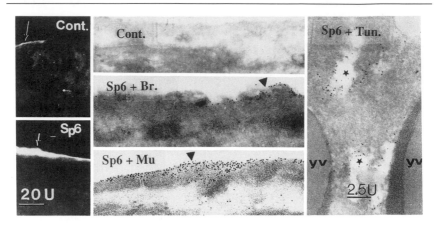

FIG. 3. Immunocytochemical detection of cholinesterases produced in *Xenopus* oocytes. Left: Surface-deposited oocyte-produced BCHE as viewed by fluorescence microscopy. Oocytes injected with synthetic BCHE mRNA alone or in conjunction with tissue RNAs were fixed, embedded in Tissue-tek OCT compound, deep frozen, and sectioned (10 μm thick). Sections were arranged on gelatin-coated slides and incubated overnight at 4° with rabbit polyclonal antibodies prepared against *Torpedo* electric organ ACHE. Sections were then incubated with either rhodamine- or fluorescein-conjugated second antibody and viewed by fluorescence light microscopy. To evaluate the nonspecific binding of second antibody, a phycoprobe-linked antibody (anti-rabbit IgG + L BIOMEDA) which fluoresces yellow when complexed with first antibody and green when adsorbed nonspecifically was employed. Middle: High magnification view of BCHE aggregates as seen by immunogold staining and electron microscopy. Following incubation with first antibody as above, sections were incubated with 5-nm colloidal gold particles conjugated to protein A, washed, and fixed in 2.5% glutaraldehyde/0.5% tannic acid. Sections were further washed, postfixed in 2% osmium tetroxide, dehydrated, and embedded in Epon. Ultrathin sections were prepared, then poststained with uranyl acetate and lead citrate. Sections from oocytes injected with Barth's medium (top) and BCHE mRNA coinjected with poly(A)$^+$ RNA from fetal brain (middle) or muscle (bottom) are presented. Right: Electron microscopic view of nascent BCHE molecules trapped intracellularly around cytoplasmic vesicles in oocytes injected with tunicamycin 24 hr prior to mRNA injection. Sections were prepared as described above.

μm^2, the same order of magnitude as in clusters harboring the nicotinic acetylcholine receptor[107,108] and ACHE[109-111] in neuromuscular junctions and along neuronal dendrites. This observation could indicate that the organization of membrane-associated molecules at the extracellular surface of *Xenopus* oocytes reaches values which are near the ultimate limit of packing densities for such proteins. The observation that coinjection with

[107] M. Salpeter, *J. Cell Biol.* **32**, 379 (1967).
[108] A. J. Sytkowski, Z. Vogel, and M. W. Nirenberg, *Proc. Natl. Acad. Sci. U.S.A.* **70**, 270 (1973).
[109] T. Rosenberry, *Biophys. J.* **26**, 263 (1979).

brain or muscle mRNAs increased the total surface area occupied by the human BCHE[47] (Fig. 3) could indicate the existence of tissue-specific factors capable of modulating the aggregation of heterologous proteins at the oocyte surface in a manner similar to that observed in native tissues.[112,113]

A similar tact was taken by Swenson *et al.*[50] in the study of gap junction channels formed in oocyte pairs injected with synthetic mRNAs derived from cloned connexin DNAs. Defolliculated oocytes were injected near the center of the vegetal hemisphere, and after an incubation period the vitelline membrane was manually detached following shrinkage of the oocyte induced by incubation in hypertonic medium. Single and paired oocytes were frozen without fixation by immersion in Freon slush. Sections were incubated with polyclonal antisera (1 : 500) or full-strength monoclonal culture supernatant, then with fluorochrome-conjugated second antibody, and viewed as described above. This study demonstrated the primarily cytoplasmic localization of connexin proteins in single oocytes and their accumulation at the site of apposition in paired oocytes. Furthermore, here, as in the immunohistochemical analysis of synthetic BCHE, fluorescent signals were particularly intense around the site of microinjection.

D. Cell Culture Assay for Secreted Growth Factors

To investigate the possible involvement of butyrylcholinesterase in cell growth and/or proliferation, conditioned medium from oocytes injected with synthetic BCHE mRNA (BCHE-OCM) was added to murine bone marrow cultures.[114] BCHE-OCM was indeed found to stimulate megakaryocyte colony formation in semisolid media and to increase the relative proportion of megakaryocytes observed by differential cell counting. A smaller but significant effect was observed with conditioned media from mock-injected oocytes, consistent with the previously described presence of platelet-derived growth factor and other growth-stimulating factors in

---

[110] E. A. Barnard, *in* "Cholinesterases: Fundamental and Applied Aspects, Proceedings of the Second International Meeting on Cholinesterases" (M. Brzin, E. A. Barnard, and D. Sket, eds.), p. 49. de Gruyter, Berlin.

[111] R. L. Rotundo, *in* "The Vertebrate Neuromuscular Junction," p.247. Alan R. Liss, New York, 1987.

[112] B. G. Wallace, R. M. Nitkin, N. E. Reist, J. R. Fallon, N. N. Moayeri, and U. J. McMahan, *Nature (London)* **315**, 574 (1985).

[113] B. G. Wallace, *J. Cell Biol.* **102**, 783 (1986).

[114] D. Patinkin, S. Seidman, F. Eckstein, F. Benseler, H. Zakut, and H. Soreq, *Mol. Cell. Biol.* **10**, 6046 (1990).

oocyte-conditioned medium.[115,116] Interestingly, supplemental horse serum, a rich source of tetrameric BCHE, did not elicit such a response, suggesting the secretion, by BCHE-producing oocytes, of a yet unidentified growth-stimulating factor. This study introduces a highly sensitive biological assay protocol for the expression of heterologous growth factors in *Xenopus* oocytes.

V. Oocyte Biology

A. Ultrastructural Considerations

We have previously emphasized viewing the *Xenopus* oocyte not as an isolated cell, but as an organ system.[5] An occyte is a cell which remains, *in vitro*, in intimate physical contact with surrounding cellular and noncellular components. Ultrastructural analysis of oocyte sections reveals the vitelline membrane, follicle cells, theca, and epithelium encompassing the oocyte[117] (Fig. 4). Micro- and macrovilli maintain important gap–junction connections between the oocyte and its enveloping follicle cells, providing chemical and electrical communication pathways[118] (see also Fig. 4). Although the follicular layers down to the vitelline membrane can be mechanically or enzymatically disengaged, these treatments clearly violate the integrity of the oocyte plasma membrane and disrupt current responses.[2,119] Although at least some of the endogenous neurotransmitter receptor activities observed in oocytes have been associated with the peripheral follicle cells,[2] denuded oocytes apparently provide the endogenous machinery for second-messenger-mediated transduction signals induced by agonist binding to heterologous receptors.[3] For example, angiotensin II induced gap–junction-mediated calcium mobilization in native follicular but not denuded oocytes, stimulating the rate of progesterone-induced maturation in a pH-sensitive manner. Nonetheless, oocytes microinjected with poly(A)$^+$ RNA from rat adrenal gland retained their responsiveness to angiotensin II after defolliculation.[100]

An additional noteworthy feature of the *Xenopus* oocyte is its polarized nature. Curiously, the prominent black/white, animal/vegetal polar asymmetry of this colossal cell reflects a general subcellular hemispheric polar-

[115] M. Mercola, D. A. Melton, and C. D. Stiles, *Science* **241**, 1223 (1988).

[116] C. Thery, P. Jullien, and D. A. Lawrence, *Biochem. Biophys. Res. Commun.* **160**, 615 (1989).

[117] S. Wischnitzer, *in* "Advances in Morphogenesis" (M. Abercrombie and J. Brachet, eds.), Vol. 5. Academic Press, New York, 1966.

[118] C. L. Browne and W. Werner, *J. Exp. Zool.* **230**, 105 (1984).

[119] R. Miledi and R. M. Woodward, *J. Physiol.* **416**, 601 (1989).

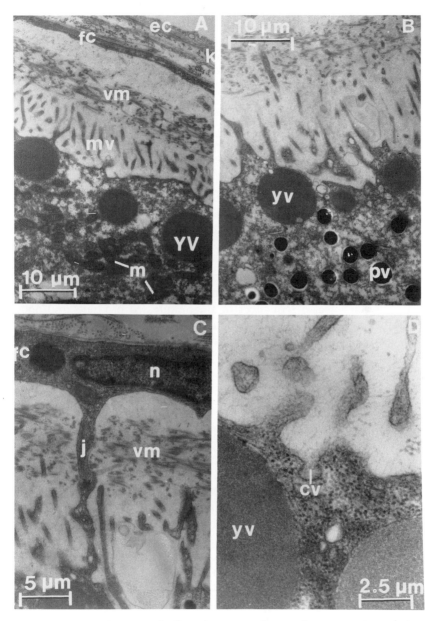

FIG. 4. Electron microscopy of salient ultrastructural oocyte features. (A) Cross-sectional view of the vegetal pole. Note the internal yolk vesicles (YV) and high concentration of mitochondria (m), protruding micro/macro villi (mv), acellular vitelline membrane (vm), follicle cells (fc), and extracellular material (ec). (B) Similar view of the animal pole. Note the pigment vesicles (pv) which are absent from the vegetal pole (A). (C) Close-up view of microvilli (j) traversing the vitelline membrane (vm) to establish gap–junction type contacts with surrounding follicle cell (fc). Note the follicle cell nucleus (n). (D) Close-up view of a coated vesicle (cv) fused to the oocyte plasma membrane. (Photographs courtesy of Dr. D. M. Phillips, The Population Council, New York.)

ization which extends to characteristics such as the distribution of yolk platelets,[120] maternal mRNA,[121] and membrane receptors/channels,[122] as well as general cytoskeletal organization[2,123] (Fig. 4). In certain instances, polarized transport of nascent heterologous proteins has also been observed.[47,123] The full implications of the asymmetric character of the oocyte with regard to expression studies is not yet clear. However, this property might be a consideration in specific instances with respect to microinjection protocols or detection methodologies.

The specialized nature of the oocyte implies that its own intrinsic character will necessarily be a factor in the production and/or detection of proteins synthesized from microinjected heterologous sequences. Thus, the translational efficiency of heterologous mRNAs in the oocytes may be affected by numerous factors, some of which are inherent to the *Xenopus* species or to the oocyte itself, whereas others depend on intrinsic properties of the injected molecules. Furthermore, specific steps in the pathway for gene expression might either be disrupted or enhanced owing to interactions between injected sequences or their expressed products and oocyte biosynthetic elements. Such interactions might eventually alter the levels of detectable "synthetic" proteins and should be taken into consideration whenever applicable. Some of these points are discussed below (see also Soreq[1]).

B. Developmental Considerations

Both the rate of protein synthesis in oocytes and the pool of elements participating in protein synthesis are prone to various modes of regulation throughout oogenesis and egg formation, and they may be further modified through the action of proteins synthesized from heterologous mRNAs. Figure 5 displays *Xenopus* oocytes at various stages of development (see Dumont[124] for a detailed description of oocyte developmental stages). Native oocyte mRNAs accumulate early in oogenesis and remain at a constant, steady-state level for the remainder of oocyte development.[125] Early in oogenesis, most of the oocyte polyadenylated mRNA is localized in the nucleus. From there it is transported to the cytoplasm where it may be detected in fully grown oocytes.[126] Within the cytoplasm of stage VI[124]

---

[120] M. V. Danilchik and J. C. Gerhart, *Dev. Biol.* **122**, 101 (1987).

[121] D. L. Weeks and D. A. Melton, *Cell (Cambridge, Mass.)* **51**, 861 (1987).

[122] Y. Oron, B. Gillo, and M. C. Gershengorn, *Proc. Natl. Acad. Sci. U.S.A.* **85**, 3820 (1988).

[123] M. W. Klymkowsky, L. A. Maynell, and A. G. Polson, *Development* **100**, 543 (1987).

[124] J. N. Dumont, *J. Morphol* **136**, 153 (1972).

[125] J. Brachet, *in* "Molecular Biology of Nucleocytoplasmic Relationships," Vol. 14, p. 189. Elsevier, Amsterdam, 1975.

[126] M.Wakahara, *J. Embryol. Exp. Morphol.* **66**, 127 (1981).

FIG. 5. Stages of *Xenopus* oocyte development. Close-up view of a representative section from a mature *Xenopus* ovary. Note the translucent stage I oocytes, opaque and white stage II oocytes, somewhat evenly pigmented stage III and IV oocytes, and larger stage V and VI oocytes with clear hemispheric differentiation. The dark pigment in stage V and VI oocytes demarcates the nucleus-containing animal hemisphere. Also visible are accessory blood vessels supplying nourishment to developing oocytes. Magnification: × 6.25.

oocytes, which are commonly used for microinjection, poly(A)$^+$ RNA is most concentrated in the vegetal subcortical region.[127]

Nucleocytoplasmic transport and mRNA mobilization in the oocytes are prerequisite to the successful expression of injected DNA sequences. These processes depend on the cytoskeletal elements of the oocytes.[128] Together with poly(A) chains, microtubule elements modulate the activity of nucleoside triphosphatase in the oocytes and affect further polyadenylation of newly transcribed mRNAs.[129] The stability and translational efficiency of oocyte mRNAs are, in turn, modulated by RNA-binding proteins which are also developmentally regulated.[130] Together with interspersed repetitive sequences of maternal origin, these RNA-binding proteins take

[127] D. G. Capco and W. R. Jefferey, *Dev. Biol.* **89**, 1 (1982).
[128] K. Shiokawa, *FEBS Lett.* **151**, 179 (1983).
[129] W. E. G. Muller, A. Bernd, and H. C. Schroder, *Mol. Cell. Biochem.* **53**, 197 (1983).
[130] J. D. Richter and L. D. Smith, *J. Biol. Chem.* **258**, 4864 (1983).

part in controlling the subcellular compartmentalization of specific mRNA pools within the oocytes.[131] Untranslatable mRNAs (~90% of total mRNA) are put into long-term storage[132] while translatable mRNA molecules are separately compartmentalized in slowly turning over pools.[133] Small nuclear RNA and tRNA, which appear not to be rate limiting, are stored in ribonucleoprotein particles from which they can be gathered for ribosome assembly.[134] Altogether, the complex transformations occurring throughout oogenesis underscore the importance of carefully preselecting the individual oocytes to be injected. These should be as uniform as possible in size and general appearance to ensure a similar stage of development.

C. Oocyte Maturation

The fully grown stage VI *Xenopus* oocytes which are commonly used for microinjection studies are naturally arrested at the end of the G_2 phase of the meiotic cell cycle. Both the *in vivo* injection with human chorionic gonadotropin (hCG) and the *in vitro* admininstration of progesterone induce, in folliculated oocytes, an AMP-dependent resumption of meiotic processes, from prophase I to metaphase II[135] through their interaction with surface membrane steroid receptors.[136] This process is termed maturation and results in the formation of metaphase-arrested fertilizable eggs. Maturation is characterized by the induction of maturation-specific transcription and translation processes[137] and a rapid and persistent decrease in cAMP.[138] Alternative agents capable of inducing oocyte maturation are drugs affecting Ca^{2+} mobilization such as trifluoperazine,[139] adenylate cyclase inhibitors which increase K^+ influx,[140] and stimulators of the Na^+/K^+ pump such as bee mellitin.[141]

Maturation is dependent on guanine nucleotides[142] and is associated

[131] D. M. Anderson, J. D. Richter, M. E. Chamberlin, D. H. Price, R. J. Britten, L. D. Smith, and E. H. Davidson, *J. Mol. Biol.* **155**, 281 (1982).

[132] M. Rosbash, *Dev. Biol.* **87**, 319 (1981).

[133] J. D. Richter, D. M. Anderson, E. H. Davidson, and L. D. Smith, *J. Mol. Biol.* **173**, 227 (1984).

[134] L. K. Dixon and P. J. Ford, *Dev. Biol.* **91**, 474 (1982).

[135] H. R. Belle, J. Boyer, and R. Ozon, *Dev. Biol.* **90**, 115 (1982).

[136] S. E. Sadler and J. L. Maller, *J. Biol. Chem.* **257**, 335 (1982).

[137] J. J. LaMarca, M. C. Srobel-Fidler, L. D. Smith, and K. Keem, *Dev. Biol.* **47**, 384 (1975).

[138] S. Schorderet-Slatkine, M. Schorderet, and E. E. Balieu, *Proc. Natl. Acad. Sci. U.S.A.* **79**, 850 (1982).

[139] T. G. Hollinger and I. M. Alvarez, *J. Exp. Zool.* **224**, 461 (1982).

[140] J. Hanocq-Quertier and E. Baltus, *Gamete Res.* **4**, 49 (1981).

[141] A. K. Deshpande and S. S. Koide, *Differentiation* **21**, 127 (1982).

[142] X. Jordana, C. C. Allende, and J. E. Allende, *Biochem. Int.* **3**, 527 (1981).

with calmodulin synthesis[143] and the subsequent activation of nucleotide phosphodiesterases.[144] Increases in active endocytosis,[145] fusion of lysosomes with yolk platelets, and release of acid hydrolases[146] induce proteolysis in hormone-treated maturing oocytes. Therefore, untreated noninduced oocytes may be advantageous for heterologous gene expression. Nonetheless, certain aspects of the maturation process may be induced by the microinjection itself and/or its resultant protein products. It should be noted that although maturation is accompanied by a considerable increase in the overall rate of protein synthesis,[147] ribosome transit time remains unchanged in mature oocytes.[148] Therefore, there is no advantage in using mature oocytes in terms of translational efficiency. Indeed, the results obtained with microinjected *Xenopus* eggs are generally similar to those reported for oocytes.[149] However, because the maturation process may specifically alter the efficiency and/or fidelity of expression of particular exogenous mRNA species and may, in fact, be stimulated by expressed foreign proteins, it merits discussion in this context.

Oocyte maturation is effected by a complex of two major proteins. One of these proteins, designated p34^{cdc2}, is a 34K phosphoprotein kinase which phosphorylates the H1 histone and is homologous to the universal eukaryotic cell cycle controller *cdc2*.[150] The p34^{cdc2} protein itself is a cell-cycle regulated substrate for tyrosine phosphorylation[151] which subsequently phosphorylates several additional nuclear and cytoplasmic proteins, all with putative roles in controlling the M phase of the cell cycle[152]; these include p$^{60c\text{-}src}$, which induces cytoskeletal rearrangements, and the elongation factors EF-1β and EF-1δ. The latter two proteins impose a block to translation which would be expected to disrupt the expression of injected mRNA. Most importantly, the *cdc2* protein kinase phosphorylates serine residues in the repetitive carboxy-terminal domain of RNA polymerase II.[153] This, in turn, may block transcriptional processes in the oocytes, a

[143] A. Cartaud and R. Ozon, *J. Biol. Chem.* **255**, 9404 (1980).

[144] F. Miot and C. Erneux, *Biochim. Biophys. Acta* **701**, 253 (1982).

[145] L. M. Tucciarrone and K. D. Lanclos, *Mol. Cell. Biochem.* **45**, 159 (1982).

[146] M. Decroly, M. Goldfinger, and N. Six-Tondeur, *Biochim. Biophys. Acta* **587**, 567 (1979).

[147] E. Younglai, F. Godeau, and E. E. Balieu, *FEBS. Lett.* **127**, 233 (1981).

[148] J. D. Richter, W. J. Wasserman, and L. D. Smith, *Dev. Biol.* **89**, 152 (1982).

[149] J. B. Gurdon, C. D. Lane, H. R. Woodland, and G. Marbaix, *Nature (London)* **233**, 177 (1971).

[150] O. Mulner-Lorillon, R. Poulhe, P. Cormier, J.-C. Labbe, M. Doree, and R. Belle, *FEBS Lett.* **251**, 219 (1989).

[151] G. Draetta, H. Piwnica-Worms, D. Morrison, B. Druker, T. Roberts, and D. Beach, *Nature (London)* **336**, 738 (1988).

[152] S. Moreno and P. Nurse, *Cell (Cambridge, Mass.)* **61**, 549 (1990).

[153] L. J. Cisek and J. L. Corden, *Nature (London)* **339**, 679 (1989).

possibility which should be seriously considered in planning DNA microinjection experiments.

Microinjection of p34^{cdc2} or of the conserved 16-amino acid peptide from its C′ terminus triggers, within 1 min, a specific 7-fold increase in the concentration of free intracellular Ca^{2+} originating from intracellular stores.[154] Therefore, microinjection of poly(A)$^+$ RNA from tissues expressing cdc2-like proteins may generate the same effects. In addition, the second-messenger pathway through which oocyte maturation may be induced involves signal transduction processes. These may also be initiated in *Xenopus* oocytes by p21^{ras} protein microinjection. In this case there is no dependence on cyclic AMP; p21^{ras} microinjection activates phosphatidylcholine metabolism in the oocytes, leading, within minutes, to diacylglycerol production and germinal vesicle breakdown (GVB).[155] Parallel consequences, although delayed, may be expected to occur following the production of p21^{ras} from microinjected mRNAs or in oocytes triggered into meiotic maturation by exogenous signals or by p34^{cdc2} derivatives, a consideration which should be taken into account when planning and evaluating microinjection experiments.

D. Environmental Considerations

Environmental conditions such as CO_2 concentration[134] and heat (i.e., $> 31°$[156]) were shown to introduce major changes in the levels and patterns of proteins synthesized in *Xenopus* oocytes. Both endogenous amphibian[100] and heterologous mammalian[50] gap junctions appear to be pH sensitive. Thus, exposure of oocytes to acidified medium gassed with CO_2 disrupts gap junction communications in both cases. The heat-shock response is characterized by the synthesis of a single major protein (M_r 70,000) and a gradual decline in the rate of protein synthesis, and it is not dependent on *de novo* transcription of RNA. The introduction of heterologous mRNAs such as that of the adenovirus E1a gene[148] may also influence oocyte metabolism. The effect of adenovirus RNA was also shown to be transcription independent and reached its peak approximately 18 hr postinjection, indicating its dependence on translation products of the microinjected RNA (reviewed by Soreq[1]). The molecular neurobiologist need particularly note that alterations in the type, number, or mean open time of ion channels in the oocyte plasma membrane may alter the uptake of

[154] A. Picard, J.-C. Cavadore, P. Lory, J.-C. Bernengo, C. Ojeda, and M. Doree, *Science* **247**, 327 (1990).
[155] J. C. Lacal, *Mol. Cell. Biol.* **10**, 333 (1990).
[156] M. Bienz and J. B. Gurdon, *Cell (Cambridge, Mass.)* **27**, 811 (1982).

amino acids into the oocytes[134,157] or affect phosphoinositide metabolism,[158] Ca^{2+} mobilization,[137] or intracellular pH,[159] and leading to altered regulation of protein synthesis.

VI. Concluding Remarks

The incalculable contribution of *Xenopus* oocyte microinjection methods to the rapid progress seen in molecular and cellular biology in the two decades since its introduction as a translation system[149,160] is profound indeed. Undoubtedly, the applications of oocyte microinjection as a heterologous expression system will continue to grow in number and scope. Yet an aspect of these remarkable cells which we have not touched on here is their potential role as a model system in which to study cellular processes interrelated with the functioning of ion channels via the microinjection of small molecules, proteins, and genetic elements encoding metabolically active polypeptides. One might therefore anticipate a renewed interest in the coming years in microinjected oocytes as cells, not merely efficient biofactories.

Acknowledgments

We are grateful to Dr. Patrick Dreyfus (INSERM, Paris) for numerous contributions to the research described in this chapter and to Dr. D. M. Phillips for electron micrographs of *Xenopus* oocytes. This work was supported by the U.S. Army Medical Research and Development Command under Grant DAMD17-90-Z-0038 (to H.S.) and by the U.S.–Israel Binational Science Foundation under Grant 89-00205/1.

[157] D. Jung and J.-P. Richter, *Cell Biol. Int. Rep.* **7,** 697 (1983).
[158] Y. Oron, N. Dascal, E. D. Nadler, and M Lupo, *Nature (London)* **313,** 141 (1985).
[159] S. C. Lee and R. A. Steinhardt, *Dev. Biol.* **85,** 358 (1981).
[160] C. D. Lane, G. Marbaix, and J. B. Gurdon, *J. Mol. Biol.* **61,** 73 (1971).

[15] Maintenance of *Xenopus laevis* and Oocyte Injection

By ALAN L. GOLDIN

Introduction

Xenopus laevis, the South African clawed frog, is a member of the anuran family Pipidae. There are six species of *Xenopus,* all indigenous to Africa, but *Xenopus laevis* is the only one commonly used in laboratory research.[1] *Xenopus laevis* frogs are distinguished by the presence of three toes with claws on each hind foot of the animal, hence the name, which means "clawed frog." There are four subspecies of *Xenopus laevis,* termed *X. laevis laevis, X. laevis petersi, X. laevis victorianus,* and *X. laevis borealis.*[1] All *Xenopus* are aerobic but entirely aquatic, so there is an absolute requirement for breathing air but no necessity for any land-based existence. *Xenopus* are very susceptible to desiccation, and they can die from being out of the water for a few hours.

Xenopus frogs are normally darkly pigmented on the dorsal side, but their skin contains chromatophores which rapidly change color according to the surrounding environment. The frog secretes both a mucouslike substance and a toxic substance with properties like epinephrine. These are most likely defense mechanisms for the frog, but they are not harmful to humans when handling the frogs.

Xenopus laevis was first used experimentally in a clinical setting as a pregnancy test. Female frogs were injected with urine from a possibly pregnant woman, which resulted in egg laying only if the urine contained human chorionic gonadotropin (hCG), an indication of pregnancy. Although the frogs are no longer used for this purpose, they are still widely used as an experimental system in developmental biology.[2] However, the most common use today of *Xenopus laevis* has been for their oocytes. Gurdon *et al.* originally demonstrated that *Xenopus* oocytes can be used to express exogenous mRNA species when microinjected into the cytoplasm.[3] Many investigators have sinced used this property to study the expression and translation of a large number of molecules (for reviews, see Refs. 4–8).

[1] J. B. Gurdon and H. R. Woodland, *in* "Handbook of Genetics" (R. C. King, ed.), p. 35. Plenum, New. York, 1975.

[2] I. B. Dawid and T. D. Sargent, *Science* **240,** 1443 (1988).

[3] J. B. Gurdon, C. D. Lane, H. R. Woodland and G. Marbaix, *Nature (London)* **233,** 177 (1971).

[4] J. B. Gurdon and M. P. Wickens, this series, Vol. 101, p. 370.

[5] A. Colman, *in* "Transcription and Translation—A Practical Approach" (B. D. Hames and S. J. Higgins, eds.), p. 271. IRL Press, Oxford, 1984.

Sources of *Xenopus* Frogs

Although *Xenopus laevis* is indigenous only to South Africa, the frogs are now available from a number of different suppliers throughout the world (see Appendix II for a listing). There are two major types of frog suppliers, those that capture and sell frogs reared in the wild and those that sell frogs bred in their own facilities. Frogs bred in the laboratory are generally more consistent and can be fed a synthetic diet available from various suppliers (see below). On the other hand, frogs captured in the wild may be less susceptible to disease than inbred frogs.[1] They will not eat a synthetic diet such as frog brittle, however, and in fact may die of starvation if kept on such a diet. They must be fed a natural diet consisting of liver, heart, or similar material (see below). Both laboratory-reared and wild-reared frogs can be used successfully for the production of oocytes.

The specifics of ordering frogs for oocyte use vary with each supplier, but it is generally a good idea to request the largest and most mature females for the purpose of obtaining stage V oocytes. These large frogs also contain a significant number of less mature oocytes, so that it is possible to obtain oocytes of all stages from them. In addition, it is advisable to request shipping with cold packs during the summer months to prevent extremes of temperature variation. To avoid this problem it may be helpful to have sufficient frogs shipped prior to the hot months of summer. Once the frogs have arrived it is a good idea to let them acclimatize for about 2 weeks before use. In addition, because of the widespread use of *Xenopus* frogs for oocyte injection experiments, there have been recent shortages, particularly from the suppliers that sell laboratory-bred frogs.

To synchronize oocyte development, the frogs can be injected with hCG, which induces ovulation and egg laying, after which oocyte development starts over approximately synchronously. An appropriate dose is 400–500 IU, which should induce egg-laying in about 12 hr.[1] Many suppliers offer frogs which have been injected with hCG. If hCG-injected frogs are to be ordered, however, be sure to request that the frogs be kept for 2–3 months after injection to allow sufficient time for mature oocytes to develop. Uninjected frogs generally are just as good a source of mature oocytes as injected ones, so that hCG injection is not usually necessary.

---

[6] H. Soreq, *Crit. Rev. Biochem.* **18**, 199 (1985).
[7] T. P. Snutch, *Trends Neurosci.* **11**, 250 (1988).
[8] N. Dascal, *Crit. Rev. Biochem.* **22**, 317 (1987).

Maintenance of Frogs

Environment for Frogs

Xenopus can tolerate a wide range of temperature fluctuations, but the quality of oocytes diminishes markedly following temperature shifts. The frogs should therefore be housed in a temperature-controlled environment at approximately 20°. Sufficient air conditioning is especially important, as heat seems to be more detrimental than cold. In addition, the frogs should be maintained in a constant light–dark cycle of 12 hr each. This helps to prevent the seasonal variability found with *Xenopus* in the wild, although oocyte quality generally does diminish during the summer months. A stable environment will help to ensure that the oocytes are still usable, however.

The containers for the frogs need not be particularly sophisticated. Large polypropylene tanks or aquarium tanks both work well. The advantage of polypropylene is that it keeps the frogs more isolated from the external environment, but aquarium glass can be painted black to achieve the same effect. This is beneficial since the frogs are easily startled by movement or light. It is important to have a tight-fitting lid with holes for air, since the frogs can easily lift a lid and escape. Frogs are obligate air breathers, and must be able to reach the surface, so the level of water should not be greater than about 6–10 inches. Suitable containers are Nalgene tanks 18 by 12 by 12 inches high (Fisher, Pittsburgh, PA, Cat. No. 14-831-330), with holes drilled in the lids. Each of these containers can hold 6–8 frogs, with 2–3 liters of water per frog.

Treatment of Water

Xenopus are freshwater frogs that prefer stationary water. If running water is used to keep the water clean, it should be provided at a slow rate to prevent development of a "red-leg" like disease.[1] It is best to keep the frogs in stationary water of a depth of 6–10 inches, and to change the water and clean the tanks twice weekly (after feeding).

Amphibians are sensitive to both chlorine and chloramine in tap water. Although many communities use only chlorine in the water supply, many others have started to add both anhydrous ammonia and chlorine, which combine to form chloramine. Chloramine is very stable, slowly decomposing back to chlorine and ammonia. Neither reverse osmosis nor deionization removes all chloramine. The method of purification required depends on the local water supply treatment. If the water supply is treated only with chlorine, it can be removed by leaving the water in open containers exposed to the air for 48 hr. This can be accomplished more rapidly by bubbling air through the tanks with an air pump designed for an aquarium. The frogs can then be safely placed in the water.

If the water supply is treated with chloramine, the above treatment will be ineffective owing to the stability of chloramine. The simplest method to remove chloramine is to use high purity carbon or Barnstead organic removal cartridge filters. Two activated charcoal filters should be used in series for safety, which will remove all detectable chloramine, chlorine, and ammonia. The water should be periodically tested at the outlet of each filter for chlorine and ammonia, and the filters replaced if either compound is present at detectable levels. This purification will result in water that can be used directly for the frogs without toxicity. An alternative method to remove chloramine is to use products which break it down to chlorine and ammonia. Many of these, such as Dechlor and Novaqua, are sold in pet stores for aquarium use. The free chlorine and ammonia must then be removed. The chlorine can be removed by allowing the water to sit exposed to the air for 48 hr, and the ammonia can be removed by filtration with zeolite. Alternatively, the ammonia can be converted to ammonium ions, which are not harmful to the frogs, by lowering the pH to 6.5.

Feeding

Frogs only need to be fed about 2 times per week. Frogs bred in the wild will eat only meat, such as beef heart, liver, or similar food. A diet of meat exclusively is deficient in calcium and vitamin D, and so either the diet should be supplemented with trout chow or the meat should be soaked in a vitamin solution. A convenient meat diet is beef heart, which is easily available from most supermarket meat departments. Frogs bred in the laboratory will subsist quite well on a synthetic diet such as frog brittle, available from NASCO (Fort Atkinson, WI; Cat. No. SA5960MP for 5 pounds). In either case about 5–10 g of food per frog is dropped into the tank, and the frogs are allowed to eat for 4–6 h. At that point the tanks are cleaned to remove residual food and waste, and the frogs are placed in fresh water.

Diseases and Disorders

Xenopus are quite hardy both in the wild and in the laboratory environment, with only a few percent mortality per year under normal laboratory conditions. The most notable disorder is red leg, in which underparts of the body and the inside of the mouth become red and swollen. This is due to an accumulation of blood. The disease can be treated by immersing the frogs in a 500 mg/ml solution of streptomycin and penicillin for a few days; otherwise, it is usually fatal.[1] Frogs which are kept in an environment with running water can suffer from a disease like red leg owing to hemolysis.

Xenopus are very sensitive to antibiotics. Sulfonamides are lethal, and penicillin or streptomycin are harmful if injected. On the other hand, the frogs are protected by a natural peptide antimicrobial, magainin.[9] Magainin is a family of closely related peptides which inhibit the growth of numerous species of bacteria and fungi and cause lysis of protozoa by osmotic disruption. In the laboratory it functions to reduce the incidence of infection, particularly following surgery for oocyte removal.

Preparation of Oocytes

Methods of Anesthesia

The most commonly used anesthetic for *Xenopus laevis* has been MS-222, the methane sulfonate salt of 3-aminobenzoic acid ethyl ester (also called tricaine, Sigma, St. Louis, MO; Cat. No. A-5040).[10] MS-222 is effective at a concentration as low as 0.15% and may be used safely to 0.35%. The compound dissolves easily in water and can be used multiple times by storing it at 4°. The frog is placed in a small container containing 0.2% MS-222 for about 10–15 min, after which the extent of anesthesia should be tested. One test is to see if the frog can right itself after being placed upside down. The effectiveness of MS-222 appears to vary with different frogs, and sometimes it can require 30 min or more for adequate anesthesia. This can be overcome by using higher levels of MS-222, but concentrations greater than 0.35% can be debilitating to the frogs. The effect can also be enhanced by using the MS-222 at 4°, but the anesthesia may wear off as the frog warms to room temperature.

An alternative anesthetic that is effective on *Xenopus* is benzocaine (ethyl *p*-aminobenzoate, Sigma, Cat. No. E-1501).[11] This compound is a reversible blocker of sodium channels and is used as a local anesthetic in mammals. It is as effective an anesthetic as MS-222 in frogs and other amphibians, and is more consistent in its effect. Benzocaine is effective on *Xenopus* at a concentration of 0.03%. Because it dissolves very slowly in water, a 100× stock (3%) should be made in 100% ethanol. The stock is then diluted 1:100 with water for immersion of the frogs, and the 1% residual ethanol causes no problems. Solutions of benzocaine are stable for at least 2 weeks at room temperature, and longer at 4°. The frog is

[9] M. Zasloff, *Proc. Natl. Acad. Sci. U.S.A.* **84,** 5449 (1987).
[10] V. Hamburger, "A Manual of Experimental Embryology." Univ. of Chicago Press, Chicago, Illinois, 1960.
[11] R. B. Borgens, M. E. McGinnis, J. W. Vanable, Jr., and E. S. Miles, *J. Exp. Zool.* **231,** 249 (1984).

anesthetized as with MS-222, which should take about 10–15 min and should be tested the same way.

Following anesthesia and surgical removal of the ovaries, the frog should be allowed to recover in a small container of water. As frogs must breathe in air, it is possible to drown them by immersion in water that is too deep while the frog is still anesthetized. When the frog has completely recovered (30–60 min), it can be safely returned to the colony.

Removal of Oocytes

When the frog is fully anesthetized, it should be placed on its back on a clean surface. A small incision about 1 cm long stretching diagonally from medial to lateral toward the head is made in the abdomen. Because the skin is extremely tough, it helps to use a sharp needle to pierce the skin, after which scissors can be used for the incision. Cut through both the skin and the underlying fascia, after which the ovaries should be visible. Remove as many oocytes as required by pulling out the lobes of the ovary with a pair of forceps, and using a pair of scissors to cut them. Place the oocytes in a 60-mm tissue culture dish containing calcium-free OR2 medium (see Appendix I). After a sufficient number of oocytes have been removed, replace the remaining ovary into the abdomen with the forceps and close the incision with one stitch using 5-0 silk.

The entire operation should be done under clean conditions, but a sterile surgical procedure is not necessary because of the magainin antimicrobial peptides secreted from the skin.[9,12] This allows a frog to be returned to the water immediately after surgery with a very low likelihood of infection developing. Healing generally occurs without gross inflammation or cellular reaction. It is possible to go back to the same frog and remove oocytes multiple times, but a deterioration in oocyte quality is sometimes seen after multiple surgeries.

To keep track of individual frogs they can be marked by various means, including toe clipping, the tying of colored threads between the toes, or tattooing.[13] The first two methods generally work poorly because it is difficult to distinguish frogs in a large tank, and the threads can easily fall off. For tattooing, use 4 N HCl to draw a number on the back of the frog while it is still anesthetized after removing the oocytes. The HCl is left on for 1–2 min, after which the frog is returned to water for recovery from anesthesia. An alternative is to use 0.5% amido black in 7% acetic acid.[1]

[12] B. A. Berkowitz, C. L. Bevins, and M. A. Zasloff, *Biochem. Pharmacol.* **39,** 625 (1990).
[13] J. B. Gurdon, *in* "Methods in Developmental Biology" (F. H. Wilt and N. K. Wessells, eds.), p. 75. Thomas Y. Crowell, New York, 1967.

While the tattooing method does theoretically provide an opportunity for infection, this is rarely seen. Frogs labeled by this method are easily distinguished in large tanks.

Removal of Follicle Cells

Oocytes can be injected with follicle cells around them. This is technically more difficult, because the follicle cells are harder to pierce with the needle and it is time-consuming to separate out individual oocytes. However, there are some responses measured in oocytes that either depend on the presence of follicle cells or are actually occurring in the follicle cells, such as the opening of K^+ channels in response to cyclic nucleotides.[14] If these responses are not important to the study, then it is best to remove the follicle cells before injection. This can be done either manually or through the use of collagenase.

Collagenase has the advantage of being able to defolliculate large numbers (thousands) of oocytes in a single step. This is essential for large-scale injections. However, overtreatment with collagenase significantly reduces oocyte viability. To avoid collagenase treatment, use a pair of forceps to pull apart the individual oocytes gently. These can then be manually defolliculated using forceps and scissors or with poly(L-lysine)-coated slides, as described by Miledi and Woodward.[14] Alternatively, the oocytes can be injected immediately with the follicle cells attached.

If collagenase is to be used, the extent of treatment should be carefully monitored for removal of follicle cells. Use collagenase at a concentration of 0.5 units/ml, which generally corresponds to 2 mg/ml for most lots of collagenase. It is a good idea to test each lot of collagenase before large-scale use, since different lots can vary in effectiveness and toxicity. For treatment, gently tease apart the oocytes while they are in calcium-free OR2 in a tissue culture dish. Use a pair of forceps, but be careful not to lyse too many oocytes. The goal is to separate the clumps of oocytes sufficiently to allow the collagenase to work. The oocytes are then incubated in a solution of collagenase in OR2 at 20°. Incubating the oocytes in a test tube on a rotator allows better mixing than using a tissue culture dish. After about 90 min replace the solution with fresh collagenase, and remove the oocytes after another 60–90 min. Monitor the treatment carefully by removal of oocytes, and stop the treatment before all of the cells have been fully defolliculated to prevent toxicity.

Collagenase is particularly damaging to oocytes if the incubation is carried out in the presence of calcium, which activates proteases. To prevent this wash the oocytes before collagenase treatment in calcium-free

[14] R. Miledi and R. M. Woodward, *J. Physiol. (London)* **416,** 601 (1989).

OR2, and then wash them extensively (4–6 times) after collagenase treatment in calcium-free OR2 before transferring them to a calcium-containing solution like ND96. Suitable sources of collagenase are Boehringer-Mannheim (Indianapolis, IN; Cat. No. 1088 793) and Sigma (type IA, Cat. No. C-9891). Some researchers use only partial collagenase treatment (minutes to an hour) to maintain oocyte viability, after which the residual follicle cells are removed manually.

Selection of Oocytes for Injection

After collagenase treatment and OR2 washing, the oocytes should be transferred to incubation solution (see below). At this point it is a good idea to select the healthiest oocytes and transfer them to a fresh dish in incubation solution. Death and lysis of many oocytes are inevitable in the original dish owing to the large number of oocytes. This quickly results in a hypertonic solution, which is harmful to the otherwise healthy oocytes. After selection the oocytes should be maintained for a few hours before injection to allow any additional oocyte death to occur. There is no disadvantage to incubating the oocytes overnight before injection.

Oocytes from *Xenopus laevis* can be classified into six stages of development on the basis of anatomy, with stage I being the earliest stage.[15] Stage I oocytes are about 50–100 μm in diameter and appear transparent. Stage II oocytes are 300–450 μm and appear either translucent or white, depending on the extent of development. Stage III oocytes are 450–600 μm and can be distinguished by the appearance of pigmentation uniformly throughout the surface. Stage IV oocytes are 600–1000 μm, with differentiated hemispheres and a very dark brown animal hemisphere. Stage V oocytes are 1000–1200 μm, with the hemispheres clearly delineated and a lightening in color of the animal hemisphere. Stage VI oocytes are the most mature, 1200–1300 μm, and can be distinguished by an unpigmented equatorial band between the two hemispheres.

The choice of which stage oocytes to inject will depend on the purposes of the experiment. For most experiments stage V oocytes are optimal because they are large, can easily be injected with up to 100 nl of solution, and have significant translational capacity. However, the large size is reflected in a very large membrane capacitance (100–200 nF), so that voltage clamp recordings from these oocytes will demonstrate a long capacitive transient (1–2 msec). Stage II or III oocytes can also be used successfully for injection and expression, and the smaller size of these oocytes will result in a faster clamp settling time.[16] However, they can only be injected with

[15] J. N. Dumont, *J. Morphol.* **136,** 153 (1972).
[16] D. S. Krafte and H. A. Lester, *J. Neurosci. Methods* **26,** 211 (1989).

about 20 nl of solution, and this must be performed using a more elaborate injection device such as those described below for nuclear injection.

Injection of Oocytes

Preparation of RNA or DNA for Injection

RNA or DNA for injection should be precipitated with ethanol at least once to remove excess salt and detergents like sodium dodecyl sulfate (SDS). These can have a markedly deleterious effect on the health of the oocytes. After precipitation the RNA should be resuspended at an appropriate concentration in either water or dilute Tris-HCl (1 mM, pH 6.5). The low pH of RNA solutions in water can be slightly more toxic to the oocytes than a buffered solution. The appropriate concentration depends on the source of the RNA and varies from 10 mg/ml for total rat brain RNA to 10 μg/ml for *in vitro* transcribed RNA. At the very low concentrations it is advisable to use carrier RNA (such as yeast tRNA) to prevent losses from RNA sticking to the glass injection needles, but this is not essential.

Cytoplasmic Injection

Cytoplasmic injections in oocytes can be performed with a very simple and inexpensive injection device.[17] This apparatus uses a Drummond 10-μl microdispenser (oocyte injector, Drummond, Broomall, PA, Cat No. 3-00-510-X) attached to a three-dimensional coarse micromanipulator such as a Brinkmann (Westbury, NY) MM-33. The attachment requires a custom-made ring that mounts onto the micromanipulator, allowing the microdispenser to be held in place by a set screw. The oocytes are placed in ND96 in a 35-mm tissue culture dish under a stereomicroscope. To keep the oocytes from moving, polypropylene mesh (Sepctra/Mesh PP, Fisher, Cat. No. 08-670-185) can be glued to the bottom of the dish. Injection needles are made by using a pipette puller to draw out the glass bores that are normally used with the Drummond microdispenser. The standard 4 inch bores work quite well with a vertical puller, but if a horizontal puller is used it requires 8 inch bores available from Drummond to make two injection needles. After pulling, the needles are broken off at a tip diameter of 20–40 μm, as measured with a reticle under a dissecting microscope. To attach the needles to the dispenser, add about 5 mm of light mineral oil into the needle near the large opening, then insert the bore onto the dispenser plunger all the way. The plunger will force the oil toward the tip

---

[17] R. Contreras, H. Cheroutre, and W. Fiers, *Anal. Biochem.* **113**, 185 (1981).

of the needle, which should result in a complete seal of oil between the metal plunger and the injection needle tip.

Before drawing the RNA into the needle the RNA solution should be centrifuged in a microcentrifuge tube for about 60 sec to pellet insoluble debris. This reduces the likelihood of needle clogging. About $2\mu l$ of RNA solution is then placed on a tissue culture dish, and the solution is drawn into the Drummond microdispenser. It is advisable to watch the RNA being drawn into the needle through the microscope to be sure that the needle has not clogged. Once the needle is filled with RNA it can be positioned over each oocyte and gently lowered until it pierces the oocyte. Up to 100 nl can then be injected into each oocyte by turning the knob on the microdispenser. By this method it is possible to inject 20 oocytes with one sample in a few minutes. Although the same needle can then be used for additional RNA samples, using a new needle for each sample prevents cross-contamination and makes needle clogging less likely.

An alternative injection device has recently been introduced by Drummond (Nanoject, Cat. No. 3-00-203-X). This consists of a motor-driven dispenser similar to the $10\text{-}\mu l$ microdispenser and includes controls for continuously drawing up solution, continuously forcing out solution, and a momentary switch for injecting 47 nl with each push of a button. This has the advantage that injections can be performed without ever looking away from the oocytes. The disadvantages are that the unit is more expensive and can malfunction more easily. An injector with adjustable volume is also available (Cat. No. 3-00-203-XV).

Nuclear Injection

Nuclear injections require a more sophisticated injection apparatus, as the maximum volumes that can be injected are about 10–20 nl. The Picopump from World Precision Instruments (Sarasota, FL) and the Picospritzer from General Valve (Fairfield, NJ) use air pressure for injection, and the Nanopump from World Precision Instruments uses a metal plunger for injection. The air pressure devices can be used for smaller volumes, but they must be calibrated for each needle. The Nanopump is slower and requires a tight seal of oil between the plunger and RNA. In either case the micromanipulator is the same, only the injection volume is controlled by either the injection pressure or the time of injection. All these instruments can, of course, be used for cytoplasmic injection also.

For nuclear injection it is necessary to align the nuclei on the top surface of the oocytes so that the needle can be inserted into the appropriate location. This can be accomplished by centrifuging the oocytes, since the nucleus is less dense than the cytoplasm and rises to the top surface. Microwell dishes from Irvine Scientific (Cat. No. 1-63118) work well for this purpose. Single oocytes are placed in each well in ND96 with the pigmented side facing up. The dish can be placed on a test tube holder in a tabletop centrifuge, such as a Beckman GPR (Fullerton, CA), for centrifugation. It is a good idea to perform a test spin with about 10 oocytes to determine the centrifugation conditions. Conditions which are a reasonable starting point are 1000 g for 13 min. Examine each oocyte for the presence of the nucleus at the top surface, which will appear as a lighter color in the pigmentation. Optimal conditions result in a clear area of decreased pigmentation in all of the oocytes without a significant number of dead ooctyes. Both the time and centrifugal force can be varied to improve oocyte viability.

The oocytes should be injected within 30–60 min after centrifugation (if they sit for too long in the microwell dish a significant number will probably die). The injection needles must be smaller in diameter than those used on the Drummond injectors for cytoplasmic injection. Suitable glass is R6, 0.63 mm OD by 0.20 mm ID by 10 cm long, also available from Drummond. The needles are pulled as usual and broken off at about 30 μm outer diameter. If a pressure injection sytem is used, the needles must then be calibrated. This is easily accomplished by injecting an aqueous dye solution (Injection Dye, see Appendix I) into glycerol, which results in a sphere of fluid. The diameter of the sphere is measured with the microscope reticle, and the volume can be directly determined using a standard graph of volume versus sphere diameter ($V = \pi d^3/6$; 10 nl is approximately equivalent to a diameter of 270 μm). The injection dye is also useful to include with the DNA solution, as this will serve as an indicator of successful nuclear injection. The DNA or RNA solution should be drawn into the injection needle with sufficient vaccum to result in about 1 μl drawn up in 60 sec. The ejection pressure or time should then be varied to achieve an appropriate injection volume. Using a Picopump, approximate conditions for injecting 10-nl volumes are a vacuum of 10–15 inches mercury and an ejection pressure of 10–14 psi, with a 1-sec injection time. It is probably best to adjust the ejection pressure and keep the time constant, as oocytes do not tolerate extremely rapid injection.

Incubation of Oocytes

After the oocytes have been defolliculated or injected they should be maintained in a relatively isoosmotic solution containing calcium. A wide variety of solutions have been used for this purpose, including Barth's,[4] OR2 with calcium,[18] normal frog Ringer's solution,[19] L-15 medium,[20] and

[18] R. A. Wallace, D. W. Jared, J. N. Dumont, and M. W. Sega, *J. Exp. Zool.* **184**, 321 (1973).

ND96[21] (recipes are given in Appendix I). ND96 is essentially OR2 which is slightly hypertonic (sodium concentration 96 compared to 82.5 mM), which may be beneficial when injecting the oocytes. In addition, the incubation medium should be supplemented with sodium pyruvate (550 mg/liter) as a carbon source[22] and gentamicin (100 μg/ml) to prevent bacterial contamination.[23] An alternative antibiotic mix that is effective is a combination of 100 U/ml penicillin and 100 μg/ml streptomycin.[24] Theophylline can be added to a concentration of 0.5 mM to inhibit phosphodiesterase and keep cAMP level high, which will prevent maturation. This is generally not necessary, however, because cAMP levels are normally high in oocytes.[25]

The oocytes should be incubated in 35-mm diameter tissue culture dishes (60-mm diameter dishes for large numbers) on a rotator at slow speed in a 20° controlled temperature incubator. Temperature fluctuations, particularly heat, can both be damaging to oocyte viability and result in erratic time courses of expression. The healthy ooctyes should be transferred to fresh dishes with new ND96 each day, and more often if there is any oocyte death. This is particularly important shortly after injection, as dying oocytes release salts which make the solution hypertonic and toxic to the remainder. The time course of expression varies with the type of RNA injected. Detectable expression of sodium channels can be observed 24 hr after injection of RNA made *in vitro* from a clone when the oocytes are incubated at 20° (lower temperatures result in a longer delay). Expression of this channel reaches maximal levels by about 2 days and does not noticeably decrease at least until 6 days postinjection.

Appendix I: Recipes

OR2—calcium-free
  82.5 mM NaCl
  2mM KCl
  1 mM MgCl$_2$
  5 mM HEPES, pH 7.5 with NaOH

---

[19] C. Methfessel, V. Witzemann, T. Takahashi, M. Mishina, S. Numa, and B. Sakmann, *Pfluegers Arch.* **407,** 577 (1986).
[20] R. A. Wallace, Z. Misulovin, D. W. Jared, and H. S. Wiley, *Gamete Res.* **1,** 269 (1978).
[21] J. P. Leonard, J. Nargeot, T. P. Snutch, N. Davidson, and H. A. Lester, *J. Neurosci.* **7,** 875 (1987).
[22] J. J. Eppig and J. N. Dumont, *In Vitro* **12,** 418 (1976).
[23] R. A. Wallace and Z. Misulovin, *Proc. Natl. Acad. Sci. U.S.A.* **75,** 5534 (1978).
[24] R. A. Laskey, *J. Cell Sci.* **7,** 653 (1970).
[25] M. F. Cicirelli and L. D. Smith, *Dev. Biol.* **108,** 254 (1985).

ND96
  96 mM NaCl
  2 mM KCl
  1.8 mM CaCl$_2$
  1 mM MgCl$_2$
  5 mM HEPES, pH 7.5 with NaOH
Modified Barth's solution
  88 mM NaCl
  1 mM KCl
  2.4 mM NaHCO$_3$
  20 mM HEPES, pH 7.5
  0.82 mM MgSO$_4$
  0.33 mM Ca(NO$_3$)$_2$
  0.41 mM CaCl$_2$
Normal frog Ringer's solution
  115 mM NaCl
  2.5 mM KCl
  1.8 mM CaCl$_2$
  10 mM HEPES, pH 7.2
L-15
  70% Leibovitz's L-15 medium (available from GIBCO, Grand Island, NY)
  10 mM HEPES, pH 7.5
100× Injection Dye
  0.2 M Tris-HCl, pH 7.5
  1 M NaCl
  4% Trypan blue
  (Use 10× for needle calibration and 1× for DNA samples)

Appendix II: Suppliers of *Xenopus laevis*

United States Suppliers

NASCO, 901 Janesville Ave., Fort Atkinson, WI 53538, (414) 563-2446.
  Ordering Information: LM535MP, adult females, 9–10.5 cm @ $20.60; LM531P, guaranteed oocyte-positive females; LM535WC, adult females, 9–10.5 cm @ $17.20 (wild-caught); LM531WC, guaranteed oocyte-positive females (wild-caught).
  Notes: The first two catalog numbers are for laboratory-reared frogs. The last two are for wild-caught, laboratory-conditioned frogs (a recent addition to the catalog). The 9–10.5 cm adult females are generally the best source of mature oocytes, as these frogs are guaranteed to be large and have always been oocyte positive.
Xenopus I, 716 Northside, Ann Arbor, MI 48105, (313) 426-2083.
  Ordering Information: 4216, adult females, oocyte positive, @ $13.90.
  Notes: Frogs captured in the wild.
Carolina Biological Supply Company, 2700 York Road, Burlington, NC 27215, (800) 334-5551, (919) 584-0381.
  Ordering Information: L1570, adult females @ $18.79.
  Notes: Laboratory-reared frogs.

European Suppliers

Xenopus Ltd., Holmesdale Nursery, Mid Street, South Nutfield, Redhill, Surrey RH1 4JY, England, 44-73-782-2687.
Ordering Information: Very large, mature females.
Notes: Either laboratory-reared or wild frogs.
Dipl. Biol.-Dipl. Ing. Horst Kähler
Institut für Entwicklungsbiologie
Kollaustrasse 113b
D-2000 Hamburg 61
Germany
40-587675
Ordering Information: *X. laevis* or *X. laevis* albinos
Notes: Either laboratory-reared or wild frogs.

Japanese Suppliers

Nippon Life Science, ordering address: 1-1-32 Shiba-Daimon, Minatoku, Tokyo 105; company address: Denmacho 19, Hamamatsu City, Shizuoka, Japan.
Seibu Department Stores
Tokyo, Japan
Notes: Xenopus can be purchased in the department stores in Japan.

South African Supplier

South African Snake Farm, P.O. Box 6, Fish Hoek, Cape Province, Republic of South Africa.
Notes: Frogs can be ordered directly from South Africa for air shipment anywhere in the world.

Acknowledgments

The author is a Lucille P. Markey Scholar, and work in his laboratory is supported by grants from the U.S. National Institutes of Health (NS-26729), the Lucille P. Markey Charitable Trust, and the March of Dimes Basil O'Connor Starter Scholar Program.

[16] Preparation of RNA for Injection into *Xenopus* Oocytes

By ALAN L. GOLDIN and KATUMI SUMIKAWA

Introduction

Xenopus oocytes are now a popular system for the expression and characterization of functional neurotransmitter receptors and ion channels. Oocytes are injected with mRNAs extracted from tissues or synthesized *in vitro* from cloned cDNAs, after which the foreign mRNAs are translated by the oocyte's own protein-synthesizing machinery, and the products are processed and incorporated into the oocyte plasma membrane. Using this method, oocytes have been induced to acquire many known neurotransmitter receptors and voltage-activated ion channels (for reviews, see Refs. 1–4). Once expressed in the large oocyte cells (over 1 mm in diameter), these receptors and ion channels may be studied with electrophysiological and biochemical techniques in a way much easier than using neuronal cells in the brain.

The most important step involved in the expression of receptors and ion channels in *Xenopus* oocytes is preparation of RNA. The quality of the mRNA dictates the number of neurotransmitter receptors and voltage-activated ion channels expressed in the oocytes. Procedures for isolation of total mRNA from the brain, size fractionation of mRNA, and synthesis of mRNA by *in vitro* transcription of cloned cDNAs are described in detail below.

Extraction of mRNA from Brain

For tissues with high RNase levels the guanidinium thiocyanate method is generally most effective in obtaining good quality of the RNA.[5] However, for fresh and frozen brains we found that the phenol–chloroform method, as a protein denaturant and as a means of separating the proteins from the nucleic acids, consistently provides us with active mRNA coding for neurotransmitter receptors and voltage-activated ion channels.[6] In our hands the RNA extracted from brains with this method is more effective in expressing functional neurotransmitter receptors and ion channels in the oocytes as compared to RNA prepared by the guanidinium thiocyanate method described by Chirgwin *et al.*[5] The drawback of the phenol–chloroform method is that it is time consuming, requiring many repetitive extraction steps involving centrifugal separation of aqueous and organic phases followed by careful collection of the aqueous layer. Thus, this method is not recommended for the simultaneous processing of a large

[1] R. Miledi, I. Parker, and K. Sumikawa, *in* "Fidia Award Lecture Series" (J. Smith, ed.), p. 57. Raven, New York, 1989.

[2] L. Kushner, J. Lerma, M. V. L. Bennett, and R. S. Zukin, *in* "Methods in Neurosciences" (P. M. Conn, ed.), Vol. 1, p. 3. Academic Press, San Diego, California, 1989.

[3] E. Sigel, *J. Membr. Biol.* **117**, 201 (1990).

[4] N. Dascal, *Crit. Rev. Biochem.* **22**, 317 (1987).

[5] J. J. Chirgwin, A. E. Przbyla, R. J. MacDonald, and W. J. Rutter, *Biochemistry* **18**, 5294 (1979).

[6] K. Sumikawa, I. Parker, and R. Miledi, *in* "Methods in Neurosciences" (P. M. Conn, ed.), Vol. 1, p. 30. Academic Press, San Diego, California, 1989.

number of brain samples for the isolation of RNA. More recently, we found that the rapid procedure combining guanidinium thiocyanate and phenol–chloroform extraction methods described by Chomczynski and Sacchi[7] results in active mRNAs that are almost comparable to those obtained by the phenol–chloroform method. The main advantage of this method is that only a single-step extraction is required, allowing the isolation of RNA from small quantities of brain tissue in a short time and simultaneous processing of a large number of brain samples.

Two different RNA extraction protocols which can be used for different experimental purposes are described. In addition, alternative techniques which are effective in isolating functional mRNA can be found in the chapter on tissue RNA as a source of ion channels ([17], this volume). It is recommended that disposable plastic gloves and eye protection should be worn when preparing or working with phenol–chloroform and guanidinium thiocyanate solutions. Disposable plastic gloves should also be worn to protect RNA from RNase during all RNA manipulations. Always use autoclaved sterile solutions, glassware, and plasticware. Plastic centrifuge tubes [Sepcor polypropylene centrifuge tubes with fluorocarbon O ring (Fisher, Pittsburgh, PA) or equivalent] are placed in 1 N HCl overnight and rinsed thoroughly with distilled water before autoclaving for 30 min. Glassware can be made RNase free by baking for at least 4 hr at 250° (or for at least 8 hr at 180°). Individually wrapped, sterile disposable glass pipettes and sterile disposable tips for pipettors are used for handling solutions.

Small-Scale Preparation of Total Brain mRNA for Injection into Oocytes by Acid Guanidinium Thiocyanate–Phenol–Chloroform Extraction

Several studies have indicated that cells in various brain areas express different receptor subtypes which can be distinguished by their structures and functional properties. It is therefore advisable to extract RNA from defined small brain areas rather than whole brains. This may also be important to reduce the potential problem of subunits from different brain areas possibly interacting to form functional receptors in oocytes in ways that do not occur in the brain.

Isolation of Total Brain RNA

Because the activity of mRNA isolated from frozen brains is not distinguishable from that of mRNA from fresh brains, dissected brain regions can be conveniently frozen in liquid nitrogen and stored at −80° until

---

[7] P. Chomczynski and N. Sacchi, *Anal. Biochem.* **162,** 156 (1987).

used. Although we describe here RNA isolation from 0.5 g of brain tissue, we have successfully used this method to isolate RNA from 50 mg of brain tissue.

1. Pipette 5 ml of solution D (4 M guanidinium thiocyanate, 25 mM sodium citrate, pH 7.0, 0.5% sarcosyl, 0.1 M 2-mercaptoethanol) into a 15-ml sterile disposable polypropylene tube (Becton Dickinson, Lincoln Park, NJ or equivalent; for smaller quantities of brain tissue 4-ml sterile disposable polypropylene tubes can be used instead).

2. Place the homogenizer probe [Polytron (Brinkmann, Westbury, NY) or its equivalent] into the tube. Drop 0.5 g of frozen brain tissue (do not thaw) into the mixing solution and homogenize for 1–2 min at room temperature.

3. Add 0.5 ml of 2 M sodium acetate (pH 4.0), 5 ml of phenol (saturated with water), and then 1 ml of chloroform–isoamyl alcohol mixture (49:1) to the homogenate with thorough mixing by inversion after the addition of each reagent.

4. Shake the suspension vigorously for 1 min and leave in ice water for 15 min.

5. Centrifuge in a swinging-bucket rotor (Beckman, Fullerton, CA, JS 13.1 rotor or equivalent; use the rubber adapter for a 15-ml tube) at 10,000 g for 20 min at 4°.

6. Use a glass pipette to collect the upper aqueous layer, avoiding white interface material, into a fresh 15-ml tube. *Keep in mind that quality of RNA is more important than quantity of RNA.*

7. Add 5 ml of 2-propanol to the collected solution and allow the RNA to precipitate at −20° for at least 1 hr.

8. Centrifuge to collect RNA in a swinging-bucket rotor (Beckman JS 13.1 or equivalent) at 10,000 g for 20 min at 4°.

9. Remove as much as possible of the supernatant and discard, saving the pellet. Dissolve the RNA pellet in 1.5 ml of solution D and transfer the RNA solution to a 4-ml sterile disposable polypropylene tube (Becton Dickinson or equivalent).

10. Add 1.5 ml of 2-propanol to the RNA solution and mix by inversion. Allow the RNA to precipitate again at −20° for at least 1 hr.

11. Centrifuge as in Step 8 (use the rubber adapter for a 4-ml tube) and discard as much as possible of the supernatant, saving the RNA pellet.

12. Wash the RNA pellet with 3 ml of 75% ethanol. For washing, grind the pellet first with a baked glass rod and then add 75% ethanol.

13. Centrifuge as in Step 8 for 10 min and discard the supernatant. Wash the pellet with 3 ml of 75% ethanol as in Step 12.

14. Centrifuge as in Step 8 for 5 min and discard the supernatant. Wash the pellet with 75% ethanol as in Step 12.

15. Centrifuge as in Step 8 for 5 min and discard as much as possible of the supernatant.

16. At this stage there are several options

16a. It is possible to leave the RNA pellet in 75% ethanol at $-20°$ indefinitely.

16b. If total RNA is used for injection, RNA should be cleaned by phenol–chloroform extraction. For this purpose, dissolve the RNA pellet (it is not necessary to first dry the pellet) in 500 μl of water and add 500 μl phenol–chloroform (1:1). After vortexing, centrifuge as in Step 8 for 5 min. Remove the upper aqueous layer, avoiding the white interface material, and collect into a fresh 4-ml tube. Add 25 μl of 4 M NaCl and 1.25 ml ethanol and allow the RNA to precipitate at $-20°$ for 1 hr. Centrifuge as in Step 8 and discard the supernatant. Wash the RNA pellet with 75% ethanol and dry the pellet. Dissolve the RNA pellet in 500 μl water.

16c. Dissolve the RNA pellet (it is not necessary to first dry the pellet) in 500 μl cold 20 mM HEPES–NaOH (pH 7.5) and use immediately for isolation of poly(A)$^+$ mRNA as described below.

The concentration of RNA can be determined by measuring the OD_{260} of 10 μl of RNA solution. An OD_{260} of 1 corresponds to approximately 40 μg/ml RNA. About 0.5 mg of total RNA should be recovered when starting with 0.5 g of brain tissue. The OD_{260}/OD_{280} ratio, which reflects the degree of contamination by protein, should be about 2.0.

Purification of Poly(A)$^+$ mRNA by Chromatography on Oligo(dT)-cellulose

Because the great variety of receptors and ion channels expressed by brain mRNA in oocytes have all been induced by poly(A)$^+$ mRNA, the removal of unwanted RNA species from the poly(A)$^+$ mRNA by using oligo(dT)-cellulose is the next stage. However, if one is using a small amount of brain tissue, the amount of RNA recovered may not be sufficient for isolation of poly(A)$^+$ mRNA. In this case, total RNA can be directly injected into oocytes. All the following manipulations are carried out at room temperature.

1. Suspend 0.2 g oligo(dT)-cellulose (Sigma, St. Louis, MO, or equivalent) in autoclaved water.

2. Pour into a column (Bio-Rad, Richmond, CA, Econo-column, 0.7 × 10 cm, or equivalent) which has been autoclaved and connected to a UV monitor (Pharmacia, Piscataway, NJ, or equivalent) via autoclaved silicone tubing.

3. Wash the column and all of the connecting system successively with (i) 10 ml of 0.1 N NaOH, (ii) 10 ml of sterile water, (iii) 10 ml of 0.5 M KCl, 5 mM HEPES–NaOH (pH 7.5).

4. Add an equal volume of 1 M KCl to the RNA solution to bring it to 0.5 M KCl and slowly apply the RNA solution onto the column.

5. Collect the unbound RNA, which is eluted, and reapply it to the column.

6. Wash the column with 0.5 M KCl, 5 mM HEPES–NaOH (pH 7.5) until the OD_{260} of the effluent is close to zero (the flow rate can be increased to save time).

7. Wash the column with 0.1 M KCl, 5 mM HEPES–NaOH (pH 7.5) as in Step 6.

8. Elute the bound RNA with 5 mM HEPES–NaOH (pH 7.5), and collect the RNA peak fraction into a 15-ml sterile, siliconized Corex tube on ice. Do not try to recover all RNA fractions. The volume of the mRNA solution should be minimized to ensure good recovery of the mRNA by ethanol precipitation.

9. Add 4 M NaCl to the RNA solution to make a 0.2 M final concentration and precipitate the RNA with 2.5 volumes of ethanol at $-20°$ overnight.

10. Collect the RNA by centrifugation in a swinging-bucket rotor (Beckman JS 13.1 or equivalent) at 10,000 rpm for at least 0.5 hr at 4°.

11. Discard the supernatant and rinse the pellet carefully with 5 ml of cold ($-20°$) 75% ethanol. Discard the ethanol (at this stage the mRNA can be stored indefinitely in 75% ethanol at $-80°$).

12. Dry the RNA pellet and dissolve in 40 μl of sterile water. The OD_{260} of a small sample should be measured to determine the concentration of RNA. The recovery of RNA is normally 2–5% of the RNA applied onto the column. The mRNA solution should be divided into small aliquots in sterile microcentrifuge tubes and stored at $-80°$. This is to avoid repeated freezing and thawing of the RNA solution which may result in RNA degradation and a gradual loss of some receptor and ion channel expression.[8]

Large-Scale Preparation of Brain mRNA for Injection of Size-Fractionated mRNA into Oocytes

If receptor or ion channel molecules being studied consist of a single subunit or several subunits encoded by mRNAs of very similar sizes, the use of fractionated rather than whole mRNA offers an important advantage for the study of receptors and ion channels expressed in oocytes. Size fractionation produces a partial purification of receptor and ion channel

---

[8] E. Sigel, *J. Physiol. (London)* **386,** 73 (1987).

mRNAs,[9,10] so that larger numbers of receptors and ion channels can be expressed in oocytes for characterization. However, it should be borne in mind that size fractionation could result in the complete loss of expression of functional receptors and ion channels if the functional expression requires two or more subunits encoded by mRNAs of very different sizes. Size fractionation of mRNA is also useful to examine the presence of other mRNA species of very different sizes encoding other subunits or enzymes that can modulate the properties of receptors and ion channels.[11]

Large-Scale Isolation of total RNA from Brains

All the following procedures are carried out at room temperature unless stated otherwise.

1. Place the homogenizer probe (Polytron or equivalent) in 0.1% (v/v) diethyl pyrocarbonate (DPC; Sigma) for 30 min and rinse thoroughly with sterile water.

2. Dissolve 500 g phenol (loose crystal, Fisher) in 55 ml of homogenization buffer [200 mM Tris-HCl (pH 9.0), 50 mM NaCl, 10 mM EDTA, 0.5% (w/v) sodium dodecyl sulfate (SDS)] at 37° just before use and add 0.55 g 8-hydroxyquinoline.

3. Pipette 100 ml of homogenization buffer into an autoclaved 250-ml centrifuge tube. Add 0.1 g heparin (Sigma) and 100 ml phenol.

4. Place the DPC-treated homogenizer probe into the 250-ml centrifuge tube and start mixing the solution. Drop 10 g of frozen brain tissue (do not thaw) into the mixing solution and homogenize for 2–3 min.

5. Cap the tube and shake vigorously for 5 min.

6. Centrifuge in a rotor (Beckman JA-14 rotor or equivalent) at 18,000 g for 15 min at about 15°.

7. Remove the bottom phenol layer with a glass pipette and discard (save the top layer including the fluffy white material).

8. Add 100 ml chloroform and shake vigorously for 5 min. Centrifuge as in Step 6.

9. Remove and discard the bottom chloroform layer, saving the upper layer and the fluffy white material.

10. Repeat Step 8.

[9] K. Sumikawa, I. Parker, and R. Miledi, *Proc. Natl. Acad. Sci. U.S.A.* **81,** 7994 (1984).

[10] A. L. Goldin, T. Snutch, H. Lubbert, A. Dowsett, J. Marshall, W. Auld, W. Downey, L. C. Fritz, H. A. Lester, R. Dunn, W. A. Catterall, and N. Davidson, *Proc. Natl. Acad. Sci. U.S.A.* **83,** 7503 (1986).

[11] V. J. Auld, A. L. Goldin, D. S. Krafte, J. Marshall, J. M. Dunn, W. A. Catterall, H. A. Lester, N. Davidson, and R. J. Dunn, *Neuron* **1,** 449 (1988).

11a. If the fluffy white material is packed into a narrow band forming a small interface layer, collect the upper aqueous layer with a glass pipette (avoid taking the white fluffy material forming the interface layer) into a fresh 250-ml centrifuge tube (Sepcor) containing 100 ml phenol–chloroform (1:1). Shake vigorously for 5 min.

11b. If there is a large interface, remove the chloroform layer and repeat Step 8. The fluffy white material must be packed into a narrow interface band before collecting the aqueous layer.

12. Centrifuge as in Step 6. With a glass pipette transfer the upper aqueous layer, without taking any of the white interface material, to a fresh 250-ml centrifuge tube (Sepcor) containing 100 ml of phenol–chloroform (1:1). Shake hard for 5 min and centrifuge as Step 6.

13. Repeat Step 12 until the interface is gone (usually 2–3 extractions).

14. Collect the upper aqueous layer into a fresh 250-ml centrifuge tube (Sepcor) containing 100 ml chloroform. Shake vigorously for 5 min and centrifuge as Step 6.

15a. If there is a white interface, repeat Step 14.

15b. If there is no white interface, collect the upper aqueous layer with a glass pipette into two fresh centrifuge tubes (Sepcor) kept on ice. Add 4 M NaCl to make the final concentration 0.2 M, and then add 2.5 volumes of cold ($-20°$) absolute ethanol. Allow the DNA and RNA to precipitate at $-20°$ for at least 2 hr or at $-80°$ for at least 0.5 hr (it is possible to leave the preparation indefinitely at this stage).

16. Centrifuge to collect the DNA and RNA in a rotor (Beckman JA-14 or equivalent) at 18,000 g for 30 min at $4°$.

17. Pour off the ethanol supernatant and aspirate the remaining ethanol carefully.

18. Dissolve the DNA and RNA pellet in one centrifuge tube (it is not necessary to first dry the pellet) in 20 ml of cold 20 mM HEPES–NaOH (pH 7.5) and transfer the solution to another centrifuge tube to dissolve the pellet. With the solution on ice, slowly add 3.6 g of NaCl (use a baked spatula) to bring the solution to 3 M. Transfer the solution to a sterile 30-ml Corex tube and cover it with Parafilm. Leave overnight at $-15°$ or on ice (this step precipitates the RNA, leaving most of the DNA in solution).

19. Centrifuge to collect the RNA in a swinging-bucket rotor (Beckman JS 13.1 or equivalent) at 9000 rpm for 30 min at $4°$. Make pinholes in the Parafilm before centrifugation.

20. Discard the supernatant, saving the RNA pellet. Wash the pellet with 10 ml of cold 3 M sodium acetate (pH 6.0). For this purpose the pellet is first ground well with a baked glass rod, then sodium acetate is added and the tube is vortexed.

21. Centrifuge as in Step 19 for 5 min and discard the supernatant. Wash the RNA pellet again as in Step 20.

22. Repeat Step 21.

23. Centrifuge as in Step 19 for 5 min and remove most of the supernatant. Wash the pellet with cold ($-20°$) 75% ethanol as in Step 20.

24. Repeat Step 23.

25. Centrifuge as in Step 19 for 5 min and discard the supernatant. (If the RNA is not used immediately, store the RNA in 75% ethanol at $-80°$).

26. Dry the pellet under reduced pressure and dissolve the RNA pellet in 1 ml cold 20 mM HEPES – NaOH (pH 7.5). Measure the concentration of RNA as before. About 8 – 10 mg of total RNA should be recovered when starting with 10 g of brain tissue.

Large-Scale Purification of Poly(A)$^+$ mRNA by Chromatography on Oligo(dT)-cellulose

About 8 – 10 mg of total RNA can be processed with 1 g of oligo(dT)-cellulose in a column (Bio-Rad Econo-column, 1.0×10 cm, or equivalent) as described for small-scale isolation of mRNA. Ten milligrams of total RNA will yield about 400 μg of poly(A)$^+$ mRNA, which is sufficient for two size fractionations of mRNA by sucrose gradient centrifugation.

Preparation of Size-Fractionated mRNA by Sucrose Gradient Centrifugation

The resolving power of sucrose gradient centrifugation is much less than that of gel electrophoresis. However, this method offers an important advantage over gel electrophoresis, namely, fractionated mRNA can be recovered easily and efficiently without losing translational activity.[9,10] The sucrose gradients may be prepared with a gradient maker or using frozen step gradients by the method of Luthe.[12] The latter method is easier and more reproducible. The following procedure is the modification of Luthe (kindly provided by M. M. Panicker) and provides equally good fractions as compared to sucrose gradients prepared with a gradient maker (M. M. Panicker and A. Morales, unpublished results, 1988, and also see Ref. 13).

1. Treat centrifuge tubes for the Beckman SW 40 Ti (or equivalent) with 0.1% (v/v) DPC overnight, and rinse thoroughly with autoclaved water before use.

[12] D. S. Luthe, *Anal. Biochem.* **135**, 230 (1983).

[13] J. A. Williams, D. J. McChesney, M. C. Calayag, V. R. Lingappa, and C. D. Logsdon, *Proc. Natl. Acad. Sci. U.S.A.* **85**, 4939 (1988).

2. To make 10–30% sucrose gradients, prepare 10% (w/w), 14% (w/w), 18% (w/w), 22% (w/w), 26% (w/w), and 30% (w/w) sucrose solutions in 10 mM HEPES–NaOH (pH 7.5), 1 mM EDTA 0.1% (w/v) lithium dodecyl sulfate (LDS) and treat with 0.05% (v/v) DPC overnight. Autoclave the DPC-treated sucrose solutions for 15 min (autoclaving helps to remove the DPC, which can interfere with translation of the mRNA).

3. Place the required number of centrifuge tubes in powdered dry ice (use an ice crusher).

4. Pipette 2 ml of 30% sucrose solution into the bottom of each tube (do not add the solution along the sides). Once the solution freezes (this takes a few minutes), pipette 2 ml of the next sucrose solution (26%) onto it.

5. Pipette 2 ml of the sucrose solutions in order of decreasing concentration as in Step 4. The final step (10%) is omitted, because the volume of the solution increases when frozen. The gradients are then stored at −80° until used.

6. Prior to using the gradients, thaw the sucrose gradients in centrifuge tubes at 4° and overlay with 10% sucrose solution (leave enough space for the mRNA samples, 200 μl).

7. Leave the sucrose gradients at 4° for at least 12–14 hr, which allows formation of continuous sucrose gradients.

8. Heat the poly(A)$^+$ mRNA solution (180–200 μg in 200 μl) to 65° for 5 min, cool on ice, and load onto the sucrose gradient.

9. Centrifuge immediately for 15–19 hr at 2°–4° in a Beckman SW 40 Ti rotor at 39,000 rpm.

10. Collect fractions (about 0.4 ml) in sterile microcentrifuge tubes. Add 20 μl of 4M NaCl and 1 ml of ethanol. Allow the mRNA to precipitate at −20° overnight.

11. Centrifuge in a swinging-bucket rotor (Beckman JS 13.1 or equivalent) at 11,000 rpm for 1 hr at 2°. Discard the supernatant and rinse the mRNA pellets twice with 1 ml of cold (−20°) 75% ethanol.

12. Dry the mRNA pellets under reduced pressure and dissolve in 10–20 μl of autoclaved water. Store at −80° until used for injection.

Preparation of Size-Fractionated mRNA by Methylmercury Gel Electrophoresis

Gel electrophoresis allows the greatest resolution of RNA size fractions.[14] However, the isolation of translationally active mRNA from denaturing gels is significantly less reproducible than the isolation of active

[14] H. Lübbert, B. J. Hoffman, T. P. Snutch, T. van Dyke, A. J. Levine, P. R. Hartig, H. A. Lester, and N. Davidson, *Proc. Natl. Acad. Sci. U.S.A.* **84**, 4332 (1987).

RNA from sucrose gradients. There are two major problems: reversal of the denaturation (necessary to get accurate size separation) and removal of residual agarose. Although a variety of procedures can be used successfully to denature RNA for gel electrophoresis (glyoxal, formaldehyde, formamide, methylmercury), methylmercury denaturation is the one that can be most reproducibly reversed. Electroelution is probably the most effective procedure for removal of agarose, but more recent techniques involving reversible binding of the RNA, such as RNaid from BIO 101 (La Jolla, CA), may work well. The following procedure was developed by Terry Snutch in the laboratories of Norman Davidson and Henry Lester at the California Institute of Technology (Pasadena, CA). While effective, it still does not result in active mRNA in all cases.

1. To pour an 8 cm 1% agarose gel, mix the following: 0.4 g agarose, 2 ml of 20 × running buffer (see below), 0.4 ml of 10% LDS, and 37.4 ml water. Boil until the agarose is completely melted, then add 0.2 ml of 1 M methylmercury.

2. Prerun the gel at 60 V for 1 hr at 4° in 1 × running buffer, then change the running buffer.

3. Prepare the RNA samples as follows:

Diagnostic sample	Preparative sample
12 µl RNA	100 µl RNA
0.6 µl of 20× running buffer	5 µl of 20× running buffer
0.2 µl of 10% LDS	1 µl of 10% LDS
Heat to 60° for 2 min	Heat to 60° for 2 min
Add 4 µl loading buffer	Add 33 µl loading buffer

4. Electrophorese at 40 V at 4° until the samples enter the gel, then increase the voltage to 60 V. Manually circulate the buffer every 30 min, and stop the electrophoresis when the bromphenol blue is 2 cm from the bottom (about 3.5 hr).

5. It is probably a good idea to stain a portion of the gel to determine the effectiveness of the fractionation. To do this cut off the portion to be stained and treat in the following series of solutions: 0.1 M ammonium acetate for 1 hr, 0.1 M ammonium acetate with 1–2 µg/ml ethidium bromide for 1 hr, and 0.1 M ammonium acetate for 1 hr. The remainder of the gel (not treated with ammonium acetate or ethidium bromide) should be used for elution as follows (see Fig. 1).

6. If the horizontal tube device is to be used, fill the bottom of the tube with 1% agarose in 0.1% sarkosyl, 0.02% sodium azide. The chamber can

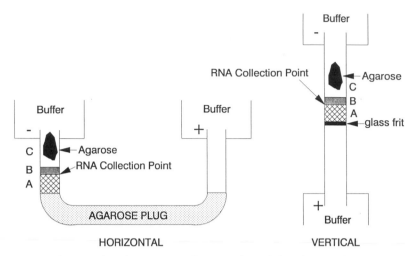

FIG. 1. Diagram of elution chambers for extraction of biologically active mRNA from methylmercury gels.

be stored with the agarose plug in this solution. Before use prerun the bridge in $1 \times$ Tris–glycine buffer at 200 V for 1 hr, then change the buffer. The vertical gel chamber requires no pretreatment, but it should be made free of RNase by either baking or storage in 0.1% sarkosyl, 0.02% sodium azide.

7. Overlay the agarose plug in the U-shaped tube or the glass frit in the vertical tube on the elution side (negative side) as follows (also see Fig. 1): (A) 2 ml of high sucrose buffer; (B) 0.25 ml of $1 \times$ Tris–glycine buffer with 10% sucrose, 1 mM dithiothreitol (DTT); (C) $1 \times$ Tris–glycine buffer with 1 mM DTT (to the top).

8. The agarose gel slice should be placed in solution C in the chamber (it should float in the solution, not touching solution B). Add 10 μl of 0.5% bromphenol blue near the gel slice. This will serve as a marker for successful electroelution.

9. Run at 2 mA per tube for 0.5 hr, then at 4 mA per tube for 2.5 hr, all at 4°. The RNA should electroelute out of the agarose and through solution B. It will be trapped at the interface between solutions B and A.

10. Remove the RNA solution from the interface between B and A, precipitate, and use.

Notes. Different lots of methylmercury can vary in quality, so each lot should be tested before use. In addition, methylmercury is a potent neurotoxin and should be handled with care. All of the electrophoresis and electroelution steps are carried out at 4° to minimize RNA degradation.

The electroelution devices described here were custom made, but Bio-Rad sells a vertical electroeluter which should work well.

Reagents

20× running buffer
    400 m*M* Boric acid, 6.2 g
    4 m*M* EDTA, 0.372 g
    20 m*M* Lithium hydroxide, 0.12 g
    Adjust the pH of a 1 : 20 dilution to 7.8 and add 0.1% LDS to the 1× running buffer
Loading buffer
    50 m*M* Methylmercury, 50 μl of 1 *M*
    40% Glycerol, 0.4 ml
    0.05% Bromphenol blue, 0.1 ml of 0.5%
    0.05% Xylene cyanol, 0.1 ml of 0.5%
    1× Running buffer, 50 μl of 20× running buffer
    0.1% LDS, 10 μl of 10% LDS
    Water, 0.3 ml
10× Tris–glycine buffer
    0.25 *M* Tris, 15.12 g
    0.75 *M* Glycine, 28.12 g
    1% Sarkosyl, 5 g
    Concentrated HCl, 14 ml (to pH about 7.4)
    Water, to 500 ml total volume
High sucrose buffer
    2 *M* Sodium chloride, 5.84 g
    20% Sucrose, 10 g
    0.1% Sarkosyl, 0.5 ml of 10% solution
    1 m*M* DTT, 50 μl of 1 *M* solution
    Note: After storage at 4° some crystals (probably sarkosyl) may appear. Therefore, warm the high sucrose buffer to room temperature before use to clear the solution, then cool to 4° again.

Preparation of Synthetic mRNA for Injection into Oocytes

Functional mRNA can be prepared by *in vitro* transcription of cloned cDNAs with bacteriophage RNA polymerase in the presence of 5′ cap analog for expression in *Xenopus* oocytes.[11,15,16] Addition of the 5′ cap

---

[15] D. A. Melton, this series, Vol. 152, p. 288.
[16] M. M. White, K. M. Mayne, H. A. Lester, and N. Davidson, *Proc. Natl. Acad. Sci. U.S.A.* **82,** 4852 (1985).

structure appears to be essential for RNA stability and efficient expression of transcribed RNA in oocytes. Using this approach, many different receptors and ion channels have been expressed in *Xenopus* oocytes (for review, see Ref. 2). However, it should be noted that functional expression in oocytes does not necessarily mean that the same receptors or ion channels are also expressed in the brain. Many different subunit subtypes exist in brains, and single subunits appear to be capable of forming functional receptors and ion channels in oocytes, leading to many combinatorial possibilities for the formation of functional receptor and ion channel structures. Furthermore, the subunit compositions of receptors and ion channels which operate in different brain areas are not well defined. Therefore, it is necessary to express different subunit combinations in oocytes and to compare the functional properties with those observed in cultured neurons to determine the biologically important forms.

In Vitro Synthesis of Capped RNA from Cloned cDNAs

Once a full-length cDNA for a neural receptor or ion channel has been cloned, it is easy to synthesize biologically active RNA from the clone. The coding region should be cloned 3' to a bacteriophage promoter, either that from SP6, T7, or T3 bacteriophage. All three bacteriophage polymerases work well to transcribe RNA uniquely from the appropriate promoter. However, the T7 polymerase has been cloned and so is significantly less expensive than SP6 polymerase. For many receptors the coding region can simply be inserted after the promoter for successful transcription and translation. In some cases, however, the resulting message will not translate well in oocytes, because of either poor stability or poor ribosome binding. In those cases the coding region can be cloned in a plasmid designed for expression in *Xenopus* oocytes, such as pSP64T[17] or pBSTA.[18] These plasmids contain the *Xenopus* β-globin 5' and 3' untranslated mRNA regions, with either an SP6 promoter (pSP64T) or a T7 promoter (pBSTA) on the 5' end and a poly(A) tail on the 3'end. This enhances both stability and translatability of some messages and can result in very significant increases in level of expression.

The plasmid DNA containing the insert should be cut with a restriction enzyme which linearizes the plasmid past the 3' end of the coding region. Both pSP64T and pBSTA have polylinker regions after the poly(A) tail which can be used for this purpose. For linearization it is best to use a restriction enzyme that leaves either a 5' overhang or a blunt end, since a 3' overhang can function as a primer for synthesis in the wrong direction,

---

[17] P. A. Krieg and D. A. Melton, *Nucleic Acids Res.* **12,** 7057 (1984).
[18] K. J. Kontis and A. L. Goldin, unpublished (1990).

making antisense RNA which has the potential to interfere with translation of the mRNA. After linearization the DNA should be extracted with phenol – chloroform – isoamyl alcohol, precipitated with ethanol, and resuspended in RNase-free water for use as a transcription template. The transcription reaction is carried out as described below to make RNA for use either as a probe (highly radioactive) or for translation (full-length).

1. The reactions are set up at room temperature as shown in the following tabulation.

	Volume (μl)	
Reagent	Probe	Full length
10× SP6/T7 buffer	2	5
10× ATP, CTP, UTP (5 mM each)	2	5
10× GTP (5 mM)	2	1
10× GpppG (5 mM)	—	5
1 M DTT	0.2	0.5
RNasin (40 U/μl)	0.5	1.25
[^{32}P]CTP (100 μCi/10 μl)	10	—
[^{32}P]CTP (diluted 1:10)	—	1
H$_2$O	1.3	26
Linearized DNA (1 μg/μl)	1	1–5
SP6 or T7 polymerase (20 U/μl)	1	2–5
Total reaction volume	20	50

2. Incubate the reaction at 37° for 1 hr for the probe reaction, for 2–3 hr for full-length transcripts. Additional enzyme can be added after 1 hr to the full-length reaction, which may increase the yield.
3. Add RNasin to 1 unit/μl.
4. Add 1 μl DNase (RNase free, either 1 U/μl or 1 mg/ml).
5. Digest at 37° for 10 min.
6. Add water to 75 μl total volume.
7. Add 1 μl of yeast carrier RNA (10 mg/ml) to the probe reaction (none to the full-length reaction).
8. Extract once with phenol – chloroform – isoamyl alcohol.
9. Purify the transcripts on a spun column[19] and freeze the eluate. The spun column can be either Sephadex G-50 or Sephacryl S-400, which also removes low molecular weight RNA [less than about 400 base pairs (bp)]. The full-length capped transcript should also be precipitated with ethanol 1–2 times before injection. After the first precipitation, resuspend the RNA in 1 mM Tris-HCl, pH 6.5. The buffered pH appears to be beneficial to the health of the oocytes following injection. The concentration will

depend on the activity of the mRNA and can vary at least from 1 μg/μl to 1 ng/μl.

Reagents

10 $\times$ SP6/T7 butter
    400 mM Tris-HCl, pH 7.5
    60 mM MgCl$_2$
    20 mM Spermidine
    100 mM NaCl

Notes. The unlabeled nucleotides should be dissolved as concentrated stocks (100 mM) in RNase-free 0.1 M Tris-HCl, pH 7.5, since the acidic nucleotides are unstable in water. The labeled CTP from Amersham (Arlington Heights, IL) or New England Nuclear (Boston, MA) at a specific activity of 400 Ci/mmol works well. However, old labeled CTP can inhibit the reaction owing to breakdown products, so it is a good idea to use fresh [32P]CTP (within a few days for a probe reaction, or within 4 weeks for a full-length reaction). SP6 and T7 polymerase from New England Biolabs (Beverly, MA), U.S. Biochemicals (Cleveland, OH), Pharmacia, and Promega (Madison, WI) have all been satisfactory. RNasin is obtained from Promega Biotec and is used to inhibit RNase activity. GpppG can be obtained from New England Biolabs or from Pharmacia. It need not be used in the methylated form, as it will become methylated in either the cell or the oocyte, but the methylated cap analog from New England Biolabs is the same price as the nonmethylated form. RNase-free DNase is available from Promega Biotec. It is important to set up the reaction at room temperature (not on ice), as it is possible for the DNA and spermidine to precipitate out at 4°.

The specific activity of the probe should be approximately 6.7 $\times$ 10^8 disintegrations/min (dpm)/μg of RNA, assuming [32P]CTP at a specific activity of 400 Ci/mmol. The yield of the full-length transcription reaction can be calculated by counting a 1-μl aliquot of the reaction mix before and after the spun column. Approximately 0.33 μg of RNA is synthesized for each 1% incorporation of the label, assuming a 50-μl reaction contains 0.5 mM CTP. If it is desirable not to have any radioactivity in the full-length transcript, the reaction can be performed without [32P]CTP. In this case the final yield of product can be quantitated by absorbance at 260 nm (be sure to use RNase-free cuvettes) or by comparing the ethidium bromide fluorescence of spots of diluted RNA with spots of diluted RNA standards.[19]

[19] J. Sambrook, E. F. Fritsch, and T. Maniatis, "Molecular Cloning: A Laboratory Manual" 2nd Ed. Cold Spring Harbor Laboratory, Cold Spring Harbor, New York, 1989.

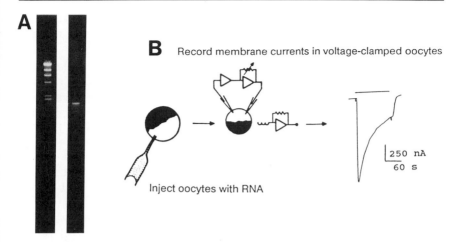

FIG. 2. Expression of GABA$_A$ receptors in *Xenopus* oocytes. GABA$_A$ receptor α_1, β_1, γ_2, and δ subunit cDNAs were cloned by the polymerase chain reaction and used for *in vitro* synthesis of mRNAs. Oocytes were injected with a mixture of mRNAs and examined using electrophysiological techniques. (A) Agarose gel electrophoresis of γ_2 subunit cDNA synthesized by the PCR (right lane) and DNA–HindIII digest (left lane), visualized by ethidium bromide fluorescence. (B) GABA-activated membrane current recorded in oocytes. Oocytes were voltage clamped at -60 mV and exposed to GABA for the length of time shown by the bar.

Cloning Receptor cDNA by Polymerase Chain Reaction

Many cDNAs coding for neurotransmitter receptors and voltage-operated ion channels have been cloned, and their DNA sequences are now available. Thus, cDNA clones encoding receptors and ion channels can be obtained relatively easily by the polymerase chain reaction (PCR).[20,21] The following procedures describe a method for obtaining a cDNA clone coding for the γ_2 subunit of the γ-aminobutyric acid (GABA) receptor by the PCR (Fig. 2). A similar approach can be used to isolate other receptor and ion channel cDNAs.

Although the standard condition used here works well for different templates and oligonucleotide primers, it may not be optimal for any particular combination. We normally synthesize two oligonucleotides identical to the 5' end sequences and two oligonucleotides complementary to the 3' end sequences. At least one of four combinations usually gives

[20] R. K. Saiki, D. H. Gelfand, S. J. Stoffel, R. Higuchi, G. T. Horn, K. B. Mullis, and H. A. Erlich, *Science* **239**, 487 (1988).

[21] C. M. Gall, K. Sumikawa, and G. Lynch, *Eur. J. Pharmacol. Mol. Pharmacol. Sect.* **189**, 217 (1990).

sufficient amounts of amplified target cDNA for cloning under the standard conditions described here. For methods to optimize the reaction conditions, see Ref. 19.

These conditions work well for a target sequence less than 2 kilobases (kb) in length. Although we have successfully cloned the kainate receptor cDNA, which is 2.8 kb in length,[21] the amplification of target sequences over 2 kb in length is normally inefficient (at least in our hands). For the cloning of longer target sequences, the sequence can be amplified in two parts having overlapping sequences which contain a unique restriction site for ligation.

1. The first step is kinasing oligonucleotide primers, oligo 4 (5'-TCT TCT GCA ACC AGA GGC G-3') and oligo 6 (5'-CTA GGT GAG ACG GAG AAA CA-3'). Oligonucleotides used for priming the PCR and cDNA synthesis are 19–22 nucleotides in length [$T_d = 2°$(number of A + T residues) + 4°(number of G + C residues) = $55° - 60°$]. Sequences containing internal complementarity should be avoided.

Mix the following in an RNase-free tube: 1 μl (1 μg) oligo 4 or oligo 6, 1 μl of 10× kinase buffer (0.5 M Tris-Cl, pH 7.6, 0.1 M MgCl$_2$, 50 mM DTT), 1 μl of 1 mM ATP, 6 μl water, and 1 μl (10 units) T4 polynucleotide kinase. Incubate at 37° for 30 min and at 65° for 10 min.

2. For cDNA synthesis with oligo 6 primers, brain mRNA isolated by the methods described above can be used for the efficient synthesis of first strand cDNA.

Mix together the following in an RNase-free tube: 2 μg mRNA, 2 μl of 50 mM Tris, pH 7.5, and water to 6.5 μl. Heat to 65° for 5 min. Place on ice. Add 4 μl of 5 × RT buffer (250 mM Tris-Cl, pH 8.3, 40 mM MgCl$_2$, 250 mM KCl), 1 μl of 6 mM DTT, 2 μl of 20 mM deoxynucleotide triphosphate (dNTP) mix, 0.5 μl RNasin (40 U/μl), 0.5 μg kinased oligo 6 (5 μl kinase reaction mix), and 1 μl reverse transcriptase (20 U/μl). Incubate at 37° for 30 min and then at 65° for 10 min.

3. For the PCR reaction, add 1.7 μl kinased oligo 4 (reaction mix), 1.7 μl kinased oligo 6 (reaction mix), 10 μl of 10× reaction buffer (100 mM Tris-Cl, pH 8.3, 500 mM KCl, 15 mM MgCl$_2$ 0.1% (w/w) gelatin, 16 μl of 1.25 mM dNTP mix, 5 μl cDNA synthesis reaction mixture, and water to 100 μl. Heat to 94° for 5 min. Cool to 72°. Add 0.5 μl (2.5 U) Taq polymerase, and mix briefly. Overlay with 50 μl mineral oil. Run the PCR for 30 cycles with the following cycle profile: 95° for 1 min, 50° for 2 min, and 72° for 3 min. Extract with 100 μl chloroform and save the top phase. Add 2 μl of 0.5 M EDTA and vortex. Add 1 volume of 4 M ammonium acetate. Add 4 original volumes ethanol. Precipitate overnight at $-20°$.

Methods for cloning PCR products in transcription vectors using linkers or adaptors can be found in Ref. 19. A more versatile system, namely, the TA cloning kit from Invitrogen (San Diego, CA) has recently been introduced which may be an easier alternative to methods using linkers and adapters. This system utilizes the fact that thermophilic enzymes used for PCR amplification have inherent terminal transferase activity, resulting in the generation of PCR products containing 3' overhangs composed of a single deoxyadenylate residue which can be directly ligated to a vector containing single 5'-deoxythymidylate overhangs. Alternatively, oligonucleotide primers containing unique restriction sites can be used for amplification of target sequences,[22] eliminating the ligation step with linkers or adapters.

Acknowledgments

The authors wish to thank Drs. Terry Snutch, Jeff Hoger, Henry Lester, and Norman Davidson for the methylmercury gel electrophoresis and electroelution procedure. A.L.G. is a Lucille P. Markey Scholar. Work in the authors' laboratories is supported by grants from the U.S. National Institutes of Health (NS-25928, NS-27341, and NS-26729), and Muscular Dystrophy Association, the Lucille P. Markey Charitable Trust, and the March of Dimes Basil O'Connor Starter Scholar Program.

[22] M. A. Frohman, M. K. Dush, and G. R. Martin, *Proc. Natl. Acad. Sci. U.S.A.* **86,** 8998 (1988).

[17] Tissue RNA as Source of Ion Channels and Receptors

By Terry P. Snutch and Gail Mandel

Introduction

The many attributes of the *Xenopus* oocyte expression system have already been outlined by Soreq and Seidman ([14] in this volume). In addition to healthy oocytes, the successful expression of exogenous excitability proteins requires intact, biologically active RNA. The RNA may be synthesized *in vitro* from a clone (see [16] and [18], this volume), or cellular RNA may be purified from a cell line or tissue known to express the ion channel or receptor of interest. If the molecule under study is a multisubunit complex, then coexpression of the complete complement of mRNAs is likely to be required to reconstitute all functional properties.[1,2]

Many excitable proteins are encoded by large RNAs [>5 kilobases (kb)], and special care must be taken to avoid degradation of the RNA. In addition, the health of the oocytes is impaired when they are microinjected with RNA samples contaminated by salts and detergents which were not removed during purification. This chapter outlines the conditions required to isolate high molecular weight cellular RNA suitable for exogenous expression. We also describe several protocols (gel electrophoresis, Northern blot analysis, RNase protection) useful for the initial characterization of purified RNA.

Enzymatic and Chemical Considerations

RNA is highly susceptible to both enzymatic and chemical degradation. During the initial homogenization ribonuclease is likely to be introduced by lysis of cells and organelles. In the later stages of purification ribonucleases are most likely to be introduced from exogenous sources. Chemical degradation of RNA can occur at all stages of RNA purification, handling, and storage. Because the initial homogenization is carried out in the presence of strong ribonuclease inhibitors, there is little concern for further precautions at this stage. However, once the partially purified RNA is removed from these strong denaturing agents, any further introduction of ribonuclease is detrimental to success. The most serious problems are ribonucleases which can contaminate glassware, solutions, equipment, and laboratory workers. A number of precautions can be taken to minimize ribonuclease contamination[3]:

1. Use only the highest quality reagents, for example, ultrapure molecular biology grade.

2. Laboratory glassware should be baked overnight at 180°. Alternatively, glassware can be treated by soaking with 0.1% diethyl pyrocarbonate (DEPC) for 60 min, rinsing several times with distilled water, and then autoclaving for 30 min.

3. Laboratory utensils for making solutions (e.g., stir bars and weighing spatulas) should be routinely baked overnight at 180°.

4. Ribonucleases are derived from living organisms, and they are

---

[1] M. M. White, K. Mixter-Mayne, H. A. Lester, and N. Davidson, *Proc. Natl. Acad. Sci. U.S.A.* **82,** 4852 (1985).

[2] D. B. Pritchett, H. Sontheimer, B. D. Shivers, S. Ymer, H. Kettenmann, P. R. Schofield, and P. H. Seeburg, *Nature (London)* **338,** 582 (1989).

[3] D. D. Blumberg, this series, Vol. 152, p. 20.

found on the skin and in the sweat of humans. To prevent introducing ribonuclease contamination, always wear disposable gloves when handling RNA samples and making stock solutions.

5. When ribonuclease-free glassware is not available, use disposable, sterile plasticware.

6. The pH meter is a likely source of ribonuclease and should not come into direct contact with solutions which are to be used after the initial homogenization of the tissue or cell line.

7. Keep a separate set of pipetting devices and an agarose gel system (combs, spacers, and gel box) for RNA use only.

8. We have not had problems with ribonuclease contamination when all laboratory glassware is routinely baked and when using autoclaved glass-distilled water and molecular biology grade reagents. However, if ribonuclease contamination is a problem, then solutions should be treated with DEPC. Using a stirrer, dissolve DEPC into the solution to a final concentration of 0.1% and let stand for 30 to 60 min. Although DEPC is a potent inhibitor of ribonucleases, it also causes the carboxymethylation of adenine bases and can result in inactivation of RNA.[4] Thus, any remaining traces of DEPC must be removed by autoclaving (DEPC decomposes into ethanol and CO_2). The extremely short half-life of DEPC in Tris buffers renders treatment of these solutions ineffective.[5]

Often overlooked is the fact that RNA is also highly susceptible to chemical degradation. It is important to avoid extremes in pH, high temperatures, and oxidizing agents.[6] At less than pH 5 some ribonucleases are highly active, whereas at high pH (>pH 9) RNA is subject to hydrolysis. Generally, RNA solutions should be maintained between pH 6.5 and 8.0, unless the RNA is being precipitated with ethanol (decrease to pH 5.2). It should be noted that at pH 7.6 poly(A)$^+$ RNA preferentially fractionates to the organic phase during phenol extractions.[7] This effect can be avoided by always extracting RNA with a 1:1 mixture of both phenol and chloroform.[8] Also to be avoided when handling RNA are polyvalent cations (such as manganese and copper) which can cause hydrolysis of RNA. If pure water is not available, RNA should be stored in a solution of 1–2 mM EDTA. The rates of both chemical and enzymatic degradation are in-

[4] L. Ehrenberg, I. Fedorcak, and F. Solymosy, *Prog. Nucleic Acid Res. Mol. Biol.* **16,** 185 (1974).
[5] S. L. Berger, *Anal. Biochem.* **67,** 428 (1975).
[6] R. M. Bock, this series, Vol. 12A, p. 218.
[7] G. Brawerman, J. Mendecki, and S. Y. Lee, *Biochemistry* **11,** 637 (1972).
[8] R. P. Perry, J. La Torre, D. E. Kelley, and J. R. Greenberg, *Biochim. Biophys. Acta* **262,** 220 (1972).

creased at elevated temperatures. Thus, thawed RNA samples should always be kept on ice, and stocks should be stored frozen at $-80°$.

Phenol must be equilibrated with aqueous buffer prior to use. Thaw a 100-g bottle of redistilled phenol [Bethesda Research Laboratories (BRL), Gaithersburg, MD, 5509UA] in a water bath at $65°$. Add a baked stir bar and approximately 100 ml of ribonuclease-free water. Add 20 ml of 2 M Tris (pH 7.5) and mix at room temperature for 20 min. Let the phases separate and then check the pH of the upper aqueous layer. If necessary adjust the pH to 7.5 with 5 M sodium hydroxide. The saturated phenol is stable at $4°$ for 3 to 4 weeks. For longer periods it should be stored in aliquots at $-20°$. If the phenol ever becomes pink or yellow it should be discarded. It is sometimes recommended that 8-hydroxyquinoline be added to a concentration of 0.1% to block oxidation of the phenol. In fact, this rather noxious agent only slows the oxidation process and, given that it is yellow in color, makes it impossible to determine when the phenol has become oxidized. Formamide should be deionized with a mixed bed resin (AG 501-X8, BioRad, Richmond, CA). Briefly, using a baked stir bar and beaker, mix 5 g of resin into 100 ml of formamide. Let stir for 30 to 60 min and then filter through Whatman (Clifton, NJ) No. 1 paper. Store the deionized formamide in aliquots at $-20°$.

Isolation of Biologically Active RNA

All cellular RNA purification protocols basically consist of two major steps: (1) initial homogenization in a solution which is designed both to disrupt cellular constituents, including RNA–protein complexes, and to inactivate ribonucleases rapidly, and (2) separation of cellular RNA from DNA, proteins, and polysaccharides. The homogenization solutions used in various procedures include strong chaotropic agents such as guanidinium thiocyanate and guanidine hydrochloride, moderately denaturing agents such as phenol and urea, and detergents such as sodium dodecyl sulfate (SDS) and sarkosyl. Historically, the denaturant was chosen to reflect the level of ribonuclease endogenous to the source tissue. In practice, however, it is prudent to use strong ribonuclease inhibitors regardless of the level of ribonuclease expected. Total cellular RNA can be separated from other cellular constituents by selective precipitation of RNA or by density gradient centrifugation. Although both separation techniques are effective, it should be noted that methods which rely on selective precipitation can result in varying yields, depending on such factors as the nature of the starting material and the degree of initial homogenization and subsequent shearing of cellular DNA.

In our experience, although may procedures yield RNA which is suitable for *in vitro* manipulations, they do not always result in RNA which is translatable in *Xenopus* oocytes. The following two protocols give good yields of high molecular weight RNA that is translatable in oocytes,[9,10] It is best to use fresh tissue; however, satisfactory results have also been obtained using samples quick frozen in liquid nitrogen and then stored at $-80°$ until required.

Guanidinium Thiocyanate/Cesium Chloride Method

The following method is modified from the protocol of Chirgwin *et al.*[11] and works well for RNA isolation from a variety of tissues and cell lines. The method is also useful for processing large numbers of samples.

1. Prepare 100 ml of GTC solution [$4M$ guanidinium thiocyanate, 25 mM sodium citrate (pH 7.0), 0.5% *N*-laurylsarcosine]. Add 0.7 ml of 2-mercaptoethanol just prior to use (final concentration is 100 mM).

2. Prepare a solution of 5.7 M CsCl containing 25 mM sodium acetate (pH 6.0) and 5 mM EDTA. Sterilize by filtration.

3. Homogenize the tissue or cells at room temperature in 10–15 ml of GTC solution per gram of sample. A Dounce homogenizer can be used for soft tissues and for cell lines, whereas a Polytron is often required for homogenization of fibrous tissues such as skeletal muscle. After homogenization, transfer the sample to a sterile centrifuge tube and remove any particulate material by centrifugation at 4000 rpm for 10 min at room temperature.

4. Pipette 3.5 ml of CsCl solution into a 14×89 mm polyallomer ultracentrifuge tube. Carefully layer the GTC supernatant onto the CsCl cushion and balance opposing tubes.

5. Centrifuge the SW41 rotor at 32,000 rpm for 20 hr at $20°$.

6. Carefully aspirate off the GTC and CsCl until just near the bottom of the tube. Decant off the remainder of the solution, and, keeping the tube inverted, cut off the tube below the CsCl cushion interface.

7. Resuspend the pellet in 250 μl of TE–SDS (10 mM Tris, pH 7.5, 1 mM EDTA, 0.1% SDS) and transfer to a sterile 1.5-ml microcentrifuge tube. Rinse the untracentrifuge tube with 150 μl TE–SDS and pool.

8. Precipitate the RNA by addition of 50 μl of 3 M sodium acetate (pH

[9] J. S. Trimmer, S. S. Cooperman, S. A. Tomika, J. Zhou, S. M. Crean, M. B. Boyle, R. G. Kallen, Z. Sheng, R.L. Barchi, F. J. Sigworth, R. H. Goodman, W. S. Agnew, and G. Mandel, *Neuron* **3,** 33 (1989).

[10] T. P. Snutch, J. P. Leonard, M. M. Gilbert, H. A. Lester, and N. Davidson, *Proc. Natl. Acad. Sci U.S.A.* **87,** 3391 (1990).

[11] J. M. Chirgwin, A. E. Przygyla, R. J. MacDonald, and W. J. Rutter, *Biochemistry* **18,** 5294 (1979).

5.2) and 2.5 to 3 volumes of ethanol. Incubate at $-20°$ overnight or on dry ice for 15 min.

9. Pellet the RNA in a microfuge for 15 min at $4°$. Remove the ethanol, wash once with 80% ethanol, and dry the pellet *only* enough to remove the last traces of ethanol. Dissolve the pellet in ribonuclease-free water at a concentration of 1 to 2 mg/ml. Store in aliquots at $-80°$.

Lithium Chloride/Urea Method

The following method is modified from the procedure of Auffray and Rougeon[12] and has the advantages that it does not involve a lengthy centrifugation through CsCl and that it can conveniently be used for large-scale RNA preparations (up to 50–60 g tissue). Total RNA isolated using this method may be contaminated by small amounts of genomic DNA (1–2%). If the RNA is to be utilized for polymerase chain reaction (PCR) analysis and primers which could distinguish between RNA and DNA products are not available, then the RNA should first be treated with ribonuclease-free DNase.

1. Prepare a solution of 6 M urea, 3 M lithium chloride, 0.5% N-laurylsarcosine, and 10 mM sodium acetate (pH 5.2). Cool to $4°$ prior to use.

2. Homogenize the tissue or cells with a precooled Dounce homogenizer in a volume of 8 to 10 ml per gram of tissue or per milliliter of packed cells. The solution will become quite viscous, and it is important to shear the DNA in order to maximize the amount of RNA recovered. Approximately 35 to 45 strokes is sufficient.

3. Place the mixture on ice in the cold overnight. Pellet the RNA in a high-speed centrifuge using baked Corex tubes at 8000 rpm for 30 min at $4°$.

4. Carefully decant off the supernatant and, with the tube inverted, wipe the sides of the tube with a tissue. Dissolve the pellet in 1–2 ml of cold 10 mM Tris (pH 7.5), 1 mM EDTA, 0.5% sarkosyl and transfer to a sterile disposable polypropylene tube. Pipetting the sample up and down several times is usually sufficient to dissolve the pellet.

5. Extract the sample with 1 volume of buffer-saturated pheno–Sevag (1:1; Sevag is chloroform–isoamyl alcohol, 24:1). For large-scale RNA preparations allow the sample to mix on a shaker in the cold for 15 min. For small-scale RNA preparations it is sufficient to add cold phenol–Sevag and mix by hand for 1–2 min. Separate out the phases by centrifugation in a $4°$ centrifuge for 5 min at 5000 rpm. Small-scale preparations can be accommodated in a microfuge.

[12] C. Auffray and F. Rougeon, *Eur. J. Biochem.* **107**, 303 (1980).

6. Transfer the upper aqueous layer to a clean tube and extract twice more with phenol–Sevag.

7. Extract the aqueous sample with 1 volume of Sevag and centrifuge as in step 5.

8. To the aqueous layer add 1/10 volume of 2.5 M sodium acetate (pH 5.2) and 2.5 volumes of ethanol. Store at $-20°$ overnight and then recover the RNA by centrifugation at 8000 rpm for 10 min.

9. Wash the RNA pellet once with 80% ethanol and centrifuge as in Step 8. Discard the ethanol, dry the pellet briefly, and dissolve the RNA at a concentration of 1 to 2 mg/ml in ribonuclease-free water. At this point the RNA is pure enough to perform Northern blots and to isolate poly(A)$^+$ RNA using affinity chromatography. For *in vitro* translation, PCR analysis, or injection into *Xenopus* oocytes, the total RNA should be reprecipitated as in Steps 8 and 9.

Prior to using the total RNA for oocyte injections the quality of the RNA should be examined. The absorbance ratio A_{260}/A_{280} of the purified RNA should be between 1.9 and 2.1. Lower ratios are usually indicative of either protein contaminants or incompletely removed reagents (e.g., sarkosyl and phenol both absorb light in the 260–280 nm range). The RNA should also be checked visually on an 0.8 to 1.0% nondenaturing agarose gel containing 1 μg/ml ethidium bromide. Alternatively, if an RNase-free gel box is not available, the RNA can be separated through a 0.8 to 1.1% agarose gel containing 1.1 M formaldehyde (see below). The total RNA should show two prominent bands representing the 28 and 18 S ribosomal RNAs. The relative intensity of the 28 to 18 S bands should be approximately 2:1, and there should be minimal smearing of lower molecular weight RNAs.

Isolation of Poly(A)$^+$ RNA

Most eukaryotic mRNAs contain stretches of 50 to 200 adenylic acid residues at the 3′ end [poly(A)$^+$ mRNA]. Commercially available cellulose resins which have covalently attached poly(dT) residues [oligo(dT)-cellulose type 3, Collaborative Research, Bedford, MA] can be used to isolate selectively mRNA from total cellular RNA.[13] With a small number of samples poly(A)$^+$ RNA can be isolated using a column chromatography protocol [columns either can be homemade, consisting of a sterile syringe plugged with baked glass wool, or they can be purchased already packed

[13] J. Sambrook, E. F. Fritsch, and T. Maniatis, "Molecular Cloning: A Laboratory Manual," 2nd Ed. Cold Spring Harbor Laboratory, Cold Spring Harbor, New York, 1989.

with oligo(dT)-cellulose; Sigma, St. Louis, MO]. For multiple RNA samples or when starting from small amounts of total RNA (<2 mg) it is practical to use a batch isolation protocol.[13]

Poly(A)$^+$ RNA can also be purified using poly(U)-Sepharose (Pharmacia, Piscataway, NJ). Owing to the collapsible nature of the matrix, only column chromatography can be performed with poly(U)-Sepharose.[14] Poly(U)-Sepharose columns have a much quicker flow rate than oligo(dT) columns and, additionally, the matrix binds poly(A)$^+$ RNA more efficiently than oligo(dT)-cellulose.

1. Equilibrate 1 g of poly(U)-Sepharose in 10 ml of $1\times$ binding buffer ($1\times$ binding buffer is 0.4 M NaCl, 10 mM Tris, pH 7.5, 1 mM EDTA, 0.1% SDS).

2. Prepare a column using a 10-ml disposable pipette in which the bottom has been plugged with baked glass wool. Form a column with the equilibrated poly(U)-Sepharose and then wash through a further 15 ml of $1\times$ binding buffer.

3. Heat the RNA sample (in sterile water or 1 mM EDTA at a concentration of 0.5 to 1 mg/ml) to 80° for 1 min. Cool on ice and then add 1 volume of $2\times$ binding buffer.

4. Pass the RNA mixture over the column and collect the elutant in a ribonuclease-free tube. Pass the elutant over the column once more and then wash the column with 10 ml of $1\times$ binding buffer.

5. Elute nonspecifically bound RNA by passing 20 ml of low salt buffer over the column (low salt buffer is 10 mM Tris, pH 7.5, 1 mM EDTA, 0.1% SDS).

6. Elute the poly(A)$^+$ RNA by passing 4 ml of room temperature 80% formamide (dionized), 10 mM Tris, pH 7.5, 1 mM EDTA through the column and collect the elutant in a sterile tube. Dilute the formamide by addition of 2 volumes of sterile water and precipitate the RNA by addition of 1/10 volume of 2.5 M sodium acetate (pH 5.2) and 2.5 volumes of cold ethanol. Incubate at $-20°$ overnight and then recover the poly(A)$^+$ RNA by centrifugation at 10,000 rpm for 20 min. Wash the RNA once with 80% ethanol, dry briefly, and then dissolve in ribonuclease-free water at a concentration of 1 mg/ml. Typically, 1.5–2% of the total RNA will be recovered as poly(A)$^+$.

The microinjection of exogenous RNA into *Xenopus* oocytes is described in [14] (in this volume). Under most circumstances robust responses are obtained by injection of approximately 100 to 200 ng of total RNA per oocyte (50 nl of a 2–4 μg/μl RNA solution). Injection of 50

[14] A. Jacobson, this series, Vol. 152, p. 254.

ng of poly(A)$^+$ RNA is sufficient for detection of many ion channels and neurotransmitter receptors. For reasons that are at present unclear, the injection of poly(A)$^+$ RNA results in responses that are only 3 to 4 fold greater than that obtained from injection of a similar amount of total RNA [a 10 to 12-fold increase in signal would be expected by poly(A)$^+$ enrichment]. However, in many instances the increase in signal obtained with poly(A)$^+$ RNA justifies the additional effort required for purification.

Denaturing Gel Electrophoresis and Northern Blot Analysis

Prior to injection into oocytes the quality of the purified RNA should be examined. First, the RNA should be examined visually for degradation using formaldehyde gel electrophoresis (Steps 1 to 5 below). Second, the RNA should be transferred to a nylon membrane and then hybridized either to a probe encoding the molecule under study (if available) or to a probe which is known to be expressed in the source tissue and is encoded by a relatively large RNA transcript (Northern blot[15]). For example, hybridization of a rat brain sodium channel cDNA probe to intact brain RNA should result in strong hybridization to 9- to 10-kb transcripts and show little evidence of a low molecular weight smear. If a probe is available for the gene under study, Northern blot analysis can also provide important information with respect to the size of the mRNA(s) encoding the gene product, the number of related mRNAs (usually an indication of gene families or alternative splicing), and the relative abundance of the transcript. In addition, Northern blot analysis using RNAs isolated from various tissues and developmental stages can give an indication as to which source will provide the most robust signal on injection of RNA into oocytes.

The following is a recipe for 150 ml of a 1% agarose gel containing 1.1 M formaldehyde.

1. Prepare 500 ml of 10× MOPS buffer (10× MOPS is 0.2 M MOPS, pH 7.0, 80 mM sodium acetate, 10 mM EDTA). Filter sterilize and store in a dark bottle at 4°.

2. Add 1.5 g of ultrapure agarose to 122 ml of distilled water and dissolve by heating in a microwave. Add 15 ml of 10× MOPS buffer and cool to 60° in a water bath. After adding a baked stir bar to the flask, pipette in 13.4 ml of formaldehyde and stir for 15–20 sec. Pour the gel in a fumehood and let set for 1 hr to overnight.

3. While the gel is setting prepare the RNA samples for loading. For

[15] P. Thomas, *Proc. Natl. Acad. Sci. U.S.A.* **77**, 5201 (1980).

each RNA sample mix 5 μl of 10× MOPS, 8.75 μl of formaldehyde, 25 μl of deionized formamide, and the RNA sample (in a volume of 11.25 μl). Use 25–30 μg of total RNA or 1–5 μg of poly(A)$^+$ RNA per lane. Heat the sample to 65° for 15 min and place on ice for 2 min. Add 2 μl of RNA loading buffer (50% glycerol, 10% Ficoll, 1 mM EDTA, 0.4% bromphenol blue, 0.4% xylene cyanol). For markers use 5 μg of synthetic RNA ladder (0.24 to 9.5 kb; BRL 5620SA). Treat the markers as above except that prior to heating add 1 μl of 1 mg/ml ethidium bromide. This obviates the need to stain the gel after running.

4. RNA bands will be fuzzy if, prior to loading the gel, the wells are not rinsed with 1× MOPS running buffer. Load the samples and electrophorese at 75–80 V until the bromphenol blue just reaches the bottom of the gel (4–5 hr for a 15-cm gel).

5. Carefully remove the gel and take a picture of the RNA markers with a ruler laid alongside as a reference.

6. Soak the gel in 1× TAE (40 mM Tris–acetate, 1 mM EDTA) for 15 min to remove most of the formaldehyde. Transfer the gel to a nylon membrane [Durlon, Strategene (La Jolla, CA); Nytran, Schleicher & Schuell (Keene, NH); Hybond-N, Amersham (Arlington Heights, IL)] by either capillary blot overnight or by electroblot using a commercial apparatus (Hoefer, San Francisco, CA). Fix the RNA to the membrane by UV cross-linking as described by the supplier.

7. Hybridization and washing conditions will depend on the characteristics of the probe and target sequences.[16] Generally, DNA probes are labeled by nick-translation or random priming, and hybridization is carried out at 42° in 50% formamide and an aqueous mixture consisting of 5× SSPE (1× SSPE is 0.18 M NaCl, 10 mM sodium phosphate, pH 7.4, 1 mM EDTA), 0.5% SDS, 0.2 mg/ml sheared, denatured salmon sperm DNA, 0.1 mg/ml yeast tRNA, and 2.5× Denhardt's [1× Denhardt's is 0.02% bovine serum albumin (BSA), 0.02% Ficoll, 0.02% polyvinylpyrrolidone]. Alternatively, hybridizations can be carried out at 68° to 70° in aqueous hybridization buffer alone. For synthetic RNA probes the high stability of RNA–RNA hybrids requires that hybridization be done under more stringent conditions, for example, in 50% formamide at 68°. Blots should be washed at 10° to 12° below the calculated T_m of the probe. At lower stringency hybridization and wash conditions nonspecific bands can appear, making the results unclear or misleading. In general, a low signal-to-noise ratio is obtained using short (<50 bases) oligonucleotide probes.

[16] G. H. Keller and M. M. Manak, "DNA Probes." Stockton Press, New York, 1989.

Ribonuclease Protection

In addition to Northern blot hybridization, another method of analyzing low abundance mRNAs, characteristic of most mammalian ion channels and receptor mRNAs, is by RNase protection.[17] This method is an extremely sensitive measure of mismatch produced between two distinct mRNAs and is therefore quite useful for discriminating between mRNAs produced by members of a multigene family. It is also useful for detection of the transcription start sites of mRNAs or for determination of intron/exon boundaries. The assay is based on the differential sensitivities of single-stranded and double-stranded nucleic acids to cleavage by certain ribonucleases. Specifically, "exactly matched" RNA hybrids are protected from digestion with ribonucleases A and T1, whereas regions of mismatch in imperfect hybrids are digested by these same ribonuclease activities. When the sequence of the RNase protection probe is known precisely, the size of the resulting "protected" fragments clearly reveals, by fractionation on denaturing sequencing gels, whether the probe and test mRNAs are identical or not. The primary advantage of the RNase protection technique over Northern blot analysis is that the former is much more sensitive. On the other hand, only Northern blot analysis can reveal the size of the mature test mRNA.

1. Antisense RNA probes are synthesized *in vitro* using either a commercial kit (Riboprobe, Promega, Madison, WI) or using the following protocol. Mix the following at room temperature: 2 μl of 10$\times$ transcription buffer, 2 μl of 5 mM ATP, 2 μl of 5 mM GTP, 2 μl of 5 mM CTP, 0.5 μl of 0.5 M dithiothreitol, 0.5 μl RNasin (30–40 U/μl), 10 μCi [α-^{32}P]UTP (400–800 Ci/mmol), 1 μl linearized plasmid DNA (1 μg/μl), and 1 μl of SP6, T3, or T7 RNA polymerase (30–40 U/μl). Incubate the sample at 37° for 1 hr. The 10$\times$ transcription buffer is 200 mM Tris (pH 7.5), 60 mM MgCl$_2$, 20 mM spermidine-HCl, and 100 mM NaCl. The cold nucleotide stocks are made up in sterile 70 mM Tris (pH 7.5).

2. Add 2 μl of DNase (RNase-free; 1000 U/ml or 1 mg/ml) and incubate at 37° for 15 min.

3. Bring the volume of the reaction to 100 μl with sterile water and extract once with 40 μl of phenol–Sevag (1 : 1), Vortex briefly and separate the phases by spinning in a microfuge for 3 min.

4. Remove the unincorporated radiolabeled nucleotides by spin column chromatography and precipitate the probe by addition of 10 μg carrier tRNA (Sigma), 0.1 volumes of 2.5 M sodium acetate, and 2.5 volumes of ethanol.

[17] K. Zinn, D. Dimaio, and T. Maniatis, *Cell (Cambridge, Mass.)* **34,** 865 (1983).

5. Recover the probe by centrifugation and dissolve in 10 μl of RNA gel buffer. Heat the sample to 90° for 5 min and run on a 6% polyacrylamide sequencing gel at 40–50 mA until the bromphenol blue is near the bottom of the gel.

6. Place the gel on a glass plate and cover with Saran wrap. Place a second glass plate on top of the wrapped gel and carry the sandwich to the darkroom. Remove the top glass plate and turn the gel face down onto a piece of film. Outline the glass plate with a marker and develop the autoradiogram. An exposure of 15–25 sec is sufficient to see the RNA band. Using the autoradiograph as a guide, excise the correct band and transfer to a sterile microfuge tube. Freeze the gel piece and then mince with a baked spatula. Add 1 ml of elution buffer (0.5 M ammonium acetate, 1 mM EDTA, 0.1% SDS) and shake at 37° for 4–6 hr.

7. Recover the eluted RNA by spinning the gel mixture in a microfuge at 4° for 15 min. Transfer the supernatant to a fresh tube and repeat the centrifugation. To the supernatant, add 10 μg of carrier tRNA and precipitate the probe with 2.5 volumes of ethanol. An alternative protocol for recovering the RNA is as follows: after the incubation in gel elution buffer pass the gel mixture through a prewetted Elutip-r prefilter (Schleicher & Schuell) and subsequently recover the RNA by ethanol precipitation.

8. Resuspend the probe in 25 to 50 μl of 5× hybridization buffer (200 mM PIPES, pH 6.4, 2 M NaCl, 5 mM EDTA) and count an aliquot in a scintillation counter. Adjust the volume of the probe so that there is 5×10^5 counts/min (cpm) per 6 μl of hybridization buffer. Add 24 μl of deionized formamide for each 6 μl of probe to yield the final hybridization mixture.

9. For each RNA sample to be analyzed, precipitate 10–50 μg of total or 1–10 μg of poly(A)$^+$ RNA and resuspend in 10 μl of hybridization mixture. Heat the samples at 90° for 5 min and then transfer to the final hybridization temperature (between 45° and 55°) for 8–16 hr.

10. To each sample add 350 μl of RNase solution (40 μg/ml RNase A, 2 μg/ml RNase T1 in 10 mM Tris, pH 7.5, 300 mM NaCl, 5 mM EDTA) and incubate at 30° for 1–2 hr. Destroy the RNase by addition of 20 μl of 10% SDS and 2.5 μl of proteinase K (20 mg/ml in water), then incubate at 37° for 15–60 min.

11. Extract the samples with 350 μl of phenol–Sevag and precipitate by addition of 10 μg of carrier tRNA and 1 ml of ethanol. Resuspend the pellet in 10 μl of gel loading buffer (80% formamide, 1 mM EDTA, 0.1% bromphenol blue, 0.1% xylene cyanol) and heat to 90° for 5 min prior to analyzing on a 6% acrylamide sequencing gel. Typical autoradiograph exposure times are 12 to 48 hr.

To demonstrate that protection has occurred, a sample of untreated probe should be run alongside the RNase-treated samples. Radiolabeled DNA size markers should also be included on the gel. A convenient way both to determine the sensitivity of the protection reaction and to quantitate the amount of target sequence in an RNA population is to generate nonlabeled sense RNA *in vitro* from the transcription plasmid containing the probe and then to titrate defined amounts of sense RNA against the radiolabeled antisense probe.

Other Considerations

Although there are numerous instances of the successful expression of ion channels and neurotransmitter receptors in *Xenopus* oocytes, it is still difficult if not impossible to detect the functional expression of certain conductances. The list of cloned receptors and channels which are poorly expressed in oocytes includes the eel sodium channel,[18] the rabbit skeletal muscle L-type calcium channel[19] and the human brain serotonin 1A receptor.[20] There are a number of possible explanations for inefficient or improper translation in oocytes. It is possible that the mRNA encoding the molecule under study is quite rare. In such instances, enrichment techniques such as poly(A)$^+$ RNA isolation and RNA fractionation methods (see [16] in this volume) may alleviate the problem. It may also be that the 5' noncoding region of the mRNA under study is actually detrimental to translation *in ovo*. Indeed, the enzymatic removal of upstream sequences has been shown to improve expression in oocytes dramatically.[21] In a number of other instances, the conductances induced in oocytes by exogenous RNA show alterations from that found in the wild-type tissue. For example, whereas the Shaker A channel is insensitive to charybdotoxin in insect cells, it is blocked by nanomolar concentrations of this toxin when expressed in oocytes.[22] Another example is the finding that the *Xenopus* oocytes do not perform the proper N-linked glycosylations of *Torpedo* nicotinic acetylcholine receptor subunits,[23] alluding to the fact that *Xenopus* oocytes do not always perform the correct posttranslational modifications[18] (see [45] in this volume).

[18] W. B. Thornhill and S. R. Levinson, *Biochemistry* **26**, 4381 (1987).
[19] E. Perez-Reyes, H. S. Kim, A. E. Lacerda, W. Horne, X. Wei, D. Rampe, K. P. Campbell, A. M. Brown, and L. Birnbaumer, *Nature (London)* **340**, 233 (1989).
[20] A. Fargin, J. R. Raymond, M. J. Lohse, B. K. Kobilka, M. C. Caron, and R. J. Lefkowitz, *Nature (London)* **335**, 358 (1988).
[21] M. M. White, L. Chen, R. Kleinfeld, R. G. Kallen, and R. L. Barchi, *Mol. Pharmacol.* **39**, 604 (1991).
[22] R. MacKinnon, P. J. Reinhart, and M. M. White, *Neuron* **1**, 997 (1988).
[23] A. L. Buller and M. M. White, *J. Membr. Biol.* **115**, 179 (1990).

[18] *In Vitro* Synthesis of RNA for Expression of Ion Channels in *Xenopus* Oocytes

By RICHARD SWANSON and KIMBERLY FOLANDER

Introduction

Xenopus oocytes are commonly used for the heterologous expression of cDNA clones encoding ion channels. The widespread use of this system is a result of the ability of the oocyte to faithfully translate RNAs encoding channels from a variety of sources, the ease with which cDNAs and RNAs can be introduced into the cells, and the suitability of the system for functional analysis by a variety of electrophysiological techniques. Expression can be effected either by nuclear injection of a cDNA itself, in an appropriate vector, or by cytoplasmic injections of RNA transcripts of a cDNA (cRNA). In this chapter, we review methods for *in vitro* synthesis of RNAs suitable for expression in *Xenopus* oocytes. Several excellent reviews on the transcription and functional expression of cRNAs have appeared in recent years.[1-5] Additional information on *in vitro* transcriptions can be found in [16] and [17] in this volume.

Template Preparation

In vitro synthesis of RNA is most easily carried out by cloning a cDNA encoding the channel into a vector containing a bacteriophage promoter and then specifically transcribing that sequence by using an RNA polymerase that recognizes the phage promoter. Many plasmid and λ vectors, containing cloning sites flanked by SP6, T7, or T3 promoters, are now commercially available. The general procedure involves linearization of the DNA template 3′ to (downstream of) the cloned cDNA, *in vitro* transcription by the addition of the appropriate RNA polymerase, and, finally, enzymatic digestion of the DNA template (Fig. 1).

Specifically, cDNAs that have been cloned into transcription vectors are digested with a restriction enzyme that cuts downstream of the end of the open reading frame encoding the channel. This digestion linearizes

[1] D. A. Melton, P. A. Krieg, M. R. Rebagliati, T. Maniatis, K. Zinn, and M. R. Green, *Nucleic Acids Res.* **12**, 7035 (1984).

[2] P. A. Krieg and D. A. Melton, *Nucleic Acids Res.* **12**, 7057 (1984).

[3] P. A. Krieg and D. A. Melton, this series, Vol. 155, p. 397.

[4] D. A. Melton, this series, Vol. 152, p. 288.

[5] J. K. Yisraeli and D. A. Melton, this series, Vol. 180, p. 42.

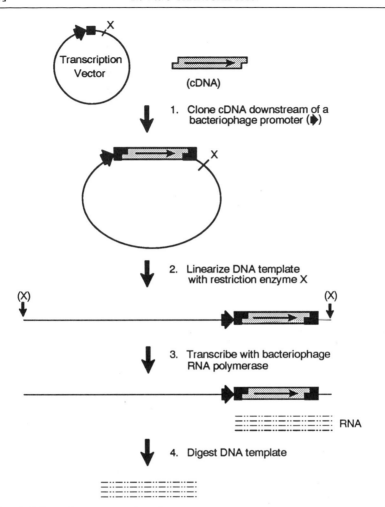

Fig. 1. Schematic representation of the steps involved in *in vitro* synthesis of RNAs. A cDNA encoding the ion channel is cloned into a vector that contains a bacteriophage promoter (▶) immediately upstream (5′) of the cloning site (■). The construct is then linearized by digestion with a restriction enzyme at a site (X) downstream (3′) of the end of the open reading frame (——→). RNA is synthesized from the template in a buffer containing nucleoside triphosphates and the bacteriophage RNA polymerase, resulting in a mixture of the newly synthesized transcripts and the original DNA template. Digestion of the DNA with RNase-free DNase leaves only the RNA transcripts.

supercoiled plasmids and eliminates long 3′ vector sequences, thereby allowing efficient (and specific) synthesis of the desired RNA. Restriction enzymes that generate 3′ protruding overhangs (e.g., *Pst*I, *Sfi*I, *Kpn*I) should be avoided if possible since the resulting templates may lead to the synthesis of extraneous RNAs, derived from transcription of the wrong (i.e., noncoding) strand.[6] These extraneous products are not produced from templates with 5' overhangs or blunt ends. If one must use an enzyme that generates 3' protruding ends, the template should be blunted, using T4 DNA polymerase,[7] prior to transcription. The linearized template is then extracted with phenol to remove the enzyme, concentrated by ethanol precipitation, dissolved in TE (10 mM Tris-HCl, pH 8.0, 1 mM EDTA) to a final concentration of about 1 mg/ml, and stored at $-20°$.

Standard Transcription Reaction Conditions

The transcription conditions outlined below are suitable for the synthesis of large amounts of RNA by SP6, T7, and T3 RNA polymerases (see also [17], this volume). Standard precautions should be taken to inactivate contaminating RNases (i.e., the use of baked glassware and solutions treated with 0.2% diethyl pyrocarbonate).[7]

Standard Transcription Reaction

40 mM	Tris-HCl, pH 7.5
8 mM	MgCl$_2$
10 mM	NaCl
2 mM	Spermidine
100 μg/ml	Nuclease-free bovine serum albumin
10 mM	Dithiothreitol
1 U/μl	RNase inhibitor
0.5 mM	ATP, CTP, and UTP
0.5 mM	7-methyldiguanosine triphosphate [mG(5')ppp(5')G]
50 μM	GTP
5 μg	DNA template
0.5–1 U/μl	SP6, T7, or T3 RNA polymerase

The reaction is typically carried out in a volume of 100 μl, at 37°, for 1.5–2 hr. The DNA template is digested with RNase-free DNase (5 U, 37°, 15 min), the proteins are removed by extraction with phenol, and the RNA product is separated from free nucleotides by ethanol precipitation and chromatography through a spun column of (RNase-free) Sephadex G-50.[7] The yield of the RNA is determined spectrophotometrically; a 1 mg/ml RNA solution has an absorbance of 25 at 260 nm.[7] Alternatively, a trace amount of one α-^{32}P-labeled nucleoside triphosphate (NTP) may be included in the transcription reaction and the yield determined from the

[6] E. T. Schenborn and R. C. Mierendorf, *Nucleic Acids Res.* **13**, 6223 (1985).

[7] J. Sambrook, E. F. Fritsch, and T. Maniatis, "Molecular Cloning: A Laboratory Manual," 2nd Ed. Cold Spring Harbor Laboratory, Cold Spring Harbor, New York, 1989.

amount of radioactivity precipitated by trichloroacetic acid.[5] The yield of RNA is typically over 25 mol/mol template.

Alternative Strategy for *In Vitro* Synthesis of Transcripts Encoding Ion Channels

Direct *in vitro* transcription of ion channel cDNAs synthesized using the polymerase chain reaction (PCR)[8] enables the rapid amplification and expression of ion channels encoded by low abundance mRNAs (Fig. 2). By incorporating a bacteriophage promoter into the 5′ PCR primer, the amplification products can be directly transcribed. This technique may, therefore, be used as a rapid functional screen to assay for the transcription of a gene encoding an ion channel in any tissue or cell line. Functional screening in this manner complements classic, structurally based screens by Northern blots and RNase protection assays. This method has been successfully used to amplify and express the I_{sK} protein from heart and uterus[9] and, by using a series of different 3′ PCR primers, has also been used to generate a series of C-terminal truncation mutants of the protein.

First-Strand cDNA Synthesis Reaction

50 mM	Tris-HCl, pH 8.3
30 mM	KCl
8 mM	$MgCl_2$
1 mM	Dithiothreitol
50 U	RNase inhibitor
2 mM	dATP, dGTP, dCTP, TTP
1 μM	Oligonucleotide primer
2 μg	Poly(A)$^+$ RNA
40 U	Avian myeloblastosis virus (AMV) reverse transcriptase

Polymerase Chain Reaction

10 mM	Tris-HCl, pH 8.3
50 mM	KCl
1.5 mM	$MgCl_2$
0.01% (w/v)	Gelatin
200 μM	dATP, dGTP, dCTP, TTP
1 μM	Each oligonucleotide primer
5 μl	First-strand cDNA reaction
2.5 U	*Taq* DNA polymerase

[8] H. A. Ehrlich, ed., "PCR Technology: Principles and Applications for DNA Amplification." Stockton Press, New York, 1989.
[9] K. Folander, J. S. Smith, J. Antanavage, C. Bennett, R. B. Stein, and R. Swanson, *Proc. Natl. Acad. Sci. U.S.A.* **87,** 2975 (1990).

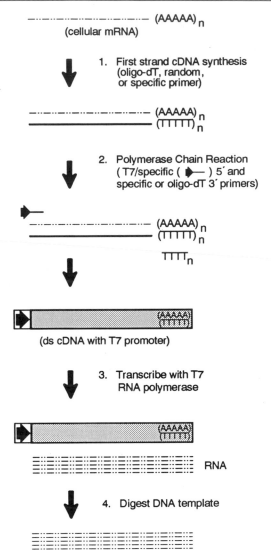

FIG. 2. Alternative strategy for the amplification and *in vitro* synthesis of transcripts encoding ion channels with known DNA sequences. Functional expression of low abundance mRNAs encoding ion channels may be rapidly accomplished by direct *in vitro* transcription of the channel cDNAs following their amplification in the polymerase chain reaction. First-strand cDNA is synthesized from an RNA pool using either oligo(dT), random, or specific primers. The cDNA encoding the ion channel is then amplified in the PCR using a 5′ PCR primer that also encodes a bacteriophage promoter. Thus, the amplified product contains the specific ion channel cDNA immediately downstream of the promoter and can, therefore, be used directly as a template for the synthesis of transcripts encoding the channel.

Messenger RNA, prepared from a tissue or cell line, is heated to disrupt secondary structure (65° for 3 min and quickly chilled on ice) and used as a template for the synthesis of first-strand cDNA with an oligo(dT), random, or a specific 3′ primer. The reaction is carried out in a volume of 40 μl, at 42°, for 60–90 min. The first-strand cDNA is then directly amplified using the polymerase chain reaction. The PCR is typically carried out in a volume of 100 μl with 25 cycles of 1 min denaturation at 94°, 2 min annealing at 56°, and 3 min extension at 72° (the optimal annealing temperature will vary with the primers used in the reaction). The 3′ PCR primer can be either oligo(dT) or, preferably, a specific sequence derived from the 3′ end of the mRNA encoding the channel. The 5′ PCR primer is a hybrid consisting of a bacteriophage promoter sequence (e.g., 5′-TAA-TACGACTCACTATAGGGAGA-3′ for a T7 promoter[10]) followed by a specific sequence, derived from the 5′ end of the channel mRNA. Amplification of the first-strand cDNA using these two primers results in the synthesis of a product consisting of the double-stranded cDNA encoding the channel directly downstream of the phage promoter (Fig. 2). The PCR product is then isolated by preparative agarose gel electrophoresis and used directly as a DNA template in an *in vitro* transcription reaction as described above.

Problems and Potential Solutions

Prior to cRNA expression in oocytes, the quality of the RNA can be monitored in two ways. An aliquot may be denatured with glyoxal, fractionated by electrophoresis, and stained with acridine orange.[11] A transcript of the expected size should be the only major species present. In addition, the RNA can be translated *in vitro*, using a rabbit reticulocyte lysate or wheat germ extract,[12] and the protein product labeled biosynthetically with [^{35}S]methionine or another ^{14}C-labeled amino acid. A protein of the predicted molecular weight should be detected following polyacrylamide gel electrophoresis and fluorography.

Transcripts prepared as described above are generally functional when injected into *Xenopus* oocytes and elicit the expression of channels that can be assayed electrophysiologically (see [19] in this volume). Problems that may result in difficulties in expression and potential solutions are discussed below.

[10] J. Dunn and F. Studier, *J. Mol. Biol.* **166,** 477 (1983).
[11] G. G. Carmichael and G. K. McMaster, this series, Vol. 65, p. 380.
[12] M. J. Clemens, *in* "Transcription and Translation: A Practical Approach" (B. D. Hames and S. J. Higgins, eds.), p. 231. IRL Press, Oxford, 1984.

Synthesis of Transcripts of Wrong Size

The product of the transcription reaction should consist of a single species of RNA with an electrophoretic mobility consistent with the size predicted from the nucleotide sequence. Minor contaminants, consisting of incomplete transcription products, are sometimes also produced, especially when transcribing extremely long templates [e.g., 8- to 10-kilobase (kb) Na^+ and Ca^{2+} channel cDNAs]. The relative abundance of these incomplete RNAs may be decreased by increasing the concentration of the nucleotides in the transcription reaction or by decreasing the temperature of the reaction from 37° to 30°.[3] (Note, however, that decreasing the temperature will also decrease the overall yield of the desired product.[3])

If only improperly sized transcripts are synthesized, the orientation of the clone in the vector, with respect to the promoter, and the site at which the template was linearized should be confirmed. The cDNA sequence should also be examined for regions with significant homology to the bacteriophage promoter or terminator sequences. Such regions might serve as spurious transcription initiation or termination signals and thereby result in the synthesis of partial transcripts. Because these signal sequences are RNA polymerase specific, problems resulting from such homologies may be prevented simply by cloning the cDNA into another vector, downstream of a different promoter.

Expression of Channel in Oocytes Cannot Be Detected

Occasionally, functional expression of a channel cannot be detected in oocytes even though the RNA is properly sized and readily translated *in vitro*. If possible, the synthesis of the protein by the oocyte should be confirmed biochemically. If an antibody to the protein is available, expression can be assayed by immunoprecipitation or Western blotting. If no antibody is available, expression can sometimes be demonstrated simply by comparison of the electrophoretic patterns of biosynthetically labeled proteins synthesized by injected and control cells. This type of biochemical analysis may help to determine whether the lack of functional expression results from impaired protein synthesis or from other defects farther downstream in the biosynthetic pathway (e.g., in the assembly of the translated proteins into functional channels or the requirement for additional subunits). If the synthesis of the protein cannot be demonstrated biochemically, the following regions of the transcripts should be examined and can be modified in attempts to increase the stability of the RNA and the expression of the protein.

5′ Cap. Transcripts must be modified with a 5′ cap structure to increase their stability in *Xenopus* oocytes.[2,13-15] Cap analogs [e.g., G(5′)ppp(5′)G, mG(5′)ppp(5′)G, or mG(5′)ppp(5′)Gm] have been demonstrated to be directly incorporated into T7[16] and SP6[17] transcripts simply by including the dinucleotide in the transcription reaction (as described above). Alternatively, the modification can be made posttranscriptionally, in a reaction catalyzed by guanylyltransferase.[14,18] The non-, mono-, and dimethyl cap analogs all increase RNA stability in *Xenopus* oocytes[2,13-15]; transcripts containing a mG(5′)ppp(5′)G cap have been reported to be most efficiently translated.[15]

Translation Initiation Site. The DNA sequence surrounding the initiating methionine codon should be examined to be sure that it conforms to the consensus for translation initiation sequences (i.e., GCCGCCRCCATGG).[19] This is generally of concern only for the expression of synthetic genes or cDNAs that have been subcloned using a restriction enzyme that might cut immediately upstream of the initiating ATG (e.g., *Nco*I). If the surrounding sequence does not represent a strong translation initiation sequence, it can be easily corrected by replacing the region with a synthetic DNA duplex that encodes the consensus sequence.

5′ and 3′ Untranslated Regions. The 5′ and 3′ untranslated regions of a transcript can also affect the level of expression of the encoded protein. Some ion channel cDNAs (e.g., K_V4, a neuronal K^+ channel[20]) contain unusually long (> 1 kb) 5′ untranslated domains. ATG codons within this region, both in and out of frame, may serve as incorrect translational initiation sites, resulting in the synthesis of nonsense proteins. Similarly, vectors containing *Sph*I or *Nco*I sites within the polylinker, upstream of the cDNA, will contain ATG codons at which translation might also begin. Extensive secondary structure in the 5′ untranslated domain has also been demonstrated to decrease translational efficiency.[21] The removal of long 5′ untranslated regions can often, therefore, increase the level of expression of the desired protein.

Conversely, in some cases, the addition of the 5′ and 3′ untranslated

[13] Y. Furuichi, A. LaFiandra, and A. J. Shatkin, *Nature (London)* **266**, 235 (1977).
[14] M. R. Green, T. Maniatis, and D. A. Melton, *Cell (Cambridge, Mass.)* **32**, 681 (1983).
[15] D. R. Drummond, J. Armstrong, and A. Colman, *Nucleic Acids Res.* **13**, 7375 (1985).
[16] D. A. Nielsen and D. J. Shapiro, *Nucleic Acids Res.* **14**, 5936 (1986).
[17] M. M. Konarska, R. A. Padgett, and P. A. Sharp, *Cell (Cambridge, Mass.)* **38**, 731 (1984).
[18] G. Monroy, E. Spencer, and J. Hurwitz, *J. Biol. Chem.* **253**, 4481 (1978).
[19] M. Kozak, *J. Cell Biol.* **108**, 229 (1989).
[20] C. J. Luneau, J. B. Williams, J. Marshall, E. S. Levitan, C. Oliva, J. S. Smith, J. Antanavage, K. Folander, R. B. Stein, R. Swanson, L. K. Kaczmarek, and S. A. Buhrow, *Proc. Natl. Acad. Sci. U.S.A.* **88**, 3932 (1991).
[21] J. Pelletier and N. Sonenberg, *Cell (Cambridge, Mass.)* **40**, 515 (1985).

domains from a mRNA that is known to be efficiently translated in oocytes may increase protein expression. Thus, the expression of the rat cardiac type I Na^+ channel[22] has been facilitated by its transcription from pSP64T, a vector in which the 5′ and 3′ untranslated domains of the *Xenopus* β-globin gene flank the cloned cDNA.[2]

3′ Poly(A) Tail. 3′-Polyadenylation may also increase RNA stability in oocytes and thereby increase the level of expression of the protein. Thus, a 3′ poly(A) tract of over 30 nucleotides has been directly demonstrated to increase the half-life of some transcripts in oocytes.[15,23] Furthermore, significant increases (10–20 fold) in the levels of expression of some proteins have been reported following polyadenylation of their transcripts.[15,24–29] In other cases, however, little or no effect was observed.[30–32] We have noted that the magnitude of the K^+ currents recorded from oocytes injected with transcripts encoding some mammalian K^+ channels are increased following polyadenylation of the RNA.[33] A 3′ poly(A) tail can be added to the RNA by cloning a synthetic dA/dT duplex into the transcription vector downstream of the cDNA[33,34] or, alternatively, can be added posttranscriptionally, in a reaction catalyzed by poly(A) polymerase.[35]

[22] L. Cribbs, J. Satin, H. Fozzard, and R. Rogart, *FEBS Lett.* **275,** 195 (1990).

[23] G. Marbaix, G. Huez, A. Burney, Y. Cleuter, E. Hubert, M. Leclercq, H. Chantrenne, H. Soreq, U. Nudel, and U.Z. Littauer, *Proc. Natl. Acad. Sci. U.S.A.* **72,** 3065 (1975).

[24] G. Huez, G. Marbaix, E. Hubert, Y. Cleuter, M. Leclercq, H. Chantrenne, R. Devos, H. Soreq, U. Nudel, and U. Z. Littauer, *Eur. J. Biochem.* **59,** 589 (1975).

[25] U. Nudel, H. Soreq, U. Z. Littauer, G. Marbaix, G. Huez, M. Leclercq, E. Hubert, and H. Chantrenne, *Eur. J. Biochem.* **64,** 115 (1976).

[26] G. Huez, G. Marbaix, A. Burney, E. Hubert, M. Leclercq, Y. Cleuter, H. Chantrenne, H. Soreq, and U. Z. Littauer, *Nature (London)* **266,** 473 (1977).

[27] G. Huez, Y. Cleuter, C. Bruck, L. Van Vloten-Doting, R. Goldbach, and B. Verduin, *Eur. J. Biochem.* **130,** 205 (1983).

[28] G. Galili, E. E. Kawata, L. D. Smith, and B. A. Larkins, *J. Biol. Chem.* **263,** 5764 (1988).

[29] H. G. Khorana, B. E. Knox, E. Nasi, R. Swanson, and D. A. Thompson, *Proc. Natl. Acad. Sci. U.S.A.* **85,** 7917 (1988).

[30] P. Sehgal, H. Soreq, and I. Tamm, *Proc. Natl. Acad. Sci. U.S.A.* **75,** 5030 (1978).

[31] A. K. Deshpande, B. Chatterjee, and A. K. Roy, *J. Biol. Chem.* **254,** 8937 (1979).

[32] H. Soreq, A. D. Sagar, and P. B. Sehgal, *Proc. Natl. Acad. Sci. U.S.A.* **78,** 1741 (1981).

[33] R. Swanson, J. Marshall, J. S. Smith, J. B. Williams, M. B. Boyle, K. Folander, C. J. Luneau, J. Antanavage, C. Oliva, S. A. Buhrow, C. Bennett, R. B. Stein, and L. K. Kaczmarek, *Neuron* **4,** 929 (1990).

[34] L. M. Hoffman and D. D. Donaldson, *Gene* **67,** 137 (1988).

[35] A. Sippel, *Eur. J. Biochem.* **37,** 31 (1973).

Posttranscriptional Polyadenylation Reaction

50 mM	Tris-HCl, pH 8.0
10 mM	$MgCl_2$
2.5 mM	$MnCl_2$
250 mM	NaCl
150 U	RNase inhibitor
25 μg	RNA
50 μg	Nuclease-free bovine serum albumin
0.1 mM	ATP (containing a trace of $[\alpha\text{-}^{32}P]ATP$)
1 U	Poly(A) polymerase

The reaction is carried out in a volume of 100 μl, at 37°, for 1 hr. The average number of moles of adenosine incorporated per mole of RNA [i.e., the average poly(A) tail length] is calculated from the fraction of $[^{32}P]ATP$ converted to an acid-insoluble form during the reaction. Typically, under the conditions given above, the average tail length is approximately 50–80 residues. The polyadenylated transcripts are extracted with phenol, to remove the protein, ethanol precipitated, and dissolved in TE at about 1 mg/ml.

[19] Electrophysiological Recording from *Xenopus* Oocytes

By WALTER STÜHMER

Introduction

The oocytes from *Xenopus laevis* have proved to be a reliable expression system for ion channels and transport systems. Electrophysiologists from all fields are increasingly making use of the oocyte expression system for their studies. Many measurements can now be carried out successfully which would have been impossible without this new technique. This chapter describes methods for electrophysiological measurements as well as some applications and limitations. Methods used to maintain *Xenopus laevis,* to microinject mRNA, to remove the follicular cell layer, and to express ion channels in *Xenopus* oocytes are described elsewhere in this volume ([14]–[16]). Therefore I only briefly describe the methods mentioned above as a general review, emphasizing some variations used in our laboratory.

Xenopus oocytes have been used successfully to translate messenger

RNAs (mRNA) into the respective proteins including posttranslational modifications.[1,2] Various receptors and ion channels have been functionally expressed after injection of poly(A) mRNA (e.g., Refs. 3-6). The development of genetic engineering techniques has made it possible to clone cDNAs coding for various receptor and/or channel proteins, to transcribe RNA from the cDNAs (cRNA), to inject the cRNAs into oocytes, and to analyze the functionally expressed proteins (e.g., Ref. 7-11).

Although other translation systems (detailed in [26]-[29], this volume) have been developed, certain experiments are only feasible in oocytes. The expression of foreign proteins in *Xenopus* oocytes has many advantages for electrophysiological measurements. The large size which easily accommodates manipulations like mRNA injections and electrode penetration is certainly one aspect. However, the fact that it is possible to obtain cell-attached patch clamp recordings[12] and, in particular, recordings from macropatches[13] is among the main advantages. These low noise recordings from a large number of channels would be difficult to obtain from other systems, and measurements, like gating current fluctuations,[14] are presently limited to the oocyte expression system.

[1] J. B. Gourdon, C. D. Lane, H. R. Woodland, and G. Marbaix, *Nature (London)* 233, 177 (1971).
[2] C. D. Lane, *Curr. Top. Dev. Biol.* 18, 89 (1983).
[3] K. Sumikawa, M. Houghton, J. S. Emtage, B. M. Richards, and E. A. Barnard, *Nature (London)* 292, 862 (1981).
[4] R. Miledi, I. Parker, and K. Sumikawa, *Proc. R. Soc. London B* 216, 509 (1982).
[5] E. A. Barnard, R. Miledi, and K. Sumikawa, *Proc. R. Soc. London B* 215, 241 (1982).
[6] C. B. Gundersen, R. Miledi, and I. Parker, *Proc. R. Soc. London B* 220, 131 (1983).
[7] M. Mishina, T. Kurosaki, T. Tobimatsu, Y. Morimoto, M. Noda, T. Yamamoto, M. Terao, J. Lindstrom, T. Takahashi, M. Kuno, and S. Numa, *Nature (London)* 307, 604 (1984).
[8] M. Noda, I. Takayuki, H. Suzuki, H. Takeshima, T. Takahashi, M. Kuno, and S. Numa, *Nature (London)* 322, 826 (1986).
[9] L. C. Timpe, T. L. Schwarz, B. L. Tempel, D. M. Papazian, Y. N. Jan, and L. Y. Jan, *Nature (London)* 331, 143 (1988).
[10] A. Mikami, K. Imoto, T. Tanabe, T. Niidome, Y. Mori, H. Takeshima, S. Narumiya, and S. Numa, *Nature (London)* 340, 230 (1989).
[11] U. B. Kaupp, T. Niidome, T. Tanabe, S. Terada, W. Bönink, W. Stühmer, N. J. Cook, K. Kangawa, H. Matsuo, T. Hirose, T. Miyata, and S. Numa, *Nature (London)* 342, 762 (1989).
[12] C. Methfessel, V. Witzemann, T. Takahashi, M. Mishina, and S. Numa, *Pfluegers Arch.* 407, 577 (1986).
[13] W. Stühmer, C. Methfessel, B. Sakmann, M. Noda, and S. Numa, *Eur. Biophys. J.* 14, 131 (1987).
[14] F. Conti and W. Stühmer, *Eur. Biophys. J.* 17, 53 (1989).

Preparation of Oocytes

Details regarding the maintenance of *Xenopus laevis* and their oocytes have been described elsewhere in this volume (see [15]). Here, a brief review on the preparation and handling of RNA is given. A precondition for electrophysiological measurements on integral membrane proteins in oocytes is the injection of a sufficient amount of mRNA. This mRNA can be from various origins: either total RNA or poly(A) mRNA extracted from tissue samples, or cDNA-derived mRNA (cRNA). A tissue sample of about 1 g yields 10 to 100 μg of poly(A) mRNA. Extraction procedures have been described for total RNA[15,16] and poly(A) mRNA.[17] The main disadvantage of injecting total RNA or poly(A) mRNA is that all possible mRNAs are translated into proteins, diluting the desired mRNA with the consequence that the desired protein will be only a fraction of the expressed proteins.

The relative abundance of the desired mRNA can be increased by size fractionation[18,19] of poly(A) mRNA. The mRNA from cellular extracts should be injected at a concentration of 1 to 10 μg/μl, and the cRNA can be injected at a concentration of 0.2 to 0.8 μg/μl, both in aqueous solution. Standard procedures to avoid and inhibit RNases should be followed, such as using baked glassware (e.g., 250° for 6 hr), wearing sterile gloves at all stages, and using diethyl pyrocarbonate-treated, autoclaved double-distilled water (DEPC–H_2O) for all solutions. Great care should be taken when handling the mRNA, not only to avoid contamination with RNases, but also to keep the solutions free of particles which could clog the injection pipettes. Thus the mRNA should be centrifuged to precipitate particles suspended in solution. Silanized Eppendorf tubes (1 or 0.5 ml) are best suited as containers for mRNA. The mRNA should be stored at −80°; however, for brief periods, storage at −20° is allowable. It is best to aliquot the mRNA into portions sufficient for the injections on a single day.

For shipment and long-term storage it is best to store the mRNA precipitated in ethanol. The precipitation of mRNA can be carried out as follows: to 1 volume of aqueous mRNA solution 0.1 volume of 20% potassium acetate and 2.5 volumes of cold 100% ethanol are added. After

[15] J. M. Chirgwin, A. E. Przybyla, R. J. MacDonnald, and W. J. Rutter, *Biochemistry* **18**, 5924 (1979).

[16] P. Dierks, A. van Ooyen, N. Mantel, and C. Weissmann, *Proc. Natl. Acad. Sci. U.S.A.* **78**, 1411 (1981).

[17] H. Aviv and P. Leder, *Proc. Natl. Acad. Sci. U.S.A.* **69**, 1408 (1972).

[18] P. Fourcroy, *Electrophoresis* **5**, 73 (1984).

[19] H. Lübbert, B. J. Hoffman, T. P. Snutch, T. van Dyke, A. J. Levine, P. R. Hartig, H. A. Lester, and N. Davidson, *Proc. Natl. Acad. Sci. U.S.A.* **84**, 4332 (1987).

mixing by inversion of the tube, the mRNA is precipitated for at least 30 min at $-70°$. The mRNA can be stored like this; however, it is more convenient to keep it in pure ethanol after decreasing the salt concentration. To do this the tube containing the mRNA is left at room temperature for about 5 min before centrifuging at 12,000 g for 15 min at 4° in a cooled centrifuge (e.g., Hettich centrifuge, Eppendorf centrifuge, or Heraeus Biofuge A). Then the supernatant is carefully removed with a sterile suction pipette, and the pellet is washed with 2 volumes of 70% ethanol. A centrifugation of 10 min will pellet the mRNA again. After removing the salt-containing supernatant the mRNA is stored in 100% ethanol. Care should be taken to avoid touching the mRNA pellet with the suction pipette and not to leave the tube open for long times to minimize contact with airborne RNases. To recover the precipitated mRNA the Eppendorf tube is brought to room temperature and centrifuged for 10 min at 12,000 g, the supernatant is removed, and the opening of the tube is covered with Parafilm into which three or four holes are poked with a sterile needle. The mRNA is then dried under reduced pressure in a desiccator for 4–10 min. Then the mRNA can slowly be resuspended in DEPC–H_2O at the desired concentration for use. The best method for drying the RNA prior to resuspending it in water is to use a Speed-Vac or any other centrifugation under vacuum, so that the alcohol evaporates at the same time that the pellet is maintained at the bottom of the Eppendorf tube.

Injection of Messenger RNA

Capillaries, such as transpipettor tubes (disposable micropipettes; Brand) are used for transferring the mRNA from the Eppendorf tube to the injection pipette. The transpipettor tubes should be made hydrophobic by flushing them in a solution consisting of 50% ether and 50% silane (dimethyldichlorosilane, Fluka, Ronkonkoma, NY; caution: health hazard). After allowing the ether to evaporate under a hood, the tubes are baked at 180° for 2 hr to inactivate any remaining RNases. Tubes treated in this way can be stored under sterile conditions for several months. A micrometer controlled syringe, filled with mineral oil to reduce the air volume and evaporation of water from the mRNA solution, is used to load the transfer pipette with a few microliters of mRNA solution.

Injection pipettes can be pulled using a standard pipette puller. They should have a long shank for estimating the volume of mRNA solution loaded and providing control that the oocytes are actually being injected. If pulled shortly before use, the injection pipettes will be RNase-free owing to the high temperature during the pulling process. The tips of the injection pipettes are broken under a microscope until they have an opening diameter of about 10 μm. Usually this will leave a rough and uneven rim, which might lesion the oocytes when penetrating them. This can be avoided by fire polishing the rim and pulling a sharp tip. This is done by approaching the tip of the injection pipette with a microfilament that has a small drop of molten glass on the tip. A sharp, needlelike protrusion can be obtained by rapidly removing the glass covered filament after touching the injection pipette. Radiation heat during this process is sufficient to smooth the ragged rim.

The injection pipettes are filled under stereomicroscopic control by suction applied through a conventional syringe. For this purpose, mRNA solution is extruded to form a droplet just outside of the transfer pipette by means of the micromanipulator-driven syringe. At this stage, the injection pipette is filled by application of suction with approximately 500 nl of mRNA solution, sufficient for injecting about 10 oocytes. After this, it is convenient to draw the remaining mRNA solution a few millimeters back into the transfer pipette by slight suction. This procedure decreases evaporation of water from the mRNA solution, which will be used for the next batch of oocytes to be injected.

A hand-driven coarse manipulator is sufficient to maneuver the injection pipette. Grooves carved into a thin Perspex plate fixed to the bottom of the injection chamber support the oocytes during injection. Alternatively, a scratched petri dish or one covered with either silicone curing agent (RTV615, General Electric, Waterford, NY) or agar can be used to fix the oocytes. The fixation of the oocytes is achieved by reducing the amount of solution in the injection chamber, so that the surface tension holds the oocytes in the grooves. Manipulation of oocytes is performed with the aid of Pasteur pipettes whose tips have been broken to enlarge the opening and then fire polished. A slight kink in the pipette tips helps in handling the oocytes during transport.

The mRNA is injected in aliquots of about 50 nl per oocyte. There are many procedures to achieve this, and very simple manual syringe systems or injection machines can be used. More sophisticated injection machines such as the Eppendorf microinjector or the Drummond oocyte injector are helpful when injecting large quantities of oocytes at a time (see also the description in [15], this volume).

Incubation

In most cases the large stage V and VI oocytes[20] can be used. The oocytes are incubated in small petri dishes in Barth's medium [in mM:

---

[20] J. N. Dumont, *J. Morphol.* **136,** 153 (1972).

84 NaCl, 1 KCl, 2.4 NaHCO$_3$, 0.82 MgSO$_4$, 0.33 Ca(NO$_3$)$_2$, 0.41 CaCl$_2$, 7.5 Tris-HCl, pH 7.4] with additions of penicillin and streptomycin (100 U/ml and 100 μg/ml, respectively). Some protocols use gentamicin (50–100 μg/ml). The optimal incubation temperature is 19°. Barth's medium should be changed every day using sterile pipettes. Oocytes having a blurred delimitation between the animal and vegetal pole should be removed.

Removal of Follicular Cell Layer

The follicular cell layer can be removed either before or after injection of the mRNA. It can be removed by purely mechanical means, but partial or total digestion of the follicular cell layer helps to prevent mechanical damage of the oocytes. Mild digestion is achieved in Ca^{2+}-free Barth's medium containing 1 mg/ml collagenase I (e.g., Sigma, St. Louis, MO, type I) for 1 hr at room temperature. The follicular cell layer is then removed mechanically using two pairs of forceps (No. 5, Dumont & Fils, Montignez, Switzerland). Alternatively, the ovaries can be incubated for 2–3 hr in the collagenase-containing solution under shaking until the oocytes are dispersed. It is important to wash (3–4 times in about 5 ml) the oocytes extensively in Barth's medium after collagenase treatment to stop the enzymatic reaction.

There seems to be no ad hoc rule to anticipate the time after injection when the various proteins will be inserted in the oocyte membrane. Therefore it is necessary to assay for protein expression from day to day. In general this will entail two-electrode voltage clamping for ion channels and electrogenic pumps. If no patch clamp experiments are intended, the vitelline envelope does not need to be removed. However, the removal of this last barrier to the plasma membrane makes insertion of electrodes easier, improves access and washout of solutes, and fixes the oocytes to the bottom of the experimental chamber (we use new 35-mm petri dishes as experimental chambers). Methods for removing the vitelline layer are described below in the section on patch clamp recording.

Electrophysiological Measurements

For most electrophysiological recordings from oocytes, the membrane potential has to be under control. This is achieved by clamping the oocyte to predefined potentials using a conventional two-electrode voltage clamp.[21,22] For this, one intracellular electrode is used to record the actual

[21] T. G. Smith, J. Lecar, S. J. Redmann, and P. W. Gage, eds., in "Voltage and Patch Clamping with Microelectrodes," American Physiological Society, Bethesda, Maryland, (Williams & Wilkins, Baltimore, Maryland), 1985.

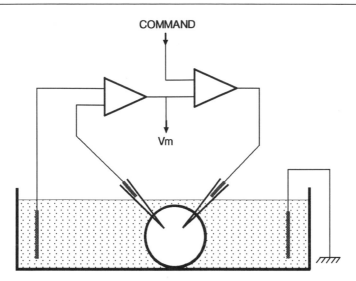

FIG. 1. Schematic diagram of the main components of a two-electrode voltage clamp. The output of the left amplifier–comparator (V_m) is the difference in potential between the bath and the oocyte electrodes. V_m is compared with the command potential, and any difference is injected into the oocyte through the current electrode. This current flows to ground through the bath ground electrode. If V_m equals the command potential, no current flows to ground.

intracellular potential, and the second electrode is used to pass current in such a way as to maintain the desired potential. This is achieved using a feedback circuit, which is the main component of the voltage clamp. The current needed to maintain a given potential is the measured parameter. There are two ways of measuring this current: as current flowing to ground through the grounding electrode (using a virtual ground amplifier) or as the current flowing through the current electrode. The intracellular potential recording electrode has such a high impedance that no current flows through it. Most voltage clamp setups use also a potential recording electrode (reference electrode) for the bath solution. This avoids polarization errors which arise from the current passing through the ground electrode. In this configuration, the transmembrane potential is taken as the difference between the intracellular potential electrode and the reference electrode. Therefore, the small error introduced because the bath is not always exactly at ground potential is corrected for. A block diagram of the main components of a voltage clamp is given in Fig. 1.

[22] N. B. Standen, P. T. A. Green, and M. J. Whitaker, eds., "Microelectrode Techniques: The Plymouth Workshop Handbook." The Company of Biologists, Cambridge, 1987.

In adition to the voltage clamp itself, a pulse generator is required for adjusting the voltage clamp amplifier and to make experiments which require measurements in response to changes in membrane potential. This pulse generator can be anything from a simple waveform generator to a computer-controlled pulse generator. If the voltage clamp does not provide a builtin filter, a Bessel low-pass filter is required to reduce the high frequency noise and as an antialiasing filter if the data are to be recorded digitally. The data recording can be a simple chart recorder (for slow events), an instrumentation tape recorder, or a computer-based data recording system. In principle, all the instrumentation will be quite similar to the one used for patch clamp recording, and details are given in [2] in this volume.

The two-electrode voltage clamp in oocytes is performed in accordance with techniques used for smaller cells.[21,22] The main differences reside in the larger size of the oocytes, which implies a higher membrane capacitance. This introduces no complication, if only slow or no changes in potential are required for the experiments. However, when measuring voltage-gated channels, the speed of the clamp and the possibility of compensating for capacitive transients are of prime importance. The time constant for charging the membrane and hence the speed of the voltage clamp depend critically on the electrode resistance through the relation $\tau = RC$, where R is the current electrode resistance and C the cell capacitance. For time-critical applications, the cell capacitance can be reduced by using smaller (stage III[20]) oocytes.[23]

Intracellular Electrodes

Standard intracellular electrodes, whose tips have been broken, have resistances below 1 MΩ and are quite appropriate for intracellular electrodes in oocytes. Electrode pipettes are made from capillary glass containing a thin filament which ensures that the electrode filling solution reaches the tip. We use glass from Clark (Reading, England), types GC120TF-10 (thin), GC150F-10 (medium), and GC200F-15 (thick). Pulling is best done using a standard intracellular electrode puller (David Kopf Instruments, Tujunga, CA) or a programmable electrode puller (see [3] in this volume). To avoid creeping of the pipette filling solution up the back end of the electrode, a small amount of dental wax or sticky wax is applied to the end after moderate heating in a small flame. Melted wax goes into the pipette along the filament for a few millimeters during application. After this, the electrode pipettes can be backfilled either with 3 M KCl or with 0.5 M

---

[23] D. S. Krafte and H. A. Lester, this volume [20].

potassium sulfate (or aspartate), containing at least 30 mM KCl. The 3 M KCl will be useful for most applications. The sulfate or aspartate solutions are to reduce the Cl$^-$ load to the oocytes, and the KCl is needed for the silver/silver chloride electrodes.

The electrode pipettes can be stored for several days sustained in a covered container which has some water at the bottom to ensure high humidity. The electrode tips are normally in the submicron range and need to be broken at the very tip to decrease the resistance to the megohm range. This is done either under a separate microscope or by simply jamming the electrode against the bottom of the recording chamber before an experiment. Typical tip diameters will be in the range of 1 to 5 μm, giving resistances in the range of 2 to 0.4 MΩ. The electrical contact to the electrode filling solution is made through a silver chloride electrode. The silver chloride electrodes can be made from chlorinated silver wire, from silver wire immersed in melted silver chloride, or from a silver/silver chloride pellet. The lifetimes of the electrodes mentioned as well as their diameters are given in ascending order.

Most voltage clamp amplifiers incorporate an electrode resistance measurement. The one from Polder (NPI Electronics, Tamm, Germany) allows electrode resistance measurements even with the electrodes positioned inside the oocyte, with a direct readout in megohms. Alternatively, the electrode resistance can be measured by applying current pulses (ΔI) through the Ag/AgCl electrode and measuring the potential jump (ΔV) induced. The electrode resistance will be $\Delta V/\Delta I$.

As long as the electrodes retain their low resistance, they can be reused for several oocytes. Clogged electrodes can be cleared by applying pneumatic back pressure with a syringe. This process can be monitored by the extrusion of the electrode filling solution (which has a different refraction index) under the microscope. A patch clamp pipette holder can be used as a electrode pipette holder. For long-term experiments, the electrode filling solution should be of such an amount that the hydrostatic pressure approximately compensates the pressure from the surface tension within the pipette. If the hydrostatic pressure in the electrode allows a large net outflow of solution, the oocytes will bulge around the site where the electrode is inserted as the oocytes accumulate KCl. If, on the contrary, there is a net inflow, the electrode resistance will increase as intracellular fluid enters the pipette. A very slight net outflow is the best choice, since this will stabilize diffusion potentials at the electrode tip.

Owing to the large size of the oocytes, the electrodes can be positioned with simple, coarse manipulators. It is important, however, that the setup be free of vibrations because otherwise oscillations will cause damage to the membrane where the electrodes penetrate the oocyte. A vibration-isolated table is usually not necessary in buildings where the floor is sufficiently quiet. This can be tested by observing the electrode tip under the set-up microscope while bouncing on the floor. A Faraday cage is usually not necessary for normal two-voltage electrode clamping, provided that standard grounding techniques are used (see [2] in this volume). The microscope can be a very simple one, either inverted or noninverted. The low magnification (5 – 10× objective) provides a comfortable working distance.

Adjusting Voltage Clamp

Once the electrodes are in solution, the offset of the potential electrode should be cancelled to within ± 1 mV. It is advisable to test this adjustment after every experiment as well, to ensure that there was no drift and that the potentials to which the oocyte had been clamped were correct. Then the electrodes are positioned with the manipulators as depicted in Fig. 2. The use of the following (or any equivalent) procedure is advisable owing to the

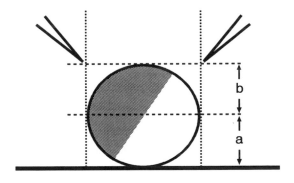

Top view :

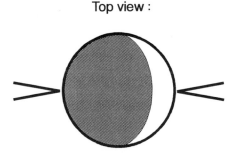

FIG. 2. Side view and top view of the oocyte with the intracellular electrodes positioned for penetration.

fact that the oocytes are opaque, so that the actual site of oocyte penetration is not visible, particularly when using an inverted microscope.

The positioning is done as follows. First, the plane even with the bottom of the oocyte is brought into focus (there are always enough particles or scratches on the bottom of the chamber). Then the rim of the oocyte is brought into focus, and the distance a can be estimated by the amount of refocusing necessary. Then the focus is brought up by about the same distance (b) so that the focal plane should be even with the top rim of the oocyte. The electrodes are brought into the positions shown in the bottom part of Fig. 2, where the electrode tips should be in focus. Axial movement of the electrodes will now ensure that the electrodes will penetrate the oocyte correctly. Obviously, for this procedure to work, it is important to have the ability to move the electrodes axially under a tilt of about 45°. To prevent oocytes from moving away during penetration, the oocyte can be supported by first advancing the current electrode until the oocyte starts to move away. Then the potential electrode is inserted. A rough bottom (polypropylene mesh, Fisher, Pittsburgh, PA, Cat. No. 08-670-185, or just simple scratches on the bottom of the experimental chamber) is helpful in holding oocytes in place during electrode penetration. The whole problem of perforating the vitelline envelope with microelectrodes and fixation of oocytes to the experimental chamber can be overcome if the vitelline envelope is removed as described below for patch clamping. This also makes the membrane more accessible to changes in the extracellular solution.

Penetration of the potential electrode can be monitored through the resting potential recorded. The resting potential of oocytes can be anywhere in the range of -20 to -80 mV. Monitoring the penetration of the current electrode is more difficult. The oocyte voltage clamp from Polder allows the measurement of the potential through the current electrode in the current clamp mode. Therefore, insertion of the current electrode is monitored in an analogous manner to the potential electrode. The Polder amplifier also features an audio monitor for the potential measured by either electrode, greatly simplifying electrode penetration while looking through the microscope. With most other amplifiers penetration of the current electrode is monitored using the current clamp mode. For this, a current pulse is passed through the current electrode, which will cause a change in potential within the oocyte as soon as the electrode penetrates the cell. This pulsed change in cell potential is monitored through the potential electrode which has to be inserted first. The amplitude of the potential response will depend on the input resistance of the oocyte, which can be obtained by evaluating $\Delta V/\Delta I$. It will be low for leaky oocytes or for oocytes expressing large quantities of exogenous channels having high probabilities of being open under the experimental conditions present.

Now that the two electrodes are inserted, the actual feedback circuit for clamping the transmembrane potential can be closed. To monitor the adjustments necessary for this step, voltage steps of about 10 mV are given as command potentials to the voltage clamp, and the current response is observed on an oscilloscope. Before closing the clamp, the gain of the feedback loop is set to a minimal value, and the holding potential is set to the desired value. Closure of the clamp will cause the membrane potential to approach the setting of the holding potential. Holding currents needed to maintain a potential of -100 mV are in the range of 100 to 200 nA, with larger currents being an indication of a leaky oocyte. The gain can now be increased carefully. Higher gain will speed up the time response of the voltage clamp, as seen from the decrease in the time constant of the current response. Too much gain, however, will cause oscillation and normally irreversible damage to the oocyte and current electrode. This can be avoided by either limiting the current output or by using an automatic oscillation shutoff feature. Both possibilities are provided in the Polder clamp, and many other commercially available clamps have provisions for current limiting.

Improving Frequency Response

The price for improved frequency response is an increase in noise. Therefore, no more gain than necessary should be used. The frequency response can be improved by higher gain, using series resistance compensation (not routinely necessary) and compensating for the potential electrode capacitance (capacitance neutralization). Electrode capacitance can be minimized by using a minimal solution level, just sufficient to cover the oocyte. Capacitive coupling between the two recording electrodes should be avoided, but the laborious procedures used for small cells like silver paint coating and subsequent isolation of the current electrode (effectively providing a grounded shield down to the very tip, see Refs. 21 and 22) are normally not required. For critical wide bandwidth applications, a simple grounded metallic shield around the current electrode reaching as close to the solution surface as possible is usually sufficient to reduce capacitive coupling between the electrodes.

Capacitance transients can be mostly compensated by adding the differentiated command voltage step with appropriate amplitude and time constant to the current trace. Two components are normally sufficient to compensate more than 90% of the capacitance transients. This compensation is only needed when recording from fast, voltage-activated channels and will in general decrease the signal-to-noise ratio.

Improving Signal-to-Noise Ratio

For slower applications, several procedures may be helpful in improving the signal-to-noise ratio. For example, the bandwidth of the feedback loop can be limited. Some amplifiers provide a feedback limited in bandwidth to 10 Hz; other units provide a setting having a high dc feedback gain. Otherwise, the lowest gain setting can be used. This, however, has the disadvantage that the clamp error will increase. The optimal solution in this case is to use an integrating feedback, which will eliminate the feedback error. This feature can even be used for fast potential-gated channels if the rise time of the pulse is adjusted accordingly. The Polder amplifier provides all the features described here.

Responses should be filtered using the lowest cutoff frequency possible. The limits of current resolution will depend on this filter setting. The range of measurable current amplitudes spans from several tens of nanoamperes for slow processes to several hundred nanoamperes for the fast voltage-activated channels. Another limitation of current resolution is the presence of endogeneous channels and carriers in the oocyte membrane. For instance, depolarization above $+30$ mV leads to development of a slow outward current, mostly owing to the presence of Ca^{2+}-activated Cl^- channels. Several other channel types are present to various degrees, and great care should be taken to avoid these currents from contaminating the desired signals.

Changes in Solution

In general, the two-electrode voltage clamp is stable for several hours, and extracellular solution changes are well tolerated. The perfusion can be simply gravity driven, with a suction pipette used for level control. While changing solutions manually with a Pasteur pipette or a syringe, the clamp feedback gain should be reduced, since otherwise some amplifiers tend to oscillate, particularly when close to a critical setting. Changes in fluid level will cause changes in electrode shunt capacitance and require readjustment of the capacitance compensation and neutralization. In some cases we have observed a dependence of current magnitude on fluid level. The reason for this is not quite clear, but it could be due to a hydrostatic access resistance dependence on the microvilli present to various degrees (depending on developmental stage) on the oocyte membranes. For critical applications, where an exact control of fluid level is mandatory, a fluid-level controller (MPCU, Adams and List, Westbury, NY) is very useful. The intracellular medium can be modified by injection of concentrated solutions. Up to 50 nl of solution can be injected while under voltage clamp with injection pipettes similar to the ones described for mRNA injection, provided that the oocyte is bare of the vitelline envelope.

Problems

In most cases, problems in recording currents from oocytes clamped with two electrodes arise from the electrodes themselves. Therefore, the possibility of measuring the electrode resistance during the experiment is very helpful. Any increase in electrode resistance deteriorates the clamp performance. This can often lead to oscillations. Also, increases (and decreases, when using capacitance neutralization) in electrode shunt capacitance can cause oscillations. This can be the case particularly when changing the solution levels. Most of the problems (and some cures) related to the electrodes and oocyte preparation have been dealt with above.

But how can the performance of the voltage clamp itself be tested? Figure 3 is a schematic diagram of the most simple equivalent circuit of an oocyte including its electrodes. It assumes intracellular electrode resistances of 1 MΩ, and the membrane resistance (1 MΩ) and capacitance

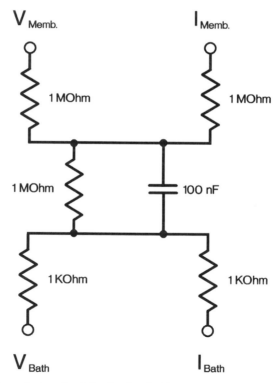

FIG. 3. Simple equivalent circuit for the electrical impedance of an oocyte. The upper part corresponds to the intracellular side.

(100 nF) correspond to average values encountered in oocytes. The 1-kΩ resistances represent mostly the bath electrode resistances. It is convenient to incorporate the circuit of Fig. 3 into a small (grounded) box with connections that can easily substitute the ones leading to the electrodes used in a real experiment. The electrical equivalent of an oocyte is also very helpful in getting acquainted with the voltage clamp amplifier being used. The equivalent circuit does not emulate the diffusion potential of 50 to 60 mV from real electrodes filled with a 3 M KCl solution.

Patch Clamp Recording from Oocytes

Prior to patch clamping on oocytes, the vitelline envelope needs to be removed. This is done mechanically after osmotic shrinkage in a solution of the following composition (in mM): 200 potassium aspartate, 20 KCl, 1 MgCl$_2$, 5 EGTA — KOH, 10 HEPES–KOH, pH 7.4. After 3 to 10 min in this solution, the transparent vitelline envelope becomes visible and can be removed mechanically with two pairs of blunt-tipped forceps. Extreme care has to be taken to avoid any damage to the cellular membrane since the bare oocytes are very fragile; they are particularly sensitive to air exposure. After a brief wash in normal frog Ringer's solution (NFR, in mM: 115 NaCl, 2.5 KCl, 1.8 CaCl$_2$, 10 HEPES, pH 7.2), the oocytes are transferred to the final experimental chamber. The bare oocytes will attach to any clean surface within minutes. Thereafter, any movement of the oocyte will cause membrane damage. More details of the procedure are described by Methfessel et al.[12]

Patch Pipettes

Pipettes for single-channel recording are similar to the ones used for small cells. However, a thick coating of a silicone (RTV615) curing agent, high enough up the pipette, is necessary because the large size of the oocytes requires deeper immersion of the pipette in the solution. The pipettes for macropatches will have tip opening diameters of 2 to 8 μm, giving resistances of 0.6 to 2 MΩ, depending also on taper shape and pipette solution. The taper should be as short as possible to avoid unnecessary access resistance. A hard aluminum silicate glass such as the one supplied by Hilgenberg (Malsfeld, Germany), or the type 7052 (e.g., A-M Systems, Everett, WA), gives smooth rims for macropipettes and low noise.

Seal Formation

The formation of a seal on oocytes is similar to seal formation in other cells for the small pipettes usually used for single-channel recording.[12] For macropatches, however, seal formation can be very slow (3–5 min). Also, the suction used to obtain seals is much smaller than when using small pipettes. It is convenient to be able to measure the pressure in the pipette. Positive pressures when entering the solution range between 10 and 40 mm H_2O; the suction during seal formation is in the range of 100 to 200 mm H_2O. The suction should be stopped when a seal resistance larger than 1 GΩ is achieved. The MPCU fluid level controller provides an adjustable pressure outlet which can be used to measure and control the pipette pressure. A slight depolarization (10–20 mV) of the patch is sometimes helpful. For macropatches seal formation becomes particularly slow when the seal resistance is in the 10 to 20 MΩ range. A few oocytes will form gigohm seals as quickly as in small cells. Seal resistances up to 100 GΩ can be obtained with small and large pipettes, having less than 200 fA root mean square (rms) of current noise. To avoid suspended particles, the solutions used in the bath and in the pipette should be filtered.

The most disturbing conductance in some oocytes is an endogenous stretch-activated channel with a conductance of about 40 pS. It can be easily identified by applying gentle suction to the patch pipette, which will cause the channel activity to increase. The open probability is also potential dependent, increasing with increasing transmembrane potential. Concentrations of 10 to 100 μM gadolinium will block these channels to a large extent.[24] These channels seem to be totally absent in more than 50% of the oocytes from the same animal.

For long depolarizations to very positive potentials, a slow outward current develops. This current is partially carried by Cl^- through Ca^{2+}-activated Cl^- channels and by a slowly activating potassium conductance. The density and number of these channels are highly variable from oocyte to oocyte. Chloride ions can be eliminated from the solution by substitution with methanesulfonic acid.[10,23] Alternatively, the current can be blocked by 0.3 mM niflumic acid or by 0.5 mM flufenamic acid.[26] Injection of EGTA or BAPTA (50 nl/oocyte at 0.1 M) also reduces the early response to inflowing Ca^{2+}. The most stringent requirements regarding the absence of any ion conductance apply when measuring gating currents. A detailed description of methods used for these particular measurements is found in [22] in this volume.

[24] X. C. Yang and F. Sachs, *Biophys. J.* **53**, 412a (1988).
[25] N. Dascal, T. P. Snutch, H. Lübbert, N. Davidson, and H. A. Lester, *Science* **231**, 1147 (1986).
[26] J. P. Leonard and S. R. Kelso, *Neuron* **4**, 53 (1990).

Other Configurations

Besides the cell-attached patch the only other easily viable configuration for macropatches is the inside-out configuration. This is obtained by abrupt withdrawal of the patch pipette from the oocyte after seal formation, in a solution normally containing high K^+ and no Ca^{2+} [e.g., (in mM): 100 KCl, 10 EGTA, 10 HEPES, pH 7.4]. Oocytes will survive for several hours in this solution as external medium, and several patches can be obtained from the same oocyte. Some batches of oocytes, usually from the same animal, do not survive the excision procedure. On rare occasions vesicles are formed during patch excision. This is in most cases a consequence of a slow excision. These vesicles are difficult to detect but are indicated by the appearance of a slow capacitive transient after excision. Therefore, before patch excision, the capacitance transient should be compensated and observed carefully during excision. With small patches, it is possible to disrupt the vesicle by brief exposure to air. With macropatches, this procedure leads to destruction of the patch in most cases. A better way of disrupting the vesicle is to carefully approach the pipette to a small, freshly made silicone (RTV615) sphere until a change in the capacitive transient is seen. This leads to rupture of the outer membrane of the vesicle and leaves the inside-out membrane intact in more than 95% of cases.

Most patches will tolerate extensive solution changes. The stability of even large patches may be due to the fact that the membrane forms an omega shape within the pipette tip. The tendency to form this omega-shaped patch is increased in oocytes after long incubation times (7 or more days after injection of the mRNA). Formation of an extensive omega patch can be detected during seal formation by the development of a slow capacitance transient, which cannot be compensated well, leaving a biphasic capacitive transient at best. These patches are not suitable for the study of kinetic effects, since solution changes as well as potential changes will not have unrestricted access to the membrane regions in close contact with the pipette wall. Also, ion accumulation effects should be considered in these cases.

Distribution of Ion Channels and Seasonal Variations

The distribution of channels is not uniform over the membrane of the oocyte. The magnitude of variation of the channel densities depends on the channel type. In our experience, sodium channels seem to be more uniformly distributed than potassium channels. Therefore, prediction of current size in a patch from the measured current size using a two-electrode voltage clamp is more accurate for the former channel type. Whole-oocyte currents of about 10 μA should yield patch currents of the order of 50 to 100 pA. When recording potassium currents from oocytes (expressing more than 10 μA of whole-cell K^+ current), some patches have practically no current whereas patches from other regions will have large currents. Therefore, several regions have to be explored with various patches in order to localize an area of high channel density. The regions of high channel density are relatively large, so that four measurements around the accessible portion of an oocyte are usually sufficient for localizing such a region. We have not found any correlation of regions of high current density with the animal or vegetal pole in oocytes, despite numerous rumors in that respect.

There are, however, clear seasonal variations. In the northern hemisphere, ion channels tend to give better results during the winter and ionic

FIG. 4. (A) "Tower" arrangement for conventional patch clamp measurements, which can also be used to make two-electrode voltage clamp experiments in oocytes.

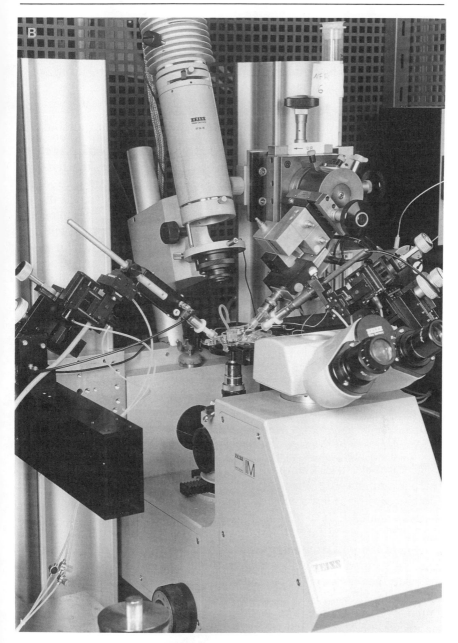

FIG. 4. (B) Close-up view of setup.

pumps seem to work better during the summer. In general, oocytes are more amenable for patch formation during the winter. The best results obtained for voltage-dependent channels are during the period from November to April. This period surprisingly cannot be shifted by using *Xenopus* acutely imported from South Africa.

Using Two-Electrode and Patch Clamp Amplifiers Simultaneously

When Na^+ channels are expressed at high densities, in particular channels that have been modified and will show no or very slow inactivation, the oocytes will spontaneously depolarize or even tend toward the sodium reversal potential. This persistent depolarization can be avoided during patch clamp experiments by holding the cell at hyperpolarizing potentials with a two-electrode voltage clamp. A high holding potential will also reduce the current required to hyperpolarize the patch, particularly with low resistance seals. However, the two-electrode voltage clamp introduces noise into the system. This noise can be drastically reduced by limiting the gain and consequently the bandwidth of the feedback amplifier of the two-electrode amplifier. Some two-electrode amplifiers have a mode that reduces the bandwidth of the feedback loop to 10 Hz (e.g., Polder), a speed sufficient to compensate for slow drifts in membrane potential. Figure 4 gives the setup for both two-electrode voltage clamp and patch clamp experiments, which can also be used for conventional patch clamp experiments on small cells.

Comparison of Recordings from Two-Electrode Voltage Clamps and from Macropatches in Oocytes

The two-electrode clamp of oocytes offers a series of advantages over patch clamp recording from macropatches. It is simpler, more stable, allows recording at lower channel densities, and the extracellular solution is easily changed. In addition, it is not so sensitive to varying oocyte conditions since the formation of gigaseals is not required. However, the kinetics of fast processes cannot be resolved owing to the high cell capacitance in combination with an upper limit of resistance of the current electrode. The temporal resolution is of the order of 200 to 1000 μsec. In addition, kinetics will depend on several parameters of the voltage-clamp amplifier (i.e., rise time limit or dc gain) as well as on the morphology of the oocyte membrane, which is extensively invaginated (microvilli). For example, extreme discrepancies can be obtained when recording from inactivating *Shaker* potassium channels in the two-electrode clamp mode. In some batches of oocytes these currents will show practically no inactivation. Currents from macropatches or ensemble averages from single-channel currents from these same oocytes will have reproducible, fast inactivating kinetics. Therefore, extreme caution has to be taken when trying to analyze channel kinetics from two-electrode data.

Macropatches achieve a high temporal resolution by electrically isolating a small area of membrane through pipettes with resistances of the order of 1 MΩ. Therefore, the temporal resolution is of the order of 50 to 200 μsec. The kinetics of fast processes are much more reproducible, probably because the microvillar structure is made more accessible in the membrane patch under the pipette. The solutions on both sides of the membrane are well defined, and exchange of the "intracellular" solution is possible in inside-out patches. Exchange of the extracellular solution is more difficult. This requires either the perfusion of the patch pipette or recording in the more difficult outside-out configuration. Recording from single channels naturally provides more information on the actual conformational changes between the open and closed states. Macropatches allow the characterization of ionic currents equivalent to the whole-cell configuration. Therefore, each of the various recording modes has its own advantages (and disadvantages), and a combination of several modes will in most cases allow a detailed characterization of the parameters under study.

[20] Use of Stage II–III *Xenopus* Oocytes to Study Voltage-Dependent Ion Channels

By Douglas S. Krafte and Henry A. Lester

Introduction

Pioneering studies on the use of *Xenopus* oocytes to express ion channels were first performed by Sumikawa *et al.*[1] on acetylcholine receptors and later by Gundersen *et al.*[2] on voltage-gated channels. Since this work, the *Xenopus* oocyte has become a convenient and robust system for expression of a variety of ion channels. Because the oocyte is able to express proteins from different species and tissues in a common membrane environment, thus removing one source of variability, an important use is to compare the detailed functional characteristics of channels induced by

[1] K. Sumikawa, M. Houghton, J. S. Emtage, B. M. Richards, and E. A. Barnard, *Nature (London)* **292,** 862 (1981).
[2] C. B. Gundersen, R. Miledi, and I. Parker, *Proc. R. Soc. London B* **220,** (1983).

METHODS IN ENZYMOLOGY, VOL. 207

distinct mRNAs. In this context, it is preferable to achieve the best possible resolution in a voltage clamp experiment.

Typically, fully grown (stage V – VI, Dumont[3]) oocytes are chosen for microinjection. Although these large cells offer the simplest arrangements for most experiments, they present limitations for the study of rapid events, particularly gating of voltage-dependent channels. To broaden the applicability of the oocyte system, we have developed techniques to express voltage-dependent channels in smaller oocytes, which offer advantages over the standard stage V – VI oocytes.

Why Stage II – III Oocytes?

Smaller oocytes offer a simple means to achieve more rapid settling of voltage clamp currents, primarily because they have less membrane capacitance. This section explains in more detail and also describes how to measure basic electrophysiological parameters. The simplest equivalent circuit of a membrane consists of a resistance R_m (comprising both active and passive components) in parallel with a capacitance C_m. With a high-quality voltage clamp, a voltage step could be achieved across such a circuit in a relatively short time (typically a few microseconds) determined by the gain and compliance of the clamp itself. The membrane gains or loses an amount of charge given by

$$Q = C_m V \tag{1}$$

However, real electrophysiological arrangements also include a resistance, R_s, in series with the membrane. R_s includes components from the external solution, the bath electrode, and other circuit elements. The membrane capacitance is now charged and discharged with a finite time constant (τ)

$$\tau = R_s C_m \tag{2}$$

The most direct way to measure Q is by integrating the voltage clamp current dQ/dt. Assuming that the voltage step is in a range where no voltage-dependent channels have opened, the current at long times after the step is simply a constant value given by V/R_m (see Fig. 1). A single-exponential fit to the capacitive currents then gives τ. Substitution of the known values for Q, τ, and V yields C_m and R_s.

The above relations are useful in determining the R_s values for any particular setup. Minimization and/or compensation for R_s is important in all voltage clamp experiments. More importantly, though, the relations illustrate two ways to improve the settling time (τ) of a voltage clamp

[3] J. N. Dumont, *J. Morphol.* **136**, 153 (1972).

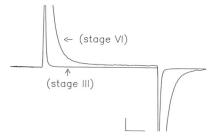

FIG. 1. Passive membrane properties of a stage III and a stage VI oocyte. Membrane currents were elicited by a voltage step from -100 to -10 mV. Membrane capacitance determined as described in the text was 49 nF for the stage III oocyte and 250 nF for the stage VI oocyte. Calibration bars: 1 μA, 2 msec. [Reprinted with permission from D. S. Krafte and H. A. Lester, *J. Neurosci. Methods* **26,** 211 (1989).]

experiment: (1) reduce the series resistance (R_s) value and (2) reduce the membrane capacitance (C_m). Using smaller oocytes for voltage clamp experiments exploits the latter. Up to 8-fold reductions in capacitance, and therefore settling times, can be achieved by using stage II–III oocytes as opposed to stage V–VI. Figure 1 illustrates the difference in settling times following an identical voltage step in a stage VI oocyte versus a stage III oocyte.

Identification of Different Stage Oocytes

Table I presents hallmarks of different stage oocytes for use in expression studies. As a practical consideration, the stage III oocytes are more easily manipulated than stage II oocytes. Whereas one achieves large de-

TABLE I
RULES FOR IDENTIFYING DIFFERENT STAGES OF *Xenopus* OOCYTES[a]

Stage	Animal/vegetal pole definition	Vitelline membrane
VI	Well-defined, marginal band between poles	Yes
V	Well-defined, no marginal band	Yes
IV	Well-defined, but vegetal pole can have significant pigment	Yes
III	No clear animal or vegetal pole, pigment granules throughout oocyte	Yes
II	No animal or vegetal pole, overall whitish appearance	Yes
I	No animal or vegetal pole, clear, nucleus visible	No

[a] See Dumont.[3]

creases in capacitance by going from stage V–VI to stage II–III, the difference between stages II and III is not so great (2-fold); the added difficulty of working with stage II oocytes is probably justified only for the most demanding applications. Finally, we have not successfully expressed ion channels in stage I oocytes, so it remains to be seen whether such oocytes can be used in expression studies. Stage I oocytes are much more fragile than those at later stages, presumably owing to the lack of a vitelline membrane.

Early stage oocytes can be isolated in the same manner as late stage oocytes using standard collagenase dispersion techniques (e.g., Leonard *et al.*[4]). Care must be taken, though, not to lose the smaller oocytes as they settle more slowly during solution changes and washes.

Mechanics of Injection Apparatus and Equipment

Stage II–III oocytes can be injected with optics and chambers similar to those used for stage V–VI oocytes. A dissecting microscope capable of 20–40× magnification is used to visualize the process. An injection chamber is made from a 35-mm tissue culture dish [Falcon (Becton Dickinson, Lincoln Park, NJ) 3001] by scratching the bottom of the dish with a hypodermic needle. The score marks from the needle prevent the oocytes from sliding during the injection process. The actual injection needle is mounted on a WPI Model A1400 Nanopump (World Precision Instruments, New Haven, CT) which in turn is mounted on a miniature micromanipulator [Narishige (Narishige USA, Greenvale, NY) MM-3-R or similar]. One microliter or less of RNA solution [1 mg/ml for poly(A$^+$)] is drawn into the needle using the reverse drive of the pump. The oocyte is penetrated with the fine drive of the miniature micromanipulator, and the pump is then used to deliver 5–10 nl of RNA to the oocyte at flow rates of 30 nl/min.

A standard oocyte injection system consists of a microdispenser mounted on a miniature micromanipulator. This arrangement is poorly suited for the stage II–III oocytes since one has to turn a knob on the microdispenser to inject RNA solutions. The physical vibrations during this process were not well tolerated by the smaller oocytes. Other equipment (voltage-clamp circuits, etc.) should be identical to standard setups. A high compliance voltage clamp (± 80 V) is necessary even for early stage oocytes.

[4] J. Leonard, T. Snutch, H. Lubbert, N. Davidson, and H. A. Lester, *J. Neurosci.* **7**, 875 (1987).

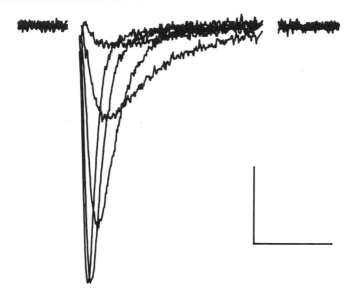

FIG. 2. Voltage-dependent Na^+ currents in a stage III oocyte injected with 5 ng of rabbit brain poly(A)$^+$ RNA. Membrane currents were elicited by voltage steps from -100 mV to -50, -40, -30, -20, and -10 mV. Passive currents have been eliminated by subtracting records in 500 nM tetrodotoxin. Calibration bars: 200 nA, 5 msec.

Examples of Expressed Voltage-Dependent Currents

Examples of expression of voltage-dependent currents are shown here and illustrate two different advantages of electrophysiological recording in stage II–III oocytes. The first advantage is improved temporal resolution. In Fig. 2, from an oocyte injected with rabbit brain RNA, Na^+ currents are elicited by voltage steps from -100 mV to test potentials more positive than -50 mV. Because the voltage change across the membrane settles much more rapidly in these cells than in stage V–VI oocytes, one achieves better resolution of the kinetic properties of the channels. Krafte and Lester[5] demonstrated that macroscopic inactivation of these channels is characterized by two exponential components.

A second advantage of recording in stage II–III oocytes is that there are fewer endogenous currents. Figure 3 shows the voltage dependence of activation and tail currents for potassium channels from an oocyte injected with rabbit brain RNA. In stage V–VI oocytes tested under the same conditions, recorded potassium currents are poorly resolved owing to overlapping Ca^{2+}-activated Cl^- currents which contaminate the records.

[5] D. S. Krafte and H. A. Lester, *J. Neurosci. Methods* **26,** 21 (1989).

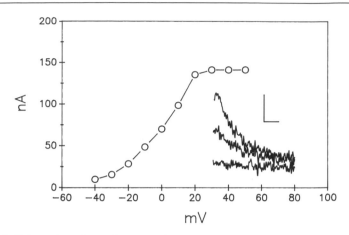

Fig. 3. Voltage-dependent K^+ currents in a stage III oocyte injected with 5 ng of rabbit brain poly(A)$^+$ RNA. Currents were elicited by voltage steps from -100 mV to test potentials more positive than -40 mV in 10-mV increments. The tail current amplitude at -100 mV versus the test potential is plotted. The inset shows actual tail currents following voltage steps to -40, 0, and $+50$ mV. Calibration bars: 50 nA, 5 msec.

Pharmacological dissection of the K^+ current from the Cl^- current remains tedious[6] although some newer blockers of the Cl^- channel offer promise.[7] The records in Fig. 3 show that the Cl^- current is much less of a problem in the stage II–III oocytes. Voltage steps from a holding potential of -100 mV to test potentials more positive than -20 mV activated K^+ currents. When the potential was returned to -100 mV, outward tail currents were observed. An instantaneous $I-V$ curve (data not shown) indicated that the reversal potential for these tail currents was approximately -80 mV.

Summary

Stage II–III *Xenopus* oocytes represent a useful extension of the standard *Xenopus* oocyte expression system. The oocytes are smaller; the reduction in membrane area, and therefore capacitance, leads to a faster settling time during a voltage clamp step. In addition, there appears to be less Ca^{2+}-activated chloride current; this may render the stage II–III oocyte a useful system for studying K^+ channels, where chloride currents can be a significant problem. Conversely, though, these early stage oocytes may not

[6] B. Rudy, J. Hoger, H. A. Lester, and N. Davidson, *Neuron* **1,** 649 (1988).
[7] M. M. White and M. Aylwin, *Mol. Pharmacol.* **37,** 720 (1990).

be useful for expression of neurotransmitter receptors coupled to phospholipase C, for such receptors are often monitored by activation of the Cl^- current.

The injections of RNA are technically more difficult in stage II–III oocytes. This can, however, be overcome with some simple modifications of the injection apparatus, mainly inclusion of a pump or similar device for actual injections.

[21] Intracellular Perfusion of *Xenopus* Oocytes

By NATHAN DASCAL, GAVIN CHILCOTT, and HENRY A. LESTER

Introduction

This chapter describes a method for recording macroscopic voltage clamp currents through the membrane of intracellularly perfused *Xenopus* oocytes. The method is similar to other internal perfusion methods described for ascidian eggs,[1-3] with several modifications, notably the addition of a perfusion tube that allows exchange of the intracellular solution within a few seconds. The extracellular solution can also be rapidly exchanged. Thus, this method allows full control of the solutions bathing the membrane. Moreover, sequential addition of transmitters, second messengers, regulatory proteins, and other agents becomes possible. The internal perfusion of large volumes of solution assures homogeneous distribution of the intracellular environment, in contrast to the spatial gradients that occur when the substances are injected (by pressure or iontophoresis) into intact oocytes. The technique has been applied in a limited number of experiments,[4] but it will presumably be exploited in the future for detailed studies of (1) ion channel permeation and (2) modulation of ion channels by second messengers and intracellular proteins. The potential importance of the internal perfusion method is enhanced by the widespread use of *Xenopus* oocytes for the heterologous expression of wild-type and mutated ion channels.[5-7]

[1] K. Takahashi and M. Yoshii, *J. Physiol. (London)* **279**, 519 (1978).

[2] S. Hagiwara and M. Yoshii, *J. Physiol. (London)* **279**, 251 (1979).

[3] M. Yoshii and K. Takahashi, *in* "Intracellular Perfusion of Excitable Cells" (P. G. Kostyuk and O. A. Krishtal, eds.), p. 77. Wiley, New York, 1984.

[4] N. Dascal, G. Chilcott, and H. A. Lester, *J. Neurosci. Meth.* **39**, 29 (1991).

[5] N. Dascal, *Crit. Rev. Biochem.* **22**, 317 (1987).

[6] T. P. Snutch, *Trends Neurosci.* **11**, 250 (1988).

[7] H. A. Lester, *Science* **241**, 1057 (1988).

Intracellular Perfusion of Oocytes

General Description of Method

The setup is presented schematically in Fig. 1. The oocyte is placed in a fire-polished Pasteur pipette (upper chamber), filled with the extracellular solution. The upper chamber is lowered into the "internal solution" chamber. The recording employs a two-electrode voltage clamp configuration. The ground electrode is in the extracellular solution; current and voltage electrodes are in the intracellular solution. Not shown in Fig. 1 are the oocyte internal perfusion tube and the perfusion apparatus of the upper chamber (see Fig. 3).

Internal Perfusion Setup

The internal solution chamber is made of transparent polystyrene (e.g., the lower third of a 25-cm² tissue culture flask) and should be large enough to accommodate the upper chamber, voltage and current electrodes, the perfusion tubes used to change the solution in this chamber (Fig. 1), and the internal perfusion tube (Fig. 3). The experiment is performed under visual observation using a standard stereomicroscope. If the microscope is on a vertical mount, a mirror is placed behind the chamber to redirect the line of sight (Fig. 2); if the microscope can be mounted horizontally, the mirror is unnecessary. The upper chamber, current and voltage electrodes, and the perfusion tubes are permanently mounted on steady rods with crocodile holders. The upper chamber holder should allow adjustment along the vertical axis.

The upper chamber is made of a Pasteur pipette cut at both ends. The

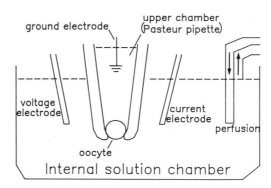

FIG. 1. Schematic presentation of the experimental arrangement for internal perfusion of an oocyte.

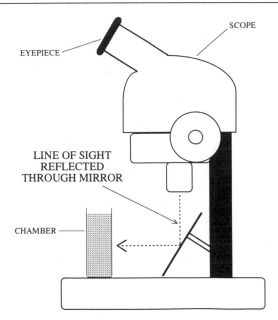

FIG. 2. Side view of the microscope and the lower chamber.

wide part is cut to a length of about 4 cm; the narrow part is cut where the outer diameter tapers to 1.2–1.8 mm. The narrow edge is then fire polished until the desired inner diameter (300–400 μm) is obtained (the oocyte passes through larger openings). The pipette serving as upper chamber can be used for only one oocyte. Both soft and borosilicate glass provides a good seal.

The perfusion tubes of the upper chamber and the ground electrode are arranged as shown in Fig. 3. The perfusion tubes must be provided with vertical movement, allowing change of the extracellular solution level. Note that the edge of the infusion tube protrudes 2–3 mm below the level of the meniscus, assuring that the perfusate will not drop on the meniscus and agitate the oocyte. The actual position of the ground electrode is of little importance as long as it comes in contact with the solution; therefore, it can be mounted separately from the perfusion tubes.

All perfusion tubes, as well as the voltage and current electrodes, are made of 1.2–2 mm (outer diameter) glass tubes. The internal perfusion tube is made as follows (Fig. 3). First, the tip is pulled manually or with a microelectrode puller, then broken to an outer diameter of 50–100 μm. Then the tube is bent, in a Bunsen burner, to give the shape shown in Fig. 3. The tube is mounted on a micromanipulator, to allow precise position-

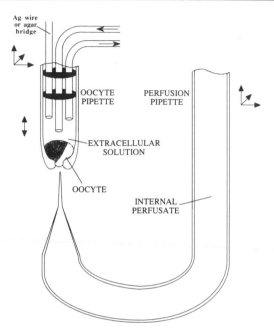

FIG. 3. Upper chamber perfusion devices and the internal perfusion pipette.

ing under the oocyte and upward–downward movement afterward. The wide end of the perfusion pipette is connected through polyethylene tubing to a syringe containing the desired internal perfusion solution.

The electrodes are filled with 3 M KCl in 1.5% agar. The ground electrode is made in the same way, or a simple Ag/AgCl wire can be used if no changes of Cl^- concentration in the extracellular solution are planned. The electrical resistance of the KCl/agar electrodes is a few tens of kiloohms (at most). The electrodes are connected to a negative feedback amplifier in the usual two-electrode voltage clamp configuration. Note that most commercially available voltage clamp amplifiers are designed for stability with electrodes of megohm resistances and may not perform optimally with low-resistance electrodes.

Preparation of Oocytes

Frogs are maintained and dissected as described.[8] The oocytes are defolliculated by treatment with 1.5–2 mg/ml collagenase in a Ca^{2+}-free

---

[8] A. L. Goldin, this volume [15].

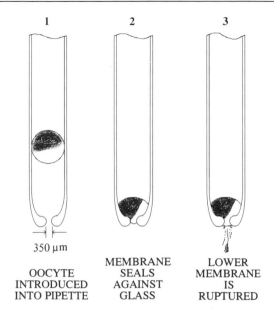

1   2   3

350 µm

	MEMBRANE	LOWER
OOCYTE	SEALS	MEMBRANE
INTRODUCED	AGAINST	IS
INTO PIPETTE	GLASS	RUPTURED

FIG. 4. *Xenopus* oocyte in a pipette.

solution,[9] transferred to a normal physiological solution (e.g., ND 96[9]), and treated as desired (injected with RNA, incubated under different conditions, etc.). Just before the internal perfusion, the vitelline membrane is stripped, essentially as described.[10] In brief, the oocyte is placed for 20–30 min into a high-osmolarity solution (e.g., ND 96 with the addition of 50 mM NaCl or 100 mM sucrose). The oocyte shrinks, and the vitelline membrane detaches from the surface at some points and can be mechanically removed with a pair of fine forceps. Do not expose the devitellinized oocyte to air as it will explode immediately; transfer it inside a horizontally positioned Pasteur pipette (to prevent sliding to the tip and exposure to air).

Succession of Events in Internal Perfusion

The succession of events in internal perfusion is shown schematically in Fig. 4. The lower and upper chambers are filled with a normal physiological solution (all solutions used in the study should be filtered through 0.2-

[9] N. Dascal, T. P. Snutch, H. Lubbert, N. Davidson, and H. A. Lester, *Science* **231**, 1147 (1986).
[10] C. Methfessel, V. Witzemann, B. Sakmann, T. Takahashi, M. Mishina, and S. Numa, *Pfluegers Arch.* **407**, 577 (1986).

or 0.45-μm filters). Solutions containing Ca^{2+} appear to promote the formation of a seal between the glass and the membrane and are therefore recommended. The oocyte is dropped into the upper chamber and settles over the bottom opening. From this moment on, the oocyte should not be moved or mechanically disturbed until the seal has formed completely, for movements would permanently destroy the seal. The solution level in the upper chamber is set 4–8 mm higher than that in the lower chamber; this positive pressure on the oocyte helps the formation of the seal. At this stage, there is no potential difference between the two chambers. The resistance of the oocyte and the seal (connected in parallel) should be continuously monitored. To do so, the feedback loop is closed, the voltage between the chambers is clamped at 0 mV, and rectangular pulses of 10 to 20-mV amplitude are applied periodically at intervals of several seconds.[4] The development of the seal is indicated by a steady increase of the measured resistance from 50–200 kΩ to a constant level of 2–10 MΩ over 15–30 min (Fig. 5). Resistances significantly lower than 2 MΩ usu-

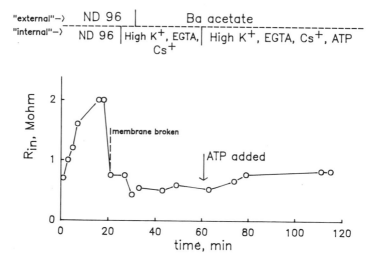

FIG. 5. Time course of changes in R_{in} as a function of time, membrane rupture, solution changes, and the addition of 1.5 mM ATP in an oocyte injected with cardiac RNA 3 days before the recording (see Ref. 9 for details). The two rows at the top indicate the solutions used: "external," the solution in the Pasteur pipette; "internal," the solution in the internal solution chamber. The vertical bars show the times of solution changes. The composition of the solutions was as follows: ND 96, 96 mM NaCl, 2 mM KCl, 1 mM MgCl$_2$, 1.8 mM CaCl$_2$, 5 mM HEPES (pH 7.5); Ba acetate solution, 40 mM barium acetate, 2 mM potassium acetate, 60 mM N-methyl-D-glucamine (acetate salt), 5 mM HEPES (pH 7.5); high K$^+$ solution, 86 mM KCl, 12 mM NaCl, 5 mM HEPES (pH 7.4), 1 mM EGTA, and 5 mM CsCl.

ally suggest low seal quality; such cells should be abandoned. The formation of a stable seal can be verified by the fact that changes in the upper chamber solution level do not affect the measured resistance. The actual resistance of the seal cannot be measured directly, but calculations show that it is at least an order of magnitude or more higher than the resistance of half of the oocyte membrane.[4,11]; the latter is usually between 0.5 and 2 MΩ in normal physiological solution.

Following the formation of a seal, the upper chamber solution level is reduced until it exceeds that of the lower chamber by no more than 1–3 mm[12] (excessive positive pressure may cause collapse of the oocyte after the lower membrane is ruptured). The lower membrane (facing the internal solution chamber) is ruptured by the tip of the internal perfusion pipette, which can be inserted a few hundred microns into the oocyte and moved in and out, to promote efflux of yolk. The yolk, however, still leaks out of the oocyte for tens of minutes even after extensive internal washings.

At this time, the internal chamber solution can be changed to the desired intracellular solution. The solutions used are designed in the same way as for experiments done in the whole-cell configuration of the patch clamp,[13,14] for example, containing a high concentration of K^+, low Na^+, Cs^+, and EGTA (Fig. 5). The voltage protocol is changed to one designed to measure the ionic currents under study, and the extracellular solution can be changed to one specially designed to measure currents through a particular class of channel; for example, currents through Ca^{2+} channels can be recorded with Ba^{2+} as the charge carrier (see Fig. 5).

Constituents of the internal chamber solution diffuse into the oocyte extremely slowly.[1–4] However, the internal perfusion tube allows one to introduce various substances and, apparently, change the actual ionic composition of the intracellular solution in just tens of seconds. The perfusion is accomplished by applying positive pressure from the syringe to the oocyte perfusion tube and blowing the solution into the oocyte. The oocyte expands visibly under the pressure of the inflowing solution and

---

[11] Assuming an average resistance of the seal formed between a normal patch pipette of 1 μm diameter with a cell membrane to be 3 GΩ, the seal of a 300-μm pipette with the oocyte membrane will be approximately 10 MΩ, assuming a similar contact between the cell surface and the glass. However, because the surface area of the contact *per unit diameter* is much larger than in the usual patch (see, e.g., Fig. 3, which presents quite a realistic picture of such a contact), the seal is probably much better.

[12] This represents the equilibrium level as modified by forces due to capillary action in the upper chamber; thus, no strain is placed on the seal.

[13] O. P. Hamill, A Marty, E. Neher, B. Sakmann, and F. J. Sigworth, *Pfluegers Arch.* **391,** 85 (1981).

[14] B. Sakmann and E. Neher, eds., "Single-Channel Recording." Plenum, New York, 1983.

then relaxes again as the perfusion stops. Note that excessive pressure may break the seal or the membrane.

To exchange the internally perfused solution, perform the following steps. Withdraw the perfusion tube from the oocyte and empty its contents, replace the syringe with one containing the new solution, fill the perfusion tube with the new solution, and insert the tube again into the oocyte and wash in the new solution. If a simple addition of a drug to the interior of the oocyte is desired, it is not necessary to exchange the solution in the entire internal solution chamber; if a more drastic change (such as replacement of an ion by another) takes place, the solution in the lower chamber should be replaced with the new one.

The resistance of the oocyte "runs down" (decreases) during the experiment[4] (Fig. 5), although currents through Cl^- and Ca^{2+} channels can still be recorded for many minutes. Our preliminary results (Ref. 4 and Fig. 5) indicate that addition of 1.5 mM MgATP to the interior of the oocyte may prevent and even reverse this rundown. In the experiment shown in Fig. 5, after the addition of ATP, Ba^{2+} current through the voltage-dependent Ca^{2+} channels was recorded without decrement for 1 hr after the addition of ATP (see Ref. 15).

Possible Drawbacks

A major problem in the internal perfusion method is the lack of an accurate estimate of the seal resistance. Current flowing through the shunt resistance of the seal may influence or even distort the recording of the current flowing through the membrane, if seal resistance is insufficiently high.[1,3] However, as noted above, the seal between devitellinized *Xenopus* oocytes and glass seems to be very good, and the errors in current amplitude estimation should be less than 10%.[4] If there is only a loose seal at the edges of the contact of the glass with the membrane, the voltage control there will be poor, and this will distort the recorded current. A rigorous test of the validity of the method for measuring fast currents has not yet been performed.

Acknowledgments

We are grateful to Dr. I. Lotan for critical reading of the manuscript. During preparation of this chapter, the authors were supported in part by grants from the Muscular Dystrophy Association (N.D.), the Israel Academy of Sciences and Humanities (N.D.), the USA–Israel Binational Fund (N.D. and H.A.L.), National Institutes of Health Grant GM 29836 (H.A.L.), and a grant from the Markey Charitable Trust (H.A.L.).

[15] J. E. Chad and R. Eckert, *J. Physiol. (London)* **378**, 31 (1986).

[22] Recording of Gating Currents from *Xenopus* Oocytes and Gating Noise Analysis

By STEFAN H. HEINEMANN, FRANCO CONTI, and WALTER STÜHMER

Introduction

Gating currents, also termed "asymmetric charge displacements" and denoted here I_g, arise from charge movements or dipole reorientations that accompany the conformational transitions of voltage-gated ion channels and are directly responsible for their voltage sensitivity. The study of gating currents therefore provides valuable information about the mechanisms by which ion channels are controlled by voltage. The first I_g measurements were made on sodium (Na^+) channels from nerve fibers.[1-3] It was found that approximately 2–3 electron charges per channel subunit are effectively translocated across the membrane before the sodium channel opens. This means that for 10,000 channels a total gating charge of approximately 10^{-14} coulomb (10 fC) has to be translocated within a typical time of some 100 μsec, yielding a mean I_g of approximately 50 pA. This rough figure sets essential conditions for I_g recordings, requiring low-noise performance at sufficiently high time resolution. For this reason, the patch clamp technique[4] using tight seals[5] is well suited for this type of measurement, provided only small cells or membrane patches are used. On the other hand, because the signal increases linearly with the number of channels, very high channel densities are also required.

In this chapter we describe methods for recording gating currents from ion channels expressed in *Xenopus* oocytes in the inside-out and cell-attached recording configuration.[5] This preparation allows combining the whole realm of genetic engineering and ion channel expression with the detailed study of voltage-dependent channel gating, a task that could not be accomplished by recording ionic currents from native excitable cells. We also demonstrate how I_g fluctuations can be used to derive information on the elementary molecular events associated with ion channel gating.

[1] C. M. Armstrong and F. Bezanilla, *Nature (London)* **242,** 459 (1973).
[2] R. D. Keynes and E. Rojas, *J. Physiol. (London)* **239,** 393 (1974).
[3] H. Meves, *J. Physiol. (London)* **243,** 847 (1974).
[4] E. Neher and B. Sakmann, *Nature (London)* **260,** 799 (1976).
[5] O. P. Hamill, A. Marty, E. Neher, B. Sakmann, and F. J. Sigworth, *Pfluegers Arch.* **391,** 85 (1981).

Oocyte Expression System for Gating Current Measurements

The oocyte expression system[6] in combination with patch clamping[7-9] offers all the advantages of the latter technique for I_g measurements in addition to the possibility of expressing specific (and mutated) channel types at a high density.

Expression of Specific Ion Channels

An obvious advantage of *Xenopus* oocytes is that the density of endogenous ion channels in these cells is very low[6] so that contaminations in the recordings by exogenous currents are usually negligible. In the worst case, the endogenous currents account for 100–400 nA in the whole oocyte,[10] whereas, for example, more than 100 μA of K^+ current can be obtained from exogenous channels. Normally even such a small contamination would be prohibitive for I_g measurements. However, oocytes with even much lower densities of endogenous channels can be frequently found, and larger membrane patches devoid of endogenous channels are not uncommon.

The exogenous channels expressed in oocytes correspond exclusively to the type encoded by the injected cRNA. This feature is of particular importance for the recording of gating currents from K^+ channels, which coexist in natural excitable preparations with Na^+ channels. In response to step depolarizations, gating currents from both these sources are elicited concurrently and are difficult to separate. So far, components arising from K^+ channels were identified in the "on" phase by their slower kinetics (the "off" responses could be separated even less reliably because their time courses are comparable) or on the basis of their different pharmacological properties.[11] Similar problems (although quantitatively less serious) exist for the measurements of Na^+ channel gating currents, possibly contaminated by contributions from K^+ channels, which may be overlooked as "baseline tilts."

Herein we discuss results obtained for one type of Na^+ channel (rat

[6] R. Miledi, this series.

[7] C. Methfessel, V. Witzemann, T. Takahashi, M. Mishina, and S. Numa, *Pfluegers Arch.* **407,** 577 (1986).

[8] W. Stühmer, C. Methfessel, B. Sakmann, M. Noda, and S. Numa, *Eur. Biophys. J.* **14,** 131 (1987).

[9] W. Stühmer, this volume [19].

[10] L. Lu, C Montrose-Rafizadeh, T.-C. Hwang, and W. B. Guggino, *Biophys. J.* **57,** 1117 (1990).

[11] S. Spires and T. Begenisich, *J. Gen. Physiol.* **93,** 263 (1989).

brain II)[12] and two types of inactivating K^+ channels (*Shaker*[5] A2[13] and a construct with the C- and N-terminal ends from *Shaker* A2 and the core domain from RCK1 termed SRS[14]). However, the methods as described here are applicable to any kind of cloned voltage-activated channel that expresses in oocytes.

High Channel Density

Because gating currents are rather small, the number of channels is the most important parameter for equal conditions of seal resistance and background noise. In this sense oocytes are ideal, because fairly high channel densities can be obtained while isolating so-called macropatches of membrane with very high seal resistances.[8] Because not all oocytes have the same degree of channel expression, we usually employ a conventional two-electrode voltage clamp for screening the channel density and select those oocytes with large currents. This procedure is rather fast and ensures high success rates in the following search for macropatches with high activity in the selected oocytes. Indeed, in particular K^+ channels seem to be clustered in "hot spots." Therefore, patches with high densities of channels were searched by screening the oocyte surface with patch clamp electrodes until such a hot spot was found. With experience the criterion for identifying a hot spot becomes obvious, given a certain level of whole-oocyte current.

Performance of Macropatches

Recording currents from macropatches as described by Stühmer *et al.*[8] has the advantage of a high time resolution, which is determined practically only by the bandwidth of the patch clamp amplifier (> 10 kHz). Indeed, despite the relatively large area, macropatches have still a small input capacitance (~ 1 pF for 100 μm^2), which, together with the fairly low access resistance (0.6 – 1.0 MΩ) yields time constants of the order of 1 μsec. On the other hand, the noise due to the input voltage fluctuations of the amplifier dropping across the load impedance is still low. In patches obtained with 0.6-MΩ pipettes root mean square (rms) noise levels as low as 160 fA in the band of 0.05 – 3.0 kHz can be obtained, whereas the mem-

[12] M. Noda, T. Ikeda, H. Suzuki, H. Takeshima, T. Tahashi, T. Kuno, and S. Numa, *Nature (London)* **322,** 826 (1986).

[13] R. Lichtinghagen, M. Stocker, R. Wittka, G. Boheim, W. Stühmer, A. Ferrus, and O. Pongs, *EMBO J.* **9,** 4399 (1990).

[14] M. Stocker, O. Pongs, M. Hoth, S. H. Heinemann, W. Stühmer, K.-H. Schröter, and J .P. Ruppersberg, *Proc. R. Soc. London B* **245,** 101 (1991).

brane surface can be large enough to yield peak ionic currents of the order of 10 nA. In such patches peak gating currents can be of the order of 100 pA so that the signal-to-noise ratio can reach values of 500 under ideal conditions.

Another advantage of the macropatch method applied to *Xenopus* oocytes is that there is no complicated preparation necessary as in the case for nerve preparations (e.g., node of Ranvier). Also, because both sides of the membrane are accessible, the cis and trans solutions can be selected freely so as to block the ionic currents completely.

Removal of Ionic Currents

To resolve gating currents all the ionic currents have to be abolished. As mentioned earlier the ratio of ionic currents to peak gating currents is of the order of 100, which implies that the ionic currents have to be avoided very efficiently.

In the case of the Na^+ channel this is accomplished by blocking the channel with $1\mu M$ tetrodotoxin (TTX). In addition, the permeant sodium ion can be substituted by Tris, which is impermeant. Thus, for cell-attached recordings the pipette solution may have the following composition (in mM): 115 Tris-Cl, 2.5 KCl, 1.8 $CaCl_2$, 10 HEPES (N-2-hydroxyl-ethylpiperazine-N'-2-ethanesulfonic acid), 1 μM TTX, pH 7.2. This solution is most effective in removing the Na^+ currents.

Potassium gating currents are best recorded from inside-out patches that allow the control of the solution on the cytoplasmic side. All permeant cations have to be replaced on both sides of the membrane by impermeants such as Tris (Tris[hydroxymethyl]aminomethane), N-methyl-glucamine or TEA (tetraethylammonium). Besides these major cations (or combinations of them) as chloride salts in 130 mM concentration the solutions contain in mM 10 HEPES and 1.8 EGTA (ethyleneglycol-bis-[β-amino-ethyl ether] N,N'-tetra acetic acid) for the intracellular side and 1.8 $CaCl_2$ for the extracellular side, respectively. Care has to be taken when channel blocking cations are used since they might cause additional gating charge immobilization.[15]

The absence of ionic currents was routinely tested by verification that the measured current signal reaches the baseline level for long enough depolarizations ($\sim$ 10 msec). As discussed in a later section, an even more accurate test is to analyze the autocovariance functions of the traces.

---

[15] F. Bezanilla, E. Perozo, D. M. Papazian, and E. Stefani, *Science* **254,** 679 (1991).

Hardware-Oriented Considerations

In this section we describe some technical details which proved to be important in recording gating currents from patches for optimizing signal-to-noise ratios and for improving the voltage clamp properties.

Low-Noise Patch Recordings

For macropatch pipettes we mostly use aluminum glass (Hilgenberg, Malsfeld, Germany). Another glass type which proved suitable is type 7052 of A-M Systems, Inc. (Everett, WA). Great care is taken to coat the pipettes with several layers of silicone (RTV615, General Electrics, Waterford, NY) as far down to the tip as possible. The pipette holder is kept as dry as possible by blowing nitrogen through the suction line while changing the pipettes. This procedure dries the inside of the holder and prevents any remaining solution from creeping up the internal electrode and thereby creating excess noise. After excision of a patch, the pipettes are lifted up as much as possible, leaving only the very tip immersed in the bath saline, in order to reduce the shunt capacitance.

Current Amplification

For our recordings we use a commercial patch clamp amplifier (EPC7, List Electronics, Darmstadt, Germany). Experiments are controlled by either a PDP 11/73 computer (Digital Inc., Marlborough, MA) or a VME-bus-based computer (MVME147, Motorola, Tempe, AZ). Both computers are equipped with 12-bit analog–digital interfaces. The output of the voltage templates generated by the computer is set larger by a factor of 40 than the final size of the command voltage in order to minimize the digitalization errors. This results in a 10-V step for a desired 250-mV change in membrane potential. For this either a 4:1 reduction in signal is made before feeding the stimulus into the patch clamp amplifier (which normally is designed to scale signals by powers of 10) or the internal scaling factor of the amplifier has to be changed.

Since even after careful analog compensation the residual capacitive artifact is often larger than the signal to be measured, great care has to be devoted to the optimization of the performance of the circuits involved in current signal amplification and capacitance compensation. If possible, these circuits should be selected according to the highest linearity. Because the current should be measured with the largest possible amplification, the effective dynamic range is *de facto* limited by the quality of the capacitive transient cancellation. In some cases, for example, when attempting to measure gating current noise, the amplitude of the capacitive transients may be further reduced by introducing a finite rise time in the stimulus input (we have used up to 100 μsec). Both this artifice and filtering the measured current (e.g., above 5 kHz) very efficiently avoid the loss of data points contaminated by capacitive artifacts whenever the highest time resolution is not mandatory. In any case, avoiding the saturation of amplifiers is essential for a meaningful and correct application of off-line procedures for digital compensation of linear capacitive and leakage currents.

Linear Correction Procedures

In I_g measurements a proper cancellation of linear capacitive transients and leak currents is mandatory because of the small size of the nonlinear signals to be detected. In addition to the considerations of the previous section, it is important to take into account that parameters like the seal resistance and the input capacitance may change during the recording of the many consecutive traces required for signal averaging. In the following we describe various strategies for extracting the nonlinear currents arising from voltage-dependent activation of channels. Although especially for I_g measurements capacitive artifacts are very important, we shall conveniently refer to these procedures as "leak correction." The first group includes those which can be applied on-line, allowing immediate qualitative inspection of the results and quick decisions on how to proceed further. This is very helpful during the actual measurements, when not all experimental parameters have been determined yet. However, a more detailed correction should be performed using off-line techniques belonging to the second group and described later. The linear procedures described here are of general use and not only applicable to I_g measurements.

On-Line Correction

Standard P/4 Methods. For the investigation of voltage-activated ion channels the so-called $P/4$ method is routinely used.[16] The method is based on the assumption that channel activity, and the associated, saturable gating charge displacements, take place only in a limited voltage range. Therefore, all the other electrical properties of the membrane, assumed to be described by a linear admittance, can be measured outside this range. $P/4$ correction pulses, which are copies of the test pulses of only one-quarter in amplitude, are normally applied around a leak holding poten-

[16] C. M. Armstrong and F. Bezanilla, *J. Gen. Physiol.* **70,** 567 (1977).

tial, V_{LH}, of, for example, -130 mV. The responses to four such leak correction pulses are accumulated and subtracted from the test pulse response. They lead to a corrected current trace representing the pure response of the voltage-gated channels. This assumes a completely linear behavior of the rest of the preparation around V_{LH}, of the amplifier, and of the recording system.

For the measurement of gating currents the test pulse should come after the leak correction pulses. This will prevent any long-lasting changes produced by the test pulse from having an effect on the leak pulse responses elicited shortly after the test pulse. Such changes could be a slow recovery of the gating charge or a long-lived activation of other permeabilities, like Ca^{2+}-activated Cl^- channels.

Alternating Leak Records and Variable P/n. A very negative (or positive, as discussed later) V_{LH} is desirable, but inherent instabilities of cell membranes at high voltages discourage in most preparations the use of potentials more negative than -150 mV and periods longer than strictly necessary solely for the measurement of leak pulse responses. The holding potential, V_H, from which the test responses are elicited is usually much less negative, also in order to reduce capacitive transients (see previous discussion). A drawback of having V_H exceed V_{LH} is often the appearance of a drooping baseline in the corrected signal. This is an artifact arising from slow transients produced by the step from V_H to V_{LH} which are erroreously subtracted from the test response. It should be noted that this slow transient is independent of the leak pulse polarity. If an even number of pulses is used, for example, for signal averaging purposes, the systematic error introduced by the step from V_H to V_{LH} can be eliminated by using trains of leak correction pulses of alternating sign. Because leak responses for negative pulses are subtracted from those for positive pulses, this procedure effectively eliminates the tilting baseline artifact.[17]

The subtraction of linear components increases the noise of the corrected record by the background noise of the leak records. This issue requires special attention and is discussed here. In the $P/4$ procedure the correction amounts to multiplying the variance of the background noise, σ_B^2, by a factor 5 (σ_B^2 from the test record and $4\sigma_B^2$ from four control records). A reasonable averaging can produce nice-looking I_g records for large test pulses for which the signal-to-noise ratio is fairly large even without averaging. However, because σ_B^2 if fairly independent of voltage, the small I_g produced by moderate depolarizations is very badly affected by this problem.

[17] F. Conti, I. Inoue, F. Kukita, and W. Stühmer, *Eur. Biophys. J.* **11**, 137 (1984).

A simple way to optimize signal-to-noise ratios is to tailor the size and the number of leak correction pulses so that the measured signals have a signal-to-noise ratio fairly independent of the stimulation voltage. For this purpose, responses to test pulses of small amplitude can be corrected using leak pulses much larger than those used in the $P/4$ procedure, with the only condition of not invading the voltage range of I_g activation. The advantage of this procedure is that a much lower noise level is obtained using the same number of leak correction pulses. For example, test pulses from a V_H of -100 mV to a test potential, V_T, of -40 mV could be corrected by four leak correction pulses from a V_{LH} of -130 to -100 mV and/or -160 mV. This causes 4 times less background noise variance to be introduced by the leak correction as compared with a $P/4$ method. At high V_T the amplitude of the leak correction pulses has to be reduced in order to remain within the linear regime, and more leak correction pulses would have to be recorded in order to obtain the same absolute noise performance; however, the more favorable signal-to-noise ratio allows limiting such an increase in the number of leak pulses to reasonable levels.

The variable signal size is taken into account by the number of averages taken for each stimulation voltage: more averages for small depolarizations, less for high voltages. All this can be done on-line and provides accurately corrected current traces; the signal quality is only limited by the time required for the recording and hence by the lifetime and stability of the preparation under investigation.

Off-Line Correction

The on-line averaging of current traces is possible and valuable for judging the quality of the recordings during an experiment. However, it is convenient to have all the individual records available before signal averaging and leakage correction for an off-line analysis. This allows, for instance, one to discard traces which have obvious extraneous electrical artifacts before averaging. Furthermore, all leak records (not only those associated with a particular test pulse) can be used to correct any test response, provided the preparation is stable. For this reason pulse protocols most convenient for off-line analysis are tailored according to strategies quite different from $P/4$.

Stimulation Protocol. Our standard stimulation protocol for I_g recordings consist of a block of eight leak correction pulses (LCP) alternating with two or three blocks of depolarizations to V_T. The LCP block consists of four pulses each from a V_{LH} of -130 mV to either -160 or -100 mV. The number of test pulses varies depending on the size of V_T. The final sequence is as follows (the upper index of V_T indicates the test potential):

LCP, V_T^{20} ($N = 4$), V_T^{-20} ($N = 8$), V_T^{-40} ($N = 12$), LCP, V_T^0 ($N = 6$), V_T^{-60} ($N = 25$), V_T^{-80} ($N = 25$), LCP, V_T^{40} ($N = 4$), V_T^{-30} ($N = 8$), LCP. This order of V_T was chosen according to the "importance" of the data points and putting the pulses to the largest V_T toward the end of the sequence. It has the advantage that the most important data will be recorded in case the patch would be lost before completion of the series.

Sodium Channels. Figure 1 shows gating currents of sodium channels recorded from a cell-attached patch. They were elicited by step depolarizations between -100 and $+40$ mV in steps of 20 mV. The stimulation protocol was the one described in the preceding section with a V_H of -120 mV and a V_{LH} potential of -140 mV. Notice that the total gating charge (the time integral of I_g) of the "on" phase, Q^{on}, does not equal the Q^{off} measured within a comparable time interval; that is, the gating charge (up to 2/3 for large and long depolarizations) becomes temporarily immobilized during inactivation as reported for other Na^+ channels.[16] It is noteworthy that, besides the above-described linear correction, no further manipulation of the current traces was performed, because the protocol of analysis does not produce any extraneous "baseline tilt."

Nonlinear Correction Procedures

In our recordings of Na^+ channel gating currents, the linear leak correction method as described above worked very well because the preparation was stable even at very negative potentials, and there was no appreciable gating charge movement in the voltage range explored by the leak correction pulses. In some instances the situation is not so favorable, and

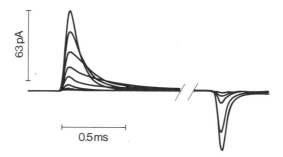

FIG. 1. Sodium gating currents in response to step depolarizations from -100 to 40 mV in steps of 20 mV after linear leak correction. The leak pulses were steps to -160 and -120 mV from a leak holding potential of -140 mV. The break in the traces corresponds to an interval of approximately 1 msec. Bandwidth: 8 kHz. Channel type: Rat brain II. [From F. Conti and W. Stühmer, *Eur. Biophys. J.* **17,** 53 (1989).]

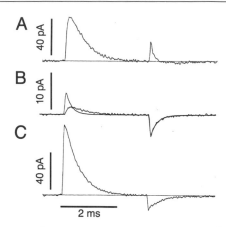

FIG. 2. Gating currents of potassium channels. (A) Gating current trace in response to a test pulse from -100 to $+20$ mV after first-order, linear correction using negative leak pulses (L^-) from -130 to -160 mV. The positive peak in the "off" phase is an artifact arising from the presence of residual gating charge movements in the range from -160 to -130 mV where the leak pulses were recorded. (B) Sum of the responses to leak pulses to -160 and to -100 mV (L^+) starting from a leak holding potential of -130 mV. Notice the change of scale from (A) by a factor of 4. In a completely linear system this tract should be flat; the "on" phase of the trace is well fitted by a double exponential with time constants $\tau(-160)$ and $\tau(-100)$, whereas the "off" phase is fitted by a single exponential with $\tau(-130)$. This allows an estimate of I_g in response to L^- pulses, which caused an error in the correction of the trace shown in (A). To reduce noise the second-order correction was performed by adding to traces (A) the noiseless estimate of $I_g(L^-)$ after filtering and appropriate scaling ($\times 4$ in this case). This procedure yields the final corrected trace in (C). Channel type: SRS. Adapted from Stühmer et al.[18]

other pulse protocols or second-order correction methods are needed as described in this section.

Nonlinear Off-Line Correction

Although the membranes of *Xenopus* oocytes are quite stable if currents are recorded from macropatches, the recording of K^+ gating currents is complicated by the fact that gating charge redistributions occur in a wider range of membrane potential that extends to very negative voltages.[18] Figure 2A shows an I_g record obtained for test pulses from a V_H of -100 mV to a V_T of 20 mV after the leak correction described before using a V_{LH} value of 130 mV and negative leak pulses to -160 mV (L^-). The positive "off" current that appears to follow the repolarization of the membrane from 20 to -100 mV is clearly an artifact. It is the result of

[18] W. Stühmer, F. Conti, M. Stocker, O. Pongs, and S. H. Heinemann, *Pfluegers Arch.* **418**, 423 (1991).

residual nonlinear current components in the voltage range where leak correction pulses where recorded. This is demonstrated by the trace of Fig. 2B which is the sum of the response to positive leak pulses, L^+ (from -130 to -100 mV), and that to L^- pulses. If the membrane patch behaved linearly in the range from -160 to -100 mV this trace would be flat. Therefore, we interpret the "on" and "off" responses in Fig. 2B as being due to gating transitions occurring in the range of -160 to -100 mV. Accordingly, the tail represents gating charge relaxation at -130 mV (both L^+ and L^- end stepping back to -130 mV), whereas the "on" phase is the sum of $I_g(-130$ to $-100)$ and $I_g(-130$ to $-160)$ and is composed of relaxations both at -160 and at -100 mV. Indeed, the "on" response is well fitted by the difference of two exponentials, one yielding a total charge ΔQ with a time constant $\tau(-100)$ of approximately 400 μsec and the other yielding a charge $\Delta Q/2$ with a time constant $\tau(-160)$ of about 200 μsec.

In general, to correct the "overcompensation" of test responses, we estimate from $(L^+ + L^-)$ records and from the first approximation of the test record to -80 mV the following quantities (1) the gating charge associated with steps from -130 to -160 mV and from -130 to 100 mV; (2) $\tau(-130)$ from the tails of $(L^+ + L^-)$ records; and (3) $\tau(-80)$ and $\tau(-100)$ from the fit of $I_g^{on}(-80)$ and $I_g^{off}(-80$ to $-100)$ with single exponentials [$I_g(-80)$ has very little artifact]. The relaxation constant at -160 mV, $\tau(-160)$, can be obtained both from (1) and by extrapolation of the values obtained for -80, -100, and -130 mV. With these data theoretical traces representing the charge movement caused by voltage steps from -130 to -160 mV and back to -130 mV were constructed, filtered, and properly scaled. The result was added to the test traces which had been corrected for linear components using negative-going leak traces only. The corrected trace is shown in Fig. 2C, where it is clearly seen that the inverted "off" tail has vanished; also the rising time course of the "on" phase is restored to the expected shape. Although having a profound effect on the initial time course of the "on" and "off" responses, particularly at large test potentials, this second-order correction never exceeded 20% in terms of absolute gating charge.

P/n at Very Positive Potentials

Not all preparations are stable enough to withstand a V_{LH} of -130 mV. For the investigation of Na^+ channel gating currents a V_{LH} at which all channels are inactivated and the gating charges reach staturation can be chosen. In fact, Bezanilla *et al.*[19] used $+60$ mV for V_{LH} for the investiga-

[19] F. Bezanilla, R. E. Taylor, and J. M. Fernandez, *J. Gen. Physiol.* **79,** 21 (1982).

tion of a slowly inactivating component of the Na^+ gating current in squid giant axons.

Charge Immobilization: Triple-Pulse Protocol

As already anticipated by Fig. 2C, the gating charge of some K^+ channels does not remobilize quickly after large depolarizations. This is deduced from the fact that Q^{on} is not equal to Q^{off}. In general, after test pulses to positive potentials lasting longer than approximately 10 msec, the gating charge can become unavailable for several milliseconds, an effect which we call "complete charge immobilization."

This effect can be used to perform a leak correction with a high signal-to-noise ratio and without introduced assumptions concerning linearity of the system or specific voltage dependencies of nonlinear gating charge components. The method employs two identical test pulses separated by a conditioning pulse sufficiently large to immobilize the gating charge. The second test pulse, which is used as a leak correction pulse, has to follow the conditioning pulse as quickly as possible (limited by the linear capacitive relaxation) so that as little as possible gating charge recovers before the leak correction pulse is applied. An example of such a "triple pulse" and the corrected trace is illustrated in Fig. 3. An estimate of the amount of remobilized charge as a function of time is obtained by comparing the integral of the "on" phase of the third pulse for different intervals between the conditioning pulse and leak correction pulse. Traces corrected according to the triple-pulse protocol using sufficiently short intervals between

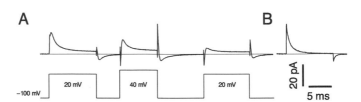

FIG. 3. Because the gating charge of potassium channels may immobilize completely at positive potentials held for more than 10 msec, the triple-pulse protocol (A) can be applied for a direct low-noise leak and capacitance correction. The center pulse immobilizes the charge so that the third pulse contains only the linear response, which is subtracted from the response to the first test pulse, yielding the corrected trace as shown in (B). Proper application of this method requires that the center pulse be large and long enough to immobilize the gating charge completely and that the interval between the end of the center pulse and the onset of the third be so short that no significant amount of charge is remobilized. Notice the fast rising phase of the trace in (B), in agreement with the trace of Fig. 2C, which was obtained with a totally different correction procedure. Channel type: SRS. Adapted from Stühmer et al.[18]

conditioning and leak pulses give not only a fairly accurate measure for the gating charge moved during the first pulse but also, in particular, a non-biased estimate of the rising phase of the gating current.

Comparison of Sodium and Potassium Channels

The study of exogenous ion channels expressed in oocytes allows the unambiguous attribution of a measured signal to a certain ion channel species. Therefore, it is possible to compare directly Na^+ and K^+ channels in terms of their gating current kinetics. The gating currents of both channel types show a very fast rise time and a single- or double-exponential decay of the "on" phase. For both channel types, Q^{on} does not equal Q^{off}; however, whereas only about two-thirds of the gating charge becomes transiently immobilized in Na^+ channels, the "off" phase of the K^+ channel may appear to vanish completely.

In Fig. 4 the steady-state activation curves, namely, a plot of Q^{on} versus membrane potential for sodium (open circles) and potassium (filled circles) channels, are compared. The mid-potentials are very similar, but it is clearly seen that the voltage dependence of the gating charge at negative potentials is much less steep for K^+ channels than for Na^+ channels. Although the measured Q^{on}_{Na} is consistent with three independent gating units per channel, where each unit effectively transports the charge of $2-3$

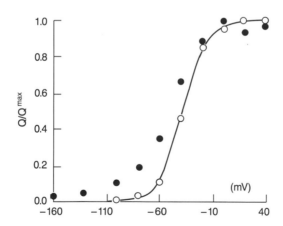

FIG. 4. Q^{on} versus membrane potential for sodium channels (open circles) and potassium channels (filled circles). The cure represents a fit to the sodium data by a Boltzmann function with an apparent gating charge of 2.5 elementary charges and a mid-potential of -38 mV. Q^{on}_K has approximately the same mid-potential but is much less steep at negative potentials. This is the reason for the need of a rather complicated correction method, because even at potentials as negative as -160 mV nonlinear charge relaxations are still appreciable.

electron charges, this description fails for the potassium channel. In the latter case, a voltage-independent final opening step as anticipated from other pieces of evidence[17,20] may be a major cause of a steepening of the $Q^{on} - V$ curve, whereas the foot at negative potentials indicates that a considerably smaller charge movement is involved in the voltage-sensitive gating transitions. Therefore, the "apparent" gating charge as estimated from the steepness of the $Q^{on} - V$ curve seems in this case grossly overestimated.

Measurement of Gating Noise

Because the measurement of mean gating currents does not provide model-independent information about the charge of the unitary gating particles (see previous section), it is desirable to acquire independent estimates of this parameter. The method of nonstationary noise analysis, as described elsewhere,[21,22] has been employed successfully for this purpose. Here we consider the specific requirements posed by the very small size of the signals which have to be measured.

Noise Considerations

In the first part of this section an estimate is made of how much noise is expected to arise from the gating process. Assume that individual gating transitions cause an equivalent translocation of z electron charges, e_0, across the membrane in a statistically independent manner. The noise generated by such a shot process is expressed by the white spectral density, $S = 2ze_0 I_g.$[23] With a peak gating current of 50 pA and $z = 2$, this gives a variance σ_g^2 of 0.32 pA^2 in the frequency band from 0 to 10 kHz. This figure is comparable with the background noise level of 0.5 pA^2 from good patches over the same frequency band. Thus, the ratio of gating noise to background noise is about 2 orders of magnitude smaller than the signal-to-noise ratio of I_g measurements. Therefore, several hundred sweeps need to be averaged for measurement of gating current variance.

A protocol that proved to be convenient consists of a block containing 100 test records to a fixed potential followed by a block with 10 leak correction pulses from a V_{LH} of -140 to -160 and -120 mV at a rate of 5/sec. The test pulse and the leak correction blocks alternate, and at least 600 test records are required to obtain a low enough standard deviation of

[20] W. N. Zagotta and R. Aldrich, *J. Gen. Physiol.* **95**, 29 (1990).

[21] F. J. Sigworth, *J. Physiol. (London)* **307**, 97 (1980).

[22] S. H. Heinemann and F. Conti, this volume [7].

[23] W. Schottky, *Ann. Phys. (Leipzig)* **57**, 541 (1918).

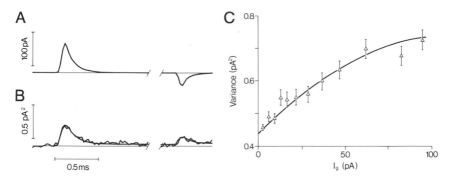

FIG. 5. Analysis of sodium channel gating noise experiments. (A) Mean current and (B) mean ensemble variance as function of time. (C) Variance, σ_g^2, as function of mean current I_g with a fit according to $\sigma_g^2 = iI_g + I_g^2/N$ where the apparent unitary current, i, is given by $2Bze_0$. B is the bandwith, z is the valence of the gating charge, and N is the number of gating units given $Q/ze_0 = (\int I_g \, dt)/ze_0$. Notice that the fitting equation contains z as the only parameter. In this case the least-squares fit yields an estimate of 2.2 for the valence of the unitary gating charge. Bandwidth: 8 kHz. Channel type: rat brain II. (From Conti and Stühmer.[24])

the ensemble variance allowing a clear separation of the component due to I_g fluctuations (roughly proportional to I_g) from the constant background.

As an example, averaged gating currents and the ensemble variance in response to 1730 test depolarizations to 20 mV are shown in Fig. 5A,B.[24] In Fig. 5C a plot of the variance versus the mean gating current of Fig. 5A is shown. The theoretical fit also takes into account the limited bandwidth of the recording, introducing a quadratic term in the I_g–σ_g^2 relationship, arising from the transient character of I_g. It should also be noted that the graph is offset by about the baseline variance which is, as the simple example showed, of the order of the gating noise variance.

Very small residual ionic currents are able to contaminate with their associated channel noise the measured variance significantly. Therefore, it is mandatory to verify the absence of ionic conductances. This can be done by calculating the autocovariance function of the baseline traces and sections of the test pulse segment containing gating currents. The correlation time of the baseline will be short and will depend only on the bandpass characteristics of the recording system. It gives an estimate of the apparent correlation, between originally unrelated events, introduced by the measuring apparatus. The correlation time of I_g fluctuations is expected to be comparable to that of the baseline noise if individual gating charge translocations are statistically independent shot events. Any other resolvable component of the autocovariance function would indicate either a viola-

[24] F. Conti and W. Stühmer, *Eur. Biophys. J.* **17**, 53 (1989).

tion of the latter hypothesis or the presence of ionic currents whose fluctuations would have correlation times corresponding to their macroscopic relaxations.[22] In the case of Na^+ channel I_g measurements recorded from oocytes, the autocovariance function has been shown to drop to zero within one sample interval, clearly indicating the absence of such components.[24]

Charge Carried by Na^+ Channel Gate

The analysis as shown in Fig. 5 is consistent with the interpretation that the unitary gating charge translocated by each of the "independent gates" of the Na^+ channel corresponds to 2.3 electron charges, which is also consistent to what has been derived from the analysis of the voltage dependence of Q^{on}. At this point it should be noted, however, that this value does not reveal the molecular mechanism of the gating process insofar as multiple cooperative steps, each involving a fraction of this charge and occurring in quick succession (faster than our time resolution would allow to detect), would lead to the same result.

Conclusions

The recording of gating currents from patches in combination with the oocyte expression system and molecular engineering, although coming at the very beginning of applications of the method, has already proved to be a powerful tool for the investigation of voltage-activated ion channels.

[23] Ligand-Binding Assays in *Xenopus* Oocytes

By AMY L. BULLER and MICHAEL M. WHITE

Introduction

Xenopus oocytes have proved to be a useful tool for the study and cloning of ion channels and receptors. Functional channels and receptors have been expressed from messenger RNAs isolated from various tissues as well as from transcripts prepared *in vitro* from cDNAs.[1] Electrophysiological characterization of these oocyte-expressed receptors and ion channels has led to measurements of current flowing through the channel itself or, in the case of receptors that cause the mobilization of intracellular Ca^{2+},

---

[1] T. P. Snutch, *Trends Neurosci.* **11**, 250 (1988).

indirect measurement of activation by studying the endogenous Ca^{2+} activated Cl^- channels in oocytes. In general, receptors and channels expressed in oocytes are pharmacologically and functionally identical to their native counterparts.

Although electrophysiological assays offer an unparalleled approach for the study of the functional properties of channels expressed in oocytes, in many cases it is desirable or even necessary to monitor expression by some other means. For example, in some instances it is useful to know how many channels are actually expressed on the surface of the oocyte in order to correlate the number of channels with the measured currents. This would be of particular importance in site-directed mutagenesis experiments if a mutation resulted in a marked diminution in the size of the macroscopic currents. Knowing the number of channels would help determine if this reduction in current was due to a change in the efficiency of expression rather than a change in the intrinsic properties of the channel. As another example, consider the biosynthesis of an ion channel that is synthesized and processed in the intracellular membranes of cells. Although electrophysiological assays can be used to detect channels that are present on the cell surface, they cannot detect those channels that are already synthesized, posttranslationally processed, and in transit to the cell surface. Finally, although electrophysiological assays have been used for expression cloning of a number of receptors and channels,[2,3] there assays can be rather labor intensive.

In the above examples, an alternative way of detecting the presence of the channel or receptor of interest is required. Ligand-binding assays offer a fairly straightforward method applicable to these and other types of questions. In this chapter we discuss the applications of this approach to ion channels and receptors expressed in *Xenopus* oocytes. We first discuss some general considerations for ligand-binding assays using oocytes, and then consider specific approaches used in our laboratory.

Basic Principles of Ligand-Binding Assays

A number of molecules have been shown to interact reversibly with high affinity with various ion channels or receptors. For example, Na^+ channels are blocked by tetrodotoxin (TTX) and saxitoxin (STX), L-type Ca^{2+} channels are blocked by dihydropyridines (DHP), and the nicotinic acetylcholine receptor is blocked by snake venom α-neurotoxins such as

---

[2] G. C. Frech, A. M. J. Van Dongen, G. Shuster, A. M. Brown, and R. H. Joho, *Nature (London)* **340**, 642 (1989).

[3] D. Julius, A. B. MacDermott, R. Axel, and T. M. Jessel, *Science* **241**, 558 (1988).

α-bungarotoxin. In these and other cases, a small molecule interacts with a target site on the channel or receptor of interest in a reversible fashion:

$$L + R \rightleftharpoons LR$$

$$K_{eq} = [L][R]/[LR] \tag{1}$$

where L is the ligand, R is the receptor, and K_{eq} is the equilibrium constant for the reaction. The dependence of [LR] on [L] is described by a rectangular hyperbola defined by the following equation:

$$[LR] = [LR]_{max}[L]/([L] + K_{eq}) \tag{2}$$

where $[LR]_{max}$ is the total number of binding sites (i.e., receptors or channels). When [L] equals K_{eq}, half of the binding sites are occupied by the ligand. Determination of K_{eq} and $[LR]_{max}$ can provide useful information; the value of K_{eq} for a series of ligands can provide a pharmacological fingerprint for the receptor of interest, whereas $[LR]_{max}$ gives the total number of receptors present in a given tissue or cell type.

Operationally, ligand-binding assays are performed by incubating intact cells, cell extracts, or purified membranes with a radiolabeled ligand for a period of time sufficient for equilibrium to be attained. The receptor–ligand complex is seperated from the unbound ligand, and the amount bound is quantified by liquid scintillation or γ counting. In this chapter, we describe how we perform ligand-binding assays using *Xenopus* oocytes under a variety of conditions. Although the precise conditions that should be used will depend on the ligand and the receptor of interest, the procedures described here should be applicable with only slight modifications.

General Hints for Carrying Out Binding Assays

Choice of Radioligand

In general, one would like to perform binding assays with as little tissue as possible. In the case of *Xenopus* oocytes, this is of particular importance since a great deal of effort is involved in the isolation and injection of occytes. Therefore, to achieve a high level of sensitivity (i.e., use the smallest amount of tissue) it is necessary to use a radioligand of very high specific activity. Although radiolabeled compounds containing ^{3}H, ^{14}C, and ^{125}I are available, the specific activity of ^{3}H and ^{14}C compounds are usually too low to be useful in most cases. Tritium-containing compounds have specific activities of 20–50 Ci/mmol [1 Ci = 2.22×10^{12} disintegrations/min (dpm)], and ^{14}C-containing compounds have specific activities of the order of 0.01 Ci/mmol. In contrast, ligands containing ^{125}I are available with specific activities of 1000–2200 Ci/mmol.

As an example, a *Xenopus* oocyte injected with *in vivo* transcribed RNA for the four subunits of the *Torpedo* acetylcholine receptor will express approximately 1–5 fmol of receptor after 48 hr.[4] If all the sites are occupied by an iodinated ligand, then a radioactivity of the order of 5000 counts/min (cpm) per oocyte would be obtained, whereas a ^{3}H-containing ligand would provide approximately 100 cpm/oocyte and a ^{14}C-containing ligand, 0.01 cpm/oocyte. Thus, a single oocyte would be sufficient to provide enough material to easily detect the receptors if an iodinated ligand were used, several might be required if a ^{3}H-containing ligand were used, and thousands would be required if the ligand contained ^{14}C. It must be emphasized that, with regard to the number of sites expressed per oocyte, the expression of a receptor from *in vitro* transcripts probably provides a best-case example. When oocytes injected with RNA isolated from a tissue (in which the mRNA that codes for the protein of interest is present at much lower levels) are used, the level of expression would be much lower. For example, injection of RNA isolated from N1E-115 neuroblastoma cells into oocytes results in the appearance of angiotensin II receptors at levels of the order of 10 amol (0.01 fmol) per oocyte.[5] In this case, even with the use of an iodinated ligand with a specific activity of 2200 Ci/mmol, it was necessary to use groups of 5–10 oocytes to obtain a detectable signal.

Nonspecific Binding

In addition to binding to the site of interest, ligands are also capable of interacting with sites other than the receptor or channel of interest. Nonspecific binding is usually defined as that binding that remains in the presence of a high concentration (usually of the order of 100–1000 K_d) of a nonradioactive ligand. In addition, nonspecific binding is not saturable, and it increases with increasing ligand concentration. Although an obvious choice of competing ligand would seem to be the nonradioactive version of the radioligand, in practice it is better to use some other compound that binds to the receptor with high affinity in order to define true nonspecific binding.[6]

Nonspecific binding can be either tissue or non-tissue related. An example of a tissue-related source of nonspecific binding would be the natural tendency for hydrophobic ligands to partition into the cell membrane, which would then add an extra component of binding. However,

[4] A. L. Buller and M. M. White, *Proc. Natl. Acad. Sci. U.S.A.* **85**, 8717 (1988).
[5] S. J. Fluharty, L. P. Reagan, and M. M. White, *J. Neurochem.* **56**, 1307 (1991).
[6] G. A. Weiland and P. B. Molinoff, *Life Sci.* **29**, 313 (1981).

because this interaction is neither of high affinity nor saturable, it can easily be determined using a high concentration of a competing ligand, which would block only the specific binding but not the nonspecific binding. An example of a non-tissue-related source of nonspecific binding would be binding of the radioligand to glass fiber filters used to separate bound from free ligand by filtration.

In addition to this type of nonspecific binding which all who use radioligand binding assays must consider, the use of heterologous expression systems can introduce another source of "nonspecific" binding, namely, that due to any endogenous channels or receptors that may be present in the cells used for expression. In the case of *Xenopus* oocytes, this is not a serious problem if one is dealing with oocytes proper, as there are essentially no endogenous receptors that can be detected by any of a number of techniques. Likewise, there are very few types of endogenous ion channels present in the oocyte membrane. However, whereas oocytes are quite barren with respect to potential types of ligand-binding sites, the surrounding follicular layer contains a number of different types of receptors, including adenosine,[7] β-adrenergic,[8] angiotensin II,[5] and chorionic gonadotropin, follicle-stimulating hormone (FSH), and lutenizing hormone (LH) receptors.[9] In this type of situation, the follicle cell layer provides an extra source of endogenous specific binding sites that would be added to those arising from the expression of the exogenous RNA, and in some cases this can be a signficant contribution. For example, injection of RNA isolated from N1E-115 neuroblastoma cells into oocytes leads to the expression of angiotensin II receptors at a level of approximately 10 amol/oocyte, whereas uninjected oocytes contain no detectable sites. However, the surrounding follicular layer of uninjected oocytes contains 35 amol angiotensin II receptors per oocyte, which, if present, would tend to obscure those receptors expressed from the exogenous RNA. In general, then, it is wise to make sure that all of the follicular cell layer has been removed prior to using the oocytes.

In general, when using peptides or proteins as the ligand, nonspecific binding can be markedly reduced by the inclusion of 1 mg/ml bovine serum albumin (BSA) in all preincubation, incubation, and wash buffers. The high concentration of BSA should saturate any nonspecific protein binding sites. In addition, if the ligand is a basic protein (as are many neurotoxins), cytochrome c (a small basic protein) can be added to all

[7] L. J. Greenfield, J. Linden, and J. T. Hackett, *Fed. Proc.* **46,** 1133 (1987).

[8] C. Van Rentergehm, J. Penit-Soria, and J. Stinnakre, *Proc. R. Soc. London B* **223,** 389 (1985).

[9] R. M. Woodward and R. Miledi, *Proc. Natl. Acad. Sci. U.S.A.* **84,** 4135 (1987).

solutions at a concentration of 0.1 mg/ml. Finally, if one is using glass fiber filters as the means of separating the oocytes from the incubation medium, the filters should be presoaked with 0.1% polyethyleneimine (PEI).

In addition to these general approaches to minimize nonspecific binding, there may be assay- or ligand-specific means of reducing the nonspecific binding. For example, ^{125}I-labeled α-bungarotoxin does not bind to Whatman (Clifton, NJ) DE-81 anion-exchange filters, whereas nicotinic acetylcholine receptors (AChRs) and AChR – ^{125}I-labeled α-bungarotoxin complexes do. This forms the basis of one way of separating bound toxin from unbound toxin.[10] Presoaking the filters with a buffer containing 1 nM unlabeled α-bungarotoxin markedly reduces the level of nonspecific binding from 50–70% of the total counts on the filter to 10–15% of the total.

Specific Protocols

We describe two different protocols for binding assays as performed in our laboratory. Some modifications may be required to adapt these protocols to a particular ligand–receptor interaction. For example, if the ligand is a peptide that is quite susceptible to proteolysis, appropriate protease inhibitors should be included. This is of particular importance when using cell extracts for the binding, which do contain proteases.

Example 1: Cell Surface Receptor Binding of [^{125}I]SARILE to Angiotensin II Receptors

^{125}I-Labeled [Sarc^1,Ile^8]angiotensin II (SARILE) is a commercially available, high specific activity (2200 Ci/mmol), high affinity angiotensin II (Ang II) receptor antagonist. [^{125}I]SARILE has been used to characterize Ang II receptors in a number of systems because it identifies a homogeneous population of Ang II receptors with high affinity, stability, and a slow dissociation rate. The latter quality is important because it allows extensive washing of the samples to reduce nonspecific binding of this ligand. In oocytes, we have used this ligand to detect Ang II receptors present both in the follicular cell layer and in follicle-free oocytes injected with RNA isolated from N1E-115 neuroblastoma cells. [^{125}I]SARILE binding to either the follicle cell Ang II receptor or to the expressed receptor is saturable and of high affinity, with a K_d of 0.7 nM, and exhibits the proper specificity when one uses a number of competing ligands.[5] The binding can be detected using a single oocyte and displays a linear increase in assays using increasing numbers of oocytes per sample at least up to 16 oocytes.

---

[10] J. Schmidt and M. A. Raftery, *Anal. Biochem.* **52**, 349 (1973).

1. Oocytes are preincubated for 15 min at room temperature with gentle shaking in 12-well tissue culture dishes in 0.5 ml SOS (100 mM NaCl, 2 mM KCl, 1.8 mM CaCl$_2$, 1 mM MgCl$_2$, 5 mM HEPES, pH 7.6) containing 1 mg/ml heat-inactivated BSA, 0.1 mg/ml cytochrome c, 0.3 U/ml aprotinin, and 0.1 mg/ml 1,10-phenanthroline (the last two components are protease inhibitors).

2. The binding reaction is initiated by adding [125I]SARILE to a final concentration of 0.5–1 nM, and the samples are incubated with gentle shaking for 60–120 min. Nonspecific binding is determined by including 1 μM Ang II in the incubation mixture.

3. The incubation medium is rapidly aspirated, and then 3 ml SOS is added to each well.

4. The oocytes are then transferred to glass fiber filters presoaked with 0.1% PEI and washed 3 times with 5 ml of SOS under reduced pressure. The oocytes frequenty disintegrate at this step, but this is not a problem.

5. The filters are then placed in test tubes and counted in a γ counter.

Example 2: 125I-Labeled α-Bungarotoxin Binding to Acetylcholine Receptors in Oocyte Extracts

We have used the following protocol to determine the total number of acetylcholine receptors expressed in oocytes after injection of *in vitro* transcripts for the four receptor subunits. In conjunction with binding assays using intact oocytes, this allows one to determine the number of assembled receptors present in the internal membranes of the oocyte in transit to the cell surface.[4] This protocol is suitable for use either with freshly injected oocytes (i.e., injected, incubated for up to several days to allow expression, and then assayed) or with oocytes that have been frozen after several days of expression. To freeze the oocytes, transfer them to microcentrifuge tubes, remove all but a very thin layer of buffer, and freeze at $-20°$. Frozen oocytes should be assayed within 2 weeks.

1. Homogenize oocytes (5–10 oocytes/500 μl) using a Dounce homogenizer in binding buffer [50 mM sodium phosphate, pH 7.2, 1% Triton X-100, 1 mM EDTA, 1 mM EGTA, 0.1 mM phenylmethylsulfonyl fluoride (PMSF), 0.1 U/ml aprotinin, 1 mg/ml BSA, 0.1 mg/ml cytochrome c]. The PMSF is added from a 100 mM stock in ethanol immediately prior to the homogenization.

2. Transfer the homogenates to 1.5-ml microcentrifuge tubes and incubate for 10 min at room temperature.

3. Remove yolk proteins by centrifuging for 15 min in a microcentrifuge. Transfer the supernatant to a new tube.

4. Preincubate the samples for 30 min at room temperature with gentle shaking. Add unlabeled α-bungarotoxin to a final concentration of 100 nM to tubes for nonspecific binding.

5. Add ^{125}I-labeled α-bungarotoxin (100–2000 Ci/mmol) to a final concentration of 1 nM to each tube. Incubate for 90 min with gentle shaking.

6. Terminate the reaction by filtering the reaction mixture through a Whatman DE-81 filter disk (preincubated for 30 min at room temperature in binding buffer containing 1 nM unlabeled α-bungarotoxin).

7. Wash filters 5 times with 5 ml binding buffer. Air-dry the filters and count in a γ counter.

The removal of the yolk proteins by centrifugation should markedly reduce the levels of nonspecific binding. However, this method does not completely eliminate yolk proteins, and if unacceptable levels of nonspecific binding are still present it may be necessary to purify oocyte membranes prior to the assay using the procedure described below. This procedure is a modification of the method of Ohlsson *et al.*[11] and does not separate surface from internal membranes.

Example 3: Xenopus Oocyte Cell Membrane Isolation

1. Homogenize oocytes (25 oocytes/0.5 ml buffer) in buffer A containing 10% (w/v) sucrose (buffer A is 150 mM NaCl, 10 mM magnesium acetate, 20 mM Tris-HCl, pH 7.6, 0.1 mM PMSF).

2. Carefully layer the homogenate on a discontinuous sucrose gradient made up of a cushion of 50% sucrose in buffer A and a small layer of 20% sucrose in buffer A.

3. Centrifuge at 15,000 g for 30 min in a SW40Ti rotor.

4. The membranes should be at the 20–50% sucrose interface. Remove them with a Pasteur pipette and transfer to a new centrifuge tube.

5. Dilute the sample with 5 volumes of buffer A.

6. Centrifuge at 100,000 g for 30 min to pellet the cell membranes.

7. Rehomogenize the pellet in the appropriate volume of binding buffer from Example 2.

[11] R. Ohlsson, C. Lane, and F. Guengerich, *Eur. J. Biochem.* **115,** 367 (1981).

[24] Expression of Gap Junctional Proteins in *Xenopus* Oocyte Pairs

By LISA EBIHARA

Introduction

The study of gap junctional channels poses a unique problem not encountered with other ionic channels in that the formation of functional gap junctional channels requires the close apposition of two cell membranes. To overcome this difficulty, Dahl and associates[1,2] studied electrical coupling between pairs of "stripped" *Xenopus* oocytes which had been injected with mRNA isolated from estrogen-induced rat myometrium or mRNA synthesized *in vitro* from gap junction cDNA. The main advantage of this system is that it provides a rapid method for assessing the function of normal and altered junctional proteins. In addition, it is possible to visualize the distribution of the junctional protein within the cell as a function of time following mRNA injection using antibodies for the protein.[3] One disadvantage is that it does not allow the study of single gap junctional channels. For such studies, it would be advantageous to stably integrate the gene into a mammalian cell line which has few or no endogenous gap junctional channels such as the human cell line SKHep1.[4] Alternatively, the channel protein could be overexpressed in bacteria and reconstituted into planar bilayers.

Methods

In Vitro Synthesis of Messenger RNA

The connexin cDNAs were separately subcloned into the transcription vector SP64T.[5] This vector contains a cloning site between the 5' and 3' noncoding of regions of β-hemoglobin and has been reported to increase the efficiency of translation of some cloned cDNAs for gap junctional proteins.[3] The DNA is linearized using an appropriate restriction enzyme

[1] G. Dahl, T. Miller, D. Paul, R. Voellmy, and R. Werner, *Science* **236**, 1290 (1987).
[2] R. Werner, T. Miller, R. Azarnia, and G. Dahl, *J. Membr. Biol.* **87**, 253 (1985).
[3] K. I. Swenson, J. R. Jordan, E. C. Beyer, and D. L. Paul, *Cell (Cambridge, Mass.)* **53**, 145 (1989).
[4] A. P. Moreno, B. Eghabali, and D. C. Spray, *Biophys. J.* **236**, 243a (1990).
[5] P. A. Krieg and D. A. Melton, *Nucleic Acids Res.* **12**, 7057 (1984).

and transcribed with SP6 RNA polymerase.[6] The methylated cap ^{7}mGppNP is added to the transcription buffer to produce capped mRNAs.[7]

Preparation of Oocytes

Female *Xenopus laevis* frogs are anesthetized with tricaine methyl sulfonate and small pieces of ovary removed through an incision in the abdomen. The pieces of ovary are stored in sterile modified Barth's solution (MB) containing (in mM) 88 NaCl, 1 KCl, 2.4 NaHCO$_3$, 15 Hepes, 0.3 CaNO$_3$, 0.41 CaCl$_2$, 0.82 MgSO$_4$, 50 mg/l Gentamicin, pH 7.4. Only stage V–VI oocytes are used. The follicular cell layer is removed by incubating small clumps of oocytes in MB without calcium or magnesium and containing 20 mg/ml collagenase (Worthington Freehold, NJ) for 30–45 min. The oocytes are then transferred to another dish containing MB without calcium or magnesium, and the follicular cell envelope is manually removed. After defolliculation, the oocytes are washed 3 times in MB with calcium and injected with 40–50 nl of mRNA dissolved in autoclaved water (20 ng/ml). The oocytes are then incubated overnight in MB solution at 18° in order to allow time for protein synthesis. The time course of protein accumulation has been measured by Swenson *et al.*,[3] who were able to detect induced junctional proteins above the background of endogenous proteins within 6 hr after injection.

The following day the oocytes are devitellinized by first incubating the oocytes in a hypertonic stripping solution (200 mM potassium aspartate, 20 mM KCl, 1 mM MgCl$_2$, 10 mM EGTA, 10 mM HEPES, pH 7.4) for 5–30 min until the vitelline membrane comes away from plasma membrane.[8] The vitelline membrane is then removed with a pair of fine forceps. The devitellinized oocytes are fragile and will lyse if exposed to an air–water interface. They are gently transferred to another dish and manipulated together into pairs. The transfer pipettes consist of either a Pipetman with a large orifice tip (USA Scientific, Ocala, FL) or a glass capillary tube with an inner diameter just larger than an oocyte in a pipetting device. After pairing, the oocytes tend to roll apart unless they are mechanically held in contact with each other. To keep them together, oocyte pairs are wedged between two strips of Parafilm separated by a 2 mm space, which are affixed onto the bottom of the dish. Gap junctions will typically form

---

[6] D. A. Melton, P. A. Krieg, M. R. Rebagliatti, T. Maniatis, K. Zinn, and M. R. Green, *Nucleic Acids Res.* **12**, 7035 (1984).

[7] M. M. Konarska, R. A. Padgett, and P. A. Sharp, *Cell (Cambridge, Mass.)* **38**, 731 (1984).

[8] C. Methfessel, V. Witzemann, T. Takahashi, M. Mishina, S. Numa, and B. Sakmann, *Pfluegers Arch.* **407**, 577 (1986).

after several hours of incubation in MB at 18°. The oocyte pairs can be stored at 18° in MB solution for several days. The membrane potential of the devitellinized oocyte pairs measured 1 day after devitellinization ranges between −40 and −90 mV. Oocytes which have sustained minor, initially undetectable degrees of damage during this procedure will have lower membrane potentials and usually lyse within 1–2 days.

Electrophysiological experiments are performed using a dual, two-microelectrode voltage clamp technique. The microelectrodes are filled with 3 M KCl and have resistances between 1 and 4 MΩ. Each oocyte is impaled with two microelectrodes, one passing current and one for measuring voltage. Both oocytes are then voltage clamped and a voltage clamp step applied to cell 1 while cell 2 is maintained at the holding potential. Under these conditions, the change in current observed in cell 2 in response to a voltage clamp step applied to cell 1 will be entirely due to current flowing through the gap junction.

Results

We chose to study initially the properties of a gap junctional protein isolated from rat heart called connexin 43.[9] Injection of connexin 43 mRNA induced the formation of gap junctional channels. The junctional conductance in the mRNA-injected cell pairs was approximately 2 orders of magnitude larger than that observed in noninjected or water-injected controls. Figure 1 shows a family of junctional current traces recorded in response to a series of depolarizing and hyperpolarizing voltage clamp steps. The junctional conductance was time and voltage independent for potentials between −60 and +60 mV. For larger voltage clamp steps, a small, slowly decaying current component was sometimes observed.

Potential Problems

One problem with this system is that the oocyte pairs have low levels of background coupling which has been attributed to the presence of an endogenous gap junctional protein called *Xenopus* connexin 38 (alpha 2).[10,11] The level of background coupling varies among frogs but is usually similar in oocytes obtained from the same frog. Factors that increase the level of endogenous coupling include incubating the cell pairs at tempera-

[9] E. C. Beyer, D. L. Paul, and D. A. Goodenough, *J. Cell Biol.* **105**, 2621 (1987).
[10] L. Ebihara, E. C. Beyer, K. I. Swenson, D. L. Paul, and D. A. Goodenough, *Science* **243**, 1194 (1989).
[11] R. L. Gimlich, N. M. Kumar, and N. B. Gilula, *J. Cell Biol.* **110**, 597 (1990).

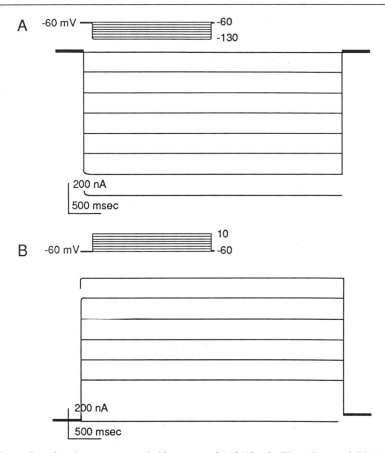

FIG. 1. Junctional currents recorded in a connexin 43/43 pair. The cells were initially held at a holding potential of −60 mV, and a series of hyperpolarizing (A) or depolarizing (B) voltage clamp steps were applied to cell 1. Junctional current was recorded in cell 2.

tures above 25° and allowing the oocytes to remain in contact with each other for more than 24 hr. To control for endogenous coupling we always perform similar experiments on water-injected or noninjected control pairs which are matched with respect to ovary and stage of development. The endogenous junctional channels can also be distinguished from mRNA-induced channels because of their striking dependence on transjunctional voltage. Figure 2 shows a family of junctional current traces recorded from a control oocyte pair which displayed unusually high levels of endogenous coupling.

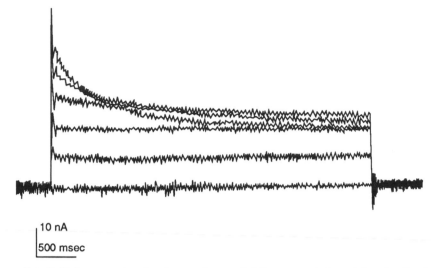

FIG. 2. Endogenous junctional currents recorded from a control noninjected cell pair during a series of depolarizing transjunctional voltage clamp steps between 0 and +50 mV in 10-mV increments. The holding potential was −33 mV.

Conclusions

Expression of gap junctional proteins in *Xenopus* oocyte pairs allows the direct measurement of cell-to-cell coupling. This technique can be used to confirm that a cloned gene encodes a gap junction protein and to study the electrophysiological properties of the different connexin subtypes. It should also be a valuable tool in studying the relationship between structure and function.

Acknowledgments

This work was supported by National Institutes of Health Grants HL45377 and HL28958, a New York Heart Grant-in-Aid, and an Irma Hirschl Career Scientist Award.

[25] Regulation of Intracellular Calcium Activity in *Xenopus* Oocytes

By YORAM ORON and NATHAN DASCAL

Introduction

Xenopus laevis oocytes serve as a useful model system for investigating the mechanism of signal transduction both of intrinsic responses in native oocytes and of acquired responses in oocytes injected with appropriate mRNAs from other tissues.[1] Native oocytes express several cell membrane receptors, stimulation of which leads to the mobilization of calcium both from cellular stores and from the medium (Table I). Following the injection of mRNAs, a number of foreign receptors are expressed in oocytes that also use calcium mobilization to produce the physiologic response, namely, activation of chloride channels (for partial list, see Table I).

Several lines of evidence indicate that cellular calcium is indeed recruited to produce the physiological responses in *Xenopus* oocytes: (1) depletion of cellular calcium by injection of calcium chelators (e.g., EGTA), by repeated resonses in calcium-free medium, or by exposure to vanadate or divalent cation ionophores in calcium-free medium abolishes responses; (2) introduction of calcium into the oocyte either by microinjection or by pretreatment with divalent cation ionophores mimics responses; (3) injection of inositol 1,4,5-trisphosphate (IP$_3$) mimics responses; (4) challenge with the physiological signals (hormones, neurotransmitters) or injection of IP$_3$ result in an increase in ^{45}Ca^{2+} efflux from oocytes that, presumably, reflects an increase in cytosolic calcium concentration. This can be directly monitored by optical methods (fluorescence of Quin-2 or Fura-2 or luminescence of aequorin). In addition, evidence points to recruitment of extracellular calcium: (1) electrophysiologic responses are blunted in the absence of extracellular calcium; (2) metal cations that block calcium entry (i.e., Mn^{2+}) partially inhibit responses to signal or microinjection of IP$_3$; (3) the second component of IP$_3$-induced current is largely dependent on extracellular calcium; (4) following calcium depletion, receptor stimulation or IP$_3$ injection produce chloride current in response to challenge by extracellular calcium.

---

[1] N. Dascal, *Crit. Rev. Biochem.* **4**, 317 (1987).

TABLE I

INTRINSIC AND ACQUIRED RESPONSES THAT UTILIZE CALCIUM

Stimulus	Re
Intrinsic	
Muscarinic (M3)	a, b
Muscarinic (M1)	c, d
Purinergic (P2?)	f, g
Angiotensin	h, i
Acquired	
Muscarinic (M1, M3, M5)	$j-m$
5-Hydroxytryptamine (5-HT)	k, n
Excitatory amino acids	$p-s$
Tachykinins	t
Vasopressin	u
Thyrotropin-releasing hormone (TRH)	v, w
Gonadotropin-releasing hormone (GnRH)	x
Cholecystokinin	y

[a] K. Kusano, R. Miledi, and J. Stinnakre, *Nature (London)* **270**, 739 (1977).
[b] N. Dascal and E. M. Landau, *Life Sci.* **27**, 1423 (1980).
[c] E. Nadler, B. Gillo, Y. Lass, and Y. Oron, *FEBS Lett.* **199**, 208 (1986).
[d] Y. Oron, B. Gillo, and M. C. Gershengorn, *Proc. Natl. Acad. Sci. U.S.A.* **85**, 3820 (1988).
[e] A. Davidson, G. Mengod, N. Matus-Leibovitch, and Y. Oron, *FEBS Lett.* **284**, 252 (1991).
[f] I. Lotan, N. Dascal, S. Cohen, and Y. Lass, *Pfluegers Arch.* **406**, 158 (1986).
[g] S. Gellerstein, H. Shapira, N. Dascal, R. Yekuel, and Y. Oron, *Dev. Biol.* **127**, 25 (1988).
[h] E. M. Landau, personal communication.
[i] M. Lupu-Meiri and Y. Oron, unpublished.
[j] Y. Nomura, S. Kaneko, K. Kato, S. Yamagishi, and H. Sugiyama, *Mol. Brain Res.* **2**, 113 (1987).
[k] N. Dascal, C. Ifune, R. Hopkins, T. P. Snutch, H. Lubbert, N. Davidson, M. Simon, and H. Lest, *Mol. Brain Res.* **1**, 201 (1986).
[l] K. Fukuda, T. Kubo, I. Akiba, A. Maeda, M. Mishina, and S. Numa, *Nature (London)* **327**, 6 (1987).
[m] H. Bujo, J. Nakai, T. Kubo, K. Fukuda, I. Akiba, A. Maeda, M. Mishina, and S. Numa, *FEBS Le* **240**, 95 (1988).
[n] C. B. Gundersen, R. Miledi, and I. Parker, *Proc. R. Soc. London B* **219**, 103 (1983).
[o] D. Julius, A. B. MacDermont, R. Axel, and T. Jessel, *Science* **241**, 558 (1988).
[p] C. B. Gundersen, R. Miledi, and I. Parker, *Proc. R. Soc. London B* **221**, 127 (1984).
[q] C. B. Gundersen, R. Miledi, and I. Parker, *Proc. R. Soc. London B* **221**, 235 (1984).
[r] K. M. Houamed, G. Bilbe, T. G. Smart, A. Constanti, D. A. Brown, E. A. Barnard, and B. N Richard, *Nature (London)* **310**, 318 (1984).
[s] T. A. Verdoorn, N. W. Kleckner, and R. Dingledine, *Mol. Pharmacol.* **35**, 360 (1989).
[t] Y. Harada, T. Takahashi, M. Kuno, K. Nakayama, Y. Masu, and S. Nakanishi, *J. Neurosci.* **7**, 32 (1987).
[u] T. M. Moriarty, S. C. Sealfon, D. J. Carty, J. L. Roberts, R. Iyengar, and E. M. Landau, *J. Bi Chem.* **264**, 13524 (1989).
[v] Y. Oron, B. Gillo, R. E. Straub, and M. C. Gershengorn, *Mol. Endocrinol.* **1**, 918 (1987).
[w] Y. Oron, R. E. Straub, P. Traktman, and M. C. Gershengorn, *Science* **238**, 1406 (1987).
[x] S. Yoshida, S. Plant, P. L. Taylor, and K. A. Eidne, *Mol. Endocrinol.* **3**, 1953 (1989).
[y] J. A. Williams, D. J. McChesney, C. Calayag, V. R. Lingappa, and C. D. Logsdon, *Proc. Natl. Aca Sci. U.S.A.* **85**, 4939 (1988).

Methods

Injection of Active Substances

All intracellular injections may be performed by two methods. When the injection is performed while the oocyte is voltage clamped, to observe the immediate effect of the injected substance, injections are done using a third pipette attached to regulated pressure source (1–3 psi) via a pneumatic valve controlled electronically by a time delay relay device. The micropipette is manually broken (tip diameter 2–5 μm) and backfilled with the desired solution. The volume of injection should not exceed 0.5% of the cell volume (i.e., ~5 nl). The values of drop diameter, volume, and nominal concentration of injected substances are given in Table II. When long-term effects are to be tested, the oocyte can be injected using an adjustable automatic micropipette (e.g., Drummond type, total dispensed volume 5–10 μl, can be adjusted to deliver 10–50 nl). The glass capillary is pulled with an electrode puller and manually broken (diameter 10–25

TABLE II
MICROINJECTION PARAMETERS[a]

Drop diameter (μm)	Drop volume (nl)	% Oocyte volume	Amount injected (pmol)	Concentration (nM)
20	0.004	0.0005	0.004	9.3
40	0.034	0.0037	0.034	74
60	0.113	0.0125	0.113	250
80	0.27	0.030	0.27	593
100	0.52	0.058	0.52	1157
120	0.9	0.1	0.9	2000
140	1.44	0.16	1.44	3176
160	2.14	0.24	2.14	4740
180	3.05	0.34	3.05	6750
200	4.19	0.46	4.19	9260
				(μM)
267	10	1.1	10	22
337	20	2.2	20	44
386	30	3.3	30	66
424	40	4.4	40	88
457	50	5.5	50	111
486	60	6.6	60	133

[a] Amounts refer to microinjected solution of any compound at 1 mM concentration in the pipette. Concentrations are nominal, assuming complete diffusion of the compound in the aqueous space of the cell (taken as 50% of cell volume for a 1.2-mm-diameter oocyte).

TABLE III
AMOUNTS AND EFFECTS OF INJECTED SUBSTANCES

Substance	Injected amount	Effect
EGTA	50 pmol	Inhibition of spontaneous fluctuations
	200–300 pmol	Inhibition of responses
	100–300 pmol	Inhibition of cellular calcium release to facilitate measurements of Ba^{2+} currents
$CaCl_2$	0.1 pmol	Threshold depolarizing fluctuations
	5–50 pmol	Medium monophasic responses
	100–300 pmol	Large responses, possibly biphasic, late fluctuations
IP_3	1–5 fmol	Threshold responses (5–10 nA, fluctuations)
	5–10 fmol	Small biphasic responses, fluctuations
	50–150 fmol	Large biphasic responses, fluctuations
	300–500 fmol	Close to saturating responses (0.5–1 μA), late fluctuations
	2–10 pmol	Desensitizing responses, $^{45}Ca^{2+}$ efflux induction

μm). The capillary is then pressure filled with light mineral or silicone oil and mounted on the wire plunger. It is back-filled with the desired solution. The volume of injected solution can be monitored by identical injection into oil under a microscope (10× objective) equipped with a reticule. The concentrations of substances needed to achieve the desired effects are given in Table III.

Remarks. The effect of intracellular injections depends to a large extent on the site and the depth of injection. To obtain a sharp and immediate effect, calcium or IP_3 should be injected as shallowly as possible (~50 μm deep). This can be performed only in manually or collagenase-defolliculated oocytes and using pipettes that have been broken to a small tip diameter. On deep injections, the resulting currents are delayed and much less sharp. The second component of responses to IP_3 is better observed using deep injections (150–300 μm).[2]

Responses obtained by injection near the animal (pigmented) pole of the oocyte differ from those observed near the vegetal pole. Chloride currents obtained by the injection of calcium are usually much sharper near the animal pole and decay faster than currents of similar amplitudes produced by injections near the vegetal pole. Injection of IP_3 at the two poles results in similar kinetics, except that the amplitude of responses

[2] B. Gillo, Y. Lass, E. Nadler, and Y. Oron, *J. Physiol. (London)* **392**, 349 (1987).

obtained near the animal pole is approximately 5 times greater than those obtained at the vegetal pole.[3]

Injections of EGTA are more effective in suppressing spontaneous current fluctuations and the rapid component of the responses to various signals than the slow component of the response elicited by calcium influx.[4] This may be because of the relatively slow kinetics of calcium chelation by EGTA. Once injected, EGTA is effective for a prolonged time (100–250 pmol will block spontaneous fluctuations for at least 30 hr).

Injections of as little as 0.5 pmol of calcium produce measurable responses. These can be repeated several times without either desensitization or potentiation. A number of consecutive threshold injections results in delayed small depolarizing current with pronounced fluctuations.[2] This may be due to activation of a separate population of chloride channels or to calcium-induced calcium release. Deep injections of large quantities of calcium (100–300 pmol) result in large currents (~1 μA, lasting 20–30 sec) which are potentiated by subsequent identical injection (possibly owing to the partial saturation of calcium stores[5]). Interestingly, unlike calcium introduced by ionophore, injected calcium does not inactivate chloride channels and is not subject to regulation by protein kinase C, indicating a possibly different site of action in the cell.[4-6]

Measurement of $^{45}Ca^{2+}$ Fluxes

Efflux Measurements. Ideally, to quantitate calcium movement, efflux studies should be performed at isotopic equilibrium. Practically, however, oocytes do not equilibrate their calcium even after 48 hr, and the studies are performed taking into account only the more rapidly exchangeable calcium pools.

To measure calcium efflux in individual cells, oocytes are incubated overnight in a small volume of NDE96[4] solution with 200 μCi/ml of $^{45}Ca^{2+}$ (specific radioactivity 100 μCi/μmol). The cells are washed several times in ND96 to remove adhering label. Efflux is monitored either by changing the solution every 0.5–5 min or by holding the oocyte in a perfusion cell and perfusing the desired solution with a peristaltic pump at a rate of 1.0–5.0 ml/min. Fractions (0.5–2.0 ml) are collected with a fraction collector and counted in 4 ml of scintillant. After a certain period needed to determine the basal rate of efflux, oocytes are stimulated with a desired stimulus (hormone, IP$_3$, etc.). At the end of the experiment, the oocyte is homoge-

[3] M. Lupu-Meiri, H. Shapira, and Y. Oron, *FEBS Lett.* **240,** 83 (1988).
[4] R. Boton, N. Dascal, B. Gillo, and Y. Lass, *J. Physiol. (London)* **408,** 511 (1989).
[5] N. Dascal and R. Boton, *FEBS Lett.* **267,** 22 (1990).
[6] M. Lupu-Meiri, H. Shapira, and Y. Oron, *Pfluegers Arch.* **413,** 498 (1989).

TABLE IV
STIMULATED EFFLUX OF $^{45}Ca^{2+}$

Stimulus	Concentration	Maximal fractional rate (%/min)[a]	Time of increased efflux (min)	Total stimulated efflux (%)
ACh-M3	0.1 mM	1.0	4.6	2.2
ACh-M1	0.1 mM	6.4	12.2	35.1
Ade	10 μM	0.5	1.9	2.0
TRH	1 μM	11.7	15.0	45.6
5-HT	1 μM	14.0	10.0	41.0
IP$_3$	>2 pmol	2.9	9.9	32.2
GTP$_\gamma$S	>20 pmol	3.3	13.7	40.7

[a] Maximal fractional rate denotes the net increase in the rate of efflux over the basal rate (0.2–1.0% of residual label/min).

nized in an identical volume of perfusion solution and evaluated for residual counts. Both methods can be used on a voltage-clamped oocyte. In this way, parallel records of electrophysiologic responses and calcium efflux may be obtained. To normalize the data from different oocytes, data can be plotted as log(% residual cpm) versus time (first-order plot) or as percent of counts per minute (cpm) out of total counts per minute in the oocyte at that point (fractional rate of efflux).

On short incubations, $^{45}Ca^{2+}$ efflux appears to follow first-order kinetics. In prolonged incubations, however, the fractional rate of basal efflux steadily decreases, indicating participation of another, slower pool. These methods are described in detail in Refs. 7 and 8. Reference values for various stimulations are given in Table IV. Oocytes take up 2000–10,000 cpm/cell, and basal efflux rates are 0.2–1.0%/min. Oocytes that exhibit much higher efflux rates and/or take up much more $^{45}Ca^{2+}$ should be discarded.

Influx Measurements. Oocytes are incubated in ND96 solution with 200 μCi/ml $^{45}Ca^{2+}$ for a desired period, then washed thoroughly of adhering label and counted. Previous stimulation with a hormone or a neurotransmitter in calcium-free medium results in an enhanced $^{45}Ca^{2+}$ influx. To demonstrate this phenomenon, oocytes are incubated for 10 min with the agonist and washed free of it for an additional 20–30 min with calcium-free medium. They are then transferred to a solution that contains the label with or without the stimulus. Reference values are given in Table IV.

[7] E. Nadler, B. Gillo, Y. Lass, and Y. Oron, *FEBS Lett.* **199**, 208 (1986).
[8] H. Shapira, M. Lupu-Meiri, M. C. Gershengorn, and Y. Oron, *Biophys. J.* **57**, 1281 (1990).

Depletion of Cellular Calcium Stores

Depletion of signal-sensitive calcium stores can be effected in calcium-free medium in three different ways: (1) by repeated exposures to the signal[8,9]; (2) by exposure to divalent cation ionophores[2]; (3) by exposure to Ca^{2+}-ATPase inhibitors (B. Gillo and Y. Oron, unpublished, 1985). It should be noted that only the first method is reversible. The effect can be measured indirectly by observing the physiological response (i.e., depolarizing chloride currents) or directly by following ^{45}Ca^{2+} efflux. These methods are briefly described below.

In depletion experiments, any suitable medium (i.e., ND96 or OR2[10]) may be used, except that calcium addition is omitted and EGTA is added (usually 0.1 mM is enough). It is our experience that maintaining calcium below 50 μM is sufficient to cause calcium depletion. Depletion of calcium stores by repeated exposures is effected by challenging the oocyte with the signal for 1–2 min at 30-min intervals. This results in a complete disappearance of the response and of the associated ^{45}Ca^{2+} efflux by the fourth exposure. Subsequent incubation of the oocyte for 10 min or more with a normal calcium concentration is sufficient to fully restore the response.[8,9]

Depletion by ionophore can be effected by preincubation of oocytes in the presence of 0.2–1 μM of A23187 [in 0.1% v/v ethanol or dimethyl sulfoxide (DMSO)] for 10–15 min. The cells are then washed free of the ionophore solution. It is notable that partial depletion (i.e., in the presence of low concentrations of A23187) affects the rapid component of the response to a larger extent than the second, slow component.[2,11] This is in agreement with our findings that the second component of the responses requires a significant contribution of extracellular calcium.[12]

Depletion with ATPase inhibitors (e.g., vanadate) is effected by adding 0.5–5.0 mM of the agent at pH 8.5–9.0. The initial response was a slowly developing depolarizing current with superimposed current fluctuations, accompanied by modest ^{45}Ca^{2+} efflux. The cell is subsequently insensitive to stimulation (B. Gillo, E. Nadler, Y. Lass, and Y. Oron, unpublished results, 1985).

Remarks. Oocytes maintained in calcium-free medium often deteriorate rapidly owing to the activation of unspecified channels.[9,13] Addition of high magnesium (20 mM) or DMSO (0.1% v/v) sometimes stabilizes the

[9] N. Dascal, B. Gillo, and Y. Lass, *J. Physiol. (London)* **366**, 299 (1985).
[10] N. Dascal, T. P. Snutch, H. Lubbert, N. Davidson, and H. A. Lester, *Science* **231**, 1147 (1986).
[11] D. Singer, R. Boton, O. Moran, and N. Dascal, *Pfluegers Arch.* **41**, 7 (1990).
[12] M. Lupu-Meiri, H. Shapira, and Y. Oron, *FEBS Lett.* **262**, 165 (1990).
[13] N. Dascal, E. M. Landau, and Y. Lass, *J. Physiol. (London)* **352**, 551 (1984).

cells. It is our experience, however, that stability to low calcium is better in oocytes of some donors than in others. In oocytes of those donors the addition of high magnesium can be avoided. Also, oocytes kept in calcium-free medium develop pronounced oscillatory currents that can interfere with responses. These eventually disappear on prolonged incubation in the absence of calcium.

A23187 is light-sensitive, and all operations should be conducted under diminished illumination and in light-protected vessels. Incubation of oocytes in the presence of calcium and A23187 results in slowly developing chloride current, reflecting most probably the kinetics of incorporation of the ionophore. Large responses can be obtained by incorporation of oocytes with A23187 in calcium-free medium, washing of the ionophore, and exposing the cells to short pulses of calcium.[4,6] The large currents can be elicited for at least 2 hr after washing out the ionophore, suggesting that it is incorporated into the membrane in a quasi-irreversible manner. The currents are biphasic, indicating two different conductances sensitive to different concentrations of extracellular calcium. The rapid conductance is subject to strong calcium-dependent inactivation.[4,14]

Calcium-Induced Chloride Currents

Depletion of calcium stores greatly facilitates the detection of chloride currents evoked by entry of extracellular calcium. Oocytes challenged in the absence of calcium by an agonist or IP_3 respond to addition of calcium (20 sec after the challenge) by a small depolarizing current. This current increases on continuous incubation in calcium-free medium (3 min), indicating that calcium depletion potentiates the subsequent response to calcium. This phenomenon is often potentiated by inclusion of 0.1% v/v DMSO in the medium (M. Lupu-Meiri and Y. Oron, unpublished, 1991). This concentration of DMSO does not affect either the holding current or the magnitude of responses. The reason for these effects of DMSO is yet to be determined. Using DMSO-containing ND96, we have adopted a method which uses repeated challenges with the signal in calcium-free medium. On signficant depletion of the stores, large calcium-evoked chloride currents are observed (>100 nA). The addition of a nonspecific calcium entry blocker (e.g., Mn^{2+}, 1 mM) completely abolishes the calcium-induced chloride current.[12]

Demonstration of Separate Calcium Stores

Signal-evoked calcium depletion can be used to demonstrate the existence (or absence) of separate, dedicated calcium stores in the oocyte.

[14] R. Boton, D. Singer, and N. Dascal, *Pfluegers Arch.* **41**, 1 (1990).

Briefly, oocytes possessing the intrinsic muscarinic response and an acquired response [e.g., to thyrotropin-releasing hormone (TRH) after injection with mRNA for TRH receptors] are repeatedly challenged with acetylcholine (ACh). In oocytes preloaded with $^{45}Ca^{2+}$ in the presence of calcium, the label in ACh-specific store is lost, whereas that in the TRH-specific store remains intact. In the absence of extracellular calcium, the response to ACh disappears, while that to TRH remains. The response to ACh can be restored by short incubation with calcium.[8]

Measuring Free Cytoplasmic Calcium

Both calcium microelectrodes and optical methods with fluorescent or luminescent indicators have been used to qualitatively follow free cytoplasmic calcium in oocytes. The main drawbacks of these methods are that they are at best semiquantitative and require specialized equipment or specific training. We have, therefore, decided to include the appropriate references without elaborating these methods here. Although measurements of changes in oocyte free calcium with ion-selective electrodes have been reported, the only well-described work is by Busa *et al.*[15] It should be stressed that electrodes measure calcium in the immediate vicinity of the electrode tip, and this may not necessarily be the compartment of interest.

We have used quin 2 to measure both resting and stimulated concentrations of free calcium in albino oocytes. The results were disappointing. The basal concentration of free calcium was indeed around 0.1 μM. However, the rises due to stimulation of native ACh receptors were negligible and implied that microinjected dye buffers the cytoplasmic calcium or that the rise occurs in a subcellular compartment which constitutes only a negligible fraction of the cytoplasmic space (Y. Oron and M. C. Gershengorn, unpublished, 1985).

Moreau *et al.* have reported using aequorin to follow calcium elevation induced by progesterone.[16] These results were difficult to duplicate.[17,18] Parker and Miledi[19] have demonstrated that injected aequorin can be used to monitor calcium elevation caused by injection of IP_3. Takahashi *et al.*[20] have described a method that used Fura-2 to monitor rises in free calcium following the stimulation of acquired 5-hydroxytryptamine (5-HT) recep-

[15] W. B. Busa, J. E. Ferguson, S. K. Joseph, J. R. Williamson, and R. Nuccitelli, *J. Cell Biol.* **101**, 677 (1985).
[16] M. Moreau, J. P. Vilain, and P. Guerrier, *Dev. Biol.* **78**, 201 (1980).
[17] K. R. Robinson, *Dev. Biol.* **109**, 504 (1985).
[18] R. J. Cork, M. F. Cicirelli, and K. R. Robinson, *Dev. Biol.* **121**, 41 (1987).
[19] I. Parker and R. Miledi, *Proc. R. Soc. London B* **228**, 307 (1986).
[20] T. Takahashi, E. Neher, and B. Sakmann, *Proc. Natl. Acad. Sci. U.S.A.* **84**, 5063 (1987).

tors. Recently, Brooker *et al.*[21] have described an elegant method to measure stimulated calcium changes in oocytes using Fura-2 fluorescence.

Recording of Currents Conducted by Plasma Membrane Calcium Channels

Native *Xenopus* oocytes exhibit low activities of intrinsic voltage-dependent calcium channels.[10] Much greater activities of voltage-dependent calcium channels are expressed in oocytes injected with mRNA from various excitable tissues.[10,22-25] Direct recording of calcium current on depolarization is impossible due to the masking effect of large calcium-activated chloride currents. It may be possible to circumvent this problem by applying the internal perfusion method described elsewhere in this volume.[26] In intact oocytes, it is possible to record currents flowing through calcium channels using barium as the charge carrier, since Ba^{2+} does not activate chloride channels.[27] The standard medium for this type of experiments contains 30–40 mM Ba,$^{2+}$ 60–70 mM Na$^+$ or N-methyl-D-glucamine, and 5 mM HEPES (pH 7.5) and substitutes methane sulfonate for chloride.[10,25] Methane sulfonate does not permeate through chloride channels and also partially blocks them.[4] Chloride channels can be further inhibited by 9-anthracenecarboxylic acid (0.5–2 mM from a stock solution of 200 mM in 0.5–1.0 M NaOH) or niflumic of fluphenamic acids (0.01–0.10 mM from a 0.1–1.0 M stock in DMSO).[4,28] Even in the presence of chloride channel blockers, large Ba^{2+} currents (>100 nA) will produce late variable inward currents, probably owing to the release of cellular calcium and activation of chloride currents. These can be completely suppressed by injections of EGTA (see above).

Note. In EGTA-injected oocytes, pure barium currents can be recorded even in solutions that contain Cl$^-$ as the major anion. EGTA, however, fails to buffer cellular calcium fully (if Ca^{2+} is present in the medium), and the resulting chloride currents will interfere with the recording (N. Dascal, unpublished, 1985).

[21] G. Brooker, T. Seki, D. Croll, and C. Wahlstedt, *Proc. Natl. Acad. Sci. U.S.A.* **87**, 2813 (1990).
[22] J. R. Moorman, Z. Zhou, G. E. Kirsch, A. E. Lacerda, J. M. Caffrey, D. M.-K. Lam, R. H. Joho, and A.M. Brown, *Am. J. Physiol.* **253**, H-985 (1987).
[23] J. P. Leonard, J. Nargeot, T. P. Snutch, N. Davidson, and H. A. Lester, *J. Neurosci.* **7**, 875 (1987).
[24] J. A. Umbach and C. B. Gundersen, *Proc. Natl. Acad. Sci. U.S.A.* **84**, 5464 (1987).
[25] I. Lotan, P. Goelet, and N. Dascal, *Science* **243**, 666 (1989).
[26] N. Dascal, G. Chilcott, and H. Lester, this volume [21].
[27] M. E. Barish, *J. Physiol. (London)* **342**, 309 (1983).
[28] M. M. White and M. Aylwin, *Mol. Pharmacol.* **37**, 720 (1990).

[26] Stable Expression of Heterologous Multisubunit Protein Complexes Established by Calcium Phosphate- or Lipid-Mediated Cotransfection

By Toni Claudio

Introduction

With most types of stable expression, foreign DNA is introduced into the genome of a recipient cell such that the foreign DNA is replicated along with the host DNA during cell division. There are several methods for introducing DNA into cells: microinjection, electroporation, fusion, viral infection, and transfection. For the latter four techniques, where large populations of cells are being treated, either populations of expressing cells ("mass" colonies) or single colonies can be isolated. For single colonies, a single "transduced" cell (one that has integrated the foreign DNA) is allowed to divide until a colony of identical cells can be harvested. This colony is then grown further to establish a new cell line that stably expresses the foreign gene product. For mass colonies, rather than isolating a single colony, large numbers of colonies are harvested and grown into a stable cell line. Every cell will not be identical to every other cell although each cell should be producing the foreign gene product. One disadvantage of stable expression systems is the length of time required to establish a new cell line, usually 6 to 8 weeks for single colony isolates. This process can be hastened by isolating mass colonies (2 to 4 weeks), but it still does not approach most transient systems (1 or 2 days).

There are several advantages of stable expression over other types of expression systems. (1) If single colonies are isolated, large quantities of identical cells can easily be produced. This feature permits many biochemical, immunological, and pharmacological techniques to be performed on the cells. (2) Experiments can be performed at any time without reestablishing expression with each new experiment. Besides simplifying the expression process, this feature provides consistency and uniformity among experiments. (3) The stable and transient expression systems in cultured cells both permit analysis of proteins in an environment with an intact surface membrane. For channels and receptors, this can sometimes be advantageous over systems in which the membrane has been disrupted, such as during the preparation of membrane fragments, reconstitution studies, or addition of material to artificial lipid bilayers.

The expression system one chooses should be dictated by the types of experiments that will be performed and the questions that will be addressed. If one needs days or weeks of continuous expression, or the ability to perform extensive biochemical or pharmacological manipulations on a population of identical proteins, then stable expression may be the best system. On the other hand, if populations of proteins are sufficient, if the initial expression system must be established within a few days rather than a few months, or if there is doubt that the protein will be functional (as might result from mutagenesis experiments or studies in which fewer than the full complement of subunits from a multisubunit protein complex are expressed), then one of the transient expression systems [such as oocytes, simian virus (SV40) in COS cells, vaccinia virus, cytomegalovirus][1] in which products can be expressed within 1 or 2 days is probably desirable. If large quantities of one subunit or fragments of polypeptides are desired, then production in one of the bacterial expression systems may be the best strategy.[1] One might also be constrained by the types of endogenous proteins expressed in the host cell. It is often easiest to detect expression of a foreign gene product in a cell not expressing the homolog of that protein.

There are many proteins which are composed of multiple, heterologous subunits, including many receptors and receptor channels. The members of the superfamily of ligand-gated receptor channels are all composed of heterologous subunits.[2,3] Musclelike acetylcholine receptors (AChRs) are composed of four different subunits (α, β, γ, δ) in the stoichiometry $\alpha_2\beta\gamma\delta$, most neuronal AChRs are composed of at least two different subunits, γ-aminobutyric acid (GABA$_A$) receptors are composed of at least three subunits (α, β, γ), and glycine receptors appear to be composed of three α subunits and two β subunits. Although some receptor and channel properties can often be reconstituted with fewer than the full complement of subunits, fully functional channels probably require all subunits. For some channels, it is easy to know when a fully functional channel has been produced, but for others it is sometimes difficult. Expression studies of various GABA$_A$ receptor subunits revealed only a few pharmacological differences between channels composed of $\alpha\beta$ subunits and those composed of $\alpha\beta\gamma$.[4] Because different combinations of these subunits can form functional channels after expression of their cDNAs, it has become very difficult to elucidate the composition and stoichiometry of a proper GABA$_A$ receptor. Establishing the identity from cloned DNAs of a proper

[1] D. V. Goeddel, ed., this series, Vol. 185.

[2] T. Claudio, in "Frontiers in Molecular Biology: Molecular Neurobiology" (D. M. Glover and B. D. Hames, eds.), p. 63. IRL Press, Oxford, 1989.

[3] H. Betz, Neuron 5, 383 (1990).

[4] D. B. Pritchett, H. Sontheimer, B. D. Shivers, S. Ymer, H. Kettenmann, P. R. Schofield, and P. H. Seeburg, Nature (London) 338, 582 (1989).

neuronal AChR also appears to suffer from problems similar to those of GABA$_A$ receptors: the true subunit composition is not known, and several *in vitro* combinations of subunits produce functional channels.[5]

There are several reasons why a subunit might form part of a functional protein complex *in vitro* but would not do so *in vivo*: (1) a subunit may be developmentally regulated such that *in vivo* it would never have the opportunity to assemble with subunits which are provided *in vitro*, (2) assembly of subunits may be regulated by tissue- or cell-specific expression, or (3) nonspecific associations may form between normally incompatible subunits because the subunits are overly expressed in the *in vitro* expression system or inappropriately regulated. Assembly of the musclelike AChRs, in general, appears to be under tighter control than that of the neuronal AChRs, but promiscuous assembly occurs with these receptors as well. In stable expression studies of the *Torpedo californica* electric organ (musclelike) AChR, only $\alpha_2\beta\gamma\delta$ pentamers are expressed on the cell surface,[6] indicating that assembly of these subunits is tightly regulated. In contrast, transient expression studies in which all four mouse AChR subunit mRNAs are mircoinjected into *Xenopus* oocytes have shown that about 33% of the cell surface AChRs are composed of only $\alpha\beta\gamma$ subunits.[7] These results would indicate that assembly of the mouse AChR subunits is not as tightly regulated as that of *Torpedo* AChR or that subunit assembly in *Xenopus* oocytes is not as tightly regulated as it is in fibroblasts. We have established stable cell lines expressing either *Torpedo* or mouse $\alpha\beta\gamma$ subunits, and cell surface expression of mouse $\alpha\beta\gamma$ subunits is achieved far more readily than that of *Torpedo*, suggesting that mouse AChR subunit assembly is not as tightly regulated as that of *Torpedo*. The problem still remains, however, concerning overexpression of subunits which leads to improper assembly or associations. For example, we introduced *Torpedo* AChR α subunit stably into rat muscle L6 cells and monitored for expression of hybrid AChRs. Although the rat muscle cell is probably optimally suited to assemble rat AChRs, with *Torpedo* α subunit being expressed at levels 10- to 20-fold greater than endogenous rat α we were able to produce *Torpedoi*–rat hybrid AChRs.[8]

Thus, there are several factors which should be entertained when deciding on an expression system. High levels of protein expression are

[5] R. L. Papke, J. Boulter, J. Patrick, and S. Heinemann, *Neuron* **3**, 589 (1989).
[6] D. S. Hartman, M.-M. Poo, W. N. Green, A. F. Ross, and T. Claudio, *J. Physiol. (Paris)* **84**, 42 (1990).
[7] R. Kullberg, J. L. Owens, P. Camacho, G. Mandel, and P. Brehm, *Proc. Natl. Acad. Sci. U.S.A.* **87**, 2067 (1990).
[8] H. L. Paulson and T. Caludio, *J. Cell Biol.* **110**, 1705 (1990).

usually desirable in order to facilitate data analysis. However, high levels may aid in the production of improper subunit associations. The recipient host cell may not have the necessary machinery required to process or assemble a particular protein complex. A stable cell line in which each cell is expressing the identical set of subunits allows the protein to be thoroughly characterized, which may be necessary for identifying subtle differences between proteins. On the other hand, if the proper combination of subunits is not known, a thorough analysis of a protein that is never expressed *in vivo* may not be very informative.

Several gene transfer protocols are described in this volume and other volumes in this series.[1] In the present chapter, I describe protocols for transfecting or lipofecting four separate cDNA constructs plus a selectable marker gene into mouse fibroblast NIH 3T3 or L cells. The cDNAs encode *Torpedo californica* electric organ AChR α, β, γ, and δ subunits[9] engineered in the vector pSV2, a vector in which transcription of the cDNA is driven by the SV40 early promoter.[10] The selectable markers used were either the herpes simplex virus-1 thymidine kinase[11] (*tk*) gene (in the vector ptk[12]) or the Tn5 aminoglycoside phosphotransferase[13] (*neo*[r]) gene (in the vector pSV2[10]); *neo*[r] confers resistance to neomycin, kanamycin, and similar compounds, such as G418. For all transfections, intact, supercoiled plasmid DNAs were used. Transfection efficiencies (the number of transfected cells that express the selectable marker gene) were compared using three different transfection protocols, and cotransfection efficiencies (either the number of cells expressing the selectable marker gene that also integrated or expressed a nonselected DNA or the number of cells that integrated and expressed two marker genes) were also determined. The transfection strategy for these studies was to mix all of the DNAs (cDNAs plus a selectable marker gene or just selectable marker genes), transfect (or lipofect) the mixture onto cells, allow time for the selectable marker gene product to be expressed, then put the cells into media such that only cells with the marker gene would survive. Surviving cells were allowed to divide until each had formed a colony of cells which could be picked and grown until enough cells were present for analysis of the expressed gene products. For some of the experiments, once colonies were formed they were merely

[9] T. Claudio, *Proc. Natl. Acad. Sci. U.S.A.* **84,** 5967 (1987).

[10] T. Claudio, W. N. Green, D. S. Hartman, D. Hayden, H. L. Paulson, F. J. Sigworth, S. M. Sine, and A. Swedlund, *Science* **238,** 1688 (1987).

[11] M. Wigler, S. Silverstein, L.-S. Lee, A. Pellicer, Y.-C. Cheng, and R. Axel, *Cell (Cambridge, Mass.)* **11,** 223 (1977).

[12] F. Colbere-Garapin, S. Chousterman, F. Horodniceanu, P. Kourilsky, and A. C. Garapin, *Proc. Natl. Acad. Sci. U.S.A.* **76,** 3755 (1979).

[13] J. Davis and A. Jimenez, *Am. J. Trop. Med. Hyg.* **29** (Suppl.), 1089 (1980).

stained and counted to determine transfection or cotransfection efficiencies.

Many transfection protocols are present in the literature. I compare and contrast two calcium phosphate-mediated (calcium phosphate transfection) protocols and one positively charged lipid-mediated (lipofection) protocol. The calcium phosphate precipitation protocols used here were further modified from Wigler et al.[11] (Protocol I) or Parker and Stark[14] (Protocol II), both of which were modifications of Graham and van der Eb.[15] The lipofection protocol (Protocol III) was a modification of Felgner et al.[16]

Methods

Protocol I: Calcium Phosphate Precipitation Transfection Procedure for Ltk⁻ Cells with tk

Perform all of the cell work in a sterile, laminar flow tissue culture hood. Spray the outsides of the containers holding DNA preparations, solutions, etc., with 70% (v/v) ethanol. It is especially important that the DNAs (also reagents and tools) used in the transfections are not contaminated with bacteria. It may be necessary to ethanol-precipitate DNA before the transfection to ensure that it is sterile. The picking of cell colonies is also performed in the tissue culture hood.

1. Twenty-four hours before the experiment, seed approximately 5×10^5 cells per 100-mm dish such that cells will number about 10^6 when the transfection is performed [a 1 : 20 dilution of a confluent (all cells are in contact with one another with no spaces remaining between them) 100-mm dish ($\sim 10^7$ cells) of L cells.]

2. Three or four hours before the transfection, replace the medium on the cells with fresh medium (referred to as fluid changing). The medium is Dulbecco's modified Eagle's (DME) medium supplemented with 10% (v/v) calf serum (some L cell lines are adapted to fetal bovine serum).

3. In sterile tube 1 (use polypropylene not plastic tubes) put 0.5 ml of $2 \times$ HBS, pH 7.1 (see below), and 10 μl of 70 mM phosphate. In sterile tube 2 put 10–20 μg of the DNA of interest, 100 ng of tk DNA, 60 μl of 2 M CaCl$_2$, and water to a final volume of 0.5 ml.

[14] B. A. Parker and G. R. Stark, J. Virol. 31, 360 (1979).
[15] R. Graham and A. van der Eb, Virology 52, 456 (1973).
[16] P. L. Felgner, T. R. Gadek, M. Holm, R. Roman, H. W. Chan, M. Wenz, J. P. Northrop, G. M. Ringold and M. Danielsen, Proc. Natl. Acad. Sci. U.S.A. 84, 7413 (1987).

4. While continuously bubbling tube 1, slowly add dropwise the contents of tube 2. Bubble for 1 to 2 min. The bubbling can be done using a disposable sterile serological pipette with cotton plug attached to the end of an automatic pipette aid. While bubbling air through the serological pipette, the contents of tube 2 can be added dropwise by dispensing it from a Pipetman.

5. Let the mixture sit at room temperature about 30 min. A very fine precipitate should form by the end of the incubation. The transfection will not work if large particulate matter is visible.

6. Add the DNA mixture (1 ml) dropwise to the top of the dish of cells and rock the medium gently to mix.

7. Incubate the cells at 37° for 5 to 8 hr (overnight is allright).

8. After 5–8 hr (or overnight) fluid change the medium (replacing it with fresh, nonselective medium).

9. Incubate at 37° for another 12–24 hr.

10. Remove the medium and put the cells into selective medium (a total of ~24–36 hr after the transfection.

11. The selection medium is changed at 3-day intervals (not every other day). The medium used in all of the remaining steps is always selection medium. Cells are maintained in selection even after colonies have been isolated and grown into stable cell lines.

12. Colonies should appear in about 9–14 days. If no colonies are visible by day 14, start again. Anything just appearing after around 20 days is probably not worth pursuing.

13. Pick colonies. This is done most easily and quickly using cotton-tipped wood applicator sticks (cotton swabs) rather than the traditional method of using cloning cylinders (also described in Step 14, below). First autoclave a batch of cotton swabs which have been placed (cotton-side down) in a beaker covered with aluminum foil. Usually, well-formed colonies can be visualized easily by holding the dish of cells up above your head, tilting it at a slight angle (taking care not to spill the medium) so that the part of the dish you are inspecting for colonies is free of medium. Often the colonies are more easily visualized if the overhead lights are not directly above the dish of cells, but rather are off to one side or the other. Take a felt-tipped pen (black or blue is better than green or red) and draw a circle around the colony, then go to an inverted microscope and, using a 4× objective, look to see that the colony has been encircled (Fig. 1A). With

FIG. 1. Colonies of L and 3T3 cells. Ltk⁻ cells were transfected with *tk* and selected in HAT medium (A, B, E, F, G, H, I). NIH 3T3 cells were transfected with *neo* and selected in G418 medium (C, D, J). Cells stained with crystal violet are shown in C, D, F, I, J. Cells were visualized using an Olympus IMT-2 inverted phase-contrast microscope and a 4 × objective (A, B, E, F, I, J) or a 10 × objective (C, D, G, H).

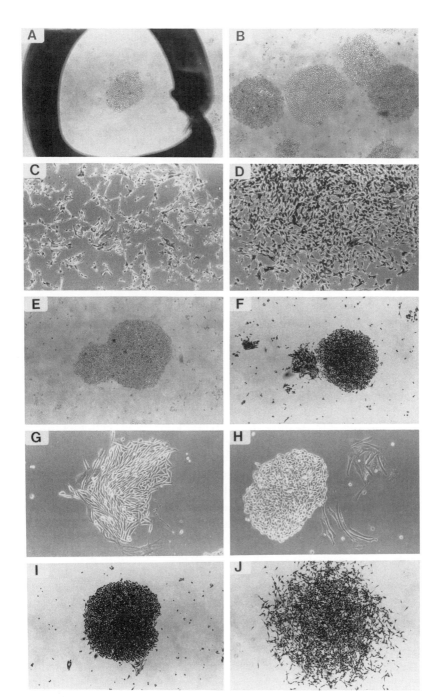

some cell types it is difficult to visualize colonies from the bottom of the culture dish. In such a case, colonies must be located by scanning the dish using a low power objective on the microscope. Mark the position of a colony with a pen, hold the dish up to the light and encircle the colony, then go back to the microscope and check to see that the colony has been encircled. Also look at the distance of other colonies from the one you wish to pick (colonies are too close in Fig. 1B) and the morphology of the colony (Fig. 1C is a very loose 3T3 colony, Fig 1D is a well-formed 3T3 colony). You want colonies that are well isolated from other colonies, that do not have "satellite" colonies around them (the very small colonies or single cells seen in Fig. 1B,E,F,I), or small colonies budding from it (Fig. 1E,F). The idea is to isolate the progeny of a single transduced cell. Different cell types produce colonies with different morphologies. For example, L colonies are usually small (1 mm in diameter) with very densely packed cells (Fig. 1A,B,E–I), whereas 3T3 colonies are much larger (5–10 mm in diameter) and the cells are often loosely packed (Fig. 1D and 1J is typical, Fig. 1C is atypically loose). Even with the same cell type, different cell morphologies can appear (L cells are shown in both Fig. 1G and H).

14. Twenty to forty evenly spaced colonies in one 10-cm dish is an optimal density for picking colonies. Have the positions of all the colonies marked before picking them. Gently rinse the dish with fresh medium, then remove it by aspiration. Have a 24-well tissue culture dish ready with 2 ml of medium in each well that will contain a colony. Take a sterile cotton swab, dip it in medium from the well the colony will go into (e.g., well 1), sweep is across the area marking a colony, and put the swab into well 1 and leave it. Take a second swab, dip it into well 2 pick another colony, and put it into well 2. Repeat the process until all the colonies of one dish have been isolated, then go back and twirl each cotton swab in its well for 2 or 3 sec, discarding the swab afterward. If more than about 5 colonies are being isolated from the same dish of cells or if the colonies are very near one another, it is safer if the dish of cells is rinsed with medium and aspirated after picking about 5 colonies before continuing to pick more colonies from the same dish. This is to prevent contaminating one colony with cells which may have been dislodged during the harvest of other colonies. After all the colonies have been picked into a 24-well dish, take the dish and gently slosh the medium forward and back 3 or 4 times, allow the medium to stop moving, then slosh it left and right. These motions will help distribute the cells evenly on the bottom of a well.

If cells do not transfer well with cotton swabs, use the cloning cylinder method of picking colonies. Cloning cylinders are small rings (6–10 mm in diameter) of chemically inert borosilicate glass (available from Bellco Biotechnology, Bellco Glass Inc., Vineland, NJ). To prepare cloning cylinders for picking colonies, take a glass petri dish and smear a thin layer of silicone grease over the inside bottom cover of the dish. Place the cloning cylinders in the grease so that the bottom of one end of a cylinder is coated with grease. Replace the top cover of the petri dish, tape the lid shut with autoclave tape, and autoclave. Be careful not to use too much grease. Only a thin layer of grease should be on the bottom surface of the cloning cylinder. For picking cell colonies, encircle the colony with a black marker pen as above. Aspirate the medium, rinse two times with phosphate-buffered saline (PBS) and aspirate, pick up a cloning cylinder in grease with a pair of forceps, and place it over the colony using the black pen markings as a guide. Gently push the cylinder down with the forceps to make a good seal with the grease between the culture dish and the cloning cylinder. Check in the microscope to be sure that the colony has been properly enclosed. Using a 1-ml glass pipette with a long tip or a disposable Pasteur pipette with rubber bulb attached, add a few drops of trypsin solution (see Step 16, below), triturate the solution inside the cloning cylinder until the cells are dislodged, and place the solution in a 24-well dish. Check in the microscope that the cells have been removed. If a significant number of cells are still attached to the culture dish, add a few more drops of trypsin solution and triturate again. If too much grease is coating the bottom of a cloning cylinder, the colony may become covered by the grease or the cylinder may skid across the colony when it is being pressed to the culture dish bottom to form a good seal. If two little grease is used or it does not contact the entire cylinder bottom, then the trypsin solution may seep underneath the cylinder before or during trituration. If the colony is left in trypsin solution too long or triturated too vigorously, cell death will occur, which may result in loss of the colony.

15. The day after picking colonies, fluid change the medium to remove bits of cotton and dead cells. The medium should not need changing before the cells become confluent. If the medium turns yellowish, it probably needs changing. If there are only a few cells in a well and the medium turns yellow, check them to be sure they are not contaminated. Inspect the wells every day or two. Cells will often grow slowly shortly after harvesting a colony, but they will begin to approach their proper doubling time after a few days (sometimes it takes a few weeks). Once confluent, the cells will by trypsinized and moved to a 35-mm dish. Sometimes cells will become densely packed in the center of a well, or around the edges, or off to one side. You may need to fluid change the well a few times while waiting for the cells to fill in as best they can. If the cells are very unevenly distributed on the well bottom, it is best to harvest them (when densely packed in some places) and move them to a 35-mm dish (or another well of a 24-well dish) even though they do not cover the entire bottom of the well.

16. Cells are moved from 24-well dishes to 6-well dishes (a 35-mm well or dish) by trypsinization. A concentration of 0.005% trypsin in 0.5 mM EDTA, PBS can be used for many cell types.

17. Once cells are confluent in a 35-mm well, trypsinize and move to a 100-mm dish. Confluent 100-mm dishes are split into more dishes which are used for making frozen stocks (see Appendix, this chapter) of the cell line, analysis, and a back dish from which cells can be plated for future experiments. We never split L cells more dilute than 1:20 (5×10^5 cells) because they tend to grow slower at low density and may grow up as colonies rather than as an even monolayer of cells.

18. It is important to make frozen stocks of cells as early after colony isolation as possible because *all* cell lines change as they are grown in culture. Any cells that lose their selectable marker gene during division will be killed because the cells are maintained in selection medium; however, changes in nonselected DNAs may go undetected.

Solutions

70 mM phosphate (100$\times$), pH 6.8: Prepare equal volumes of 70 mM NaH$_2$PO$_4$ and 70 mM Na$_2$HPO$_4$. Mix the two solutions together, and it should be pH 6.8; sterile filter. Store from $-20°$ to room temperature.

2 M CaCl$_2$: Mix in water, sterile filter. Store from $-20°$ to room temperature.

2$\times$HBS: 50 mM HEPES, 280 mM NaCl; adjust to pH 7.1 $\pm$ 0.05 with 1 M NaOH. Long-term storage is at $-20°$. The bottle in use can be stored from 4° to room temperature, and the pH should be measured (and adjusted if necessary) before each use.

Selection medium: 1$\times$ HAT in DME plus 10% calf serum (some L cells are adapted to fetal bovine serum).

200$\times$ stock of HAT: 0.3 g hypoxanthine, 20 mg aminopterin, and 103 mg thymidine in 100 ml of 50 mM NaOH. Sterile filter and store in aliquots at $-20°$. A final concentration of 1$\times$ HAT is used (15 μg/ml H, 1 μg/ml A, 5.15 μg/ml T).

Comments. The pH of the 2$\times$ HBS is critical. Transfection efficiency is also dependent on DNA concentration.[15,17] We use 10–20 μg of total DNA per 10^6 cells and have used the strategy of keeping the selectable marker gene amount low while keeping the four AChR subunit cDNAs high. For example, we typically use 5 μg each of the four subunit cDNAs with 100 ng *tk*. If 100 ng of *tk* does not give enough colonies, raise it to 250 ng or more.

[17] A. Loyter, G. A. Scangos, and F. H. Ruddle, *Proc. Natl. Acad. Sci. U.S.A.* **79,** 422 (1982).

I have not done an analysis of the cotransfection efficiency keeping the cDNAs constant and varying the selectable marker gene.

Protocol II: Calcium Phosphate Precipitation Transfection Protocol for NIH 3T3 Cells with neo

1. Twenty-four hours before the experiment, seed approximately 4×10^5 cells per 100-mm dish such that cells will number around 10^6 when the transfection is performed [a $1:15$ dilution of a confluent 100-mm dish ($\sim 5 \times 10^6$ cells) of 3T3 cells which double every ~ 18 hr].

2. In a sterile tube put 0.5 ml HBS, pH 7.0–7.1, 10–20 μg of the DNA of interest, and 1 μg of *neo* DNA. Mix and add 31 μl (final concentration 125 mM) of sterile filtered 2 M CaCl$_2$; flick tube about 20 times.

3. Incubate at room temperature 45 min (you should not see particles, just a faint bluish color).

4. Remove the medium from the cells, add the DNA mixture (~ 0.5 ml), and incubate in the hood for 20 min (rock after 10 min).

5. Add 10 ml DME plus 10% calf serum (some 3T3 cells are adapted to fetal bovine serum) to the dish and incubate at 37° for 4 hr.

6. Glycerol shock the cells. Remove medium, add 2 ml of sterile-filtered 15% glycerol in HBS. Incubate at 37° for 3.5 min. Rapidly remove medium by aspiration, add 10 ml DME plus 10% calf serum, swirl, remove medium by aspiration, and add 10 ml DME plus 10% calf serum. Return cells to the incubator and let them grow until confluent (~ 2 days.).

7. Split cells $1:10$ or $1:20$ into *neo*-selection medium, fluid change at 3-day intervals. Colonies should appear in about 9 days.

8. Pick colonies (as described above for L cells) and grow in 24-well dishes, then 6-well dishes (35 mm), then 100-mm dishes. Make frozen stocks of the new cell lines.

Solutions

HBS: 137 mM NaCl, 5 mM KCl, 0.7 mM Na$_2$HPO$_4$, 6 mM dextrose, 21 mM HEPES, pH adjusted to 7.0–7.1 exactly with NaOH.

2 M CaCl$_2$ (as described for L cells)

Selection medium: 0.6 mg/ml G418 (GIBCO Laboratories, Grand Island, NY). The concentration required will vary with the cell type and may vary with different lot numbers of G418.

20× stock of G418: 20 mg/ml G418 in 100 mM HEPES, pH 7.3; sterile filter and store at 4° or for long-term storage at −20°.

Comments. The pH of the HBS is critical. Some protocols include a glycerol shock (Step 6), others do not. See Results (below) for further discussion.

Protocol III: Lipofection Protocol for Ltk⁻ and 3T3 Cells

The following is a transfection protocol (referred to as lipofection) which uses a positively charged lipid, DOTMA {N-[1-(2,3-dioleoyloxy)-propyl]-N,N,N-trimethylammonium chloride}.[16]

1. The day before the experiment, plate cells so that they will be approximately 50% confluent when the transfection is performed. Procedures are given for lipofecting onto a 100-mm dish of cells. Procedures for lipofecting onto a 35-mm dish of cells are given in brackets.

2. Wash cells 3 times with DME (*no* serum).

3. Mix solutions in polystyrene tubes. The DNA mixture contains 1–10 μg of DNA diluted to 1.5 ml with DME (no serum) (1–2 μg DNA to 200 μl with DME for a 35-mm dish). The DOTMA (lipofectin reagent from Bethesda Research Laboratories, Gaithersburg, MD) mixture contains 10–100 μg of DOTMA diluted to 1.5 ml with DME (no serum) (10 μg DOTMA to 200 μl with DME for a 35-mm dish). Mix the two solutions gently (use a ratio of DOTMA to DNA between 4 : 1 and 10 : 1).

4. Remove the medium from the cells, add 3 ml of the DOTMA–DNA mixture (400 μl for a 35 mm dish), and incubate at 37° for 3–24 hr. I used a 5-hr incubation for experiments described in the Results. Note that some very sensitive cell lines may not survive well in serum-free medium for 24 hr.

5. Add 10 ml DME plus calf serum to dishes (1.6 ml to a 35-mm dish).

6. Incubate until confluent (1–2 days), then split into selection medium, fluid change at 3-day intervals, pick colonies, and grow as above.

Comments. Protocols and discussions of cationic liposome-mediated transfection can be found elsewhere.[16,18] Protocol III, presented here, has not been optimized for L or 3T3 cells.

Results

In a previous study, the four *Torpedo* AChR subunit cDNAs plus an adenine phosphoribosyl transferase (*aprt*) gene were cotransfected (following Protocol I) into Ltk⁻aprt⁻cells, and eleven colonies were isolated and grown into stable cell lines. Southern blot analysis of the colonies revealed that 9 of 11 colonies had integrated all four cDNAs (80% cotransfection efficiency), the copy number varied between 1 and 20 per cDNA and, of the 9 positive cell lines, half contained approximately equal numbers (10–20) of each cDNA.[9]

[18] P. L. Felgner and M. Holm, *in* "Focus" Vol. 11, No. 2, p. 21. Bethesda Research Laboratories/Life Technologies, Gaithersburg, Maryland, 1989.

Results from our laboratory in which we cotransfected one, two (33, 46, 75%), or three (40, 50%) cDNAs plus one selectable marker (Protocol I or II) indicate around 45% cotransfection efficiency when analyzing for protein products. When cotransfecting all four cDNAs and analyzing for functional cell surface AChR, the efficiency is approximately 20–25%. Newly established cell lines are analyzed for expression of functional AChR complexes or protein. Because we do not routinely analyze the DNA content, we do not know if the cotransfection efficiency at the DNA level is always 80% with only a 25% efficiency at the functional level. Thus, the best strategy when cotransfecting subunit cDNAs of a heterologous multisubunit protein in which a functional complex will be analyzed is probably to isolate about 24 colonies from a transfection to ensure that 2–5 cell lines will have the desired properties.

A few experiments were performed for this chapter in order to compare transfection efficiencies among different protocols, to determine the cotransfection efficiency using a lipofection protocol, and to determine how DNA concentration might influence liposome-mediated transfection efficiency. The protocols (or modifications of them) were performed on L and 3T3 cells in 35-mm dishes.

Transfection Efficiencies

In one set of experiments, different transfection protocols were tested on L and 3T3 cells in order to compare transfection efficiencies. In each experiment, 1 μg of pSV2-*neo* was used as the selectable marker gene. For Protocol I, around 10^5 L cells were transfected, and cells were incubated with DNA for 5 hr before fluid changing (or a glycerol shock was performed at this time) and put into selection 36 hr after the transfection. For Protocol II, approximately 10^5 L or 6×10^4 3T3 cells were transfected. For Protocol III, about 6×10^5 L or 3×10^5 3T3 cells were transfected, and the DOTMA–DNA mixture (at a ratio of 10:1) was left on the cells for 5 hr.

The results are presented in Table I. The number of colonies produced after transfecting 1 μg of marker DNA and selecting for that DNA and the transfection efficiencies (number of colonies produced divided by the number of cells transfected) are shown. The following conclusions could be drawn. (1) In L cells using Protocol I, glycerol shock did not alter the transfection efficiency. (2) In L cells using Protocol II, the transfection efficiency with a glycerol shock was approximately 10-fold better than without the shock. In 3T3 cells using Protocol II, the glycerol shock improved the efficiency by about 5-fold. (3) Lipofection (Protocol III) increased the number of colonies of L cells by about 20-fold compared with Protocol II with a glycerol shock (~200-fold over Protocol I or

TABLE I
COMPARISON OF TRANSFECTION AND LIPOFECTION PROTOCOLS

Protocol	Glycerol shock	Cell type	Number of cells transfected	Number of colonies[a]	Transfection efficiency[b]
I	−	L	10^5	130	10^{-3}
I	+	L	10^5	150	10^{-3}
II	−	L	10^5	100	10^{-3}
II	+	L	10^5	1200	10^{-2}
III	−	L	6×10^5	25,000	4×10^{-2}
II	−	3T3	6×10^4	50	10^{-3}
II	+	3T3	6×10^4	260	4×10^{-3}
III	−	3T3	3×10^5	290	10^{-3}

[a] Colonies produced after transfecting 1 μg of DNA (pSV2-*neo*) onto cells in 35-mm dishes.

[b] The number of colonies formed divided by the number of cells transfected.

Protocol II without a shock). In 3T3 cells, Protocols II with a shock and III produced the same number of colonies, and both increased the number of colonies by about 5-fold over Protocol II without the shock.

Transfection Efficiency Summary. Of the 3–5 transfection protocol variations tested on 3T3 or L cells, a liposome-mediated protocol produced the greatest number of colonies in L cells (20- to 250-fold more colonies). However, because more cells are transfected when using Protocol III, there is only a 4- to 40-fold increase in the transfection efficiency of this method over the other methods. Although the efficiency is not greatly enhanced, the number of colonies produced is, and it is this latter factor that is usually of most importance. In contrast, with 3T3 cells, lipofection did not produce an increase in colony number compared with Protocol II with a glycerol shock.

Cotransfection Efficiencies Using Lipofection

In another set of experiments, L cells were cotransfected with two selectable marker genes (*tk* and *neo*^{r}) using Protocol III (lipofection) to determine the cotransfection efficiency of this method. Cells were cotransfected with 1 μg each of pSV2-*neo* and p*tk* with 10 μg of DOTMA, then selected in medium supplemented with G418 (for *neo*^{r}), HAT (for *tk*), or G418 plus HAT. More colonies were obtained in G418 medium compared with HAT medium and so the cotransfection efficiency was determined by dividing the number of colonies produced in G418 plus HAT medium by the number of colonies produced in HAT medium. A cotransfection efficiency of approximately 50% (55% and 56% in two separate experiments) was obtained. We do not know how efficiently four nonselected DNAs are introduced into the same cell using a lipofection protocol.

Amount of DNA

It has been shown that there is an optimal amount of DNA that should be used in calcium phosphate-mediated transfections.[11,15,17] High molecular weight carrier DNA is often prepared and used as part of the calcium phosphate mixture, in part, to bring the amount of total DNA to this level. In one experiment, L cells were transfected (using Protocol III) with 0.1, 0.5, or 1 μg p*tk* (the selectable marker gene). In other cells, carrier DNA (pSV2-*neo* was used) was added when the selectable marker gene was less than 1 μg. In this experiment, 35-mm dishes containing 6 × 10^5 L cells were lipofected with 10 μg of DOTMA and the following: 0.1 μg of p*tk*, 0.5 μg of p*tk*, 1 μg of p*tk*, 0.1 μg of p*tk* plus 1 μg pSV2-*neo*, or 0.5 μg of p*tk* plus 0.5 μg pSV2-*neo*. Cells were selected in HAT medium.

The results are shown in Table II. Transfection efficiencies were improved 3-fold when 1 μg of carrier DNA was added to 0.1 μg of selectable marker DNA, and efficiencies were improved 1.4-fold when 0.5 μg of carrier was added to 0.5 μg of marker DNA. If the total amount of DNA was about 1 μg, then the number of colonies was proportional to the amount of selectable marker gene added. The actual reason why transfection efficiency was improved with carrier DNA is not known. It could be related to the total amount of DNA used or to the ratio of DOTMA to DNA used. In this experiment the ratio of DOTMA to DNA varied between 10:1 and 100:1. In other lipofection experiments, when the DOTMA to DNA ratio was 2:1, the transfection efficiency was worse than when the ratio was 4:1 or 10:1. It is not known if approximately 10:1 is

TABLE II
EFFECT OF CARRIER DNA ON LIPOFECTION

Amount of *tk* (μg)	Amount of carrier DNA[a] (μg)	Number of colonies[b]
1.0	None	2500
0.5	None	1000
0.5	0.5	1400
0.1	None	100
0.1	1.0	300

[a] Carrier was pSV2-*neo* plasmid DNA.
[b] Performed on 35-mm dishes of L cells.

an optimal ratio with reduced efficiencies being observed when this ratio is either much larger or smaller.

Summary

For most of the studies conducted in our laboratory, we were interested in expressing the AChR in a nonmuscle cell background, in part, to distinguish inherent AChR properties from those contributed by its environment. The fibroblast host cells we used do not express endogenous AChRs, and although there can be considerable daily variability, for most of the studies, approximately 80% of the cells did not express any type of endogenous channel (S. Sine, personal observations, 1988). Thus, characterization of *Torpedo* AChRs in fibroblasts was simplified by not having endogenous channel expression. We have shown that the pharmacological and electrophysiological properties of AChRs appear fully correct,[10] only $\alpha_2\beta\gamma\delta$ pentamers are expressed on the cell surface,[6] AChRs expressed in fibroblasts can be regulated as they are in muscle cells by agents that increase intracellular levels of cAMP[19] and the AChRs cluster in fibroblasts as they do in muscle cells in response to extracellularly added clustering factors.[20] We have stably expressed *Torpedo* AChRs in NIH 3T3 and L fibroblasts[21,10] as well as rat muscle L6 cells.[8] Different transfectants express different levels of AChRs, with the numbers varying between about 2.5×10^4 and 1.5×10^6 surface AChRs per cell.

In choosing a host cell for expressing a protein of interest, it is always prudent to characterize the line prior to transfection for expression of the homolog of the protein being introduced. When transfecting channel proteins, one might also wish to characterize the host cell for endogenous channel expression. Other considerations for selecting a host cell depend on the types of experiments that will be performed. For example, if fluorescent microscopic experiments will be performed, the ability of the cells to adhere to glass coverslips can be a very important consideration.

Appendix

Frozen Stocks of Cells

Wash a 100-mm dish of cells twice with PBS, trypsinize, add to 5–10 ml of medium in a round-bottomed 15-ml clear tube, and spin gently (e.g.,

[19] W. N. Green, A. F. Ross, and T. Claudio, *Proc. Natl. Acad. Sci. U.S.A.* **88,** 854 (1991).
[20] D. S. Hartman, N. S. Millar, and T. Claudio, *J. Cell Biol.* **115,** 165 (1991).
[21] S. M. Sine, T. Claudio, and F. J. Sigworth, *J. Gen. Physiol.* **96,** 395 (1990).

3 min in a tabletop clinical centrifuge at ~1000 rpm) to pellet cells. Aspirate the supernatant, add 1 ml of 10% dimethyl sulfoxide (DMSO) in medium, resuspend gently, and put 0.5 ml in each of two polypropylene cryotubes. Freeze at −70° in a freezer overnight (most cells will be stable for a month at this temperature), then transfer to a liquid nitrogen storage freezer.

When thawing cells, have a 100-mm dish containing 10 ml of medium ready in the tissue culture hood. Thaw cells quickly by removing a cryotube of cells from the liquid nitrogen storage tank and placing it in a 37° water bath. Shake the tube gently, and as soon as the cells are thawed (or almost thawed) transfer them to the medium-containing dish, distribute the cells evenly on the bottom of the dish (Protocol I, Step 14), and place in an incubator. Most cells will sit down within 1–4 hr. At this time (or wait until the next day), fluid change and place in drug-containing medium.

Comments. There are multiple reasons why low cell viability will result from freezing, including the following: cells were in trypsin too long, they were pelleted too hard or resuspended too roughly, or they were sitting around in 10% DMSO too long before freezing or after thawing. These problems can all be overcome with proper technique; however, there are some cell lines that simply do not freeze well. The day after thawing, instead of having a 1:5 to 1:3 dish of cells, it will only be around 1:100, and the cell doubling time may also be slower. For such cell lines, several vials of frozen cells should be added to the same 100-mm dish, and the dish should be fluid changed as soon as the cells are sitting down in order to remove the DMSO and cell debris. It might also be helpful to leave out the selectable marker drug until the cells are growing properly, but add it back before the cells become confluent. If the line is really sick, let the dish become confluent without adding drug then split it heavily (1:4) into drug-containing medium.

We make freezes (one or two 100-mm dishes) of all newly established cell lines *before* they have been analyzed for expression (Protocol I, Step 17). Although it is a little time consuming and can be expensive, this procedure ensures that an early freeze is made of each line and that we have a backup in case the line in culture is lost (owing to contamination, equipment failure, etc.). Once the lines are screened, we discard the unwanted cell lines and make more freezes of the lines we want to keep (four–eight 100-mm dishes worth of freezes).

Staining Colonies of Cells

The following procecure is for 100-mm dishes; modify volumes for other dish sizes.

1. Wash a dish of cells twice in PBS, then remove the PBS by aspiration.

2. Add 2 ml of 3.7% formaldehyde in PBS. Incubate at room temperature for about 1 h.

3. Rinse with distilled water (can be done under the faucet with a gentle stream of water along the side). Dishes can be dried overnight by inverting on paper, but it is not necessary for them to be dry, only that the excess water be removed (by shaking, for example).

4. Add enough 1% crystal violet in water to cover the bottom of one dish. Swirl for about 10 sec (much longer times are allowable but seconds are sufficient), then pour it into the next dish of colonies. While the second dish is sitting in crystal violet, rinse the first dish in water (done under the faucet with a gentle stream) to remove the stain, shake out the excess water, and invert on paper to dry. Pour the crystal violet from the second dish into a third dish and rinse out the second dish. Repeat until all dishes are stained and washed. Be careful that the dishes do not get mixed up during the staining procedure. Because most dishes are only labeled on their lids, it's important not to separate the dish bottoms from lids.

[27] Vaccinia Virus as Vector to Express Ion Channel Genes

By ANDREAS KARSCHIN, BARBARA A. THORNE, GARY THOMAS, and HENRY A. LESTER

Introduction

Most cDNAs for ion channels, receptors, and transporters arise from eukaryotes, particularly mammals. Heterologous expression of these cDNAs in mammalian cells has the advantage that the host cells may be expected to resemble the cells of origin with regard to biosynthesis, posttranslational modification, sorting, assembly, and targeting of the expressed proteins. The most commonly used eukaryotic expression systems are now (1) the microinjection of cDNA or mRNA into *Xenopus* oocytes (see [14]–[25], this volume) and (2) transiently or stably transfected mammalian cell lines (see [26], this volume). A number of additional approaches utilize expression plasmids and/or viruses. The viruses used as expression vectors include members of many families; yet to date only orthopox-,[1] retro-,[2] and insect baculoviruses (see [28], this volume) have been considered as vectors for the heterologous expression of ion channel genes.

We describe here the use of vaccinia virus (VV), a lytic DNA virus, to achieve the transient expression of voltage-gated ion channels in various primary cell cultures and cell lines. The origin of VV, the live vaccine used to immunize against smallpox (leading to the eventual eradication of this disease), is unclear; but vaccinia belongs to the family of the orthopoxviruses and closely resembles cowpoxvirus, which was used for the first pox immunizations 200 years ago.[3,4] The rationale of using VV as an expression vector is to introduce the foreign ion channel gene by targeted homologous recombination into a site of the VV genome without substantially reducing the ability of the virus to replicate and express proteins in the host cell. The successful transcription and translation of heterologous DNA via VV has already been demonstrated for various viral surface antigens[5,6] and soluble proteins, such as chloramphenicol acetyltransferase (CAT)[7] or neuropeptides.[8,9] Of more relevance here, heterologous expression has been achieved for hormone receptors, such as the atrial natriuretic peptide clearance receptor,[10] and for a voltage-gated ion channel,[1] implying that the expressed gene products are also properly assembled and transported to the cell surface.

Recombinant VV vectors present several advantages for eukaryotic expression technology.[11] First, VV particles have a complex morphology with an outer envelope surrounding the biconcave core and two lateral bodies of unknown function. The genome consists of a linear, double-

[1] R. J. Leonard, A. Karschin, J. Aiyar, N. Davidson, M. A. Tanouye, L. Thomas, G. Thomas, and H. A. Lester, *Proc. Natl. Acad. Sci U.S.A.* **86,** 7629 (1989).

[2] T. Claudio, H. L. Paulson, W. N. Green, A. F. Ross, D. S. Hartman, and D. Hayden, *J. Cell Biol.* **108,** 2277 (1989).

[3] F. Fenner, R. Wittek, and K. R. Dumbell, "The Orthopoxviruses." Academic Press, New York, 1989.

[4] E. Jenner, "An Inquiry into the Causes and Effects of the Variolae Vaccinae, a Disease Discovered in Some of the Western Counties of England, Particularly Gloucestershire, and Known by the Name of the Cow Pox." Sampson Low, London, 1798.

[5] G. L. Smith, M. Mackett, and B. Moss, *Nature (London)* **302,** 490, (1983).

[6] D. Panicali, S. W. Davis, R. Weinberg, and E. Paoletti, *Proc. Natl. Acad. Sci. U.S.A.* **80,** 5364 (1983).

[7] M. Mackett, G. L. Smith, and B. Moss, *J. Virol.* **49,** 857 (1984).

[8] G. Thomas, E. Herbert, and D. Hruby, *Science* **232,** 1641 (1986).

[9] M. R. MacDonald, J. Takeda, C. M. Rice, and J. E. Krause, *J. Biol. Chem.* **264,** 15578 (1989).

[10] J. G. Porter, Y. Wang, K. Schwartz, A. Arfsten, A. Loffredo, K. Spratt, D. B. Schenk, F. Fuller, R. M. Scarborough, and J. A. Lewicki, *J. Biol. Chem.* **263,** 18827 (1988).

[11] B. Moss and C. Flexner, *Annu. Rev. Immunol.* **5,** 305 (1987).

stranded DNA molecule of 187 kilobases (kb), packaged within the virus core. This large genome can be increased still further by incorporation of one or more foreign DNA sequences. At least 25 kb (severalfold more than for smaller viruses) can be introduced without compensatory deletion of VV DNA[12]; the upper limit to added DNA is not yet known.

Second, VV is able to infect 100% of a eukaryotic cell population. Vaccinia virus permits the correct processing and expression of foreign genes in cells of various species, including mammals, birds, and insects, both *in vivo* and *in vitro*. In contrast, plasmid vectors can be used with a limited number of cell types that are competent for microinjection or transection techniques; even then, only a small fraction (typically <1% for stable transfection) take up and express the desired gene. Third, VV carries out its complete life cycle in the cytoplasm of the infected cells. This property, which is unusual for DNA viruses, facilitates the integration of the foreign gene into the VV genome. More importantly, the cytoplasmic life cycle ensures that the transcription and processing of the VV DNA are carried out under the control of viral enzymes (e.g., RNA polymerase), which are packaged in the virus core. Thus, there is no dependence on transcriptional regulation and RNA processing, which complicate expression for viruses that reside in the nucleus of the host cell. However, because VV enzymes do not splice out introns, only cDNAs can be used.

Fourth, although the epidemiology of genetically altered VV has not been explored sufficiently, much information is available about the parental VV wild-type strains that are successfully used in live virus vaccines. The low rate of VV-associated postimmunizing complications[13] may be decreased even further with highly attenuated VV mutants.[14,15] Similarly, thymidine kinase-negative VV phenotypes, as described below, exhibit reduced virulence and thus may represent relatively safe recombinant VV expression vectors for laboratory research.[16] Nevertheless, careful precautions *must* be observed at all times to prevent accidental exposure of personnel, to inactivate totally all virus, and to sterilize all solutions and equipment, which could contain recombinant VV. Protective measures, such as wearing surgical gloves, should be taken by personnel working with

[12] G. L. Smith and B. Moss, *Gene* **25,** 21 (1983).
[13] C. Kaplan, *Arch. Virol.* **106,** 127 (1989).
[14] D. Rodriguez, J.-R. Rodriguez, J. F. Rodriguez, D. Trauber, and M. Esteban, *Proc. Natl. Acad. Sci. U.S.A.* **86,** 1287 (1989).
[15] R. M. L. Buller, S. Chakrabarti, J. A. Cooper, D. R. Twardzik, and B. Moss, *J. Virol.* **62,** 866 (1988).
[16] R. M. L. Buller, G. L. Smith, K. Cremer, A. L. Notkins, and B. Moss, *Nature (London)* **317,** 813 (1985).

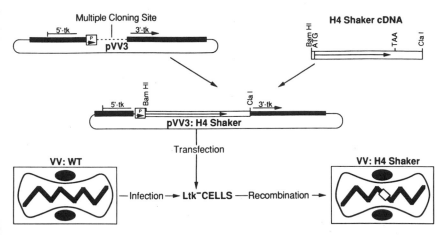

FIG. 1. Design of a recombinant vaccinia virus VV:H4 containing the *Drosophila Shaker* H4 cDNA. For details of the construction of the recombination plasmid pVV3:H4, the transfection/infection procedures, and the selection of recombinants, see text.

the virus. The authors can provide a manual of the safety precautions observed in this laboratory.

Strategy for Construction of Recombinant Vaccinia Virus

Design of Plasmid

To insert an ion channel cDNA at the desired point in the VV genome, we employ the two-step procedure previously developed to construct and isolate VV recombinants that express foreign genes (Fig. 1).[17,18] Initially, standard DNA technology is used to construct a chimeric recombination plasmid. The recombination plasmid pVV3 contains sequences for replication and antibiotic selection in *Escherichia coli* as well as the VV thymidine kinase *(tk)* gene, interrupted by a multiple cloning site downstream of the highly efficient VV 7.5-kDa promoter.[19] Thus, the plasmid permits the targeted integration of the ion channel gene downstream of this VV promoter. Insertion of the foreign cDNA into the *tk* sequence, a nonessential region for virus growth in cell culture, allows a primary enrichment of recombinant viruses in the following purification steps.

[17] M. Mackett, G. L. Smith, and B. Moss, *in* "DNA Cloning" (D. M. Glover, ed.), Vol. 2, p. 191. IRL Press, Oxford, 1985.
[18] D. E. Hruby, G. Thomas, E. Herbert, and C. A. Franke, this series, Vol. 124, p. 295.
[19] C. M. Rice, C. A. Franke, J. H. Strauss, and D. E. Hruby, *J. Virol.* **56,** 227 (1985).

Infection/Transfection and Marker Rescue

In the second step, part of the chimeric plasmid is inserted into the *tk* gene of VV. Because the large VV DNA renders *in vitro* engineering impractical with present techniques, the foreign gene is introduced into VV by homologous recombination *in vivo*. The recombination step is achieved by infecting fibroblasts with wild-type VV (VV : WT), followed by transfection with the recombination plasmid containing the cDNA insert. The VV plasmid sequences undergo homologous recombination with the VV DNA, resulting in the integration of the foreign cDNA at the *tk* locus of the VV genome. The recombinant DNA molecules are then replicated and packaged into mature virions. Because the insertion in the *tk* gene has not interrupted any essential viral function, the recombinants are viable. The fibroblasts are next incubated in the presence of 5'-bromodeoxyuridine (BrdU) to amplify virus in which the *tk* gene has been inactivated by successful homologous recombination. Recombinants are then isolated by several rounds of plaque hybridization and grown to large scale.

Techniques

Marker Rescue

The host cells used for initial infection are thymidine kinase-negative L cells (Ltk⁻), cultured in Eagle's minimal medium (MEM) plus 10% heat-inactivated fetal calf serum (FCS), TAGG supplement (16 μM thymidine, 50 μM adenosine, 50 μM guanosine, 10 μM glycine), 25 μg/ml gentamicin, and 2 mM glutamine. The cells are normally maintained in 25 μg/ml BrdU to select for the tk⁻ phenotype, but they have to be removed from the BrdU selection at least 2 days prior to performing the marker rescue. At the time of infection the cell monolayer should be 70% confluent. Two techniques for introducing plasmid DNA into the infected cells for recombination are described here: calcium phosphate precipitation and lipofection. Standard virology manipulations for performing the marker rescue procedures are described below. All recombinant DNA techniques and solutions follow the standard protocols.[20]

The DNA mixture used for standard $CaPO_4$ transfection is prepared 1 day prior to the marker rescue. The following parts are combined; 5 μg of the insertion plasmid containing the ion channel DNA sequence (e.g., pVV3:H4), 1 μg of intact VV:WT DNA, and 14 μg sonicated salmon

[20] J. Sambrook, E. F. Fritsch, and T. Maniatis, "Molecular Cloning: A Laboratory Manual," 2nd Ed. Cold Spring Harbor Laboratory, Cold Spring Harbor, New York.

sperm or calf thymus DNA as a carrier. VV:WT DNA (WR strain from ATCC, Rockville, MD) can be prepared in advance as described below and stored at $-20°$ for several months under ethanol. The total 20 μg of DNA is coprecipitated with the standard ethanol technique and resuspended in 440 μl water (using aseptic techniques) by simple diffusion at 4° overnight. Vigorous mixing is avoided to prevent shearing of genomic VV DNA.

On the day of marker rescue, Ltk$^-$ cells are infected with VV:WT at a multiplicity of infection (MOI) of 0.05 plaque-forming units (PFU)/cell. The infection procedure is as usually performed with partially purified virus on cells grown in monolayers in the absence of BrdU (perform a cell count in a control dish), and the VV:WT stock is diluted in PBS-MB (phosphate-buffered saline plus 1 mM $MgCl_2$ plus 0.1% bovine serum albumin). Briefly, the cells are washed with PBS-M (phosphate-buffered saline plus 1 mM $MgCl_2$) at 37° and the virus plated at the desired MOI in a small volume [1 ml total volume for a monolayer of BSC-40, Ltk$^-$, or HeLa (see below) cells grown in 100-mm-diameter tissue culture dishes]. Cells are incubated with the inoculum for 30 min at room temperature with intermediate rocking of the plates. Finally the virus inoculum is removed, and the cells are refed with medium and incubated for 3 hr at 37°.

Fifteen minutes prior to the end of the incubation period, a standard calcium phosphate precipitation of the DNA mixture is prepared.[21] All reagents used for precipitation must be at room temperature and in polypropylene tubes. Sixty-two microliters of 2 M $CaCl_2$ is gently added to the 440 μl VV DNA, and the mixture is slowly added to 0.5 ml of 2× HEPES-buffered saline (pH 7.1) until precipitation occurs. Be very gentle to avoid clumping of the DNA. Next, the medium is aspirated from the cells, and the DNA precipitate is carefully added dropwise over the cells. Rocking of plates is avoided to prevent aggregation of DNA. After a second incubation period of 2.5 hr at 37°, the precipitated DNA is aspirated off the cells, which are then washed once with MEM plus 10% FCS. The cells are then briefly shocked with 15% glycerol in 2× HEPES-buffered saline for 40 sec to facilitate DNA uptake. Cells are then rapidly washed twice with MEM plus 10% FCS and placed in a humidified CO_2 incubator for 2 days.

Because the percentage of cells which receive plasmid DNA using standard calcium phosphate techniques is very low, we investigated alternative methods for introducing VV insertion plasmids into VV-infected cells. We found that lipofection of the recombinant VV plasmid greatly increases the percentage of VV which undergo successful recombination yet is a much simpler procedure. Instead of Ltk$^-$ cells, lipofection tech-

---

[21] F. L. Graham and A. J. van der Eb, *Virology* **52**, 456 (1973).

niques are optimal using HeLa cells grown to 50–80% confluence on 35-mm culture dishes for the recombination step. Cell cultures are infected with VV:WT for 30 min at a MOI of 0.1–1 as described for Ltk⁻ or BSC-40 cells. During the inoculation, plasmid DNA containing VV sequences (in sterile water) is complexed with lipofectin reagent (BRL, Gaithersburg, MD, 8292SA) for transfection as follows: for each 35-mm culture dish, 20 μg Lipofectin reagent (in 50 μl sterile water) is added to 3 μg plasmid DNA (in 50 μl sterile water), gently mixed, and allowed to stand 15 min at room temperature. Because no genomic VV DNA is used, resuspension of ethanol-precipitated DNA can be accomplished by vortexing. No overnight incubation is required. Following removal of the VV inoculum, cells are rinsed 3 times with warm serum-free MEM, and 1 ml serum-free MEM is added to each culture dish. The lipofectin–DNA complex is then added dropwise to the cells, and the plates are gently swirled. Cultures are incubated for 4 hr at 37° with 5% CO_2 in these serum-free conditions. Without removing the medium containing the lipofectin–DNA, 1 ml MEM containing 20% FCS is added to each dish, and cells are cultured an additional 2 days.

Following the marker transfer by either transfection method a crude virus stock is prepared by detaching the infected cells from the culture dish using a rubber policeman and pelleting the cells at low speed in a benchtop centrifuge for 5 min. Greater than 90% of VV usually remains cell associated. Therefore, after a wash with PBS-M, the cell pellet is resuspended in 1 ml PBS-M and frozen/thawed 3 times in liquid N_2 to release the virus. The recombinant virus may be amplified by passage through Ltk⁻ cells, which are prepared 3 days ahead and grown in BrdU (25 μg/ml). Infection with the VV crude stock should then be at a MOI of 0.2 PFU/cell (a VV titer of 1×10^7 PFU/ml estimated). After another 3 days of incubation a second crude stock of VV is harvested and processed as before.

The recombinant viruses comprise an estimated 0.1% of the progeny viruses using $CaPO_4$ precipitation,[22] but typically 1–2% with the alternative lipofection method. To distinguish the successful recombinants from wild-type or parental viruses that either (1) survived the BrdU selection or (2) result from mutations of the tk^- gene, several rounds of plaque purification with a hybridizing DNA probe are performed.[12] Routinely, African green monkey BSC-40 cells, grown in MEM plus 10% heat-inactivated FCS and 25 μg/ml gentamicin at 40°, are used as confluent monolayers for virus infection. Assuming a titer of 10^7 PFU/ml from the second Ltk⁻ VV crude stock, a standard infection procedure is performed on a series of

[22] G. L. Smith and B. Moss, *BioTechniques* **Nov/Dec,** 306 (1984).

fresh BSC-40 monolayers, with 10-fold serial dilutions to yield plaque numbers between 1 and 1000 per dish.

Plaque Lift and Filter Preparation

The infected cells are grown for 2 days at 37°, then stained with neutral red (1%) until the plaques become visible. The medium is removed and a NEN (Boston, MA) or Nytran colony plaque screen filter is pressed onto the moist culture dish. For later recovery of the viable virus a replica filter is produced by firmly pressing a nitrocellulose filter against the original with a damp tissue (wetted in PBS-M). After punching orientation holes through the two filters, the replica is set on a damp filter paper, placed in a petri dish, and stored at −70°. The original filter is probed for nucleic acid sequences of the DNA insert by standard procedures for plaque hybridization[23]: the filter is successively denatured (0.5 M NaOH/1.5 M NaCl), neutralized (3 M NaCl/0.5 M Tris base), washed in 2 × SSC, and the VV DNA fixed to the filter by baking *in vacuo* for 30 min at 80°. The baked filters are then incubated at 55° for 30 min with 100 mM Tris-Cl, 150 mM NaCl, 10 mM EDTA, 0.2% (v/v) sodium dodecyl sulfate (SDS), and 50 μg/ml proteinase K to facilitate nucleic acid hybridization. The filters are then placed in prehybridization solution (50% formamide, 1% (v/v) SDS, 1 M NaCl, 10% (w/v) dextran sulfate, and 100 μg/ml denatured salmon sperm DNA) for 2 hr at 42°.

The filters are finally hybridized with a ^{32}P-labeled DNA or RNA probe (specific for the foreign gene), in the prehybridization solution at 42° overnight, washed, and the autoradiography performed. The positions of the recombinant viruses, indicated by dark labeling, are identified; and corresponding plaques are recovered from the nitrocellulose replica using a sterile paper punch or a razor blade. The viruses on the punched dots are released by freeze/thawing 3 times in 200 μl PBS-MB, vortexing, and indirect sonication (6 times, 10 sec). BSC-40 cells grown to monolayers on 100-mm plates are then infected with 5–10% of this stock (diluted in PBS-MB) and cultured for 2 days to allow plaque formation. The plaque purification step is repeated until 90–100% of the plaques are recombinant (determined by nucleic acid hybridization, usually three rounds).

For the final round of plaque purification an agar overlay is performed in which a monolayer of BSC-40 cells grown on a 100-mm plate is infected with the final VV crude stock to yield 25–200 plaques. Following virus adsorption the cells are refed with 10 ml of MEM containing 10% FCS in 0.75% agarose (prepared by mixing 5 ml of 2 × MEM plus 20% FCS with

[23] J. Meinkoth and G. Wahl, *Anal. Biochem.* **138,** 267 (1984).

5 ml 1.5% agarose, Sea Plaque, FMC, Rockland, ME). After 48 hr a second agar overlay with 5 ml of warm 1% agarose containing 200 μl neutral red is overlaid on the dishes. After the dye has stained the cells (about 2 hr), a series of well-separated plaques are isolated by aspiration into the stem of a Pasteur pipette. The agar plug is then transferred to a polypropylene tube containing 400 μl PBS-M, and the virus is released by freeze/thawing. Finally, BSC-40 cells, grown in a 24-well plate, are infected with 100 μl of this stock. Two days after infection, crude stocks are prepared from the infected 24-well cultures and used for both dot-blot hybridization (to confirm that the agar overlayed plaque contains a VV recombinant) and for growing a larger scale preparation of the VV recombinant. However, prior to making a larger scale preparation of the virus, a final plaque screen is performed to ensure purity of the sample.

Growth and Preparation of Partially Purified Vaccinia Virus

Once the VV recombinant is purified, the virus is grown to large amounts to prepare a partially purified virus stock for use in expression experiments. Typically, four 150-mm plates of BSC-40 cells are infected with the recombinant virus at a low MOI of 0.005 PFU/cell (the crude stock of recombinant virus sample from 24-well cultures is assumed to contain roughly 10^7 PFU/ml). The cells are scraped 48 to 72 hr later and pelleted in a clinical centrifuge at low speed. The cells are washed once in PBS-M. Finally the cells are broken in 8 ml of 10 mM Tris (pH 9.0) on ice with 20–25 strokes of a Dounce homogenizer. Nuclei and unbroken cells are separated by centrifugation at 750 g for 5 min at 4°. The pellet is resuspended in an additional 8 ml cold 10 mM Tris, pH 9.0, rehomogenized, and pelleted. The pooled supernatants are overlayed onto a 16-ml sucrose pad [36% (w/v) in 10 mM Tris, pH 9.0] in a 38-ml ultracentrifuge tube and virus particles sedimented through the sucrose pad at 25,000 g (Beckman SW 27 rotor, 13,500 rpm, 80 min, 4°). The visible VV pellet is resuspended in 1–2 ml of 10 mM Tris, pH 9.0, and processed in a Dounce homogenizer until a uniform suspension is obtained (7 strokes).

The final preparation is assumed to have of the order of 10^{10} PFU/ml; however, the actual titer is determined by plaque counting. Therefore, a number of BSC-40 plates are simply infected with dilutions of the virus to obtain between 10 and 1000 plaques per plate. Two days after infection, plaques are visualized by replacing the culture medium with 0.5% (w/v) methylene blue in 50% methanol. After staining 10 min at room temperature, cells are washed twice with 10% methanol, and plaques are counted. The partially purified VV preparation is stored in small aliquots at −70° and is stable for several years. Once thawed, an aliquot can be kept up to 1 month at 4° without loss of titer but should not be refrozen.

Preparation of Vaccinia Virus DNA

The preparation of VV:WT DNA, which is needed for the initial marker rescue transfection of Ltk⁻ cells by CaPO₄, is started with a VV:WT infection of BSC-40 cells using standard inoculation procedures (MOI 0.05 PFU/cell). After 2–3 days the cells are harvested, washed once with PBS-M, and brought to a final volume of 600 μl in PBS-M. Triton X-100 (30 μl of a 10% stock), 250 mM EDTA (48 μl), and 2-mercaptoethanol (1.5 μl) are added, and the sample is kept on ice for 10 min with occasional vortexing to liberate the virus from the cellular membranes. The suspension is centrifuged for 2–3 min at 3000 rpm (TOMY RC-15A microcentrifuge) to remove cellular debris, and the supernatant is spun again at 15,000 rpm for 10 min to pellet the virus particles. The virus is resuspended in 100 μl TE (10 mM Tris-Cl/1 mM EDTA) (pH 8.0) and carefully broken up. Vigorous mixing from this step forward should be avoided to prevent shearing the VV DNA. Sodium chloride (4 μl, 5M), SDS (10 μl, 10%), 2-mercaptoethanol (0.3 μl), and proteinase K (25 mg/ml, 1.5 μl), freshly made in its buffer, are added to the suspension, mixed gently by hand, and incubated at 50° for 30 min to release the virus DNA. Following phenol–chloroform and ether extraction, the VV DNA is ethanol precipitated, pelleted, and stored for further use at −20°. To confirm that the VV DNA was not sheared, it is suggested that a restriction endonuclease digestion with *Hin*dIII be performed and checked for sharp bands on an agarose gel.

DNA from the recombinant virus can be similarly prepared for a restriction endonuclease digest and Southern blot hybridization to verify the correct insertion site of the cDNA to be expressed. Vaccinia virus restriction maps for a number of standard enzymes such as *Bgl*I, *Hin*dIII, *Sal*I, or *Sma*I already exist,[24] and the enzyme of choice will depend on the restriction sites present in the foreign ion channel DNA insert. *Hin*dIII digestion generates a relatively small number of large segments, and the VV *tk* sequences are located in the 4.8-kb J fragment. If one compares *Hin*dIII digests of VV:WT and DNA of the recombinant VV with a coelectrophoresed standard molecular weight marker, the correct insertion of the inserted DNA into the *tk* site can be verified.

Expression of Potassium Channels via Vaccinia Virus Vectors

In an example of the procedures, the 2.2-kb coding region of a *Drosophila* transient K⁺ channel cDNA, *Shaker* H4,[1,25] was subcloned from pBluescript (Stratagene, La Jolla, CA) into the VV recombination plasmid

---

[24] F. M. DeFilippes, *J. Virol.* **43**, 136 (1982).
[25] L. Iverson, M. A. Tanouye, H. A. Lester, N. Davidson, and B. Rudy, *Proc. Natl. Acad. Sci. U.S.A.* **85**, 5723 (1988).

pVV3 cleaved with *Bam*HI/*Cla*I[19] to produce the plasmid pVV3:H4. In this case the initiation codon of the H4 cDNA was placed 17 base pairs (bp) downstream from the 3′ end of the VV "early" transcriptional 7.5-kDa start site. Additional ATG triplets between these two sites should be avoided to assure a correct reading frame. The homologous recombination, marker rescue, and large-scale virus growth were performed as described above to yield the recombinant virus VV:H4 over a period of 3 weeks.

Infections with VV:H4 for the electrophysiological experiments were carried out as described above for Ltk⁻ or BSC-40 cells. Varying infection rates between 1 and 5 MOI were used to infect both a number of cell types in primary culture and several cell lines grown to monolayers (see Table I) with inoculation times of 30 min. In primary cultures, which were sensitive to any change in medium, the cells were refed after infection with a mixture of the original and fresh medium at 37°. The various cell lines and primary cultures were maintained under individually optimal conditions. Voltage clamp recordings of whole-cell currents, performed 24–72 hr

TABLE I

VACCINIA VIRUS-INDUCED EXPRESSION OF *Drosophila Shaker* H4 POTASSIUM CHANNELS IN CELL LINES AND PRIMARY CELL CULTURES[a]

Host cell	Peak current (nA)	Membrane area (μm^2)	Current density ($\mu A/cm^2$)	Channels total	Density/μ
Cell lines					
PC-12 (rat)	1.2 ± 0.2 (7)	1.25 ± 4 × 10³	96	1500	1.2
CV-1 (monkey)	1.2 ± 0.4 (11)	5.5 ± 3.2 × 10³	22	1475	0.3
NIH 3T3 (mouse)	2.9 ± 1.3 (5)	8.5 ± 3.9 × 10³	34	3560	0.5
AtT-20 (mouse)	3.6 ± 1.3 (9)	1.9 ± 0.2 × 10³	189	4500	2.4
RBL-1 (rat)	1.2 ± 0.3 (12)	1.75 ± 0.3 × 10³	66	1440	0.8
Primary cells[b]					
Atrial cells	3.7 ± 2.1 (19)	5 ± 0.7 × 10³	74	4625	0.9
Ventricular cells	4.0 ± 1.3 (3)	4.6 ± 3 × 10³	88	5060	1.1
Cardiac fibroblasts	3.3 ± 0.9 (12)	10.8 ± 2.1 × 10³	30	4060	0.4
SCG cells	2.8 ± 1.7 (5)	4.5 ± 1.1 × 10³	62	3460	0.8
Pyramidal cells	5.7 ± 1.4 (7)	3.2 ± 0.7 × 10³	178	7150	2.3
Brain astroglia	7.6 ± 2.8 (8)	10.5 ± 2.3 × 10³	73	9500	0.9

[a] All infections were performed at an MOI of 1–5 per cell for 30 min at room temperature. The peak current was measured from the response of cells to a voltage step from $V_H = -90$ mV $V_C = +30$ mV at room temperature. The cell membrane surface area was determined from the capacitance of infected cells assuming a specific membrane capacity of 1 $\mu F/cm^2$. The channel density was estimated based on a single channel conductance of 16 pS and a channel open probability of 0.5. Given values are means ±S.D. (N).

[b] All primary cell cultures were obtained from neonatal rats (P2-3).

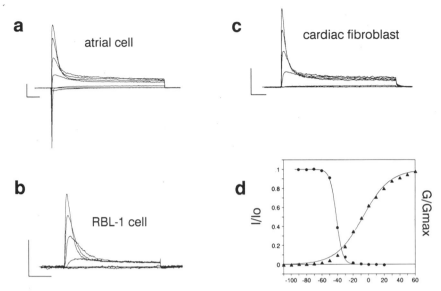

FIG. 2. Examples of VV:H4-induced transient A-type K^+ currents expressed in RBL-1 cells (b) and neonatal rat primary cultures of cardiac atrial cells (a) and fibroblasts (c). The families of current traces are elicited by voltage steps from a holding potential of -90 mV to test potentials between -60 and $+30$ mV. Scale bars are 1 nA, 20 msec. The diagram in (d) shows the steady-state current inactivation I/I_o (circles) and the relative peak conductance G/G_{max} (triangles) for the fibroblast in (c) as a function of the prepulse and pulse potential, respectively. The ratio I/I_o is derived as the peak current response to a voltage step from -120 to $+30$ mV (I_o) versus currents induced from voltage steps to $+30$ mV from decreasing holding potentials (I). The inactivation curve, the best data fit to a Boltzmann distribution, has a slope of 4 mV change for every e-fold change and a midpoint of -41 mV. The conductances for the ratio G/G_{max} were determined from $G = I/V_m - V_k$, assuming $V_k = -70$ mV and G_{max} at $+60$ mV.

postinfection, indicated that the H4 cDNA gene products were successfully assembled to functional A-type K^+ channels when expressed via VV.

Figure 2 compares the waveforms of current responses elicited by a family of depolarizing voltage steps in one cell line and in primary cultures from cardiac tissue. The VV:H4-infected cells displayed transient outward currents that were activated at membrane potentials more positive than -40 mV. In summary, these currents, which were not exhibited by noninfected or by VV:WT-infected control cells, were characterized by all the hallmarks of the native A_1 currents of embryonic *Drosophila* myotubes. The voltage dependence of activation was indistinguishable among cell types; the rate constant for activation was a single-exponential function of voltage. The time course of macroscopic inactivation was also voltage dependent. Recovery from inactivation occurred only at membrane potentials more negative than -40 mV (Fig. 2d). Both the voltage for half-inactivation (-36 mV) and the time constant for recovery from steady-state inactivation ($\tau = 38$ msec, 90% recovery in 200 msec) agree with data for *Drosophila* muscle and for H4 mRNA-injected oocytes. The currents were highly selective for K^+ ions, as indicated by the reversal potential of the tail currents. Another hallmark of the native A current, the sensitivity to 4-aminopyridine, was also demonstrated for the VV:H4-induced K^+ channels.

In cells, such as RBL-1 or fibroblasts, with a low density of intrinsic voltage-dependent ionic conductances, the reconstituted H4 currents were readily apparent in the raw records. Most primary cells from muscle or nervous tissue, however, display endogenous transient A-type K^+ channels, as well as other types of more slowly activating outward K^+ conductances, whose activation and inactivation ranges partially overlap with that of the H4 current. Nevertheless, the fast *Drosophila* currents could be unmistakably identified after pharmacological dissection or by using appropriate voltage protocols.

Coexpression experiments can be performed using multiple VV recombinants. In addition to H4 recombinant VV, we have constructed and isolated a chimeric VV carrying the gene for the rat brain "delayed rectifier-type" K^+ channel RBK-1.[26] Simultaneous coinfection of heart cells with both the VV:H4 and VV:RBK1 recombinants revealed successful expression of both gene products.

The infection rate of recombinant VV is high with most cell types ($>80\%$) but depends on several factors, such as species of origin or the multiplicity of infection. Usually, VV-infected cells after 24 hr are easily identified by distinct morphological changes: rounding up, detachment from the substrate, and cytoplamsic granular inclusions. The efficiency of VV:H4 expression, judging from the fraction of obviously infected cells expressing the transient H4 current, was estimated to be about 50% for the PC-12, RBL-1, and CV-1 cell lines and surprisingly lower ($\sim 10\%$) in the NIH 3T3 fibroblasts, but much higher percentages ($>80\%$) are routinely observed by immunofluorescence of VV proteins. The primary cardiac fibroblasts, on the other hand, appeared to express at a higher rate ($\sim 80\%$) than the heart muscle cells and neurons ($\sim 50\%$). The differences presumably arise from a varying susceptibility to VV infections, rather than from differing viral replication rates in the infected cells. If the infection of a given cell type is less efficient than found for any cells thus far, the standard infection procedure may be varied in several ways. Optimal inoculation conditions may be obtained (1) by incubation up to 2 hr at 30°, (2) by

[26] M. J. Christie, J. P. Adelman, J. Douglass, and R. A. North, *Science* **244**, 221 (1989).

superinfection a few hours after the first infection, or (3) by decreasing serum concentrations in the growth medium.

Using the A_1 single-channel parameters obtained for cultured *Drosophila* myotubes[27] we calculated the densities of VV-expressed H4 channels for each tested cell type (see Table I). Average peak channel densities up to 2–3 channels/μm^2 cell membrane should be more than sufficient to perform single-channel measurements on the infected cells for a more detailed analysis of the location and characteristics of the expressed membrane proteins.

Recent Developments

One particularly exciting application of recombinant VV vectors would be to increase the level of foreign gene expression in order to obtain milligram quantities of ion channel protein. This goal may be achieved either by altering the site of gene insertion, by using stronger promoters, or by insertion of multiple gene copies. For maximum expression, a novel strategy might be based on coinfection with two recombinant VV strains. One recombinant virus would be designed to carry the coding region of the desired ion channel gene, flanked by potent bacteriophage T7 RNA polymerase promoter and termination sequences; the second VV would be vTF7-3, which harbors the T7 polymerase gene downstream of a regular VV promoter.[28] For the heterologous expression of β-galactosidase, this dual infection procedure is roughly 300 times more efficient than simple infection with a single recombinant virus.[28,29] In addition, to improve the translatability of the largely uncapped VV transcripts, an encephalomyocarditis virus (EMCV) untranslated region (UTR), which confers a high affinity ribosome binding site to the 5′ end of the RNA, can be inserted between the T7 promoter and the foreign gene. Judged by a CAT expression assay, use of the EMCV UTR appeared to enhance the VV/T7 hybrid system 4- to 7-fold.[30] This variation of a VV expression vector, consisting of a DNA virus vector, a bacteriophage promoter, and an RNA virus ribosome binding site, has already been successfully used for the heterologous expression of serotonin $5HT_{1A}$ receptors in neonatal cardiac atrial cells.[31]

[27] C. K. Solc, W. N. Zagotta, and R. W. Aldrich, *Science* **236**, 1094 (1987).
[28] T. R. Fuerst, E. G. Niles, W. Studier, and B. Moss, *Proc. Natl. Acad. Sci. U.S.A.* **83**, 8122 (1986).
[29] T. R. Fuerst, P. Earl, and B. Moss, *Mol. Cell. Biol.* **7**, 2538 (1987).
[30] O. Elroy-Stein, T. R. Fuerst, and B. Moss, *Proc. Natl. Acad. Sci. U.S.A.* **86**, 6126 (1989).
[31] A. Karschin, B. Y. Ho, C. Labarca, O. Elroy-Stein, B. Moss, N. Davidson, and H. A. Lester, *Proc. Natl. Acad. Sci. U.S.A.* **88**, 5694 (1991).

Although this chapter gives methods for the routine construction of a recombinant VV applicable to virtually any cDNA, it is still a tedious process. Other selection procedures for recombinant plaques, for instance, by antibiotic resistance[32] or large plaque size,[33] may be more efficient than *tk* gene disruption. More importantly, expression of ion channels can still be detected in a procedure that entirely eliminates construction of recombinant viruses. In this alternative protocol, the target cells are simultaneously infected/transfected with (1) a VV harboring the T7 gene and (2) a plasmid containing the desired foreign gene downstream of the T7 promoter and the EMCV untranslated gene. The absolute level of expression per cell with such a system is expected to be lower than for the double-VV infection protocol.[30] However, a majority of the cells do express the gene. The limiting factor in a typical calcium phosphate transfection procedure is not the uptake of the precipitated DNA into the cytoplasm, but rather its transport into the nucleus. The high fraction of expressing cells using this system is due to the fact that VV transcription enzymes work in the cytoplasm and the transfected plasmid does not have to enter the nucleus for transcription.

Future Directions

Vaccinia virus infections can lead to cell death in the majority of cell lines after several days in culture. The time window between first expression of the foreign protein and lysis of the host cell, as in other transient expression systems, naturally limits the types of experiments that can be performed using the VV expression system. Significant efforts in some laboratories are directed toward addressing this shortcoming by cloning the target gene into a conditional lethal VV mutant, or by performing the experiments in the presence of drugs that decrease the virus replication rate (e.g., 5-fluorodeoxyuridine, rifampicin, interferons). For the study of ion channel proteins, however, the infected cells are usually viable longer than necessary for the electrophysiological assay. The principal advantage of recombinant VV over other transient expression systems is the broad range of mammalian cell types that can be infected at high success rates. Thus, VV may prove extremely useful in the expression of ion channel and receptor/ionophore proteins requiring cell-specific posttranslational processing, or when cell-specific subunits of these excitability proteins are required for complete functional reconstitution.

[32] C. Franke, C. M. Rice, J. H. Strauss, and D. E. Hruby, *Mol. Cell. Biol.* **5**, 1918 (1985).
[33] J. F. Rodriguez and M. Esteban, *J. Virol.* **63**, 997 (1989).

Acknowledgments

We wish to thank Drs. Jayashree Aiyar, Reid Leonard, and Laurel Thomas for help and advice with the development of the recombinant vaccinia virus technology; Drs. Mark Tanouye and John Adelman for K⁺ channel cDNAs; and Annie Gouin, Carolyn Nolan, and Mary King for excellent technical help. Work described here was supported by grants from the National Institutes of Health (GM-29836 and DK-37274), the Max-Kade Foundation, and the Scott Helping Hand Fund through the Cystic Fibrosis Foundation.

[28] Expression of Ion Channels in Cultured Cells Using Baculovirus

By ALEXANDER KAMB, JUAN I. KORENBROT, and JAN KITAJEWSKI

Introduction

Baculovirus is gaining popularity as a vehicle for production of preparative quantities of foreign proteins in insect cells. The advantages of the baculovirus system for protein expression are numerous. As higher eukaryotic cells, insect cells generally produce correctly folded foreign proteins. The yields of protein can be astonishing: up to 500 mg/liter of cells for a soluble protein and up to 150 mg/liter for a membrane protein.[1,2] The protocol for producing recombinant virus that carry a foreign gene is nearly as simple as growing λ phage in *Escherichia coli*. The infection is reliable and can be scaled up easily as needed. Because the infection is lytic, cells do not need to be grown for a long period of time constantly expressing large quantities of protein that may impair cell growth or viability.

The viral vector used for expression is *Autographa californica* nuclear polyhedrosis virus, a type of baculovirus with a large host range among lepidopteran insects. Infection by wild-type baculovirus produces two types of virus: extracellular virus that bud from the infected cell membrane and intracellular virus that form large, refractile granules in conjunction with a viral protein called polyhedrin. These granules are referred to as occlusion bodies or polyhedra. Polyhedra protect viral particles from the environment after the infected animal dies. Because the polyhedrin gene is involved only in producing the intracellular form of the virus, the gene can be deleted without hindering production of extracellular virus. This characteristic of the virus, as well as its ability to infect insect cells in culture, is

---

[1] V. A. Luckow and M. D. Summers, *Virology* **170**, 31 (1989).
[2] J. Vialard, M. Lalumiere, T. Vernet, D. Briedis, G. Alkhatib, D. Henning, D. Levin, and C. Richardson, *J. Virol.* **64**, 37 (1990).

the basis for procedures and vectors developed primarily by Summers and co-workers which have been used with tremendous success for producing a variety of mammalian proteins in bulk.[3] The strategy involves construction of transfer vectors that carry a foreign gene under the control of the polyhedrin promotor. The transfer vector is introduced along with wild-type baculovirus DNA by transfection into a cell line, Sf9, derived from tissue of the moth *Spodoptera frugiperda.* Homologous recombination results in the substitution of the polyhedrin gene on the virus with the foreign gene on the transfer vector. Recombinant viruses that carry a foreign gene under the control of the polyhedrin promotor can be distinguished from wild-type viruses using light microscopy.

We have found that baculovirus-directed expression of human K^+ channels is an effective method for producing correctly folded protein suitable for biochemical studies. In addition, Sf9 cells are superb for electrical analysis of these channels. The cells have very few endogenous channels. They can be readily patch clamped, either in the cell-attached or whole-cell mode, and they generate large, well-behaved ion currents.

Growth of Sf9 Cells

Sf9 cells are easy to grow. They require minimal tissue culture equipment: only a 28° incubator, inverted compound microscope, and spinner flasks. Most of the manipulations can be performed on the benchtop. Sf9 cells can be cultured as monolayers or in suspension, and changing between these two culture conditions is straightforward because the cells adhere only loosely to tissue culture vessels. The cells grow in Grace's insect cell medium (GIBCO, Ground Island, NY) supplemented with 10% fetal calf serum (heat inactivated at 56° for 30 min), Tc yeastolate, and Tc lactalbumin hydrolyzate each at 3.3 g/liter (Difco, Detroit, MI), and penicillin/streptomycin. This supplemented Grace's medium is referred to as complete medium. A continuous culture of cells can be maintained in spinner flasks on stir plates in a humidified incubator at 28°. Suspension cultures of Sf9 cells require adequate aeration. Therefore, cultures maintained in spinner flasks should not exceed certain volumes; for instance, cells grown in a 1-liter spinner flask (Wheaton, Millville, NJ) must be kept in no more than 200 ml of medium. Under these conditions, Sf9 cells grow rapidly, doubling approximately daily. Cells are routinely split every 2 days

---

[3] M. D. Summers and G. E. Smith, "A Manual of Methods for Baculovirus Vectors and Insect Cell Culture Procedures" (Bulletin No. 1555). Texas Agricultural Experiment Station and Texas A&M University, College Station, Texas, 1987.

to a density of 1×10^6 cells/ml. After 2 days, the maximal cell density of 4×10^6 cells/ml is achieved.

Transfection

Construction of Transfer Vectors

A number of vectors are now available to construct the initial plasmids for gene transfer into Sf9 cells. All vectors are designed to facilitate homologous recombination between the foreign gene, in this case a human K⁺ channel cDNA clone called HuKIV, and the polyhedrin gene of the viral genome.[4] We used pVL1393, a plasmid that contains a polylinker located between the polyhedrin promoter and polyadenylation site for convenient insertion of cloned genes.[5] This vector takes advantage of the observation that sequences immediately upstream and, possibly, downstream of the translational start site are important for efficient transcription of the polyhedrin gene. Mutation of the polyhedrin start codon to ATT allows insertion of cloned genes 35 nucleotides downstream of the normal polyhedrin start codon, thus retaining critical promoter regulatory elements. Translation is initiated at the first ATG in the foreign gene downstream of the mutated polyhedrin ATT triplet.

Preparation of DNA

Production of recombinant virus requires both purified transfer vector DNA and purified wild-type baculovirus DNA. The transfer vector should be purified by ultracentrifugation in a CsCl gradient. Baculovirus DNA is prepared by extraction of whole-cell DNA from Sf9 cells infected by wild-type baculovirus.

1. Split cells to a density of 1×10^6 cell/ml the day prior to infection. (If the cell density doubles by the next day, the cells are healthy and can be used for infection.)
2. Centrifuge the cells at room temperature (RT) for 10 min at 500 g and resuspend in the required amount of viral inoculum that has been increased to 20 ml with complete medium for a multiplicity of infection (MOI) of 10; that is, 10 infectious viruses per cell.
3. Incubate the suspension for 1 hr at 28°, rocking the cells every 15 min.

[4] M. Ramswami, M. Gautam, A. Kamb, B. Rudy, M. A. Tanouye, and M. K. Mathew, *Mol. Cell Neurosci.* **2**, 214 (1990).
[5] N. R. Webb and M. D. Summers, *Technique* **2**, 173 (1990).

4. Transfer the culture to a spinner flask and increase the volume to 200 ml with complete medium.
5. Harvest cells 72 hr after infection and centrifuge for 5 min at 500 g.
6. Wash once with phosphate-buffered saline (PBS) and centrifuge again.
7. Resuspend the cells in 100 ml of lysis buffer (30 mM Tris, pH 7.5, 10 mM magnesium acetate, 1.0% (v/v) Nonidet P-40) and leave on ice for 10 min.
8. Centrifuge the lysate at 1000 g for 5 min to pellet the nuclei.
9. Wash the nuclei once with PBS and centrifuge again.
10. Add 200 ml of extraction buffer (0.1 M Tris, pH 7.5, 0.1 M Na$_2$EDTA, 0.2 M KCl) to resuspend the pellet.
11. Incubate the mixture at 50° overnight in the presence of 40 μg/ml proteinase K.
12. Taking great care not to shear the DNA, extract the solution twice with phenol–choloroform–isoamyl alcohol (25:24:1).
13. Add 20 ml absolute ethanol and precipitate the DNA at −80° for 20 min.
14. Centrifuge the mixture at 2500 g for 20 min at RT, resuspend the precipitated DNA in 0.1 × TE (1 mM Tris, pH 7.5, 0.1 mM Na$_2$EDTA), and incubate at 65° for 30 min to dissolve the pellet.

Normally 4 mg of nucleic acid is obtained using this procedure. Roughly one-quarter of this nucleic acid is baculovirus DNA.

Gene Transfer into Sf9 Cells

Sf9 cells are transfected by Transfection Method 1 described in Summers and Smith.[3]

1. Seed 25-cm² flasks with 2 × 10⁶ cells per flask and allow the cells to attach.
2. Remove media from flask and replace with 0.75 ml Grace's medium plus 10% FCS and penicillin/streptomycin. Leave flask at room temperature.
3. Mix 2 μg of transfer vector and 1 μg of wild-type baculovirus DNA (i.e., 4 μg of total nucleic acid extracted from baculovirus-infected cells) in a sterile tube.
4. Add 0.75 ml of transfection buffer (25 mM HEPES, pH 7.1, 140 mM NaCl, 125 mM CaCl$_2$), vortex, and add the DNA solution dropwise to the medium in the culture flask.
5. Incubate the precipitate with the Sf9 cells for 4 hr at 28°.
6. Remove the medium and add 4 ml fresh medium.
7. After 4 days, collect the culture supernatant and store at 4°.

By 4 days, most of the cells will contain polyhedra characteristic of the intracellular form of the virus. These polyhedra are clearly visible under the light microscope. The culture medium contains predominantly wild-type virus with a small percentge of recombinant virus.

Viral Plaque Assay

The viral plaque assay is used to titer viral stocks and to screen for recombinant baculovirus. It is important that the Sf9 cells used in the assay are healthy and in logarithmic growth.

1. Seed 60-mm tissue culture dishes (LUX, Nunc, Naperville, IL) with 2×10^6 Sf9 cells per dish.
2. After allowing the cells to adhere to the dish for 1 hr, remove the medium and add 1 ml of serially diluted viral stock (see below). Leave the Sf9 cells in the inoculum for 1 hr at 28°.
3. During this time, prepare an agar overlay that consists of 1.0% agarose (Sea Plaque low-melt agarose, FMC, Rockland, ME) in complete medium. Melt 1.0 g agarose in 20 ml sterile water in a microwave and add 80 ml complete medium. Store this solution at 37° until use.
4. After the viral adsorption, remove the inoculum and slowly add 4 ml of the agar overlay to each 60-mm dish. Leave the plates at room temperature for 30 min to allow the agar to solidify before returning them to the 28° incubator.

Within 6–7 days, plaques become visible. Wild-type plaques appear as small opaque dots on the cell monolayer. Recombinant plaques are more difficult to identify. To facilitate detection of plaques, another agar overlay that contains 0.04% (w/v) trypan blue is added on day 6 (1.5 ml per 60-mm dish). By the following day, the trypan blue has penetrated the infected cells, and the plaques appear as blue dots on the monolayer. After trypan blue staining, recombinant plaques can be scored under a compound microscope (see below).

Isolation of Recombinant Virus

To identify recombinant viruses, a plaque assay is performed using the transfection culture supernatant as inoculum. The supernatant is diluted by 10^3, 10^4, and 10^5, and 1 ml of each dilution is used to infect a 60-mm dish of Sf9 cells in a standard plaque assay. Five or six days after the infection the cells are stained with trypan blue, and the following day recombinant viral plaques are identified among the numerous wild-type viral plaques. Typically, 1% or less of the plaques will lack occlusion bodies (occ⁻) and these plaques appear as pale blue clusters of cells under the microscope. In contrast, wild-type plaques with polyhedra stain more intensely blue. To confirm that putative occ⁻ plaques truly lack polyhedra, it is necessary to use high power phase-contrast microscopy.

1. Pick several occ⁻ plaques using a micropipettor by inserting the pipette tip directly into the agarose over the plaque and removing an agar plug.
2. Place the plugs in 1 ml of complete medium and leave at 4° for at least 1 h to allow dispersion of the plug.
3. Dilute 100 μl of the solution that contains the agar plug into 1 ml of complete medium and use to infect a fresh 60-mm dish of Sf9 cells in a standard plaque assay.
4. Repeat the process of plaque purification 2 more times.

The final round of purification should yield dishes that contain only occ⁻ plaques. In general, 4–6 independent occ⁻ viruses should be purified since spontaneous nonrecombinant occ⁻ viruses are sometimes isolated.

Production of Pass 1 Stock

Preparation of a stock of the plaque-purified virus for storage and small-scale protein expression studies (the pass 1 stock) is the next step in handling recombinant viruses.

1. Use 900 μl of the agar plug solution from the final round of plaque purification to inoculate 1×10^6 Sf9 cells that have been seeded in a 25-cm² flask.
2. After viral adsorption, remove the inoculum and add 4 ml of fresh medium.
3. After 3 days, collect the medium.
4. Remove 1 ml of this viral stock for long-term storage at −80°. The remainder can be used to generate more viral stock or to infect cells for small-scale analysis of protein expression.

Production of Pass 2 Stock

To generate protein on a scale suitable for biochemical studies, it is necessary first to produce a large viral stock (the pass 2 stock).

1. Centrifuge 50 ml of culture from a spinner flask (2×10^6 cells/ml) at 500 g for 5 min at RT and resuspend the pellet in 0.2 ml of pass 1 stock (MOI ~ 0.2).
2. Incubate the mixture at 28° for 1 hr and add to 100 ml of complete medium.

3. After 4 days of growth in a spinner flask, centrifuge the culture and maintain the supernatant at 4°.

The titer of the pass 2 stock can be determined by a plaque assay. Normally, it is $1-2 \times 10^8$ infectious viruses/ml.

Production of Protein

For production of human K$^+$ channel (HuKIV) protein, we typically use 100-ml cultures.

1. Harvest 100 ml of Sf9 cells during growth in log phase ($\sim 2 \times 10^6$/ml) and centrifuge at 500 g for 5 min.
2. Resuspend the cells in a volume of pass 2 stock which corresponds to 2×8^8 infectious viruses (usually 10–20 ml), an MOI of 20.
3. Leave cells at 28° for 1 hr and then dilute to a final volume of 100 ml with complete medium. The infection is allowed to progress in a spinner flask at 28° for 40–44 hr, when cells are harvested.
4. Centrifuge the infected cells at 500 g for 5 min and wash 2 times in cold PBS. (Infected cells are fragile and should be resuspended gently using a pipette.)
5. Following the washes, resuspend cells in 4.5 ml cold lysis buffer [10 mM Na$_2$HPO$_4$, pH 7.0, 1 mM Na$_2$EDTA, 0.02% NaN$_3$, 5 mM dithiothreitol, 20 μg/ml leupeptin, 50 μg/ml aprotinin, 100 μM phenylmethylsulfonyl fluoride (PMSF)] and rupture by 20 strokes in a Dounce homogenizer. Perform all subsequent manipulations at 4°.
6. Remove nuclei by centrifugation at 1000 g for 5 min.
7. Centrifuge the supernatant at 10,000 g for 10 min to remove particulate matter.
8. Centrifuge this supernatant at 160,000 g for 1 hr to pellet membrane vesicles.
9. Resuspend the pellet with the aid of a Dounce homogenizer in 2.5 ml lysis buffer with NaCl added to 0.5 M.
10. Centrifuge the solution again at 160,000 g for 1 hr and resuspend the pellet in 2 ml lysis buffer for analysis using a Dounce homogenizer.

This protocol leads to the recovery of nearly all of the HuKIV protein detectable by Western blot in the crude lysate of infected Sf9 cells. The membrane-bound protein can be solubilized by treatment with detergents.

Electrophysiology Using Sf9 Cells

To prepare Sf9 cells for electrophysiology, the cells are grown in spinner flasks as described above. Cells infected at an MOI of 20 are used 24 hr or less after infection because the cells are healthier than at later times. Cells are removed from the culture and placed directly on untreated glass coverslips. Within 30 min the cells adhere and are ready for recording.

To record membrane currents under voltage clamp, tight-seal electrodes in the whole-cell mode can be readily used with Sf9 cells that are attached to clean glass coverslips. We have successfully recorded from cells bathed in Grace's medium using electrodes filled with 135 mM MOPS, 20 mM potassium aspartate, 10 mM KCl, 10 mM NaCl, 4 mM MgCl$_2$, 0.1 mM EGTA, 3 mM Na$_3$ATP, 1 mM Na$_3$GTP, pH 6.5, 300 mOsM. Electrodes are fabricated from Corning 1724 aluminosilicate glass and can be used both with and without Sylgard coating. Tight seals form readily in over 90% of the cells whether or not the cells are infected.

On-cell recordings revealed vigorous single-channel activity in Sf9 cells infected with baculovirus carrying HuKIV at membrane potentials above 0 mV, but not in uninfected cells or in cells infected with a different recombinant virus carrying the *int-1* protooncogene. Figure 1 illustrates membrane currents measured in the whole-cell mode in both uninfected and infected cells. In uninfected cells, the whole-cell mode was attained easily after tight-seal formation by rupturing the membrane with a brief (0.2 msec) pulse of voltage. The observed membrane currents were small and not voltage dependent. In infected cells, however, large outward membrane currents were activated by depolarizing voltages. These currents had kinetics and voltage dependence of activation typical of neuronal delayed rectifier currents. The current density observed in these cells was in the same range as that seen for delayed rectifier currents in neurons. To obtain stable, whole-cell recordings in infected cells, we found it necessary to sustain a slight (4–6 cm H$_2$O) suction on the recording electrode after rupturing the membrane; otherwise the membrane tended to reseal rapidly and clog the electrode tip.

Summary

The utility of baculovirus as a vehicle for protein expression for both soluble and integral membrane proteins has been proved repeatedly. Our results suggest that baculovirus also holds promise as a means for expressing ion channel proteins. Because Sf9 cells are especially well suited to electrophysiology and because the construction of recombinant viruses that carry cloned genes is easy, baculovirus may also prove valuable for detailed functional studies of ion channels.

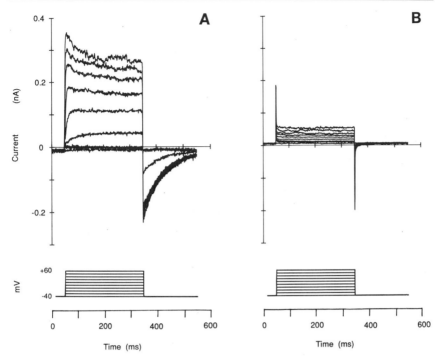

FIG. 1. (A) Membrane currents measured under voltage clamp in an Sf9 cell infected with HuKIV-bearing baculovirus; (B) currents measured in uninfected Sf9 cells. In both cells, the membrane voltage was held at −40 mV, and the holding current was near zero. The membrane voltage was changed between −40 and +60 mV in 10-mV steps. The voltage steps were 300 msec in duration and were repeated every second. Recording was done at room temperature with a bandwidth of 1 kHz. The records shown were not corrected for leakage or compensated for membrane capacitance.

Acknowledgments

We thank Drs. R. M. Stroud and H. E. Varmus for support and encouragement, Dr. S. Wang and Y. Liu for providing antiserum, and Dr. M. Shuster and Dr. D. Morgan for comments on the manuscript. A. Kamb was supported by Damon Runyon–Walter Winchell fellowship DRG-978. J. Kitajewski was a Fellow of the Jane Coffin Childs Memorial Fund for Medical Research. This work was supported by National Institutes of Health Grant GM-24485 and National Science Foundation Grant PCM 83 16401.

[29] Probing Molecular Structure and Structural Changes of Voltage-Gated Channel by Expressing Mutant Channels in Yeast and Reconstituting Them into Planar Membranes

By M. Colombini, S. Peng, E. Blachly-Dyson, and M. Forte

Introduction

It is becoming increasingly clear that no one method can yield a thorough understanding of the molecular structure and mode of action of proteins. X-Ray crystallography, although extremely powerful, yields the static structure of a protein in a crystal, which may be quite different from its functional state. This chapter presents an example of the insights that can be gleaned from molecular genetics and electrophysiology.

It has long been recognized that the study of membrane protein structure has built-in advantages brought about by the nonpolar nature of the membrane interior which places constraints on the possible structures of membrane proteins. One such constraint is the requirement, resulting from energetic considerations, that the portion of the protein forming the transmembrane domains be organized with secondary structure that maximizes the formation of hydrogen bonds by the protein backbone. Another is the requirement that a nonpolar surface face the nonpolar portion of the phospholipid bilayer of the membrane. For channel-forming membrane proteins, the structure must include a polar protein surface that will allow water to penetrate an extremely small tunnel within the protein, forming an aqueous channel. These considerations place some restrictions on the nature of the transmembrane domains of the membrane protein.

For membrane channels that form large aqueous pores, it seems inevitable that amino acid side chains form at least part of the polar wall of the aqueous pore. The nature and especially the charge of these side chains must influence if not dominate the ion selectivity of the channel. Thus, changing the appropriate side chain should change the ion selectivity of the channel. Conversely, amino acid substitutions that change the ion selectivity are good indicators of locations within the protein that form part of the wall of the aqueous pore.

Strategy for Deducing Folding Pattern for Protein Domains Forming Aqueous Pore

The use of selectivity changes to determine which portions of the protein form the water-filled pore can yield a wealth of information about the protein structure. The approach involves changing the charge at a specific location by site-directed mutagenesis. The ion selectivity of the resulting mutant channel is then assessed. In the simple case, no selectivity change should occur if the residue faces the medium on either side of the membrane. A selectivity change of the appropriate sign and magnitude should occur if the residue faces the aqueous pore. In practice, the situation can be clouded by questions concerning residues located in boundary regions and by possible conformational changes distal from the site of a specific mutation. These problems can be minimized in several ways.

Selection of Sites to Mutate

When site-directed mutagenesis is used, mutations are not generated at random. It is important to choose specific mutation sites in order to test a specific model. This model can be based on the known characteristics of the channel such as pore radius, ion selectivity, circular dichroism spectrum, and, most importantly, the primary sequence. Because we are attempting to model the wall forming the aqueous pore, a first approach is to assume the wall is composed of "sided" α helices and/or β sheets, one face forming a polar surface and the other an apolar one. For a transmembrane β sheet, this condition can be met by sequences, about 10 amino acids long, in which the residues alternate between polar and nonpolar. For an α helix, the polar/nonpolar alternation should occur on the average every 3.6 residues, and transmembrane sequences should be about 20 amino acids long. Having identified possible candidate sequences, one can decide on the location of specific mutations.

The considerations presented above for selecting protein sequences most likely to form the wall of the aqueous pore were used to work out an approximation to the nature and location of the amino acid sequences forming the pore of the mitochondrial channel, VDAC.[1] In the case of VDAC, a small amount of protein (for fungi, 283 amino acids) is used to form a large pore (about 3 nm in diameter). Thus, it is likely that a single layer of protein separates the aqueous pore from the hydrocarbon medium of the phospholipid membrane. As circular dichroism measurements indicate a high degree of β structure, a simple program was used to look for "sided" β strands. The results of this analysis for the three published VDAC protein sequences[2–4] are shown in Fig. 1. The higher the peak the

[1] E. Blachly-Dyson, S. Peng, M. Colombini, and M. Forte, *Science* 247, 1233 (1990).
[2] K. Mihara and R. Sato, *EMBO J.* 4, 769 (1985).
[3] R. Kleene, N. Pfanner R. Pfaller, T. A. Link, W. Sebald, W. Neupert, and M. Tropschung, *EMBO J.* 6, 2627 (1987).
[4] H. Kayser, H. D. Kratzin, F. P. Thinnes, H. Götz, W. E. Schmidt, K. Eckart, and N. Hilschmann, *Biol. Chem. Hoppe-Seyler* 370, 1265 (1989).

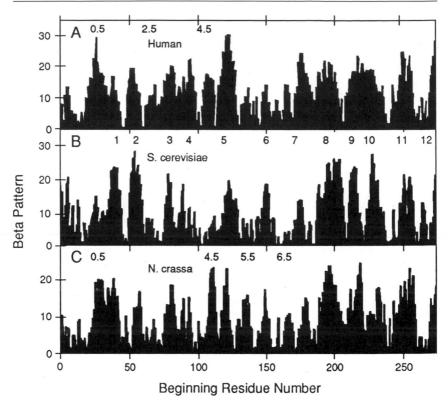

Fig. 1. Evaluation of the potential of stretches of amino acids in the VDAC sequences from *Saccharomyces cerevisiae* and *Neurospora crassa*, and of a homologous human sequence to form β strands lining the walls of a water-filled pore. The hydropathy value [J. Kyte and R. F. Doolittle, *J. Mol. Biol.* **157,** 105 (1982)] of each group of 10 adjacent amino acids was combined as follows: $\Sigma_{i=1}^{10} (-1)^{i+1}v(i)$, where $v(i)$ is the hydropathy value of the ith amino acid. The absolute value of these sums was plotted against the number of the first amino acid in the summation. The numbers in (B) refer to the locations of the major hydropathy peaks. The numbers in (A) and (C) refer to major peaks found that were minor or absent in (B).

better the alternating polar–nonpolar pattern. Hence the peaks predict locations in the primary amino acid sequence which are good candidates for β strands that might form the wall of the aqueous pore. Although a similar pattern can be seen for all the known sequences, some differences are also evident. Focusing on the yeast sequence, 12 likely β strands were identified. Obviously, there is some subjectivity in deciding which of the small peaks to include. In addition to these 12 β strands, it has been recognized[2] that the N terminus of yeast VDAC can form an almost perfect "sided" α helix (nonpolar groups every 3 or 4 residues). Thus, this potential helix is also a candidate for a part of the wall of the aqueous pore.

Type of Mutations

Of the variety of possible alterations that can be generated in the VDAC protein by molecular genetic tools (insertions, deletions, single and multiple substitutions at defined positions), we have found single substitutions to be the most informative. Subtle mutations are easiest to understand and interpret with confidence. If channel-forming activity is lost, little can be said except that "somehow" that site is important. Not only does this statement have little informative value, but it may be misleading because one can easily come up with many possible ways in which channel-forming activity could be lost even if the site of mutation had little importance to channel function (e.g., mutation caused channels to aggregate and/or not insert into the membrane; the mutation caused the protein to have a different structure at a location distal from the site of mutation; the mutation caused the protein to be misfolded; the mutation reduced an energy barrier to protein denaturation). However, if the mutant protein forms channels structurally comparable to the wild type, save for a small change in selectivity, it is easier to make an interpretation that is both informative and likely to be correct.

In generating mutants with single amino acid substitutions, we used our model not only to identify putative transmembrane strands to modify, but also to choose sites in which the mutation would cause minimal disruption of the regional structure. As we specifically planned to change charges at sites affecting channel selectivity, we avoided inserting a charge where the model indicated it would face the lipid bilayer. In general, to maximize the selectivity change, we changed positively charged residues to negative and vice versa, thereby changing the charge at these locations by two units. At times, a charged residue was not located in the region of interest so a neutral residue was replaced by a charged one, taking care that the neutral residue was one that would have faced the aqueous phase (according to the model).

Expression of Mutant Channels in Yeast

The general experimental approach is shown in Fig. 2. A yeast strain (*Saccharomyces cerevisiae*) was constructed in which the genomic copy of the VDAC gene had been deleted. This was accomplished by the single-step gene disruption procedure,[5] using the cloned VDAC gene. The result-

[5] R. J. Rothstein, this series, Vol. 101, p. 202.

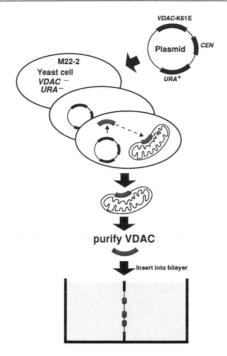

Fig. 2. Schematic of the steps for producing mutated VDAC for electrophysiological analysis. As an example, a plasmid containing the VDAC gene with a K61E mutation is shown. *URA* was used as the selection marker. M22-2 is the designation of the yeast strain lacking the chromosomal copy of the VDAC gene. (Reproduced with permission from Ref. 1.)

ing strain was viable, and, surprisingly, could grow on nonfermentable carbon sources (such as glycerol) at 30° but not at 37° (the parent strain could grow on glycerol at both temperatures). Although growth on nonfermentable carbon sources would not have been required since VDAC can be isolated from petite mitochondria whose DNA has been scrambled with ethidium bromide, this growth indicates that yeast cells contain a redundant function that can substitute for VDAC at normal growth temperatures. This yeast strain was used to express plasmid-borne VDAC genes containing mutations constructed *in vivo*.

Many techniques for oligonucleotide-directed site-specific mutagenesis have been described,[6–9] so the details of the method will not be discussed

[6] M. J. Zoller and M. Smith, *DNA* **3**, 479 (1984).

[7] W. A. Kramer, *et al., Nucleic Acids Res.* **12**, 9441 (1984).

[8] T. A. Kunkel, *Proc. Natl. Acad. Sci. U.S.A.* **82**, 488 (1985).

[9] T. A. Kunkel, J. D. Roberts, and R. A. Zakour, this series, Vol. 154, p. 367.

here. Briefly, a 1.8-kilobase (kb) DNA fragment containing the gene is subcloned into M13mp9 and subjected to site-directed mutagenesis with synthetic oligonucleotides using the "gapped duplex" method.[7] Single-stranded DNA is prepared, and oligonucleotides with single or mutiple mismatches, complementary to the mutagenesis targets, are used to prime the synthesis of mutant second strands. Mutant clones are selected[7-9] and then screened by DNA sequencing. The mutant gene is subcloned into a yeast shuttle vector containing a yeast selectable marker gene (*URA3*), a centromere, and a replication origin (*CEN4, ARS1*). The shuttle vector is carried at a single copy number in yeast cells. The resulting plasmids are introduced by transformation into the yeast strain lacking the chromosomal copy of the VDAC gene. Single-copy yeast plasmids containing the indicated VDAC genes are introduced into VDAC-deleted yeast strain M22-2 by lithium acetate transformation.[10] The mutant VDAC protein is successfully inserted into the mitochondrial outer membranes. Thus, mitochondria from the transformed yeast are used as a source of mutant VDAC protein.

VDAC, either wild type or mutant, is purified from mitochondrial membranes. The mitochondria are isolated from yeast cells essentially as previously described.[11] The mitochondria are hypotonically shocked in 1 mM KCl, 1 mM Tris-HCl, pH 7.5, and the membranes are collected by centrifugation at 27,000 g for 20 min. The membranes are solubilized with 2.5% Triton X-100, and VDAC is purified on a mixed hydroxyapatite–Celite column as previously described.[12] The solution containing the solubilized channels is supplemented with dimethyl sulfoxide to 15% (v/v) and stored frozen at $-80°$.

The ability to perform the single-step gene disruption is the distinct advantage of the yeast system that allowed the exclusive expression of yeast VDAC genes carried on shuttle plasmids. The genes on the plasmids were either wild-type or had been modified *in vitro* by site-directed mutagenesis.

Analysis of Mutants

The chosen strategy, to produce small changes in the properties of the channels, resulted in the problem of measuring these small changes in a reliable way. Because VDAC channels undergo large changes in selectivity on channel closure (the closed state still conducts small ions but does so at a lower rate), it is important to make sure that selectivity measurements are made with channels in the appropriate (open) conformation. This, and the

[10] H. Ito, Y. Kukuda, K. Murata, and A. Kimura, *J. Bacteriol.* **153**, 163 (1983).
[11] G. Daum, P. C. Böhni, and G. Schatz, *J. Biol. Chem.* **257**, 13028 (1982).
[12] H. Freitag, R. Benz, and W. Neupert, this series, Vol. 97, p. 286.

usual problem of parallel conducting elements affecting the results, are the primary reasons for making selectivity measurements on membranes containing one or a few channels. In this way, the discrete conductance changes observed during the experiment could be used to distinguish between the various possible sources of conductance, thus avoiding introducing erroneous estimates into the data pool.

The channels are inserted into planar phospholipid membranes made with soybean phospholipids by the monolayer method of Montal and Mueller[13] (as modified[14]). Because VDAC allows both cations and anions to cross the membrane, a 10-fold salt gradient (1 M versus 0.1 M KCl; each side also contained 5 mM CaCl$_2$, 1 mM MES, pH 5.8) is used as the condition under which ion selectivity is estimated. A 2- to 10-μl aliquot of a wild-type or mutant VDAC sample in the storage solution (as above) is added (with stirring) to 4 ml solution in the high salt side of the membrane. Channels insert spontaneously after few minutes of delay. The voltage needed to bring the current to zero (the reversal potential) provides a measure of channel selectivity. All experiments are performed under voltage clamp conditions as previously described.[14] To minimize the liquid junctions between the calomel electrodes and the solutions, KCl is used as the salt in the aqueous phase. No attempt is made to balance the osmotic pressure between the solutions on the two sides of the membrane since the problems of trying to correct for reflection coefficients and activity changes were judged to be worse than the effects of water flow through the channel during the experiment. In any event, the aim is not to measure channel selectivity in an absolute way, but to measure changes in selectivity induced by specific mutations.

In a sample experiment (Fig. 3), current was monitored as a function of time in the presence of a 10-mV driving force. The lack of detectable current demonstrated the highly impermeable nature of the planar phospholipid membrane prior to channel insertion. If this criterion is not met the experiment is terminated. The first downward current deflection at left corresponds to the spontaneous insertion of one channel. The change in current (I) observed is due to two driving forces: the voltage (V) and the EMF (electromotive force) of the ion gradient. Only some of the EMF is contributing (the reversal potential, E) because VDAC channels conduct both K$^+$ and Cl$^-$. Thus,

$$I = G(V - E)$$

To estimate the magnitude of E, the voltage required to bring the current,

[13] M. Montal and P. Mueller, *Proc. Natl. Acad. Sci. U.S.A.* **69**, 3561 (1972).
[14] M. Colombini, this series, Vol. 148, p. 465.

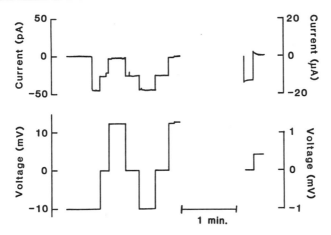

FIG. 3. Measurement of the reversal potential of a VDAC channel produced by the mutant protein T256K (see definition of mutant code in Table I). The lower tracings show the applied potential (the sign refers to the high salt side), and the upper tracings are the recorded current. Left-hand records: Insertion of one channel is indicated by the first abrupt change in current (note: no change in applied potential at this point). An applied potential of 12.6 mV was needed to bring the current to 0. This is the reversal potential for the particular channel. Right-hand records: After the membrane broke, the electrode asymmetry was estimated by placing both calomel electrodes in the high-salt side and finding the voltage that brought the current to zero, in this case 0.4 mV (note the changes of scale). Thus, the reversal potential for this channel was 12.2 mV.

I, to zero is determined by iteration. Because VDAC channels are voltage gated, care is taken to measure E at a time when the channel is open.

Analysis of Results

The results of observations made on the mutant VDAC proteins that have been generated to date are summarized in Table I. The measured parameter, the reversal potential, was subtracted from the value of the wild-type reversal potential in order to focus on changes (δ) in this value and correlate them with the charge change engineered into the protein. Because both the sign and magnitude of the change in charge varied depending on the particular mutation, the change in reversal potential had to be divided by the change in charge (z). The results allow the mutations to be placed into one of two categories: (1) charge changes that caused a significant change in selectivity and (2) charge changes that caused no significant change in selectivity. These results fit very well with a simple interpretation: category 1 represents locations lining the aqueous pore, whereas, category 2 represents locations outside the pore.

TABLE I
SELECTIVITY CHANGES INDUCED IN VDAC BY SINGLE AMINO ACID SUBSTITUTION

Species[a]	Reversal potential (mV)[b]	δ^c (mV)	δ per unit charge[d]
Group 1			
Wild type	10.3 ± 0.1 (6)	—	—
D15K	15.6 ± 0.3 (13)	5.3	2.7
K19E	−3.2 ± 0.2 (14)	−13.5	6.8
D30K	14.9 ± 0.2 (16)	4.6	2.3
K46E	4.9 ± 0.4 (9)	−5.4	2.7
K61T	4.9 ± 0.2 (2)	−5.4	5.4
K61E	1.8 ± 0.2 (7)	−8.5	4.3
K65E	5.7 ± 0.1 (6)	−4.6	2.3
K84E	3.4 ± 0.1 (22)	−6.9	3.5
K95E	3.4 ± 0.1 (18)	−6.9	3.5
R124E	5.2 ± 0.1 (3)	−5.1	2.6
E152K	12.1 ± 0.1 (3)	1.8	0.9
G179D	7.0 ± 0.2 (3)	−3.3	3.3
K234Q	6.1 ± 0.2 (3)	−4.2	4.2
K248E	6.0 ± 0.2 (12)	−4.3	2.2
T256K	12.2 ± 0.2 (4)	1.9	1.9
D282K	15.4 ± 0.3 (5)	5.1	2.6
Group 2			
D51K	10.1 ± 0.2 (3)	−0.2	−0.1
K108E	9.4 ± 0.0 (2)	−0.9	0.5
K132E	10.1 ± 0.2 (3)	−0.2	0.1
D156K	9.8 ± 0.1 (3)	−0.5	−0.3
R164D	9.2 ± 0.1 (3)	−1.1	0.6
K174E	9.4 ± 0.2 (2)	−0.9	0.5
D191K	9.6 ± 0.6 (4)	−0.7	−0.4
K205Q	10.0 ± 0.0 (2)	−0.3	0.3
K205E	9.2 ± 0.7 (4)	−1.1	0.6
K211E	11.0 ± 0.1 (3)	0.7	−0.4
E220K	10.9 ± 0.1 (2)	0.6	0.3
R252E	10.1 ± 0.6 (3)	−0.2	0.1
K267E	10.6 ± 0.2 (4)	0.3	−0.2
K274E	10.1 ± 0.2 (3)	−0.2	0.1

[a] Mutations are indicated by the one-letter amino-acid codes for the amino acid changed and the new residue at the corresponding position. For example, D15K indicated that aspartate (D) at position 15 was changed to lysine (K). Single-letter amino acid abbreviations used are A, Ala; D, Asp; E, Glu; G, Gly; K, Lys; Q, Gln; R, Arg; T, Thr.

[b] Means ± S.E. Numbers in parentheses indicate number of observations.

[c] Differences between the mutant and wild-type reversal potentials.

[d] The changes in reversal potential between mutant and wild type were divided by the change in charge induced by the mutation.

At present, one cannot rigorously eliminate alternate interpretations for mutations causing effects typical of category 1. Several of the characteristics of the induced changes indicate that alternative interpretations are unlikely. First, distal effects do not need to correlate in sign with the change in charge imposed by the mutation. However, in all cases in category 1, a perfect correlation was obtained between the direction of the selectivity change and the sign of the charge change caused by the mutation. (In category 2, this correlation did not hold, but these changes were judged nonsignificant.) Second, there was little change in the observed single-channel conductance for all mutants,[1] indicating that the mutations did not change the basic structure of the channels. Third, further evidence against complex conformational changes distorting the results and their interpretation comes from the results obtained with VDAC proteins carrying multiple mutations (Table II). The changes in the values of the reversal potential for multiple mutant channels were very close to what might be expected by simply adding the results observed with the corresponding single mutations.

Correlating Results with Model

The model used as a guide in choosing the sites for mutation must remain fluid, changing with the new experimental findings. Of the original 12 predicted β strands in the yeast protein, one was found not to form a transmembrane segment based on lack of effect of charge changes at

TABLE II
SELECTIVITY CHANGES IN VDAC INDUCED BY MULTIPLE SUBSTITUTIONS

Species	Reversal potential[a] (mV)	δ (mV) Observed	δ (mV) Expected[b]	δ (mV) Observed minus expected
K19E + K61E	-12.5 ± 0.4 (6)	-22.8	-22.0	-0.8
K19E + K95E	-10.8 ± 0.2 (6)	-21.1	-20.4	-0.7
K19E + K248E	-4.8 ± 0.3 (4)	-15.1	-17.8	2.7
K46E + K61E	-2.6 (1)	-12.9	-13.9	1.0
K46E + K95E	-2.4 ± 0.2 (10)	-12.7	-12.3	-0.4
K61T + K65E	0.4 ± 0.4 (7)	-9.9	-10.0	0.1
K61E + K65E	-2.4 ± 0.4 (4)	-12.7	-13.1	0.4
K46E + K61E + K65E	-6.1 ± 0.1 (2)	-16.4	-18.5	2.1

[a] Mean ± S.E. Numbers in parentheses indicate the number of observations.

[b] The changes observed in the corresponding single mutations (from Table I) were summed.

positions 205 and 211 (Fig. 4). Another was moved to account for the observed lack of effect of charge changes at position 156. However, a segment beginning at position 28 which appears not to form a good "sided" β strand (Fig. 1) was found to form a transmembrane strand based on the effects on the reversal potential of charge changes at position 30. Interestingly, the β patterns of the *Neurospora crassa* and human sequences show strong peaks at this position. Other regions where stronger peaks of β structure are present in these sequences but not in yeast are now being checked to see if other transmembrane strands have been missed. Finally, the mutations at positions 15 and 19 clearly demonstrate that the N-terminal α helix is also forming part of the wall of the aqueous pore (Fig. 4).

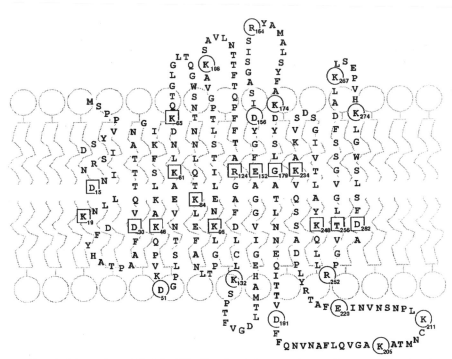

FIG. 4. Schematic model of the VDAC molecule in a membrane. The strands forming the wall of the aqueous pore are shown in a background of phospholipids. The amino-terminal α helix (left) is followed by 12 strands forming an antiparallel β sheet. Mutation of the circled residues left the selectivity unchanged. Mutation of the boxed residues altered the selectivity of the channel (see Table I).

Strategy for Working with Mutations That Seem to Perturb Channel Structure

VDAC proteins carrying lysine to glutamate changes at positions 234 and 236 failed to form channels in the phospholipid membranes. The reasons for this are unclear, but there is evidence to indicate drastic changes in conformation (changes in elution properties in the purification process). A simple solution was the construction of a less dramatic change in this position, namely conversion of lysine to glutamine rather than glutamate at position 234. The milder mutation resulted in channels with the appropriate conductance but a reversal potential shifted by an amount (Table I) typical for locations forming the wall of the aqueous pore.

Detection of Structural Changes in Channels

If one can map out the portions of the VDAC protein that form the wall of the aqueous pore in one conducting state, as described above, one should be able to determine if these same regions also perform the same function in another state. For VDAC, evidence points to a large change in pore diameter on channel closure, indicating a reduction in the surface area of protein forming the wall of the pore. If so, then some subset of the regions affecting ion selectivity in the open state should no longer affect selectivity in the closed state. The same collection of mutants can be tested in an attempt to identify the classes of sites predicted by this idea, namely, sites that affect only the open-state selectivity, sites affecting both open and closed-state selectivities, and sites that have no influence on either open- or closed-state selectivities. By placing particular sites in the molecule into one of these three categories, we should be able to develop a picture of the conformational changes that take place in the VDAC protein during the transition from the open to the closed state. These experiments are currently under way for VDAC.

Strategy for Controlling Variability in Protein Conformation

The strategy of generating mutations that minimize perturbations in protein structure results in small changes in the parameter of interest (here, the reversal potential). This is not a serious problem unless this parameter varies either as a normal property of the protein or as a result of the mutation. Naturally, any variation makes the detection of small changes in this parameter a more difficult task. One way of reducing this variation is to add an agent that favors a particular conformation, thus increasing the free energy difference between possible states.

In the case of VDAC, problems were encountered when the strategy used to probe the structure of the highest-conducting, open, state was applied to the low-conducting, "closed," states. Because VDAC is voltage gated (open state preferred at low membrane potentials and closed at high potentials) the first attempts tried to determine the selectivity of voltage-induced closed states (these are still permeable to KCl but much less so). As even single VDAC channels close to a variety of closed states, distinguishable by small differences in conductance and selectivity, attempts to detect mutation-induced selectivity changes in the presence of this variability failed. In an effort to reduce the variability, a synthetic polymer, referred to as König's polyanion,[15] was added to the medium. This agent increases the voltage dependence of VDAC but, more importantly, results in a marked reduction in the variability among the accessible closed states. Thus, under these conditions, it is possible to determine changes in the closed-state reversal potential between wild-type and mutant VDAC channels.

In at least one case, the mutant channels in the closed state showed quite a bit more variability than the wild-type channels, suggesting that the mutation allowed the structures to be more fluid. Although on average the closed-state selectivity of the mutant still differed from the wild type, the higher variability raises general questions about the reliability of information on the closed state. A solution might be to change the nature of the mutation at sites which cause such fluidity in the protein structure, hoping to obtain more reproducible results.

Limitations of Approach

The general approach outlined here for obtaining information about the molecular structure of channels is probably limited to channels that form large conducting pathways. If selectivity were the result of a well-defined selectivity filter, changing the amino acid residues in the filter would probably have drastic effects on channel conductance. In addition, if the selectivity were the rate-limiting step, little could be learned from subtle changes in the pathways leading to this filter. However, it is still unclear, for the narrow channels, how localized the selectivity mechanism really is and where the rate-limiting steps lie.

Acknowledgments

This work was supported by Grant N00014-90-J-1024 from the Office of Naval Research to M.C. and Grant GM35759 from the National Institutes of Health to M.F.

[15] M. Colombini, C. L. Yeung, J. Tung, and T. König, *Biochim. Biophys. Acta* **905**, 279 (1987).

Section III

Reconstitution of Ion Channels in Lipid Bilayers

[30] Insertion of Ion Channels into Planar Lipid Bilayers by Vesicle Fusion

By PEDRO LABARCA and RAMÓN LATORRE

Introduction

Model or "artificial" lipid bilayers have become common tools in the study of the molecular mechanisms that underly the ion-translocating properties of cell membranes. The assembly in model lipid bilayers of the molecular entities engaged in ion translocation permits their study in convenient isolation from the rest of the cell machinery. In the extreme case, a transport protein is isolated to purity from its native environment and functionally reconstituted in a model bilayer allowing for the establishment of detailed structure–function correlates. In fact, the ultimate criterion to qualify a given molecular entity as responsible for an ion-translocating function still is its functional reconstitution in a model lipid bilayer. The study of membrane transport in model lipid bilayers has been rightly defined as an extreme reductionist approach, a quality to which it owes much success. Among the variety of transport mechanisms exhibited by cell membranes, ion channels are the ones that have profited the most from model bilayers. The simplicity and current resolution capabilities of current-measuring devices available and the fantastic rates at which ion channels catalyze ion fluxes through lipid bilayers make it possible to monitor the currents flowing through a single channel molecule inserted or reconstituted in a special type of model lipid membrane, the planar lipid bilayer.

In this chapter we summarize the experimental procedures used to insert ion channels derived from cell membranes into planar lipid bilayers using fusion approaches.[1-3] An attempt has also been made to introduce the most general concepts that usually guide preliminary studies of ion channels in planar lipid bilayers. Some of the limitations of the technique as well as some less explored avenues of research are introduced.

[1] C. Miller and E. Racker, *J. Membr. Biol.* **30**, 283 (1976).
[2] F. S. Cohen, M. H. Akabas, J. Zimmemberg, and A. Finkelstein, *Science* **217**, 458 (1982).
[3] W. Hanke, H. Eibl, and G. Boheim, *Biophys. Struct. Mech.* **7**, 131 (1981).

Insertion of Ion Channels into Planar Lipid Bilayers Using Fusion
Technique

Planar Lipid Bilayers

Planar lipid bilayers amenable to ion channel insertion can be produced either by the "painting" technique[4] or from two lipid monolayers
spread at the air–water interface.[5] In the "painting" variant a small aliquot
of a solution of lipid in decane (10–50 mg/ml) is applied in the aperture
(0.1–0.5 mm diameter) of a plastic septum separating two aqueous
chambers. A short time after deposition of a drop of lipid in the aperture a
lipid bilayer develops in the center of the aperture. The lipid bilayer so
formed hangs from a solvent torus which is adsorbed to the edge of the
aperture. The resulting bilayers have capacitances of the order of 0.3–
0.6 $\mu F/cm^2$ and conductances of about 10 pS, with peak-to-peak electrical
noise of 0.5–2 pA at 1-kHz low-pass filtering. Bilayer thinning can be
monitored optically through a microscope and/or by capacitance measurements. A thin lipid bilayer looks black to the eye since it reflects little light.
For this reason the planar bilayer is often referred to as a "black membrane." After thinning, the bilayer should occupy 75–80% of the aperture
area.

It is worth noting that "painted" bilayers, in addition to lipid, contain
solvent. The presence of solvent means that painted bilayers possess electrical capacitances that are notoriously smaller than those of cell membranes.

The technique to assemble lipid bilayers from monolayers spread at the
air–water interface was developed by Montal and Mueller,[5] following the
original suggestion of Langmuir[6] and the experiences of Takagi *et al.*[7] In
this method two lipid monolayers are spread in each compartment of a
chamber separated by a septum in which a small hole 50–300 μm in
diameter has been punctured. Previous to bilayer assembly the aperture is
doped with petroleum jelly or squalene, providing the system with a torus
to support the bilayer. A few minutes after spreading the monolayers to
allow for solvent (usually pentane) evaporation, the buffer in each chamber
is sequentially raised above the aperture resulting in the formation of a flat
lipid film exhibiting capacitances of the order of 0.6–0.8 $\mu F/cm$ and high
electrical resistances. Bilayers assembled by this procedure are virtually

[4] P. Mueller, D. Rudin, H. T. Tien, and W. C. Wescott, *Circulation* **26,** 1167 (1962).
[5] M. Montal and P. Mueller, *Proc. Natl. Acad. Sci. U.S.A.* **69,** 3561 (1972).
[6] I. Langmuir, *J. Chem. Phys.* **1,** 756 (1933).
[7] M. Takagi, K. Azuma, and U. Kishimoto, *Annu. Rep. Biol. Works Fac. Sci. Osaka Univ.*
13, 107 (1965).

"solvent free,"[8] providing a lipid matrix devoid of solvent in which the insertion of ion channel molecules can be attempted. Because bilayers made from monolayers do not thin, bilayer formation in this case is followed only through capacitance measurements.

Lipid Requirements

The experimenter engaged in preliminary attempts to fuse vesicular fragments derived from cell membranes or lipid vesicles into which purified channel proteins have been reconstituted should benefit from the proper choice of lipids used in bilayer formation. Experience indicates the convenience of including an acidic lipid in the lipid mixture used to form bilayers since they increase the chances of fusion. The most widely used acidic lipid is phosphatidylserine (PS), which is mixed with a neutral lipid such as phosphatidylethanolamine (PE) at various ratios, although other acidic lipids like phosphatidic acid and cardiolipin have also been reported to yield satisfactory results. (The reader should be aware that the surface charge contributed by acidic lipid can influence ion channel properties.[9])

Table I shows the lipid composition of planar bilayers in where ion channels were inserted using the fusion method.[10-24] Vesicle fusion to planar lipid bilayers can also be obtained in planar bilayers made of neutral lipids. A variety of natural and synthetic lipids are commercially available from Avanti Polar Lipids (Birmingham, AL), or, alternatively, lipids of

[8] S. H. White, *In* "Ion Channel Reconstitution" (C. Miller, ed.), p. 3. Plenum, New York and London, 1986.

[9] R. Latorre, P. Labarca, and D. Naranjo, this volume [32].

[10] J. C. Tanaka, R. E. Furman, and R. L. Barchi, *In* "Ion Channel Reconstitution" (C. Miller, ed.), p. 277. Plenum, New York and London, 1986.

[11] R. P. Hatshorne, B. U. Keller, J. A. Talvenheimo, W. A. Caterall, and M. Montal, *Proc. Natl. Acad. Sci. U.S.A.* **82**, 240 (1985).

[12] B. K. Krueger, F. W. Jennings, and R. J. French, *Nature (London)* **303**, 172 (1983).

[13] E. Moczydlowski, S. S. Garber, and C. Miller, *J. Gen. Physiol.* **84**, 665 (1984).

[14] J. S. Smith, T. Imagawa, M. Jianjie, M. Fill, K. P. Campbell, and R. Coronado, *J. Gen. Physiol.* **92**, 1 (1986).

[15] B. A. Suarez-Isla, V. Iribarra, A. Oberhauser, L. Larralde, C. Hidalgo, and E. Jaimovich, *Biophys. J.* **54**, 737 (1988).

[16] H. Affolter and R. Coronado, *Biophys. J.* **48**, 341 (1985).

[17] A. Lievano, E. C. Vega-Saenz de Miera, and A. Darszon, *J. Gen. Physiol.* **95**, 273 (1990).

[18] M. T. Nelson, R. J. French, and K. Krueger, *Nature (London)* **308**, 77 (1984).

[19] E. Moczydlowski, O. Alvarez, C. Vergara, and R. Latorre, *J. Membr. Biol.* **82**, 273 (1985).

[20] J. Bell and C. Miller, *Biophys. J.* **45**, 279 (1984).

[21] P. Labarca, Ph.D. Thesis, Brandeis University, Waltham, Massachusetts (1980).

[22] P. Labarca, R. Coronado, and C. Miller, *J. Gen. Physiol.* **76**, 396 (1980).

[23] J. A. Hill, R. Coronado, and H. C. Strauss, *Circ. Res.* **62**, 411 (1988).

[24] M. M. White and C. Miller, *J. Biol. Chem.* **254**, 10161 (1979).

TABLE I

Lipids Most Commonly Used to Form Bilayers in Ion Channel Reconstitution
Studies

Channel (source)	Type of lipid	Ref.
Purified Na$^+$ channel (skeletal muscle)	Phosphatidylethanolamine, phosphatidylcholine	10
Purified Na$^+$ channel (brain)	Phosphatidylethanolamine, phosphatidylcholine	11
Na$^+$ channel (rat brain homogenates)	Phosphatidylserine, phosphatidylethanolamine	12
Na$^+$ channel (rat skeletal muscle plasma membrane)	Phosphatidylethanolamine, phosphatidylcholine	13
Ca^{2+} channel (purified from skeletal muscle sarcoplasmic reticulum)	Phosphatidylethanolamine, phosphatidylserine	14
Ca^{2+} channel (skeletal sarcoplasmic reticulum vesicular fragments)	Phosphatidylethanolamine, phosphatidylcholine Phosphatidylethanolamine, phosphatidylserine	15
Ca^{2+} channel (T-tubule membranes)	Phosphatidylethanolamine, phosphatidylserine	16
Ca^{2+} channel (sperm plasma membranes)	Phosphatidylethanolamine, phosphatidylserine	17
Ca^{2+} channel (brain homogenates)	Phosphatidylethanolamine, phosphatidylserine	18
Ca^{2+}-activated K$^+$ channel (T-tubule membranes)	Phosphatidylethanolamine, phosphatidylserine, phosphatidylcholine	19
K$^+$ channels (skeletal muscle sarcoplasmic reticulum membranes)	Phosphatidylserine, phosphatidylethanolamine Phosphatidylethanolamine, phosphatidylcholine Asolectin, cardiolipin Asolectin, phosphatidic acid	20 21
K$^+$ channel (cardiac muscle sarcoplasmic reticulum membranes)	Phosphatidylserine, phosphatidylethanolamine	22
Cl$^-$ channel (fish electric organ)	Phosphatidylethanolanine, phosphatidylglycerol	23
Cation-selective pore (peroxisomal membranes)	Phosphatidylethanolamine, phosphatidylcholine	24

bilayer quality can be purified in the laboratory. Stock lipid solutions are kept at $-20°$ to $-80°$ under a nitrogen atmosphere, and lipid used in bilayer work must be prepared fresh from stocks each day. Bilayer chambers and glass material used in bilayer work should be immaculate.

Fusion of Vesicular Membrane Fractions to Planar Lipid Bilayers

The main criterion that makes a given membrane preparation suitable for fusion into planar lipid bilayers is its purity, apart from the obvious interest of a researcher in studying the conductance properties of a given cell membrane at the single-channel level. Use of a highly purified membrane fraction provides the potential reproducibility that is so important in this kind of experimental work. However, crude membrane extracts have also been used in some cases.[25]

For reasons that have not been completely understood vesicle fusion in painted bilayers is more easily achieved than fusion in bilayers made from monolayers. However, the experimental conditions needed to fuse lipid vesicles are identical for both types of model planar bilayers.[21,22,26]

For operational purposes the two aqueous compartments separated by the lipid film are defined as cis and trans compartments. The cis compartment is the one to which a voltage generator is connected to the bilayer through an Ag/AgCl electrode. The trans compartment is connected to the input of the current-measuring amplifier through a second Ag/AgCl electrode. It is recommended that both electrodes be connected to the chambers through $3\ M$ KCl agar bridges that, ideally, should be made each working day to avoid bacterial and fungal contamination. Both chambers must be provided with stirring facilities. It is also necessary to provide the setup with a perfusion system.

After a bilayer is formed under the buffer solution of choice, the Ca^{2+} concentration in the cis compartment must be raised by adding an aliquot from a concentrated stock solution. From now on all operations are made under stirring in the cis compartment. (The precise divalent ion concentration in the cis side must be empirically determined; fusion requires the presence of Ca^{2+} in the $1-10$ mM range.) An aliquot of vesicles is added to the cis chamber. The membrane vesicles should be loaded with a buffer (usually a sucrose buffer) at $0.3-0.5\ M$ solute concentration prior to their use in fusion experiments. Because sucrose is an impermeable solute, the vesicles swell when added into the bilayer chamber, which helps in pro-

[25] I. B. Levitan, *in* "Ion Channel Reconstitution" (C. Miller, ed.), p. 523. Plenum, New York and London, 1986.

[26] F. S. Cohen, *in* "Ion Channel Reconstitution" (C. Miller, ed.), p. 131. Plenum New York and London, 1986.

moting fusion. The amount of vesicles to be added depends on the fusogenic properties of each particular preparation, but it ranges from 0.2 to 50 μg protein/ml (final concentration). Using bilayer chambers that hold small volumes (0.3–0.5 ml) is important when the purified membranes are scarce.

Under the experimental conditions that we presented (vesicles and Ca^{2+} in the cis side, and the same ionic buffer in both compartments), vesicle fusion to planar lipid bilayers is infrequent. However, fusion can now be induced in a controlled fashion by establishing an osmotic gradient between the cis and the trans bilayer sides. For this, an aliquot of a concentrated solute (e.g., KCl or LiCl) is added to the cis chamber. The magnitude of the osmotic gradient necessary to trigger fusion is variable, but a 3:1 gradient should be used to start. Larger osmotic gradients result in more fusion. The researcher should spend some time becoming familiar with the experimental conditions needed to achieve fusion of a particular membrane preparation in order to monitor single-channel currents. Experimental conditions must be found that are reproducible and efficient in obtaining the insertion of single ion channel molecules into the planar bilayer. Conditions leading to massive fusion of vesicles to planar bilayers should be avoided.

Bilayer conductance is followed throughout these procedures by applying a potential difference and by measuring current. At the beginning of the experiment the bare bilayer conductance should be low (<10 pS). Electrically unstable bilayers should be avoided in fusion work. It is also recommended that the experimenter occasionally check the membrane state (optically and/or by capacitance measurements). Planar lipid films sometimes become thick during the initial fusion procedures. Ion channels do not work in thick membranes. It is not uncommon that a lack of reproducibility in ion channel insertion from one planar bilayer to another is due to a lack of control of the thickness and size of the membrane during the fusion steps. This lack of control leads to unsuccessful attempts to insert ion channels via fusion in lipid films that are not of bilayer dimensions.

The order of addition of Ca^{2+} and the vesicles and the moment chosen to establish the osmotic gradient between the cis and trans compartments can be varied. The sequence followed above establishes the osmotic gradient after vesicle addition to the cis side, allowing the adequate magnitude of the gradient necessary to trigger fusion to be defined in each assay. Woodbury and Miller[26a] have demonstrated that the fusion of liposomes containing ion-channels can be promoted when the liposomes are doped

[26a] D. J. Woodbury and C. Miller, *Biophys. J.* **E8,** 833 (1990).

with nystatin, a pore-forming antibiotic. This approach demands that the liposomes are enriched with sterol, which is required for the formation of oligomeric nystatin channels.

Incorporation of ion channels into the planar bilayer as the result of a single fusion event is evidenced either by the occurrence of a conductance change, in the form of a sudden jump in the current record, followed by single-channel fluctuations if the fusing vesicle contains several channels; or by the sudden appearance of conductance fluctuations reflecting the opening and closing of a single channel molecule inserted. At this point to stop fusion it is necessary to remove the vesicles and Ca^{2+} and to reestablish osmotic equilibrium by perfusion of the cis bilayer side with an ionic buffer of choice. Usually, perfusion with eight chamber volumes of buffer suffices to remove most of the vesicles, to restore osmotic equilibrium, and to lower the calcium concentration in the cis chamber to contaminant levels. The cis chamber can also be perfused with a solution containing a calcium chelator, most commonly EGTA, to establish a well-defined calcium concentration in the cis chamber. The levels of calcium contaminating the aqueous buffers being used should be known.

If the membrane breaks during the fusion procedure it can be rapidly rebuilt and the assay started again. However, the membrane should be optically monitored or its capacitance should be checked after repainting to make sure that it thinned to a bilayer.

Fusion of Lipid Vesicles Containing Purified Ion Channels

Purified ion channels reconstituted in planar bilayers allow for detailed studies on the ion selectivity, voltage dependence chemical modulation, and pharmacological properties associated with a well-defined molecular structure. Few ion channels have been isolated to purity and functionally reconstituted in model planar bilayers. Nicotinic cholinergic receptor channels purified from fish electric organ,[27,28] tetrodotoxin-sensitive Na^+ channels,[10,11,29–31] and the ryanodine receptor from sarcoplasmic reticulum

[27] N. Nelson, R. Anholt, J. Lindstrom, and M. Montal, *Proc. Natl. Acad. Sci. U.S.A.* **77**, 3057 (1980).
[27a] G. Boheim, W. Hanke, F. J. Barrantes, H. Eibl, B. Sakmann, G. Fels, and A. Maelicke, *Proc. Natl. Acad. Sci. U.S.A.* **78**, 3586 (1981).
[28] P. Labarca, J. Lindstrom, and M. Montal, *J. Gen. Physiol.* **83**, 473 (1984).
[29] R. L. Rosenberg, S. A. Tomiko, and W. S. Agnews, *Proc. Natl. Acad. Sci. U.S.A.* **81**, 5594 (1984).
[30] W. Hanke, G. Boheim, J. Barhanin, D. Pauron, and M. Lazdunski, *EMBO J.* **3**, 509 (1984).
[31] S. Shenkel, E. C. Cooper, W. James, W. S. Agnew, and F. J. Sigworth, *Proc. Natl. Acad. Sci. U.S.A.* **86**, 9592 (1989).

skeletal muscle[14] form the small repertoire of ion channels that have been isolated to purity and functionally reconstituted in model membranes. Partially purified K^+ channels from the squid axon have also been inserted into planar lipid bilayers.[32]

The requirements to achieve fusion of lipid vesicles containing purified ion channels are identical to those found with vesicular membrane fragments. Tetrodotoxin-sensitive Na^+ channels, purified from several sources, have been functionally inserted in both "painted" and "solvent-free" bilayers.[10-13,29-31] Single-channel currents have been monitored after removal of inactivation with toxins or proteolytic enzymes.[10-13,29,31] The results obtained indicate that, with the exception of Na^+ channel inactivation, which is hard to study in a planar bilayer, the integral membrane protein associated with the tetrodotoxin receptor suffices to account for sodium channel function. A similar impression emerges from the studies of purified nicotinic receptor channels in model membranes.[28] However, because the variety of purified ion channels that have been functionally reconstituted in planar bilayers is not large, the involvement of peripheral membrane proteins or even soluble proteins in the control of ion channel behavior has not been determined. As the repertoire of purified ion channels available for planar bilayer work expands, these and other questions related to the modulation of ion channel proteins by chemical modification, second messengers, and cell metabolism will probably gain the attention of experimenters. The planar lipid bilayer provides a powerful experimental assay bridging membrane biochemistry, which scrutinizes purified ion channel molecules, and cell physiology, which attempts to unveil the way in which ion channel molecules are coupled to cell function. Furthermore, pharmacological studies of newly discovered ion channels in planar lipid bilayers are helpful in identifying new high affinity toxins that will make it easier for membrane biochemists to achieve the purification and reconstitution of new ion channel molecules.[33]

Vesicle Fusion to Planar Bilayers Made from Monolayers

There are few cases of ion channels studied in lipid bilayers made from monolayers[21,22,27,27a,28,30] which from here on will be referred to as "solvent-free" bilayers. Although, so far, the experimental conditions found to assist vesicle fusion to "painted" and "solvent-free" bilayers are identical, namely, presence of acidic lipid, millimolar concentrations of Ca^{2+} in the

[32] G. Prestipino, H. H. Valdivia, A. Lievano, A. Darszon, A. N. Ramirex, and L. D. Possani, *FEBS Lett.* **250,** 570 (1989).
[33] C. Miller, E. Moczydlowski, R. Latorre, and M. Phillipps, *Nature (London)* **313,** 316 (1985).

cis side, and osmotic gradients,[21,22,26] fusion is far more difficult to achieve in the latter type of model membrane. This, plus the fact that, according to the available evidence, the solvent contained in "painted" membranes does not seem to have major effects in ion channel behavior, has led most laboratories using fusion approaches to resort to this type of membrane. There are no reports, however, that purified acetylcholine receptor channels have been functionally reconstituted in painted bilayers, suggesting that, at least in some cases, the presence of solvent can be of relevance to ion channel reconstitution in planar bilayers. Furthermore, there are some questions regarding ion channels that can be better attacked in "solvent-free" bilayers. These questions deal with the effects of lipid asymmetry on ion channel behavior.

"Painted" bilayers cannot be easily made asymmetric in lipid composition, although they can be made asymmetric in surface charge. Thus, "solvent-free" membranes provide a more adequate experimental system in which to study the effects of lipid composition on ion channel properties. As ion channels are studied in more and more detail, questions regarding the effects of the bilayer environment on ion channel function will become relevant. It is hoped that, by then, a better understanding of the mechanisms of vesicle fusion to planar lipid bilayers will facilitate the use of "solvent-free" bilayers in ion channel work.

Choosing Appropriate Ionic Conditions to Carry Out Ion Channel Measurements

Prior to the removal of the fusogenic conditions the experimenter can gain some preliminary information regarding the selectivity of the ion channels inserted. Under the experimental conditions that have been assumed here the cis compartment displays a higher concentration of cations and anions than the opposite compartment. If a cation-selective channel inserts into the bilayer, there will be a net flow of cations from cis to trans at zero applied voltage. In this case the sign of the resulting current is defined as positive, and therefore the development of current fluctuations in the positive quadrant indicates the insertion of a cation-selective channel. On the contrary, the insertion of an anion-selective channel will result in a negative current (anions flowing from cis to trans) under zero applied voltage. Based on the above preliminary information the experimenter might wish to perfuse the cis chamber with a solution different from that bathing the trans bilayer side in order to gain some insight regarding the ion selectivity properties of a given channel. For example, assume that a channel has been inserted which is perfectly cation selective and that the trans compartment contains a KCl buffer. Perfusing the cis chamber with a

NaCl buffer makes it possible to obtain a permeability ratio between Na^+ and K^+ ions. Furthermore, the cis side of the bilayer into which a channel has been inserted can now be sequentially perfused with different cationic buffers to determine permeability ratios for several cations in the same experiment.

On the other hand, if the experimenter decides to obtain single-channel current records under symmetrical ionic conditions, it is convenient to carry out these measurements at different applied voltages. This will permit the plotting of the current–voltage relation whose slope defines the channel conductance, as well as the gathering of preliminary information on the voltage-dependent properties of a given channel. Other maneuvers can also be performed such as checking the possible effects of chemical effectors like Ca^{2+}, cyclic nucleotides, or other agents on channel behavior. They should be tested in both the trans and the cis bilayer sides since the orientation of an ion channel inserted by fusion of vesicular fragments of cell membrane into a planar bilayer cannot be predicted *a priori*. If possible, experimental records should be stored without filtering on magnetic tape for further analysis. Appropriate low-pass filtering of the data can be done prior to data analysis, from the unfiltered records stored on tape.

In some cases the insertion of ion channels into planar lipid bilayers is used to study a particular channel known to be present in a purified membrane fraction in isolation from its native environment. In other cases, the experimenter might wish to discover and study those ion channels present in membrane fractions whose conductance properties are unknown, as is the case with membrane fractions derived from cell organelles.[1,34,35] In any case, it is convenient to choose the appropriate ionic conditions in which to carry out bilayer studies, because the same membrane fraction can exhibit different types of channels. A researcher interested in studying sodium channels in planar lipid bilayers would like to avoid the inconvenience of having, for example, a Na^+- selective channel and a Cl^- channel operating simultaneously. Thus, the dissection and study of the type of ion channels present in a given membrane preparation will be easier if performed under ionic conditions in which only one type of conductance is monitored at a time. To discard the contribution of anion-selective conductances in the experimental record, it is appropriate to use an ionic buffer made of a chosen cation and a large anion. The proper choice of cations will allow monitoring of K^+-, Na^+-, or Ca^{2+}-selective

[34] P. Labarca, D. Wolff, U. Soto, C. Necochea, and F. Leighton, *J. Membr. Biol.* **94**, 285 (1986).

[35] P. Labarca, R. Anholt, and S. Simon, *Proc. Natl. Acad. Sci. U.S.A.* **85**, 544 (1988).

channels in planar bilayers. The study of anion-selective channels can be done in a solution in which, for example, choline is the cation.

The studies by Affolter and Coronado[36] of Ca^{2+}-selective channels from the T-tubule of skeletal muscle provide an example of a wise choice of experimental conditions in which to monitor a particular type of channel in planar lipid bilayers. Here planar bilayers were made of an equimolar mixture of brain PE–PS (20 mg/ml, in decane) in an aqueous buffer made of 50 mM NaCl, 1 mM EGTA, 10 mM HEPES–Tris, pH 7.0. The addition of acidic lipid to the lipid mixture increases the rate of vesicle fusion to the planar lipid bilayer. After bilayer thinning the concentration of $BaCl_2$ in the cis side was raised to 100 mM, providing an osmotic gradient and at the same time establishing conditions suitable to monitor Ca^{2+}-selective channels. These conditions are also appropriate to avoid the contribution of Ca^{2+}-activated K^+ channels to the experimental record since they are blocked by this divalent cation. T-tubule membranes were then added into the cis side ($\sim$ 30 μg protein/ml), and the use of agonists of Ca^{2+}-selective channels (Bay-K8644 or CGP-28392) permitted the monitoring of single Ca^{2+} channel currents over extended periods of time. Thus, the experimental conditions chosen, in addition to eliminating the contribution of K^+-selective channels to the experimental record, make use of the known ion-selective and pharmacological properties of the channels under study in these experiments.

Preliminary Steps in Characterization of Ion Channels

Only the most general case will be considered here, namely, the fusion to planar bilayers of purified vesicular fragments derived from cell membranes for which there is little information on the types of ion channels displayed. In such a case, it is necessary to carry out a dissection of each channel type that, as the result of fusion, inserts in the planar bilayer. This dissection is easier to perform by choosing the appropriate ionic conditions (see Ref. 36), because the most general criterion for identifing a given ion channel lies in its ionic selectivity. A minimal description of a newly discovered ion channel should include the factors described below.

Conductance. A channel can exhibit one or more open conductance states in addition to the closed, zero-conductance state. The conductance of a given open state is obtained by building the current–voltage relation for that particular state. If this relation is linear, the conductance corresponds to the slope of the straight line. If it is nonlinear, the conductance can be taken as the slope of the current–voltage relation over a voltage

[36] H. Affolter and R. Coronado, *Biophys. J.* **48**, 341 (1985).

range where it is approximately linear. A report of the open state conductances should include the ionic conditions under which the measurements were made and the voltage range explored.

Selectivity. It is necessary to determine whether a channel is cation selective or anion selective. This can be done under conditions in which the ionic buffer contains chloride salts of a chosen cation (K^+, Na^+, Ca^{2+}) and salt is more concentrated in one chamber (e.g., 0.3 M KCl cis, 0.1 M KCl trans). For a channel that is perfectly cation selective or perfectly anion selective, the current will reverse sign at a voltage, E_{rev}, whose magnitude equals the Nernst potential:

$$E_{rev} = (RT/zF) \ln([A]_{trans}/[A]_{cis})$$

The sign of this "reversal potential" will depend on whether the channel is cation or anion selective. The convention used here is that, under the experimental conditions assumed above, the Nernst potential will have a negative sign if the channel is cation selective (because the *trans* bilayer side has been defined as virtual ground). If a channel can be permeated by both cations and anions, however, the reversal potential will be less than that predicted by the Nernst equation, namely,

$$E_{rev} = (RT/F) \ln\{(P_{cation}[A^+]_{trans} + P_{anion}[A^-]_{cis})/(P_{cation}[A^+]_{cis} + P_{anion}[A^-]_{trans})\}$$

Here, P is the permeability coefficient, and the concentrations must be corrected for ion activity coefficients. The sign and the magnitude of the reversal potential will depend on the degree to which a given channel selects between one particular cation and the anion. The ion selectivity can be conveniently expressed in terms of the permeability ratio:

$$P_{cation}/P_{anion} = \{[A^-]_{cis} - [A^-]_{trans} \exp(FE_{rev}/RT)\}/\{[A^+]_{cis} \exp(FE_{rev}/RT) - [A^+]_{trans}\}$$

Most channels are perfectly cation or perfectly anion selective. Moreover, most channels found in cell membranes are cation selective.

The next step consists of determining to what degree a given ion channel is able to discriminate between ions exhibiting the same valence sign. For example, the permeability ratios for a family of monovalent or divalent cations can be easily obtained using the planar bilayer assay. For this, one bilayer chamber (usually the *cis* chamber) is sequentially perfused with different ionic buffers while the ionic buffer in the opposite chamber is left unchanged. Estimates of permeability ratios for several ions can thus be obtained in a single experiment.

In the case in which only one divalent cation A^{2+} is present in the cis chamber and only one divalent cation B^{2+} is present in the trans chamber, the permeability ratio is

$$P_A^{2+}/P_B^{2+} = ([B^{2+}]/[A^{2+}]) \exp(2FE_{rev}/RT)$$

When one side of the bilayer contains only one monovalent cation and the other side only one divalent cation, the ratio becomes

$$P_B'^{2+}/P_A^+ = ([A^+]/4[B^{2+}]) \exp(FE_{rev}/RT)$$

where $P_B'^{2+} = P_B^{2+}/[1 + \exp(FE_{rev}/RT)]$.[37]

In all cases ion concentrations should be expressed as activities. Ion channel selectivity can be most easily studied in the planar bilayer because the ionic conditions can be conveniently chosen and both bilayer chambers are accessible.

Voltage Dependence. The fraction of time that a channel spends in the open conformation(s) might depend on the applied voltage. Usually, studies of channel voltage dependence are carried out under symmetrical conditions. As a minimum, it is necessary to find out whether a given channel displays voltage dependence. If a channel exhibits only one conductance state and one closed state, a plot of the fraction of time the channel spends in the open state as a function of the applied voltage will indicate how strong this voltage dependence is and the voltage range over which it occurs under a set of particular experimental conditions. For channels exhibiting several open conductance states the fraction of time in each state should be studied as a function of the applied voltage.

Pharmacology. Ion-channel blockers, toxins, agonists, and antagonists are useful tools in a preliminary characterization of ion channels. Different ion channels display distinctive pharmacological profiles, and the experimenter involved in dissecting ion channels present in a given membrane fraction should take advantage of the pharmacological properties of previously known channels. Furthermore, pharmacological studies in planar lipid bilayers can also be useful in gathering information concerning ion channels present in cell membranes that are not amenable to direct electrophysiological inquiries. Defining the role of a given ion channel in cell function can be assisted by the use of specific blockers of such a channel previously identified in planar lipid bilayers. In addition, the identification of toxins that bind with high affinity and specificity to ion channels is important to biochemists engaged in ion channel purification and reconstitution. Planar bilayers have already been shown to be quite useful in providing new information on ion channel pharmacology.[33]

[37] C. A. Lewis, *J. Physiol. (London)* **286**, 417 (1977).

Limitations of Planar Bilayer Assay

Solvent. There are many advantages to using planar lipid bilayers in ion channel work, as well as some disadvantages. The presence of solvent in the "painted" variant of the planar bilayer has not been completely discarded as a limitation. Many ion channels inserted in painted bilayers have been derived from purified membrane fractions whose conductance properties are not very well known at the single-channel level. The bilayer assay only monitors those channels that survive the bilayer environment. On these grounds, it would seem that the bilayer assay is not a truly reliable assay to attempt the dissection of ion channels in cell membranes not accessible to the patch clamp. However, the dissection of ion channels present in the sarcoplasmic reticulum membrane using planar lipid bilayer approaches seems to indicate otherwise. The sarcoplasmic reticulum is the only cell membrane whose ion channels have been monitored by using painted bilayers over a long period of time and by different laboratories. The results derived from bilayer work agree with those obtained by other experimental approaches used to monitor the permeability properties of this cell membrane. Furthermore, the few purified ion channels that have been studied in painted bilayers display properties that faithfully mimic those displayed in their native membrane environment. In spite of this, we lack sufficient experience with purified reconstituted ion channels to disregard the presence of solvent as a limitation of the planar lipid bilayer.

Time Resolution. A second limitation of the planar lipid bilayer is its large capacitance, which limits the resolution of the system. The analysis of ion channel current records is difficult to perform above 1000-kHz low-pass filtering. The planar lipis bilayer is not the best experimental system in which to carry out detailed kinetic studies of ion channel gating. Furthermore, ion channels exhibiting small conductances and fast kinetics are difficult to monitor in planar lipid bilayers. However, the above statement must be qualified since Wonderlin *et al.*[37a] introduced some valuable technical innovations that permit achievement of low noise recording of single-channel currents within a millisecond after a voltage step. To achieve this, they used small bilayer apertures ($25-80\ \mu$m diameter) having thin edges, low stray capacitance, and a geometry that minimizes access resistance to the bilayer. Single-channel currents were followed by means of a patch-clamp headstage endowed with logic control switching between a high gain (50 GΩ) and a low gain (50 MΩ) feedback resistance, for rapid charging of the bilayer capacitance. Residual currents, produced by gain

---

[37a] W. F. Wonderlin, A. Finkel, and R. J. French, *Biophys. J.* **58**, 289 (1990).

switching and electrostrictive changes in bilayer capacitance, were digitally substracted, allowing for a steady-state baseline within 1 ms.

Lack of Control of Vesicle Fusion. When fusing vesicular fractions derived from cell membranes, the control of the type, number, and origin of the channel inserted is not necesssarily easy. The problem of the type of channel to be monitored can be partially solved by choosing the ionic buffer. However, a frequent complication occurs when more than one channel type exhibiting similar selectivity properties or two channels of the same type are simultaneously inserted into the bilayer. Insertion of single channel molecules can be obtained in a significant number of assays by choosing the adequate fusion conditions. This requires some experience. Finally, it is difficult to determine with absolute certainty, by only using the planar bilayer assay, whether a channel inserted comes, as expected, from the membrane fraction of interest or from some minor contaminant fraction that is particularly fusogenic. Use of highly purified membrane preparations and the reproducibility of the experimental results from one preparation to another help the experimenter to gain confidence in this experimental approach and the results obtained.

Artifacts

Massive Vesicle Fusion. A common artifact is the develoment of conductance increases that result from massive fusion of lipid vesicles to the planar bilayer. The establishment of fusogenic conditions in the presence of pure lipid vesicles can produce stepwise increases in bilayer conductance resembling those observed on the insertion of ion channels. Massive fusion of vesicles may result in the breakage of the bilayer. In some cases, the removal of vesicles and the osmotic gradient cause a leaky bilayer exhibiting an unpredictable behavior. The most clear evidence that conductance increases observed are due to massive vesicle fusion is the observation of increases in bilayer capacitance (P. Labarca, unpublished results, 1980), which are not expected to occur when a single or a few vesicles fuse to the planar bilayer. The fusion of one lipid vesicle having a diameter of 1 μm to a 300-μm-diameter planar bilayer having a capacitance of 0.3 μF/cm^2 will increase the bilayer capacitance by only 0.02%. A 5% increase in bilayer capacitance would require the fusion of about 250 vesicles 1 μm in diameter. As stated previously, the main criterion for discarding a conductance increase as artifactual in fusion studies is the monitoring, in a reproducible fashion, of single-ion channel currents and the reproducibility in channel behavior observed from assay to assay under a particular experimental condition.

Mitochondrial Contamination. A second type of artifact often observed in fusion experiments arises from the contamination of the membrane preparation with channel-forming material. The most common is the voltage-dependent, anion-selective pore, VDAC,[38] present in the mitochondrial outer membrane. VDAC does not denaturate when removed from its native environment, and it can spontaneously insert into a planar lipid bilayer from the aqueous buffer. Its large conductance, the presence of conductance substates, and its voltage-dependent behavior might lead to the erroneous impression that the insertion of a single VDAC molecule into the planar bilayer corresponds to a fusion event. To avoid this it is fundamental to use a clean, highly purified membrane fraction. In addition several clues might indicate that VDAC is present in a given preparation. First, in contrast to most channels VDAC does not select very well between anions and cations. Second, they can insert spontaneously into the planar bilayer without requiring Ca^{2+} or osmotic gradients. The basic properties of VDAC are described in Ref. 39.

Bacterial Contamination. Other pore-forming contaminants that usually plague bilayer work include the porins present in the outer membrane of gram-negative bacteria.[40] Like VDAC, porins insert spontaneously into bilayers from the aqueous buffer, display conductance substates and voltage-dependent behavior,[41] and discriminate poorly among cations and anions.[40] The annoyances that arise from bacterial contamination can be avoided by careful handling of solutions and by keeping all materials used for bilayer work clean and neat. For example, stock solutions kept outside the refrigerator, old agar bridges, and dirty glass material are some of the potential sources of bacterial contamination. The habits of cleanliness should be exaggerated in work with planar lipid bilayers. The researcher must also realize that the ultimate goal of ion channel studies in planar lipid bilayers is the monitoring of the small currents flowing through single channel molecules. A careful handling of all the materials required for bilayer work will help the researcher gain confidence in bilayer studies and in the experimental results obtained using the technique.

Lipids. Artifacts can also occur when using poorly handled lipids. Use of oxidized lipids, lipids that have been contaminated with aqueous buffers or membrane preparations, and crude lipid extracts sometimes makes it impossible to form stable bilayers. In other cases, lipid bilayers made with such lipids, can be formed, but they will exhibit spontaneous conductance fluctuations that, in some cases, can even resemble those contributed by ion channel molecules. Obviously, the occurrence of some of these diffi-

[38] C. Doring and M. Colombini, *J. Membr. Biol.* **83,** 81 (1985).

[39] M. Colombini, *in* "Ion Channel Reconstitution" (C. Miller, ed.), p. 533. Plenum, New York and London, 1986.

[40] R. Benz, *Curr. Top. Membr. Transp.* **21,** 199 (1984).

[41] A. Mauro, M. Blake, and P. Labarca, *Proc. Natl. Acad. Sci. U.S.A.* **85,** 1071 (1988).

culties should lead the experimenter to avoid the use of such lipids and to improve techniques in handling planar bilayers.

Other Methods to Insert Ion Channels into Planar Lipid Bilayers

Fusion of lipid vesicles to "solvent-free" planar bilayers made of 1-stearoyl-3-myristoyl/glycero-2-phosphocholine, a mixed-chain lecithin, at temperatures below the lipid phase transition has been reported by Boheim and collaborators (see e.g., Ref. 3). This method is not widely used since it requires the use of a particular lipid at restricted temperatures. The method of Schindler[42] to transfer proteins from lipid vesicles to a planar bilayer has been successfully used to insert channels into planar bilayers[27,28,43] and to form bilayers from monolayers in the tip of a patch pipette.[44-46]

Acknowledgments

This work was supported by grants from the Fondo Nacional de Investigacion (1167–1988, 451–1988), the National Institutes of Health (GM-35981), and the Tinker Foundation. R.L. is a recipient of a John S. Guggenheim Fellowship. He also wishes to thank the Dreyfus Bank (Switzerland) for generous support from a private foundation that they made available.

[42] H. Schindler, *FEBS Lett.* **122,** 77 (1980).
[43] H. Schindler and U. Quast, *Proc. Natl. Acad. Sci. U.S.A.* **77,** 3052 (1980).
[44] W. Hanke, C. Methfessel, H. U. Wilsem, and G. Boheim, *Biochem. Bioeng. J.* **12,** 329 (1984).
[45] R. Coronado and R. Latorre, *Biophys. J.* **43,** 231 (1983).
[46] B. Suarez-Isla, K. Wan, J. Lindstrom, and M. Montal, *Biochemistry* **22,** 2319 (1983).

[31] Planar Lipid Bilayers on Patch Pipettes: Bilayer Formation and Ion Channel Incorporation

By BARBARA E. EHRLICH

Introduction

This chapter describes how to make bilayers on the tip of patch-style pipettes and then incorporate channels into these bilayers. As there have been several reports describing the specific techniques needed for the formation of this type of bilayer,[1-4] and even more reports describing how

[1] R. Coronado and R. Latorre, *Biophys J.* **43,** 231 (1983).
[2] T. Schuerholz and H. Schindler, *FEBS Lett.* **152,** 187 (1983).
[3] B. A. Suarez-Isla, K. Wan, J. Lindstrom, and M. Montal, *Biochemistry* **22,** 2319 (1983).
[4] W. Hanke, C. Methfessel, H. V. Wilmsen, and G. Boheim, *Biochem. Bioeng. J.* **12,** 329 (1984).

to insert channels into membranes,[5] the emphasis of this chapter will be to transcribe some of the "oral lore and traditions" that accompany the formation of bilayers on the tip of patch-style pipettes and the study of channels in these bilayers. After a short overview, I describe the method for making these bilayers and outline the assorted techniques that have been developed to incorporate channels into the membrane. As appropriate, I shall include some of the pitfalls that need to be avoided.

Overview

Tip-dip membranes, double dip membranes, and patch membranes are bilayers whose name refers to the method of membrane formation. In other words, bilayers are made at the tip of patch-style pipettes by passing the tip of the pipette through a monolayer of lipid two times.

Tip-dip bilayers represent the combination of two other techniques, patch clamping and planar lipid bilayers. Using these techniques an artificial membrane is formed that allows for alterations in the lipid and salt composition, the ionic strength, and the presence of cofactors, agonists, and the like. In addition, the small size of the membrane means low current noise, which allows rapid changes in the voltage and good resolution of small currents, especially those from rapidly gating channels. Although tip-dip bilayers provided the best way to make small, high resolution bilayers for many years, more recently a technique that makes small, traditional, decane-containing bilayers has been described that allows high frequency resolution.[6]

An early example of making small bilayers on the tip of glass pipettes used a glass pipette "patched" onto a standard black lipid bilayer.[7] The large membrane is constructed from lipid dissolved in n-decane. Then a fire-polished, silanized glass pipette is brought up to the membrane as if it were a cell, and, after waiting 0.5 to 5 min, a small membrane is formed on the tip and the large membrane seals around the pipette. With this configuration it is possible to measure both macroscopic and single-channel currents. Disadvantages of this method are that two bilayers must remain intact and that lipid is dissolved in n-decane. Using decane probably is not a serious disadvantage, but there are some reports of differences in channel kinetics when compared in painted bilayers and in solvent-free bilayers.[8]

Subsequently, methods to make solvent-free bilayers on the tip of glass

---

[5] C. Miller, ed., "Ion Channel Reconstitution." Plenum, New York, 1986.

[6] W. Wonderlin, A. Finkel, and R. French, *Biophys. J.* **58,** 289 (1990).

[7] Anderson and Muller, *J. Gen. Physiol.* **80,** 403 (1982).

[8] W. Hanke, *in* "Ion Channel Reconstitution" (C. Miller, ed.), p. 141. Plenum, New York, 1986.

pipettes were developed.[1-4] It is this type of tip-dip bilayer that will be described in this chapter. Even though the basic concept for making bilayers at the tip of patch pipettes is simple, construction of these membranes has some requirements that differ from those used in standard bilayer formation.

Fabrication of Tip-Dip Bilayers

Starting Materials

The basic equipment needed for any bilayer setup includes an oscilloscope, a Faraday cage, a vibration-isolation table, a stimulator for pulsing the bilayer and for setting the holding potential, a signal generator for capacitance measurements if the stimulator cannot make a triangle wave, a stirrer, and chart and tape recorders.[9] To make tip-dip bilayers a patch clamp amplifier, a pipette puller, and a micromanipulator are needed in addition to the basic electronic components used to make painted or solvent-free membranes. Optional equipment includes a computer with software and an analog–digital interface for taking, storing, and analyzing the data.

High-quality lipids should be obtained from a supplier such as Avanti Polar Lipids (Birmingham, AL). I have had good success in storing the lipids in chloroform at − 70°. When ready to use, the lipid is brought to room temperature, the desired volume is put in a clean glass tube, the chloroform is removed by a stream of nitrogen, and the lipid is resuspended in the desired solvent. In all cases where I suggest hexane as the solvent, pentane can be substituted.

The final component needed is membrane vesicles or a protein to study. Numerous proteins can or have been studied in bilayers.

Fabrication of Pipettes

The membrane is literally made at the end of a glass micropipette. The pipettes such as Kimax-51 capillary tubes ($0.8-1.1 \times 100$ mm) and micro-hematocrit capillary tubes (without heparin), are constructed as if they were to be used for patch clamping a cell. Standard methods for pulling the pipettes are employed.[10] Briefly, the pipettes are pulled in two stages: the first stage thins the glass over about 10 mm and the second stage pulls the

[9] O. Alvarez, in "Ion Channel Reconstitution" (C. Miller, ed.), p. 115. Plenum, New York, 1986.
[10] O. P. Hamill, A. Marty, E. Neher, B. Sakmann, and F. J. Sigworth, *Pfluegers Arch.* **391,** 85 (1981).

recentered capillary into two pipettes. It is not necessary to coat the shank of the pipette with Sylgard, nor to heat polish the tip of the pipette. It is useful, however, to determine the size of the pipette tip by attaching the pipette shank to a 10-cm³ syring with a Touhy-Borst adaptor with the syringe plunger set at 10 cm.³ The tip of the pipette is then dipped into a scintillation vial filled with methanol. The syringe plunger is depressed until the first bubbles are formed, and the location of the plunger is noted. Pipettes that form bubbles between approximately 1.5 and 2 atm pressure (between 3.5 and 5 ml on the syringe) usually give good results.

Many pipettes (5 to 20 acceptable ones) should be made and kept in a clean, dry location before the experiment (a small, square desiccator is a good holder). New pipettes should be made each day. Solution should be added to the pipette just before use, as is the custom in patch clamp experiments. The easiest way to make a device to fill the pipette is to pull a 1-cm³ syringe to a fine tube by gently heating while turning over a small flame. This device has several advantages: it fits into the shank of the pipette, there are no metal parts to leach divalent cations into the pipette solution (which will alter the pH), and it is easy to clean or replace.

Preparation of Bilayer Chamber

A 96-well microtiter plate, preferably with flat bottoms, is used. The flat bottoms make it possible to stir the solution in each well with a 3 × 5 mm magnetic flea without disturbing the surface. To utilize the microtiter plate to the fullest extent, cover it with a piece of Parafilm to keep the unused wells clean. To begin an experiment four (or more) wells are uncovered; one well is for the ground wire, and the other wells are for experimental solutions. The wells are connected by agar bridges. Because it is important that the bridges do not move and do not interfere with the path of the patch pipette during the experiment, make the bridges from short lengths of glass tubing that have been bent in as U shape to span the wells without projecting more than a few millimeters above the level of the microtiter plate. The tubes are filled with 2% agar dissolved in 0.5 M CsCl. With this type of setup each well can have a different solution. To change solutions during the experiment the pipette is moved to the appropriate well, rather than perfusing a new solution into the well.

Making Bilayers

A buffered salt solution is added to each well (usually 0.3 ml) and a solution of lipid in hexane is layered on top of the solution. One to 2 μl of a lipid (10 mg/ml) is used even though this is much more than is needed to form a monolayer. After the hexane is evaporated, the pipette is passed through the monolayer several times, starting with the pipette in the air, then into the solution, into the air, and back into the solution. It is best to have the tip just below the surface to minimize the capacitance due to the submerged glass. Alternatively, the glass could be coated with Sylgard, but it seemed easier to just keep the tip close to the surface. During the dipping procedure it is useful to pulse the membrane with a small voltage (~ 10 mV) to monitor the formation of the membrane.

There are several common outcomes of the dipping procedure. (1) Formation of a membrane is not possible, that is, the tip resistance remains low. The only way to deal with this outcome is to discard the pipette and take another pipette. Often the tip has broken and is too large to hold a membrane. (2) The membrane resistance is extremely high ($> 10^{11}$ Ω), suggesting that there is a glob of lipid clogging the tip of the pipette. Again, discard the clogged pipette and take another pipette. (3) The membrane is formed, but the resistance is too low to be useful. Usually the membrane resistance can be increased by applying suction via a tube attached to the pipette holder as if one were patching onto a cell. Satisfactory results are usually obtained with membrane resistances of $2-10 \times 10^{10}$ Ω. Because there is no cytoskeleton, too much pressure will rupture the membrane and the procedure must be started again. It is possible to form a membrane on a pipette more than once if too much suction is applied.

Once the bilayer is formed, the membrane is ready for the experiment (Fig. 1). To change solutions lift the pipette out of the solution and move the pipette to another well with a different solution composition. Sufficient solution adheres to the pipette to maintain the bilayer, but not so much solution that there is concern of contaminating the new solution. The use of a micromanipulator makes this procedure more convenient.

Advantages and Disadvantages

The main advantage of tip-dip membranes is that they are very small ($1-5$ μm) so the drawbacks associated with larger bilayers are avoided. Fast, small events can be detected with tip-dip membranes. The main reasons many investigators have avoided tip-dip membranes are that access to the solution inside the pipette is difficult and some channel types just will not insert into this type of membrane. It may be possible to perfuse the inside of the pipette as has been demonstrated in patch-clamped cells,[11] (see also articles 10 and 48 this volume by Tang *et al.*), but we have not tried this. There are no obvious solutions to the lack of channel insertion. Another problem is that there can be cation-selective channellike events in the absence of added protein, possibly owing to an

---

[11] M. Soejima and A. Noma, *Pfluegers Arch.* **400**, 424 (1984).

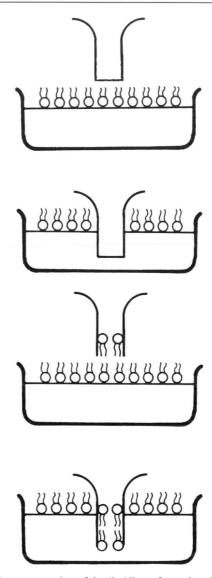

FIG. 1. Schematic representation of tip-dip bilayer formation. See text for details.

interaction between the lipid and the glass. Using high divalent cation concentrations (> 25 mM), especially in the pipette solution, and synthetic lipids (rather than extracted lipids) and taking extra care in cleaning the solutions and all associated paraphernalia often reduce the appearance of these events. As in making all types of bilayers, it is useful to be obssessive about cleanliness.

Incorporation of Channels

Although the method to construct tip-dip bilayers is different from more traditional methods used for making lipid bilayers, channel incorporation occurs by the same techniques developed previously. Of the two main techniques, the first uses proteoliposomes as the source of lipid for the solvent-free or the tip-dip membranes. The protein of interest is reconstituted into vesicles (or native membrane vesicles are used), and these vesicles are layered on top of the saline solution. Vesicles at the air–water interface will lyse and form a monolayer with protein embedded at the density found in the vesicles. When the membrane is formed the protein is already in the monolayer and will, therefore, be in the bilayer. The membrane protein at the air–water interface is subjected to the same forces that induce the lysing of the vesicles and therefore, probably denature the protein. Nonetheless, some membrane proteins have been incorporated into bilayers with this method. It is conceivable that the proteins at the interface are denatured and that the proteins incorporated into the bilayer actually get there by fusion of the vesicles below and not at the interface. This is the first technique.

In the majority of work that studies channels after they have been incorporated into bilayers, the vesicles have been fused to the bilayer. It is generally assumed that fusion occurs by a mechanism that imitates cellular exocytosis. Either native membrane vesicles or purified protein reconstituted into lipid vesicles can be used. Fusion requires addition of calcium to the vesicle-containing bath and an osmotic gradient across the bilayer, where the hyperosmotic solution is the vesicle-containing bath. Actually, if both the phospholipid vesicles and the bilayer contain no negatively charged lipids, calcium is not required.[12] In practice, however, calcium is necessary because many applications use native membrane vesicles which contain negatively charged lipids, or negatively charged lipids are added to the phospholipid vesicles to allow the control of fusion by calcium. The calcium is needed to allow the vesicles to "adhere" to the bilayer.[13] The osmotic gradient is needed to promote vesicle swelling and eventual lysis by stimulating diffusion of water across the membrane.[14] If the vesicle is very close to the bilayer ("adhered"), then swelling increases the area of contact between the vesicle and the bilayer. It is theorized that vesicles swollen to the maximum volume will interact with the bilayer, and this interaction initiates fusion between the two bilayers.

Some degree of control can be obtained by varying the magnitude of

[12] F. S. Cohen, M. H. Akabas, J. Zimmerberg, and A. Finkelstein, *J. Cell Biol.* **98,** 1054 (1984).
[13] M. H. Akabas, F. S. Cohen, and A. Finkelstein, *J. Cell Biol.* **98,** 1063 (1984).
[14] F. S. Cohen, M. H. Akabas, and A. Finkelstein, *Science* **217,** 458 (1982).

the osmotic gradient and the number of vesicles. For example, to terminate fusion, the osmotic gradient can be neutralized and the vesicles removed from the bath by perfusion. Nonetheless, additional vesicles often will fuse in the absence of an osmotic gradient. This phenomenon is probably due to the observation that vesicles completely coat the bilayer and perfusion will not remove all the "adhered" vesicles. The fact that the vesicles coat the bilayer also explains why adding more vesicles often will not increase the number of fusion events or decrease the lag between vesicle addition and the first fusion event.

Conclusion

It is important to remember that although the methods outlined here and in other reviews appear to be straightforward, there are numerous steps that cannot be explained theoretically at the present time. A better understanding of the physical processes that govern protein – lipid and lipid – lipid interactions will assist in future efforts in bilayer formation and channel reconstitution.

A final word of caution was given by Finkelstein[15] in a chapter on planar lipid bilayer formation in an earlier volume of this series. This piece of advice can be applied to tip-dip bilayers. He noted, " . . . that some manipulation of variables is required before everything is working properly. Then, one can make stable membranes quickly and reliably week after week until, as happens to everyone I know who works with this system, one day a stable membrane cannot be formed. After a few agonizing days of changing and permuting the lipid, the septa, the brush, the distilled water source, and your socks, everything works properly again. Most likely, the conditions are the same as before." I have been asked many times if it is truly necessary to change your socks. The answer is, yes.

Acknowledgments

I thank Dr. Ilya Bezprozvanny for comments on the manuscript. This work was supported by National Institutes of Health Grants HL-33026 and GM-39029. B.E.E. is a Pew Scholar in the Biomedical Sciences.

[15] A. Finkelstein, this series, Vol. 32, p. 489.

[32] Surface Charge Effects on Ion Conduction in Ion Channels

By RAMÓN LATORRE, PEDRO LABARCA, and DAVID NARANJO

Introduction

Ion channels belong to a class of integral membrane proteins which when confronted with a stimulus are able to change their conformation from a nonconductive to an ion-conductive one or vice versa. In the conductive configuration (open state) ion channels let tens of millions of ions to flow passively across the cell membrane, allowing a number of important cellular events to occur. The interface formed by an electrolyte in contact with a charged surface can have important effects in ion channel function. Thus, a charged surface in the neighborhood of the ion conduction system influences the concentration of ions at the channel entrances and hence its conductance.[1] Moreover, voltage-dependent channels, such as the sodium and potassium channels of nerve and muscle, have sensors of transmembrane electrical potential. The sensors, known as gating charges, sense the *difference* in the transmembrane potential proper. Their position in the electric field, which determines whether the channel is closed or open, is always a function of the applied potential (measured by the electrodes) *plus* the difference in surface potential induced by any of the charged interfaces that are in the proximity of the gating machinery. Therefore, changes in the surface potential, promoted, for example, by changes in the ionic strength of the solution, can have profound effects in the channel activation versus voltage curves, a fact known since the pioneering work of Frankenhaeuser and Hodgkin.[2]

In the above examples the role played by fixed charges in controlling ion permeation and gating has been emphasized. However, these electrostatic interactions can also be used to obtain valuable information about the gross architecture of ion channels. Changes in surface potential can be used to test surface charges densities in the neighborhood of the voltage sensors and the permeation sites.[3-7] On the other hand, reconstitution of

[1] H.-J. Apell, E. Bamberg, H. Alpes, and P. Lauger, *J. Membr. Biol.* **31**, 171 (1977).

[2] B. Frankenhaeuser and A. L. Hodgkin, *J. Physiol. (London)* **137**, 218 (1957).

[3] D. L. Gilbert and G. Ehrenstein, *J. Gen. Physiol.* **55**, 822 (1969).

[4] G. N. Mozhayeva and A. P. Naumov, *Nature (London)* **228**, 164 (1970).

[5] B. Hille, A. M. Woodhull, and B. I. Shapiro, *Philos. Trans. R. Soc. London, Ser. B* **270**, 301 (1975).

[6] T. Begenisich, *J. Gen. Physiol.* **66**, 47 (1975).

[7] R. MacKinnon, R. Latorre, and C. Miller, *Biochemistry* **28**, 8092 (1989).

ion channels in planar bilayers allows comparisons of channel activity in bilayers made of neutral lipids with channel activity in bilayers made of charged lipids. These studies have made it possible to know the degree of insulation of the conduction system and the channel gating machinery from the surface potential originated by the fixed charges located in the plane of the lipid bilayer membrane.[8-12]

In biological membranes fixed charges can originate in principle from three different sources: (1) the membrane lipids, (2) ionized groups in the channel-forming protein, and/or (3) nonprotein domains forming part of the molecular structure of the channel, such as acidic carbohydrate domains containing sialic acid. The most abundant charged phospholipid in biological membranes is phosphatidylserine (PS), which possesses a single negative charge from the carboxyl group in the serine residue of the lipid. Several studies indicate that the lipids forming cell membranes are not randomly distributed between the two monolayers, but rather occur in a asymmetric arrangement.[13] In particular, the studies made with erythrocyte membranes revealed that the outer monolayer of the erythrocyte membrane consists mainly of neutral lipid, whereas the inner monolayer contains most of the PS. A lipid distribution of this nature would promote a difference in surface potential across the lipid bilayer *of about 67 mV* if we considered the red cell to be filled with a 120 mM NaCl solution. Asymmetry of the charged PS between membrane monolayers has also been determined in isolated disc membranes of rod outer segment. In this case about 72% of the PS is found on the external monolayer.[14] This case is of interest because the charged lipid composition was determined by estimating the surface potentials of the inner and outer monolayer.

Charged amino acid side chains of the channel protein (e.g., lysine, arginine, and glutamine residues) can originate surface potentials. In this case assessing the electrostatic influence of these charges on ion transport or channel gating is difficult as, in general, molecular structure, charge location, and distribution of the charged residues are not known. A notable exception to this situation is the case of the acetylcholine receptor channel, where studies using site-directed mutagenesis have shown that a ring of glutamine residues located in a known segment of the protein called M2

[8] H.-J. Apell, E. Bamberg, and P. Lauger, *Biochim. Biophys. Acta* **552**, 369 (1979).

[9] X. Cecchi, O. Alvarez, and R. Latorre, *J. Gen. Physiol.* **66**, 535 (1981).

[10] J. Bell and C. Miller, *Biophys. J.* **45**, 279 (1984).

[11] E. Moczydlowski, O. Alvarez, C. Vergara, and R. Latorre, *J. Membr. Biol.* **83**, 273 (1985).

[12] R. Coronado and H. Affolter, *J. Gen. Physiol.* **87**, 933 (1986).

[13] J. Rothman and J. Lenard, *Science* **195**, 743 (1977).

[14] F. C. Tsui, S. A. Sundberg, and W. L. Hubbell, *Biophys. J.* **57**, 85 (1990).

controls the rate of ion transport through this channel.[15] The importance of the surface charge associated with carbohydrates in modulating channel ion permeation and gating is less clear despite the fact that in some cases, like the sodium channel, the channel protein is heavily glycosylated and some 40% of the total carbohydrates is sialic acid.[16]

When trying to elucidate the role of fixed charges in ion channel modulation, what is wanted finally is quantification of the effects of surface charges on channel conductance and gating, a difficult task, especially since in most cases the structure of the molecular domain where the fixed charges reside is unknown. Most authors have prefered to explain their results in terms of a relatively simple and elegant theory: the Gouy–Chapman theory of the electrical double layer.[17,18] The principles underlying this theory are very similar to those involved in the Debye–Hückel theory for strong electrolytes.[19] The only difference between the two theories is that in the Gouy–Chapman treatment the equations are those for planar rather than spherical symmetry, which makes the theoretical treatment simpler mathematically. The Gouy–Chapman theory was improved by Stern,[20] who brought it more in agreement with physical reality by suggesting that ions cannot be treated as point charges. The Stern model extends the Gouy–Chapman theory by including specific ion binding at the interface. It is amazing how well this theory has stood the test of time. For example, the ion distribution in an electrolyte solution in contact with a charged phospholipid membrane has recently been measured directly, and the results qualitatively agree with the Gouy–Chapman–Stern theory.[21] In the words of McLaughlin,[22] "the unexpected feature of the Gouy–Chapman theory is that it works. . . . " The Interested reader who wishes a detailed historical and mathematical account of this theory is referred to Refs. 23 and 24. For general treatments relating this theory and its applicability to membranes, the reader should consult Refs. 22 and 25.

[15] K. Imoto, C. Bush, B. Sakmann, M. Mishina, T. Konno, J. Nakai, H. Bujo, Y. Mori, K. Fukuda, and S. Numa, *Nature (London)* **335,** 645 (1988).
[16] S. R. Levinson, W. B. Thornhill, D. S. Duch, E. Recio-Pinto, and B. W. Urban, *in* "Ion Channel" (T. Narahashi, ed.), Vol. 2, p. 33. Plenum, New York, 1990.
[17] M. Gouy, *J. Phys. Radium* **9,** 457 (1910).
[18] D. L. Chapman, *Philos. Mag.* **25,** 475 (1913).
[19] V. P. Debye and E. Hückel, *Phys. Z.* **24,** 185 (1923).
[20] O. Stern, *Z. Elektrochem.* **30,** 508 (1924).
[21] M. J. Bedzyk, G. M. Bommarito, M. Caffrey, and T. L. Penner, *Science* **248,** 52 (1990).
[22] S. McLaughlin, *Annu. Rev. Biophys. Biophys. Chem.* **18,** 113 (1989).
[23] J. O. Bockriss and A. K. N. Reddy, "Modern Electrochemistry," Vol. 2, p. 623. Plenum, New York, 1970.
[24] R. Aveyard and D. A. Haydon, "An Introduction to the Principles of Surface Chemistry." Cambridge Univ. Press, London, 1973.
[25] S. G. A McLaughlin, *Curr. Top. Memb. Transp.* **9,** 71 (1977).

Gouy–Chapman Theory of Electrical Double Layers

Here the problem is reduced to calculating the surface potential, ϕ_s, given a surface charge density, σ, or vice versa. Consider then an infinite plane surface in which the surface charge is smeared in contact with an electrolyte solution in which ions are considered as point charges. Let us assume, just to illustrate the important points of the system, that the surface is charged with negative charges (Fig. 1). At an infinite distance from the surface the potential will be that of the bulk solution (zero in the absence of an applied potential). However, as the surface is approached the potential will change until it reaches its maximum (absolute) value at the surface itself ($x = 0$). It would be useful to answer the following questions: What is the relation between ϕ_s and σ? What is the relation between ϕ_s and the ion concentration, and how does ϕ_s change with distance?

Case of Mixed Electrolytes

To find an expression to relate surface charge density with $\sigma_s(x = 0) \equiv \phi_o$, let us recall the Poisson equation since it relates charge density and electrostatic potential. In one dimension the Poisson equation is

$$d^2\phi(x)/dx^2 = -\mathbf{R}(x)/E_oE_s \tag{1}$$

where $\mathbf{R}$ is the charge density (C/m³), E_o the permittivity of the free space (C²/Nm²), and E_s the dielectric constant of the solution (water has a

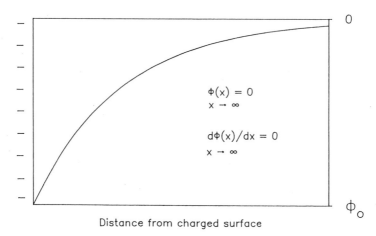

$\phi(x) = 0$
$x \rightarrow \infty$

$d\phi(x)/dx = 0$
$x \rightarrow \infty$

Distance from charged surface

FIG. 1. Properties of the electrostatic potential generated by a charged surface. The potential falls to zero as the distance from the surface is large, and its first derivative is also zero. These are the necessary boundary conditions to integrate the Poisson–Boltzmann equation. At the surface itself the potential reaches a maximum value equal to ϕ_o.

dielectric constant of 80.36 at 20°). On the other hand, the Boltzmann distribution law relates ion concentration and potential:

$$C_i(x) = C_i^{\text{bulk}} \exp[-z_i F\phi(x)/RT] = C_i(\infty) \exp[-z_i F\phi(x)/RT] \quad (2)$$

where C_i is the concentration of the ionic species i, z_i is the valence, and F, R, and T have their usual meanings. In the present case, **R** is the excess charge promoted by the presence of the fixed surface charge at any given point x in the solution, therefore,

$$\mathbf{R}(x) = \Sigma\, z_i F C_i(x) \quad (3)$$

which combined with Eqs. (1) and (2) yields

$$d^2\phi(x)/dx^2 = -(F/E_oE_s)\{\Sigma\, z_i C_i(\infty) \exp[z_i F\phi(x)/RT]\} \quad (4)$$

For the sake of compactness, the following definitions will be used: $\psi(x) \equiv F\phi(x)/RT$ and $C_i(\infty) \equiv C_i$, $\psi(x)$ is usually called the reduced potential. With these definitions Eq. (4) becomes

$$d^2\psi(x)/dx^2 = -(F^2/RTE_oE_s)\{\Sigma\, z_i C_i \exp[z_i\psi(x)]\} \quad (5)$$

Equation (5) can be integrated multiplying both sides of Eq. (5) by the integrating factor $d\psi/dx$. Using the boundary conditions given above yields

$$(1/2)(d\psi/dx)^2_{x=0} = (F^2\, \Sigma\, z_i C_i/RTE_oE_s)[\exp(-z_i\psi_o) - 1] \quad (6)$$

Because electroneutrality must hold in the double layer, the excess of charge in the solution must be equal with opposite sign to that in the surface. This implies that

$$\sigma = -\mathbf{R}(x)\, dx \quad (7)$$

Combination of Eqs. (1) and (7) gives

$$d[d\psi(x)/dx] = -(F/RTE_oE_s)\mathbf{R}(x)\, dx = (F/RTE_oE_w)\sigma \quad (8)$$

which on integration and recalling the boundary conditions becomes

$$[d\psi(x)/dx]^2_{x=0} = -(F/RTE_oE_s)\sigma^2 \quad (9)$$

Combination of Eqs. (6) and (9) then gives

$$(F/RTE_oE_w)^2\sigma^2 = (2F^2\, \Sigma\, C_i/RTE_oE_s)[\exp(-z_i\psi_o) - 1] \quad (10)$$

Recalling the definition of reduced potential and taking the square root of the left- and the right-hand sides of Eq. (10), the so-called Gouy–Chapman equation is immediately obtained:[26]

$$\sigma = \pm\{2RTE_oE_s\, \Sigma\, C_i[\exp(z_i F\psi_o/RT) - 1]\}^{1/2} \quad (11)$$

[26] D. C. Grahame, *Chem. Rev.* **41**, 441 (1947).

Equation (11) must be used for solutions of mixed electrolytes, which is the case of biological solutions containing both monovalent and divalent cations. Notice that if RT is given in joules/mole, the term $(RTE_oE_s \Sigma C_i)^{1/2}$, and therefore σ, is given in coulombs/m^2.

Case of One Symmetrical Electrolyte of $z:z$ Type

If for simplicity only a symmetrical electrolyte (e.g., NaCl) is considered, the Poisson equation is given by the relation

$$d^2\psi(x)/dx^2 = -(F^2/RTE_oE_s)[C^+(x) - C^-(x)] \tag{12}$$

where C^+ and C^- are the concentrations of the cation and the anion, respectively, which in turn are functions of the distance x from the charged surface. Using the identity $\sinh x \equiv (e^x - e^{-x})/2$, Eq. (12) becomes

$$d^2\psi(x)/dx^2 = (2F^2C_i/RTE_oE_s) \sinh z_i\psi(x) \tag{13}$$

which on a first integration, following a procedure similar to the one used to obtain Eq. (6), gives

$$[d\psi(x)/dx]^2_{x=0} = (4F^2C_i/RTE_oE_s)[\cosh z_i\psi_o - 1] \tag{14}$$

Equation (14) together with Eq. (9), the identity $[(\cosh x - 1)/2] \equiv \sinh(x/2)$, and recalling the definition of reduced potential leads to

$$\sigma = (8RTE_oE_sC_i)^{1/2} \sinh(z_iF\phi_o/2RT) \tag{15}$$

where it is assumed that $C^+ = C^- = C_i$ and $z^+ = z^- = z$. Equation (15) is considerably simpler than Eq. (11).

Both Eqs. (11) and (15) have been used profusely in membrane transport physiology (see Refs. 22 and 25 for reviews), and they illustrate that the surface potential produced by a given surface charge density depends on ion concentration and valence (see Fig. 2A,B). Multivalent ions are more effective in screening the surface charge than monovalent ions. The higher the concentration the lower the surface potential. At very low electrolyte concentrations high potentials are expected. In this case the Gouy–Chapman theory predicts that the concentration of counterions at the membrane surface becomes a constant independent of the bulk electrolyte concentration. This is easily shown taking as a starting point Eq. (15), which at high potentials (assuming a negatively charged surface) becomes[22]

$$-[\exp(-zF\phi_o/2RT)] = 2\sigma/(8RTE_oE_sC_i)^{1/2} \tag{15a}$$

Rearranging and squaring both sides of Eq. (15a) yield

$$C_o = C_i \exp(-zF\phi_o/RT) = \sigma^2/2RTE_oE_s \tag{15b}$$

Thus, in the limit of very low concentrations the electrolyte concentration at the surface is proportional to the square of the charge density and is independent of both valence and bulk electrolyte concentration. This result has important consequences in the interpretation of the values for ion channel conductance obtained at low salt concentrations as discussed below.

For practical purposes a convenient way to write Eq. (15) is

$$\sigma = [(C_i)^{1/2}/136] \sinh(z_i F \phi_o/2RT) \tag{16}$$

since in this case the concentration is given in moles/liter, σ in $-e$/Ångstrom^2, and ϕ_o in millivolts. All these are units used routinely by electrophysiologists. Moreover, making use of the identity $\sinh^{-1} x \equiv \ln[x + (x^2 + 1)^{1/2}]$, it is possible to obtain ϕ_o from the equation

$$\phi_o = (2RT/z_i F) \ln[X + (X^2 + 1)^{1/2}] \tag{17}$$

where $X = 136\sigma/(C_i)^{1/2}$.

Surface Potential as Function of Distance

Equation (12) can be written as

$$d^2\psi(x)/dx^2 = -(F^2 C_i/RTE_o E_s)\{\exp[-\psi(x)] - \exp[\psi(x)]\} \tag{18}$$

For the case of small potentials {when $\psi(x) \ll kT$, $\exp[\pm\psi(x)] = 1 \pm \psi(x)$) Eq. (18) becomes

$$d^2\psi(x)/dx^2 = (2F^2 C_i/RTE_o E_s)\psi(x) \tag{19}$$

Defining $K^2 \equiv 2F^2 C_i/RTE_o E_s$, the general solution of the differential equation [Eq. (19)] is

$$\psi(x) = A \exp(-Kx) + B \exp(Kx) \tag{20}$$

Given the boundary conditions of the problem, $\psi(x) \to 0$ as $x \to \infty$, B must be zero, otherwise the electrostatic potential would grow indefinitely with distance, a result with no physical meaning. On the other hand, the constant A must satisfy any particular solution of Eq. (20), in particular when $x \to 0$, $A = \psi(0) = \psi_o$. Therefore,

$$\psi(x) = \psi_o \exp(-Kx) \tag{21}$$

Thus, for small surface potentials the Gouy–Chapman theory predicts that the potential falls exponentially from the charged surface. From Eq. (21) it can be seen that the potentials falls to ψ_o/e when $x = 1/K$, which is referred to as the thickness of the diffuse double layer. The parameter K is identical to the reciprocal of the Debye length of the Debye–Hückel theory for strong electrolytes. The thickness of the diffuse double layer decreases as the ion concentration is increased, and the larger the ionic strength the faster the double layer potential falls with distance. For example, for a $1:1$ salt $1/K = 0.96$ nm at 0.1 M and 3.04 nm at 0.01 M. The Debye length depends also on the type of electrolyte. For a $2:2$ electrolyte $1/K = 0.48$ nm at 0.1 M and 1.52 nm at 0.01 M, and, as expected, $\phi(x)$ decays faster with distance than in the presence of a $1:1$ electrolyte.

From Eq. (21) and the boundary conditions we have

$$-(d\psi/dx)_{x=0} = \psi_o K \qquad (22)$$

On the other hand, from the Poisson equation [Eq. (1)] and the fact that the double layer must be, as a whole, electrically neutral [Eq. (7)], we obtain

$$-(d\psi/dx)_{x=0} = F\sigma/RTKE_oE_s \qquad (23)$$

Combining Eqs. (22) and (23) and remembering that $\psi = F\phi/RT$ we arrive as the expression

$$\phi_o = \sigma/E_oE_sK \qquad (23a)$$

Equation (23a) is the equation for a parallel plate capacitor in which ϕ_o is the potential difference between plates separated by distance $1/K$ when the charge density is σ.

A function $\phi(x)$, applicable to interfacial potentials of arbitrary magnitude, can be obtained by integration of Eq. (13) with the result[24,25]

$$\phi(x) = (2/z)\ln\{[1 + \alpha\exp(-Kx)]/[1 - \alpha\exp(-Kx)]\} \qquad (24)$$

where $\alpha = [\exp(z\phi_o/2) - 1]/[\exp(z\phi_o/2) + 1]$

Figure 2A shows the double layer potential as a function of distance from the charged surface at various different univalent electrolyte concentrations. Figure 2A was constructed using Eq. (24) with a fixed charge density of -0.2 e/nm^2. It should be noted that, as demanded by Eq. (15), the potential decreases with the square root of the electrolyte concentration. For small potentials, this decay from the surface toward the bulk solution is exponential, as predicted by Eq. (21) (see inset, Fig. 2A). Notice also that the lower the electrolyte concentration the slower the potential decay. The effect of electrolyte valence is shown in Fig. 2B, where Eq. (24) has been used for a $1:1$ and a $2:2$ electrolyte.

Shortcomings of Diffuse Double Layer Theory

The Gouy–Chapman theory suffers from several limitations:[22,24,25] (1) The dielectric constant is supposed to be independent of the distance from the charged surface. (2) The charge is assumed to be smeared over the

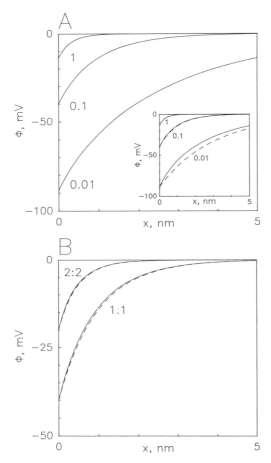

FIG. 2. (A) Potential profile predicted by the Gouy–Chapman theory [Eq. (24)] for different concentrations (in M) of monovalent electrolyte in the bulk solution. As the concentration is raised the value of the potential is smaller and decays faster (the value of the Debye length decreases as the concentration increases). The potential follows an exponential decay when the double layer potential is small, but at high potentials the decay is faster [solid line in the inset is drawn according to Eq. (24)] than that expected from an exponential function [Eq. (21), dashed line in the inset]. (B) Surface potential as a function of distance according to Eq. (24) for symmetrical 1:1 and 2:2 electrolytes. In both cases the electrolyte concentration is 0.1 M. Debye lengths are 0.96 and 0.48 nm for the monovalent and the divalent electrolyte solutions, respectively. Note that the divalent cation is much more effective in screening the potential than the monovalent cation. Solid lines were drawn according to Eq. (24). Dashed lines were drawn according to Eq. (21). $\sigma = -0.2$ e/nm^2.

TABLE I
MOST USED EQUATIONS WHEN PLAYING WITH SURFACE CHARGE

Equation	Eq. no.	Name
$d^2\phi/dx^2 = -\rho(x)/\epsilon_0\epsilon_s$	(1)	Poisson's equation
$C_i(x) = C_i(\text{bulk}) \exp[-z_iF\,\phi(x)/RT]$	(2)	Boltzmann distribution law
$\sigma = \pm \{2RT\,\epsilon_0\epsilon_s\Sigma C_i \exp(z_iF\,\phi_0/RT) - 1]\}^{1/2}$	(11)	Gouy-Chapman also known as Grahame equation (Must be used when ionic solutions contain mixed electrolytes)[24,25,26]
$\sigma = (8RT\epsilon_0\epsilon_s C_i)^{1/2} \sinh(z_iF\,\phi_0/2RT)$	(15)	Simple form of the Gouy-Chapman equation in the case of symmetrical electrolytes[24,25]
$C_o = C_i \exp(-zF\phi_0/RT) = \sigma^2/2RT\,\epsilon_0\epsilon_s$	(15b)	Surface concentration limit at very low bulk concentrations
$\phi(x) = \phi_0 \exp(-Kx)$	(21)	A useful equation to calculate ϕ at a distance when ϕ_0 is small (<25 mV)
$\phi(x) = (2/z) \ln\{[1 + \alpha \exp(-Kx)]/[1 - \alpha \exp(-Kx)]\}$	(24)	The relation between ϕ and x for surface potentials of arbitrary magnitude[24,25]

surface. (3) Image charge effects are ignored. (4) Ions have been assumed to be point charges. Experimental tests and experimental observed limitations of the Gouy-Chapman theory as applied to membranes have been discussed in detail in Ref. 22.

It is convenient when using the double layer theory to know exactly its range of application. First, direct measurements have demonstrated that the theory predicts well the potential profile and the ion distribution in contact with a charged membrane.[21,22] Second, the assumption that the charges are smeared over the surface in the form of discrete ions appears to be reasonable when working with monovalent lipids [PS, phosphatidyl-glycerol (PG), and phosphatidylinositol (PI)].[27] However, when charges are embedded in a medium of low dielectric constant[28] or when the lipids are multivalent the assumption breaks down, and more sophisticated theories are needed.[29] Finally, one of the weakest points of the Gouy–Chapman theory is the treatment of ions as point charges. A model more close to physical reality in which ions are considered to have a finite size was proposed by Stern[20] (discussed in Refs. 23–25 and 30). This model is briefly discussed below.

[27] A. P. Winiski, A. C. McLaughlin, R. V. McDaniel, M. Eisenberg, and S. McLaughlin, *Biochemistry* **25**, 8206 (1986).

[28] O. S. Andersen, S. Feldberg, H. Nakadomari, S. Levy, and S. McLaughlin, *Biophys. J.* **21**, 35 (1988).

[29] M. Langner, D. Cafiso, S. Marcelja, and S. McLaughlin, *Biophys. J.* **57**, 335 (1990).

[30] S. McLaughlin, N. Mulrine, T. Gresalfi, G. Vaio, and A. McLaughlin, *J. Gen. Physiol.* **77**, 445 (1981).

Stern Correction of Gouy–Chapman Theory

In the Stern model ions in the first layer adjacent to the surface are adsorbed, building up the so-called Stern layer (Fig. 3). This adsorption of ions is sufficient to decrease the surface potential to an extent compatible with the experimental results. The Stern model makes use of a treatment first proposed by Langmuir in which ions in solution and those in the Stern layer are in equilibrium. Thus, the correction implies a combination of the double layer theory with a Langmuir isotherm to account for specific ion binding at the interface. Figure 3 shows that the total charge σ_t (assumed to be negative) is balanced by the charge σ_s of ions in the Stern layer *and* the charge σ_{gc} of ions in the diffuse layer. In other words, the condition of electroneutrality demands that $\sigma_t = -(\sigma_s + \sigma_{gc})$.

The Stern correction becomes important when dealing with biological membranes since they are bathed by solutions containing alkali metal and alkali earth cations. Both types of cations are known to bind to some extent to negatively charged lipids[30] and probably to the negatively charged sites that modulate channel gating.[2-5] Ions in the bound layer reduce the surface charge density and, hence, the surface potential predicted by the Gouy–Chapman theory.

To illustrate how the Stern correction works, a case in which the solution is in contact with a negatively charged surface containing both

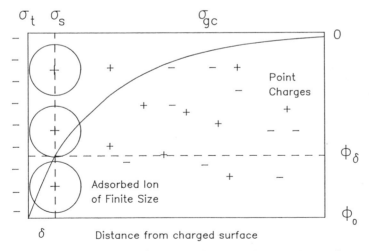

FIG. 3. Schematic representation of the Gouy–Chapman–Stern model. In this model ions with a finite size (radius δ) bind to the charged surface, decreasing the effective surface charge density and hence the surface potential. Ions beyond the Stern layer are considered as point charges. The model predicts that in the Stern layer the potential falls linearly with distance, whereas from δ to ∞ the potential falls off approximately exponentially.

monovalent and divalent cations is used. The negative charges are able to form $1:1$ complexes with the cations according to the reactions

$$L^- + C^+(0) \overset{K_1}{\rightleftharpoons} L^-C^+$$
$$L^- + C^{2+}(0) \overset{K_2}{\rightleftharpoons} L^-C^{2+}$$

where L^-, the negative fixed charge, is able to bind monovalent $[C^+(0)]$ or divalent cations $[C^{2+}(0)]$ (the 0 in parentheses means ions right at the membrane surface but in solution) with intrinsic association constants K_1 and K_2 (M^{-1}), respectively. Assuming that both reactions are in equilibrium it is clear that

$$\{L^-\}[C^+(0)]K_1 = \{L^-C^+\} \tag{25}$$
$$\{L^-\}[C^{2+}(0)] = \{L^-C^{2+}\} \tag{26}$$

where braces denote surface concentrations (e.g., moles/m^2). The total surface concentration of negative charges $\{L^-\}_{total}$ is the sum of free and bound surface concentrations:

$$\{L^-\}_{total} = \{L^-\} + \{L^-C^+\} + \{L^-C^{2+}\} \tag{27}$$

If $K_1 \ll K_2$, combination of Eqs. (25)–(27) gives an expression that relates the surface charge density, σ', after considering binding with the initial surface charge density, σ_t,

$$\sigma' = \sigma_t + \sigma_{st} = -\{L^-\}_{total}[1 - K_2C^{2+}(0)]/[1 + K_1C^+(0) + K_2C^{2+}(0)] \tag{28}$$

TABLE II

AFFINITY OF DIFFERENT BIOLOGICALLY
RELEVANT CATIONS FOR
PHOSPHATIDYLSERINE[a]

Ion	$K_A(M^{-1})$
Li$^+$	0.8
Na$^+$	0.6
K$^+$	0.15
Mg^{2+}	8.0
Ca^{2+}	12.0
Ba^{2+}	20.0

[a] Data from McLaughlin et al.[30]

where $C^+(0)$ and $C^{2+}(0)$ are given by Eq. (2) expressed for monovalent and divalent cations, respectively,

$$C^+(0) = C^+(\infty) \exp(-F\phi_o/RT) \tag{29}$$

$$C^{2+}(0) = C^{2+}(\infty) \exp(-2F\phi_o/RT) \tag{30}$$

The set of Eqs. (11), (28), (29), and (30) is solved by substituting Eqs. (29) and (30) into Eq. (28) and then setting Eq. (28) equal to Eq. (11), yielding an implicit function for surface potential which can be obtained numerically, for example, with the Newton–Raphson method. Binding constants for monovalent and divalent ions to phosphatidylserine bilayer membranes are given in Table II.

Digression on Interfacial Potentials

The effect of double layer potentials arising from charged lipids on channel conduction and gating have been used to elucidate structural and functional characteristics of ion channels. It is important, therefore, to digress briefly on how these potentials are measured and what are the main assumptions involved in these measurements. Both lipids and proteins are molecular dipoles. If they are charged the total potential at the interface is the sum of the aqueous double layer potential due to the fixed charges located at the surface *and* dipole potentials associated with the oriented dipolar molecules.[24,25] Figure 4A shows a schematic representation of the electrostatic potential profile across a lipid bilayer composed of charged lipids, considering the interior of the lipid bilayer to be continuous with a low dielectric constant, for example, that of a hydrocarbon, which is about 2. In Fig. 4B a schematic representation of the electrical dipole potential profile is shown, and Fig. 4C illustrates the resultant potential profile on adding those of Fig. 4A,B. Below some hints on how to estimate these potentials are given.

Recalling that the membrane conductance (g_m) is proportional to the mobility of the ion in the membrane (u_m) and its concentration inside it (C_m), g_m can be written

$$g_m \approx u_m C_m \approx u_m C_{solution} \exp(-ze\Gamma/kT) \tag{31}$$

where $C_{solution}$ is the concentration of the ion in the bulk solution surrounding the membrane and Γ is the sum of the double layer potential and the dipole potential. Equation (31) provides a method for estimating Γ. In particular the dipole potential (ϕ_D) can be obtained in a neutral bilayer or charged bilayers at very high ionic strength so that the contribution of the double layer potential is negligible. For these cases a comparison

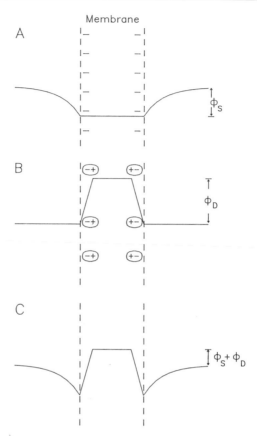

FIG. 4. Schematic representation of the variation of the electrostatic potential across a lipid bilayer membrane. (A) The lipids in both monolayers are negatively charged, giving rise to a negative surface potential ϕ_S. (B) Owing to the fact that phospholipids are dipoles, a dipole potential ϕ_D is generated which is positive in the bilayer interior. (C) The total potential difference between the interior of the membrane and the aqueous solution is the sum of the ionic double layer, ϕ_S, and the dipole contribution, ϕ_D.

between the membrane conductance induced by hydrophobic ions of the same size, valence, and shape but different sign, like tetraphenylarsonium and tetraphenylborate ions, yields[31,32]

$$g_- \approx u_- C_- \exp(e\phi_D/kT) \tag{32}$$
$$g_+ \approx u_+ C_+ \exp(-e\phi_D/kT) \tag{33}$$

[31] O. H. Leblanc, *Biophys. J.* **10** 94a (1970).
[32] E. A. Liberman and V. P. Topaly, *Biofizika* **14**, 452 (1969).

Given the molecular characteristics of these two ions it is safe to assume that $u_- = u_+$ and the ratio between Eqs. (32) and (33) yields

$$g_-/g_+ = (C_-/C_+) \exp(2e\phi_D/kT) \tag{34}$$

Because the concentrations of the hydrophobic ions are known, it is possible to obtain the dipole potential from the measured conductance ratio. Conductance ratios of 10^6 to 10^9 have been obtained for phosphatidylethanolamine and lecithin–cholesterol membranes, repectively, which implies dipole potentials of the order of 190 to 288 mV positive inside the membrane.[31,33] The interested reader should consult Ref. 33 for a more "realistic" estimation of the membrane dipole potential. It is clear that because of the presence of dipole potentials lipid bilayers are intrinsically more permeable to negative than to positive ions.

A similar principle can be used to calculate the double layer potential, because at low ionic strength Eq. (31) is given by

$$g_m \approx u_m C_{solution} \exp[ze(\phi_D + \phi_S)/kT] \tag{35}$$

where g_m is the nonactin- or valinomycin-induced membrane conductances.[25,34] On the other hand, at very high ionic strength $\phi_S \approx 0$. The ratio of the bilayer conductance g_m obtained at high and low ionic strength then provides an estimation of ϕ_S. High ionic strength can be obtained by adding NaCl or LiCl to the solution bathing the bilayer since Na^+ and Li^+ are transported by the carriers very poorly. A more accurate method to estimate surface charge densities of charged membranes, which does not rely on an estimation of the conductance in the uncharged membrane, is to measure g_m as a function of salt concentration. These data are then fit using Eqs. (17) or (11) and (35).[34] In all these cases it is assumed that ϕ_D does not change with ionic strength. This is a reasonable assumption.[35] Another alternative to estimate ϕ_S is to compare the conductance induced by the carriers in charged and neutral bilayers. However, in this case we are forced to assume that ϕ_D is the same in both types of membranes.

An independent method of obtaining the magnitude of the interfacial potentials is to measure the potential across a monolayer, made of the same type of lipid as the bilayer, spread on the desired buffer solution. The air/solution potential difference is measured using a ^{210}Po electrode placed right above the monolayer surface and a Ag/AgCl electrode placed in the solution connected to a high impedance electrometer.[36,37] As above, ϕ_S is

[33] O. S. Andersen and M. Fuchs, *Biophys. J.* **15**, 795 (1975).
[34] S. McLaughlin, G. Szabo, G. Eisenman, and S. M. Ciani, *Proc. Natl. Acad. Sci. U.S.A.* **67**, 1268 (1970).
[35] R. Latorre and J. E. Hall, *Nature (London)* **264**, 361 (1976).
[36] S. Ohki and R. Sauve, *Biochim. Biophys. Acta* **511**, 377 (1978).
[37] J. Reyes and R. Latorre, *Biophys. J.* **28**, 259 (1978).

taken as the difference between the interfacial potential measured at low and high ionic strength. The limitation of this method is the choice of the area per phospholipid molecule in the monolayer that best represents that of the lipid in the bilayer. Areas per phospholipid molecule ranging from 0.53 to 0.65 nm^2 have been reported.[38]

Several other experimental approaches can be utilized to measure the electrostatic potential adjacent to negatively charged phospholipid bilayer membranes.[14,39] However, they need a more sophisticated experimental setup and may be less familiar to the researcher.

Surface Charge Modulation of Channel Conductance

Fixed Charge and Ion Channel Conductance

In this section the effect that fixed charges located at the entrances of ion channels may have in determining ion channel conductance is examined. The discussion will be centered on estimations of the surface charge density using single-channel conductance measurements inasmuch as these are devoid of the uncertainty present in most macroscopic experiments, namely, that the current comes about from a single class of channel. The conductance of a channel to permeable ions depends on the number of ions near the pore entrance. This number is determined by the density, location, and sign of the fixed charges located in the neighborhood of the ion channel conduction system. Only in those cases in where the studies are done with reconstituted channels inserted into neutral planar lipid bilayers can one be certain that the fixed charges which modulate channel conductance are located in the channel-forming protein itself. In this regard, bilayers provide a less ambiguous assay than patch clamp studies. To facilitate the analysis of results that follows, we have classified ion channels into wide aqueous pores, one-ion pores, and multi-ion pores. For the sake of simplicity the analysis is restricted to cation-selective pores.

Wide Pores

Some wide pores show a certain degree of cation selectivity.[40,41] A wide pore with fixed negative charges at its entrance or in the pore walls can explain this behavior. They are like wide-pore fixed-negative ion ex-

[38] R. C. MacDonald and S. Simon, *Proc. Natl. Acad. Sci. U.S.A.* **84,** 4089 (1987).
[39] O. Alvarez, M. Brodwick, R. Latorre, A. McLaughlin, S. McLaughlin, and G. Szabo, *Biophys. J.* **44,** 333 (1983).
[40] P. Labarca, D. Wolff, U. Soto, C. Necochea, and F. Leighton, *J. Membr. Biol.* **94,** 285 (1986).
[41] R. Latorre, H. Lecar, and G. Ehrenstein, *J. Gen. Physiol.* **60,** 72 (1972).

changers.[42] In this type of pore it is expected that ions retain their hydration and that the channel permeability is proportional to the ion mobilities in free solution. Ion transport for this class of channels can be interpreted in terms of classic electrodiffusion theory.[43] In symmetrical solutions this theory predicts that single-channel conductance, g_{pore}, is proportional to the first power of the surface concentration, C_{pore}, of the permeant ion according to the relation

$$g_{pore} = AC_{pore} = AC_{bulk} \exp(-zF\phi_S/RT) \tag{36}$$

where A is a constant and ϕ_S can be calculated numerically by giving values to the surface charge density until the best fit to the g_{pore} versus C_{bulk} concentration curve is obtained. Wide pores containing a fixed negative charge at their entrances have larger conductances at relatively low ionic concentration than unchartged pores (i.e., $\phi_S = 0$), and when the permeant ion concentration is very low, channel conductance does not vanish but tends to a lower limit because the concentration at the surface, in the limit of very low ion concentration, is constant and predicted by Eq. (15b).

Simulated results for this type of pore are shown in Fig. 5A. We have assumed here that the pore behaves according to Eq. (36), and the cases of the neutral and the charged pore at two different surface charge densities are considered. In all cases changes in permeant ion concentration are made symmetrically. This channel behavior has been found for the excitability-inducing material (EIM) channel incorporated into planar bilayers. EIM behaves as porins which show cation selectivity but discriminate very poorly between monovalent cations.[41,44] In this case the best fit to the data using Eq. (36) was obtained with $\sigma = -0.1$ e/nm^2.[45] This net charge may be the sole cause of the cation versus anion selectivity found in these channels.[46]

Ion-Selective Channels with Linear g_{pore} versus C_{pore} Relations

Fixed charges at the pore mouth give rise to a surface potential which in turns depends on the ionic strength of the medium as predicted by the Gouy–Chapman theory. If the conductance of a given K^+ channel is modified when the concentration of an impermeant cation is altered but the concentration of the permeant ion is kept constant, then the results can be interpreted as changes in the *local* K^+ concentration.[47] Hidden in the

[42] G. Eisenman, J. P. Sandblom, and J. I. Walker, *Science* **155**, 965 (1967).

[43] A. L. Hodgkin and B. Katz, *J. Physiol. (London)* **108**, 37 (1949).

[44] R. Bean, *in* "Membranes: A Series of Advances" (G. Eisenman, ed.), Vol. 2, p. 409. Dekker, New York, 1973.

[45] R. Latorre and O. Alvarez, *Comments Mol. Cell. Biophys.* **5**, 193 (1988).

[46] J. A. Dani, *Biophys. J.* **49**, 607 (1986).

[47] M. J. Kell and L. J. DeFelice, *J. Membr. Biol.* **102**, 1 (1988).

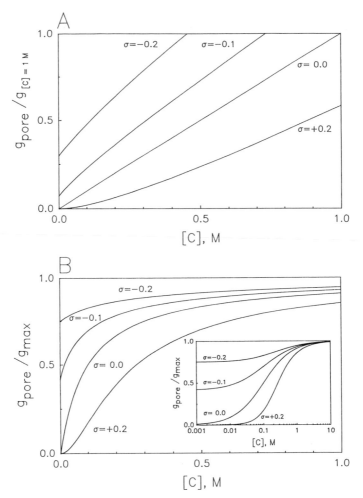

FIG. 5. (A) Simulated relative conductance versus concentration curves for a wide cationic pore containing different surface charge densities at its entrance. σ is given in $-e/\text{nm}^2$, and the curves are drawn according to Eq. (36). (B) The one-ion pore case. Curves were drawn according to Eq. (40) for different surface charge densities at the pore entrances. K is taken to be 0.1 M. Notice that for large negative surface charge densities the pore is essentially saturated at low concentrations of the permeant ion, and when σ becomes positive the curve has an S shape. In the inset the abscissa is logarithmic to stress that the pore conductance reaches limiting values at very low and very high concentrations.

relationship between g_{pore} and impermeant ion concentration is the value of the charge density near the pore, which can be extracted using the same principles discussed above. Assuming that Eq. (36) holds it follows that

$$g_{\text{pore}}/g_{\text{pore}}^{\circ} = \exp(-zF\phi_{\text{S}}/RT) \qquad (37)$$

where g_{pore}° is the conductance of the uncharged pore, which is obtained by using concentrations of impermeant monovalent or divalent cation such that a further addition of the impermeant ion does not change the channel conductance. Taking the logarithm of both sides of Eq. (37) and solving for ϕ_S gives

$$\phi_S = -(RT/zF) \ln(g_{pore}/g_{pore}^{\circ}) \tag{38}$$

The estimate of ϕ_S assumes that the pore conductance is proportional to the first power of the surface concentration of K^+ [Eq. (36)]. It should always be kept in mind that experiments where the screening ions are added to one side of the membrane are difficult to interpret since, if the local concentration varies as described above, not only the conductance, but also the electrical driving force will change. To avoid this problem it is always more convenient to change the impermeant ion concentration equally at both sides of the membrane.[48,49]

Equation (38) and the procedure described above (concentration of the impermeant ion was changed only in the external side of the channel) have been used to obtain the local charge near an inward rectifier channel from chick ventricular cells.[47] The analysis of the results must be taken with the caveat that the impermeant ion must be completely inert. In other words, the decrease should occur solely through a screening effect and not by other mechanisms such as channel blockade. In the case of the cardiac inward rectifier, the patch clamp studies showed that channel conductance decreases to the same extent whether Mg^{2+}, Ca^{2+}, or Ba^{2+} is used as the impermeant cation. Because all these divalent cations have the same effect on conductance, it is reasonable to assume that ion blockade or binding to the fixed charges (Stern effect) are not the causes of the observed decrease in conductance. Furthermore, the surface potentials calculated from conductance measurements using Na^+ as the impermeant cation are essentially the same as those obtained using divalent cations. Using the method described above a surface charge density near the open inward rectifier pore yielded a value of $-0.25 \ e/nm^2$.

One-Ion Pores

Structural data on the acetylcholine receptor channel,[50] experiments with organic ions that block potassium channels,[51-54] and measurements of

[48] S. G. A. McLaughlin, G. Szabo, and G. Eisenman, *J. Gen. Physiol.* **58**, 667 (1971).
[49] W. N. Green, L. B. Weiss, and O. S. Andersen, *J. Gen. Physiol.* **89**, 841 (1987).
[50] C. Toyoshima and N. Unwin, *Nature (London)* **336**, 247 (1988).
[51] C. M. Armstrong, *in* "Membrances: A Series of Advances" (G. Eisenman, ed.), Vol. 3, p. 325. Dekker, New York, 1975.
[52] R. Latorre and C. Miller, *J. Membr. Biol.* **71**, 11 (1983).
[53] C. Miller, *J. Gen. Physiol.* **79**, 869 (1982).
[54] A. Villarroel, O. Alvarez, A. Oberhauser, and R. Latorre, *Pfluegers Arch.* **418**, 118 (1988).

streaming potentials[55,56] have lead to identifying common features in ion-selective channels. The entrances are fairly large mouths which are connected by means of a narrow tunnel, giving the pore a hourglass shape. Some of the complexities in the ion conductance data are explained if the vestibules are assumed to contain a net charge as descussed below.

One-ion pores are those that do not allow the presence of more than one ion at a time in their conduction system. This situation may arise in pores which possess short tunnels connecting the vestibules and with low binding energies, as electrostatic repulsion prevents a second ion from entering the channel. Ion conduction in one-ion pores has been analyzed using reaction-rate theory.[57–59] In this case the zero voltage (small signal) conductance versus ion concentration relation is described by a Langmuir isotherm of the type

$$g_{pore} = g_{max}[C]/(K + [C]) \tag{39}$$

where g_{max} is the maximum channel conductance and [C] is the permeant ion concentration in the vestibules. For an uncharged pore this is equal to the bulk concentration. K is the [C] for half-maximal conductance, which for a one-ion pore corresponds exactly to the equilibrium dissociation constant for ion binding in the channel. Criteria used to define a channel as a one-ion pore can be found in Refs. 58, 60, and 61. Suffice it to say here that the fact that conductance reaches a maximum value (g_m) as the concentration of the permeant ion is raised provides one of the most direct tests that the channel cannot be occupied by more than one ion. However, this test has its limitations. One of these limitations is of concern in the case when changes in [C] are made without keeping the ionic strength constant. In this case surface charge effects can distort the shape of the g_{pore} versus [C] curve (Fig. 5B). Distortions of this type can arise as a consequence of the net charge of phospholipids or from charges contained in the channel-forming protein. The possible contribution of fixed charges located in the protein itself can be unveiled when channels are incorporated into neutral lipid bilayer membranes.[49,62]

In the lack of structural information regarding the location and distri-

[55] C. Miller, *Biophys. J.* **38**, 227 (1982).

[56] C. Alcayaga, X. Cecchi, O. Alvarez, and R. Latorre, *Biophys. J.* **55**, 367 (1989).

[57] S. Glasstone, K. J. Laidler, and H. Eyring, "The Theory of Rate Processes." McGraw-Hill, New York, 1941.

[58] P. Lauger, *Biochim, Biophys. Acta* **311**, 423 (1973).

[59] O. S. Andersen, this series, Vol. 171, p. 62.

[60] B. Hille, *in* "Membranes: A Series of Advances" (G. Eisenman, ed.), Vol. 3, p. 255. Dekker, New York, 1975.

[61] R. Coronado, R. L. Rosenberg, and C. Miller, *J. Gen. Physiol.* **76**, 425 (1980).

[62] G. Eisenman, J. Sandblom, and E. Neher, *Biophys. J.* **22**, 307 (1978).

bution of the fixed charges, the most simple approach to obtain σ near the pore entrances is to combine Eqs. (2) and (38), yielding

$$g_{pore} = g_{max}/\{1 + K/[C]_{bulk} \exp(-zF\phi_S/RT)\} \qquad (40)$$

where ϕ_S is calculated using Eq. (17) as was done for wide pores. Equation (40) assumes that in symmetrical salt solutions, ϕ_S at the internal and external mouths of the pore are similar. In one-ion pores with charged vestibules conductance approaches a finite limit both as $[C] \rightarrow \infty$ [Eq. (38)] and as $[C] \rightarrow 0$ [Eq. (15b)]. Therefore, it is expected that in charged one-ion pores single-channel conductance at low concentrations is higher than expected. Simulated results using Eq. (40) and different surface charge densities are shown in Fig. 5B. As expected Fig. 5B shows that for a positively charged vestibule pore conductance is smaller than the conductance obtained for the neutral pore. Moreover, owing to the particular manner in which the concentration increases in the pore entrance (see Fig. 5A), the g_{pore} versus [C] curve has an S shape reminiscent of cooperative processes and can be wrongly interpreted in such a way. Other causes can be the origin of the deviation of the simple behavior given by Eq. (39), for example, a multi-ion channel with two or more sites that can be simultaneously occupied.[63,64] Distortions of the simple Langmuir isotherm can also be expected if the mechanism of ion transport is given by an energy barrier of the fluctuating type.[65] However, the predictions of the Gouy–Chapman theory (or its modifications) are always more easily tested by modifying the ionic strength of the electrolyte medium with impermeant cations and measuring its effect on the shape of the current versus voltage and conductance versus [C] relationships. Deviations of Eq. (39) arising from multi-ion occupancy are discussed in the next section.

A study of the shape of the current–voltage relationships for the open pore can give information about possible asymmetries in the surface charge density in the internal and external vestibules.[46,65,66] Asymmetries in the surface charge density give rise to rectification in the current–voltage relationship for the open pore, but in order to quantify the effect a specific model for ion transport through the one-ion pore must be given.[46,66,67] In this case it is customary to describe ion transport in the channel with Eyring transition state theory. In Eyring's theory the narrow region of the channel is viewed as a series of energy barriers and wells, and the rate, k_{ij}, of an ion moving from site i to j is of the form $k_{ij} = (kT/h) \exp(-G^*/kT)$,

[63] B. Hille and W. Schwarz, *J. Gen. Physiol.* **72**, 409 (1978).
[64] P. Lauger, W. Stephan, and E. Fredhland, *Biochim. Biophys. Acta* **602**, 167 (1980).
[65] J. Dani and G. Eisenman, *J. Gen. Physiol.* **89**, 959 (1987).
[66] A. Villarroel, Ph.D. Thesis, University of California, Los Angeles, California (1989).
[67] A. M. Correa, R. Latorre, and F. Bezanilla, *J. Gen. Physiol.* in press (1991).

where h is Planck's constant and G^* denotes the height of the free energy barrier the ion must jump to go from i to j (see Refs. 59 and 68 for recent reviews of Eyring and alternative theories to describe ion transport through channels).

To illustrate the effect of an asymmetric distribution of surface charge we have chosen a model consisting of three barriers and two sites, the 3B2S model (Fig. 6A). The model includes the effect the charged vestibules has on ion transport through the channel, and, for demonstrative purposes, we have assumed that the inner vestibule is uncharged and the outer vestibule has a σ value of -0.2 e/nm^2. The strategy followed here to obtain the predicted open channel current–voltage relationships is the same as that described in Ref. 69, and no further details will be given. In this approach the surface potential is computed according to Eq. (17), and the surface charge density is expressed in terms of a circle of radius R containing one charge. Figure 6B shows the current versus voltage relationships for the open pore at two different permeant ion concentrations. At these concentrations the inward going current is larger than the outward one because the outer (charged) vestibule is able to concentrate cations in a position for them to be transported inward. Moreover, the model predicts that, at all but large concentrations, the conductance measured at high negative potentials is larger than that measured at high positive potentials.

Another alternative to approach the problem of charged vestibules is to assume a given distribution for the fixed charges and a given geometry for the vestibule.[46,66,70,71] In this case, the Poisson–Boltzmann equation needs to be solved for the particular geometry chosen for the vestibule, and the integration is always done numerically. Eventually all models that take into account surface charge effect must consider the actual structure of the channel mouths. However, lacking detailed structural information regarding channel architecture, and because both vestibule geometry and charge location influence the magnitude of the surface potential, it is preferable to use the less realistic, but more simple Gouy–Chapman theory. Up to now the only case in which vestibule dimensions can be estimated is the acetylcholine receptor channel, but the exact location of the fixed charges is unknown.

It must be mentioned that estimating the surface potential on the basis of the nonlinear Poisson–Boltzmann equation solved for cylindrically symmetrical systems led to a surprising result.[71] In contrast to the prediction of the Gouy–Chapman theory that the concentration of permeant ion

[68] D. G. Levitt, *Annu. Rev. Biophys. Biophys. Chem.* **15,** 29 (1986).
[69] O. Alvarez, A. Villarroel, and G. Eisenman, this volume [56].
[70] P. C. Jordan, *Biophys. J.* **51,** 297 (1987).
[71] M. Cai and P. C. Jordan, *Biophys. J.* **57,** 883 (1990).

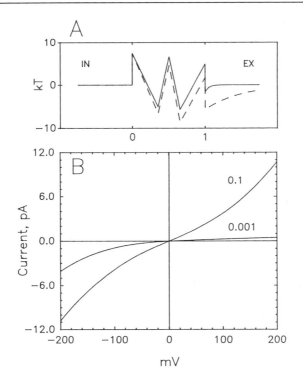

FIG. 6. Asymmetric ion transport through a single-ion channel with a net charge in the external vestibule. (A) Energy profile for a monovalent cation in the narrow part of the channel. The profile is calculated for 2 (dashed line) or 100 mM (solid line) electrolyte and a $\sigma_{external}$ value of -0.2 e/nm^2. (B) Single-channel current versus potential curves predicted by the energy barrier model given in (A) in 2 and 100 mM monovalent electrolyte. Note the large inward rectification revealed at low electrolyte concentrations. It is assumed that the potential does not fall in the vestibules. The energy barrier is symmetrical, and from left to right the first well is located at an electrical distance of 0.35 and the central peak at an electrical distance of 0.5. Energy for barriers are set to 7.4kT units for all the peaks and to $-4.6kT$ units for the two wells. A large "repulsion factor" equivalent to 20kT units makes entrance of a second ion to the channel extremely improbable.

in the charged vestibule tends to a lower limit as the concentration in the bulk tends to zero [Eq. (15b)], a charge vestibule model, with defined geometry and surface charge density, predicts that the local concentration falls to zero when C_{bulk} tends to zero. This is a prediction that needs to be confronted with channel conductance data obtained over a wide range of ion concentrations and especially in the low concentration end.

Moreover, in a recent reivew, Green and Anderson[71a] confronted the

[71a] W. N. Green and O. S. Andersen, *Annu. Rev. Physiol.* **53,** 341 (1991).

predictions of Gouy–Chapman and Debye-Hückel theories regarding the surface potential behavior in the limit of very low ionic strength. As we discussed in an early section (Eq. 17; Fig. 2A), Gouy–Chapman theory demands that in this limit ϕ_s will approach $\pm \infty$. In the Debye-Hückel theory, which assumes a spherical symmetry, the potential at the surface of the sphere when the ionic strength approaches zero is given by $z \cdot e/4\pi\epsilon_0\epsilon_s r$ where e is the electronic charge and r is the radius of the sphere; that is ϕ_s approaches a finite value and hence the interfacial concentration will approach zero.[71a] This clearly exemplifies the fact that the surface potential at the channel entrances, in the limit of low concentrations, will obtain a value highly dependent on channel geometry.

Multi-Ion Pores

Multi-ion pores are the class of channels that can contain several ions simultaneously in their narrow region.[64,72,73] For this class of channels it is expected to find g_{pore} versus [C] curves that deviate significantly from Eq. (39). In particular, conductance may increase up to a certain concentration, reach a maximum, and then decrease at higher concentrations. Hille and Schwarz[63] have discussed in detail the criteria that define multi-ion pores. A practical way to generate current–voltage relationships for the open pore and predicting channel conductance as a function of [C] using Eyring formalism is given in Ref. 69. This methodology also incorporates the effects on channel conductance of fixed charges in the vestibules.

The charged vestibules 3B2S model is taken to exemplify the case of a channel in which two ions can reside simultaneously in its conduction system (both binding sites loaded). Figure 7A shows the g_{pore} versus [C] curve generated by the barrier model shown in the inset, when the vestibules do not have a negative charge or when they are symmetrically charged with a σ value of -0.2 e/nm^2. Although multi-ion occupancy produces as a result a g_{pore} versus [C] curve that does not comply with Eq. (39), at odds with the charged vestibule pore, it cannot give rise to a limiting conductance at very low [C]. For purposes of comparison, in Fig. 7B we show the curves generated by a single-ion pore having an energy barrier equal to that shown in the inset of Fig. 7A. Notice that for the one-ion pore channel conductances are too low and saturate very rapidly. The 3B2S two-ion pore generates larger conductances[45,74] than the single-ion channel, but at low concentrations of the permeant ion the limiting

[72] A. Finkelstein and O. S. Andersen, *J. Membr. Biol.* **59**, 155 (1981).

[73] R. W. Tsien, P. Hess, E. McCleskey, and R. L. Rosenberg, *Annu. Rev. Biophys. Biophys. Chem.* **16**, 265 (1987).

[74] R. Latorre, *in* "Ion Channel Reconstitution" (C. Miller, ed.), p. 431. Plenum, New York, 1986.

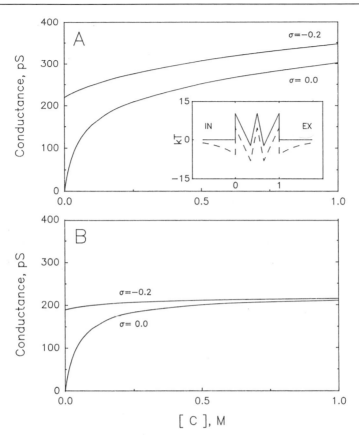

FIG. 7. (A) Channel conductance as a function of monovalent cation concentration for a two-ion channel described by the 3B2S energy profile given in the inset. The vestibules are symmetrically charged with a σ value of 0.2 e/nm² (dashed line; barrier given for a 1 mM electrolyte concentration) or neutral (solid line). (B) Channel conductances as a function of monovalent electrolyte concentration for a one-ion channel described by the same energy profile shown in the inset of Fig. 7A. The one-ion pore condition is set by imposing a large repulsion factor between ions inside the pore. Energy parameters for the two-ion channel are as follows: energy for barriers, 10.3kT; energy for wells, −2.3kT; repulsion factor, 1.5kT. Electrical distances are as in Fig. 6.

conductance is the same because [C] in both cases is given by Eq. (15b) (but see Ref. 71).

Surface Charge Effects Revealed by Channel Modification with Trimethyloxonium

Trimethyloxonium (TMO) is a cationic methyl group donor, and TMO reagents are known to react specifically with carboxylate residues in several enzymes.[75,76] In particular, TMO has been shown to alter profoundly the properties of the Na^+ channel (e.g., Refs 77 and 78) and of a Ca^{2+}-activated K^+ channel.[7,79] In both cases treatment of the channel with TMO led to a reduction in the single-ion conductance without affecting ion selectivity. There is a difference between these two kind of channels, however. In sodium channels treatment with the modifying reagent led to a irreversible decrease in channel conductance. Channel conductance was not further modified when the sodium channel was exposed again to TMO.[78] In contrast, Ca^{2+}-activated K^+ channels are subject to a progressive reduction in single-channel conductance on repeated TMO treatments.[79] Modification of single Na^+ channels is then all or none, suggesting a single carboxyl group involved in conduction. The simplest explanation for the results obtained with the Ca^{2+}-activated K^+ channel is, therefore, that more than one site is modified in this channel. Characterization of the TMO effect on channel conductance has provided strong evidence that these kind of channels carry ionized carboxyl groups near the pore entrance. For the Ca^{2+}-activated K^+ channel the g_{pore} versus $[K^+]$ curve is similar to that shown in Fig. 7A for the charged-vestibule 3B2S model, but after channel treatment with TMO added to the external side the curve looks like that generated by the plain 3B2S model. In other words, the TMO effect is ionic strength dependent. The lack of TMO effect at high $[K^+]$ indicates that TMO does not affect sites located deep in the channel, but rather operates on more peripherally located carboxylate groups.

Removal of an externally located fixed charge in the Ca^{2+}-activated K^+ channel has shown to affect both inward and outward currents. The reduction in inward current can be rationalized in terms of a decrease of the local $[K^+]$. On the other hand, the decrease in outward current is explained in terms of a decrease of the electical gradient since the presence of a fixed negative charge only in one side of the channel induces a difference in potential across the membrane. This difference produces a tilt of the energy barrier profile. Thus, in the presence of a negative charge in the external face of the channel, *outward* K^+ flux will be *kinetically* favored.[7]

Modification of ion channels by TMO revealed another important role of fixed charges at the channel entrances. The high conductance Ca^{2+}-activated K^+ channel is blocked in a bimolecular fashion by nanomolar con-

[75] H. Nakayama, K. Tanizawa, and Y. Kanaoka, *Biochem. Biophys. Res. Commun.* **40**, 537 (1970).

[76] S. M. Parsons, L. Jao, F. W. Dalhquist, C. L. Borders, T. Groff, J. Racs, and M. A. Raftery, *Biochemistry* **8**, 700 (1969).

[77] F. J Sigworth and B. C. Spalding, *Nature (London)* **283**, 293 (1980).

[78] J. F. Worley, R. J. French, and B. K. Krueger, *J. Gen. Physiol.* **87**, 327 (1986).

[79] R. MacKinnon and C. Miller, *Biochemistry* **28**, 8087 (1989).

centrations of a scorpion toxin called charybdotoxin (CTX).[80] Furthermore, a kinetic analysis of the channel blockade by this toxin led to the conclusion that a single CTX molecule is positioned at an external ion entryway, hindering ion flux. After treatment with TMO the affinity of CTX for the modified channel is much weaker.[79] This decrease in apparent affinity after carboxyl groups in the channel protein are esterified is in agreement with the known structure of CTX, which is a basic protein having a net charge of $+5$ at neutral pH.[81,82] Apparently, the carboxyl groups involved in CTX binding are the same ones involved in modulating channel conductance since the presence of CTX during TMO treatment protects the channel from being modified by the latter compound. Tetramethyloxonium applied on the external face of the Na^+ channel eliminates saxitoxin sensitivity, and in turn saxitoxin protects Na^+ channels from being modified. However, in this last case if appears that the saxitoxin binding site is separated from the channel entrance.[83] Thus, it appears that carboxyl groups influence ionic conduction and participate in toxin block.

Lipid Effects

The topic in this section is an evaluation of the role of the electrical charge in the bilayer in the observed conduction characteristics of ion channels incorporated into planar lipid bilayer membranes.

Some Experimental Caveats

In cases where the experiment needs to be done in the presence of mixed electrolytes and where binding can be appreciable (see Table II), the set of Eqs. (11), (28), (29), and (30) need to be used to obtain the surface potential from the known surface charge density. Keep in mind that because of the large binding constant of divalent cations to negatively charge phospholipids (e.g., PS and PG) even trace amounts of divalent ions are important in determining the surface potential. In particular, special caution must be taken when determining conductances at very low concentrations of the permeant ion.[8] To exclude the effects of multivalent cations ethylenediaminetetraacetic acid (EDTA) can be added to the electrolyte. Direct methods to obtain the bilayer surface potentials under a

[80] C. Miller, E. Moczydlowski, R. Latorre, and M. Phillips, *Nature (London)* **313**, 316 (1985).

[81] C. Smith, M. Phillips, and C. Miller, *J. Biol. Chem.* **261**, 14607 (1986).

[82] G. Gimenez-Gallego, M. A. Navia, J. P. Reuben, G. M. Katz, G. J. Kaczorowski, and M. L. Garcia, *Proc. Natl. Acad. Sci. U.S.A.* **85**, 3329 (1988).

[83] W. N. Green, L. B. Weiss, and O. S. Andersen, *J. Gen. Physiol.* **89**, 873 (1987).

given set of experimental conditions were described above (see Disgression on Interfacial Potentials).

Most authors have used planar bilayers made of phosphatidylethanolamine (PE) or mixtures of PE and phosphatidylcholine (PC) when wanting neutral bilayers. However, PE bears an appreciable amount of charge at pH values below 6 or above 8.[34] Charged membranes can be made of PS or PG, although PS has been more widely used. Mixtures of PS and PE can be used when different surface charge densities are required. The bilayer surface can be made positively charged using the synthetic lipid 1,2-dihexadecylphosphatidylcholine (DHPDC) mixed with PE, but the membranes are very unstable.[84]

Charged Lipids and Channel Conductance

As stated above channel conductance can be modulated by fixed charges present in the channel protein itself or by the charged lipid surrounding the channel. Channel proteins are large polypeptides. For example, the acetylcholine receptor channel protrudes well into the external and internal solutions, and it is expected that the conduction system of this channel is insulated from the potential arising from the phospholipid fixed charges. We would like to answer the following questions: How can we determine the extent of insulation of a channel from the lipid bilayer? What kind of conclusions can we extract from the results regarding channel structure?

To accomplish these goals it becomes essential to be able to incorporate the channel in membranes of well-known composition, as discussed in Ref. 85. The channel needs to be incorporated into neutral membranes in order to obtain the particular behavior of the g_{pore} versus [C] curve without interferences from the bilayer electrostatics. The second step is to incorporate the channel into a charged membrane and to determine again the g_{pore} versus [C] curve, comparing it with that obtained in the neutral membrane. We assume, then, that any distortion in the g_{pore} versus [C] curve was promoted for unspecific electrostatic reasons and not by specific lipid–protein interactions owing to differences in either the acyl chains or the head groups of the lipids. If the channel is well insulated from the surface of the bilayer membrane, no differences between the conductance obtained in the neutral and the charged bilayers are expected. On the other hand, if the ion conduction system is sensing part of the surface potential generated in the bilayer surface, channel conductance will be larger in the charged membrane. This result can be interpreted assuming that the ion

[84] J. Bell, *in* "Ion Channel Reconstitution" (C. Miller, ed.), p. 469. Plenum, New York, 1986.
[85] P. Labarca and R. Latorre, this volume [30].

concentration involved in conduction is not the bulk concentration, but the concentration at the channel entrances separated by some distance from the lipid bilayer surface, that is,

$$C_{pore} = C_{bulk} \exp[-zF\phi(x)/RT] \tag{41}$$

where $\phi(x)$ can be obtained using Eq. (24) and the known ϕ_s of the lipid bilayer. In the absence of structural information the distance x can be interpreted as the distance separating the center of the channel mouth from the lipid bilayer surface. For one-ion pores that follow Eq. (39) (the entrances are neutral) g_{pore} versus [C] curves in neutral and charged membranes would look exactly as those shown in Fig. 5B, in which σ is replaced by distance.

For batrachotoxin-modified Na^+ channels the surface charge effects are very small.[49,86] The Bay K-8644-activated Ca^{2+} channel is also well insulated from the bilayer surface.[12] For the sarcoplasmic reticulum channel and the Ca^{2+}-activated K^+ channel conductances are 2- to 2.5-fold larger in PS than in PE at low permeant ion concentration $(10-20 \text{ m}M)$.[10,11] These values imply distances of the order of 1 to 2 nm between the pore mouth and the surface of the lipid bilayer. It is not surprising that the conduction system of the pores does not sense the full value of the surface potential in the charged phospholipid. The dimensions of the integral protein can be larger than the 4 nm thickness of the lipid bilayer, and the sites could be insulated from the bilayer surface by distances lateral and/or normal to the membrane.

Warning Concerning Protons

If the channel mouths are charged or if the channel is not completely insulated from charged phospholipids, the permeant ion concentration as well as the proton concentration at the channel entrances will be higher than those in the bulk solution. This is because the concentration in the pore of both H^+ and K^+ are governed by Eq. (2). If protons are able to block the channel,[87-89] it is expected that, at the same K^+ concentration, the conductance will be larger at the higher pH. Therefore, proton block will become a problem in the interpretation of the single-channel conductance obtained in charged membranes since surface pH values as low as 4.5 can be calculated in certain experimental conditions.[89] The strategy to

[86] R. Coronado, *Annu. Rev. Biophys. Biophys. Chem.* **15**, 259 (1986).
[87] A. Woodhull, *J. Gen. Physiol.* **61**, 687 (1973).
[88] P. Labarca, Ph.D. Thesis, Brandeis University, Waltham, Massachusetts (1980).
[89] J. Bell, *Biophys. J.* **48**, 349 (1985).

avoid misinterpretation of results owing to proton block is to characterize the effect of protons on channel conductance in neutral membranes. Once this is done it is possible to work with charged membranes at pH values well above the inhibition constant for protons, K_i, and/or to include a proton block explicity in Eq. (39). In this case K is given by the expression

$$K = K_{app}/\{1 + [H] \exp(-F\phi_S/RT)/K_i\} \tag{42}$$

where K_{app} is the apparent K for K^+ and [H] is the bulk proton concentration. Note that if the fixed charge is located in the channel protein other types of problems in data interpretation arise. For example, the pK_i for proton block will be overestimated, and particular chemical groups can be mistakenly identified as forming part of the proton site or participating in proton binding.

Ion Blockade Used to Advantage

Channel blockade by ions can be used to obtain an independent estimation of the surface charge of the vestibules.[7,62,90] It is easy to show that the probability of opening, P_o, of a given channel in the presence of blocker is given by

$$P_o = \langle i \rangle/i_o = \{1 + [B]/K_d\}^{-1} \tag{43}$$

where $\langle i \rangle$ is the average current passing through the channel in the presence of the blocker B, i_o is the single-channel current in the absence of B, and K_d is the apparent dissociation constant for the blocking reaction. So, the simplest blocking model predicts that the fractional current as a function of the blocker concentration should be a hyperbola given by Eq. (42). A fixed charge near the blocking site will act to concentrate the blocker, and hence it will potentiate block since

$$[B] = [B]_{bulk} \exp(z_B F\phi_S/RT) \tag{44}$$

Therefore, it is expected that the blocking potency will be a function of ionic strength. Increasing ionic strength will decrease ϕ_S and hence [B]. Equation (44) shows that multivalent blockers will be more affected that monovalent ones.[90] The surface potential can be calculated by fitting the curve $\langle i \rangle/i_o$ versus ionic strength to Eq. (11) or (15) using different values of σ until the best fit is obtained.

The double layer potential can also be estimated from experiments in which the fractional current is plotted against $[B]_{bulk}$. In this case it is expected that large deviations from a hyperbola will be found because at

[90] C. Smith-Maxwell and T. Begenisich, *J. Gen. Physiol.* **90**, 361 (1987).

low $[B]_{bulk}$ blockade will be stronger than that predicted by Eq. (43). Of course deviations will be larger when the ionic strength of the medium is low. At high ionic strength deviations from a hyperbolic shape will be small because ϕ_S is small. Fitting the experimental data to Eqs. (42), (43), and the Gouy–Chapman theory will give an estimate of σ near the blocking site.[7,62] For ions that block channels, K_d has usually been found to be voltage dependent, and, therefore, these calculations assume that changes in ϕ_S will not affect K_d. If changes in ionic strength are made with ions that compete for the blocking site [e.g., Eq. (42)] or enhance blockade, these effects should be taken into account in the analysis.

Acknowledgments

This work was supported by grants from the Fondo Nacional de Investigacion (451-1988, 1167-1988, and 961-1991), the National Institutes of Health (GM-3598), and the Tinker Foundation. R.L. is a recipient of a John S. Guggenheim Fellowship. He also wishes to thank the Dreyfus Bank (Switzerland) for generous support from a private foundation that they made available.

[33] Determination of Ion Permeability by Fluorescence Quenching

By ANA MARIA GARCIA

Introduction

Movement of ions across membranes is generally studied by either radioisotope flux or current (e.g., planar lipid bilayer, voltage clamp or patch clamp) measurements. Here, an alternative method is described which can be used to study both electrogenic and electrically neutral fluxes. This technique is based on the quenching of the fluorescence of a fluorophore by an ionic quencher. The fluorophore is first trapped inside membrane vesicles or cells. On mixing the vesicles or cells with a solution containing the quencher ion, a decrease in the fluorescence is observed as quencher ions diffuse into the vesicles. The rate of the quenching reaction is very fast ($> 10^{-3}$ sec), and the extent of quenching is proportional to the concentration of quencher. Therefore, the rate of fluorescence change reflects the rate of quencher movement across the membrane, that is, the membrane permeability of the ionic quencher.

The fluorescence quenching technique has the advantage that it can be used to measure ion fluxes on a millisecond to second time scale. It is applicable to ion channels as well as ion exchangers and carriers. It was first used to describe ion fluxes through the acetylcholine receptor in membrane vesicles[1,2] and in cells.[3] Its use has been extended to potassium channels[4] and ion exchangers.[5,6] A recent review[7] describes the use of several fluorescent indicators such as fluorescent substrates, complexation indicators, potentiometric dyes, as well as collisional quenchers to measure ion fluxes across biological membranes. Readers are referred to that review[7] for a detailed analysis of the first three techniques. This chapter describes in detail the methodology pertaining the measurement of monovalent ion fluxes through membrane vesicles by the use of collisional quenchers, with some reference to studies in whole cells.

Theory

The fluorescence of a fluorophore is decreased on collision or direct interaction with quencher molecules. This is mainly due to a transfer of energy from the fluorophore in the excited state to the quencher. The extent of quenching depends on the concentration of quencher and is described by the Stern–Volmer relationship:

$$F_o/F = 1 + K_{SV}[Q] \tag{1}$$

where F_o and F correspond to the fluorescence in the absence and presence of the quencher, respectively, and $[Q]$ is the quencher concentration. A plot of F_o/F versus $[Q]$ yields a straight line with a slope corresponding to the Stern–Volmer constant, K_{SV}. Equation (1) takes into account only dynamic or collisional quenching, which is the dominant component of the quenching reaction. However, deviations from linearity are often observed when contributions by static quenching become significant, or when the quencher is not efficient. This seems to be particularly true in the cases described here, in which a highly charged fluorophore is being used at a relatively high ionic strength (for a detailed discussion on factors which affect quenching, see Eftink and Ghiron[8]).

Table I shows examples of quenching constants under different conditions for the fluorophore 1,3,6,8-pyrene tetrasulfonate (PTS). It is obvious

[1] J. W. Karpen, A. B. Sachs, D. J. Cash, E. B. Pasquale, and G. P. Hess, *Anal. Biochem.* **135**, 83 (1983).
[2] H.-P. H. Moore and M. A. Raftery, *Proc. Natl. Acad. Sci. U.S.A.* **77**, 4509 (1980).
[3] J. W. Karpen, A. B. Sachs, E. B. Pasquale, and G. P. Hess, *Anal. Biochem.* **157**, 353 (1986).
[4] A. M. Garcia and C. Miller, *J. Gen. Physiol.* **83**, 819 (1984).
[5] N. P. Illsley and A. S. Verkman, *Biochemistry* **26**, 1215 (1987).
[6] P. Y. Chen and A. S. Verkman, *Biochemistry* **27**, 655 (1988).
[7] O. Eidelman and Z. I. Cabantchik, *Biochim. Biophys. Acta* **988**, 319 (1989).
[8] M. R. Eftink and C. A. Ghiron, *Anal. Biochem.* **114**, 199 (1981).

TABLE I
QUENCHING CONSTANTS FOR 1,3,6,8-PYRENE TETRASULFONATE

Quencher	K_{SV} (M^{-1})	Conditions
TlNO$_3$	68	0.1 M KCl
Tl glucuronate	120	0.1 M Glucose
Tl glucuronate	80	0.1 M Choline glucuronate
Tl glucuronate	50	0.1 M Choline glucuronate, 0.2 mg/ml SR vesicles
Cs glucuronate	4	0.1 M Choline glucuronate
Methionine	20	0.15 M Glucose, 20 mM KCl
Methionine	12	0.15 M Glucose, 20 mM KCl, 0.15 mg/ml SR vesicles

that the ionic strength as well as the ions present in the reaction affect the quenching constant. Therefore, the first experimental step when measuring ion fluxes by collisional quenchers is the determination of K_{SV} under exactly the same experimental conditions as the flux measurements.

Equipment

For studies of permeability of membrane vesicles, a stopped-flow apparatus with a fluorescence detection and data acquisition system appropriate to the time scale of the experiments is necessary. If measurements are only in the time scale of seconds, a spectrofluorometer with stirred cuvette and injection port (to facilitate quencher additions) might suffice. For whole-cell determination, a fluorescence microscope equipped for fluorescence-digital imaging microscopy or other appropriate means of digitizing the data is necessary.[9]

Choosing a Fluorophore

Perhaps the first property to look for in a fluorophore is whether it has a high quantum yield. However, in the cases described here, several other parameters, like quenching constant and membrane permeability, require close consideration as well. If it is necessary to use a relatively high concentration of the fluorophore to obtain enough signal from the internal volume of the vesicles, care should be taken that self-quenching and other artifacts do not occur.

A common feature of the fluorescent probes used to measure ion fluxes is the presence of one or more rings with charged substituent groups. Thus,

[9] S. J. Ram and K. L. Kirk, *Proc. Natl. Acad. Sci. U.S.A.* **86,** 10166 (1989).

a fluorescent probe with a negative charge is quenched by cations, and, vice versa, a positively charged ring structure is quenched by anions. For example, sulfonate groups are found in the following anionic probes: 8-amino-1,3,6-naphthalene trisulfonate (ANTS, used to determine the permeability of *Torpedo* synaptic membrane vesicles[2]), 1,3,6,8-pyrenetetrasulfonic acid [PTS, used in the determination of cation permeability of sarcoplasmic reticulum (SR) vesicles[4]], and anthracene-1,5-disulfonic acid (ADS, for acetylcholine receptor-mediated cation fluxes[1]). In the case of cationic probes, the detailed study of Krapf *et al.*[10] on the properties of quinoline and acridine derivatives suggests that the best probe to be used with anion quenchers like chloride or bromide is 6-methoxy-N-(3-sulfopropyl)quinolinium (SPQ), a quinoline ring with electron donor substituents.

When choosing a fluorescent probe to determine ion permeability, consideration should also be given to the permeability of the probe across the membrane. The permeability properties of different probes can be partially predicted by looking at the octanol–water partition coefficient: the higher the partition coefficient, the higher the lipid permeability of the probe and, therefore, the faster it will move across the membrane. However, variations exist among membranes, and the final decision should be made based on performance in the system under study. This is better illustrated by the example of ANTS: whereas ANTS is too permeable across sarcoplasmic reticulum membrane vesicles, it was successfully used to measure cation permeability across *Torpedo* membrane vesicles. This difference is probably due to the different lipid composition of the two membranes. In the case of the cationic probe SPQ, the apparent half-time for efflux across cell membranes at 37° is 10–30 min, whereas at 23° it is 30–90 min.[10] Therefore, at least for SPQ, conditions such as temperature can be chosen to decrease its membrane permeability. As a general rule, it is advisable to determine the amount of probe that leaks out during the experiment. This is done by using a membrane-impermeant quencher (see section on determination of fluorescent probe leakage).

In general, the fluorescent probes are available as salts. It might be desirable to change the counterion to one appropriate to the experimental conditions chosen (see Experimental Design). This can be simply done by passing the fluorophore through an ion-exchange column equilibrated with the counterion of choice.

In summary, when choosing a fluorophore, the following considerations must be taken: high quantum yield, low membrane permeability, and

[10] R. Krapf, N. P. Illsley, H. C. Tseng, and A. S. Verkman, *Anal. Biochem.* **169,** 142 (1988).

good quenching by the ion of interest. The reader is advised to check recent catalogs, for example, Molecular Probes (Eugene, OR), for suggestions of fluorescent probes, since new probes are constantly becoming available.

Quencher

For Anionic Probes

The most frequently used quenchers of anionic probes like ANTS or PTS are the heavy metals Tl^+ and Cs^+. Of the two, Tl^+ is in general more efficient than Cs^+. However, this varies from probe to probe. These two ions can often substitute for other monovalent cations through cation channels. Another example of a cationic quencher is Ag^+; however, this ion is in general not permeable through cation channels.

Problems often encountered are unavailability of the desired salts of the quencher ion and poor solubility of the salts (especially Tl^+) that are available. Also, their purity might not be adequate (as in the case for most gluconate salts). An alternative is to start with the hydroxide of the cationic quencher, followed by neutralization with the acid of the anion of choice. This provides the experimenter with a wider variety of counterions and with more confidence about their purity. This same approach is valid for anionic quenchers (see below).

For Cationic Probes

The most common quenchers for cationic probes are halides, whose quenching efficiencies are $I^- > Br^- > Cl^-$. Other examples of quenchers in this category are SCN^- and citrate, but both are much poorer quenchers of fluorophores of the SPQ family.

Nonpermeant Quenchers

It is important to control the amount of fluorophore leakage during the experiment by measuring the amount of probe outside of the vesicles during and at the end of the experiment. This is done by mixing the vesicles with a quencher that permeates the membrane slowly. Some examples of slowly permeant quenchers are methionine, acrylamide, pyridine, and succinimide. Depending on the particular fluorophore being used, the appropriate quencher with the highest quenching efficiency (i.e., highest K_{SV}) should be chosen. Also, the permeability across the membrane under study should be determined independently.

Loading of Probes

The most obvious way of loading the fluorescent probe is by passive diffusion. In general, a 2- to 3-hr incubation at room temperature or overnight on ice is enough to reach equilibrium. Different systems have different permeability properties and might be amenable to different manipulations such as osmotic shock or freeze–thawing. The choice must depend on the system under study.

After loading, the external probe needs to be removed. When working with vesicles, they are passed through a 1-ml ion-exchange resin or a small size-exclusion column. Centrifugation of the vesicles and resuspending them in the appropriate medium are not advisable if the probe can leak out in less than the time required to complete the procedure. For experiments involving attached cells, washing with a solution without the fluorophore is sufficient to remove the external probe.

Experimental

The fluorescent probe can be loaded in the presence or absence of quencher ion. In the first case, the efflux of quencher will result in a fluorescence increase, whereas in the second case, the influx of quencher will decrease the fluorescence. Furthermore, the experiment can be done with a permeable or nonpermeable counterion. The possibilities of varying parameters keep increasing according to the ingenuity of the experimenter. To provide the reader with a set of basic conditions, only one experimental protocol will be described. This is the case of vesicles loaded with fluorophore and mixed at time $t = 0$ with a solution containing quencher ion. Following the description of the protocol, there will be a discussion on the experimental design itself, namely, how to set up the ion distribution across the membrane to determine the permeability of the ion of interest.

Fluorescence Quenching

Immediately after removal of the external fluorescent probe, the vesicles are loaded into syringe A of a stopped-flow apparatus. Syringe B is loaded with a solution of the same composition but without vesicles. The solutions are mixed and F_0 recorded. Syringe B is now changed for one containing quencher ion. Normally, the highest quencher concentration which gives a linear response is used. On mixing, the decrease in fluorescence, F, is recorded until no further change is observed. The experiment can be repeated with the same or different quencher solution. At this point, the effect of membrane potential, ion channel blockers, and transport inhibitors can be tested. The additional components to be tested either can be added to syringe B, together with the quencher, or they can be preincubated with the vesicles and loaded into syringe A. Care must be taken to keep the osmolarity and ionic strength constant throughout the experiment. The experiment can be similarly done in a regular spectrofluorometer, where the vesicles are mixed with quencher ion and immediately loaded in the cuvette at time $t = 0$.

Determination of Fluorescent Probe Leakage

Depending on the length of the experiment and on the permeability of the fluorescent probe, it is important to include a control in which the quencher ion is replaced by a membrane-impermeable quencher and mixed with the vesicles. Under these conditions, the fluorescent probe that has leaked out during the experiment will be instantaneously quenched. If the leakage of fluorescent probe is significant, it should be considered in the final calculations. In this case

$$F_o = F_{in} + F_{out} \tag{2}$$

where F_o is the total fluorescence and F_{in} and F_{out} correspond to internal and external probe, respectively. F_{out} is estimated from the instantaneous quenching by rearranging the Stern–Volmer equation [Eq. (1)]:

$$F_{out} = (1 + K'_{SV}[Q']) F_{inst} \tag{3}$$

where K'_{SV} and $[Q']$ refer to the nonpermeant quencher and F_{inst} is the amount of fluorescence instantaneously quenched. F_{in} can now be calculated from Eq. (2).

Channel Distribution

When working with isolated membrane vesicles, the distribution of ion channels or transporters as well as the vesicle size are very likely heterogeneous. Although it will be impossible to determine an average number of channels per vesicle with this method, it is possible to at least determine the fraction of vesicles without transporters. For this, the membrane vesicles are incubated with an ionophore prior to mixing with quencher ion. The ionophore used should increase the permeability of the membrane to the quencher and/or to the counterions present, like gramicidin A for cations. It is expected that, in this case, the quenching of the fluorescence probe will occur within the mixing time of the apparatus. If all the vesicles have channels, the final F obtained from this experiment should agree with the F obtained by long-term reaction in the absence of ionophore. A difference in the F values suggests the presence of a population of vesicles lacking the transporter under study and therefore, presenting a higher permeability barrier to ion movement.

Calculation

After finishing the data collection, the fluorescence values are converted to quencher concentration by the Stern–Volmer equation described above. Considering fluorescent probe leakage, the concentration of quencher inside the vesicles at time t will be given by

$$[Q_i]_t = [F_{in}(1 + K_{SV}Q_o)/F + K_{SV}Q_o - F_{out}]/K_{SV} \tag{4}$$

where F_{in} and F_{out} are calculated by Eqs. (2) and (3). K_{SV} and Q_o refer to the Stern–Volmer constant and external concentration of permeable quencher. Under the conditions of the experiment, Q_o is a constant.

Subsequently, the data are fitted by a single exponential

$$Q_i = Q_o(1 - e^{-kt}) \tag{5}$$

where k is the rate constant of the flux. However, if it has been determined that a fraction of the vesicles do not carry the transporter, at least two exponentials will be needed to fit the data. The permeability (P_n) of the vesicles to the ion n will then be given by

$$P_n = kV/A \tag{6}$$

where V and A are the volume and area of the vesicle, respectively. Because the vesicles are heterogeneous in size, an average value should be considered.

Experimental Design

Regardless of whether one starts with the quencher ion and the fluorophore in the same or opposite compartments, the experiment can be designed in either of two ways: (1) the quencher ion is the least permeable of the ionic species present, or (2) the quencher ion is the most permeable. In the first case, the quenching of fluorescence will be directly related to the permeability of the quencher across the membrane. In the second case, the conditions of the medium across the membrane can be adjusted so that the fluorescence quenching is a function of the permeability of a counterion. In both cases, the mobility of the counterions is very important: whether the experiment is designed to study the properties of an ion channel or an ion exchanger, the movement of one ion across the membrane is always accompanied by the movement of a counterion. If no counterion movement is allowed, a membrane potential will be generated which will stop further quencher ion flux across the membrane. The counterion can be of the same or opposite charge to the quencher ion being used. However, the presence of a counterion of opposite charge to the quencher will result in the migration of two ions in the same direction, with the consequent creation of an osmotic gradient.

Quencher Ion as Least Permeable Species

In the case of the quencher ion being the least permeable species, the membrane is made permeable to counterions with the use of ionophores. For example, in a study to examine the anion permeability of placental microvillous vesicles, Illsley and Verkman[5] used valinomycin to increase the membrane permeability to K^+. Every time one molecule of quencher ion Cl^- moves inside of the vesicles, one K^+ ion moves along with it. Thus, the decrease in fluorescence due to the influx of quencher is determined exclusively by the permeability of the quencher. This approach allows the possibility of creating a potential across the membrane: because the most permeable species is the ion permeable through the ionophore, its concentration at both sides of the membrane determines a Nernst potential across the membrane. However, the movement of quencher ion eventually collapses this membrane potential.

Quencher Ion as Most Permeable Species

The case of the quencher ion being the most permeable species is best illustrated by the study on the membrane permeability of sarcoplasmic reticulum vesicles.[4] These vesicles have a K^+ channel, whose properties have been studied in detail in artificial planar bilayers.[11] Of the most used cation quenchers, Cs^+ is a blocker of this channel, but Tl^+ is very permeable. A study of membrane permeability was designed the following way: the vesicles were loaded with the fluorescent probe choline-PTS, in the presence of a salt of the cation under study (i.e., K^+, Na^+, Li^+, choline). The anion of the salt was nonpermeant (glutamate or gluconate). After removal of the external probe, the vesicles were mixed with a solution containing thallium glutamate or thallium gluconate. When necessary, the ionic strength and osmolarity were kept constant with a salt of a nonpermeant cation (like N-methylglucamine) and a nonpermeant anion. On mixing and because of its concentration gradient, Tl^+ moves into the vesicles. Because Tl^+ is the most permeable species, a membrane potential is created which would prevent further Tl^+ movement. However, the

[11] C. Miller, *J. Membr. Biol.* **40**, 1 (1978).

cation (the only other permeable species) trapped inside the vesicles moves out, collapsing the membrane potential and allowing Tl$^+$ influx to continue. Depending on the cation trapped inside, different rates of fluorescence quenching were obtained. From the rate constants, the permeability of the different cations was calculated [Eq. (5)] and when compared to reported or expected values was found to correlate very well. Thus, this approach is useful to determine the membrane permeability to different ions.

In conclusion, the fluorescence quenching method allows the determination of membrane permeability to monovalent ions on a time scale of milliseconds to seconds. The advantages of this technique are its low invasiveness to the system and the possibility of using it with electrogenic as well as nonelectrogenic systems.

[34] Synthetic Peptides and Proteins as Models for Pore-Forming Structure of Channel Proteins

By ANNE GROVE, TAKEO IWAMOTO, MYRTA S. MONTAL,
JOHN M. TOMICH, and MAURICIO MONTAL

Introduction

In this chapter, we describe a strategy which aims to identify, in the primary structure of channel proteins, segments that determine the functional characteristics of the protein.[1,2] Primary structures of several superfamilies of ligand-gated and voltage-gated channel proteins have been elucidated[3]; however, high resolution structural information is not available. Extensive sequence similarity among members of superfamilies of channel proteins, the occurrence of homologous subunits organized around a central pore, and the identification of segments capable of forming α-helical transmembrane structures suggest a common structural motif for the ionic pore: a cluster of amphipathic α helices, gathered such that nonpolar residues are in contact with the membrane interior or hydrophobic parts of the protein, and charged or polar residues are facing the central hydrophilic, ion-conducting pore.[1,2,4]

[1] R. E. Greenblatt, Y. Blatt, and M. Montal, *FEBS Lett.* **193**, 125 (1985).
[2] M. Montal, *FASEB J.* **4**, 2623 (1990).
[3] S. Numa, *Harvey Lect.* **83**, 121 (1989).
[4] M. Montal, *in* "Ion Channels" (T. Narahashi, ed.), Vol. 2, p. 1. Plenum, New York, 1990.

Selection of Pore-Forming Segments

Amphipathic α-helical segments are identified in the primary structure using secondary structure prediction algorithms.[5-7] Segments are selected based on sequence similarity between members of a superfamily of channel proteins; helical modules must be of sufficient length to span the lipid bilayer (i.e., greater than 20 amino acids), and the sequence should contain functional residues, compatible with the observed ionic selectivity of the protein.[1,2,4-7]

Evaluation of Design

Peptides with sequences representing such functional elements are synthesized, and their ability to emulate the pore properties of the target protein is tested by incorporating the peptides into lipid bilayers. It is conjectured that the peptides self-assemble in the membrane to form conductive oligomers of different sizes.[8,9] To design oligomeric proteins with predetermined conformation, the identified channel-forming peptides are covalently attached to a carrier molecule (template).[10-12] Such nonlinear protein molecules exhibit template-induced secondary structure, and the strategy has been successfully applied to the generation of four-helix bundle molecules.[12-14]

Nicotinic Cholinergic Receptor Channel: A Test Case

Synthesis and functional characterization of monomeric peptides and nonlinear proteins representing the presumed pore-forming structure of the nicotinic acetylcholine receptor (AChR) from *Torpedo californica* are

[5] D. Eisenberg, R. M. Weiss, T. C. Terwilliger, and W. Wilcox, *Faraday Symp. Chem. Soc.* **17,** 109 (1982).

[6] D. Eisenberg, *Annu. Rev. Biochem.* **53,** 595 (1984).

[7] J. Finer-Moore, J. F. Bazan, J. Rubin, and R. M. Stroud, *in* "Prediction of Protein Structure and the Principles of Protein Conformation" (G. D. Fasman, ed.), p. 719. Plenum, New York, 1989.

[8] S. Oiki, W. Danho, and M. Montal, *Proc. Natl. Acad. Sci. U.S.A.* **85,** 2393 (1988).

[9] S. Oiki, W. Danho, V. Madison, and M. Montal, *Proc. Natl. Acad. Sci. U.S.A.* **85,** 8703 (1988).

[10] M. Mutter, E. Altmann, K.-H. Altmann, R. Hersperger, P. Koziej, K. Nebel, G. Tuchsherer, S. Vuilleumier, H.-U. Gremlich, and K. Müller, *Helv. Chim. Acta* **71,** 835 (1988).

[11] M. Mutter, K.-H. Altmann, G. Tuchscherer, and S. Vuilleumier, *Tetrahedron* **44,** 771 (1988).

[12] M. Mutter, R. Hersperger, K. Gubernator, and K. Müller, *Proteins: Struct. Funct. Genet.* **5,** 13 (1989).

[13] J. Rivier, C. Miller, M. Spicer, J. Andrews, J. Porter, G. Tuchscherer, and M. Mutter, *Proc. Int. Symp. Solid Phase Synthesis* p. 39 (1991).

[14] M. Montal, M. S. Montal, and J. M. Tomich, *Proc. Natl. Acad. Sci. U.S.A.* **87,** 6929 (1990).

described. The AChR is composed of four subunits with a stoichiometry of $\alpha_2\beta\gamma\delta$, assembled with the ion channel in the center.[15-17] Four hydrophobic transmembrane segments, M1–M4, have been identified in each subunit.[3,17] Labeling of the channel with noncompetititve inhibitors indicates that the M2 segment is involved in forming the AChR channel.[18-20] A 23-residue segment corresponding to the M2 segment of the δ subunit, EKMSTAISVLLAQAVFLLLTSQR (amino acids 255–277[3]), is selected.[9] Assembly of four M2δ peptides on the multifunctional template molecule, KKKPGKEKG, generates a four-helix bundle structure, $T_4M2\delta$.[14]

Synthesis of Peptides and Oligomeric Proteins

Solid-Phase Synthesis

Peptides and proteins are synthesized by solid-phase methods according to general principles[21-23] using an Applied Biosystems Model 431 peptide synthesizer (ABI, Foster City, CA). N^α-*tert*-Boc-N^γ-tosyl-L-arginine-4-oxymethylphenylacetamidomethyl resin (capacity 0.62 mmol/g) and *tert*-Boc-glycine-4-oxymethylphenylacetamidomethyl resin (PAM resin, capacity 0.75 mmol/g) are also from ABI. Amino acid derivatives having the L-configuration are used (The Peptide Institute, Osaka, Japan). Peptides are built up using the acid-labile *tert*-butyloxycarbonyl (*t*-Boc) protecting groups for N^α-protection and dicyclohexylcarbodiimide (DCC)/hydroxybenzotriazole (HOBt) for coupling: after preactivation for 43 min at room temperature in 1 M DCC/1 M HOBt in N-methylpyrrolidone (NMP), couplings are performed in NMP containing dimethyl sulfoxide (DMSO) for 32 min using a 10-fold excess of the preformed HOBt esters. Coupling efficiencies at each step are monitored by the quantitative ninhydrin test.[24]

[15] J. Reynolds and A. Karlin, *Biochemistry* **17**, 2035 (1978).

[16] C. Toyoshima and N. Unwin, *J. Cell Biol.* **111**, 2623 (1990).

[17] J. P. Changeux, *Trends Neurosci.* **11**, 485 (1990).

[18] J. Giraudat, M. Dennis, T. Heidmann, J.-Y. Chang, and J.-P. Changeux, *Proc. Natl. Acad. Sci. U.S.A.* **33**, 2719 (1986).

[19] F. Hucho, W. Oberthür, and F. Lottspeich, *FEBS Lett.* **205**, 137 (1986).

[20] F. Revah, J.-L. Galzi, J. Giraudat, P.-Y. Haumont, F. Lederer, and J.-P. Changeux, *Proc. Natl. Acad. Sci. U.S.A.* **87**, 4675 (1990).

[21] R. B. Merrifield, *J. Am. Chem. Soc.* **85**, 2149 (1963).

[22] R. B. Merrifield, *Biosci. Rep.* **5**, 353 (1985).

[23] S. B. H. Kent, *Annu. Rev. Biochem.* **57**, 957 (1988).

[24] V. K. Sarin, S. B. H. Kent, J. P. Tam, and R. B. Merrifield, *Anal. Biochem.* **117**, 147 (1981).

Loading Capacity of Resin

Synthesis is performed using high accessibility/low substitution resins in order to improve coupling efficiencies. Reduced loading capacity of the preloaded resins is accomplished during coupling of the first amino acid. A reduced amount of the *t*-Boc amino acid derivative is allowed to react with the resin for 1 hr. Remaining sites are capped using acetic anhydride, by treatment with 10% (v/v) acetic anhydride/5% diisopropylethylamine in NMP for 5 min. This strategy is designed to allow coupling to only the most accessible sites on the resin. The substitution is measured as follows: capped resin containing the first two amino acids is deprotected by treatment with 65% trifluoroacetic acid in dichloromethane (DCM) for 10 min at room temperature. The acid-treated resin is washed extensively with DCM and an aliquot removed and weighed. The moles of residual free N terminus are calculated using the spectrophotometric, quantitative ninhydrin assay.[24] The reduced substitution resin (0.1 mmol) is returned to the synthesizer.

Synthesis of Linear Peptide

Preloaded N^{α}-*t*-Boc-N^{γ}-tosylarginine-PAM resin is treated with 0.50 equivalent of N^{α}-*t*-Boc-L-glutamine for 1 hr and remaining sites capped with acetic anhydride to achieve a reduced loading capacity of 0.3–0.4 mmol/g (Fig. 1A,B). Preformed HOBt esters of N^{α}-*t*-Boc-protected amino acids are used to assemble the full-length sequence (Fig. 1C,D). The side chains of trifunctional amino acids are protected as follows: E(OBzl) (benzyl ester), S(Bzl) (benzyl ether), T(Bzl), R(Tos) (tosylate). Multiple couplings are performed to achieve coupling efficiencies exceeding 99.5% per site. Capping with acetic anhydride (Fig. 1E) follows the final coupling step and is included to produce shorter failed sequences and thereby aid in the purification of the full-length peptide. The N terminus of the peptide is not acetylated.

Template Synthesis

Synthesis of the four-helix bundle protein is accomplished by a two-step procedure: the template molecule is synthesized using orthogonal lysine side-chain protection followed by the simultaneous assembly of peptide blocks. Steric constraints of the nonlinear, four-helix bundle proteins require the use of low substitution resins for optimal synthetic yield. The preloaded *t*-Boc-glycine-PAM resin (capacity 0.75 mmol/g) is reacted with 0.25 equivalents of N^{α}-*t*-Boc N^{ϵ}-9-fluorenylmethoxycarbonyl-(fmoc) lysine for 1 hr (Fig. 2A,B). Using this lower substitution (0.1–0.2 mmol/g),

A

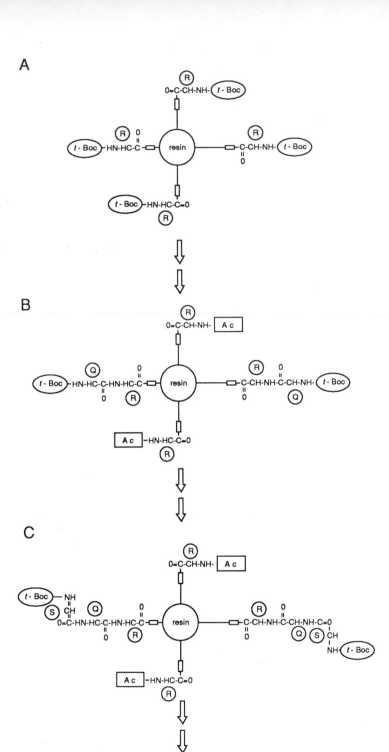

B

C

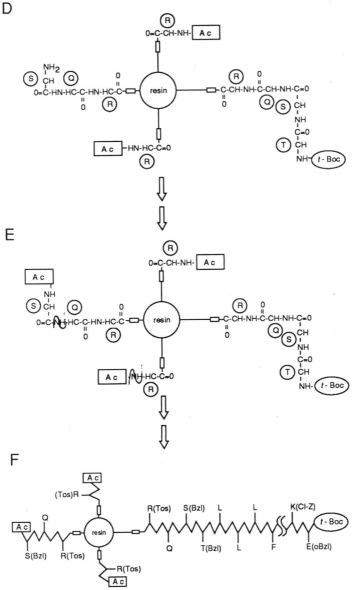

D

E

F

FIG. 1. Flow chart diagram of the protocol used to synthesize peptides of defined amino acid sequence. (A) Monomer synthesis begins with the highly substituted polystyrene resin, 0.62 mmol/g, containing N^{α}-t-Boc-N^{γ}-tosyl-L-arginine (R) attached through a PAM linker (open box). (B) The loading capacity is decreased in the first step by adding 0.5 equivalents of t-Boc-L-glutamine (Q) as the preactivated HOBt ester. All unreacted sites are capped using an excess of acetic anhydride (Ac). (C, D) Amino acids are added as the HOBt esters to the growing peptide chain. (E) Failed sequences are capped with acetic anhydride. (F) The completed peptide, assembled as described.

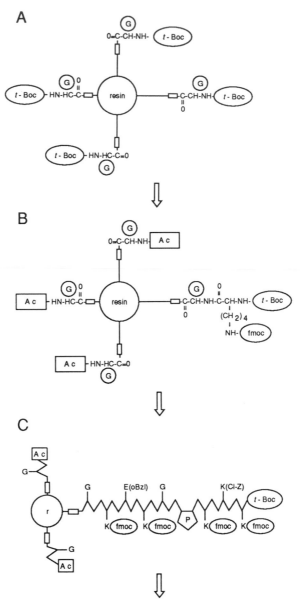

Fig. 2. Synthesis of the tethered four-helix bundle molecule. (A) The highly substituted resin (0.75 mmol/g) containing *t*-Boc-glycine (G) attached through a PAM linker (open box) is shown. (B) The most accessible sites on the resin are selected by coupling with 0.25 equivalents of the N^{α}-*t*-Boc-N^{ε}-fmoc-lysine. The remaining unreacted sites are blocked using an excess of acetic anhydride (Ac). (C) The template KKKPGKEKG is completed. (D) After blocking of the free N terminus of the template with excess acetic anhydride, the fmoc groups are removed by treatment with 20% piperidine in DMF (E). (F, G) Single amino acids are added to assemble sequentially the peptide modules of the synthetic proteins.

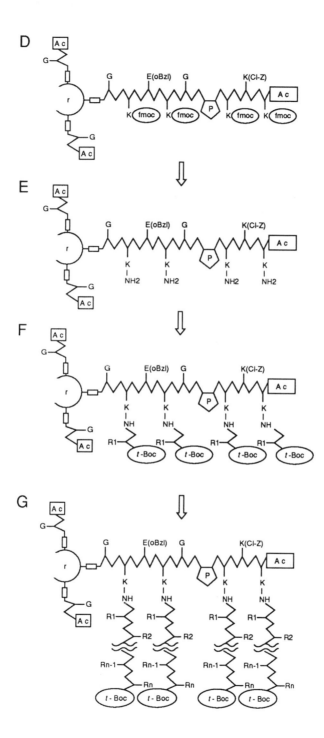

the number of recouplings required to attain coupling efficiencies greater than 99.5% are reduced and increased weight gains obtained. The 9-amino acid template Ac-K(N^ϵ-fmoc)-K(N^ϵ-Cl-Z)-K(N^ϵ-fmoc)-P-G-K(N^ϵ-fmoc)-E(γ-OBzl)-K(N^ϵ-fmoc)-G-PAM resin is synthesized (Cl-Z is 2-chlorobenzyloxycarbonyl). Preformed HOBt esters of N^α-t-Boc-protected amino acids are used to obtain quantitative coupling of peptide blocks to the ϵ-amino groups of lysine. Failed sequences produced during synthesis of the template, as well as the N terminus, are capped using acetic anhydride (Fig. 2C-D). The sequence is confirmed using automated Edman degradation on an ABI Model 477A peptide sequencer and the N-terminal lysine acetylated by treatment with acetic anhydride.

Parallel Assembly of Peptide Blocks

The four peptide blocks are built up in parallel on the template molecule. Assembly of peptide blocks typically starts with 50 mmol of template – resin (Fig. 2D). The base-labile N^ϵ-fmoc protecting groups are removed by incubation with 20% piperidine (v/v) in dimethylformamide (DMF) for 20 min (Fig. 2E). The four peptide blocks are assembled simultaneously (Fig. 2F,G) by stepwise synthesis using HOBt-activated esters of N^α-t-Boc amino acid derivatives on an ABI Model 431 synthesizer. Side chains of the trifunctional amino acids are protected as follows: E(OBzl), S(Bzl), T(Bzl), R(Tos). Multiple couplings (3 – 5 per residue) are performed to ensure high coupling efficiencies (> 99.5%) for each step. Capping of failed sequences with acetic anhydride may be included. Tracer amounts of ^{3}H-labeled t-Boc-leucine are added to determine concentrations in solution. Amino termini are not capped. Weight gains of over 80% are typically obtained for the channel proteins.

Cleavage and Deprotection

Cleavage and deprotection[25] are performed in anhydrous HF for 30 min at $-10°$ and for 60 min at $0°$ in the presence of p-cresol and p-thiocresol, 1.4 ml/g of resin. After the HF is removed, the resulting peptide – resin mixture is washed with anhydrous diethyl ether and dried overnight under reduced pressure over KOH pellets. Cleaved protein – resin mixtures are stored under reduced pressure in the dark at room temperature.

[25] S. Sakakibara, Y. Shimonishi, Y. Kishida, M. Okada, and H. Sugihara, *Bull. Chem. Soc. Jpn.* **40**, 2164 (1967).

Purification and Characterization

Purification

Peptides and proteins are purified by reversed-phase high-performance liquid chromatography (HPLC) as shown in Fig. 3. M2δ and T$_4$M2δ are extracted from the resin with trifluoroethanol (TFE, 99+% pure, Aldrich, Milwaukee, WI) and purified by multiple HPLC runs on a Vydac C$_4$ (semi-prep) 214 TP 1010 RP column. HPLC-grade trifluoroacetic acid (TFA) and acetonitrile is used (Pierce, Rockford, IL).

For M2δ, the column is equilibrated in 90% solvent A (deionized/distilled water containing 0.1% TFA) and 10% solvent B [80% (v/v) acetonitrile/water containing 0.1% TFA]. The peptide is eluted from the column using a linear gradient to 100% solvent B in 90 min. Eluted peptide is reinjected onto a narrow-bore Vydac C$_4$ RP column TP 214 54 equilibrated as described above. The peptide is eluted using the gradient shown (Fig. 3A).

T$_4$M2δ is purified by multiple runs on a Vydac C$_4$ (semi-prep) 214 TP 1010 RP column equilibrated in 75% solvent A and 25% solvent B. The protein is purified through a series of gradient steps followed by 30-min isocratic periods at 55, 62, and 75% of solvent B. Fractions are collected and analyzed by sodium dodecyl sulfate–polyacrylamide gel electrophore-

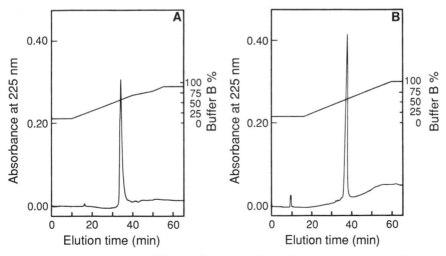

FIG. 3. Reversed-phase HPLC of synthetic channel peptides and proteins. (A) M2δ and (B) T$_4$M2δ eluted from the column described using the indicated gradient.

sis (SDS-PAGE).[26] Purified protein is reinjected onto a narrow-bore Vydac C_4 RP column TP 214 54 equilibrated as decribed above. Protein is eluted using the gradient shown (Fig. 3B). M2δ and T_4M2δ elute from the column as well-resolved peaks.

Amino Acid Analysis and Microsequencing

Peptide composition and sequences are confirmed by amino acid analysis on an ABI Model 420 PTC derivatizer/analyzer after hydrolysis with 6 M HCl for 1.5 and 3 hr at 165° *in vacuo,* confirming the stoichiometry of four oligopeptides per template, as well as automated Edman degradation on an ABI Model 477A peptide sequencer.[14] Homogeneity of the preparations is confirmed by capillary zone electrophoresis[27] conducted in 20 mM citrate buffer (pH 2.5) at 38° on an ABI Model 270A instrument and SDS-PAGE[26] on 16% tricine gels (Novex, Encinitas, CA). Molecular weights are determined using low range M_r markers (Diversified Biotech, Newton Centre, MA). The M_r of T_4M2δ is approximately 11,000, consistent with a protein containing 101 residues. M2δ migrates with an apparent M_r of 2000. The yield of monomeric peptide is 5 – 10%; yields of T_4M2δ are low, less than about 1% of cleaved protein – resin mixture.

Reconstitution in Planar Lipid Bilayers

Formation of Bilayers at the Tip of Patch Pipettes

Bilayers are formed by apposition of two monolayers, initially formed at the air–water interface, at the tip of patch pipettes.[28] This method is based on the technique of bilayer formation across an aperture in a Teflon partition that separates two aqueous compartments[29] and has proved to be superior in terms of time resolution and sensitivity.[30] Pipettes are fabricated from microhematocrit capillary tubes (Fisher, Pittsburgh, PA) using an automated pipette puller (Sutter Instrument Co., San Rafael, CA, Model P-87).[28,31] The tip size is adjusted to yield 5 – 10 MΩ open resistance when the pipette is filled and immersed in the buffer described. Mono-

---

[26] U. K. Laemmli, *Nature (London)* **227,** 680 (1970).

[27] M. A. Firestone, J. F. Michaud, R. H. Carter, and W. Thormann, *J. Chromatogr.* **407,** 363 (1987).

[28] T. Hamamoto and M. Montal, this series, Vol. 126, p. 123.

[29] M. Montal, this series, Vol. 32, p. 545.

[30] M. Montal, R. Anholt, and P. Labarca, *in* "Ion Channel Reconstitution" (C. Miller, ed.), p. 157. Plenum, New York, 1986.

[31] B. A. Suarez-Isla, K. Wan, J. Lindstrom, and M. Montal, *Biochemistry* **22,** 2319 (1983).

layers are formed from solutions of lipid (Avanti Biochemicals, Alabaster, AL): PC (1,2-diphytanoyl-*sn*-glycero-3-phosphocholine), 5 mg/ml in hexane (Sigma, St. Louis, MO), or POPE/POPC [1-palmitoyl-2-oleoyl-*sn*-glycero-3-phosphoethanolamine (POPE) and 1-palmitoyl-2-oleoyl-*sn*-glycero-3-phosphocholine (POPC)], 4:1, 5 mg/ml in hexane. Planar bilayer experiments are performed at $24° \pm 2°$. Other conditions are as decribed.[28,29]

Buffer Solutions

The aqueous compartments separated by the lipid bilayer contain 500 mM NaCl, 1 mM CaCl$_2$, and 10 mM HEPES (Sigma), adjusted to pH 7.3 with NaOH. Chambers for bilayer formation are single wells cut out from a multiwell Disposo-tray (Linbro, Flow Laboratories, Inc., McLean, VA). Each tray has 48 flat-bottomed wells (0.8×0.6 cm), and the well capacity is approximately 400 μl. Each well is used only once.

Reconstitution of Protein in Planar Bilayers

Peptide or protein is incorporated into bilayers by either of two approaches: purified protein is extracted with lipid, either PC or POPE/POPC, 4:1, 5 mg/ml in hexane,[14] by vortexing 1 min and sonicating for 10 sec in a water bath sonicator (Laboratory Supplies Co., Inc., Hicksville, NY) to achieve final protein/lipid ratios in the range of 1:1000. Lipid–protein mixtures are spread into monolayers at the air–water interface and bilayers formed as described above. Alternatively, peptide or protein may be dissolved in TFE; TFE extracts are added to the aqueous subphase following formation of lipid bilayers (final concentration 300–600 nM). Spontaneous insertion into bilayers is observed after 5–15 min. Both methods of incorporation lead to similar channel recordings.[14]

Single-Channel Characterization of Synthetic Peptide and Protein

Electrical Recordings

Electrical recordings are carried out as described[14,31] using a patch clamp system (List L-M EPC-7, Medical Systems Corp., Greenvale, NY). Constant voltage is applied from a dc source (Omnical 2001, World Precision Instruments, New Haven, CT). The signal output from the clamp is stored on videocassette (Sony Betamax) equipped with a modified[32] digital

---

[32] F. Bezanilla, *Biophys. J.* **47**, 437 (1985).

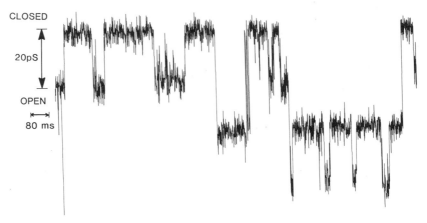

FIG. 4. Single-channel current recordings from a PC membrane containing M2δ in symmetric 0.5 *M* NaCl. Currents were recorded at 100 mV. The records were filtered at 2 kHz and sampled at 0.1 msec. Single-channel conductances are 20 and 40 pS.

audioprocessor (Sony PCM 501ES, Unitrade, Philadelphia, PA). The two aqueous compartments are connected to the amplifier by means of two Ag/AgCl electrodes (In Vivo Metric Systems, Healdsburg, CA).

Data Processing

Records are filtered at 2 kHz with an 8-pole Bessel filter (Frequency Devices, Haverhill, MA) and digitized at 100 μsec per point using an INDEC-L-11/73-70 microcomputer system (INDEC, Sunnyvale, CA). Single-channel open and closed conductance levels are discriminated using a pattern recognition algorithm.[33] Conductance values are calculated from current histograms best fitted by the sum of two Gaussian distributions. Channel open and closed lifetimes are determined by exponential fits to probability density distributions of dwell times in the open and closed states.[30,31,33]

Single-Channel Properties of M2δ

Figure 4 shows that the synthetic peptide M2δ forms discrete channels in lipid bilayers. The most frequent conductance event has a single-channel conductance, γ, of 40 pS in symmetric 0.5 *M* NaCl. A distinct feature of recordings obtained with M2δ is the occurrence of events with smaller

[33] P. Labarca, J. Lindstrom, and M. Montal, *J. Gen. Physiol.* **83,** 473 (1984).

($\gamma = 20$ pS, Fig. 4) or larger conductance, detected with lower frequency than the 40-pS conductance event. The channels recorded with M2δ are the result of the peptides self-assembling in the bilayer, indicated by the heterogeneity of γ, the observation that smaller conductances show longer mean open times, and the finding that the frequency of occurrence of large conductances is low.[9,14]

Characteristics of Pore-Forming Protein, $T_4M2\delta$

A single-channel recording from a PC membrane containing the tethered tetrameric protein, $T_4M2\delta$, is shown in Fig. 5. Single-channel current histograms indicate that γ in symmetric 0.5 M NaCl is 19 pS. The homogeneity of the unitary conductance events and the frequent occurrence of openings lasting on the order of seconds are conspicuous.

The 40-pS conductance event observed with the M2δ monomer matches the conductance observed with authentic *Torpedo* AChR, identified as a pentameric structure.[15,16] The pore dimensions of a pentameric cluster of α helices would account for the 40-pS conductance, whereas the 20-pS channel is consistent with a tetrameric array.[2]

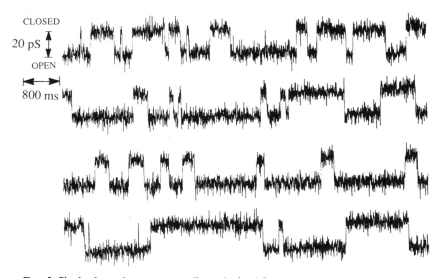

CLOSED

20 pS

OPEN

800 ms

FIG. 5. Single-channel current recordings obtained from a PC membrane containing the synthetic $T_4M2\delta$ protein in symmetric 0.5 M NaCl. Currents were recorded at 100 mV. Records were filtered at 2 kHz and sampled at 0.1 msec. The calculated single-channel conductance is 19 pS.

Advantages and Limitations

Design of Pore-Forming Proteins

Oligomeric clusters of amphipathic α helices, whether self-assembled in the lipid bilayer or covalently attached to a template molecule in a four-helix bundle configuration, form ionic channels in bilayers. The functional characteristics of these bundles of α helices match those of the authentic AChR. Both monomeric peptides and tethered, tetrameric proteins are readily reconstituted in lipid bilayers.

The evidence that fundamental pore properties may be reproduced within a bundle of α helices representing selected sequences from the primary structure of a channel protein lends credence to the notion that a cluster of amphipathic α helices constitutes a general pore-forming motif for channel proteins. Additional support is furnished by the findings that synthetic monomeric peptides representing the S3 segment of the brain sodium channel form cation-selective channels[8] and a four-helix bundle molecule representing a suggested pore-lining, transmembrane segment of the dihydropyridine-sensitive calcium channel mimics the pore properties of the target protein.[34] Work in progress corroborates the general validity of this approach in identifying pore-lining elements of a voltage-gated potassium channel and the glycine receptor. The fidelity of the strategy is asserted by the fact that peptide segments which correspond to sequences of hydrophobic transmembrane segments postulated to occur at interstices between helices or at the protein–lipid boundary but not to contribute to the channel lining do not form channels, either as monomeric peptides or tetrameric proteins. These include the monomeric peptides M1δ (LFTVINFITPCVLISFLASLAFY)[9] of the AChR, IS6 (IFFVLVIFLGS-FYLINLILAVV)[4] of the rat brain sodium channel, and the tetrameric bundle T$_4$CaIVS5, where IVS5 (YVALLIVMLFFIYAVIGMQMFGK)[34] is from the dihydropyridine-sensitive calcium channel.

Synthetic Strategy

Peptides 21–23 residues in length, representing a single transmembrane segment, are quite accessible to chemical synthesis and purification. Monomeric peptides provide a tool to assay readily the involvement of such segments in forming an ionic channel. The functional significance of individual residues in lining the lumen of the pore may be evaluated by specific residue replacement. However, the propensity of peptides to aggre-

[34] A. Grove, J. M. Tomich, and M. Montal, *Proc. Natl. Acad. Sci. U.S.A.* **88**, 6418 (1991).

gate in lipid bilayers to form conductive oligomers yields a heterogeneous population of conductance events.

The design of tethered oligomeric proteins offers the pronounced benefit of predefined oligomeric number and orientation of peptide blocks relative to one another, features which greatly facilitate interpretation of the displayed pore-properties. However, the branched structure of these proteins requires specific attention to possible steric hindrance during assembly of peptide blocks (such as the use of resins with low substitution) and attainment of high coupling yields for each residue, as separation of closely related impurities is intricate. Tethered, heterooligomeric synthetic proteins would more closely mimic the proposed heterooligomeric pore structure of authentic proteins.[2] The realization of such molecules will greatly improve the potential of this strategy in investigating the structure and functional properties of the pore-forming element of channel proteins.

This strategy outlines steps toward the reconstruction of the structure of the whole protein from its component parts. The fact that the identified structural motif represents a functional domain of a channel protein validates its use as a model for the pore-forming structure. This reduction in complexity with the added feasibility of selectively labeling specific residues with NMR-active isotopes[35] makes the synthetic channel proteins ideal objects for structure determination using two-dimensional NMR spectroscopy. This opens the prospects for detailed structural analysis of synthetic channels in a lipid bilayer, the physiological environment of channel proteins.

Acknowledgments

Supported by grants from the U.S. Public Health Service (GM-42340 and MH-44638 to M.M., and GM-43617 to J.M.T.), the Office of Naval Research (N00014-89-J-1469 to M.M.), the Department of the Army Medical Research (DAMD17-89-C-9032 to M.M.), and by a Research Scientist Award to M.M. from the Alcohol, Drug Abuse and Mental Health Administration's National Institute of Mental Health (MH-00778).

[35] K. Wütrich, *Science* **243**, 45 (1989).

Section IV

Purification of Ion Channel Proteins and Genes

[35] Purification and Reconstitution of Skeletal Muscle Calcium Channels

By VINCENT FLORIO, JÖRG STRIESSNIG, and WILLIAM A. CATTERALL

Introduction

Voltage-sensitive calcium channels mediate increases in calcium influx into cells in response to changes in membrane potential. Their calcium conductance is responsible for the plateau depolarization of the action potential in nerve and muscle cells. Calcium which enters the cell through voltage-gated calcium channels serves as the intracellular second messenger of the action potential, initiating excitation–contraction coupling in cardiac and smooth muscle, secretion of neurotransmitters and hormones in neural and endocrine cells, and activation of intracellular protein phosphorylation and gene expression in many cell types. Physiological and pharmacological studies have defined at least four classes of calcium channels.[1] L-type calcium channels mediate long-lasting calcium currents in response to strong depolarizations. They are the principal calcium channels in muscle tissues and are also present in neurons and many other cell types. They are distinguished by their sensitivity to organic calcium channel modulators including the dihydropyridines (DHP), phenylalkylamines, and benzothiazepines. N-type calcium channels are neurospecific. They mediate both transient and sustained components of calcium current in response to strong depolarizations and are blocked by ω-conotoxins. T-type calcium channels mediate transient calcium currents in response to small depolarizations. P-type calcium channels, first described in cerebellar Purkinje neurons, have physiological properties similar to N-type channels but are blocked by funnel web spider toxins and not by ω-conotoxins. At present, only L-type calcium channels have been investigated thoroughly at the molecular level.

In skeletal muscle, L-type calcium channels mediate a slowly activated, long-lasting calcium current in the transverse tubules which is blocked by DHP calcium channel antagonists.[2,3] They also serve as voltage sensors for initiation of excitation–contraction coupling through an unknown mechanism that does not require their calcium conductance activity.[4] These

[1] B. P. Bean, *Annu. Rev. Physiol.* **51**, 367 (1989).
[2] J. A. Sanchez and E. Stefani, *J. Physiol. (London)* **283**, 197 (1978).
[3] W. Almers, R. Fink, and P. T. Palade, *J. Physiol. (London)* **312**, 177 (1981).
[4] E. Rios and G. Brum, *Nature (London)* **325**, 717 (1987).

calcium channels are concentrated in the transverse tubule membrane fraction of muscle tissue, which contains the highest density of high affinity receptor sites for DHPs and other organic calcium channel modulators of any membrane system studied.[5,6] The DHPs, phenylalkylamines (e.g., verapamil), and benzothiazepines (e.g., diltiazem) bind at three separate, allosterically coupled receptor sites on the L-type calcium channels.[7] The availability of high affinity ligands for the different receptor sites on L-type calcium channels and the high density of these calcium channels in T-tubule membranes have made the L-type calcium channel from skeletal muscle the principal molecular model for studies of calcium channels. Therefore, this chapter will focus exclusively on purification and reconstitution of the L-type calcium channel from skeletal muscle. Some progress has been made in identifying and purifying calcium channels in brain, heart, and smooth muscle, but no widely accepted method for purification and reconstitution of these channels has been developed as yet.

Isolation of Transverse Tubule Membranes

Rationale

Owing to the selective distribution of DHP-sensitive calcium channels in the T-tubules of skeletal muscle, dihydropyridine binding activity can be enriched from homogenates about 7-fold in crude T-tubule fractions and about 50-fold in purified T-tubule preparations.[6,8] Standard procedures for complete T-tubule membrane purification employ differential centrifugation followed by sucrose density gradient centrifugation and calcium loading to selectively remove contaminating vesicles of sarcoplasmic reticulum (SR) origin.[9,10] The sucrose gradient and calcium loading steps are more time consuming and result in a lower yield of dihydropyridine receptors per gram wet weight of starting muscle tissue.[11] Fully purified T-tubules are preferred for functional[12] or biochemical[13] studies of the calcium channel

[5] M. Fossett, E. Jaimovich, E. Delpont, and M. Lazdunski, *J. Biol. Chem.* **258**, 6086 (1983).

[6] H. Glossmann, D. R. Ferry, and C. B. Boschek, *Naunyn-Schmiedebergs Arch. Pharmacol.* **323**, 1 (1983).

[7] H. Glossmann, D. Ferry, A. Goll, J. Striessnig, and M. Schober, *J. Cardiovasc. Pharmacol.* **7**, 520 (1985).

[8] J.-P. Galizzi, M. Fosset, and M. Lazdunski, *Eur. J. Biochem.* **144**, 211 (1984).

[9] C. Hidalgo, M. E. Gonzales, and R. Lagos, *J. Biol. Chem.* **258**, 13937 (1983).

[10] M. Rosemblatt, C. Hidalgo, C. Vergara, and N. Ikemoto, *J. Biol. Chem.* **256**, 8140 (1981).

[11] F. Flockerzi, H.-J. Oeken, and F. Hofmann, *Eur. J. Biochem.* **161**, 217 (1986).

[12] H. Affolter and R. Coronado, *Biophys. J.* **48**, 341 (1985).

[13] B. M. Curtis and W. A. Catterall, *Proc. Natl. Acad. Sci. U.S.A.* **82**, 2528 (1985).

in the membrane-bound state. In contrast, the rationale for preparative calcium channel purification is to use a membrane isolation procedure which allows only partial purification of T-tubule membranes but results in a high recovery of dihydropyridine receptors from the homogenate. SR vesicles contain almost no glycoproteins, and contaminating proteins from them are therefore easily removed by lectin affinity chromatography of the glycosylated calcium channel complex. All membrane preparations should start with freshly obtained skeletal muscle tissue. The use of previously frozen and thawed muscle has been recommended to increase the yield of T-tubule membranes.[8] However, this procedure has been shown to activate endogenous proteases, causing selective degradation of the $\alpha 1$ and β subunits.[14,15]

The following procedure describes the preparation of crude T-tubule membranes. Highly purified T-tubules can be obtained by further sucrose-density gradient centrifugation and calcium loading.[9]

Purification of Crude T-Tubule Membranes from Rabbit Skeletal Muscle

Reagents and Buffers

All chemicals are from Sigma (St. Louis, MO), except calpain inhibitors, which are from Calbiochem (LaJolla, CA).

Buffer MA: 100 mM Tris–maleate, pH 7.0, 0.3 M sucrose
Buffer MB: 20 mM Tris–maleate, pH 7.0, 0.3 M sucrose

Buffers contain the following final concentrations of protease inhibitors: 1 μg/ml leupeptin, 200 μg/ml o-phenanthroline, 200 μg/ml iodoacetamide, 20 μg/ml phenylmethylsulfonyl fluoride (PMSF), 1 μM pepstain A, 2 μg/ml aprotinin, 16 μg/ml benzamidine. Iodoacetamide can be omitted if the sulfhydryl reagent interferes with the calcium channel function to be studied. Stock solutions of protease inhibitors are prepared freshly in distilled water, dry acetone (PMSF), or ethanol (pepstatin A).

To increase the yield of the full-length form of the $\alpha 1$ subunit ($\alpha 1_{212}$, see section on subunit composition and properties), specific inhibitors of the Ca^{2+}-activated protease calpain (calpain inhibitors I and II) should be added to the buffers at final concentrations of 8.5 μg/ml. In our hands, these specific inhibitors prevent cleavage of $\alpha 1_{212}$ more effectively than our cocktail of other protease inhibitors or chelation of Ca^{2+} with EGTA.

---

[14] P. L. Vaghy, J. Striessnig, K. Miwa, H. G. Knaus, K. Itagaki, E. McKenna, H. Glossmann, and A. Schwartz, *J. Biol. Chem.* **262**, 14337 (1987).
[15] M. Takahashi, M. J. Seagar, J. F. Jones, B. F. Reber, and W. A. Catterall, *Proc. Natl. Acad. Sci. U.S.A.* **84**, 5478 (1987).

Procedure

1. A 2.5-kg New Zealand White rabbit is killed by an intravenous overdose of pentobarbital. Skeletal muscle tissue from back and leg muscles is rapidly removed, immediately put on ice, and cleaned of nonmuscle tissues.

2. All following steps are carried out at 4° or on ice. The muscle is ground in a clean meat grinder and its weight determined. Routinely, 400 g muscle is obtained from one rabbit. Five hundred milliliters of buffer MA is added to the ground muscle followed by homogenization in a Waring blender for 20–30 sec. An additional 500 ml of buffer MA is added followed by homogenization for 20 sec. After adding an additional 500 ml of buffer MB and homogenizing for 15 sec, the mixture is allowed to cool on ice for 5 min. Homogenization for 30 sec followed by cooling is repeated 7 times.

3. The homogenate is centrifuged at 3600 rpm in a Beckman J6 rotor for 20 min. The top layer and the pellet are discarded, and the supernatant is filtered through a double layer of cheesecloth.

4. The filtered supernatant is centrifuged at 10,000 g for 20 min.

5. To the supernatant, solid KCl is added to final concentration of 0.5 M (37.3 g/liter), and the solution is stirred for 30 min.

6. Crude T-tubule membranes are collected by centrifugation at 150,000 g for 45 min.

7. The pellets are washed once in buffer MB and finally resuspended in buffer MB at a membrane concentration of 10 mg/ml. The membranes are homogenized 6 times in a Potter–Elvehjem homogenizer (tight pestle), quickly frozen in liquid nitrogen, and stored in aliquots at −85° until use.

The concentration of membrane protein is measured with the Peterson protein assay using bovine serum albumin as a standard. One gram of membranes is typically obtained from 400 g wet weight of muscle. The specific activity of DHP receptors in the membranes is determined using (+)-[^{3}H]PN200-110 as a radioligand as described previously.[16]

Solubilization and Purification

Rationale

The presence of a high density of L-type calcium channels having high affinity DHP receptor sites in transverse tubule membrane preparations provides an enriched preparation for biochemical isolation of these channels. The typical specific binding capacity of approximately 10 to 20 pmol/mg for the partially purified T-tubule preparations currently used in

---

[16] B. M. Curtis and W. A. Catterall, *Biochemistry* **23**, 2113 (1984).

purifications corresponds to 0.4 to 0.8% of the theoretical value of 2500 pmol/mg for a protein of 400 kDa having one high affinity DHP receptor site. Therefore, it was anticipated in development of a purification scheme that conventional purification methods would be sufficient to yield a highly purified preparation.

The purification procedures for skeletal muscle calcium channels were developed based on previous studies of brain calcium channels.[17,18] These studies showed that high affinity DHP receptor sites could be solubilized in good yield from rat brain membranes with the nonionic detergent digitonin. The previously bound DHP label, [^{3}H]nitrendipine, remained bound after solubilization and dissociated very slowly. Analysis of the solubilized DHP receptor showed that it was a glycoprotein which bound to WGA (wheat germ agglutinin)-Sepharose and that the detergent–DHP receptor complex was a large structure with a sedimentation coefficient of 19 S. These characteristics of the brain DHP receptor formed the basis for purification of the more abundant skeletal muscle calcium channel to near homogeneity. The transverse tubules of skeletal muscle form an intracellular invagination of the plasma membrane and contain few glycoproteins. Affinity chromatography on WGA-Sepharose therefore is a powerful purification method. In addition, the large size of the digitonin-solubilized DHP receptor sites results in clear resolution from other glycoproteins of the transverse tubules by sucrose gradient sedimentation. Sequential application of these purification methods yields a nearly homogeneous calcium channel preparation.[15,16,19]

Reagents. (+)-[^{3}H]PN200-110 (70–80 Ci/mmol) was obtained from New England Nuclear (Boston, MA), BAY K8644 was kindly provided by Dr. Alexander Scriabine (Miles, New Haven, CT), digitonin was from ICN, wheat germ agglutinin was from Sigma, and CNBr-activated Sepharose 4B and DEAE-Sepharose were from Pharmacia (Piscataway, NJ). Wheat germ agglutinin-Sepharose was prepared as described[20] at a substitution level of 10 mg lectin/ml swollen gel.

Buffers

Buffer A: 10 mM HEPES–Tris, pH 7.4, 185 mM KCl, 0.5 mM CaCl$_2$, 1% (w/v) digitonin
Buffer B: Buffer A containing 0.1% (w/v) digitonin
Buffer C: 20 mM Tris-HCl, pH 6.9, 20 mM NaCl, 0.5 mM CaCl$_2$
Buffer D: 9 volumes of buffer C, 1 volume of buffer A

[17] H. Glossmann and D. R. Ferry, *Naunyn-Schmiedebergs Arch. Pharmacol.* **323**, 279 (1983).
[18] B. M. Curtis and W. A. Catterall, *J. Biol. Chem.* **258**, 7280 (1983).
[19] J. Striessnig, K. Moosburger, A. Goll, D. R. Ferry, and H. Glossmann, *Eur. J. Biochem.* **161**, 603 (1986).
[20] R. P. Hartshorne and W. A. Catterall, *Proc. Natl. Acad. Sci. U.S.A.* **78**, 4620 (1981).

Buffer E: 9 volumes of buffer C, 1 volume of buffer A, 280 mM NaCl
Buffer F: 5 mM MOPS, pH 7.4, 0.1% (w/v) digitonin

Protease inhibitors are added from stock solutions to the same concentrations employed for membrane isolation.

Procedures

Prelabeling with (+)-[^{3}H]PN200-110. To follow the calcium channel throughout the purification procedure its DHP receptor site is reversibly labeled with the calcium channel blocker (+)-[^{3}H]PN200-110 before solubilization. Membranes (0.5–1 g) suspended in buffer MB are thawed on ice, and (+)-[^{3}H]PN200-110 is then directly added from the 1 μCi/μl stock solution. The mixture is incubated in the dark for 90 min on ice. The prelabeled membranes are collected by centrifugation at 150,000 g for 35 min. Typically, these conditions result in labeling of 4 to 8% of the DHP receptor sites. The (+)-[^{3}H]PN200-100–DHP receptor complex is very stable at 4° ($t_{1/2} > 5$ hr). Provided all steps are carried out at a temperature of 4° or less, only a small fraction of the calcium channels must be labeled to recover sufficient radiolabeled calcium channel after four steps of purification. For reconstitution studies, the DHP calcium channel activator BAY K8644 is bound to the remaining DHP receptor sites not occupied with (+)-[^{3}H]PN200-110 by incubation with a final concentration of 3μM BAY K8644 for 15 min on ice. Three micromolar BAY K8644 is then included in all buffers through the remainder of the purification steps.

Owing to the light sensitivity of the dihydropyridines, samples are protected from UV light during all further manipulations. Typical fluorescent lighting contains little UV and does not cause significant inactivation of PN200-110 or BAY K8644.

Solubilization. The prelabeled T-tubule membrane pellet is homogenized in buffer A by gently drawing it up and down in a 5-ml plastic pipette tip to give a final digitonin to protein ratio of 5:1 (w/w) (100 ml of buffer A/0.2 g of membrane protein). After incubation for 30 min on ice with stirring, insoluble protein is removed by centrifugation as above for 45 min.

Affinity Chromatography on WGA-Sepharose. The solubilized DHP receptor is passed through a column containing 20 ml of WGA-Sepharose equilibrated in buffer B at a flow rate of 8–10 ml/min. The column is then washed at maximal flow rate with 10–20 bed volumes of buffer B to remove unbound protein. Bound receptor is biospecifically eluted with buffer B containing 3.8% (w/v) N-acetyl-D-glucosamine at a flow rate of 0.3 ml/min. DHP receptor-bound radioactivity in 50-μl aliquots of the eluted fractions is determined by liquid scintillation counting.

DEAE-Sephadex Ion-Exchange Chromatography. Fractions containing DHP binding activity are pooled (5–10 ml) and diluted with 0.1% (w/v) digitonin in distilled water to a conductivity equivalent to that of buffer D. The diluted mixture is batch incubated for 30 min with 10 ml of DEAE-Sephadex equilibrated in buffer D. After transfer of the mixture to a column, the binding of the DHP receptor to the resin is verified by quantification of unbound radioactivity in the flow-through fractions. The resin is washed with 10 bed volumes of buffer D at maximum flow rate. Bound protein is eluted by a linear gradient from buffer D to buffer E at a flow rate of 1 ml/min. Eluted fractions are collected and counted for radioactivity as above.

Concentration by WGA-Affinity Chromatography. The most active fractions are pooled and batch incubated for 30 min with 10 ml of WGA-Sepharose equilibrated in buffer B. The mixture is transferred to a column and washed quickly with 10 bed volumes of buffer B. Bound protein is eluted with 4.4% (w/v) N-acetyl-D-glucosamine as above and collected in 50-μl fractions.

Sucrose Density Gradient Centrifugation. Sucrose gradients from 5 to 20% (w/w) sucrose are prepared in Beckman (Fullerton, CA) Quickseal tubes in buffer G. Two to four milliliters of the pooled DHP receptor eluted from the second WGA column is loaded on each sucrose gradient. After centrifugation for 90 min in a Beckman VTi50 rotor at 210,000 g (50,000 rpm), 1.5 to 2-ml fractions are collected at a flow rate of 2 ml/min from the bottom of the tubes. Active fractions are pooled, quickly frozen, and stored in liquid nitrogen until use.

Partial Purification Procedures

For studies which do not require completely purified calcium channels or which require maximum retention of calcium conductance activity, a rapid two-step purification employing WGA affinity chromatography and ion-exchange chromatography may be substituted for the complete four-step procedure. Such preparations contain contaminating polypeptides migrating with apparent molecular weights of 110K and 300–400K and exhibit maximal binding activities of 600–1000 pmol/mg of protein.[16]

Subunit Composition and Properties

The purified L-type calcium channel from skeletal muscle is a complex of several subunits (reviewed in Catterall *et al.*[21] Figure 1A is a collage of

---

[21] W. A. Catterall, M. J. Seagar, and M. Takahashi, *J. Biol. Chem.* **263**, 3535 (1988).

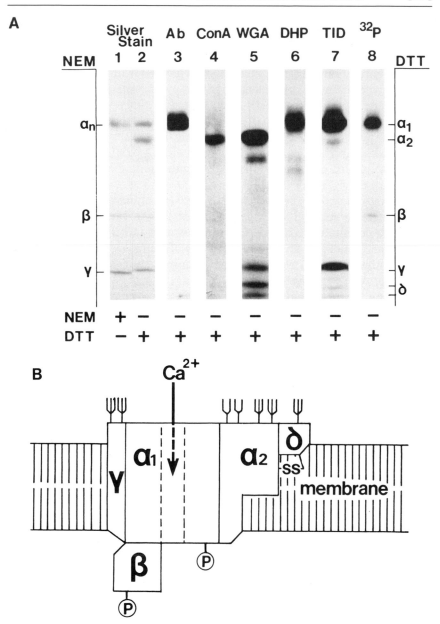

Fig. 1. (A) SDS gel analysis of calcium channel subunits. (B) Transmembrane complex of subunits.

several sodium dodecyl sulfate (SDS) gel analyses which summarize the biochemical properties of the calcium channel subunits.[15] Silver staining of the polypeptide components of the purified calcium channel after SDS gel electrophoresis without reduction of disulfide bonds [i.e., in the presence of N-ethylmaleimide (NEM) but absence of dithiothreitol (DTT)] reveals three major size classes of protein subunits: α subunits of 175K, β subunits of 54K, and γ subunits of 32K (Fig. 1A, lane 1). Similar analysis after reduction of disulfide bonds reveals two high molecular weight polypeptide components, α1 of 175K and α2 of 143K, which comigrate with an apparent molecular weight of 170K before reduction of disulfide bonds (Fig. 1A, lane 2). This unusual behavior of the α2 subunit results from its disulfide linkage to δ subunits of 24K and 27K. These two proteins do not stain well with silver (Fig. 1A, lane 2), but they are easily visualized with glycoprotein labeling methods such as binding of ^{125}I-labeled wheat germ agglutinin (Fig. 1A, lane 5). The α1, α2, β, γ, and δ subunits are coimmunoprecipitated as a complex by antibodies against each of the individual subunits and therefore are specifically associated polypeptide components of the L-type calcium channel.

The different biochemical properties of the calcium channel subunits can be revealed by specific labeling methods. The lectin concanavalin A, which binds to high mannose carbohydrate chains on glycoproteins, labels only the α2 subunit (Fig. 1A, lane 4). Wheat germ agglutinin, which binds to complex carbohydrate chains, labels the α2, γ, and δ subunits (Fig. 1A, lane 5). Thus, the α2, γ, and δ subunits are glycoproteins; approximately 30% of their apparent mass is due to carbohydrate residues.[15] The DHP receptor sites of L-type calcium channels can be specifically labeled by the photoreactive DHP azidopine. Only the α1 subunit is labeled by azidopine (Fig. 1A, lane 6), placing the DHP receptor site on that subunit. Similar experiments show that the receptor sites for phenylalkylamines and benzothiazepines are also located on α1. The transmembrane segments of membrane proteins can be specifically labeled with the hydrophobic photoaffinity probe 3-(trifluoromethyl)-3-(m-iodophenyl)diazirine (TID). TID is incorporated at high levels into the α1 and γ subunits and at lower level into the α2 subunit (Fig. 1A, lane 7). These results indicate that the α1 subunit is the principal transmembrane subunit in the calcium channel complex and that the α2 and γ subunits also are integral membrane proteins.

The activation of L-type calcium channels in cardiac and skeletal muscle is enhanced by activation of cAMP-dependent protein kinase in intact muscle cells. The α1 and β subunits of the purified calcium channel complex are both substrates for cAMP-dependent protein kinase (Fig. 1A, lane 8). The relative phosphorylation of these two polypeptides varies with experimental conditions; incorporation into the β subunits ranges from 20 to 80% of that of the α1 subunit. Evidently, phosphorylation of one or both of these two subunits is important in the regulation of calcium channel activation through the cAMP second-messenger pathway.

These biochemical properties of the subunits of the purified calcium channel outlined above indicate that they are arranged in a transmembrane complex as illustrated in Fig. 1B. The α1 subunit is the central transmembrane subunit of the complex and interacts with a disulfide-linked glycoprotein dimer of α2 and δ subunits, an intracellularly disposed, phosphorylated β subunit, and a transmembrane glycosylated γ subunit.

The conclusions of these biochemical studies are supported by the primary structures of the subunits deduced from cDNA cloning and sequencing experiments. The predicted primary structure of the α1 subunit reveals a protein of 212K with four homologous domains which each contain six predicted transmembrane segments.[22] Its primary structure is homologous to that of the sodium channel α subunit, and expression of the α1 subunit alone is sufficient to direct the synthesis of functional calcium channels in mammalian cells.[23] Antipeptide antibodies directed against amino acid sequences in the carboxy-terminal domain of the α1 subunit show that there are two forms of the α1 subunit in purified calcium channel preparations, isolated T-tubule membranes, and intact skeletal muscle cells in culture.[24,25] The major 175K form of the α1 subunit (Fig. 1, lane 2) does not contain the carboxy terminus encoded by the α1 gene.[24] A minor 212K form is recognized by antibodies directed against the carboxy terminus and comprises no more than 5% of the calcium channel α1 subunits in purified preparations. The functional significance of the two forms of the α1 subunit is not known at present, but they differ in potential sites of phosphorylation by cAMP-dependent protein kinase, suggesting that they may have different regulatory properties.

The gene encoding the α2 subunit encodes a protein of 1106 amino acid residues with multiple hydrophobic segments, multiple sites of N-linked glycosylation, and no homology to other known proteins.[26] Amino acid sequence analyses and studies with antipeptide antibodies show that

[22] T. Tanabe, H. Takeshima, A. Mikami, V. Flockerzi, H. Takahashi, K. Kangawa, M. Kojima, H. Matsuo, T. Hirose, and S. Numa, *Nature (London)* **328**, 331 (1987).

[23] E. Perez-Reyes, H. S. Kim, A. E. Lacerda, W. Horne, X. Y. Wei, D. Rampe, K. P. Campbell, A. M. Brown, and L. Birnbaumer, *Nature (London)* **340**, 233 (1989).

[24] K. S. De Jongh, D. K. Merrick, and W. A. Catterall, *Proc. Natl. Acad. Sci. U.S.A.* **86**, 8585 (1989).

[25] Y. Lai, M. J. Seagar, M. Takahashi, and W. A. Catterall, *J. Biol. Chem.* **265**, 20839 (1990).

[26] S. B. Ellis, M. E. Williams, N. R. Ways, R. Brenner, A. H. Sharp, A. T. Leung, K. P. Campbell, E. McKenna, W. J. Koch, A. Hui, A. Schwartz, and M. M. Harpold, *Science* **241**, 1661 (1988).

the δ subunits are encoded by the same gene as $\alpha2$.[27] The $\alpha2$ subunit contains residues 1 to 934, whereas the δ subunit contains residues 935 to the carboxy terminus of the $\alpha2\delta$ precursor. The 24K and 27K forms of the δ subunit have the same amino terminus and are likely to differ in their extent of glycosylation. Evidently, the $\alpha2\delta$ precursor is disulfide linked and is cleaved to yield the $\alpha2$ and δ subunits during biosynthesis and assembly of the calcium channel complex.

The β and γ subunits are encoded by two additional genes.[28,29] The β subunit is a uniformly hydrophilic protein of 524 amino acid residues with multiple potential sites for protein phosphorylation. The γ subunit is a protein of 222 amino acid residues with three predicted transmembrane segments and multiple potential sites of N-linked glycosylation. The functional roles of the $\alpha2$, β, γ, and δ subunits of the calcium channel complex are not defined at present, but the phosphorylation of the β subunits of the calcium channel suggest that they may be involved in channel regulation.

Reconstitution

Rationale

The purified skeletal muscle calcium channel binds the major classes of calcium channel ligands including DHPs, verapamil, and diltiazem, but other aspects of calcium channel function cannot be examined with solubilized channels in detergent solution. To assess the calcium conductance activity of purified calcium channels and their regulation by membrane voltage, calcium channel modulating drugs, and protein phosphorylation, they must be returned to a phospholipid bilayer membrane across which calcium conductance can be measured.

The ion conductance activity of purified and reconstituted calcium channels has been studied in two ways. Purified calcium channels reconstituted in phospholipid vesicles have been studied by measurement of $^{45}Ca^{2+}$ influx into vesicles containing calcium channels activated by binding of the DHP calcium channel activator BAY K8644.[30,31] Because skeletal muscle calcium channels mediate calcium currents that last for seconds, their activity can be effectively measured by relatively slow ion flux meth-

[27] K. S. De Jongh, C. Warner, and W. A. Catterall, *J. Biol. Chem.* **265**, 14738 (1990).
[28] P. Ruth, A. Röhrkasten, M. Biel, E. Bosse, S. Regulla, H. E. Meyer, V. Flockerzi, and F. Hofmann, *Science* **245**, 1115 (1989).
[29] S. D. Jay, S. B. Ellis, A. F. McCue, W. E. Williams, T. S. Vedvick, M. M. Harpold, and K. P. Campbell, *Science* **248**, 490 (1990).
[30] B. M. Curtis and W. A. Catterall, *Biochemistry* **25**, 3077 (1986).
[31] K. Nunoki, V. Florio, and W. A. Catterall, *Proc. Natl. Acad. Sci. U.S.A.* **86**, 6816 (1989).

ods. The ion flux method has low time resolution and limited voltage control, but it has the advantage that the functional properties of all of the purified calcium channels can be assessed. Purified calcium channels have also been incorporated into planar phospholipid bilayers by fusion of reconstituted phospholipid vesicles or by direct incorporation from detergent solution.[32,33] Calcium currents mediated by single calcium channels can then be recorded at different membrane potentials. This method provides improved time resolution, excellent voltage control, and experimental access to both sides of the calcium channel protein, but correlation of the functional properties of the recorded single channels with the biochemical properties of the purified calcium channel population as a whole is very difficult. Thus, the ion flux and single-channel recording methods provide complementary information on the functional properties of purified calcium channels.

Preactivation with BAY K8644

Calcium channels are more stable during detergent solubilization and purification if their DHP receptor site is occupied. To increase the probability of purification of the calcium channel in an active state for ion conductance, calcium channel preparations to be used for reconstitution are usually preactivated by binding of BAY K8644 and then purified in the continuous presence of BAY K8644 or a similar DHP calcium channel activator.[30] This procedure is necessary for consistent recovery of functional purified calcium channels.

Reconstitution into Phospholipid Vesicles

Calcium Channel Preparations. Following elution from WGA-Sepharose, skeletal muscle calcium channels are approximately 40% pure, as judged by silver staining of polyacrylamide gels. The channels are highly concentrated at this stage (0.2 mg/ml channel protein) and can be reconstituted to yield a vesicle preparation with a relatively large specific calcium flux activity. However, attempts to measure calcium influx after phosphorylation by cAMP-dependent protein kinase result in marked stimulation by ATP alone, possibly reflecting contamination with an endogenous kinase or the ATP-sensitive calcium release channel. Further purification of the WGA eluate on DEAE-Sephadex eliminates this problem and yields a

[32] V. Flockerzi, H. Oeken, F. Hofmann, D. Pelzer, A. Cavalié, and W. Trautwein, *Nature (London)* **323,** 66 (1986).
[33] L. Hymel, J. Striessnig, H. Glossmann, and H. Schindler, *Proc. Natl. Acad. Sci. U.S.A.* **85,** 4290 (1988).

preparation that is approximately 80% pure[31] but less concentrated (~30 μg/ml). We have used these preparations for most of our reconstitution studies because further purification by sucrose gradient centrifugation causes a large dilution, introduction of a high sucrose concentration which interferes with reconstitution and ion flux measurement, and loss of calcium flux activity.

Lipid Preparation. A number of different lipid combinations and sources have been examined in the reconstitution procedure described below. The best results have been obtained with egg phosphatidylcholine from Avanti Polar Lipids (Birmingham, AL). Cholesterol, cardiolipin, soybean asolectin, and phosphatidylserine should be avoided (in any proportion) as they do not improve reconstitution but cause high nonspecific calcium binding or flux. CHAPS is used as the solubilizing detergent because it efficiently dissolves the phosphatidylcholine without significant inactivation of the calcium channel.[30] To prepare the lipid–detergent mixture, 100 mg of phosphatidylcholine is dried to a fine powder under nitrogen with stirring (3–6 hr). 150 mg of CHAPS dissolved in 1 ml of 0.1 M triethylammonium (TEA)–MOPS, pH 7.0, are added, and the mixture is sonicated to clarity under nitrogen. Then 6.8 ml of 20 mM $Ca(CH_3SO_3)_2$, pH 7.0, and 2.2 ml of 50% sucrose are added, and the aliquots are stored at $-20°$ under nitrogen.

Buffers

Reconstitution buffer: 10 mM TEA–MOPS (pH 7.0), 10 mM $Ca(CH_3SO_3)_2$ (pH 7.0), 0.32 M sucrose, 3 μM BAY K8644. Adjust to about 400 mOsmol/liter.

Sucrose buffer: 10 mM TEA–MOPS (pH 7.0), 0.34 M sucrose, 3 μM BAY K8644. Adjust to same osmolality as reconstitution buffer.

Procedure. Reconstitution is performed at 4° in a darkened room by removal of detergent via gel filtration on a column (1 × 25 cm) of Sephadex G-50–150. The column is first preequilibrated by applying a mixture of 2.2 ml of the above lipid solution plus 0.8 ml of 0.1% digitonin (ICN, Cleveland, OH), eluting with 25 ml reconstitution buffer, and repeating this procedure twice. The column is then washed with an additional 50 ml of reconstitution buffer. It is only necessary to preequilibrate the column once; column lifetimes for repetitive uses are generally 1 to 2 months, or 20–50 runs. The column should be washed with at least 50 ml of reconstitution buffer between each run and should be left in 0.02% NaN_3 after each experiment.

Purified calcium channels (0.8 ml; 40–400 pmol) are gently mixed with 2.2 ml of lipid–detergent solution, to which 3 μM BAY K8644 has been added, and applied to the column. The column is eluted with reconstitution buffer at 0.5 ml/min, and 1-ml fractions are collected in polystyrene tubes. The void volume is marked by three turbid fractions. These three fractions contain the calcium channel, as detected by specifically bound [³H]PN200-110, and the phospholipid vesicles, as shown by light scattering, inorganic phosphate measurement, and the internal volume measurement described below. The yield of reconstituted calcium channels is generally 50–75%, and their subunit composition is the same as that of the applied detergent-solubilized channels.[31] The extent of incorporation of calcium channels into vesicles can be measured by retention of bound [³H]PN200-110 on GF/F filters, cosedimentation of channels with phospholipid vesicles on sucrose gradients,[30] and by the inability of the reconstituted calcium channels to bind to WGA-Sepharose (V. Florio, and W. Catterall, 1989, unpublished results). The reconstituted vesicles cannot be frozen and thawed with retention of specific ion flux activity, but they have a functional half-life of about 18 hr on ice.

Calcium Influx Measurement

Reagents and Materials

Chelex buffer: 10 mM TEA–MOPS (pH 7.0), 0.34 M sucrose, 1 mg/ml bovine serum albumin (Sigma, Fraction V; do not use the crystallized product). Adjust to same osmolality as reconstitution buffer.

$^{45}CaCl_2$: 10 mCi/ml from New England Nuclear

Chelex columns: The Chelex columns are used in calcium flux measurements to bind free $^{45}Ca^{2+}$ while allowing vesicles to elute in the excluded volume. They are made from 5.75-inch Pasteur pipettes to which are attached the cut ends of 1-ml pipette tips. A small amount of glass wool is added as a flow screen, and wet Chelex 100 resin (Bio-Rad, Richmond, CA; 100–200 mesh) is added to a volume of 1.4 ml. The columns are equilibrated with 5 ml of Chelex buffer at room temperature. It is possible to regenerate these columns by washing immediately after use with 4 ml of 0.2 M EDTA followed by 12 ml of deionized water. These columns bind approximately 99.95% of free $^{45}Ca^{2+}$ under the assay conditions described below.

To obtain measurable uptake of $^{45}Ca^{2+}$ into vesicles, it is necessary to reconstitute in the presence of a high concentration (10 mM) of unlabeled calcium and then remove most of the extravesicular calcium. The resultant outward gradient of calcium ion concentration causes the added $^{45}Ca^{2+}$ to concentrate inside vesicles as it exchanges for cold calcium and isotopic equilibrium is approached.[30,34,35] Extravesicular calcium is removed by rapid gel filtration on prespun columns. The columns consist of standard 3-ml syringes with nylon mesh disks (4 layers of Nitex) as flow screens. Sephadex G-50–150 is added to a packed volume of 2.5 ml, and the columns are equilibrated with 10 ml of sucrose buffer. Just before use, the columns are centrifuged at 1000 g for 90 sec. Aliquots of vesicles (0.2 ml) are applied to each column, and the columns are inserted into 3-ml Beckman Bio-vials and centrifuged as above. Dilution of the vesicles in this procedure is approximately 25%.

To provide a measure of nonspecific calcium influx and binding to the reconstituted vesicles, some samples are incubated for 90 min at 4° with a calcium channel blocker such as verapamil prior to the ion flux measurement. Following this incubation period, 0.4-ml aliquots of each vesicle sample are prewarmed to 37° for 90 sec, and 0.4 ml of prewarmed sucrose buffer containing 8 μCi of $^{45}Ca^{2+}$ is added. Aliquots (190 μl) are removed at 15, 30, 60, and 120 sec, applied to small columns of Chelex resin prepared as described above, and eluted with 1 ml Chelex buffer into 20-ml scintillation vials. Ten milliliters of scintillation fluid are added, and samples are left in the dark for 1 hr before counting. Blanks obtained from incubations in the absence of reconstituted vesicles are subtracted from all data. Nonspecific uptake is determined in the presence of a saturating concentration of an L-type calcium channel blocker such as verapamil or on equivalent samples of vesicles reconstituted in the absence of calcium channels. A relative measurement of the total internal vesicular volume accessible to $^{45}Ca^{2+}$ is obtained by performing an uptake assay for 5 min in the presence of 5 μM A23187. Uptake into different batches of reconstituted vesicles is then corrected for small differences in internal volume.

Phosphorylation of Vesicles

Activation of L-type calcium channels in cardiac and skeletal muscle cells is increased when cAMP-dependent protein kinase is activated. This regulation is due to direct phosphorylation of the $\alpha 1$ and β subunits of the calcium channel complex and can be studied by phosphorylation of reconstituted preparations of calcium channels with cAMP-dependent protein kinase.[31–33] Because high concentrations of calcium inhibit cyclic AMP-dependent protein kinase (PKA), extravesicular calcium must be removed before phosphorylation. A rapid means of accomplishing this is to apply

[34] J. A. Talvenheimo, M. M. Tamkun, and W. A. Catterall, *J. Biol. Chem.* **257**, 11868 (1982).
[35] H. Garty, B. Rudy, and S. J. D. Karlish, *J. Biol. Chem.* **258**, 13094 (1983).

the vesicles (3 ml) to a 2-ml column of Chelex that has been extensively equilibrated with sucrose buffer. The column is then centrifuged at 500 g for 10 sec. The eluted vesicles are phosphorylated for 1 hr at 4° in a reaction mixture containing 5–50 pmol of channel, 5 mM MgCl$_2$, 0.28 μM catalytic subunit of PKA, 1 mM ATP, and 0.1% 2-mercaptoethanol. A high concentration of ATP is required, possibly because the phosphorylation must be carried at low temperature to preserve calcium channel activity. The reaction is terminated by gel filtration on spun columns as described above. Control samples with kinase alone, ATP alone, and no additions should be included in the experimental design. With impure calcium channel fractions, ATP alone can have a marked effect on calcium influx owing to endogenous kinase activity or to ATP-gated channels. With more highly purified calcium channel preparations, samples treated with ATP alone typically display slightly higher flux than buffer-only or kinase-only controls.[31]

Functional Properties of Reconstituted Calcium Channels

Figure 2 illustrates the uptake of ^{45}Ca^{2+} into vesicles containing purified calcium channels from skeletal muscle. Calcium influx is low when the reconstituted vesicles are not phosphorylated or when PKA is added without ATP (Fig. 2A, squares). This low level of calcium influx is generally 2- to 3-fold higher than that in protein-free vesicles and is blocked almost completely by verapamil and other L-type calcium channel blockers (Ref. 30 and W. A. Catterall, unpublished experiments). A slight increase in calcium influx is observed with ATP alone (Fig. 2A, triangles), but a large increase is obtained by treatment with both ATP and PKA (open circles). The high level of calcium influx observed after phosphorylation by PKA is inhibited nearly 90% by the membrane-impermeant phenylalkylamine D890 (Fig. 2A, filled circles), indicating that the phosphorylated calcium channels are oriented inside-out in the vesicle bilayer as expected. Phosphorylation of these reconstituted inside-out channels can increase calcium influx up to 10-fold in favorable preparations of calcium channels (Fig. 2 and Ref. 31), but a range of stimulation from 1.5- to 10-fold is observed in different calcium channel preparations. The factors causing this variability in results have not been clearly defined.

Calcium influx is inhibited by agents known to block L-type calcium channels *in vivo* at concentrations similar to those that are active in intact muscle cells. The inhibition of calcium flux by organic calcium channel antagonists of the DHP and phenylalkylamine classes shows the same order of potency (PN200-110 > D890 ≈ D600 > verapamil) observed in intact preparations (Fig. 2B). Specific calcium flux into channel-containing

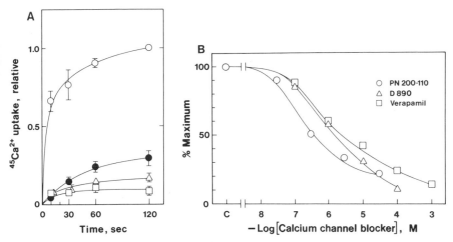

FIG. 2. Uptake of $^{45}Ca^{2+}$ in vesicles. (A) Uptake versus time. (B) Inhibition of calcium flux.

vesicles is also blocked by inorganic cations with the same order potency ($Cd^{2+} > Ni^{2+} > Mg^{2+}$) that is seen in intact muscle fibers.[36]

Reconstitution for Single-Channel Analysis

L-type calcium channel activity has been observed at the single-channel level after reconstitution of purified channel preparations into planar bilayers. Flockerzi et al.[32] added a phosphatidylethanolamine–phosphatidylserine–cholesterol mixture to calcium channels purified in digitonin and dialyzed the mixture against calcium-containing buffer for 16 hr and calcium-free buffer for 34 hr. The resulting vesicles were added to a solution on which a phosphatidylethanolamine–phosphatidylserine–cholesterol–n-hexane monolayer had been formed, and bilayers were formed on the tips of patch micropipettes placed in contact with the surface of the bath. Single calcium channels with a conductance of 20 pS were observed that were inhibited by organic calcium channel antagonists and activated more completely after phosphorylation by PKA.

In a more recent study,[33] vesicles were prepared by diluting purified calcium channels in 0.1% digitonin 400-fold into a dilute (10 μg/ml) suspension of crude soybean lipid with sonication. The suspension was then swirled with glass beads over a dried film of soybean lipid and cholesterol (6:1). The resulting vesicles were incorporated into planar

[36] P. T. Palade and W. Almers, *Pfluegers Arch.* **405**, 91 (1985).

bilayers by the septum-supported, vesicle-derived bilayer technique, which has been reviewed by Schindler elsewhere in this series.[37] In the planar bilayer configuration, the ion conductance activity of purified calcium channels is modulated by calcium channel activators and inhibitors as expected. Activation of the channel by Bay K8644 required phosphorylation by cAMP-dependent protein kinase, and the level of channel activity was greatly increased by phosphorylation as observed in ion flux experiments. However, in the planar bilayer system, individual ion conductance events ranged in size from 0.9 to 60 pS, in contrast to the 20-pS values observed with patch pipette methods. Nevertheless, the results of ion flux, patch pipette, and macroscopic bilayer methods together show that purified skeletal muscle calcium channels retain selective ion conductance activity that is modulated appropriately by calcium channel-modulating drugs and protein phosphorylation.

[37] H. Schindler, this series, Vol 171, p. 225.

[36] Purification and Reconstitution of Nicotinic Acetylcholine Receptor

By AMITABH CHAK and ARTHUR KARLIN

Introduction

The nicotinic acetylcholine receptor (AChR) was the first receptor, and certainly the first ligand-gated channel, to be purified (see Ref. 1 for review). This success during 1973 and 1974 was due largely to the availability of very rich sources, namely, the electric tissues of electric rays (family Torpedinidae) and the electric eel *(Electrophorus electricus),* and to the availability of reversible and irreversible ligands for the acetylcholine binding sites. Also, at that time there was increasing awareness of the properties of amphipathic, integral membrane proteins, of the hydrophobic effect, and of detergents.[2] It became clear that the techniques used to purify water-soluble proteins could be applied to membrane proteins in detergent complexes. The minimal criteria for the choice of detergent were that the protein was efficiently solubilized and that at least one assayable property, binding in the case of receptors, was preserved.

For receptors like the AChR that are ligand-gated channels, although

[1] A. Karlin, *in* "The Cell Surface and Neuronal Function" (C. W. Cotman, G. Poste, and G. L. Nicolson, eds.), p. 191. Elsevier/North-Holland Biomedical, Amsterdam, 1980.
[2] C. Tanford, "The Hydrophobic Effect," 2nd Ed. Wiley, New York, 1980.

binding is preserved in detergent solution, the critical functional properties of gating and ion conduction are observed only in membranes. Therefore to assay these functions in purified ligand-gated channels, one must reincorporate them into artificial membranes. Also, studies of receptor topology and interactions with other membrane proteins and lipids require either native membranes or reconstituted membranes. We describe the preparation, starting with *Torpedo californica* electric tissue, of AChR-rich native membranes, solubilized and purified AChR, and AChR reconstituted in liposomes. Acetycholine receptor subtypes have also been purified from other tissues (e.g., muscle and neuronal tissue) and from many other species (see Ref. 3 for review). Recently, many subtypes have been characterized genetically rather than biochemically (see Ref. 4 for review).

Electric Tissue

Live rays[5] are dissected, and the tissue is stored in liquid nitrogen. Alternatively, tissue can be purchased already frozen. In the former case, working in the cold room and wearing rubber gloves, we remove the ray from its container and immediately pith it. The two electric organs fill the "wings." A scalpel is used to cut around the edges of the kidney-shaped organs. The dermal layer on the dorsal side is peeled off with a surgical clamp, following which the organ is shelled off the ventral surface by blunt dissection. The electric organs are weighed, placed on an enameled metal tray, and cut into approximately 2 cm chunks, which are dropped one by one into liquid nitrogen for quick freezing. We have stored tissue under liquid nitrogen for several months. A medium-sized ray yields 600 to 1200 g of electric tissue.

Preparation of Membranes Rich in Nicotinic Acetylcholine Receptor

For the preparation of AChR-rich membranes,[6] approximately 120 g of frozen *Torpedo* tissue is thawed in 300 ml of PE buffer.[7] Subsequent

---

[3] J. Lindstrom, R. Schoepfer, and P. Whiting, *Mol. Neurobiol.* **1**, 281 (1987).
[4] T. Claudio, *in* "Frontiers of Neurobiology: Molecular Neurobiology" (D. Glover and D. Hames, eds.), p. 63. IRL, London, 1989.
[5] Pacific Biomarine (Venice, CA) or Marinus (Long Beach, CA).
[6] C. Czajkowski, M. DiPaola, M. Bodkin, G. Salazar-Jimenez, E. Holtzman, and A. Karlin, *Arch. Biochem. Biophys.* **272**, 412 (1989).
[7] Abbreviations for solutions: PE buffer is 10 mM sodium phosphate, 1 mM EDTA, pH 7; NPxxx is xxx mM NaCl, 10 mM sodium phosphate, 1 mM EDTA, 0.02% NaN$_3$, pH 7.0, unless otherwise specified in parentheses, as, for example, NP100 (pH 8.0); TNPxxx is 0.2% Triton X-100 in NPxxx; 2CNPxxx (pH 7.4) is 2% sodium cholate in NPxxx (pH 7.4); 0.5 CNPxxx (pH 7.4) is 0.5% sodium cholate in NPxxx (pH 7.4); B buffer is 0.2% Triton X-100, 10 mM NaCl, 10 mM MOPS, pH 7.4.

operations are carried out at 4°. The mixture containing nearly thawed tissue is placed in a food processor[8] and minced by four bursts of 30 sec each. Most of the foam is allowed to settle. One-third of the suspension at a time is homogenized at full speed, twice, for 1 min, in a 500-ml flask of a homogenizer[9] fitted with two sharpened blades. The resulting suspension (H1) is centrifuged in two JA-14 rotor bottles[10] at 6000 rpm for 10 min. The supernatants are pooled and saved on ice. The pellets are suspended in 100 ml of PE buffer and rehomogenized in a 250-ml flask,[9] at full speed, twice, for 1 min. This supernatant is added to the saved pool, which is then filtered through eight layers of cheesecloth. The pellets (P1) are discarded. The supernatant (S1) is centrifuged in 26-ml polycarbonate bottles in a 60 Ti rotor[10] at 45,000 rpm for 40 min. The resulting supernatant (S2) is discarded. The pellets (P2) are suspended in 100 ml of 40% (w/v) sucrose in PE buffer and homogenized in a 250-ml flask at full speed twice for 1 min. This homogenate (H2) is centrifuged in polycarbonate bottles in a 60 Ti rotor at 60,000 rpm for 140 min. The resulting supernatants (S3) are discarded. The pellets (P3) are suspended in 40 ml PE buffer and homogenized in a 100-ml flask,[9] with one blade on the shaft, at full speed for 1 min. Homogenized P3 is frozen and stored in liquid nitrogen in 4-ml aliquots.

Discontinuous sucrose density gradients with 12 ml of 44% (w/v) sucrose in PE buffer under 10 ml of 40% (w/v) sucrose are made in 60 Ti rotor polycarbonate bottles. P3 is removed from liquid nitrogen and thawed. Eighteen milliliters of P3 is mixed with 3.6 ml of 3 M NaCl in PE buffer to give a final concentration of 0.5 M NaCl. This suspension is passed in and out 10 times through a blunted 22-gauge, 3.5-inch spinal needle attached to a 30-ml plastic syringe. A 3-ml aliquot of this suspension (H3) is layered over each of six discontinuous gradients, and the gradients are centrifuged in a 60 Ti rotor at 60,000 rpm for 140 min. After centrifugation, two bands and a pellet are visible. The second band at the interface of the 40 and 44% sucrose layers is collected by suction from the top. These second bands are pooled and diluted with 4 to 5 volumes of PE buffer to a volume of 150 ml and centrifuged at 60,000 rpm in a 60 Ti rotor for 80 min. The resulting pellets in 4 ml of PE buffer are passed in and out, 10 times, through a 22-gauge needle, and this suspended AChR-rich membrane is stored in liquid nitrogen.

The yield is 1 to 2 mg of protein at a specific activity of about 3 nmol of α-bungarotoxin binding sites per milligram of protein. The protein of this membrane therefore is about 40% AChR. Further purification can be

---

[8] Cuisinart Model DLC-10 Plus.
[9] Homogenizer Model Virtis-45, The Virtis Co., Inc. (Gardiner, NY).
[10] Rotors and centrifuges from Beckman Instruments Inc. (Palo Alto, CA).

obtained by extraction of peripheral membrane proteins with dilute NaOH (pH 11)[11] or with lithium diiodosalicylate,[12] but with consequent increased disordering of the membrane. Also, affinity partitioning has given high specific activities.[13]

Purification of Detergent-Solubilized Receptor

The detergent used to solubilize the AChR should give a good yield while maintaining structural and functional integrity. Triton X-100 efficiently solubilizes the AChR and maintains its snake curarimimetic toxin-binding activity and, with some alteration, its binding of acetylcholine (ACh) and other small ligands.[1] Ligand-gated channel activity of the AChR cannot be recovered once it has been solubilized in Triton X-100, even though the solubilized AChR can be reincorporated into liposomes.[14] For reconstitution of channel activity, solubilization and purification in cholate and phospholipid mixtures are required.[15] For structural studies, the higher yield and purity obtained by purification in Triton X-100 is preferred. We describe solubilization and purification of AChR both in Triton X-100 and in a cholate and phospholipid mixture.

Preparation of Affinity Gel

Bromoacetylcholine Bromide (BAC)[16]. To 18.4 g of choline bromide is added dropwise, over 40 min, with stirring, 24.2 g of bromoacetyl bromide. The viscous mixture is stirred for an additional 75 min in an ice bath. To the mixture is slowly added 75 ml of absolute ethanol. The white crystalline product is recrystallized twice from 400 ml of 2-propanol. The yield is 67% and the product melts at $135° - 136°$. BAC is stored desiccated at $4°$ and is not opened for weighing until at room temperature.

Affinity gel[17]. Affi-Gel 10,[18] 25 ml in isopropanol as supplied, is washed

[11] R. R. Neubig, E. K. Krodel, N. D. Boyd, and J. B. Cohen, *Proc. Nat. Acad. Sci. U.S.A.* **76**, 690 (1979).

[12] J. Elliot, S. G. Blanchard, W. Wu, J. Miller, C. D. Strader, P. Hartig, H.-P. Moore, J. Racs, and M. A. Raftery, *Biochem. J.* **185**, 667 (1980).

[13] G. Johansson, R. Gysin, and S. D. Flanagan, *J. Biol. Chem.* **256**, 9126 (1981).

[14] M. G. McNamee, C. L. Weill, and A. Karlin, *in* "Protein–Ligand Interactions" (H. Sund and G. Blauer, eds.), p. 316. deGruyter, Berlin, 1975.

[15] M. Epstein and E. Racker, *J. Biol. Chem.* **253**, 6660 (1978).

[16] V. N. Damle, M. McLaughlin, and A. Karlin, *Biochem. Biophys. Res. Commun.* **84**, 845 (1978).

[17] We previously used Affi-Gel 401,[18] a crosslinked agarose gel with attached homocysteinyl residues at about 10 μmol per ml packed gel, which we reacted with BAC to form the cholinocarboxymethyl adduct [A. Reynolds and A. Karlin, *Biochem.* **17**, 2035 (1978)]. Bio-Rad recently discontinued manufacture of Affi-Gel 401. We now use Affi-Gel 10, a N-hydroxysuccinimide ester linked to crosslinked agarose by a neutral 10 atom spacer.

[18] Bio-Rad Laboratories, Inc. (Richmond, CA).

in a Buchner funnel with 250 ml of cold water. The gel should not be allowed to dry during this or any subsequent step. The wet gel is transferred to a beaker with 75 ml 13.3 mM cystamine in 100 mM HEPES (pH 8.0). This slurry is shaken for 1 hr at room temperature. The gel is transferred to a Buchner and washed with 50 ml HEPES buffer and 4 × 50 ml water. The gel is mixed in a beaker with 50 ml 20 mM dithiothreitol in 200 mM TRIS buffer (pH 8.0) and shaken for 1 hr. The gel is washed with 600 ml water. Aliquots (100 μl) of the gel and of the supernatant are saved for sulfhydryl assay. The gel is mixed with 50 ml 20 mM BAC in 200 mM NaCl, 50 mM sodium phosphate buffer (pH 7.0) and shaken for 1 hr. The gel is washed with 500 ml water. Before the reaction with BAC, the gel contains about 10 μmol sulfhydryl per ml gel, and after BAC it contains no free sulfhydryls. The affinity gel is made before the membrane extract. The gel can be stored at 4° in 0.02% NaN$_3$ for a few days.

Assay for Sulfhydryl Groups in Gel. A gel sample is pipetted into a tared 12-ml conical glass centrifuge tube and sedimented for 1 min. The supernatant is carefully removed with a pulled out Pasteur pipette, and the packed gel is weighed. The gel is suspended by vortexing in 1 ml of 1 mM 5,5'-dithiobis(2-nitrobenzoic acid) in 200 mM Tris-Cl (pH 8.0) and after 1 min is sedimented. To 100 μl of the supernatant is added 1 ml of 200 mM Tris-Cl (pH 8.0), and the absorbance at 412 nm of this mixture is determined. A control without gel is treated similarly. A molar extinction coefficient of 13,600 is used to determine the sulfhydryl concentration in the cuvette, from which the sulfhydryl titer of the gel can be calculated. If after treatment with BAC the gel contains unreacted sulfhydryl groups, these can be blocked with N-ethylmaleimide.

Crude Membrane Preparation

For the preparation of crude membranes,[19] 600 g of *Torpedo* tissue is partially thawed and added to 2400 ml of 1 mM EDTA, pH 7.4. (If preservation of *Torpedo* AChR dimers is desired, 2 mM N-ethylmaleimide is added here to block free sulfhydryls, preventing the reduction of the disulfide linking the monomers in dimer.) All subsequent steps are carried out at about 4°. Two milliliters of 0.5 M phenylmethylsulfonyl fluoride (PMSF) in ethanol is added, and the tissue is homogenized for 60 sec in a 1-gallon commercial blender[20] set at "low." This homogenate is filtered through four layers of cheesecloth and centrifuged in about 10 JA-14 polycarbonate bottles at 14,000 rpm for 90 min. The supernatant is discarded. The pellets are collected in 1200 ml of the EDTA buffer and

---

[19] C. L. Weill, M. G. McNamee, and A. Karlin, *Biochem. Biophys. Res. Commun.* **61**, 997 (1974).

[20] Waring Products Division (New Hartford, CT).

homogenized in the blender set at "high" for 20 sec. This homogenate is centrifuged in the JA-14 rotor at 14,000 rpm for 90 min. The supernatant is removed carefully from the loose pellets and discarded. The pellets are pooled in a 250-ml graduated cylinder and brought up to a volume of 173 ml with EDTA buffer; 21 ml of 500 mM NaCl, 100 mM sodium phosphate, 10 mM EDTA, 0.2% sodium azide, pH 7, is added. This suspension is homogenized in the 1-liter container for the blender set at "medium" for 20 sec.

Extraction of Membrane in Triton X-100

For Triton X-100 extraction,[19] the crude membrane preparation is stirred vigorously in a 250-ml Erlenmeyer flask, while 50 μl of 0.5 M PMSF in ethanol and 16 ml of 40% Triton X-100 in water is added. The final medium contains 3% Triton X-100, 50 mM NaCl, 10 mM sodium phosphate, 1 mM EDTA, 0.02% azide pH 7. After being stirred for 1 hr at 4°, the mixture is transferred to eight 60 Ti rotor tubes and centrifuged at 60,000 rpm for 30 min. The extract is carefully removed from the pellets and pooled. It can be stored overnight in ice *and* in the refrigerator, or it can be loaded slowly onto the affinity column, for elution the next day.

Affinity Chromatography in Triton X-100

The extract can be adsorbed onto the affinity gel either prior to pouring the column or after pouring the column.[17,19] All operations are carried out in a cold room. In the first case, 24 ml of affinity gel is washed 3 times with 25 ml of TNP50.[7] The affinity gel and extract are stirred together with a magnetic bar at slow speed in a 250-ml beaker for 2 hr. The suspension is transferred to four 50-ml centrifuge tubes and sedimented for 4 min. The supernatant is carefully removed. The gel is washed twice with TNP50. About 25 ml of TNP50 is added to the gel, and the slurry is transferred with a broken-off Pasteur pipette into a 1.5-cm diameter glass column, using more TNP50 to complete the transfer as needed. During this process, eluate is collected at 1 drop/sec at a drop size of 15 μl.

In the second case, the affinity gel is poured into a column in TNP50. The extract is loaded onto this column overnight at a flow rate of 10 to 15 ml/hr, controlled by a peristaltic pump. After the extract is loaded, the column is washed with 200 ml of TNP50 at 60 ml/min.

After loading the extract by either method, the column is washed with 200 ml of TNP100 at 60 ml/hr. Finally, the AChR is eluted with 50 ml of 10 mM carbamylcholine in TNP90 at 30 ml/hr. This eluate is collected in 5-ml fractions, and the AChR elutes in fractions 2 through 7. These fractions in 1-ml aliquots are frozen in liquid nitrogen for assay and use. A typical yield is about 30 mg of AChR with specific activities in the different fractions ranging from 5 to 8 nmol curarimimetic toxin binding sites per milligram of protein. (The theoretical maximum specific activity is 7 nmol sites/mg, but there are inaccuracies in both the binding site and protein assays.)

The above procedure results in pure or nearly pure AChR from *Torpedo californica*. In variations of the above method, slight additional purification and, in addition, resolution of monomeric and dimeric forms of AChR were obtained by sucrose density gradient centrifugation.[17,21] In purifying AChR from less concentrated sources than *Torpedo*, other laboratories have used, in addition to affinity chromatography, chromatography on DEAE-cellulose,[22] on immobilized lectin,[23] and on immobilized antibody.[24] Also, affinity chromatography has been carried out on agarose linked either to other small ligands[21] or to various elapid snake toxins.[25]

Purification in Cholate and Phospholipid Mixture

The following variations of the above procedures preserve the ligand-gated channel function of the AChR, which one can assay on reincorporation of the AChR into membrane.[26] A crude membrane fraction is prepared as above, up to the last sedimentation, except typically at one-sixth the scale, starting with 100 g of tissue. The 1-liter blender container is used in place of the 1-gallon container. The crude membrane pellet is suspended in a final volume of 60 ml of 0.5% asolectin[27] in 2CNP100 (pH 7.4)[7] by homogenization at "low" for 20 sec. The mixture is stirred for 1 hr and then centrifuged at 60,000 rpm for 30 min. The supernatant is used immediately or stored in ice overnight. The BAC–Affi-Gel 10 affinity gel described above is washed with 0.0625% asolectin in 0.5CNP50 (pH 7.4), and 6 ml is poured into a 0.9-cm-diameter glass column. The supernatant is pumped onto the column at 30 ml/hr. The column is washed at the same rate successively with 25 ml of 0.5CNP50 (pH 7.4) and 50 ml of 0.5CNP100 (pH 7.4), and the AChR is eluted with 15 ml of 10 mM carbamylcholine in 0.5CNP90 (pH 7.4). The elution of protein can be followed in this case by its absorbance at 280 nm. Carbamylcholine-eluted fractions containing appreciable protein are stored in liquid nitrogen.

[21] A. Karlin, M. G. McNamee, C. L. Weill, and R. Valderrama, *in* "Methods in Receptor Research" (M. BLecher, ed.), p. 1. Dekker, New York, 1976.

[22] G. Biesecker, *Biochemistry* **12**, 4403 (1973).

[23] S. Froehner, C. G. Reiness, and Z. Hall, *J. Biol. Chem.* **252**, 8589 (1977).

[24] P. J. Whiting and J. Lindstrom, *Biochemistry* **25**, 2082 (1986).

[25] J. Lindstrom, B. Einarson, and S. Tzartos, this series, Vol. 74, p. 452.

[26] R. L. Huganir, M. A. Schell, and E. Racker, *FEBS Lett.* **108**, 155 (1979).

[27] Associated Concentrates (Woodside, NY).

Assays

Acetylcholine Binding Sites

Elapid snake venom contains basic polypeptide toxins which bind with high specificity and affinity to the ACh binding sites. Radioactive derivatives of these toxins are available. For example, the widely used α-bungarotoxin (Bgtx) is commercially available as an ^{125}I derivative,[28] and *Naja naja siamensis* toxin 3 is readily tritiated.[29] In the former case, the iodinated toxin (initially about 140 Ci/mmol) is diluted 50 to 100 times with unlabeled α-bungarotoxin[30] prior to use.[31] We henceforth refer to the high-specific activity ^{125}I-labeled bungarotoxin diluted with unlabeled Bgtx simply as [^{125}I]Bgtx.

On the day of use, [^{125}I]Bgtx is diluted to about 140 pmol/ml in 100 mM NaCl, 10 mM MOPS, pH 7.4, 0.2% Triton X-100, and 0.1 mg/ml bovine serum albumin. Two 25-mm DE-81 filters[32] are placed on each screen of a filtration manifold and are washed with 4 ml of buffer B. Fifty microliters of diluted [^{125}I]Bgtx is added to 50 μl of AChR containing about 2 pmol of sites and mixed well. After 30 min at room temperature, the mixture is diluted with 5 ml of cold B buffer[7] and poured onto the doubled DE-81 filters.[29,33] Mild suction at 50 to 100 mm Hg pressure difference is applied, the tube is washed onto the filter with another 5 ml of B buffer, and the filter is washed with 15 ml more of B buffer, now with a pressure difference of 200 mm Hg for faster filtration. After the filters are placed in the bottom of a scintillation vial, 250 μl of 1 M acetic acid is pipetted onto

---

[28] New England Nuclear Corp. (Boston, MA).

[29] V. N. Damle and A. Karlin, *Biochemistry* **17**, 2039 (1978).

[30] Sigma Chemical Co. (St. Louis, MO).

[31] One should determine that [^{125}I]Bgtx and Bgtx bind equivalently to AChR as follows. A fixed quantity of AChR is titrated with the diluted [^{125}I]Bgtx. If unlabeled Bgtx bound with higher affinity than [^{125}I]Bgtx, then the quantity of ^{125}I bound would reach a maximum at equivalence, when all Bgtx plus [^{125}I]Bgtx were bound and all toxin binding sites were occupied, and then the ^{125}I bound would decrease with increasing total Bgtx concentration to a new steady level as the excess unlabeled Bgtx competed with the [^{125}I]Bgtx for the toxin binding sites; that is, there would be a cusp in the binding curve around equivalence. If no cusp is observed, one may conclude that labeled and unlabeled Bgtx bind equivalently for the purposes of this assay. In addition, one should routinely determine the fraction of ^{125}I in the [^{125}I]Bgtx preparation that is bound by excess AChR. This fraction is typically about 85%. One determines the factor (cpm/mol) for converting radioactivity to the quantity of Bgtx by counting a known quantity of [^{125}I]Bgtx plus Bgtx on filters, by multiplying these counts by the fraction of ^{125}I that could be bound by AChR, and by dividing by the quantity of Bgtx plus [^{125}I]Bgtx in the sample.

[32] Whatman, Inc. (Clifton, NJ).

[33] R. P. Klett, B. W. Fulpius, D. Cooper, M. Smith, E. Reich, and L. D. Possani, *J. Biol. Chem.* **252**, 4811 (1973).

the filters. After 10 to 20 min at room temperature, 5 ml of Scintisol[34] or equivalent counting cocktail is added, and the vial is placed in a scintillation counter.

Composition by Sodium Dodecyl Sulfate Gel Electrophoresis

Standard Laemmli[35] conditions are used with a 7.5% acrylamide resolving gel. The sample is prepared initially by dissolving in 2% sodium dodecyl sulfate (SDS), 10 mM Tris, pH 8.0, 10 mM DTT, and holding at 50° for 30 min. (Higher temperatures can lead to degradation of subunits.) Thirty millimolar N-ethylmaleimide is added, and after an additional 15 min 9 volumes of acetone is added. After a few minutes, the precipitate is sedimented in a clinical centrifuge and washed with 5 volumes of acetone. The pellet is dried in a vacuum desiccator and dissolved in Laemmli sample buffer. Reduced and alkylated AChR yields four bands (α, β, γ, and δ) with apparent masses of about 40, 48, 58, and 64 kDa.[19] If DTT and 2-mercapto-ethanol in the Laemmli sample buffer are omitted, the 64-kDa band (δ) is diminished, and a 130-kDa band (δ dimer) is seen instead. Purity is indicated by the absence of other bands on the gel.

Reconstitution and Flux Assay[15,36,37]

For reconstitution and flux assays,[15,36,37] ligand-gated channel activity is more effectively reconstituted in a mixture of phosphatidylethanolamine (PE),[38] phosphatidylserine (PS),[38] and cholesterol than in asolectin. A PE–PS–cholesterol mixture, 6 : 3 : 1 by weight, is made at 100 mg/ml in chloroform. Two hundred microliters of this mixture is dried to a film with a stream of N_2 and then dried further under reduced pressure for 3 hr. To the dried lipid is added 570 μl of 2CNP100 (pH 8.0).[7] The final concentration of lipid is 3.5%. The mixture is vortexed and sonicated under N_2 in a bath sonicator[39] until the mixture is opalescent (about 5 min).

Native *Torpedo* membrane vesicles are quite leaky. The ligand-gated channel activity of the AChR can be more reproducibly assayed after solubilization and reconstitution into liposomes. Membrane, containing

[34] Isolab, Inc. (Akron, OH).

[35] U. K. Laemmli, *Nature (London)* **227,** 680 (1979).

[36] M. Montal, R. Anholt, and P. Labarca, *in* "Iron Channel Reconstitution" (C. Miller, ed.), p. 157. Plenum, New York, 1986.

[37] M. G. McNamee, O. T. Jones, and T. M. Fong, *in* "Ion Channel Reconstitution" (C. Miller, ed.), p. 231. Plenum, New York, 1986.

[38] Avanti Polar Lipids (Birmingham, AL).

[39] Laboratory Supplies Company, Inc. (Hicksville, NY).

about 1 nmol of toxin binding sites, is sedimented and suspended in 620 μl of 0.5% lipid in 2CNP100(pH 8.0), made by diluting the above 3.5% lipid solution with 2CNP100 (pH 8.0). All steps are at 4°. After being stirred for 1 hr, the suspension is centrifuged 30 min in a microcentrifuge. The supernatant is mixed with an equal volume of the 3.5% lipid solution and agitated for 30 min. This mixture is dialyzed successively against 500 ml of NP100 (pH 8.0) for 14 to 16 hr, against 500 ml of 180 mM sucrose, 10 mM Tris-Cl, pH 8.0, 0.02% NaN$_3$ for 10 to 12 hr, and against a fresh 500 ml of the same sucrose solution for about 18 hr.

To assay ligand-gated channel activity in purified AChR, AChR containing about 1 nmol of toxin binding sites in the cholate–asolectin mixture (see above) is made up to a final concentration of 2% lipid in 2CNP100 (pH 8.0) with the 3.5% lipid mixture and 2CNP100 (pH 8.0) and dialyzed as above.

Prior to reconstitution, ion-exchange resin for the flux assay is prepared as follows. One hundred grams of AG 50W-X8 (100–200 mesh)[18] in the hydrogen form is washed twice with 500 ml of methanol and twice with 1 liter of deionized water. The resin is equilibrated with 200 ml of 1 M Tris base by stirring gently for 30 min and is washed 3 times with 1 liter of deionized water. The resin is similarly equilibrated with 200 ml of 1 M HCl and washed with water. It is reequilibrated with 200 ml of 1 M Tris base, allowed to settle, suspended in 1 liter of water, and allowed to settle. The equilibration with Tris is repeated until the supernatant pH is 8.0. Sodium azide is added to 0.02%, and the resin is stored at 4°.

Prior to the flux assay, 1.4 ml of resin is added to disposable columns and is washed first with 2 ml of 1 mg/ml bovine serum albumin, 0.02% NaN$_3$ and second with 2 ml of 180 mM sucrose, 10 mM Tris-Cl (pH 8.0), 0.02% NaN$_3$. The latter solution is used below as the eluant.

In the flux assay, ^{22}Na is diluted with the above eluant to contain about 145,000 counts/min (cpm) per 10 μl. Ten microliters of this ^{22}Na and 10 μl of either 700 μM carbamylcholine in eluant or eluant alone are added to the bottom of a tube. Fifty microliters of reconstituted AChR is rapidly added and mixed with an Eppendorf pipette. With a second pipette, the mixture is immediately transferred to the top of the column of resin. This is followed immediately with 1.5 ml of eluant. The entire output from the column, approximately 1.5 ml, contains the vesicles and the included ^{22}Na but not free ^{22}Na. This is collected in a 7-ml scintillation vial; 5 ml of Scintisol is added, and the vial is counted. Although a white gel forms, more Scintisol does not result in higher efficiency counting. Under the above conditions, approximately 4 times as many counts (about 600 cpm) are taken up by the reconstituted vesicles in the presence of carbamylcholine than in its absence (about 150 cpm).

[37] Purification, Affinity Labeling, and Reconstitution of Voltage-Sensitive Potassium Channels

By Hubert Rehm and Michel Lazdunski

Introduction

Voltage-dependent K^+ channels are integral membrane proteins which allow, in a voltage-dependent manner, the passage of K^+ ions across the cell membrane. They play a decisive part in the repolarization of the neuron and in the maintenance of the membrane potential. Voltage-dependent K^+ channels can be subdivided into three main categories: delayed rectifiers, A-type channels, and inward rectifiers. On depolarization, delayed rectifiers activate slowly but do not inactivate, whereas A-type channels activate and inactivate quickly. The inward rectifiers open when the membrane is hyperpolarized.[1]

A breakthrough in the protein chemistry of voltage-dependent K^+ channels was the discovery that certain snake venoms contain highly specific ligands for these channels.[2,3] Potassium channel toxins are also found in the venoms of insects and scorpions. To date four classes of toxins are known which bind to and affect the activity of voltage-dependent K^+ channels (Table I).

The venoms of *Dendroaspis angusticeps* and *Dendroaspis polylepis* contain a variety of basic peptides having a molecular weight of about 7000 that are called dendrotoxins.[4-6] These peptides are homologous to Kunitz-type protease inhibitors although they do not inhibit proteases. Dendrotoxins facilitate neurotransmitter release at the neuromuscular junction and induce epileptic crises on intracerebroventricular (icv) injection.[4,7] Prominent members of the dendrotoxin family are dendrotoxin (DTX) and dendrotoxin I (DTX_I). DTX_I from *Dendroaspis polylepus* venom is the most toxic dendrotoxin known.

The β-bungarotoxins are a family of basic proteins with a molecular weight of about 21K from the venom of the snake *Bungarus multicinctus*

[1] B. Rudy, *Neuroscience* **25**, 729 (1988).
[2] E. Moczydlowski, K. Lucchesi, and A. Ravindran, *J. Membr. Biol.* **105**, 95 (1988).
[3] P. N. Strong, *Pharmacol Ther.* **46**, 137 (1990).
[4] A. L. Harvey and E. Karlsson, *Br. J. Pharmacol.* **77**, 153 (1982).
[5] A. L. Harvey and A. J. Anderson, *Pharmacol Ther.* **31**, 33 (1985).
[6] H. Schweitz, J. N. Bidard, and M. Lazdunski, *Toxicon* **28**, 847 (1990).
[7] G. Gandolfo, C. Gottesmann, J. N. Bidard, and M. Lazdunski, *Eur. J. Pharmacol.* **160**, 173 (1989).

TABLE I
POTASSIUM CHANNEL TOXINS

Toxin	Source	Molecular weight	Structure	Dissociation constants of binding (K_D in nM)[a]	Type of K⁺ current blocked (preparation)	Ref.
Dendrotoxins	Snake venom (*Dendroaspis*)	7000	Protease inhibitor homologs, 3 internal disulfide bridges	0.02–0.05 (dendrotoxin I) 0.4–0.8 (dendrotoxin)	Slow inactivating (sensory A neurons of nodose ganglion) Noninactivating (dorsal root ganglion neurons) Slow-inactivating (node of Ranvier)	b, c c, d c, e
β-Bungarotoxin	Snake venom (*Bungarus multicinctus*)	21000	Phospholipase A₂ enzyme with associated protease inhibitor homolog	0.2–0.4	Noninactivating (dorsal root ganglion neurons)	c, f
Mast cell degranulating peptide	Bee venom	2500	22-amino acid peptide with weak sequence similarity to phospholipase A₂ subunit of β-BTX, 2 internal disulfide bridges	0.1–0.2	Slow inactivating (sensory A neurons of nodose ganglion)	c, g
Charybdotoxin	Scorpion venom (*Leiurus quinquestriatus hebraeus*)	4500	37-amino acid peptide with 3 internal disulfide bridges	0.7	Noninactivating (dorsal root ganglion neurons)	c, h
Noxiustoxin	Scorpion venom (*Centruroides noxius Hoffmann*)	4700	39-amino acid peptide with 3 internal disulfide bridges, homologous to charybdotoxin	Not determined	Noninactivating (squid giant axon)	c, i

[a] High affinity binding to rat brain membranes. Lower affinity binding sites are also identified.
[b] C. Stansfeld, S. Marsh, J. Halliwell, and D. Brown, *Neurosci. Lett.* **64**, 299 (1986).
[c] F. Dreyer, *Rev. Physiol. Biochem. Pharmacol.* **115**, 93 (1990).
[d] C. Stansfeld and A. Feltz, *Neurosci. Lett.* **93**, 49 (1988).
[e] E. Benoît and J. M. Dubois, *Brain Res.* **377**, 374 (1986).
[f] M. Petersen, R. Penner, F. Pierau, and F. Dreyer, *Neurosci. Lett.* **68**, 141 (1986).
[g] C. Stansfeld, S. Marsh, D. Parcy, J. O. Dolly, and D. Brown, *Neuroscience* **23**, 983 (1987).
[h] H. Schweitz, C. Stansfeld, J. N. Bidard, L. Faqui, P. Maes, and M. Lazdunski, *FEBS Lett.* **250**, 519 (1989).
[i] E. Moczydlowski, K. Lucchesi, and A. Ravindran, *J. Membr. Biol.* **105**, 95 (1988).

(Table I).[8] They consist of two subunits (14K and 7K) which are linked by a disulfide bridge. The 14K subunit is a phospholipase A_2; the 7K subunit is homologous to dendrotoxins. β-Bungarotoxins block neuromuscular transmission in a triphasic course with an intermediate facilitatory phase. In addition, they are cytotoxic for cholinergic and γ-aminobutyric acid (GABA) ergic neurons.[9]

Mast cell degranulating peptide (MCD) from bee venom is a basic peptide of 22 amino acids with two intramolecular disulfide bridges.[10] MCD, which releases histamine from mast cells, is relatively nontoxic on peripheral injection, but its toxicity becomes exceptionally high when it is injected icv.[11,12] In sublethal doses MCD induces hippocampal theta rhythms and long-term potentiation.[13] These central effects are due to the K^+ channel blocking activity of MCD. Very high affinity binding sites for MCD are present in the brain, but lower affinity binding sites are present as well.

Charybdotoxin (CTX) is a basic peptide of 37 amino acids from the venom of the scorpion *Leiurus quinquestriatus hebraeus*.[2,14] In peripheral tissues, CTX blocks a class of Ca^{2+}-dependent K^+ channels[2]. Charybdotoxin is also a potent convulsant when injected intraventricularly, and it releases GABA from preloaded synaptosomes.[15] A near relative of CTX is noxiustoxin, a basic peptide of 39 amino acids from the venom of the scorpion *Centruroides noxius Hoffmann*. It inhibits a delayed rectifier K^+ current in the squid giant axon.[2,16]

In brain, the dendrotoxins, β-bungarotoxins, MCD, and CTX bind with high affinity to proteinaceous sites that are allosterically related, as each toxin inhibits the binding of the others.[3,15,17-22] Solubilization with

[8] B. D. Howard and C. B. Gundersen, *Annu. Rev. Pharmacol. Toxicol.* **20**, 307 (1980).

[9] H. Rehm, T. Schäfer, and H. Betz, *Brain Res.* **250**, 309 (1982).

[10] E. Habermann, *Science* **177**, 314 (1972).

[11] E. Habermann, *Naunyn-Schmiedebergs Arch. Pharmacol.* **300**, 189 (1977).

[12] J. N. Bidard, G. Gandolfo, C. Mourre, C. Gottesmann, and M. Lazdunski, *Brain Res.* **418**, 235 (1987).

[13] E. Cherubini, Y. Ben Ari, M. Gho, J. N. Bidard, and M. Lazdunski, *Nature (London)* **328**, 70 (1987).

[14] C. Miller, E. Moczydlowski, R. Latorre, and M. Philipps, *Nature (London)* **313**, 316 (1985).

[15] H. Schweitz, J. N. Bidard, P. Maes, and M. Lazdunski, *Biochemistry* **28**, 9708 (1989).

[16] E. Carbone, G. Prestipino, L. Spadavecchia, F. Franciolini, and L. D. Possani, *Pfluegers Arch.* **408**, 423 (1987).

[17] H. Rehm and H. Betz, *J. Biol. Chem* **257**, 10015 (1982).

[18] J. N. Bidard, C. Mourre, and M. Lazdunski, *Biochem. Biophys. Res. Commun.* **143**, 383 (1987).

[19] I. B. Othman J. W. Spokes, and J. O. Dolly, *Eur. J. Biochem.* **128**, 267 (1982).

[20] R. Schmidt, H. Betz, and H. Rehm, *Biochemistry* **27**, 963 (1988).

[21] A. Breeze and J. O. Dolly, *Eur. J. Biochem.* **178**, 771 (1989).

detergent does not destroy this mutual interaction because the different toxin binding sites copurify.[23,24] Thus, in brain, the membrane protein which binds dendrotoxins with high affinity is also the target for β-bungarotoxin, MCD, and CTX. This protein is called the DMB protein (from D, DTX; M, MCD; B, binding).

The DMB protein was purified from rat brain membranes.[24] This oligomeric protein complex has a molecular weight of 450K and consists of α subunits of molecular weight 76K–80K and β subunits of molecular weight 42K and 38K. The glycosylated α subunit (76K–80K) bears the different toxin binding sites since in cross-linking experiments the iodinated toxins (DTX$_I$, MCD, CTX) label a band of similar molecular weight.[15,23,25]

The DMB protein appears to be a voltage-dependent K$^+$ channel of the delayed rectifier type.[26–32] Antibodies raised against partial peptide sequences derived from mammalian delayed rectifier K$^+$ channel clones recognize specifically the toxin-binding α subunit of the DMB protein.[33] After reconstitution, the purified DMB protein behaves as a K$^+$ channel. Its opening probability is regulated by phosphorylation of the toxin-binding α subunit.[34]

Purification of Dendrotoxin I

Dendrotoxin I is the most potent voltage-dependent K$^+$ channel blocker known till now and has been essential in biochemical studies involving this channel. *Dendroaspis polylepis* venom (2 g) is dissolved in

---

[22] J. N. Bidard, C. Mourre, G. Gandolfo, H. Schweitz, C. Widmann, C. Gottesmann, and M. Lazdunski, *Brain Res.* **495,** 45 (1989).

[23] H. Rehm, J. N. Bidard, H. Schweitz, and M. Lazdunski, *Biochemistry* **27,** 1827 (1988).

[24] H. Rehm and M. Lazdunski, *Proc. Natl. Acad. Sci. U.S.A.* **85,** 4919 (1988).

[25] H. Rehm and H. Betz, *EMBO J.* **2,** 1119 (1983).

[26] E. Benoît and J. M. Dubois, *Brain Res.* **377,** 374 (1986).

[27] C. Stansfeld, S. Marsh, J. Halliwell, and D. Brown, *Neurosci. Lett.* **64,** 299 (1986).

[28] C. Stansfeld, S. Marsh, D. Parcey, J. O. Dolly, and D. Brown, *Neuroscience* **23,** 893 (1987).

[29] C. Stansfeld and A. Feltz, *Neurosci. Lett.* **93,** 49 (1988).

[30] F. Dreyer and R. Penner, *J. Physiol. (London)* **386,** 455 (1987).

[31] H. Schweitz, C. Stansfeld, J. N. Bidard, L. Faqui, P. Maes, and M. Lazdunski, *FEBS Lett.* **250,** 519 (1989).

[32] W. Stühmer, M. Stocker, B. Sakmann, P. Seeburg, A. Baumann, A. Grupe, and O. Pongs, *FEBS Lett.* **242,** 199 (1988).

[33] H. Rehm, R. Newitt, and B. Tempel, *FEBS Lett.* **249,** 224 (1989).

[34] H. Rehm, S. Pelzer, C. Cochet, B. Tempel, E. Chambaz, W. Trautwein, D. Pelzer, and M. Lazdunski, *Biochemistry* **28,** 6455 (1989).

10 ml of 0.1 M ammonium acetate, pH 6.8, and centrifuged for 1 hr at 100,000 g. The supernatant is loaded onto a column of Sephadex G-50 fine (l, 80 cm; d, 2.5 cm) which has been equilibrated with 100 mM ammonium acetate, pH 6.8.[35] The column is eluted at a speed of 36 ml/hr. Fractions are monitored at 280 nm. Usually, three peaks of UV-absorbing material are obtained; the middle peak contains the different dendrotoxins which are present in the venom and is taken for further purification.

The dendrotoxin peak from the Sephadex column is then loaded (41 ml/hr) onto a column of Bio-Rex 70 (Bio-Rad, Richmond, CA; l, 25 cm; d, 2.5 cm) which had been equilibrated with 0.1 M ammonium acetate, pH 7.0. The column is eluted with a 2-liter linear gradient from 0.1 to 1.4 M ammonium acetate, pH 7.0, and the fractions in the eluate are monitored for absorption at 280 nm.[35] Dendrotoxin I is in a major peak which elutes at higher concentrations of ammonium acetate. Peaks containing DTX_I are identified by their ability to inhibit [^{125}I]DTX_I binding to rat brain membranes (see assay described in Ref. 24).

After Bio-Rex 70 chromatography DTX_I is already about 90% pure and can be used for affinity chromatography. Near to complete (>99%) purification is achieved by an additional reversed-phase chromatography on a Lichrosorb RP-18 column from Merck (Darmstadt, Germany) with a linear gradient of 10–40% acetonitrile.[23] Techniques used to purify other members of the dendrotoxin family can be found in Ref. 6.

Iodination of Dendrotoxin I

Solutions

I: 1.0–1.3 mCi Na^{125}I (NEN, Du Pout de Nemours, Paris; NEZ 033L, pH 8–10, 16–18 Ci/mg)

II: 0.5 M Sodium phosphate buffer, pH 7.4

III: DTX_I (1 mg/ml) in 0.1 M sodium phosphate buffer, pH 7.4

IV: 0.5 mM Chloramine-T in 0.1 M sodium phosphate buffer, pH 7.4 (prepared immediately before use)

V: 10 mM Sodium phosphate, pH 7.4, 0.02% Triton X-100

VI: 10 mM Sodium phosphate, pH 7.4, 1 mg/ml bovine serum albumin (BSA)

VII: 10 mM Sodium phosphate, pH 7.4, 1 mg/ml BSA, 0.66 M NaCl

Procedure. The reaction is performed in the Na^{125}I-containing glass vial delivered by the manufacturer. To the Na^{125}I (in 2.5–3.0 μl) are added 10 μl of solution II and 7 μl of solution III. The reaction is started by

[35] A. L. Harvey and E. Karlsson, *Naunyn-Schmiedebergs Arch. Pharmacol.* **312**, 1 (1980).

addition of 5 μl of solution IV (chloramine-T) and mixing. Higher concentrations of chloramine-T lead to inactivation of DTX_I. After 30 sec at room temperature the reaction is stopped with 1 ml of solution V.

The diluted reaction mixture is loaded onto a column of Bio-Rex 70 (in a 2-ml syringe, gel volume 1.4 ml) which had been equilibrated in solution VI. The column is washed with 7–10 ml of solution VI, and $[^{125}I]DTX_I$ is eluted with solution VII. Fractions with a volume of 1 ml are taken and 2-μl aliquots of the fractions are counted. The specific activity of the iodinated DTN_I is 420–460 Ci/mmol. Similar procedures can be used for other dendrotoxins.

Affinity Labeling of DMB Protein with $[^{125}I]DTX_I$

Buffers

I: 50 mM Tris-Cl, pH 7.5, 140 mM NaCl, 1.3 mM MgSO$_4$, 5 mM KCl, 1 mg/ml BSA
II: 50 mM Sodium phosphate, pH 8.5
III: 100 mM Tris-Cl, pH 7.4

Procedure. $[^{125}I]DTX_I$ (0.2–1.0 nM, 400–2000 Ci/mmol) and 300–500 μg rat brain membranes are incubated in 0.5 ml of buffer I for 30 min at 4° in the absence or presence of unlabeled DTX_I or other antagonists. The membranes are washed twice with cold buffer II (centrifugation at 10,000 g for 5 min at 4°), to remove free $[^{125}I]DTX_I$. Thereafter the labeled membranes are resuspended in 0.5 ml of buffer II containing 0.1–1 mg/ml of the cross-linker dimethyl suberimidate. Membranes and cross-linker are incubated for 10–20 min at 4°. The reaction is stopped by washing the membranes twice with 1 ml of 100 mM Tris-Cl, pH 7.4. For analysis the membranes are solubilized with Laemmli sample buffer and electrophoresed on 8 or 10% sodium dodecyl sulfate (SDS)-polyacrylamide gels. The gels are dried for autoradiography and exposed for 4–8 days to Kodak (Rochester, NY) X-Omat AR film.

Preparation of Dendrotoxin I Affinity Column

Solutions

I: 0.5 M Potassium phosphate buffer, pH 7.6
II: 0.1 M Ethanolamine, pH 7.9
III: DTX_I in 0.5 M potassium phosphate buffer, pH 7.6 (protein concentration 8–10 mg/ml). When DTX_I is purified with a Bio-Rex 70 column and ammonium acetate buffers, residual NH$_4^+$ ions in the DTX_I solution have to be removed by dialysis in Spectrapor 3 mem-

branes (Spectrum, Los Angeles, CA; molecular weight cutoff 3500) against 0.5 M potassium phosphate buffer, pH 7.6.

IV: 20 mM Na–HEPES, pH 7.2, 1 mM NaEDTA, 120 mM KCl, 10% (v/v) glycerol, 0.05% (w/v) Triton X-100, 0.01% (w/v) phosphatidylcholine (crude soybean)

Procedure. For the preparation of the DTX$_I$ AcA 22 column, glutaraldehyde-activated Ultrogel AcA 22 (15 ml) is washed on a glass funnel with 230 ml of water and thereafter with 100 ml of solution I. The wet AcA 22 cake is associated to 30–35 ml of DTX$_I$ solution (III) and incubated for 20–25 hr at room temperature on an end-over-end shaker. The supernatant is removed and the gel incubated for 3–4 hr at 4° with 20 ml of solution II. The gel is washed with 0.5 liters of solution I and equilibrated in buffer IV.

The column may be prepared with pure DTX$_I$ (after reversed-phase high-performance liquid chromatography) or with only 90% pure DTX$_I$ (after Bio-Rex 70). Both preparations give satisfactory results in affinity chromatography. The [^{125}I]DTX$_I$ AcA 22 column can be used repeatedly for at least 1 year, provided the use of reducing agents like dithiothreitol is avoided.

Purification of DMB Protein

Buffers

I: 20 mM Na–HEPES, pH 7.2, 1 mM NaEDTA, 120 mM KCl, 10% (v/v) glycerol, 1% (w/v) Triton X-100, 0.2% (w/v) phosphatidylcholine (crude soybean)

II: 20 mM Na–HEPES, pH 7.2, 1 mM NaEDTA, 202 mM KCl, 10% (v/v) glycerol, 1% (w/v) Triton X-100, 0.2% (w/v) phosphatidylcholine (crude soybean)

III: 20 mM Na–HEPES, pH 7.2, 1 mM NaEDTA, 120 mM KCl, 10% (v/v) glycerol, 0.05% (w/v) Triton X-100, 0.01% (w/v) phosphatidylcholine (crude soybean)

IV: 20 mM Na–HEPES, pH 7.2, 1 mM NaEDTA, 820 mM KCl, 10% (v/v) glycerol, 0.05% (w/v) Triton X-100, 0.01% (w/v) phosphatidylcholine (crude soybean)

Procedure. To 320 ml of rat brain membranes (13 mg/ml; 4.16 g of protein; preparation described in Ref. 36), 0.5 ml of 100 mM phenylmethylsulfonyl fluoride (PMSF) in dimethyl sulfoxide (DMSO) is added, and the mixture is incubated under stirring for 10 min at room temperature.

[36] J. W. Taylor, J. N. Bidard, and M. Lazdunski, *Nature (London)* **328,** 70 (1987).

The membranes are mixed with 2200 ml of cold buffer I and stirred at 4° for 60 min. After centrifugation for 45 min at 35,000 rpm and 4° the pooled supernatants are put (72 ml/hr) onto a DEAE-Affi-Gel blue column (diameter 5 cm, gel volume 300 ml) which had been equilibrated in buffer I.

The column is washed with 1 liter of buffer I (rate 72 ml/hr) and eluted with buffer II at a rate of 50 ml/hr. Fractions of 5–6 ml are collected. The fractions may be screened for [^{125}I]DTX$_I$ binding activity, but because eluted binding activity parallels the protein concentration of the eluate it is sufficient to determine protein and to pool the eluted protein peak.

The pool from the ion-exchange column is applied at a speed of 16 ml/hr onto the DTX$_I$ AcA 22 column (diameter 1.0–1.6 cm, gel volume 6–10 ml) which had been equilibrated with buffer III. The affinity column is washed first with 50 ml of buffer I then with 30 ml of buffer III at a rate of 28 ml/hr. The column is eluted with buffer IV at a speed of 16 ml/hr; 2.5-ml fractions are collected. Fractions containing the DMB protein are identified either from their binding activity or from the increase in buffer conductivity. The binding fractions (or fractions showing a conductivity increase) are pooled. After each run the DTX$_I$ AcA 22 column should be washed with buffer I containing 1 M KCl and thereafter equilibrated in buffer III. The pool from the affinity chromatography step is applied at speed of 9 ml/hr onto a column of wheat germ agglutinin (WGA)-Sepharose (diameter 0.9 cm, gel volume 8–10 ml) which had been equilibrated in buffer III. The WGA column is washed with 10 ml of buffer III at 18 ml/hr and eluted with buffer III containing 50 mM N-acetylglucosamine at a speed of 9 ml/hr; 1.5-ml fractions are collected.

N-Acetylglucosamine gives an intense staining in the Pierce (Rockford, IL) protein assay. This reaction may be used to determine the position of the eluted glycoproteins. Alternatively the binding assay with [^{125}I]DTX$_I$ can be used for this purpose. The purification produces 20–30 μg of the DMB protein, which is at about 90% purity. The purified DMB protein is stored in liquid nitrogen. Storage at −20° leads to loss of binding activity.

Reconstitution of DMB Protein

Buffers

I: 20 mM Na–HEPES, pH 7.2, 120 mM KCl, 1 mM EDTA, 10% (v/v) glycerol, 1% (w/v) octylglucoside, 1.3 mg/ml phosphatidylcholine, 0.6 mg/ml phosphatidylethanolamine, and 1 mg/ml phosphatidylserine. All phospholipids are from Avanti Polar Lipids (Birmingham, AL).

II: 20 mM Na–HEPES, pH 7.2, 1 mM EDTA, 120 mM KCl

Procedure. The purified DMB protein (4–10 μg/ml) in 20 mM Na–HEPES, pH 7.2, 1 mM NaEDTA, 120 mM KC1, 10% (v/v) glycerol, 0.05% (w/v) Triton X-100, 0.01% (w/v) phosphatidylcholine (crude soybean), 20–50 mM N-acetylglucosamine is diluted 10-fold with buffer I and dialyzed for 48 hr at 4° against three changes of buffer II. After dialysis, 70% of the DTX$_I$ binding acitivity which had been present in the soluble DMB protein preparation is recovered. The affinity of [^{125}I]DTX$_I$ for the reconstituted channel is similar to the affinity of the toxin for the channel in native brain membranes. The reconstituted channel preparation can then be used for electrophysiological experiments as described by Rehm *et al.*[34]

The phospholipid composition of buffer I influences the reconstitution of [^{125}I]DTX$_I$ binding activity. The recovery of binding activity is significantly lower when buffer I contains only phosphatidylethanolamine and phosphatidylserine, whereas replacement of phosphatidylserine or phosphatidylethanolamine by phosphatidylcholine does not change the results. Reconstitution of [^{125}I]DTX$_I$ binding activity from Triton X-100–phospholipid mixtures with the help of SM-2 beads (Bio-Rad, Richmond, CA) was not successful.

Acknowledgments

This work was supported by the Centre National de la Recherche Scientifique, the Deutsche Forschungsgemeinschaft, and DRET (Grant 90/192).

[38] Affinity Purification and Reconstitution of Calcium-Activated Potassium Channels

By DAN A. KLÆRKE and PETER L. JØRGENSEN

Introduction

Calcium-activated potassium channels are ubiquitous, and their properties differ with respect to calcium sensitivity, single-channel conductance, voltage dependence, and sensitivity to high affinity inhibitors.[1,2] In the thick ascending limb of Henle's loop (TAL) in the mammalian kidney, the active reabsorption of NaCl depends on the function of K$^+$ channels and the Na,K,Cl-cotransport system in the luminal membrane together with the Na,K-pump and a net Cl$^-$ conductance in the basolateral mem-

[1] O. H. Petersen and Y. Maruyama, *Nature (London)* **307,** 693 (1984).
[2] E. Moczydlowski, K. Lucchesi, and K. Ravindran, *J. Membr. Biol.* **255,** 4087 (1988).

brane.[3,4] In luminal plasma membrane vesicles from TAL we have used flux studies to characterize K⁺ channels, which are regulated by physiological concentrations of Ca^{2+}, calmodulin, pH, and phosphorylation.[5,6] A charybdotoxin-sensitive, Ca^{2+}-activated K⁺ channel with a single-channel conductance of 127 pS has been studied in cultured cells from this tissue.[7] In intact tubuli, Ca^{2+}-insensitive, ATP-sensitive K⁺ channels with single-channel conductances of 20 pS[8] and 60–70 pS[9] have been identified in the luminal membrane of TAL using the patch clamp technique.

As estimated on the basis of ion fluxes, the number of K⁺ channels per cell in TAL is very low, relative to the number of Na,K-pumps and Na,K,Cl-cotransporters.[10] The sparse distribution of K⁺ channels explains the problems encountered in attempts to purify and characterize their proteins. As an approach to solve these problems we have used calmodulin affinity chromatography based on the observation that a portion of the Ca^{2+}-activated K⁺ channel activity from TAL is stimulated by calmodulin from the cytoplasmic face with a very high affinity.[6] The initial steps in the purification procedure are preparation of a plasma membrane fraction enriched in vesicles from the luminal membrane of TAL and subsequent detergent solubilization of the membrane proteins. To monitor K⁺ channel activity after purification procedures, the protein is reconstituted into phospholipid vesicles, where the K⁺ channel activity can be measured using a very sensitive flux assay.

Purification Procedures

A flow diagram of the steps in the purification procedure is shown in Fig. 1. The starting material is the macroscopic red inner stripe of outer medulla of the pig kidney, where the thick ascending limb of Henle's loop is the predominant structure. A crude plasma membrane fraction is prepared by differential centrifugation,[11,12] and luminal and basolateral plasma membrane vesicles are partially separated on a metrizamide density gradient. Luminal plasma membrane vesicles form the starting mate-

[3] R. Greger and E. Schlatter, *Pfluegers Arch.* **396,** 325 (1983).
[4] S. C. Hebert, P. A. Friedman and T. E. Andreoli, *J. Membr. Biol.* **80,** 201 (1984).
[5] D. A. Klærke, S. J. D. Karlish, and P. L. Jørgensen, *J. Membr. Biol.* **95,** 105 (1987).
[6] D. A. Klærke, J. Petersen, and P. L. Jørgensen, *FEBS Lett.* **216,** 211 (1987).
[7] S. E. Guggino, W. B. Guggino, N. Green, and B. Sacktor, *Am. J. Physiol.* **252,** C128 (1987).
[8] W. Wang, S. White, J. Geibel, and G. Giebisch, *Am. J. Physiol.* **258,** F244 (1990).
[9] M. Bleich, E. Schlatter, and R. Greger, *Pfluegers Arch.* **415,** 449 (1990).
[10] D. A. Klærke and P. L. Jørgensen, *Comp. Biochem. Physiol. A: Comp. Physiol.* **90A,** 757 (1988).
[11] P. L. Jørgensen, this series, Vol. 32, p. 277.
[12] P. L. Jørgensen, this series, Vol. 156, p. 29.

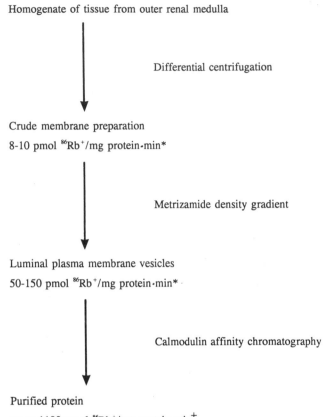

Homogenate of tissue from outer renal medulla

Differential centrifugation

Crude membrane preparation
8-10 pmol $^{86}Rb^+$/mg protein·min*

Metrizamide density gradient

Luminal plasma membrane vesicles
50-150 pmol $^{86}Rb^+$/mg protein·min*

Calmodulin affinity chromatography

Purified protein
up to 1100 pmol $^{86}Rb^+$/mg protein·min†

FIG. 1. Flow diagram of purification procedures for the Ca^{2+}-activated K^+ channel from outer renal medulla. The K^+ channel activity is measured in plasma membrane vesicles (*) or reconstituted vesicles (†) as described in text.

rial for calmodulin affinity chromatography. They are characterized by a high Ca^{2+}-activated K^+ channel activity and a high content of the Na,K,Cl-cotransport system identified by furosemide-sensitive ion fluxes, whereas the Na,K—ATPase is a marker for the basolateral membrane.[13]

Materials

3-[(3-Cholamidopropyl)dimethylammonio]-l-propane   sulfonate (CHAPS) was obtained from Aldrich, Steinheim, Germany; metrizamide

[13] C. Burnham, S. J. D. Karlish, and P. L. Jørgensen, *Biochim. Biophys. Acta* **821,** 461 (1985).

from Nycomed, Oslo, Norway; Sephadex G-50 coarse and calmodulin-Sepharose 4B from Pharmacia, Uppsala, Sweden; ouabain and sodium monovanadate were from Merck, Darmstadt, Germany; Dowex (50W-X2 or 50W-X8, 50–100 mesh, H⁺ form) from Fluka, Buchs, Switzerland; ^{86}RbCl from Amersham, England; dodecyloctaethylene glycol monoether ($C_{12}E_8$) from Nikko Chemicals, Tokyo, Japan; ethylene glycol bis(β-aminoethyl ether)-N,N,N',N'-tetraacetic acid (EGTA), phosphatidylcholine (soybean, type II-S), DL-dithiothreitol (DTT), phenylmethylsulfonyl fluoride (PMSF), 3-(N-morpholino)propanesulfonic acid (MOPS), and all other chemicals were from Sigma (St. Louis, MO)

Preparation of Luminal Plasma Membrane Vesicles

In a local slaughterhouse pig kidneys are taken out as soon as possible after desanguination (maximum 30 min) and transferred to ice-cold P buffer (see Table I for buffer compositions). The kidneys are stored in ice-cold P buffer for a maximum of 6 hr until further preparation. The red inner stripe of outer medulla is obtained by dissection,[12] and the tissue is homogenized using 10 ml of P buffer per gram of tissue by 5 strokes in a tightly fitting Teflon–glass homogenizer (Braun, Melsungen, Germany). It is important that all preparation procedures are carried out at 0°–4°.

The homogenate is centrifuged at 7000 rpm (5000 g) for 15 min in a Sorvall high-speed centrifuge (Dupont, Wilmington, Delaware), SS-34 rotor. The supernatant is collected, and the sediment is resuspended in 5 ml of P buffer per gram of original tissue and centrifuged again for 15 min at 7000 rpm. The two supernatants are mixed and centrifuged for 30 min at 20,000 rpm (40,000 g). The pellet is resuspended by homogenization (10 strokes) in 1.5 ml of P buffer per gram of original tissue to a protein concentration of 8–12 mg/ml.

The resuspended vesicles (1.5 ml) are placed on either (a) a linear

TABLE I
COMPOSITION OF BUFFERS

Buffer	Composition (mM)							pH
	Sucrose	KCl	EGTA	CaCl₂	PMSF	DTT	MOPS–Tris	
Preparation (P)	250	50	1	—	—	—	10	7.2
Dilution (Di)	—	50	—	—	—	—	10	7.2
Reconstitution (R)	250	50	1	—	—	1	10	7.2
Dowex (Do)	350	—	—	—	—	—	10	7.2
Calcium (C)	250	50	—	0.1	0.02	—	20	7.2
EGTA (E)	250	50	5	—	0.02	—	20	7.2

density gradient or (b) a step gradient of Metrizamide. (a) The linear gradient (8 ml) is made of 5–15% (w/v) Metrizamide dissolved in P buffer on top of a 1-ml cushion of 30% (w/v) Metrizamide. After centrifugation at 19,000 rpm (40,000 g) for 16 hr in a TST 41.14 swinging-bucket rotor (Kontron, Zurich, Switzerland), the gradient is separated into fractions of 1 ml. Luminal plasma membrane vesicles characterized by a high K^+ channel activity are harvested in fractions 3–5 from the top at Metrizamide concentrations of 4–8%. (b) The step gradient consists of 9 ml of 12% (w/v) Metrizamide in P buffer. It is centrifuged at 35,000 rpm (150,000 g) for 1 hr or at 19,000 rpm for 3.5 hr in the TST 41.14 rotor. Two fractions of 1 ml are recovered from the top of the tube. The floating luminal vesicles are found in the second fraction from the top at a protein concentration of 5–6 mg/ml.

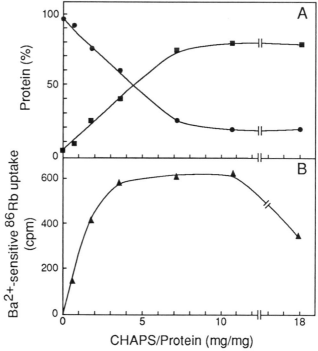

FIG. 2. (A) Solubilization by CHAPS of plasma membrane vesicles (2.8 mg protein/ml) prepared on a linear Metrizamide density gradient. Soluble protein (■) and nonsoluble protein (●) were separated by centrifugation. (B) Reconstitution of K^+ channel activity at increasing CHAPS concentrations. Aliquots (200 μl) of the supernatant from the experiment in (A) were mixed with solubilized lipid, reconstituted vesicles were eluted from the Sephadex G-50 column, and K^+ channel activity was measured as described in the text.

For assay of K⁺ channel activity, vesicles are diluted in P buffer to a protein concentration of 1–2 mg/ml, and ^{86}Rb⁺ uptake is measured as described below (Assay for Potassium Channel Activity).

Solubilization and Reconstitution

The K⁺ channel protein is further purified by fractionation after solubilization in detergent. To measure the K⁺ channel activity after solubilization and purification steps, we have developed methods for flux assay after reconstitution of the K⁺ channel into phospholipid vesicles.[5,13] To find the optimum procedure, a number of different detergents, including CHAPS, $C_{12}E_8$, octylglucoside, and cholate, and reconstitution procedures, including freeze–thaw sonication and dialysis, were examined. Figure 2 shows that the optimal conditions for solubilization of the K⁺ channel protein are found at a CHAPS/protein weight ratio of 4:1 to 10:1. It is seen that the K⁺ channel protein is solubilized at a lower detergent concentration than the total protein (Fig. 2). Reconstitution into phospholipid vesicles is optimal at a protein/lipid ratio of 1:10 (Fig. 3). The sensitivity of the K⁺ channels to Ca²⁺, pH, and various inhibitors is preserved after solubilization and reconstitution, indicating that the K⁺ channel protein is not denatured.[5]

Procedure. Prior to solubilization, luminal plasma membrane vesicles are collected following 30- to 100-fold dilution in Di buffer (Table I) and then centrifuged for 90 min at 40,000 rpm (4°). The vesicles are resus-

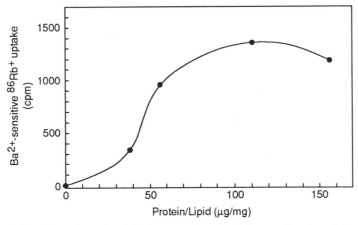

FIG. 3. Potassium channel activity in reconstituted vesicles as a function of the protein/lipid ratio in the reconstitution mixture. Membranes were solubilized at a CHAPS/protein weight ratio of 7:1. For reconstitution the amount of lipid was kept constant, whereas the amount of protein was varied to give the indicated ratio.

pended in 200 μl of R buffer at a protein concentration of 4 mg/ml by homogenization in a tightly fitting Teflon–glass homogenizer. For solubilization, an aliquot of 20 μl of 0.5 M CHAPS in R buffer is added to obtain a CHAPS/protein weight ratio of 7 : 1. The mixture is incubated for 1 min at room temperature and centrifuged for 10 min at 95,000 rpm (180,000 g) in a Beckman (Palo Alto, CA) Airfuge to sediment the insoluble residue.

Phosphatidylcholine is dissolved in R buffer to a final concentration of 50 mg/ml and sonicated for several minutes at room temperature to translucency in a Bransonic bath sonicator with a drop of CHAPS solution added to the bath. An aliquot of 200 μl sonicated lipid is solubilized by addition of 40 μl of 0.5 M CHAPS in R buffer.

The solubilized protein and lipid are mixed at a ratio of 80 μg/mg at room temperature by dragging the mixture back and forth through the constriction of a Carlsberg pipette 5–10 times. For reconstitution by removal of CHAPS, the mixture is applied to a Sephadex G-50 course column (1 $\times$ 30 cm), which has been preequilibrated at 20° with R buffer. The column is eluted with R buffer at a rate of 0.5–1 ml/min, and the reconstituted vesicles are harvested in the void volume in 2–3 fractions of 1 ml.

Assay for Potassium Channel Activity

In both plasma membrane vesicles and reconstituted vesicles K^+ channel activity is determined in a very sensitive assay, measuring $^{86}Rb^+$ uptake into the vesicles against a large opposing gradient of K^+.[14]

Procedure. Dowex 50W-X2 or 50W-X8 is equilibrated with Tris-Cl and poured into Pasteur pipettes, which are plugged with a small ball of glass wool, to make columns with a void volume of 650 μl. The columns are flushed twice with 500 μl ice-cold Do buffer (Table I) and kept on ice until use. To block Na,K-ATPase, the vesicles (1–2 mg protein/ml) are incubated for 10 min at 20° in P buffer with 0.5 mM ouabain, 2 mM MgCl$_2$, 100 μM sodium monovanadate. To deplete Ca^{2+}, the vesicles are incubated for 10–30 min at room temperature with 3.75 mM EGTA–Tris and 10 μM of the Ca^{2+} ionophore A23187 from a stock solution (2 mM) in ethanol.

Prior to assay, external K^+ is exchanged for Tris$^+$ by adding 180 μl of the preincubated vesicles to ice-cold Dowex columns. The vesicles are eluted with 650 μl Do buffer and kept on ice until use. To assay for stimulation by Ca^{2+}, 540 μl of the Ca^{2+}-depleted vesicles are incubated for 1 min at 20° with 100 μl Do buffer containing EGTA to give a final concentration of 3 mM and 0.1–4.0 mM Ca^{2+} to give the desired concen-

---

[14] H. Garty, B. Rudy, and S. J. D. Karlish, *J. Biol. Chem.* **258,** 13094 (1983).

tration of free Ca^{2+}. In solutions containing Ca^{2+}/EGTA buffer systems, it is important to ensure that pH is maintained. The free Ca^{2+} concentrations are calculated according to Pershadsingh and McDonald.[15]

For initiation of the assay, 270 μl of the vesicles is mixed at 20° with 60 μl Do buffer containing reagents to obtain the following final concentrations: 1 mM Furosemide, 100 μM ^{86}RbCl [5 × 10^5 counts/min (cpm)/ 100 μl], 3 mM EGTA, and CaCl$_2$ to give the desired concentration of free Ca^{2+}. The mixture is then incubated for 10 min at room temperature. To stop ^{86}Rb$^+$ uptake, 270 μl of the medium is transferred to an ice-cold Dowex column for removal of external ^{86}Rb$^+$. The ^{86}Rb$^+$ trapped inside the vesicles is eluted into counting vials with 3 times 500 μl ice-cold Do buffer, and activity is measured by Cerenkov counting in the tritium channel of a Packard scintillation counter. The assay is usually done in the absence and presence of an inhibitor or activator, and the K$^+$ channel activity is expressed as the ^{86}Rb$^+$ uptake into the vesicles in picomoles per milligram protein.

Affinity Chromatography

Calmodulin ($K_{0.5}$ ~0.1 nM) doubled the Ca^{2+} stimulation of the K$^+$ channel activity in the reconstituted vesicles, where the cytoplasmic aspect of the K$^+$ channels is exposed to the medium, whereas there was no effect of addition of calmodulin in absence of Ca^{2+}. This observation forms the basis for attempts toward purification of the K$^+$ channel proteins by calmodulin affinity chromatography.[6,16] Solubilized protein from luminal plasma membrane vesicles is added to a calmodulin affinity column in presence of Ca^{2+}. More than 99% of the protein does not bind to the column and is eluted in peak 1 (Fig. 4). The bound protein can be eluted after addition of EGTA (peak 2, Fig. 4), and this protein shows very high Ca^{2+}-activated K$^+$ channel activity after reconstitution into phospholipid vesicles.

Procedure. Luminal plasma membrane vesicles are collected by centrifugation as described above for reconstitution. The vesicles are resuspended in 6 ml of C buffer (Table I) to a concentration of 3–4 mg/ml and homogenized by 10 strokes in a Teflon–glass homogenizer. CHAPS is added from a stock solution (0.5 M in C buffer) to a final concentration of 25 mg/ml to obtain a CHAPS/protein ratio of 6–8. After incubation for 1 min at room temperature, the insoluble residue is sedimented by centrifugation for 10 min in a Beckman Airfuge at 95,000 rpm (180,000 g) or 70,000 rpm (200,000 g) in a Beckman TL-100 ultracentrifuge.

[15] H. A. Pershadsingh and J. M. McDonald, *J. Biol. Chem.* **255**, 4087 (1980).
[16] V. A. Niggli, J. T. Penniston, and E. Carafoli, *J. Biol. Chem.* **254**, 9955 (1979).

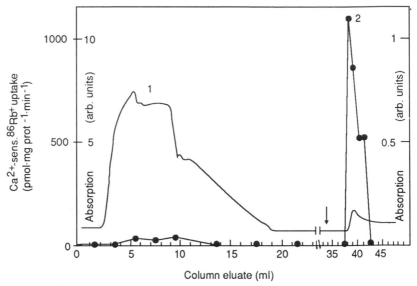

FIG. 4. Affinity chromatography on the calmodulin-Sephadex column. The sample was added at zero, and the change from C to E buffer is indicated by an arrow. Protein from the fractions was reconstituted into phospholipid vesicles, and the Ca^{2+}-sensitive $^{86}Rb^+$ uptake was measured as the difference in uptake in the presence and absence of 1 μM free Ca^{2+} (●). The unbroken line shows the UV absorption with different scales for peak 1 (fractions 2–19) and peak 2 (fractions 37–42).

A calmodulin affinity column (1 × 12 cm) is set up and equilibrated at room temperature with C buffer. It is important to degas the buffer. The soluble protein is loaded on the column at a speed of 0.1 ml/min, and the nonbound protein is eluted from the column with C buffer. The column is then washed with 20 ml of C buffer. The bound protein is eluted from the column after changing to E buffer. The column eluent is continuously monitored by UV absorption at 280 nm, and the protein is collected in Minisorp tubes (Nunc, Copenhagen, Denmark).

To measure K^+ channel activity in the eluate, an aliquot of 200 μl is mixed with 240 μl solubilized lipid and reconstituted into vesicles by removal of the detergent on the Sephadex G-50 column as described above. Calcium-activated K^+ channel activity is measured as described under Assay for Potassium Channel Activity.

Discussion

The procedure described here purifies the calmodulin-binding Ca^{2+}-stimulated K^+ channel more than 100-fold, from 8–10 pmol $^{86}Rb^+$/ mg protein · min in crude plasma membrane vesicles up to 1100 pmol

^{86}RB$^+$/mg protein · min in the reconstituted vesicles after affinity chromatography (Fig. 1). The amount of purified protein is less than 0.7% of the total solubilized protein from the luminal plasma membrane vesicles applied on the affinity column. The purified K$^+$ channel is activated by Ca^{2+} in the same range of concentration as in the native system, and the K$^+$ channel activity is stimulated by phosphorylation from a cAMP-dependent protein kinase[6] as previously described for K$^+$ channels from snail helix ganglia.[17] However, it seems that the K$^+$ channel has lost its sensitivity to Ba^{2+} after the purifying procedure. The procedure has been reproduced in colon epithelium to purify a high conductance Ca^{2+}-activated K$^+$ channel.[18] From Ehrlich ascites tumor cells a Ba^{2+}-sensitive K$^+$ channel can be partially purified.[19]

Gel electrophoresis of the purified protein shows two major bands of 36K and 51K, a minor band at 43K and minor bands at several other positions.[6] After transfer to an Immobilon (Millipore, Bedford, MA) membrane, N-terminal amino acid sequencing identified the band at 51K as β_2-glycoprotein I,[20] and the 43K band has been identified as actin. Repeated attempts at sequencing the 36K band have been unsuccessful.

[17] D. A. Ewald, A. Williams, and I. B. Levitan, *Nature (London)* **315**, 503 (1986).
[18] S. Lin, W. P. Dubinsky, M. K. Haddox, and S. G. Schultz, *Am. J. Physiol.* **261**, C713–C717 (1991).
[19] F. Jessen, E. K. Hofmann, and B. Aabin, *Acta Physiol. Scand.* **136**, 10A, C14 (1989).
[20] J. Lozier, N. Takahashi, and F. W. Putnam, *Proc. Natl. Acad. Sci. U.S.A.* **81**, 3640 (1984).

[39] Assay and Purification of Neuronal Receptors for Inositol 1,4,5-Trisphosphate

By SUNIL R. HINGORANI and WILLIAM S. AGNEW

Introduction

Receptors for inositol 1,4,5-trisphosphate (IP$_3$) serve as ligand-activated, Ca^{2+}-release channels found in intracellular membrane compartments. The initiation of the phosphoinositide cascade, in response to a variety of extracellular signals, results in the formation of IP$_3$ and the subsequent mobilization of Ca^{2+} from intracellular stores.[1] The response

[1] M. J. Berridge and R. F. Irvine, *Nature (London)* **341**, 197 (1989).

can exhibit complex spatial and temporal patterns of Ca^{2+} release[2,3] such as oscillating spikes, intracellular longitudinal and spiral waves, and even intercellular waves spead via gap junctions.[4,5] These signals underly a number of important cellular functions ranging from the stimulation of protein kinases and proteases to the activation of ion channels to the generation of polarity in developing oocytes. Calcium ion itself participates in the modulation of these signals via the mechanism of Ca^{2+}-induced Ca^{2+} release and by regulating IP_3 binding to its receptors, which may occur in a biphasic manner[6] perhaps involving some combination of direct binding of Ca^{2+} by the receptor[7] and/or interaction of the receptor with a dissociable membrane protein that mediates Ca^{2+} sensitivity.[8]

Recent studies have established that mammalian brain, cerebellum in particular, is an extraordinarily rich source of IP_3 receptor proteins, facilitating development of binding assays, methods of purification and reconstitution, as well as molecular cloning. We here describe procedures for rapid purification of the IP_3 receptor complex and an assay to measure quantitatively the binding properties of $[^3H]IP_3$ to both native and purified preparations. These methods should have general utility in the characterization of receptors from a variety of tissues, leading to better understanding of the intracellular receptor–structure mechanism. An adequate binding assay should also help elucidate the regulatory role of accessory proteins such as that involved in Ca^{2+} inhibition of IP_3 binding.

Preparation of Cerebellar Membranes

The techniques described below have been applied to both bovine and rat preparations with qualitatively similar results.

Solutions

Buffer E: 20 mM EPPS [4-(2-hydroxyethyl)-1-piperazinepropanesulfonic acid], pH 8.5 at around 4°, 20 mM NaCl, 100 mM KCl, 1 mM EDTA, 0.02% NaN_3, 1× protease cocktail (added just prior to each step)

Protease cocktail (100×): 50 mM o-phenanthroline, 50 mM L-1-chloro-

[2] M. J. Berridge, *J. Biol. Chem* **265,** 9583 (1990).

[3] T. Meyer, *Cell (Cambridge, Mass.)* **64,** 675 (1991).

[4] A. H. Cornell-Bell, S. M. Finkbeiner, M. S. Cooper, and S. J. Smith, *Science* **247,** 470 (1990).

[5] J. Lechleiter, S. Girard, E. Peralta, and D. Clapham, *Science* **252,** 123 (1991).

[6] E. A. Finch, T. J. Turner, and S. M. Goldin, *Science* **252,** 443 (1991).

[7] R. F. Irvine, *FEBS Lett.* **263,** 5 (1990).

[8] S. K. Danoff, S. Suppatapone, and S. H. Snyder, *Biochem. J.* **254,** 701 (1988).

3-(4-tosylamido)-4-phenyl-2-butanone (TPCK), 100 mM phenyl-methylsulfonyl fluoride (PMSF) made up in acetone

Buffer S: 20 mM EPPS, pH 8.5, 320 mM sucrose, 1 mM EDTA

Membrane Preparation

For bovine preparations, typically two cerebella (35–45 g total wet weight) are obtained from freshly decapitated calves at an abattoir, placed immediately into ice-cold buffer E, and transported back to the laboratory (30 min, door to door). Rat cerebella (~5–6 g total wet weight) are obtained from 15–20 decapitated animals and similarly placed immediately into ice-cold buffer. After weighing, the tissue is transferred to 10 volumes of fresh cold buffer and homogenized with a Polytron tissue homogenizer (Brinkman Instruments, Westbury, NY) with 2 bursts at setting 35 for 15–20 sec each. The homogenate is then sedimented at 100,000 g (35,000 rpm, Beckman Ti 35 rotor) for 30 min at 2°–4°. The supernatant is discarded and the pellet(s) homogenized as above in half the original volume of buffer and sedimented as above. The supernatant is again discarded and the final pellets resuspended in buffer E at a final concentration of 4 mg (wet weight) tissue/ml buffer with the aid of a motor-driven pestle in a Potter–Elvejhem (Thomas Scientific, Swedesboro, NJ) glass homogenizer. This final suspension is stored in 1-ml aliquots at −70° and can be used within 6 months to 1 year without appreciable loss in binding activity. The protein concentration (Bio-Rad, Richmond, CA) of these cerebellar membranes is typically around 20 mg/ml.

Differential sedimentation protocols have also been attempted, and although such an approach does result in fractions with increased specific activity, we generally use the procedure described above to maximize total receptor yield. Nonetheless, Fig. 1 illustrates a scheme involving differential centrifugation. As can be seen in Table I, the final pellet (P4) is enriched in binding activity compared with other fractions, but the yield is quite low.

Protocol. Rat cerebella are homogenized in pairs immediately after decapitation in 20 ml buffer S in a 30 ml Potter–Elvejhem glass homogenizer and stored on ice until 10 cerebella are so collected. Homogenization is performed with 5–6 strokes of a motor-driven pestle. The homogenates are pooled and spun at 30 g (1000 rpm, Sorvall SS-34 rotor) for 10 min. The supernatant is sedimented at 12,000 g (10,000 rpm, Sorvall SS-34) for 20 min and the resulting supernatant (S2) saved; the pellet is resuspended in 100 ml buffer S and spun at 12,000 g (10,000 rpm, Sorvall SS-34) for 20 min. This pellet (P3) is weighed, resuspended in 4 volumes buffer S, and

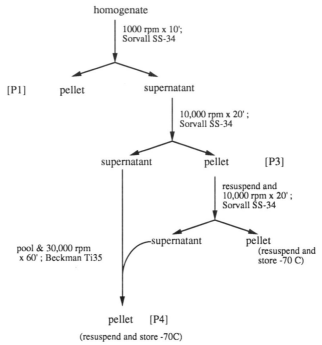

FIG. 1. Scheme for fractionation of cerebellar membranes by differential sedimentation.

stored in 1-ml aliquots at $-70°$. The supernatant (S3) is pooled with S2 and then sedimented at 100,000 g (30,000 rpm, Beckman Ti 35 rotor). This final pellet (P3) appears as a translucent, yellowish smear; 1.5–2.0 ml buffer S is used to recover it with scraping and pipetting repeatedly. The suspension is stored in 200-μl aliquots at $-70°$.

TABLE I

FRACTIONS OBTAINED FROM DIFFERENTIAL CENTRIFUGATION[a]

| Fraction | Protein concentration | | Yield (pmol) |
	mg/ml	pmol/mg[b]	
P1	2.4	1.5	126
P3	7.8	1.1	69
P4	2.6	7.6	36

[a] See text and Fig. 1 for description of fractions.
[b] Nonsaturating conditions.

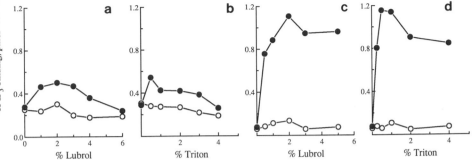

FIG. 2. Solubilization of [^{3}H]IP$_3$ binding sites by Lubrol-PX and Triton X-100 as measured with a gel filtration spun-column assay (a, b) and the syringe assay (c, d). Membrane aliquots prepared as described in the text were incubated for 10 min with varying detergent concentrations and sedimented at 200,000 g (35 Psi, Beckman Airfuge A 18/100 rotor) for 5 min. Supernatants were then assayed for total (filled symbols) and nonspecific (open symbols) [^{3}H]IP$_3$ binding by either assay. Note the high background (nonspecific) and low specific signals observed with the gel filtration, spun column assay as compared with the syringe assay.

Solubilization of Binding Sites

Our initial investigations of receptor binding properties employed a gel filtration, spun-column assay developed for the purification of the tetrodotoxin-binding sodium channel from the Amazonian eel, *Electrophoricus electricus*. The assay separates bound from unbound toxin on the basis of molecular size. Although we were able to compare the relative abilities of various detergents to solubilize IP$_3$ receptor sites from cerebellar membrane fragments using this assay (Fig. 2), the method proved unsatisfactory for more quantitative studies, as is discussed below.

Although another laboratory has reported difficulty in solubilizing IP$_3$ receptors from mouse cerebellum using nonionic detergents (Ref. 9, but see also Ref. 10), we found that IP$_3$ receptors from both rat (Fig. 2) and bovine cerebellar membranes could be readily released into solution by a number of such detergents. In particular, Lubrol PX and Triton X-100 were found to be most effective, whereas sodium cholate, an anionic detergent, was a poor solubilization agent.

Thus, for standard binding experiments and for subsequent receptor purification, we prepare a 2% Lubrol extract in the following manner. One milliliter of thawed cerebellar membranes is added to a 10 ml Potter–Elvejhem glass homogenizer (on ice) containing 3.5 ml buffer E and 2% (w/v) Lubrol PX (0.5 ml of 20%, w/v, stocks stored at −20°). The mixture

^{9} N. Maeda, M. Niinobe, K. Nakahira, and K. Mikoshiba, *J. Neurochem.* **51**, 1724 (1988).
^{10} N. Maeda, M. Niinobe, and K. Mikoshiba, *EMBO J.* **9**, 61 (1990).

is homogenized by hand with 15 strokes, poured into a chilled centrifuge tube, and inverted at 4° for 20 min. The sample is then sedimented at 100,000 g (35,000 rpm, Beckman Ti 40 rotor) for 40 min. The supernatant is referred to as detergent-solubilized extract.

Ion-Exchange Resin Binding Assay

The profiles in Fig. 2 also illustrate some of the inherent limitations with the gel filtration, spun-column assay, namely, the high and variable background ("breakthrough") signals observed. For reasons that remain unclear, the Sephadex G-50 resin inadequately retains unbound [^{3}H-]IP$_3$. Experimenting with different resin buffer conditions (buffer type, ionic strength, detergent concentration), resin type (including BioGel P-10), and column height failed to resolve these problems. However, the highly charged nature of IP$_3$ suggested a basis for another assay involving ion exchange to adsorb unbound ligand and size exclusion to exclude ligand bound to receptor. Finally, dissociation rate experiments indicated that ligand unbinding occurred extremely rapidly compared with the time required for completion of a spun-column assay (45–60 sec), emphasizing the need for methods with more rapidly separate free from bound ligand. A comparatively rapid syringe method was established.

Solutions/Materials

Buffer C: 50 mM Tris, pH 8.5
3-cm^3 Disposable syringes with plungers
SpectraMesh (Spectrum Medical Industries, Los Angeles, CA) nylon
  filters (10 μm pore size)
Whatman (Clifton, NJ) filter paper
Dowex 1-X2, chloride form, anion-exchange resin (Bio-Rad)

Resin Preparation

Fifty grams of Dowex 1-X2 anion-exchange resin is equilibrated with 2 × 100 ml of double-distilled water and excess fluid decanted after resin settling. To the swollen resin is added 100 ml buffer C containing 15 mg/ml bovine serum albumin (BSA), and the slurry is agitated for 2 hr at 4°. The BSA prevents nonspecific protein binding by the resin without affecting its ability to retain free IP$_3$.

Syringe Preparation

Small disks are punched out of nylon mesh (SpectraMesh; 10 μm pore size) using a hammer and a disk cutter. Similar disks are also cut out of Whatman paper. A nylon disk is placed into the barrel of a 3-cm^3 syringe followed by a Whatman disk and then a drop of water to ensure that the disks remain flat when adding the resin. The resin slurry is then added to give a settled bed volume of 0.6 ml. The resin is retained by the Whatman paper disk, while the nylon disk serves to support the former against the high pressures developed during assay. Without the underlying nylon disk, a hole is blown through the paper during the assay and the resin simply passes through. Prepared syringes are typically used within 30 min of pouring.

Assay Mechanics

Although spun-column assays using anion-exchange resin in place of gel filtration resin gave dramatically reduced breakthrough counts, increased net binding, and highly improved reproducibility (Table II[11]), elution by centrifugation was still slow compared with the dissociation rate of receptor–ligand complexes, leading to underestimation of total receptor sites and perhaps distorting estimates of binding affinities. To decrease sample transit time, we employed the plunger to manually expel the sample into a waiting scintillation vial. Figure 3 illustrates the steps in the process.

After resin settling, the tip of the column is gently touched to a Kimwipe to drain a small amount of more fluid. This prevents fluid from flowing up the syringe wall as it is turned horizontally during the following manipulations. With the syringe held horizontally, sample containing equilibrated receptor–ligand complexes is applied to the side wall but not allowed to drain onto the resin bed (Fig. 3a). The plunger is then gently inserted as the syringe is turned vertically (Fig. 3b) and then forced through smoothly and rapidly over a scintillation vial clamped to a ring stand (Fig. 3c). The movement is rapid but not abrupt; an explosive plunge results in high background and variability. However, with a small amount of practice the method becomes remarkably reproducible. The actual transit time of the sample through the resin bed is no greater than 1 sec, resulting in 50% increase in specific binding versus centrifugation using the same resin, and 160% increase compared with gel filtration assays (Table II; Ref. 7). Background counts remain extremely low.

After assay, the nylon filters are retrieved and cleaned by placing them in a small beaker of distilled water and briefly (20–30 sec) sonicating the beaker in a bath sonicator (Laboratory Industries, Hicksville, NY).

With this method, 4- to 5-fold higher receptor abundance than previously reported was found for rat cerebellum together with a previously undetected second class of binding sites (Fig. 4 and Ref. 7). In

[11] S. R. Hingorani and W. S. Agnew, *Anal. Biochem.* **194,** 204 (1991).

TABLE II
COMPARISONS AMONG SYRINGE ASSAY AND SPUN-COLUMN PROCEDURES[a]

Assay	Column volume (ml)	Total ± SD (cpm)	Nonspecific ± SD (cpm)	Specific ± SD (cpm)	Recovery %	n
Sephadex G-50 spun column	1.5[b]	522 ± 98	158 ± 126	394 ± 114	100	4
	1.1[c]	1192 ± 126	495 ± 2	697 ± 128	177	3
Dowex spun column	0.4[d]	1279 ± 96	71 ± 16	1208 ± 112	307	3
Dowex syringe	0.6	1890 ± 36	104 ± 27	1786 ± 63	453	3

[a] Parallel assays of soluble receptor preparations were performed with optimized conditions for the ion-exchange syringe assay and the spun-column gel filtration and ion-exchange protocols. Further increasing centrifugal flow rates for the gel filtration assays did not yield any increase in net (specific) binding. These and similar experiments indicate that (1) the syringe assay detected specific binding 2.5-fold higher than gel filtration spun columns and 1.5-fold higher than ion-exchange spun columns; (2) ion-exchange methods give lower backgrounds than gel filtration; and (3) ion-exchange spun-column procedures are sensitive to rotational velocity (speed of separation), as discussed in the text. Although not evident in these data, gel filtration backgrounds often varied by 2-fold or more. From Hingorani and Agnew.[11]

[b] 2000 rpm × 60 sec.
[c] 1500 rpm × 60 sec.
[d] 2800 rpm × 60 sec.

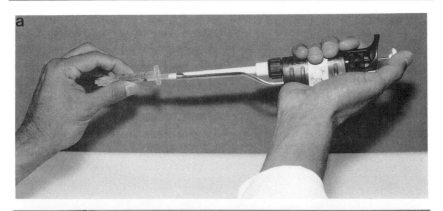

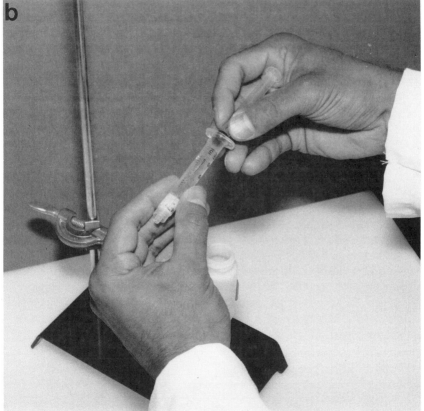

FIG. 3. Steps in executing the plunge assay. (a) Sample addition to the syringe; (b) insertion of the plunger; (c) completion of the assay with sample extrusion into a scintillation vial. *(Figure 3 continues on p. 582)*

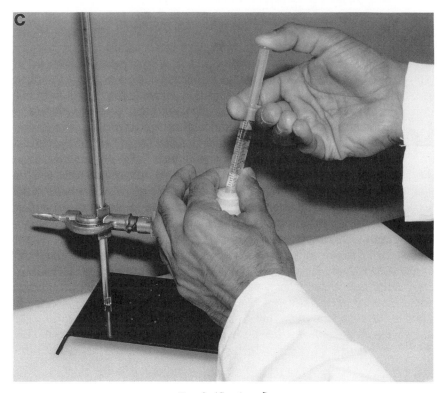

FIG. 3. *(Continued)*

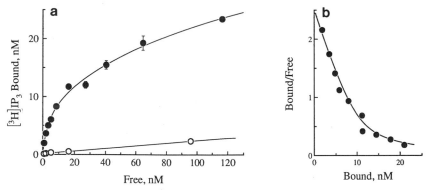

FIG. 4. Saturation binding of [^{3}H]IP$_3$ to detergent-solubilized extract. (a) Binding in the absence (filled symbols) and presence (open symbols) of 12.5 μM unlabeled IP$_3$ and various concentrations of labeled IP$_3$. (b) Scatchard replot of data in (a) fit to a two-binding site model with $K_{d1} = 4$ nM *(B_{max} 8 pmol/mg protein)* and $K_{d2} = 120$ nM *(B_{max} 22 pmol/mg protein)*.

addition, manual sample extrusion confers an added degree of flexibility on the experimenter. For example, by performing the extrusion in two steps it was possible to obtain more accurate estimates of dissociation kinetics.[11] In this procedure, the sample was added to the syringe and the plunger inserted only far enough to force the sample into (but not through) the resin. After waiting variable but measured amounts of time, the sample extrusion was completed by forcing the plunger through the rest of the way. In this manner, rapid dissociation time points were obtainable that were impossible to acquire by more conventional means.

Performing Saturation Isotherms

The extremely high specific activity of commercially available [^{3}H]IP$_3$ (17 Ci/mmol; NEN–Du Pont, Boston, MA) raises practical and economic barriers against performing conventional saturation binding isotherms. In many tissue preparations, receptor abundance is low. At concentrations of ligand required for saturation, nonspecific backgrounds become large compared to specific binding. Also, the expense of saturating receptors with high specific activity ligand is prohibitive.. Thus, many workers have resorted to more indirect methods of measuring binding parameters in which nonsaturating levels of high specific activity ligand are displaced by competition with unlabeled ligand or inhibitors. A discussion of the potential pitfalls and limitations of such methods is given elsewhere.[11] Saturation binding isotherms cna be achieved by using an assay with a high signal-to-noise ratio and decreasing the specific activity of commercial [^{3}H]IP$_3$ with high purity unlabeled IP$_3$ (Boerhinger-Mannheim, Germany).

Preparation of [^{3}H]IP$_3$ Stocks

Stocks of unlabeled IP$_3$ (100 or 200 μM) are prepared in buffer E (minus protease inhibitors) and stored in 20-μl aliquots at $-70°$. From these aliquots, 8 μM stocks, stored at 4°, are prepared and used within 1 week to make final [^{3}H]IP$_3$ stocks of 1.0 or 1.8 Ci/mmol, depending on the aims of the experiment. These final stocks are made fresh for each experiment. To measure nonspecific binding, we use less expensive and somewhat less pure ($\sim$85%) unlabeled IP$_3$ from Sigma (St. Louis, MO); 500 μM stocks are stored in 50-μl aliquots at $-70°$.

Assay Conditions

Assays are typically performed in final volumes of 200 μl including the appropriate volume of radiolabel mix, 140 μl detergent solubilized extract, buffer E, with or without $\pm$5 μl of 500 μM unlabeled Sigma IP$_3$. The samples are vortexed gently, incubated for 5 min at 4°, and then assayed as described.

Figure 4 shows the results of isotherms performed on solubilized receptors from rat cerebellar membranes; similar profiles are seen for bovine preparations.[12] In each case, Scatchard replots are nonlinear, suggesting the presence of either distinct classes of sites or a single class of sites with modifiable affinity states. Curves are fit to a two-binding site model, giving affinities of 4–8 and 100–150 nM for the two states, respectively, with the high affinity state present at 25–35% the abundance of the low affinity state.

Purification of Cerebellar Inositol 1,4,5-Trisphosphate Receptors

The purification protocols discussed exploit some unusual biochemical properties of cerebellar IP$_3$ receptors. These include a complex interaction of the receptor with wheat germ agglutinin and the tendency of the receptor to associate in homotetramers, a property reminiscent of its skeletal muscle homolog, the ryanodine receptor/Ca^{2+} release channel.[13]

Lectin Affinity Purification

Receptors for IP$_3$ interact with a number of lectins, some of which have been used in other purification protocols.[10,15] The complex nature of the heterogeneous oligosaccharide chains on this intracellular receptor is intriguing and at present not understood. In the course of our investigations with lectin–receptor interactions (Table III), we discovered an unusually avid binding of receptors to wheat germ agglutinin (WGA)-coupled resins of high capacity (lower capacity resins gave negligible uptake; Table III). In analytical experiments, adsorption of [^{3}H]IP$_3$ binding activity could not be prevented by preincubating resin aliquots with even 200 mM N-acetylglucosamine (NAG), the competing glycoside (often concentrations of 100 mM NAG are used to desorb proteins from WGA resins specifically). However, a commercially available trimeric form of this sugar, N,N,N-triacetylchitotriose (NAG$_3$), was exceptionally effective at blocking such uptake. The affinity of NAG$_3$ for WGA is approximately 3 orders of magnitude higher than that of its monomeric form, NAG. This property led to a simplified purification scheme that can be performed on a preparative scale.

[12] S. R. Hingorani, K. Ondrias, B. E. Ehrlich, and W. S. Agnew, submitted for publication.
[13] F. L. Lai, H. P. Erickson, L. Rousseau, Q. Y. Liu, and G. Meissner, *Nature (London)* **331**, 315 (1988).
[14] Porath, this series, Vol. 34, p. 13.

TABLE III
LECTIN–RECEPTOR INTERACTIONS

Lectin	mg Lectin/ml resin	Uptake	Sugar specificity[a]
Triticum vulgaris (wheat germ agglutinin)	0.5	0	NAG$_3$ > NAG
	1.6	0	
	>8.0	+++	
Lens culinaris (lentil)	7.5	+	MMP
Pisum sativum (pea)	2.0	+	MMP
Limax flavus agglutinin (slug)	5.0	0	NeuNAc
Solanum tuberosum (potato)	3.8	0	NAG$_3$

[a] NAG; *N*-Acetylglucosamine; NAG$_3$, *N,N,N*-triacetylglucosamine; MMP, methylmannopyranoside; and NeuNAc, *N*-acetylneuraminic acid.

Solutions

Buffer ELPC: 20 m*M* EPPS, pH 8.5, 20 m*M* NaCl, 100 m*M* KCl, 1 m*M* EDTA, 0.02% NaN$_3$, 0.1% Lubrol, 0.0184% phosphatidylcholine (Sigma). Phosphatidylcholine (PC) is dried down in a Pyrex tube under a stream of argon gas to a waxy consistency. Two milliliters buffer E is added to the dried lipid and then sonicated, sealed under argon, to a milky white suspension. The total amount of detergent required to give 0.1% in the final buffer is then added and the suspension sonicated to clarity. This mixture is then added to the appropriate volume of buffer E to create ELPC.

WGA-Sepharose 4B-CL: WGA is purchased from E-Y Labs, Inc. (San Mateo, CA) coupled to Sepharose 4B-CL as described by Porath.[14] Resins are prepared at approximately 9 mg WGA/ml settled resin.

Sucrose stocks: 5% and 20% stocks prepared in ELPC 1 × 3.5 inch., quick-seal polyallomer centrifuge tubes (Beckman)

Procedure. Three to five milliliters of rat cerebellar membranes (or 10 ml bovine cerebellar membranes) are solubilized as described as above. To 20–22 ml of the recovered supernatant (~30 ml for bovine solubilization) NAG is added to a final concentration of 100 m*M*. This is then mixed with 5 ml of WGA-resin preincubated with 100 m*M* NAG and the slurry incubated batchwise with rotation for 3 hr at 4°. As shown in Fig. 5, the majority (~80%) of sites are adsorbed extremely rapidly ($t_{1/2}$ ~5 mins); a subsequent 15–20% of sites are adsorbed more slowly ($t_{1/2}$ ~100 min).

[15] S. Supattapone, P. F. Worley, J. M. Baraban, and S. H. Snyder, *J. Biol. Chem.* **263**, 1530 (1988).

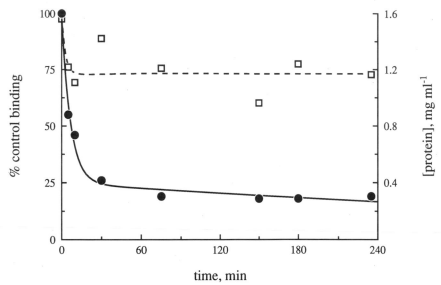

FIG. 5. Rate of uptake of protein (open symbols) and specific [³H]IP₃ binding (filled symbols) by lectin affinity resin. Binding was measured in the presence of 25 nM [³H]IP₃. The curve for the binding profile is a double-exponential fit with the majority of sites (81%) adsorbed rapidly ($t_{1/2}$ ~5 min) and the remainder (19%) adsorbed much more slowly($t_{1/2}$ ~100 min). The double-exponential fit was substantially better than the best single-exponential fit.

From here, the procedure can be performed either in a batchwise or column fashion depending on the subsequent aims of the experiment.

Batchwise Elution. Batchwise elution allows rapid (few hours) preparation of highly purified, though not rigorously homogeneous, receptor. After adsorption of receptors, the resin is washed batchwise with 4 × 40 ml ELPC; low-speed centrifugation (500 rpm, Dynac II desktop centrifuge) in between washes speeds resin setting. After resin washing, an equal volume of ELPC containing 2 mM NAG₃ is added and the slurry incubated with rocking for 10 min at 4°. The eluant, containig enriched IP₃ receptors, is

TABLE IV
SPECIFIC ACTIVITY OF PEAK BINDING FRACTIONS

Preparation	Binding affinity (mM)	B_{max} (pmol/mg)	Yield (%)
Solubilized extract	4 (I)	8–10	
	120 (II)	22–26	100
WGA (batchwise)	5	400–800	40–50
WGA (column pool)	5	325	27
Sucrose (one step)	—	3000–3500	—
WGA/sucrose	26	4100	10

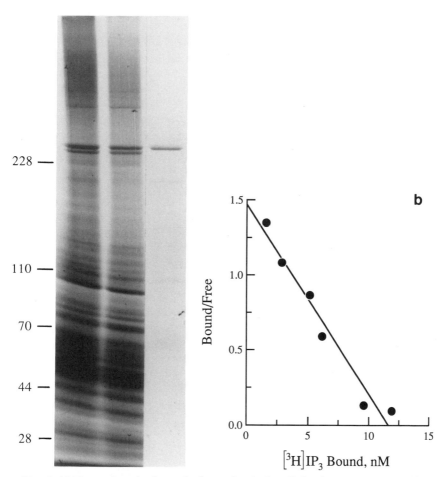

FIG. 6. (a) Electrophoresis of samples from a batchwise affinity chromatography purification. Lane 1, Detergent-solubilized extract; lane 2, extract minus material adsorbed to lectin-resin; lane 3, batchwise eluant of resin. A single prominent peptide of M_r 243,000 is revealed. (b) Scatchard plot of IP$_3$ receptors purified as in (a). The fit reveals a single class of sites with $K_d = 8$ nM. Affinities of purified material ranged between 4 and 8 nM.

then drained and the resin washed with another volume of ELPC to recover more receptor. Receptor isolated in this fashion is of high specific activity (Table IV) and essentially devoid of contaminants (Fig. 6a). Scatchard replots of saturation isotherms (Fig. 6b) reveal a single class of binding sites of high affinity (4–8 nM). In addition, the sensitivity of IP$_3$

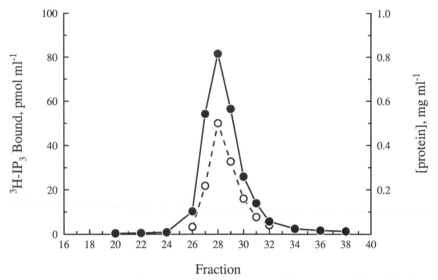

FIG. 7. [³H]IP₃ binding (filled symbols) and protein profiles (open symbols) of fractions collected from columnwise elution of the lectin-affinity resin.

binding to inhibition by Ca^{2+} is lost, a property that can be reconferred by addition of residual extract from which IP_3 binding activity has been removed (i.e., the column "flow-through"). This preparation has been reconstituted into planar lipid bilayers and demonstrated to form an IP_3-activatable, Ca^{2+}-permeable, heparin-sensitive channel.

Columnwise Elution. To obtain a completely homogeneous preparation, receptors are eluted from the lectin-resin in a columnwise fashion, peak fractions pooled, and the pool centrifuged through a linear 5–20% sucrose gradient. After uptake, the WGA-resin is poured into a 15-ml polypropylene column (Bio-Rad) and eluted with 10 ml of 1 mM NAG₃ at a flow rate of 4 ml/hr using a peristaltic pump to control flow. Significantly faster flow rates result in little or no eluted receptor. The binding activity (under nonsaturating conditions) and protein profiles of eluted fractions are shown in Fig. 7.

Sucrose Gradient Centrifugation. Thirty-eight-milliliter 5–20% linear sucrose gradients in ELPC are poured using a gradient maker. Approximately 2.5 ml of pooled fractions from the lectin affinity elution is added to the top of the gradients, and the tubes are heat sealed and run in a vertical reorienting rotor at 150,000 g (45,000 rpm, Beckman VTi 50 rotor) for 2 hr at 2–4°. Parallel gradients can be run containing molecular weight standards, 1 mg each of BSA and thyroglobulin, in 2.5 ml ELPC.

Fractions (90 drops, ~1.4 ml) are collected by puncturing the tube bottom. Recovered [^{3}H]IP$_3$ binding activity migrates at 22 S, implying a molecular weight for the complex of around 1,000,000 (Fig. 8). By sodium dodecyl sulfate–polyacrylamide gel electrophoresis (SDS-PAGE), however, peak fractions run as 243,000 peptides (Fig. 9). Peak binding fractions are of extremely high specific activity (Table IV) and appear pure by SDS-PAGE. Isotherms of purified material again reveal a single class of binding sites with high affinity.

The tendency of the receptor to migrate as a large molecular weight complex enables a substantial purification in a different single-step procedure than described above. Solubilized extract (2–2.5 ml) can be applied directly to a 38-ml 5–20% sucrose gradient and centrifuged as described. The resulting peak fractions (fractions 10 and 11) are highly purified, exhibiting specific activities of approximately 3000–3500 pmol/mg protein (Table IV), and are essentially devoid of contaminants (Fig. 10). It appears that beginning with the P4 pellet of cerebellar membranes with this procedure gives essentially homogeneous material in a single step of su-

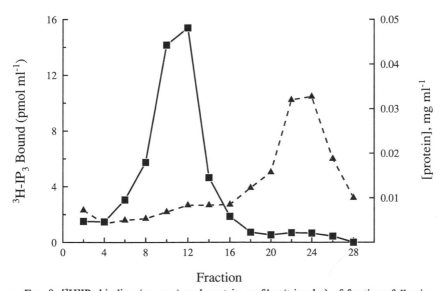

FIG. 8. [^{3}H]IP$_3$ binding (squares) and protein profiles (triangles) of fractions following sucrose gradient centrifugation of pooled fractions from lectin affinity column chromatography.

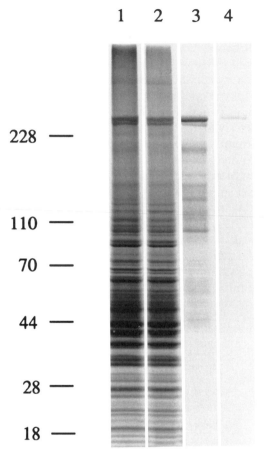

FIG. 9. Electrophoresis of samples from complete, two-step purification of IP$_3$ receptors. Lane 1, Detergent-soulblized extract; lane 2, lectin column flow-through; lane 3, pool of peak fractions from lectin affinity column (see Fig 8); lane 4, 0.8 μg protein of the peak fraction from sucrose gradient centrifugation.

crose gradient centrifugation. The ryanodine-sensitive Ca^{2+} release channel from skeletal muscle has been purified to homogeneity with this approach beginning with T-tubule membranes enriched in ryanodine binding activity.[13]

Receptor purified in this fashion also retains its sensitivity to Ca^{2+}, suggesting that an associated membrane protein that is lost during chromatographic methods is instead retained during sucrose gradient centrifugation. Thus, it is possible in a few hours to obtain highly purified receptor

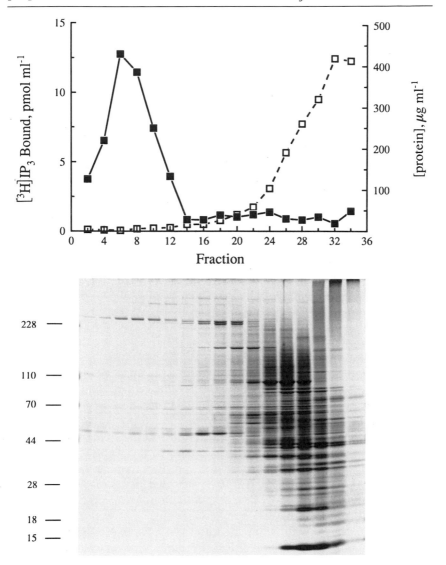

FIG. 10. (a) [^{3}H]IP$_3$ binding (filled symbols) and protein profiles (open symbols) from sucrose gradient centrifugation of detergent-solubilized extract; SDS-PAGE of same.

preparations with or without associated sensitivity to Ca^{2+}, using either the batchwise lectin chromatography or one-step sucrose gradient procedures, respectively. Such preparations should prove useful in investigations of the complex nature of Ca^{2+} interactions with IP$_3$ receptors.

[40] Isolation of Ion Channel Genes by Expression Cloning in *Xenopus* Oocytes

By GEORGES C. FRECH and ROLF H. JOHO

Introduction

The classic method to determine the molecular, primary structure of a protein begins with its biochemical purification. The purified protein serves as a basis for either generating antibodies or obtaining a partial amino acid sequence. Antibodies or oligonucleotide probes derived from the amino acid sequence are then used as the tools to screen cDNA libraries. Purification of ion channel proteins is hindered by the fact that most ion channels are low abundance, integral membrane proteins. Furthermore, selective, high-affinity ligands, useful tools for the purification of low abundance proteins via affinity chromatography, are not available in some instances. Alternative methods that circumvent the purification of such ion channel proteins are therefore needed.

The approach of expression cloning described in this chapter offers such an alternative. It combines directional cDNA cloning in a transcription-competent vector with a functional electrophysiological assay in cRNA-injected *Xenopus* oocytes. By taking this approach, three cDNA clones encoding two members of a class of voltage-gated ion channels and one member of a class of ligand-gated ion channels have been successfully isolated, namely, a slowly activating voltage-gated potassium channel,[1] a delayed rectifier-type potassium channel,[2] and a kainate receptor.[3]

The strategy described here is not limited to the identification of ion channels but may be applied to any gene product that can be assayed by oocyte expression. In fact, this strategy was used first to isolate a cDNA clone encoding a lymphokine, the IgG$_1$ induction factor, now called interleukin 4 (IL-4).[4] Microinjected oocytes secreting IL-4 were identified by collecting the oocyte incubation medium and assaying its IgG$_1$-inducing activity on lipopolysaccharide (LPS)-stimulated spleen cells. A cDNA clone encoding a Na$^+$/glucose-cotransporter has also been isolated by

---

[1] T. Takumi, H. Ohkubo, and S. Nakanishi, *Science* **242,** 1042 (1988).

[2] G. C. Frech, A. M. J. VanDongen, G. Schuster, A. M. Brown, and R. H. Joho, *Nature (London)* **340,** 642 (1989).

[3] M. Hollmann, A. O'Shea-Greenfield, S. W. Rogers, and S. Heinemann, *Nature (London)* **342,** 643 (1989).

[4] Y. Noma, P. Sideras, T. Naito, S. Bergstedt-Lindquist, C. Azuma, E. Severinson, T. Tanabe, T. Kinashi, F. Matsuda, Y Yaoita, and T. Honjo, *Nature (London)* **319,** 640 (1986).

expression cloning in *Xenopus* oocytes.[5] In this case the bioassay involved the measurement of Na^+-dependent ^{14}C-labeled sugar uptake into microinjected oocytes. Furthermore, expression cloning of the enzyme steroid 5α-reductase was achieved by assaying steroid 5α-reductase activity by thin-layer chromatography using [^{14}C]testosterone as a substrate.[6] Finally, several G protein-coupled neurotransmitter receptor cDNAs have been isolated by expression cloning, namely, the substance K receptor,[7] the serotonin 1c (5HT1$_c$) receptor,[8] the substance P receptor,[9] the neurotensin receptor,[10] the thyrotropin-releasing hormone (TRH) receptor,[11] the bombesin/gastrin-releasing peptide (GRP) receptor,[12] an endothelin (ET) receptor,[13] the platelet-activating factor (PAF) receptor,[14] and a metabotropic glutamate receptor.[15] When expressed in the oocyte, these receptors couple to an endogenously present G protein on stimulation with the appropriate ligand. Receptor activation leads to phosphatidylinositol-4,5 biphosphate (PIP$_2$) turnover, elevation of intracellular Ca^{2+} concentration through release from intracellular, inositol 1,4,5-trisphosphate (IP$_3$)-sensitive stores, and finally activation of a Ca^{2+}-sensitive Cl$^-$ conductance. It is the Ca^{2+}-induced Cl$^-$ current that allows one to screen for expression of these receptors electrophysiologically.

Principle of Method

The core steps involved in the expression cloning approach are illustrated in Fig. 1. Poly(A)$^+$ RNA from a tissue or a cell line is microinjected into *Xenopus* oocytes, and the expression of functional ion channels of

[5] M. A. Hediger, M. J. Coady, T. S. Ikeda, and E. M. Wright, *Nature (London)* **330**, 379 (1987).

[6] S. Andersson, R. W. Bishop, and D. W. Russell, *J. Biol. Chem.* **264**, 16249 (1989).

[7] Y. Masu, K. Nakayama, H. Tamaki, Y. Harada, M. Kuno, and S. Nakanishi, *Nature (London)* **329**, 836 (1987).

[8] D. Julius, A. B. MacDermott, R. Axel, and T. M. Jessell, *Science* **241**, 558 (1988).

[9] Y. Yokota, Y. Sasai, K. Tanaka, T. Fujiwara, K. Tsuchida, R. Shigemoto, A. Kakizuka, H. Ohkubo, and S. Nakanishi, *J. Biol. Chem.* **264**, 17649 (1989).

[10] K. Tanaka, M. Masu, and S. Nakanishi, *Neuron* **4**, 847 (1990).

[11] R. E. Straub, G. C. Frech, R. H. Joho, and M. C. Gershengorn, *Proc. Natl. Acad. Sci. U.S.A.* **87**, 9514 (1990).

[12] E. R. Spindel, E. Giladi, P. Brehm, R. H. Goodman, and T. P. Segerson, *Mol. Endocrinol.* **4**, 1956 (1990).

[13] H. Arai, S. Hori, I. Aramori, H. Ohkubo, and S. Nakanishi, *Nature (London)* **348**, 730 (1990).

[14] Z. Honda, M. Nakamura, I. Miki, M. Minami, T. Watanabe, Y. Seyama, H. Okado, H. Toh, K. Ito, T. Miyamoto, and T. Shimuzu, *Nature (London)* **349**, 342 (1991).

[15] M. Masu, Y. Tanabe, K. Tsuchida, R. Shigemoto, and S. Nakanishi, *Nature (London)* **349**, 760 (1991).

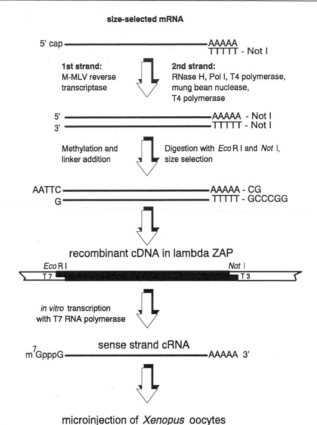

FIG. 1. Construction of cDNA libraries for expression cloning in *Xenopus* oocytes. The individual steps are described in the text.

interest is assayed by electrophysiological recording techniques. The poly-(A)$^+$ RNA is size fractionated on a sucrose gradient, and the size fraction giving rise to peak responses in the oocytes is used to construct a cDNA library in the transcription-competent bacteriophage vector λZAP.[16] First-strand cDNA synthesis is primed with an oligo(dT) primer/linker containing a *Not*I recognition sequence at its 5′ end. Double-stranded cDNA is synthesized basically according to the method described by Gubler and Hoffman.[17] The double-stranded cDNA is methylated at all *Eco*RI recognition sites before *Eco*RI linkers are ligated to the cDNA. Double restric-

[16] J. M. Short, J. M. Fernandez, J. A. Sorge, and W. D. Huse, *Nucleic Acids Res.* **16,** 7583 (1988).
[17] U. Gubler and B. J. Hoffman, *Gene* **25,** 263 (1983).

tion digestion with *Eco*RI and *Not*I creates cDNA molecules with asymmetric ends. The entire cDNA population is size separated by agarose gel electrophoresis, and only that size range is selected corresponding to the mRNA fraction it is derived from. After electroelution the cDNA is ligated directionally into λZAP and subsequently packaged *in vitro* to create a phage cDNA library. The cDNA library is amplified as independent sublibraries, and phage DNA is prepared from each sublibrary. Sense complementary RNA (cRNA) is synthesized *in vitro* from phage DNA derived from individual sublibraries. After microinjection of cRNA into *Xenopus* oocytes, sublibraries containing cDNA clones encoding functional ion channel proteins are identified by electrophysiological recording techniques. A positively responding sublibrary is subdivided into smaller pools through a process of sib selection, until a single cDNA clone is identified.

In the case of an ion channel consisting of several different subunits, that is, with more than one type of subunit being required for function, the response will be lost while going through the process of sib selection. Therefore, at the pool size level where the response is lost, cRNAs derived from different pools need to be coinjected into *Xenopus* oocytes to identify two (or more) different pools which restore a response. One such positive pool can then be taken through the process of sib selection at a time, by always coinjecting with the other positive pool(s).

Materials

The cloning vector λZAP, its *Escherichia coli* host stain XL1-Blue, and the Gigapack Gold and Gigapack Plus *in vitro* packaging extracts are available from Stratagene (La Jolla, CA). Phosphorylated *Eco*RI linkers can be purchased from New England Biolabs (Beverly, MA). The agarose Seakem GTG is available from FMC (Rockland, ME). The electroelution device Elutrap is obtained from Schleicher and Schuell (Keene, NH). The cap analog $m^7G(5')ppp(5')G$ can be purchased from Pharmacia (Piscataway, NJ). The RNAase inhibitor RNasin is available from Promega (Madison, WI).

Buffers and Solutions

TE: 10 mM Tris-HCl (pH 7.5), 1 mM EDTA
5× First-strand cDNA synthesis buffer: 0.25 M Tris-HCl (pH 8.3), 0.375 M KCl, 50 mM dithiothreitol (DTT), 15 mM MgCl$_2$, 5 mM of each dNTP (N = A, C, G, T), 0.5 mg/ml bovine serum albumin (BSA)
5× Second-strand cDNA synthesis buffer: 0.2 M Tris-HCl (pH 8.0), 45 mM MgCl$_2$, 40 mM DTT

5× Mung bean nuclease buffer: 0.15 M Sodium acetate (pH 5.0), 0.25 M NaCl, 5 mM ZnCl$_2$, 25% glycerol

10× T4 DNA polymerase buffer: 0.5 M Tris-HCl (pH 8.0), 0.15 M (NH$_4$)$_2$SO$_4$, 0.1 M MgCl$_2$, 1 mM EDTA, 0.1 M DTT, 2 mM of each dNTP, 1 mg/ml BSA

5× EcoRI methylase buffer: 0.5 M Tris-HCl (pH 8.0), 0.5 M NaCl, 5mM EDTA, 50 mM DTT, 0.4 mM S-adenosylmethionine (SAM), 0.5 mg/ml BSA

10× Ligation buffer: 0.5 M Tris-HCl (pH 7.5), 0.1 M MgCl$_2$, 0.1 M DTT, 10 mM ATP, 0.5 mg/ml BSA

10× High salt buffer: 0.5 M Tris-HCl (pH 7.5), 1M NaCl, 0.1 M MgCl$_2$, 10 mM dithioerythritol (DTE)

SM: 50 mM Tris-HCl (pH 7.5), 0.1 M NaCl, 10 mM MgCl$_2$, 0.01% (w/v) gelatin

Methods

Standard techniques for isolating and characterizing nucleic acids, such as phenol–chloroform extractions, alcohol precipitations, agarose gel electrophoresis, quantification of ^{32}P-labeled nucleic acids, and selection of poly(A)$^+$ RNA by oligo(dT) cellulose chromatography, are described in detail elsewhere.[18,19] Also, the following procedures are described in detail elsewhere in this volume: purification of RNA as a source of ion channels ([16] and [17]), maintenance of *Xenopus laevis* for oocyte production [15], microinjection of *Xenopus* oocytes [14], electrophysiological recording from *Xenopus* oocytes [19].

Size Fractionation of Poly(A)$^+$ RNA

1. Using a standard gradient mixer, prepare a linear 10–30% sucrose gradient containing 10 mM Tris-HCl (pH 7.5), 1 mM EDTA, 0.5% sodium sarcosinate in a 38.5-ml open-top centrifuge tube (Beckman, Palo Alto, CA, Cat. No. 344058).

2. Heat poly(A)$^+$ RNA in TE for 5 min at 70°, chill it on ice, and load it on top of the sucrose gradient (up to 1 mg of RNA in 0.5 ml TE per tube). Do not add sodium sarcosinate to the RNA because this may lead to the formation of a precipitate.

3. Centrifuge the RNA in a SW-28 Beckman rotor for 20 hr at 25,000 rpm (110,000g_{max}) at 4°.

[18] S. L. Berger and A. R. Kimmel, eds., this series, Vol. 152.

[19] J. Sambrook, E. F. Fritsch, and T. Maniatis, "Molecular Cloning: A Laboratory Manual," 2nd Ed. Cold Spring Harbor Laboratory, Cold Spring Harbor, New York, 1989.

4. Puncture the bottom of the tube with a hypodermic needle, collect 38 fractions (1 ml each) and determine the RNA concentration of each fraction by measuring the OD_{260} (consider that the varying sucrose concentration gives rise to a different baseline for each fraction).

5. Adjust the sodium acetate concentration to 0.3 M and precipitate the RNA with 2.5 volumes of ethanol. Dissolve the RNA in 350 μl TE and extract with phenol–chloroform. Reprecipitate the RNA with sodium acetate and ethanol.

6. Dissolve the RNA from each fraction in 100 μl water. The size-fractionated RNA is now ready to be microinjected into oocytes and assayed for the expression of ion channels. Store the RNA at −80°.

Usually, after one passage over oligo(dT)-cellulose, poly (A)$^+$ RNA still contains sufficient 28 S and 18 S rRNAs to serve as convenient internal size markers on the sucrose gradient profile. The size distributions of the different RNA fractions may be determined more accurately by running a small aliquot from each fraction on a denaturing agarose gel, transferring it to a hybridization filter, and hybridizing to it a ^{32}P-labeled oligo(dT) probe.[20]

cDNA Synthesis

To enrich for full-length cDNA it is wise to use the best reverse transcriptase available. We have compared different lots of Moloney murine leukemia virus (M-MLV) reverse transcriptase from three different vendors (Fig. 2). The enzymes obtained from Bethesda Research Laboratories (BRL, Gaithersburg, MD) performed best, in terms of both yield and size distribution of first-strand cDNA, although there were considerable differences between individual lots. These differences were especially apparent when long RNA was used as template, but they became negligible when using shorter RNA (Fig. 2). In any case, we recommend testing any reverse transcriptase, as well as all other enzymes involved in library construction, before use on a precious RNA sample.

1. Mix 10–20 μg of fractionated poly(A)$^+$ RNA with 5 μg primer/linker in 40 μl of 10 mM Tris-HCl (pH 7.5). Heat the mixture for 1 min at 70° and let it cool slowly to room temperature.

2. Add water, 20 μl of 5× first-strand cDNA synthesis buffer, 160 units RNasin, and 10 μl M-MLV reverse transcriptase (200 units/μl), to obtain a total volume of 100 μl. Incubate for 2 hr at 37°.

---

[20] A. L. Goldin, T. Snutch, H. Luebbert, A. Dowsett, J. Marshall, V. Auld, W. Downey, L. C. Fritz, H. A. Lester, R. Dunn, W. A. Catterall, and N. Davidson, *Proc. Natl. Acad. Sci. U.S.A.* **83**, 7503 (1986).

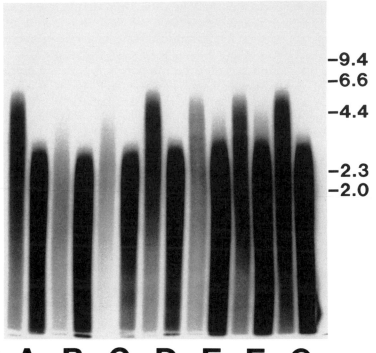

A B C D E F G

Fig. 2. Size analysis of first-strand cDNA: Comparison of different M-MLV reverse transcriptases. Two different mRNA size fractions (~5.5–6.5 and ~2.5–3.0 kb) obtained from a sucrose gradient were used to synthesize first-strand cDNA as described in the text. The reactions were scaled down to 10-μl volumes, and 5 μCi [α-^{32}P]dCTP was present in each sample. Each enzyme was tested on both mRNA size fractions (left-hand side, 5.5–6.5 kb; right-hand side, 2.5–3.0 kb). The enzymes tested and their unit concentrations as supplied by the manufacturers were as follows. (A, D–G) Various lots from BRL (200 U/μl); (B) Pharmacia (15 U/μl); (C) Stratagene (15 U/μl). The samples were electrophoresed through a 1.0% alkaline agarose gel using *Hin*dIII-cut λ DNA as a size marker.

3. Add water, 80 μl of 5× second-strand cDNA synthesis buffer, 1–2 μCi fresh [α-^{32}P]dCTP (*note:* the radioactive label may be added to either the first-strand or the second-strand reaction), 8 units RNase H, and 200 units *E. coli* DNA polymerase I, to obtain a total volume of 400 μl. Incubate sequentially for 1 hr at 12° and 1 hr at room temperature.

4. Heat the mixture for 10 min at 70°.

5. Add 20 units of T4 DNA polymerase and incubate for 15 min at 37°.

6. Extract with phenol–chloroform, adjust the NaCl concentration to 0.1 *M*, and precipitate the cDNA with 2 volumes of ethanol.

The primer/linker we have used successfully contained 50 dT residues preceded by a *Not*I recognition sequence and an extra 10 nucleotides at its 5' end.[2,11,21] The radioactive label [α-^{32}P]dCTP, may be added to either the first-strand or the second-strand reaction.

Opening of Hairpin Structures, Linker Addition, and Restriction Digestion

As discussed elsewhere,[21] for a portion of the cDNA population, second-strand priming occurs via hairpin loop formation. These molecules are not available for further manipulations unless the hairpins are opened up by treatment with mung bean nuclease.

1. Incubate the cDNA in 200 μl of 1$\times$ mung bean nuclease buffer containing the appropriate concentration of mung bean nuclease for 1 hr at 30°. (The exact amount of nuclease needs to be determined empirically to assure mild conditions that only open hairpin structures without shortening the double-stranded cDNA.[21])

2. Extract with phenol–chloroform, adjust the ammonium acetate concentration to 1 *M*, and precipitate the cDNA with 2 volumes of ethanol.

3. Incubate the cDNA in 100 μl of 1$\times$ T4 DNA polymerase buffer containing 10 units T4 DNA polymerase for 30 min at 37°. (This serves to assure blunt end formation, a requirement for successful linker addition.)

4. Extract with phenol–chloroform, adjust the ammonium acetate concentration to 2 *M*, and precipitate the cDNA with 2 volumes of ethanol.

5. Incubate the cDNA in 40 μl of 1$\times$ *Eco*RI methylase buffer containing 80 units *Eco*RI methylase for 1 hr at 37°. Add an additional 80 units of enzyme and incubate for another hour at 37°.

6. Extract with phenol–chloroform, adjust the ammonium acetate concentration to 2 *M*, and precipitate the cDNA with 2 volumes of ethanol.

7. Incubate the cDNA in 20 μl of 1$\times$ ligation buffer containing 15% polyethylene glycol (PEG 8000), 2 μg phosphorylated *Eco*RI linkers (dodecamers), and 2 units T4 DNA ligase at 16° overnight.

8. Adjust the NaCl concentration to 0.1 *M* and heat to 65° for 15 min.

9. Add water, 33 μl of 10$\times$ high salt buffer, and 70 units of *Not*I (high concentration, > 35 U/μl), to obtain a total volume of 350 μl. Incubate for 2 hr at 37°. Add an additional 70 units of *Not*I and continue the incubation at 37° for 2 hr longer.

10. Add 250 units of *Eco*RI (high concentration, > 70 U/μl) and incu-

---

[21] G. C. Frech and R. H. Joho, *Gene Anal. Tech.* **6**, 33 (1989).

bate for 2 hr at 37°. Add an additional 250 units of *Eco*RI and continue the incubation at 37° for 2 hr longer.

11. Extract with phenol–chloroform; precipitate the DNA with 2 volumes of ethanol.

*Eco*RI adaptors may be used instead of *Eco*RI linkers. In that case, two enzymatic manipulations, methylation of internal *Eco*RI sites and restriction digestion with *Eco*RI, are traded for one enzymatic manipulation, phosphorylation of the adaptors already ligated to the cDNA. However, it has been reported that the cloning efficiency with linkers can be as much as 5-fold higher than that with adaptors.[18]

Size Selection and Electroelution

1. Prepare a 1% Seakem GTG agarose gel, using a slot former with teeth 2 cm wide. Load the DNA into one 2-cm well, flanked on each side by appropriate size markers.

2. Electrophorese at low, constant voltage (~ 2 V/cm) for about 3 hr at room temperature.

3. Excise the cDNA which corresponds in size to the mRNA it is derived from. (This may represent only a small percentage of the total cDNA synthesized, depending on the size of the mRNA.)

4. Electroelute the cDNA in the Elutrap electroseparation system, according to the manufacturer's instructions. Electroelute until no more radioactivity remains in the gel slice.

5. Extract with phenol–chloroform, adjust the NaCl concentration to 0.1 M, and precipitate the cDNA with 2 volumes of ethanol.

6. Dissolve the cDNA in 10 mM Tris-HCl (pH 7.5) at a concentration of 10–50 ng/μl. The cDNA is now ready to be ligated into the cloning vector.

Preparation of Cloning Vector

1. Incubate 10 μg of undigested λZAP DNA in 100 μl of 1× ligation buffer containing 10 units T4 DNA ligase overnight at 4° to ligate the cohesive ends.

2. Adjust the NaCl concentration to 0.1 M and heat to 65° for 15 min.

3. Add water, 60 μl of 10× high salt buffer, and 20 units of *Not*I to obtain a total volume of 700 μl. Incubate for 2 hr at 37°. Run an aliquot of the reaction mixture on an analytical 0.4% agarose gel to ensure that *Not*I has digested to completion.

4. Add 200 units of *Eco*RI (high concentration, > 70 U/μl) and incubate for 2 hr at 37°.

5. Add 0.1 unit of calf alkaline phosphatase and incubate for an additional 30 min at 37°.

6. Heat to 65° for 45 min and then extract with phenol–chloroform. Precipitate the phage DNA with 1 volume of 2-propanol.

7. Dissolve the phage DNA in 10 mM Tris-HCl (pH 7.5) at a concentration of 0.2–0.4 μg/μl. The ligation and packaging efficiency of the prepared vector DNA should now be examined prior to cDNA library construction. (This step is especially important if the cDNA is precious because the starting material, the mRNA template, is not easily obtained in large quantities).

8. Mix 200 ng of λ vector DNA with 0.6 μl of 10× ligation buffer and adjust the volume to 4 μl with water. Split the mix into two tubes (A and B). To tube A add 0.3 μl T4 DNA ligase (1 U/μl) and 0.7 μl water; to tube B add 0.3 μl ligase and 0.7 μl of a test insert (containing *Eco*RI/*Not*I asymmetric ends). Aim for a molar ratio of test insert to vector DNA of 2:1. Incubate for 2 hr or longer at 14°.

9. For the *in vitro* packaging procedure, follow the manufacturer's instructions. Package 1 μl from tube A and 1 μl from tube B, by splitting one Gigapack Plus packaging extract. (A Gigapack Plus packaging extract may be used at this step. However, for the final library construction the highest efficiency packaging extract, Gigapack Gold, should be used.)

10. Titrate the diluted phage suspensions on *E. coli* XL1-Blue in the presence of isopropyl-β-D-thiogalactopyranoside (IPTG) and 5-bromo-4-chloro-3-indolyl-β-D-galactoside (X-Gal). Using a control *Eco*RI/*Not*I fragment of 2 kilobases (kb), we routinely obtain over 5×10^7 plaque-forming units (pfu) per microgram of λ vector DNA with a background of less than 0.2% (white pfu).

Ligation, Packaging, and Amplification of Library

Once the cloning vector has been tested, we recommend determining the molar ratio of cDNA to vector DNA that maximizes the number of recombinants and the ratio of recombinant to background. The bulk of the cDNA material can then be ligated under the optimal conditions, packaging enough to obtain the desired amount of cloning events.

1. Set up ligation reactions as described above (Preparation of λ Cloning Vector, Step 8). Keep the vector DNA constant (e.g., 20 ng per reaction) and vary the cDNA concentration. Test at least three different molar ratios of cDNA to vector DNA (e.g., 1:1, 3:1, 9:1).

2. Package 1 μl from each ligation reaction with Gigapack Gold (highest efficiency) packaging extracts (one packaging extract can be split and used for up to four samples).

3. Titrate the cDNA libraries on *E. coli* XL1-Blue in the presence of IPTG and X-Gal.

4. Ligate an appropriate amount of cDNA with vector DNA at the most favorable molar ratio (as determined in Steps 1–3). Package with Gigapack Gold and titrate the cDNA library to determine the amount of cloning events obtained.

5. Amplify the phage as independent sublibraries, by growing them on agar plates in the absence of IPTG and X-Gal for 6–7 hr at 37°. Do not overgrow the phage because this may lead to underrepresentation of more slowly growing phage clones in the amplified libraries.

6. Overlay the plates with 10 ml SM and 0.5 ml chloroform, and leave them at 4° overnight. Store the amplified sublibraries in tightly capped tubes over a 1/20 volume of chloroform at 4°.

The amount of phage to be amplified per sublibrary should depend on two considerations. (1) If the size of the sublibrary is very small, many sublibraries may have to be screened before finding the one which contains a functional clone. (2) If the sublibrary comprises too many independent cloning events, a functional clone may not be detectable.

Phage DNA Preparation, in Vitro Transcription, and Sib Selection

1. Determine the phage titers of the amplified sublibraries. Plate about 10^6 phage per 90-mm agar plate to obtain optimal DNA yields.

2. Prepare phage DNA according to one of several standard procedures [e.g., the polyethylene glycol (PEG)/CsCl method[19] or the cetyltrimethylammonium bromide (CTAB)–DNA precipitation method[22]].

3. Digest the phage DNA with *Not*I to completion, then incubate in the presence of 50 μg/ml proteinase K and 0.5% sodium dodecyl sulfate (SDS) for 30 min at 37°. Extract with phenol–chloroform and precipitate the phage DNA with ethanol.

4. Synthesize cRNA *in vitro* for microinjection into *Xenopus* oocytes as described elsewhere in this volume (see [18]). We have determined that, in order to obtain quantitatively capped cRNA, a 5× molar excess of the cap analog m^{7}G(5')ppp(5')G over GTP is required, when using T3 or T7 RNA polymerase.

5. Identify a sublibrary which gives a positive signal after microinjection of cRNA into *Xenopus* oocytes.

6. Assuming that a "positive" sublibrary consisted of 10,000 independent cloning events, containing one positive cDNA clone, then plate out

[22] G. Del Sal, G. Manfioletti, and C. Schneider, *BioTechniques* **7**, 514 (1989).

the sublibrary on, for example, 20 (or 30) agar plates at a density of 1000 phage per plate. [Assuming a Poisson distribution, the probability that the clone of interest will be present at least once on 20 (30) plates is 86.5% (95%).] Grow the phage for 6–7 hr at 37°.

7. Cut the agar from each plate into eight equally sized pieces, each one containing approximately 125 phage clones. Transfer each agar piece to a fresh tube, add SM–chloroform, and let the phage diffuse out of the agar overnight. Each supernatant is referred to as a pool of 125.

8. Combine small aliquots from eight pools of 125 to obtain pools of 1000. Identify a pool of 1000 that gives rise to a positive signal after microinjection of cRNA into *Xenopus* oocytes.

9. Test the corresponding eight pools of 125 to identify the one which is positive.

10. Plate the positive pool of 125 at low density and pick single phage clones. Suspend each clone in 1 ml SM.

11. Combine small aliquots from, for example, 20 individual clones to obtain pools of 20. Identify a positive pool of 20.

12. Test the corresponding 20 clones to identify a single, functional clone.

Discussion

We would like to emphasize several points that are important for the construction of an expression library maximally enriched for full-length cDNAs of interest. If the RNA for the ion channel of interest does not give large currents in the oocyte, enrichment is of particular importance, in order to limit the number of pools screened. Optimal enrichment is of somewhat less concern in the case of, for example, a G protein-coupled receptor where amplification of the signal occurs (one activated receptor leads to the opening of many Cl⁻ channels).

(1) If possible, a tissue (e.g., a defined area of the brain) or a cell line should be identified which is enriched for the protein of interest. (2) The poly(A)⁺ RNA should be size selected and fractions should be identified for cDNA library construction. (3) A directional cDNA library should be constructed. This reduces the number of recombinants to be screened by a factor of 2, as compared to a nondirectional library. (4) The synthesized cDNA should be size-selected by agarose gel electrophoresis. Selection of cDNA corresponding in size to the mRNA it is derived from dramatically enriches for full-length inserts. We do not recommend fractionating the cDNA by column chromatography, a method often recommended to separate cDNA from linkers, since this method usually does not allow accurate size fractionation.

The successful application of the method described in this chapter is somewhat dependent on the length of the mRNA or, more precisely, on the distance from the initiation codon of the coding region to the poly(A) tail. (Note: A cDNA clone may encode a functional ion channel protein without containing the entire coding region.[23]) Theoretically, the limiting factor for cDNA cloning is the maximal insert size in the expression vector (~ 10 kb for λZAP). However, the longer the mRNA, the more technically demanding the undertaking becomes, mainly owing to the inefficient performance of reverse transcriptases on long mRNA templates. The lengths of the cDNA clones isolated by expression cloning in *Xenopus* oocytes are as follows: 0.6 kb (slow K^+ channel),[1] 3.4 kb (delayed rectifier-type K^+ channel),[2] 3.0 kb (kainate receptor),[3] 0.75 kb (IL-4),[4] 2.2 kb (Na^+,glucose-cotransporter),[5] 2.5 kb (steroid 5α-reductase),[6] 2.5 kb (substance K receptor),[7] 3.0 kb ($5HT1_c$ receptor),[8] 3.4 kb (substance P receptor),[9] 3.6 kb (neurotensin receptor),[10] 3.8 kb (TRH receptor),[11] 1.8 kb (bombesin/GRP receptor),[12] 3.2 kb (ET receptor),[13] 3.0 kb (PAF receptor),[14] and 4.3 kb (metabotropic glutamate receptor).[15]

In mammalian genes, *Not*I recognition sites (5'-GCGGCCGC-3') occur, on average, only once in 10^6 base pairs owing to the significant underrepresentation of CG dinucleotides.[24] This rare occurrence of the *Not*I recognition sequence makes it clearly the best-suited restriction enzyme recognition sequence in the primer/linker used to initiate cDNA synthesis. Nevertheless, the possibility exists that an internal *Not*I site may be present in the gene of interest. Recently, a method has been developed which eliminates this risk.[25] Synthesis of the cDNA is primed by a primer/linker containing a *Xho*I recognition site, and 5-methyl dCTP is incorporated into the first-strand cDNA to obtain hemimethylated cDNA. The restriction enzyme *Xho*I does not cleave internal, hemimethylated recognition sites but does cleave the unmethylated primer/linker-derived *Xho*I site. *Eco*RI linkers or adapters are added to allow creation of cDNA molecules with asymmetric *Eco*RI/*Xho*I ends that can be directionally cloned into λZAP cut with *Eco*RI and *Xho*I.

Acknowledgments

We wish to thank Dr. J. A. Drewe for critically reading the manuscript. This work was supported by an Advanced Technology Program Award from the State of Texas and by grant NS28407 (R.H.J.).

[23] A. M. J. VanDongen, G. C. Frech, J. Drewe, R. H. Joho, and A. M. Brown, *Neuron* **5,** 433 (1990).
[24] M. N. Swartz, T. A. Trautner, and A. Kornberg, *J. Biol. Chem.* **237,** 1961 (1962).
[25] W. D. Huse and C. Hansen, *Strategies* **1,** 1 (1988).

[41] Hybrid Arrest Technique to Test for Functional Roles of Cloned cDNAs and to Identify Homologies among Ion Channel Genes

By Ilana Lotan

Introduction

Recombinant DNA techniques have enabled the isolation of cDNA clones that code for receptor- and voltage-operated ion channel proteins. The identity of each of the cDNA clones has to be confirmed by a functional assay, which includes the expression of the clone in an expression system and measurement of the resulting current. One of the expression systems of choice, whenever possible, has been the *Xenopus* oocyte. This approach, however, requires the isolation of clones containing all the sequence needed for proper translation and, in the case of multisubunit channels, the corresponding sequences for all the subunits (e.g., the nicotinic receptor). The approach based on hybrid or antisense arrest of mRNA expression in *Xenopus* oocytes overcomes this limitation. The idea is derived from works of many laboratories demonstrating that DNA and RNA antisense sequences, complementary to a target mRNA, can block selectively the expression of this mRNA in prokaryotic and eukaryotic cells.[1]

The utility of antisense RNA in blocking gene expression, however, turned out to be limited in some respects. The degree of inhibition is effective only if the antisense sequence covers the 5' terminus and codon of the mRNA, presumably by interfering with ribosome binding. In some expression systems it is either partial or reversible, perhaps because these cells contain an endogenous helicase, demonstrated in *Xenopus* fertilized eggs and to some extent in oocytes,[2,3] which unwinds RNA–RNA hybrids and thereby relieves the inhibition of translation achieved by the antisense RNA.

Antisense DNA sequences turned out to be more successful in inhibiting the expression of target mRNAs in many systems. In *Xenopus* oocytes it has recently been shown[4,5] that short complementary oligodeoxynucleo-

[1] P. J. Green, O. Pines, and M. Inouye, *Annu. Rev. Biochem.* **55,** 569 (1986).
[2] M. R. Rebagliati and D. A. Melton, *Cell (Cambridge, Mass.)* **48,** 599 (1987).
[3] B. L. Bass and H. Weintraub, *Cell (Cambridge, Mass.)* **48,** 607 (1987).
[4] C. Cazenave, N. Loreau, N. T. Thuong, J. Toulme, and C. Hélène, *Nucleic Acids Res.* **15,** 12 (1987).
[5] P. Dash, I. Lotan, M. Knapp, E. R. Kandel, and P. Goelet, *Proc. Natl. Acad. Sci. U.S.A.* **84,** 7896 (1989).

METHODS IN ENZYMOLOGY, VOL. 207

tides (antisense oligonucleotides) are effective in sequence-specific inactivation of target mRNA by an RNase H-like activity (an enzyme that selectively degrades the RNA moiety of RNA–DNA hybrids). This antisense DNA arrest has an inherent advantage over antisense RNA arrest as it is irreversible, so that it requires relatively much shorter complementary sequences and usually works equally well with probes directed against the 3' and 5' end of the gene (the last advantage is more pronounced with the modified method mentioned below).

Here we describe a modified method for antisense arrest of expression with oligo nucleotides, applicable for messages coding for channel proteins contained in tissue-derived total RNA. This provides a powerful method to test for functional roles of cloned cDNAs, which by themselves do not direct expression of currents when injected into oocytes.[6,7] It also enables the identification of potential homologies between sequenced mRNAs coding for channel proteins and mRNAs than can be assayed in the oocyte.[8-11]

General Considerations

The hybrid arrest experiments are performed by introducing into the oocyte total RNA together with an oligonucleotide complementary to a stretch of mRNA which codes for a channel protein (target mRNA) present in this heterologous population. The injected oocytes are kept in culture medium for several days and electrophysiologically assayed for induced currents.[12] The amplitude of the current through the target channels should be reduced or completely abolished (in an oligonucleotide dose-dependent manner) as compared with the same current expressed in oocytes injected with RNA alone. The expression of all other mRNA species should be unaffected, provided they do not contain a sequence of marked similarity with the chosen stretch in the target mRNA. In the oocyte cytoplasm the oligonucleotide hybridizes (forms hydrogen bonds between appropriate base pairs) with the complementary stretch in the

[6] I. Lotan, P. Goelet, A. Gigi, and N. Dascal, *Science* **243**, 666 (1989).
[7] D. F. Slish, D. B. Engle, G. Varadi, I. Lotan, D. Zinger, N. Dascal, and A. Schwartz, *FEBS Lett* **250**, 509 (1989).
[8] I. Lotan, A. Volterra, P. Dash, S. A. Siegelbaum, and P. Goelet, *Neuron* **1**, 963 (1988).
[9] K. Folander, J. S. Smith, J. Antanrvage, C. Bennett, R. B. Stein, and R. Swanson, *Proc. Natl. Acad. Sci. U.S.A.* **87**, 2975 (1990).
[10] H. Akagi, D. E. Patton, and R. Miledi, *Proc. Natl. Acad. Sci. U.S.A.* **86**, 8103 (1989).
[11] K. Sumikawa and R. Miledi, *Proc Natl. Acad. Sci. U.S.A.* **85**, 1032 (1988).
[12] This volume, [14]–[25].

target mRNA and induces a rapid[13] irreversible RNA degradation via RNase H activity (which is the main factor responsible for the inhibition of translation in the oocyte). Initially, the target mRNA is cleaved at the oligonucleotide hybridization site, and then after several hours its fragments can no longer be detected in the cytosol, probably because of degradation by cellular nucleases.[5] As the oligonucleotides are short lived inside the oocyte with a half-life of less than 30 min,[14] the blockade of the target mRNA translation is completed, in fact, within several hours, provided that appropriate concentrations of oligonucleotide are employed.[13]

In the arrest of expression of pure message injected into oocytes, oligonucleotides about 15 to 30 bases long are effective when coinjected with the mRNA. The molar concentration of oligonucleotide required for effective inhibition is up to 200 times greater than that of the mRNA. In the study of ion channels, however, the target channel mRNA is diluted in a heterologous population (total RNA) so that its expression arrest is more complex. It is conceivable that the association rate between the oligonucleotide and its specific site is reduced, therefore we increase the oligonucleotide concentration and allow for more effective hybridization by including, prior to injection, heat denaturalization of the mRNA and oligonucleotide followed by *in vitro* hybridization. To lessen nonspecific interactions between the oligonucleotide and the irrelevant mRNA species, we use longer oligonucleotides, 40 to 90 bases, which have increased affinity for the target mRNA and reduced affinity for the irrelevant mRNAs. The increased affinity increases the association time between the RNA and the oligonucleotide, thereby serving for better chances of cleavage by RNase H. The incision of denaturation is based on a study (reported in Ref. 8; see Table I) which demonstrates that the blockade is enhanced when the RNA is denatured prior to hybridization. In fact denaturation causes a severalfold increase in the potency of a complementary oligonucleotide in blocking the expression of tetrodotoxin (TTX)-sensitive Na^+ current, suggesting that secondary structure and aggregation of mRNAs can attenuate effective hybridization between the target mRNA and the oligonucleotide. The importance of the *in vitro* hybridization has never been checked by us.

Under these modified conditions, hybrid arrest is selective for the target channel mRNA, as it would definitely not affect the expression of other channel mRNAs which contain sequence of up to 50% sequence similarity with the target mRNA[6]; the complementary sequence of the target site of

[13] C. Jessus, C. Cazenave, R. Ozon, and C. Hélène, *Nucleic Acids Res.* **16**, 2225 (1988).
[14] T. M. Woolf, C. G. B. Jennings, M. Rebagliati, and D. A. Melton, *Nucleic Acids Res.* **18**, 1763 (1990).

TABLE I
DOSE-DEPENDENT INHIBITION OF SODIUM CURRENT EXPRESSION

Nucleotide concentration (ng/oocyte)	Inhibition[a]	(%)
	native RNA	denatured RNA
2.5	99.8 ± 0 (2)	99.7 (1)
0.25	84.3 ± 3.4 (4)	98.6 ± 1 (3)
0.085	—	92.6 (1)
0.025	$-4^b \pm 18$ (2)	72.1 ± 4 (2)
0	0 ± 5.6 (6)	0 ± 4.6 (2)

[a] Without and with denaturation of the RNA prior to *in vitro* hybridization with oligonucleotide. Fifty nanoliters of 8.7 μg/μl chick brain total RNA injected into oocytes expressed, among others, voltage-dependent Na^+ and A-type K^+ currents. The RNA was hybridized to different concentrations of 81-mer oligonucleotide complementary to a portion of the rat brain Na^+ channel II sequence. Numbers are means $\pm$ SEM (number of frogs given in parentheses; 2–10 oocytes were tested per frog). For each group, the mean ratio between the amplitudes of both currents (I_{NA}/I_A) was measured and compared with the mean ratio measured in the group injected with RNA alone, with the mean percentage of inhibition of Na^+ current being calculated. I_A served as a reference to minimize the effects of variable expression in oocytes and of variations in the technical handling.
[b] The mean ratio of this group was higher than in the control.

the mRNA, however, need not be identical to the oligonucleotide sequence. Thus, we used 50-base-long oligonucleotides, targeted at a rabbit skeletal muscle Ca^{2+} channel mRNA stretch, to suppress the expression of Ca^{2+} channel mRNA (98% reduction in current amplitude) from rabbit heart of which the complementary sequence similarity with the corresponding stretch in the rabbit mRNA is 77%. We also could inhibit selectively the expression of Na^+ channel mRNA across species (see Ref. 8; e.g., Table I).

Oligonucleotides targeted at different portions of the coding region of the mRNA have similar potency; the extent of inhibition is dependent on the concentration of the oligonucleotide. Increasing amounts cause increased inhibition, with significant inhibition occurring when the molar ratio of oligonucleotide to the target mRNA is 1000–10,000.[8] The high molar excess of oligonucleotide may be due to nonspecific binding and the fact that the effective intracellular concentration of oligonucleotide is reduced because of its rapid degradation in the oocyte. It should be noted that some investigators have succeeded in effectively arresting the expression of target channel mRNA contained in a heterologous population [poly(A⁺) mRNA, however] by coinjecting relatively short oligonucleotides (30-mer) without predenaturation and *in vitro* hybridization.[10]

Modified oligonucleotides have been developed in several laboratories in order to improve their antisense performance. In the oocyte, only two derivatives are relevant for better hybrid arrest since they are chemically modified in a way that does not interfere with RNA cleavage by RNase H, on hybridization of the oligonucleotide with the RNA. Indeed, these derivatives were shown to be more efficient agents for hybrid arrest of expression of pure message injected into oocytes, and therefore they are of potential value of hybrid arrest of ion channel expression in oocytes. Phosphorothioate derivatives, having a sulfur atom substituted for the free oxygen group in the phosphodiester bond, are less vulnerable to nucleases (with half-lives greater than 3 hr) and can therefore remain at antisense-effective concentrations inside the oocyte.[14] Thus they are effective in total blockade of target RNA at severalfold lower concentrations than those needed for unmodified oligonucleotides.[15] The second modification proven to be useful in the oocyte system is intercalating agents terminally linked to oligonucleotides, which increase binding affinity of short oligonucleotides. Thus a specific and significant inhibition was reached with an 11-mer acridine-linked oligonucleotide, whereas the unmodified homologous oligonucleotide had no significant effect over the same concentration range.[4] It should be pointed out that both modifications render the oligonucleotides more toxic, as they may be sequence-unspecific inhibitors of protein synthesis in the concentration range used with unmodified oligonucleotides of the same length. Thus these derivatives should be used for specific arrest at lower doses.

Antisense RNAs have also been used successfully by several investigators to inhibit the expression of ion channel genes. Full-length antisense RNAs were transcribed *in vitro* from the cDNA, subcloned in an expression vector, and coinjected with the mRNA.[11,16] In our hands, however, when comparing antisense RNA translation arrest of α- and β-globin mRNA (pure messages) with antisense DNA arrest, the antisense RNA led to inconsistent arrest, whereas antisense DNA caused an effective and reproducible block of globin translation when assayed by [³⁵S]methionine pulse labeling in oocytes.[17]

[15] C. Cazenave, C. A. Stein, N. Loreau, N. T. Thuong, L. M. Neckers, C. Subasighe, C. Hélène, J. S. Cohen, and J. J. Toulme, *Nucleic Acids Res.* 17, 4255 (1989).
[16] G. Dahl, T. Miller, D. Paul, R. Voellmy, and R. Werner, *Science* 236, 1291 (1987).
[17] I. Lotan and P. Goelet, unpublished results (1986).

Experimental

Reagents and Solutions

Oligonucleotides 40–80 bases long are synthesized in an automatic DNA synthesizer and purified by electrophoresis through a denaturing polyacrylamide gel or by reversed-phase chromatography (see Sambrook *et al.*)[18] They should be then purified from any traces of RNAse A by a series of two phenol–chloroform extractions followed by two chloroform extractions to get rid of any traces of phenol, which can be toxic to the oocyte.[19] The purified oligonucleotides are recovered by ethanol–salt precipitation and dissolved in double-distilled autoclaved water in several concentrations, such as 1 mg/ml, 300 μg/ml, and 30 μg/ml (calculated to result 8.3, 2.5, and 0.25 ng/oocyte, respectively, when injected according to the following procedure).

Total RNA which expresses the ion channel under study in oocytes, is purified from the appropriate tissue[20] and is dissolved in water. The final concentration of the RNA should be around 10 μg/μl. In the hybridization mixture its concentration is only 2/3 and its potency to induce currents in oocytes is reduced by about 20 to 50% (depending on the quality of the purified RNA) because of the high temperature incubation.

6× Hybridization buffer: 600 mM NaCl, 60 mM Tris (pH 7.6)

Pure paraffin oil

Double-distilled autoclaved water

Autoclaved Eppendorf tubes (preferentially 0.5 ml)

Procedure

In one experiment there should be a set of, at least, two reactions: one control reaction which contains the RNA alone in the hybridization medium and the second experimental reaction which contains RNA and oligonucleotide. Both reactions are treated in the same way. It is advisable to have more reactions with different concentrations of the oligonucleotide. Before setting up the reactions, RNA and oligonucleotide are denatured separately by 2-min incubations at 65° and subsequent cooling on ice.

The following is an experimentally convenient setting for handling and for injecting enough oocytes to permit statistical analysis (with limiting quantities, the volumes can be scaled down). Set the reactions on ice.

---

[18] J. Sambrook, E. F. Fritch, and T. Maniatis, "Molecular Cloning: A Laboratory Manual," 2nd Ed., pp. 11.23, 11.29. Cold Spring Harbor Laboratory, Cold Spring Harbor, New York, 1989.

[19] I. Lotan and P. Goelet, unpublished results (1987).

[20] B. P. Bean, this volume [11].

Reaction 1 (control) contains 2 μl denatured RNA solution, 0.5 μl of 6× hybridization buffer, and 0.5 μl water. Reaction 2 (experiment) contains 2 μl denatured RNA solution, 0.5 μl of 6× hybridization buffer, and 0.5 μl oligonucleotide solution (e.g., 300 μg/μl). Add few drops of paraffin oil to each tube to cover the water phase, spin down, and incubate for 2 min at 65°, followed by 0.5 to 3 hr of incubation at 37°; place on ice. Take up the contents of the tube (including the paraffin oil) and place on a petri dish; the aqueous phase is centered and is easily taken up by a microinjector. Inject 50 nl into each oocyte. At least 25–30 oocytes can be injected per reaction.

Evaluation of Results

If there is 100% sequence complementarity between the oligonucleotide and the target sequence, the 30 μg/ml oligonucleotide should inhibit the target current significantly (more than 70%). If oocytes injected with the experimental reaction containing 1 mg/ml still express the target current and it is not significantly reduced as compared to that in the control reaction, then there is definitely no significant complementarity. This high concentration of oligonucleotide usually causes nonselective suppression (20–50%) of all the currents expressed; thus to ensure that the inhibition is indeed specific to the target, one should compare the effects of the oligonucleotide on other currents preferably expressed in the same oocytes injected with the same RNA. This is recommended for lower oligonucleotide concentrations as well, to ensure that the oligonucleotide is not contaminated by impurities which can cause unspecific inhibition and also to overcome effects of variable expression among different oocytes. The concentration of 300 μg/μl of oligonucleotide is high enough to block (more than 90% inhibition) the target channel (of more than 77% sequence complementarity) and is low enough to have no unspecific effects on irrelevant channels. It should be pointed out that oligonucleotides complementary to a channel mRNA often enhance significantly the expression of other channels.[6,8]

Discussion

The described hybrid arrest method identifies sequence similarities of channel genes and may thereby permit the isolation of homologs by screening relevant libraries with the effective oligonucleotide sequences. Moreover, this method should be applicable for isolating genes of unknown sequence that code for channel proteins whose expression in mRNA-injected oocytes can be assayed. We could show (see Fig. 1)[19] that the expression of α-globin mRNA was arrested specifically by its complementary DNA sequences generated from an orientation-specific construct of the α-globin clone in an expression vector. (In the construct the α-globin cDNA is inserted in specific orientation with respect to the promoter on

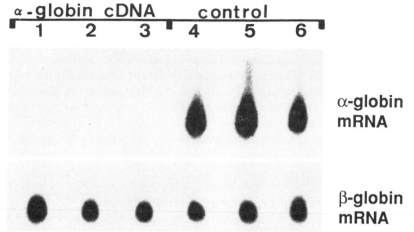

FIG. 1. Specific degradation of α- but not β-globin mRNA in oocytes by α-globin antisense DNA, as revealed by Northern blot analysis. Approximately 10 μg of total oocyte RNA extracted from a set of injected oocytes was run on each track of a formaldehyde agarose gel. The gel was transferred to nitrocellulose and probed with radioactive α- and β-antisense DNAs. Oocytes were injected either with 5 ng of globin mRNA (α, β) alone (control, lanes 4–6) or with a mixture of the mRNA and a 3-fold molar excess of α-globin antisense DNA (lanes 1–3). The reagents were either injected directly (lanes 1, 4), preincubated at 70° for 2 min (lanes 2, 5), or preincubated at 70° for 2 min followed by incubation at 37° for 30 min (lanes 3, 6). α-Globin antisense DNA sequences were reverse transcripts generated from *in vitro* transcripts of the sense orientation (first-strand cDNAs).

the expression vector and permits the synthesis of sense RNA by *in vitro* transcription and antisense single stranded DNA by reverse transcription.) Thus, hybridization of a heterologous mRNA population, which expresses a target channel in oocytes, with antisense DNA sequences generated from an *orientation*-specific cDNA library constructed from the mRNA species should arrest significantly the expression (or at least change the kinetics in the case of multisubunit protein) of the target mRNA in oocytes. By testing successively smaller subpools of the DNA antisense sequences, the specific cloned DNA which blocks the channel expression may be identified. This cDNA isolation procedure should circumvent the possible difficulties which the common "positive selection" cloning technique in oocytes (e.g., Ref.[21]) would encounter, in the case of a multisubunit (e.g., nicotinic channel) or a large single-subunit protein (e.g., Na⁺ channel). Indeed a similar technique was partly used in the cloning of the serotinin (5-HT1c) receptor.[22]

[21] G. C. Frech, A. M. J. Van Dongen, G. Schuster, A. M. Brown, and R. H. Joho, *Nature (London)* **340**, 642 (1989).
[22] H. Lubert, T. P. Snutch, T. Van Dyke, A. J. Levine, P. R. Hartig, H. A. Lester, and N. Davidson, *Proc. Natl. Acad. Sci. U.S.A.* **84**, 4332 (1987).

[42] Cloning of Ion Channel Gene Families Using the Polymerase Chain Reaction

By ELEAZAR C. VEGA-SAENZ DE MIERA and JEN-WEI LIN

Introduction

The polymerase chain reaction (PCR), developed by Mullis *et al.*,[1] is a method to amplify a fragment of DNA sequence by using specific primers that flank a region of interest. Thermostable polymerases are used for repeated DNA synthesis from a given template. This method has been used in various stages of ion channel cloning.[2-6] When properly used, it is an extremely powerful technique, and one may obtain more than 10^6-fold amplification. Many reviews on the theory and applications of PCR methods are available[7,8]; interested readers may obtain basic information from these sources. We describe here the use of the PCR to identify putative members of an ion channel gene family. In comparison to low-stringency hybridization, we have found the PCR to be a faster and more efficient method to discover new members of the family. For this purpose, the template for the PCR can be either genomic DNA[9] or cDNA. The cDNA may be derived from a cDNA library[10] or, as in the example described here, may be single-stranded cDNA synthesized from the mRNA of a tissue of interest. In addition, if cDNA is used as template, the amplified

[1] K. Mullis, F. Faloona, S. Scharf, R. Saiki, G. Horn, and H. Erlich, *Cold Spring Harbor Symp. Quant. Biol.* **51,** 263 (1986).
[2] G. I. Fishman, D. C. Spray, and L. A. Leinwand, *J. Cell Biol.* **111,** 589 (1990).
[3] K. Folander, J. S. Smith, J. Antanavage, C. Bennett, R. B. Stein, and R. Swanson, *Proc. Natl. Acad. Sci. U.S.A.* **87,** 2975 (1990).
[4] K. Otsu, H. F. Willard, V. K. Khanna, F. Zorzato, N. M. Green, and D. H. MacLennan, *J. Biol. Chem.* **265,** 13472 (1990).
[5] J. C. L. Tseng-Crank, G.-Y. Tseng, A. Schwartz, and M. A. Tanouye, *FEBS Lett.* **268,** 63 (1990).
[6] Z-Y. Zhao and R. H. Joho, *Biochem. Biophys. Res. Commun.* **167,** 174 (1990).
[7] H. A. Erlich, ed., "PCR Technology: Principles and Applications for DNA Amplification." Stockton, New York, 1989.
[8] M. A. Innis, D. H. Gelfand, J. J. Sninsky, and T. J. White eds., "PCR Protocols: A Guide to Methods and Applications." Academic Press, San Diego, California, 1990.
[9] A. Kamb, M. Weir, B. Rudy, H. Varmus, and C. Kenyon, *Proc. Natl. Acad. Sci. U.S.A.* **86,** 4372 (1989).
[10] B. F. O'Dowd, T. Nguyen, A. Tirpak, K. R. Jarvie, Y. Israel, P. Seeman, and H. B. Niznik, *FEBS Lett.* **262,** 8 (1990).

METHODS IN ENZYMOLOGY, VOL. 207

PCR fragments provide a profile of gene expression in a specific tixsue.[11] In principle, this can be done with RNA from a very small amount of tissue.[12]

In this chapter, we discuss the choice and design of PCR primers. Detailed PCR protocols provided here have worked for us and are generally applicable in most circumstances.

Design of Primers

The choice and design of primers constitute perhaps the most important variable for a successful PCR. To optimize the identification of distinct members of a gene family, one should amplify a region of a gene that is variable within the family. This variable region should be flanked by conserved areas where the PCR primers will be located. Therefore, the design of primers depends on previous knowledge of the sequence of some member(s) of the family. For general considerations of primer design, good sources of references are available.[7,8] Some special considerations for the amplification of members of a gene family are listed below.

Selecting Regions for Amplification. The first step is to determine the regions of the channel to be amplified. One should start with collecting as many examples of existing amino acid sequences of members of a gene family as possible so that conserved regions can be identified by sequence comparison. In case that only one or a few members of the family are known, and doubts exist about the likelihood of conservation of certain regions, knowledge or assumptions of amino acid sequence domains with specific functional roles can be used as a guide. For example, based on the speculations that the H5 region of *Shaker* potassium channels is involved in the selectivity for potassium ions,[9] one may design a primer from this region. (A hybrid arrest experiment may be used to evaluate the validity of the assumptions; see [41] in this volume.)

In a typical PCR two conserved regions are necessary so that the sequence flanked by two primers is amplified. However, it is not necessary that both primers hybridize to sequences in the channel cDNA. For example, to amplify from library cDNA,[10] one may use a primer that hybridizes to the sequence of interest and a second primer that hybridizes to sequences on a host cloning vector.

Minimizing Codon Degeneracy. After conserved areas are identified, the amino acid sequence of the regions should be analyzed to locate a stretch of amino acid sequence with minimal codon degeneracy. (PCR primers typically have a length of 18 to 30 nucleotides.[8]) Given the as-

[11] E. Vega-Saenz de Miera, N. Chiu, K. Sen, K. Lau, J. W. Lin, and B. Rudy, *Biophys. J.* **59**, 197a (1991).

[12] H. H. Li, U. B. Gyllensten, X. F. Cui, R. K. Saiki, and H. A. Erlich, *Nature (London)* **335**, 414 (1988).

sumption that only amino acid sequence is conserved in the course of evolution, it is necessary to use degenerate primers to include all the possible nucleotide sequences which code the same amino acid sequence. It is important to minimize primer degeneracy since a high primer concentration is an essential requirement of a successful PCR, and a highly degenerative primer effectively reduces the concentration of individual primer sequence. In addition, a highly degenerative primer increases nonspecific amplification. We have succeeded in isolating sequences of interest with primers that contain less than 512 possible combinations. Other laboratories may have had a different experience for the maximal amount of degeneracy allowed. For example, a primer degeneracy up to 13,824 has been used, but additional screening steps to eliminate nonspecific products were needed.[6]

The length of the region to be amplified should be within the limitation of the PCR, typically less than 3 kilobases (kb), although best results are obtained when the amplified product is shorter than 1 kb. Knowledge of the expected length of the PCR product(s) is helpful in the identification of the sequence of interest since PCR products may contain more than one band.

Perfecting 3' Base Pairing. Because polymerase extension starts at the 3' end of the primer, it is extremely important to obtain perfect base pairing in this region in order to ensure proper priming. We find that the last 10 bases of the primers at the 3' end must have a perfect base matching. Thus, we typically include all possible combinations of the codons in this region of the primer. More mismatches are tolerated as one moves away from the last 10 bases. It may also be helpful to use codon usage to decrease degeneracy away from this region.[13]

One may add linkers to the 5' end of the primers to facilitate subsequent subcloning of amplified products. However, this is not essential, and it is not desirable if the degeneracy of the primers is high. End filling by Klenow and blunt-end ligation of amplified fragments works well for subcloning.

These considerations apply to the use of both cDNA and genomic DNA. However, owing to the possibility of the existence of intervening introns, the amplification from genomic DNA may not work if the length of the intron is longer than that permissible for PCR amplification.

Contamination

Because the PCR is so powerful in amplifying DNA sequences, a contamination of a few molecules may be amplified and become a significant fraction of the products. This is a particularly serious problem in a laboratory that handles many different types of DNAs, especially cloned

DNAs. The few molecules of the cDNAs that may exist on the bench, the hands of experimenters, and particularly the tips of pipettes can be catastrophic. Therefore, one takes precautions similar to those necessary for RNase-free conditions.[7,8] For example, we keep a separate set of tips, tubes, and pipettors that are used exclusively for PCR experiments. Several companies sell pipette tips designed especially for the PCR. We use long (sequencing type) pipette tips.

Control Experiments

Given all the precautions one may take to avoid contamination, a series of control reactions should always be performed in each experiment to evaluate the possibility of contamination. We routinely perform one reaction with both primers in the absence of DNA template in order to evaluate possible contamination of water, reagents, or pipette tips.

In the case of amplification from single-stranded cDNA, one may have contamination from genomic DNA present in the original mRNA samples. In addition to treating the mRNA with RNase-free DNase , special controls are necessary to eliminate possible contamination of this type. We routinely carry out a control PCR with a sample containing all the reagents, including the mRNA, but in which cDNA was not synthesized. One may also destroy the mRNA RNase A. Enzymes and reagents of the highest quality are required to avoid amplification of contaminating DNAs. Furthermore, when exploring for amplified products in analytical DNA gels, we apply 5–20 times more reaction sample in the case of the control experiments.

Experimental Protocols

General Considerations

The GeneAmp PCR kit (Perkin Elmers Cetus, Norwalk, CT) which utilizes Taq polymerase is used for all of our experiments. Other thermostable polymerases are also available. The reagents and the polymerase are aliquoted immediately after receipt of the kit to minimize contamination. Separate incubation tubes and pipettors are set aside for the same reason. The PCR reaction is carried out in a DNA thermal cycler (Perkin Elmers Cetus). The experiments are carried out in the step cycle mode of the DNA thermal cycler. Typically 30–35 cycles are used; additional cycles increase only primer–dimer and background.

The most sensitive parameter that may affect the specificity of amplification is the annealing temperature (T_{an}). Higher T_{an} preferentially allows for perfectly matched base pairing and leads to more specific amplification.

From this point on, as we discuss the stringency of the PCR, T_{an} will be the parameter under consideration. However, if one is interested in a family of related genes, lowering T_{an} may allow genes that have less perfect matching with primers to be amplified. In addition, when a degenerate primer is used, individual primers may have different melting temperatures owing to different nucleotide compositions. For example, we had an experience where certain genes showing a perfect base match with primers were only amplified under lower stringency conditions (see below). As expected, lower stringency leads to more nonspecific amplifications. In general, it is necessary to go through a few trials in order to identify the lowest stringency conditions where a well-defined band(s) of an expected size is obtained. Another variable of importance is the concentration of magnesium ions in the reaction. Magnesium ions affect enzymatic activity as well as the annealing of primers, the strand dissociation temperature, and the formation of primer dimers, among others.[8] As the importance of these factors depends on the specific reaction, it is recommended to test the PCR with any new primers or DNA at various Mg^{2+} concentrations in the range of 0.5 to 5 mM.

After the amplification, 5 – 10% of the product is used to run an analytical gel to determine if bands of the appropriate size appear. PCR products can be sequenced directly[8]; however, if several different products of similar length are expected (as in the case discussed here), it is necessary to clone the amplified products to isolate individual sequences for analysis. We have used one of two options. One may use the entire PCR product to run a preparative gel and purify the bands of interest for cloning, or one may reamplify the initial reaction products. (If one is interested in genes expressed in low abundance, reamplification may not be a good idea. This is because the procedure may amplify abundant genes or nonspecific products preferentially, and the relative quantities of the rare genes will be further reduced.) Two approaches were used to amplify further the product of interest. One was to dilute the PCR product, 1/1000, and repeat the same amplification reaction. Alternatively, one can gel purify the band(s) of interest and reamplify the purified fragments.

Protocols

cDNA Synthesis. The following conditions are for the synthesis of first-strand cDNA starting with 1 μg of mRNA. All the reagents should be RNase free (see [17] in this volume).

Mix, on ice, 1 μg of poly(A)$^+$ mRNA with 0.5 μg of oligo(dT)$_{12-18}$ (Pharmacia, Piscataway, NJ) and adjust the volume to 12 μl with water that has been treated with DEPC (diethyl pyrocarbonate). Incubate the mixture at 70° for 10 min, chill on ice, and add the following: (1) 4 μl of 5×

reaction buffer [5× is 250 mM Tris-HCl (pH 8.3), 375 mM KCl, 15 mM MgCl₂]; (2) 2 μl of 0.1 M dithiothreitol (DTT) (Boehringer-Mannheim, Indianapolis, IN); (3) 1 μl of deoxynucleoside triphosphate (dNTP) mix [1 μl of each 100 mM dNTP (Pharmacia) with 6 μl of water].

Incubate for 2 min at 37°. Then add 1 μl (200 units) of RNase H⁻ reverse transcriptase [Bethesda Research Laboratories (BRL), Gaithersburg, MD; Cat. No. 8053SA]. Quickly remove 5 μl and mix with 1 μl of [α-³²P]dCTP in a separate tube. Incubate both tubes for 1 hr at 37°, boil for 2 min, and freeze until needed. The reaction mix can be used as PCR templates directly without precipitation of the cDNA or removing the mRNA.

To estimate the yield, take 1 μl from the tube that has the radioactive dCTP and precipitate with trichloroacetic acid (TCA). Compare the TCA-precipitated counts with the total counts in 1 μl. A yield of cDNA equivalent to 5 to 15% of the starting mRNA should provide a good PCR template. To evaluate the size of the products, use 2 μl to run an alkaline agarose gel and check by autoradiography.[14]

Polymerase Chain Reaction. All the reagents are from the GeneAmp DNA amplification reagent kit. As explained earlier the kit is aliquoted on arrival. The initial aliquots are done by mixing 5 volumes of 10× reaction buffer with 1 volume of each dNTP (10 mM). Eighteen microliters of this mixture is added to 0.5-ml microcentrifuge tubes (Robin Scientific). Each tube is good for one reaction in a final volume of 100 μl. The aliquots can be stored at −20° for up to a month.

To the prealiquoted microfuge tubes add the following: (1) primers (the PCR requires high primer concentrations, which should be adjusted empirically, and excess primer can lead to nonspecific amplification; we usually get good results with primer concentrations in the range of 100 ng for a nondegenerate 20-mer, or 1–1.5 μg for degenerate 20-mers) and (2) 1–3 ng of cDNA, estimated from the radioactivity incorporation mentioned earlier [use 100 ng for genomic DNA and use DNA equivalent to 10⁷ plaque-forming units (pfu) if library cDNA is used]. Adjust the volume to 99.5 μl with water. Then add Taq polymerase, 2.5 units in 0.5 μl. (Alternatively, one can dilute the enzyme slightly, 2- to 4-fold, for the accuracy of volume measurement.) Add mineral oil, 50 μl or 3 to 4 drops, on top to prevent evaporation of the aqueous phase during the reaction. The order of adding the reagents is to minimize possible contamination.

[13] K. Wada, Y. Wada, H. Doi, F. Ishibashi, T. Gojobori, and T. Ikemura, *Nucleic Acids Res.* **19,** 1981 (1991).

[14] J. Sambrook, E. F. Fritsch, and T. Maniatis, eds., "Molecular Cloning: A Laboratory Manual," 2nd Ed. Cold Spring Harbor Laboratory, Cold Spring Harbor, New York, 1989.

The thermocycling profile depends on the expected length of the fragment of interest and the stringency one would like to use. It is almost always necessary to go through a few trials in order to find optimal conditions for each experiment. A typical high-stringency reaction, where a fragment of 230–500 base pairs (bp) is expected, has the following profile: 94°, 1 min; 55°, 1 min; 72°, 1 min. Typically 30–35 cycles is used. Sometimes, a 3-min denaturing period at 94° is applied before the cycles start to ensure complete melting, and an extension period of 7 min at 72° is used after the cycles are terminated to complete the extension of any incomplete products amplified during the reaction. To facilitate the removal of the oil we freeze the aqueous phase at $-20°$. The reaction products are transferred to a clean tube.

Take 10 μl of the reaction product to run an analytic gel. PCR products are purified, if desired (see above). For cloning, the purified PCR products are kinased with T4 kinase. (If the kinase is shared in the laboratory, it is better to kinase the PCR fragment rather than the primers to minimize contamination.) If blunt-end ligation is used the kinased products are filled with Klenow before ligation to the desired vector. If the primers are designed with cloning sites, the products are digested with the appropriate restriction enzyme prior to ligation.

Results and Discussion

The following example shows the number of fragments of interest obtained from a particular experiment. By lowering the stringency of the reactions, the number of related fragments increased dramatically. In a high-stringency amplification (i.e., 94° for 1 min, 55° for 1 min, 72° for 1 min), two different potassium channel sequences were identified in cDNA from PC12 cells. With the same set of primers and cDNA, a low-stringency reaction (i.e., 94° for 1 min, 45° for 1 min, 55° for 1 min) produced seven different sequences. It is obvious from this example that low-stringency conditions are useful for more extensive detection of expressed genes. Whether the characterization of the expression profile is exhaustive depends on the design of primers and how much work one is willing to put into the sequencing of the subclones.

Each fragment thus identified was subsequently used to isolate clones from cDNA libraries utilizing high-stringency hybridization conditions. Close to 100% of the isolated clones corresponded to the expected ones. In this regard, PCR-generated probes reduce, to a great extent, the amount of work one devotes to the characterization of false positives during low-stringency screening of libraries.

[43] Overview of Toxins and Drugs as Tools to Study Excitable Membrane Ion Channels: I. Voltage-Activated Channels

By Toshio Narahashi and Martin D. Herman

Sodium Channels

Chemicals Acting on Sodium Channels

A variety of chemicals are now known to act on voltage-activated sodium channels, and some of them are being widely used as useful tools to characterize sodium and other channels. The use of chemicals as tools for the study of ion channels was ignited by the discovery of the puffer fish poison tetrodotoxin (TTX) as a specific and potent blocker of the sodium channels.[1,2] This discovery not only led a widespread use of TTX as a tool but also paved the avenue to the use of other specific toxins and chemicals as tools, the concept almost unthinkable in the early 1960s.[3,4] Since that time, a variety of chemicals have been studied along this line, and several of them are now being employed as useful tools in the laboratory because of their specific or unique actions on ion channels.

Chemicals acting on sodium channels may be classified into two large groups, namely, blockers and modulators. Modulators could be subdivided into several groups based on the mechanism of action, but this sometimes leads to misconception. Therefore, modulators are simply pooled in one group, and the mechanism of action of each chemical is described.

Chemicals That Block Sodium Channels

Tetrodotoxin and Saxitoxin. Tetrodotoxin is a potent neurotoxin contained in the ovary and liver of puffer fish. Tetrodotoxin has also been discovered in a variety of animal species not related to each other, and recent studies have clearly indicated that it is biosynthesized by certain species of bacteria and reaches the various animals via food chain.[5] Saxi-

---

[1] T. Narahashi, T. Deguchi, N. Urakawa, and Y. Ohkubo, *Am. J. Physiol.* **198,** 934 (1960).
[2] T. Narahashi, J. W. Moore, and W. R. Scott, *J. Gen. Physiol.* **47,** 965 (1964).
[3] T. Narahashi, *Physiol. Rev.* **54,** 813 (1974).
[4] W. A. Catterall, *Annu. Rev. Pharmacol. Toxicol.* **20,** 15 (1980).
[5] T. Yasumoto, D. Yasumura, T. Yotsu, T. Michishita, A. Endo, and Y. Kotaki, *Agric. Biol. Chem.* **50,** 793 (1986).

toxin (STX) is contained in the toxic Alaska butter clam *Saxidoma giganteus,* but the toxin actually derives from the dinoflagellate *Gonyaulax catanella.*[6] Both TTX and STX exert an identical effect on the nerve membrane. The following description about TTX also applies to STX unless otherwise stated.

Tetrodotoxin reversibly blocks the voltage-activated sodium channels at nanomolar concentrations without any effect on potassium and other channels.[3] The apparent dissociation constant to block the sodium channels is estimated to be 3 nM for squid giant axons[7] and 2 nM for mouse neuroblastoma cells (N1E-115 line).[8] Block occurs on a one-to-one stoichiometric basis. It is effective only when applied to the external membrane surface, causing no effect when perfused intracellularly.[9] Because of its highly specific and potent blocking action on the sodium channels, TTX has become a valuable tool for the study of various ion channels. For example, in the presence of TTX in the external perfusate, potassium channel currents only can be analyzed without any contamination with sodium channel currents. Tetrodotoxin has no effect on the neurotransmitter release nor on the postsynaptic membrane. Therefore, TTX-treated neuromuscular or synaptic preparations are suitable for the study of membrane depolarization–secretion coupling.[10,11] Tetrodotoxin has been used to estimate the density of sodium channels in excitable membranes since the first such study using TTX bioassay techniques.[12] Most studies since that time utilized tritiated TTX or STX, especially the latter.[13] Tetrodotoxin has been routinely used to isolate the sodium channels from excitable membranes.

Local Anesthetics and Other Blockers. Local anesthetics have been studied extensively in their mechanism of action to block nerve conduction since the first pioneering voltage clamp studies by Taylor[14] and Shanes et al.[15] Procaine and cocaine were found to block both sodium and potassium channels of squid giant axons. Studies along this line were facilitated

[6] E. J. Schantz, J. M. Lynch, G. Vayvada, K. Matsumoto, and H. Rapoport, *Biochemistry* **5,** 1191 (1966).

[7] L. A. Cuervo and W. J. Adelman, Jr., *J. Gen. Physiol.* **55,** 309 (1970).

[8] F. N. Quandt, J. Z. Yeh, and T. Narahashi, *Neurosci. Lett.* **54,** 77 (1985).

[9] T. Narahashi, N. C. Anderson, and J. W. Moore, *J. Gen. Physiol.* **50,** 1413 (1967).

[10] J. Bloedel, P. W. Gage, R. Llinás, and D. M. J. Quastel, *Nature (London)* **212,** 49 (1966).

[11] B. Katz and R. Miledi, *J. Physiol. (London)* **192,** 407 (1967).

[12] J. W. Moore, T. Narahashi, and T. I. Shaw, *J. Physiol. (London)* **188,** 99 (1967).

[13] T. Narahashi, *in* "Handbook of Natural Toxins, Volume 3: Marine Toxins and Venoms" (A. Tu, ed.), p. 185. Dekker, New York and Basel, 1988.

[14] R. E. Taylor, *Am. J. Physiol.* **196,** 1071 (1959).

[15] A. M. Shanes, W. H. Freygang, H. Grundfest, and E. Amatniek, *J. Gen. Physiol.* **42,** 793 (1959).

by using various analogs of local anesthetics. For example, both tertiary and quaternary derivatives of lidocaine and other local anesthetics were effectively utilized based on their pK_a values to determine the site of action and active form of local anesthetics.[16,17] Local anesthetics block the sodium and potassium channels from inside of the nerve membrane in their charged cationic form. The frequency- or use-dependent nature of channel block has been investigated extensively since its first study by Strichartz.[18] The modulation hypothesis that calls for variable affinities of a blocking agent for various states of ion channels has also been developed.[19] Several other chemicals are also known to block the sodium channels in a manner similar to local anesthetics.[20,21] Thus, local anesthetics, their derivatives, and related blocking agents have been used extensively to elucidate the mechanisms underlying the sodium channel block. However, it should be emphasized that these compounds are not selective for the sodium channels, and they are not very potent in their blocking action, usually requiring above micromolar concentrations.

Conotoxins. The venom of the marine snail *Conus geographus* contains several conotoxins. One group of conotoxins, μ-conotoxins, inhibit the binding of TTX.[22] Different voltage-activated sodium channels can be distinguished by sensitivities to μ-conotoxin GIIIA and STX.[23] The sodium channel of skeletal muscle is very sensitive to both TTX and conotoxins (μ-conotoxin and geographutoxin II), the sodium channels of mouse neuroblastoma cells and crayfish giant axons are sensitive to TTX but insensitive to μ-conotoxin, and the sodium channels of heart and denervated skeletal muscle are insensitive to both toxins.[23,24] Geographutoxin II binds to the same sodium channel site as TTX and STX.[25]

Chemicals That Modulate Sodium Channels

The kinetics of sodium channel current are complex owing to the involvement of inactivation and therefore will be simplified if the inactiva-

[16] T. Narahashi, D. T. Frazier, and M. Yamada, *J. Pharmacol. Exp. Ther.* **171,** 32 (1970).

[17] D. T. Frazier, T. Narahashi, and M. Yamada, *J. Pharmacol. Exp. Ther.* **171,** 45 (1970).

[18] G. Strichartz, *J. Gen. Physiol.* **62,** 37 (1973).

[19] B. Hille, *J. Gen. Physiol.* **69,** 497 (1977).

[20] J. Z. Yeh, *J. Gen. Physiol.* **73,** 1 (1979).

[21] J. Z. Yeh and T. Narahashi, *J. Gen. Physiol.* **69,** 293 (1977).

[22] Y. Yanagawa, T. Abe, and M. Satake, *J. Neurosci,* **7,** 1498 (1987).

[23] E. Moczydlowski, B. M. Olivera, W. R. Gray, and G. R. Strichartz, *Proc. Natl. Acad. Sci. U.S.A.* **83,** 5321 (1986).

[24] M. Kobayashi, C. H. Wu, M. Yoshii, T. Narahashi, H. Nakamura, J. Kobayashi, and Y. Ohizumi, *Pfluegers Arch.* **407,** 241 (1986).

[25] Y. Ohizumi, H. Nakamura, J. Kobayashi, and W. A. Catterall, *J. Biol. Chem.* **261,** 6149 (1986).

tion is selectively removed. For this reason, much attention has been paid to finding out sodium inactivation blockers.

Pronase. The enzyme pronase is a classic example of a blocker of sodium channel inactivation.[26,27] When perfused internally, pronase eliminates the sodium channel inactivation in squid axons without much effect on the sodium channel activation. Thus pronase has been used extensively to analyze the kinetics of sodium channel activation.

N-Bromoacetamide. *N*-bromoacetamide and *N*-bromosuccinimide also block the sodium channel inactivation when perfused intracellularly.[28]

Sea Anemone Toxins. Certain polypeptide neurotoxins have been found to be useful tools because they block the sodium channel inactivation selectively, and also because they bind to a receptor irreversibly. The selective block of the sodium channel inactivation was first shown by Narahashi *et al.*[29] for a toxin isolated from the sea anemone *Condylactis gigantea.* It is effective on crayfish and lobster giant axons but ineffective on squid giant axons. Since then several other sea anemone toxins have been found to exert the same effect as *Condylactis* toxin on the sodium channels, including *Anemonia sulcata* toxin II[30] and anthopleurin A.[31] The amino acid sequences of these and other sea anemone toxins have been determined.[32]

Scorpion Toxins. Scorpion toxins are known to modify the kinetics of sodium channel current. However, the situation is complex because there are many polypeptide components in a venom from one species of scorpion, and also because different components exert different effects on sodium channels. There are at least three classes of scorpion toxins based on the physiological effect on the sodium channel. The first group includes the toxins that inhibit the sodium channel inactivation: toxins V and varI-3 from *Centruroides sculpturatus,* toxin IIα from *Leiurus quinquestriatus,* toxins M_7 and 2001 from *Buthus eupeus,* toxins V and XII from *Buthus tamulus,* and toxins I and II from *Androctonus australis.*[33] The second group of toxins that modify the sodium channel activation kinetics, causing a shift of the voltage dependence in the hyperpolarizing direction and a suppression of the sodium current, includes toxins I, III, IV, VI, and VII

[26] C. M. Armstrong, F. Bezanilla, and E. Rojas, *J. Gen. Physiol.* **62**, 375 (1973).

[27] E. Rojas and B. Rudy, *J. Physiol. (London)* **262**, 501 (1976).

[28] G. S. Oxford, C. H. Wu, and T. Narahashi, *J. Gen. Physiol.* **71**, 227 (1978).

[29] T. Narahashi, J. W. Moore, and B. I. Shapiro, *Science* **163**, 680 (1969).

[30] G. Romey, J.-P. Abita, H. Schweitz, G. Wunderer, and M. Lazdunski, *Proc. Natl. Acad. Sci. U.S.A.* **73**, 4055 (1976).

[31] P. A. Low, C. H. Wu, and T. Narahashi, *J. Pharmacol. Exp. Ther.* **210**, 417 (1979).

[32] L. Beress, *Pure Appl. Chem.* **54**, 1981 (1982).

[33] H. Meves, J. M. Simard, and D. D. Watt, *Ann. N.Y. Acad. Sci.* **479**, 113 (1986).

from *Centruroides sculpturatus,* toxin II from *Centruroides suffusus,* and toxin γ from *Tityus serrulatus.*[33] The third group of toxins inhibits the sodium channel inactivation and shifts both the steady-state sodium channel activation and inactivation curves in the hyperpolarizing direction; it includes TsIV-5 from *Tityus serrulatus.*[34]

Chloramine-T. Chloramine-T inhibits the sodium channel inactivation, causing a prolonged sodium current flow in the nodes of Ranvier of the frog and toad.[35,36] The steady-state sodium inactivation curve is shifted in the depolarizing direction. Furthermore, chloramine-T when kept in solution appears to contain at least two active forms, one to inhibit the sodium inactivation and the other to block the sodium activation.[37]

Batrachotoxin. Batrachotoxin (BTX) is contained in the skin secretion of the Colombian poison arrow frog *Phyllobates aurotaenia.*[3,38] It causes a large and irreversible membrane depolarization arising from opening of sodium channels.[39] The sodium channel inactivation is removed, the single-channel open time is prolonged, and the sodium channel activation–voltage curve is greatly shifted in the hyperpolarizing direction.[40,41] Batrachotoxin preferentially binds to the open sodium channel.[38] Because of the prolonged opening of single sodium channels, BTX and its analog, batrachotoxin A 20α-benzoate, have been used extensively for experiments in which the sodium channels isolated from various excitable membranes were incorporated into planar phospholipid bilayers.[42,43] Batrachotoxin appears to bind to the same intrasodium channel site as that of local anesthetics.[44,45]

Grayanotoxins. Grayanotoxins (GTXs) are the toxic principles obtained from the leaves of various plants *(Leucothoe, Rhododendron, Andromeda, Kalmia)* belonging to the family Ericaceae.[3] Several GTXs, including grayanotoxin I and α-dihydrograyanotoxin II, are known to cause a large depolarization of the membrane through opening of the sodium

---

[34] G. E. Kirsch, A. Skattebl, L. D. Possani, and A. M. Brown, *J. Gen. Physiol.* **93,** 67 (1989).
[35] W. Ulbricht and M. Stoge-Herzog, *Pfluegers Arch.* **402,** 439 (1984).
[36] G. K. Wang, *J. Physiol. (London)* **346,** 127 (1984).
[37] J. M. Huang, J. Tanguy, and J. Z. Yeh, *Biophys. J.* **52,** 155 (1987).
[38] B. I. Khodorov, *Prog. Biophys. Mol. Biol.* **45,** 57 (1985).
[39] T. Narahashi, E. X. Albuquerque, and T. Deguchi, *J. Gen. Physiol.* **58,** 54 (1971).
[40] B. I. Khodorov and S. V. Revenko, *Neuroscience* **4,** 1315 (1979).
[41] F. N. Quandt and T. Narahashi, *Proc. Natl. Acad. Sci. U.S.A.* **79,** 6732 (1982).
[42] E. Moczydlowski, S. S. Garber, and C. Miller, *J. Gen. Physiol.* **84,** 665 (1984).
[43] E. Moczydlowski, A. Uehara, X. Guo, and J. Heiny, *Ann. N.Y. Acad. Sci.* **479,** 269 (1986).
[44] E. X. Albuquerque, I. Seyama, and T. Narahashi, *J. Pharmacol. Exp. Ther.* **184,** 308 (1973).
[45] B. I. Khodorov, E. M. Peganov, S. V. Revenko, and L. D. Shishkova, *Brain Res.* **84,** 541 (1975).

channels.[46,47] Several moieties in the GTX molecule have been identified as being responsible for the sodium channel opening action.[48] Grayanotoxins have also been used as chemical tools in the same manner as BTX, albeit less frequently owing to the limited availability of toxins.

Veratridine. Veratridine is one of the veratrum alkaloids contained in plants that belong to the tribe Veratreae and the family Liliaceae.[3] They are found in various genera such as *Veratrum, Schoenocaulon,* and *Zygadenus.* Veratridine causes a membrane depolarization as a result of opening of sodium channels.[49] The mean open time of sodium channels is prolonged by veratridine,[50] and the single-channel conductance is decreased.[51,52] Like BTX, veratridine binds preferentially to the open sodium channel.[51,53,54] Veratridine has been used to keep the sodium channel open.

Aconitine. Aconitine is a toxic component contained in the plant *Aconitum napellus.* It causes an inhibition of sodium channel inactivation and a shift of sodium channel activation voltage in the hyperpolarizing direction.[55-57] However, the degrees of these changes differ among different preparations, including the node of Ranvier, skeletal muscle, and neuroblastoma cells. Bursts of channel openings and closings underlie the slow inactivation in the presence of aconitine.[58] The ionic selectivity of sodium channels is decreased by aconitine.[56,57,59]

Brevetoxins. Brevetoxins are produced by the dinoflagellate *Ptychodiscus brevis* (formerly *Gymnodinium breve)* which causes red tides that kill fish.[60,61] Several components of brevetoxins have been identified, yet their nomenclature is confusing. A notation system has been proposed to designate the various brevetoxins isolated and identified by various groups. Eight components are now known, and they are called PbTX-1 through

[46] I. Seyama and T. Narahashi, *J. Pharmacol. Exp. Ther.* **184**, 299 (1973).
[47] T. Narahashi and I. Seyama, *J. Physiol. (London)* **242**, 471 (1974).
[48] T. Masutani, I. Seyama, T. Narahashi, and J. Iwasa, *J. Pharmacol. Exp. Ther.* **217**, 812 (1981).
[49] M. Ohta, T. Narahashi, and R. F. Keeler, *J. Pharmacol. Exp. Ther.* **184**, 143 (1973).
[50] S. S. Garber and C. Miller, *J. Gen. Physiol.* **89**, 459 (1987).
[51] S. Barnes and B. Hille, *J. Gen. Physiol.* **91**, 421 (1988).
[52] M. Yoshii and T. Narahashi, *Biophys. J.* **45**, 184a (1984).
[53] J. B. Sutro, *J. Gen. Physiol.* **87**, 1 (1986).
[54] M. D. Leibowitz, J. R. Schwarz, G. Holan, and B. Hille, *J. Gen. Physiol.* **90**, 75 (1987).
[55] H. Schmidt and O. Schmitt, *Pfluegers Arch.* **349**, 133 (1974).
[56] D. T. Campbell, *J. Gen. Physiol.* **80**, 713 (1982).
[57] I. I. Grishchenko, A. P. Naumov, and A. N. Zubov, *Neuroscience* **9**, 549 (1983).
[58] B. Nilius, K. Benndorf, and F. Markwardt, *Pfluegers Arch.* **407**, 691 (1986).
[59] G. N. Mozhayeva, A. P. Naumov, Y. A. Negulayev, and E. D. Nosyreva, *Biochim. Biophys. Acta* **466**, 461 (1977).
[60] G. Strichartz, T. Rando, and G. K. Wang, *Annu. Rev. Neurosci.* **10**, 237 (1987).
[61] C. H. Wu and T. Narahashi, *Annu. Rev. Pharmacol. Toxicol.* **28**, 141 (1988).

PbTX-8.[61-63] The potency and efficacy to affect the sodium channels are different among the brevetoxins. Their effects on the sodium channels resemble those of batrachotoxin: brevetoxins inhibit the sodium channel inactivation, causing a prolonged sodium current to flow during a step depolarization; brevetoxins shift the sodium channel activation voltage in the hyperpolarizing direction; brevetoxins appear to bind the local anesthetic site in the sodium channel; the effects of brevetoxins are irreversible after washing with toxin-free medium; and brevetoxins cause a membrane depolarization.[64-67]

Goniopora Toxins. The stony corals (*Goniopora* species) contain polypeptide toxins. One of them, called GPT, prolongs the cardiac action potential, causes a positive inotropic effect, and increases the transmitter release from nerve terminals. These effects are ascribed to inhibition of the sodium channel inactivation, an increase in leakage conductance, and membrane depolarization.[68] Another polypeptide toxin has also been isolated from a *Goniopora* species.[69] This component appears to be a calcium channel activator.

Ciguatoxin. Ciguatoxin is the toxic component that causes poisoning from the consumption of tropical fish which are ordinarily nontoxic but at times become toxic.[61] However, the toxin is produced by a toxic dinoflagellate, *Gambrierdiscus toxicus.* Ciguatoxin inhibits the sodium channel inactivation and shifts the sodium channel activation voltage in the hyperpolarizing direction.[70,71] These modifications of the sodium channel gating kinetics lead to a membrane depolarization. Thus, the effects of ciguatoxin on the sodium channel resembles those of batrachotoxin and brevetoxins. In fact ciguatoxin and brevetoxins have been found to bind to the same site.[72]

Palytoxin. Palytoxin is contained in coelenterate species that belong to the genus *Palythoa.* It has a unique and complex structure with a molecu-

[62] Y. Shimizu, *Pure Appl. Chem.* **54,** 1973 (1982).

[63] M. A. Poli, T. J. Mende, and D. G. Baden, *Mol. Pharmacol.* **30,** 129 (1986).

[64] J. M. C. Huang, C. H. Wu, and D. G. Baden, *J. Pharmacol. Exp. Ther.* **229,** 615 (1984).

[65] C. H. Wu, J. M. C. Huang, S. M. Vogel, V. S. Luke, W. D. Atchison, and T. Narahashi, *Toxicon* **23,** 481 (1985).

[66] W. D. Atchison, V. S. Luke, T. Narahashi, and S. M. Vogel, *Br. J. Pharmacol.* **89,** 731 (1986).

[67] R. E. Sheridan and M. Adler, *FEBS Lett.* **247,** 448 (1989).

[68] I. Muramatsu, M. Fujiwara, A. Miura, and T. Narahashi, *J. Pharmacol. Exp. Ther.* **234,** 307 (1985).

[69] J. Qar, H. Schweitz, A. Schmid, and M. Lazdunski, *FEBS Lett.* **202,** 331 (1986).

[70] J.-N. Bidard, H. P. M. Vijverberg, C. Frelin, E. Chungue, A. M. Legrand, R. Bagnis, and M. Lazdunski, *J. Biol. Chem.* **259,** 8353 (1984).

[71] E. Benoit, A. M. Legrand, and J. M. Dubois, *Toxicon* **24,** 357 (1986).

[72] A. Lombet, J. N. Bidard, and M. Lazdunski, *FEBS Lett.* **219,** 355 (1987).

lar weight of 2680 ($C_{129}H_{223}N_3O_{54}$) and 64 chiral centers.[73-75] The total synthesis of palytoxin has recently been accomplished by Professor Y. Kishi of Harvard University after 8 years of work. The toxin has more than 10^{21} possible isomers.[76,77]

Palytoxin causes a large depolarization of squid nerve membranes which is reversed slowly after washing. It is effective only from outside the membrane. The depolarization is due to an increase in membrane permeability to sodium, but it is only partially reversed by a high concentration (1 μM) of TTX. The ionic selectively in palytoxin-poisoned membrane is considerably different from that of normal membrane. However, the kinetics of sodium current undergo little or no change in palytoxin. These observations have led to the conclusion that palytoxin creates a new channel in the membrane which is permeable to sodium but resistant to TTX.[78] Palytoxin also depolarizes every membrane examined, including cardiac muscle, skeletal muscle, smooth muscle, and myelinated and nonmyelinated nerves.[61] Single-channel recording experiments have clearly shown that palytoxin elicits openings of individual channels in ventricular myocytes with a conductance of 9.5 pS.[79] These currents are dependent on the presence of Na^+ ions, are observed in Na^+-free NH_4^+, Li^+ or Cs^+ solution, and are resistant to TTX and Co^{2+}. Palytoxin also causes K^+ efflux in erythrocytes which is prevented by ouabain. In the crayfish axon, the palytoxin-induced depolarization requires ATP intracellularly and is inhibited by ouabain.[80] These and other observations have led to the hypothesis that palytoxin converts the Na^+,K^+-ATPase-dependent ion pump to an ion channel.[81-83]

[73] R. E. Moore and G. Bartolini, *J. Am. Chem. Soc.* **103**, 2491 (1981).

[74] D. Uemura, K. Ueda, and Y. Hirata, *Tetrahedron Lett.* **22**, 2781 (1981).

[75] R. E. Moore, G. Bartolini, J. Barchi, A. A. Bothner-By, J. Dadok, and J. Ford, *J. Am. Chem. Soc.* **104**, 3776 (1982).

[76] R. W. Armstrong, J.-M. Beau, S. H. Cheon, W. J. Christ, H. Fujioka, W.-H. Ham, L. D. Hawkins, H. Jin, S. H. Kang, Y. Kishi, M. J. Martinelli, W. W. McWhorter, Jr., M. Mizuno, M. Nakata, A. E. Stutz, F. X. Talamas, M. Taniguchi, J. A. Tino, K. Ueda, J. Uenishi, J. B. White, and M. Yonaga, *J. Am. Chem. Soc.* **111**, 7525 (1989).

[77] R. W. Armstrong, J.-M. Beau, S. H. Cheon, W. J. Christ, H. Fujioka, W.-H. Ham, L. D. Hawkins, H. Jin, S. H. Kang, Y. Kishi, M. J. Martinelli, W. W. McWhorter, Jr., M. Mizuno, M. Nakata, A. E. Stutz, F. X. Talamas, M. Taniguchi, J. A. Tino, K. Ueda, J. Uenishi, J. B. White, and M. Yonaga, *J. Am. Chem. Soc.* **111**, 7530 (1989).

[78] I. Muramatsu, D. Uemura, M. Fujiwara, and T. Narahashi, *J. Pharmacol. Exp. Ther.* **231**, 488 (1984).

[79] I. Muramatsu, M. Nishio, S. Kigoshi, and D. Uemura, *Br. J. Pharmacol.* **93**, 811 (1988).

[80] C. H. Wu and K. Marx, *Biophys. J.* **51**, 387a (1987).

[81] E. Habermann and G. S. Chhatwal, *Naunyn-Schmiedebergs Arch. Pharmacol.* **319**, 101 (1982).

[82] E. Habermann, *Naunyn-Schmiedebergs Arch. Pharmacol.* **323**, 269 (1983).

[83] G. S. Chhatwal, H.-J. Hessler, and E. Habermann, *Naunyn-Schmiedebergs Arch. Pharmacol.* **323**, 261 (1983).

Pyrethroids. Pyrethroids are synthetic derivatives of pyrethrins which are contained in the flowers of *Chrysanthemum cinerariaefolim.* The pyrethrum insecticide had been used extensively until World War II. However, development of a variety of synthetic and long-lasting insecticides such as DDT and parathion during the early postwar period made the pyrethrum insecticide obsolete. Beginning in the mid-1960s, serious concerns over the possible long-term effects of insecticides on humans and environmental contaminations have been voiced, and a large number of derivatives of pyrethrins were synthesized and tested. Some two dozen of them have proved useful, owing to high insecticidal potencies, low mammalian toxicities, and biodegradability, and are being used extensively as insecticides. The mechanisms of action of pyrethroids on nerve membranes have been reviewed extensively.[84-91] It has now been well established and accepted that the major target site of pyrethroids is the voltage-activated sodium channel. Only those studies which are relevant to the use of pyrethroids as tools are described here.

Pyrethroids may be divided into two large groups based primarily on their structures. Those which do not have a cyano group at the α position are called type I, and those which do have an α-cyano group are called type II. Both modify the sodium channel in a similar manner. Pyrethroids cause a marked prolongation of the mean open time of the sodium channel. The open time is normally of the order of a few milliseconds, but ranges from several hundred milliseconds in the presence of tetramethrin, a type I pyrethroid,[92] to several seconds in fenvalerate and deltamethrin, type II pyrethroids.[93,94] The sodium channel inactivation is inhibited, the activa-

[84] T. Narahashi, *in* "Advances in Insect Physiology" (J. W. L. Beament, J. E. Treherne, and V. B. Wigglesworth, eds.), Vol. 8, p. 1. Academic Press, London and New York, 1971.

[85] T. Narahashi, *in* "Insecticide Biochemistry and Physiology" (C. F. Wilkinson, ed.), p. 327. Plenum, New York, 1976.

[86] T. Narahashi, *Neurotoxicology* 6(2), 3 (1985).

[87] T. Narahashi, *in* "Sites of Action for Neurotoxic Pesticides" (R. M. Hollingworth and M. B. Green, eds.), American Chemical Society Symposium Series, No. 356, p. 226. American Chemical Society, Washington, D.C., 1987.

[88] T. Narahashi, *in* "Neurotox '88: Molecular Basis of Drug and Pesticide Action" (G. G. Lunt, ed.), p. 269. Elsevier, Amsterdam, 1988.

[89] T. Narahashi, *in* "Insecticide Action: From Molecule to Organism" (T. Narahashi and J. E. Chambers, eds.), p. 55. Plenum, New York, 1989.

[90] G. S. F. Ruigt, *in* "Comprehensive Insect Physiology, Biochemistry and Pharmacology" (G. A. Kerkut and L. I. Gilbert, eds.), Vol. 12, p. 183. Pergamon, Oxford, 1984.

[91] W. Wouters and J. van den Bercken, *Gen. Pharmacol.* **9.** 387 (1978).

[92] D. Yamamoto, F. N. Quandt, and T. Narahashi, *Brain Res.* **274,** 344 (1983).

[93] K. Chinn and T. Narahashi, *J. Physiol. (London)* **380,** 191 (1986).

[94] S. F. Holloway, T. Narahashi, V. L. Salgado, and C. H. Wu, *Pfluegers Arch.* **414,** 613 (1989).

tion kinetics are slowed, and the activation voltage is shifted in the hyper-polarizing direction. Thus, the sodium current is greatly prolonged during and after a depolarizing pulse, and the membrane is depolarized. These changes in sodium channel kinetics lead to repetitive afterdischarges in nerve fibers, massive discharges from sensory neurons, and synaptic and neuromuscular disturbances, which form the basis for the symptoms of poisoning in mammals and insects.

Pyrethroids are useful chemical tools because of their action to keep the sodium channels open for an extremely long period of time. Holloway *et al.*[94] took advantage of this action and measured a current–voltage curve of a single sodium channel modified by fenvalerate. The curve shows a marked rectification at large negative potentials, the characteristic bending commonly observed by recording the sodium currents arising from many channels.

Potassium Channels

Whereas the delayed rectifying potassium channel was the only potassium channel that had been studied extensively until 10–15 years ago, more recent developments in electrophysiological techniques such as patch clamp recording have made it possible to disclose a variety of potassium channels in diverse tissues including both excitable and inexcitable cells. We now know the gating mechanism, the properties, and blocking agents related to the various potassium channels. The developments along this line are not limited to academic interest; a variety of drugs are being studied for potential clinical uses for cardiovascular and other diseases. Several reviews have been published recently.[95–100]

Voltage-Activated Potassium Channels

Delayed Rectifying Potassium Channels. Delayed rectifying potassium channels are present in various excitable cells and are responsible for the falling phase of the action potential. The channels are opened and are kept open by membrane depolarization. Under normal physiological condi-

[95] G. Yellen, *Annu. Rev. Biophys. Chem.* **16,** 227 (1987).
[96] N. S. Cook, *Trends Pharmacol. Sci.* **9,** 21 (1988).
[97] N. A. Castle, D. G. Haylett, and D. H. Jenkinson, *Trends Neurosci.* **12,** 59 (1989).
[98] D. J. Adams and W. Nonner, *in* "Potassium Channels: Structure, Classification, Function and Therapeutic Potential" (N. S. Cook, ed.), p. 40. Horwood, Chichester, 1990.
[99] J. R. de Weille and M. Lazdunski, *in* "Ion Channels" (T. Narahashi, ed.), Vol. 2, p. 205. Plenum, New York, 1990.
[100] F. Dreyer, *Rev. Physiol. Biochem. Pharmacol.* **115,** 93 (1990).

tions, potassium ions flow in the outward direction through the open channels.

The agent that was first discovered to block the delayed rectifying potassium channels is tetraethylammonium (TEA),[101] and its interaction with potassium channels was reviewed by Armstrong,[102] Stanfield,[103] and Adams and Nonner.[98] When applied internally to the squid giant axon, TEA and its longer chain derivatives such as nonyltriethylammonium block the outward potassium currents without affecting the inward potassium current.[104-106] The potassium channel is blocked while opening and in a manner dependent on current direction. However, TEA is not potent, requiring 20–40 mM internally to block the potassium current completely in squid axons. It is not highly selective for the potassium channels, suppressing the sodium current to some extent (by 10–30%) at 20–40 mM. In the node of Ranvier and skeletal muscle but not in squid giant axon, TEA is effective in blocking the potassium current from either side of the nerve membrane.[107] Tetraethylammonium has been used extensively as a tool to block the delayed rectifying potassium channel.

4-Aminopyridine (4-AP) was discovered by Pelhate and Pichon[108] to block the delayed rectifying potassium current. 2-, 3-, and 4-Aminopyridines block the potassium channel from either side of the nerve membrane in a manner dependent on the membrane potential and the duration of depolarization.[109] The block becomes less pronounced with larger depolarizations and is slowly relieved during a depolarizing pulse. The apparent K_d value is approximately 30 μM. 3,4-Diaminopyridine is more potent than 4-AP, with an apparent dissociation constant of 5.8 and 0.7 μM for external and internal application, respectively.[110] The aminopyridines have been used often as tools because of their potent and selective action on the potassium channels, but caution must be exercised in using them because of their voltage- and time-dependent nature of action.

Cesium ions block the delayed rectifying potassium channels in a manner dependent on the direction of current flow. Internal Cs^+ blocks the

[101] I. Tasaki and S. Hagiwara, *J. Gen. Physiol.* **40**, 859 (1957).
[102] C. M. Armstrong, *in* "Membranes: A Series of Advances" (G. Eisenman, ed.), Vol. 3, p. 325. Dekker, New York, 1975.
[103] P. R. Stanfield, *Rev. Physiol. Biochem. Pharmacol.* **97**, 1 (1983).
[104] C. M. Armstrong and L. Binstock, *J. Gen. Physiol.* **48**, 859 (1965).
[105] C. M. Armstrong, *J. Gen. Physiol.* **54**, 553 (1969).
[106] C. M. Armstrong, *J. Gen. Physiol.* **58**, 413 (1971).
[107] B. Hille, *J. Gen. Physiol.* **50**, 1287 (1967).
[108] M. Pelhate and Y. Pichon, *J. Physiol. (London)* **242**, 90P (1974).
[109] J. Z. Yeh, G. S. Oxford, C. H. Wu, and T. Narahashi, *J. Gen. Physiol.* **68**, 519 (1976).
[110] G. E. Kirsch and T. Narahashi, *Biophys. J.* **22**, 507 (1978).

outward current, whereas external Cs^+ blocks the inward current.[111,112] High concentrations are required for these effects. Barium ions block the delayed rectifying potassium channels from both sides of the membrane on a one-to-one stoichiometric basis.[113,114]

Several toxins are known to block the delayed rectifying potassium channels. Scorpion toxins have been shown to affect the delayed rectifying potassium channels. Early studies were performed using venoms of *Leiurus quinquestriatus* and *Buthus tamulus;* both sodium and potassium channels were affected, the latter being suppressed.[115,116] Noxiustoxin, isolated and purified from the venom of the Mexican scorpion *Centruroides noxius,* blocks the delayed rectifying potassium channels in their open configuration selectively and reversibly, with a K_d of 400 nM.[117,118] Although recognized primarily as a specific blocker of Ca^{2+}-activated potassium channels,[119] charybdotoxin isolated from *Leiurus quinquestriatus* var. *hebraeus* also blocks Ca^{2+}-insensitive potassium channels from murine T lymphocytes,[120] the potassium channel that generates A-current in *Drosophila,*[121] and the component f1 of the delayed rectifying potassium current in frog nodal membranes.[122]

Toxins isolated from bee venom have received much attention. Although apamin primarily blocks the Ca^{2+}-activated potassium channel, mast cell degranulating (MCD) peptide blocks the f1 component of the delayed rectifying potassium current of frog nodes[122] and the slowly activating outward potassium current at mammalian motor nerve terminals.[100] MCD peptide appears to be similar to dendrotoxins in its blocking action on the delayed rectifying potassium current.

Snake venoms contain many components, some of which block the

---

[111] W. K. Chandler and H. Meves, *J. Physiol. (London)* **180,** 798 (1965).

[112] W. J. Adelman and J. P. Senft, *J. Gen. Physiol.* **51,** 102S (1968).

[113] C. M. Armstrong and S. R. Taylor, *Biophys. J.* **30,** 473 (1980).

[114] D. C. Eaton and M. S. Brodwick, *J. Gen. Physiol.* **75,** 727 (1980).

[115] E. Koppenhöfer and H. Schmidt, *Pfluegers Arch.* **303,** 133 (1968).

[116] T. Narahashi, B. I. Shapiro, T. Deguchi, M. Scuka, and C. M. Wang, *Am. J. Physiol.* **222,** 850 (1972).

[117] E. Carbone, E. Wanke, G. Prestipino, L. D. Possani, and A. Maelicke, *Nature (London)* **296,** 90 (1982).

[118] E. Carbone, G. Prestipino, L. Spadavecchia, F. Franciolini, and L. D. Possani, *Pfluegers Arch.* **408,** 423 (1987).

[119] C. Miller, E. Moczydlowski, R. Latorre, and M. Philipps, *Nature (London)* **313,** 316 (1985).

[120] R. S. Lewis and M. D. Cahalan, *Science* **239,** 771 (1988).

[121] R. MacKinnon, P. H. Reinhart, and M. M. White, *Neuron* **1,** 997 (1988).

[122] M. E. Bräu, F. Dreyer, P. Jonas, H. Repp, and W. Vogel, *J. Physiol. (London)* **420,** 365 (1990).

delayed rectifying potassium channels. Dendrotoxin isolated from the green mamba *Dendroaspis angusticeps* and dendrotoxin I (toxin I) isolated from the black mamba *Dendroaspis polylepis* are known to block the transient A-current and voltage-activated transient Ca^{2+}-dependent potassium current,[123,124] but the delayed rectifying potassium current is more sensitive than these transient potassium currents. For instance, the f1 component of the delayed rectifying potassium current of myelinated nerve fibers is blocked by dendrotoxin I with an apparent K_d of 0.4 nM,[125] a portion of the delayed rectifying potassium current of the guinea pig dorsal root ganglion cells is blocked by dendrotoxin at concentrations of 0.14 to 1.4 nM,[126] and a portion of the delayed rectifying potassium current of visceral sensory neurons of rat nodose ganglia is blocked.[127,128] Dendrotoxin also blocks the delayed rectifying potassium current of *Myxicola* giant axons, albeit with a relatively low potency (K_d 150 nM).[129]

Inward Rectifying Potassium Channels. Inward rectification or anomalous rectification was first discovered in frog muscle.[130] The inward rectifying channel is permeable to potassium and has been studied, in addition to skeletal muscle, in egg cells and cardiac muscle. The agents that block the inward rectifying potassium channels include Ba^{2+} ions,[131,132] Cs^+ ions,[133] and H^+ ions.[131,134]

Transient Potassium Channels (A-Currents). The transient potassium channel is opened by membrane depolarization, causing an outward potassium current which attains a peak and decays. It is often referred to as A-current. This current was first discovered in molluscan neurons by Hagiwara *et al.*,[135] and it was named A-current by Conner and Stevens.[136] It appears that A-current modulates the firing rate in neurons.

[123] J. V. Halliwell, I. B. Othman, A. Pelchen-Matthews, and J. O. Dolly, *Proc. Natl. Acad. Sci. U.S.A.* **83**, 493 (1986).
[124] C. W. Bourque, *J. Physiol. (London)* **397**, 331 (1988).
[125] E. Benoit and J.-M. Dubois, *Brain Res.* **377**, 374 (1986).
[126] R. Penner, M. Petersen, F.-K. Pierau, and F. Dreyer, *Pfluegers Arch.* **407**, 365 (1986).
[127] C. E. Stanfeld, S. J. Marsh, J. V. Halliwell, and D. A. Brown, *Neurosci. Lett.* **64**, 299 (1986).
[128] C. E. Stanfeld, S. J. Marsh, D. N. Parcej, J. O. Dolly, and D. A. Brown, *Neuroscience* **23**, 893 (1987).
[129] C. L. Schauf, *J. Pharmacol. Exp. Ther.* **241**, 793 (1987).
[130] B. Katz, *Arch. Sci. Physiol.* **3**, 285 (1949).
[131] S. Hagiwara, S. Miyazaki, W. Moody, and J. Patlak, *J. Physiol. (London)* **279**, 167 (1978).
[132] H. Ohmori, S. Yoshida, and S. Hagiwara, *Proc. Natl. Acad. Sci. U.S.A.* **78**, 4960 (1981).
[133] S. Hagiwara, S. Miyazaki, and N. P. Rosenthal, *J. Gen. Physiol.* **67**, 621 (1976).
[134] W. J. Moody and S. Hagiwara, *J. Gen. Physiol.* **79**, 115 (1982).
[135] S. Hagiwara, K. Kusano, and N. Saito, *J. Physiol. (London)* **155**, 470 (1961).
[136] J. A. Connor and C. F. Stevens, *J. Physiol. (London)* **213**, 21 (1971).

Aminopyridines block the transient potassium channels in molluscan neurons[137,138] and mammalian neurons.[139] Charybdotoxin, a toxin contained in the venom of the Mideastern scorpion *Leiurus quinquestriatus* var. *hebraeus,* is widely known as a specific blocker of the Ca^{2+}-activated potassium channels, but it also blocks A-current in *Drosophila.*[121] Dendrotoxin isolated from the green mamba *Dendroaspis angusticeps* is a specific blocking agent acting on the delayed rectifying potassium channels, but it has been found to block the transient potassium channel in hippocampal CA1 neurons, albeit slowly and not so potently.[123,140]

Transient Calcium-Dependent Potassium Channels. Magnocellular neurosecretory cells of the rat supraoptic nucleus generate a transient outward potassium current on depolarization.[124] This current is dependent on intracellular calcium and is abolished by removal of Ca^{2+} from the external perfusate by application of Co^{2+}, Mn^{2+}, or Cd^{2+} or by replacement of Ca^{2+} by Ba^{2+}. It is not affected by TEA but is blocked by 4-aminopyridine and dendrotoxin.

Sarcoplasmic Reticulum Potassium Channels. Sarcoplasmic reticulum (SR) contains a voltage-activated channel permeable to potassium. Both Cs^+ and TEA are slightly permeant to the channel, yet they block the flow of K^+ ions through the channel.[141-143] The channel is blocked by 1,3-bis[tris(hydroxymethyl)methylamino]propane.[142]

M-Currents. M-Current was first identified in the large B cells of frog lumbar sympathetic ganglia.[144] It is inhibited by muscarinic acetylcholine (ACh) receptor agonists. M-Current was also found in other preparations, including rat and rabbit superior sympathetic ganglia, spinal cord neurons, hippocampal pyramidal cells, cortical neurons, and neuroblastoma cells.[145] M-Current does not generate action potentials by itself, but appears to modulate excitability such as the frequency and pattern of discharges.[146] M-Current is inhibited by a variety of agents: muscarinic ACh receptor agonists such as ACh, muscarine, methacholine, and oxotremorine; pep-

[137] S. H. Thompson, *J. Physiol. (London)* **265,** 465 (1977).
[138] S. Thompson, *J. Gen. Physiol.* **80,** 1 (1982).
[139] B. Hustafsson, M. Galvan, P. Grafe, and H. Wigstroem, *Nature (London)* **299,** 252 (1982).
[140] J. O. Dolly, J. V. Halliwell, J. D. Black, R. S. Williams, P. Pelchen-Matthews, A. L. Breeze, F. Mahraban, I. B. Othman, and A. R. Black, *J. Physiol. (Paris)* **79,** 280 (1984).
[141] R. Coronado, R. L. Rosenberg, and C. Miller, *J. Gen. Physiol.* **76,** 425 (1980).
[142] R. Coronado and C. Miller, *J. Gen. Physiol.* **79,** 529 (1982).
[143] S. Cukierman, G. Yellen, and C. Miller, *Biophys. J.* **48,** 477 (1985).
[144] D. A. Brown and P. R. Adams, *Nature (London)* **283,** 673 (1980).
[145] D. A. Brown, *in* "Ion Channels" (T. Narahashi, ed.), Vol. 1, p. 55. Plenum, New York, 1988.
[146] P. R. Adams, *Trends Neurosci.* **5,** 116 (1982).

tides such as t-luteinizing hormone-releasing hormone (t-LHRH), substance P, substance K, eledoisin, kassinin, and physalaemin; and Ba^{2+} ions.[145]

Ligand-Activated Potassium Channels

Calcium-Activated Potassium Channels. Calcium-activated potassium channels are modulated by an increase in intracellular calcium concentration. Three types of Ca^{2+}-activated potassium channels can be distinguished based on single-channel conductance: (1) "Big" or "maxi" potassium channels which exhibit a high single-channel conductance of 100–250 pS; (2) "intermediate" potassium channels which have an 18–50 pS single-channel conductance; and (3) "small" potassium channels with a 10–14 pS single-channel conductance.[96,100] Big and intermediate channels are opened by both an increase in intracellular Ca^{2+} concentration to 0.1–10 μM *and* membrane depolarization, whereas the small channels are opened by an increase in intracellular Ca^{2+} concentration only.

Apamin, one of the components in the venom of the honeybee *Apis mellifera,* blocks the small Ca^{2+}-activated potassium channels in mammalian neurons and skeletal muscle.[147–149] The apamin-sensitive channel is not blocked by TEA, whereas other types, presumably large channels, are blocked by TEA and charybdotoxin but not by apamin.[119,150–152] The apamin-sensitive Ca^{2+}-activated potassium channels are also blocked by quinine, tubocurarine, and pancuronium.[153,154]

Charybdotoxin, isolated from the scorpion *Leiurus quinquestriatus* var. *hebraeus,* is a potent and specific blocker acting on the big Ca^{2+}-activated potassium channels.[119] Block occurs at the external orifice of the channel on a one-to-one stoichiometric basis. Charybdotoxin also blocks the intermediate Ca^{2+}-activated potassium channel.[155] The venom of this scorpion species contains another toxic component, leiurotoxin I, which blocks the small Ca^{2+} activated potassium channel.[156]

[147] M. Hugues, H. Schmid, G. Romey, D. Duval, C. Frelin, and M. Lazdunski, *EMBO J.* **1,** 1039 (1982).
[148] M. J. Seagar, C. Granier, and F. Couraud, *J. Biol. Chem.* **259,** 1491 (1984).
[149] A. L. Blatz and K. L. Magleby, *Nature (London)* **323,** 718 (1986).
[150] G. Romey and M. Lazdunski, *Biochem. Biophys. Res. Commun.* **118,** 669 (1984).
[151] G. Romey, M. Hugues, H. Schmid-Antomarchi, and M. Lazdunski, *J. Physiol. (Paris)* **79,** 259 (1984).
[152] P. Pennefather, B. Lancaster, P. R. Adams, and R. A. Nicoll, *Proc. Natl. Acad. Sci. U.S.A.* **82,** 3040 (1985).
[153] D. H. Jenkinson, D. G. Haylett, and N. S. Cook, *Cell Calcium* **4,** 429 (1983).
[154] N. S. Cook and D. G. Haylett, *J. Physiol. (London)* **358,** 373 (1985).
[155] A. Hermann and C. Erxleben, *J. Gen. Physiol.* **90.** 27 (1987).

Other Potassium Channels

ATP-Sensitive Potassium Channels. A new type of potassium channel was discovered in cardiac membranes which is inhibited by an increase in intracellular ATP concentration.[157] Similar potassium channels sensitive to ATP were subsequently found in skeletal muscle, pancreatic β-cells, arterial smooth muscle, and cortical neurons.[99,158-162]

The ATP-sensitive potassium channel plays an important role in the physiological function of cells. In pancreatic β-cells, glucose metabolism produces ATP which closes the ATP-sensitive potassium channels. This leads to a membrane depolarization which in turn opens the voltage-activated calcium channels, causing an influx of Ca^{2+} ions. The increase in intracellular Ca^{2+} concentration releases insulin.[159] In cardiac muscle, a decrease in ATP level such as that resulting from ischemia causes openings of the ATP-sensitive potassium channels which in turn cause a membrane hyperpolarization and a shortening of the action potential. The resultant decrease in excitability prevents further consumption of the high-energy phosphate, thereby preventing the cell from irreversible damage.[158] The role of the ATP-sensitive potassium channels in skeletal muscle is less clear, but a situation similar to that in cardiac muscle could occur.

Hypoglycemic sulfonylureas such as tolbutamide and glibenclamide have been used in the treatment of diabetes for many years. It has been found recently that they block the ATP-sensitive potassium channel, thereby eliciting insulin secretion in a manner similar to an increase in intracellular ATP concentration.[161,163]

Potassium Channel Openers. Recent developments of potassium channel openers have aroused widespread interest, as these drugs could open an avenue to a new type of drug in the treatment of cardiovascular disorders.[163,164] Several types of chemicals have been shown to open potassium

[156] G. G. Chicchi, G. Gimenez-Gallego, E. Ber, M. L. Garcia, R. Winquist, and M. A. Cascieri, *J. Biol. Chem.* **263**, 10192 (1988).

[157] A. Noma, *Nature (London)* **305**, 147 (1983).

[158] A. Noma and T. Shibasaki, *in* "Ion Channels" (T. Narahashi, ed.), Vol. 1, p. 183. Plenum, New York, 1988.

[159] F. M. Ashcroft, *Annu. Rev. Neurosci.* **11**, 97 (1988).

[160] P. Rorsman and G. Trube, *in* "Potassium Channels: Structure, Classification and Therapeutic Potentials" (N. S. Cook, ed.), p. 96. Horwood, Chichester, 1990.

[161] P. Rorsman, P.-O. Berggren, K. Bokvist, and S. Efendic, *News Physiol. Sci.* **5**, 143 (1990).

[162] B. P. Bean and D. D. Friel, *in* "Ion Channels" (T. Narahashi, ed.), Vol. 2, p. 169. Plenum, New York, 1990.

[163] U. Quast and N. S. Cook, *Trends Pharmacol. Sci.* **10**, 431 (1989).

[164] G. Edwards and A. H. Weston, *Trends Pharmacol. Sci.* **11**, 417 (1990).

channels: the benzopyrans as represented by cromakalim, the first agent to be designated a potassium channel opener; guanidine/thiourea derivatives as represented by pinacidil; pyridine derivatives as represented by nicorandil; pyrimidines as represented by minoxidil; benzothiadizines as represented by diazoxide; and thioformamides as represented by RP52891.[164]

Opening of potassium channels would lead to relaxation of blood vessels. However, it could elicit dysrhythmias in the heart and release of excitatory amino acids in the brain owing to accumulation of K^+ in the interstitial space, which in turn would cause cell death. Therefore, it would be highly desirable to develop potassium channel openers having a selective action on certain types of cells. A question had remained until recently whether the target for these potassium channel openers was the ATP-sensitive potassium channel. A recent single-channel study has clearly shown that cromakalim opens the ATP-sensitive potassium channel, the action that is inhibited by glibenclamide.[165] Furthermore, the relaxations of norepinephrine-induced contractions of mesenteric artery by diazoxide, cromakalim, and pinacidil were reversed by Ba^{2+} and glibenclamide.[165] Thus, it has become clear that these vasodilators work through openings of the ATP-sensitive potassium channels. It should be pointed out, however, that pinacidil appears to have a blocking action on the voltage-activated calcium channel and Ca^{2+}-activated potassium channel in addition to an opening action on the ATP-sensitive potassium channel. Because of potential clinical applications, potassium channel openers have now become a very hot subject of investigation.

Calcium Channels

Classification and Properties of Calcium Channels

Voltage-activated calcium channels play important roles in a variety of functions of excitable cells. These include generation of action potentials, control of bursting and pacemaker activity, secretion of neurotransmitters and hormones, and excitation–contraction coupling. Therefore, drugs that specifically modulate the activity of calcium channels could potentially be used in the treatment of certain diseases. For this reason, calcium channel blocking agents or calcium antagonists have long been studied for potential applications to cardiovascular disorders, and some of them have been successfully developed into useful therapeutic agents.[166–168] Examples in-

[165] N. B. Standen, J. M. Quayle, N. W. Davies, J. E. Brayden, Y. Huang, and M. T. Nelson, *Science* **245,** 177 (1989).
[166] R. A. Janis, P. J. Silver, and D. J. Triggle, *Adv. Drug. Res.* **16,** 309, (1987).

clude verapamil, nifedipine, and diltiazem which are being used in the treatment of angina, paroxysmal supraventricular tachyarrhythmias, atrial fibrillation, hypertension, and/or cardioplegia.

Development and wide applications of patch clamp techniques[169] in the early 1980s have made it possible to record and characterize the calcium channel activity of neuronal and cardiac membranes. We now know that there are several types of calcium channels in various excitable cells and that each type of calcium channel exhibits specific physiological and pharmacological characteristics.[170-172] This has aroused interest in seeking blocking agents acting selectively on one of the several types of calcium channels. Studies along this line have been successful to some extent, yet most such chemical agents exhibit a specificity for channel type only to a limited extent. One difficulty that has arisen along this line of investigation is the failure to establish a clear-cut definition for classification of channel types. Thus a chemical agent that has been shown to block specifically one calcium channel type in a cell may or may not exhibit the same specificity for a similar but not identical calcium channel type in another kind of cell.

The classification of calcium channels should not be taken as representing a rigid and clear-cut definition. Although there have been many discussions about how to classify the calcium channels and whether any particular calcium channel type found in a cell is the same as or different from a similar type of calcium channel in another cell type, the conventional classification described below is a matter of convenience.

T-Type Calcium Channels. T-Type or type I calcium channels are commonly defined by two characteristics: (1) the current is inactivated during a step depolarization, and (2) the channel is opened at relatively large negative potentials or with a small depolarization (sometimes called low-threshold current). T-Type calcium channels are distributed widely, including certain neuroblastoma cells (N1E-115 line),[173] neuroblastoma-glioma hybrid cells (NG108-15 line),[174] dorsal root ganglion neurons,[175]

[167] A. Fleckenstein, "Calcium Antagonism in Heart and Smooth Muscle: Experimental Facts and Therapeutic Prospects." Wiley, New York, 1983.

[168] F. Urthaler, *Am. J. Med. Sci.* **292,** 217 (1986).

[169] O. Hamill, A. Marty, E. Neher, B. Sakmann, and F. J. Sigworth, *Pfluegers Arch.* **391,** 85 (1981).

[170] P. Hess, *Annu. Rev. Neurosci.* **13,** 337 (1990).

[171] M. M. Hosey and M. Lazdunski, *J. Membr. Biol.* **104,** 81 (1988).

[172] B. P. Bean, *Annu. Rev. Physiol.* **51,** 367 (1989).

[173] T. Narahashi, A. Tsunoo, and M. Yoshii, *J. Physiol. (London)* **383,** 231 (1987).

[174] A. Tsunoo, M. Yoshii, and T. Narahashi, *Proc. Natl. Acad. Sci. U.S.A.* **83,** 9832 (1986).

[175] M. C. Nowycky, A. P. Fox, and R. W. Tsien, *Nature (London)* **316,** 440 (1985).

endocrine cells (GH_3),[176] cardiac cells (sinoatrial and Purkinje cells, pacemaker cells of the atria and ventricles),[177] and vascular smooth muscle.[178] T-Type calcium channels appear to play an important role in eliciting a burst of action potentials in certain neurons, in pacemaker activity in cardiac cells, and in aldosterone secretion.

N-Type Calcium Channels. N-Type calcium channels are usually defined by two characteristics: (1) the current is slowly inactivated during a step depolarization, and (2) the channel is opened at relatively small negative potentials or with large depolarizations. The N-type of calcium channel is usually found in neuronal membranes, and it appears to play a role in neurotransmitter release.[179,180]

L-Type Calcium Channels. L-Type or type II calcium channels are defined by two characteristics: (1) the current is not significantly inactivated during a step depolarization, and (2) the channel is opened at relatively small negative potentials or with large depolarizations (sometimes called high-threshold current). L-Type calcium channels are present in neuroblastoma cells (N1E-115 line),[173] neuroblastoma–glioma hybrid cells (NG108-15 line),[174] skeletal muscle,[181] cardiac cells,[182] vascular smooth muscle,[183] dorsal rot ganglion neurons,[184] and endocrine GH_3 cells.[176] The role of L-type calcium channels appears to involve excitation–contraction coupling in muscles, release of neurotransmitters and hormones, and modulation of excitability.

P-Type Calcium Channels. The P-type class of calcium channels was first discovered in Purkinje cells[185,186] and is characterized by being activated at small negative potentials (high thresholds), by insensitivity to dihydropyridines and ω-conotoxin, and by sensitivity to the blocking ac-

[176] C. J. Cohen and R. T. McCarthy, *J. Physiol. (London)* **387**, 195 (1987).
[177] G.-N. Tseng and P. A. Boyden, *Circ. Res.* **65**, 1735 (1989).
[178] G. Loirand, C. Mironneau, J. Mironneau, and P. Pacand, *J. Physiol. (London)* **412**, 333 (1989).
[179] L. D. Hirning, A. P. Fox, E. W. McCleskey, B. M. Olivera, S. A. Thayer, R. J. Miller, and R. W. Tsien, *Science* **239**, 57 (1988).
[180] R. W. Tsien, D. Lipscombe, D. V. Madison, K. R. Bley, and A. P. Fox, *Trends Neurosci.* **11**, 431 (1988).
[181] C. Cognard, G. Romey, J. P. Galizzi, M. Fossert, and M. Lazdunski, *Proc. Natl. Acad. Sci. U.S.A.* **83**, 1518 (1986).
[182] B. P. Bean, *Proc. Natl. Acad. Sci. U.S.A.* **81**, 6388 (1984).
[183] M. Nelson and J. F. Worley, *J. Physiol. (London)* **412**, 65 (1989).
[184] R. T. McCarthy, *in* "Nimodipine and Central Nervous System Function: New Vistas" (J. Traber and W. Gispen, eds.), p. 35. Schattauer, Stuttgart and New York, 1990.
[185] R. Llinás, M. Sugimori, J.-W. Lin, and B. Cherksey, *Proc. Natl. Acad. Sci. U.S.A.* **86**, 1689 (1989).
[186] R. R. Llinás, M. Sugimori, and B. Cherksey, *Ann. N.Y. Acad. Sci.* **560**, 103 (1989).

tion of certain spider toxins.[187-189] P-Type calcium channels have not been found in thalamic or inferior olivary cells.

Calcium Channel Blockers

L-Type Calcium Channels. A variety of chemical agents are known to block L-type calcium-channels.[166] These include phenylalkylamines (verapamil, D600, or gallopamil), dihydropyridines (felodipine, nifedipine, nimodipine, nisoldipine, nitrendipine), benzothiazepines (diltiazem), opioids and other endogenous peptides (Leu-enkephalin, somatostatin),[174] ω-conotoxin GIVA, benzodiazepines (diazepam, chlordiazepoxide),[190,191] barbiturates (phenobarbital, pentobarbital),[192] ethanol,[193] and polyvalent inorganic cations (Co^{2+}, Ni^{2+}, Cd^{2+}, Pb^{2+}, La^{3+}).[173,194]

The potencies of these agents in blocking L-type calcium channels vary greatly depending on the kinds of agents and cells. Some examples of effective doses (ED_{50} values) are given in Table I. Some of the agents listed deserve comments. Dihydropyridines are generally potent in selectively blocking L-type calcium channels of cardiac, smooth, and skeletal muscles, with subnanomolar ED_{50} values. They are less potent on neuronal L-type calcium channels. Dihydropyridine block is voltage dependent, the potency increasing with depolarization. This is due to a higher affinity of dihydropyridines for the inactivated channel than the resting channel. A higher affinity for the inactivated state relative to the resting state is also the case for chlorpromazine block of L-type calcium channels.[195] Leu-enkephalin is a potent and selective L-type channel blocker.[174] In this respect, it is almost equivalent to tetrodotoxin on sodium channels. Several polyvalent cations block L-type calcium channels; they also block T-type calcium channels albeit with different potency profiles among several cations tested. For L-type channels, La^{3+} and Pb^{2+} are most potent, and Co^{2+} is least potent.[173,194] ω-Conotoxin GVIA blocks L-type calcium channels in

[187] V. P. Bindokas and M. E. Adams, *J. Neurobiol.* **20,** 171 (1989).
[188] B. M. Salzberg, A. L. Obaid, K. Staley, J.-W. Lin, M. Sugimori, B. D. Cherksey, and R. Llinás, *Biophys. J.* **57,** 305a (1990).
[189] B. Cherksey, J.-W. Lin, M. Sugimori, and R. Llinás, *Biophys. J.* **57,** 305a (1990).
[190] S. Watabe, M. Yoshii, N. Ogata, and T. Narahashi, *Soc. Neurosci. Abstr.* **12,** 1193 (1986).
[191] E. Reuveny and T. Narahashi, *Soc. Neurosci. Abstr.* **15,** 355 (1989).
[192] D. A. Twombly, M. D. Herman, and T. Narahashi, *Soc. Neurosci. Abstr.* **13,** 102 (1987).
[193] D. A. Twombly, M. D. Herman, C. H. Kye, and T. Narahashi, *J. Pharmacol. Exp. Ther.* **254,** 1029 (1990).
[194] E. Reuveny and T. Narahashi, *Brain Res.* **545,** 312 (1991).
[195] N. Ogata and T. Narahashi, *J. Pharmacol. Exp. Ther.* **252,** 1142 (1990).

TABLE I
POTENCY OF VARIOUS AGENTS IN BLOCKING L-TYPE CALCIUM CHANNELS

Agent	Cell	ED_{50}	Ref.
Dihydropyridines			
Nitrendipine	Ventricular muscle	0.36 nM (I)	a
		700 nM (R)	a
Felodipine	Atrial muscle	0.67 nM	b
(+)-Isradipine	Vein	0.02 nM	c
(+)-Isradipine	Skeletal myoball	0.1 nM	d
Nimodipine	Pituitary cell	0.5 nM	e
Nimodipine	Dorsal root ganglion	1.3 nM	f
Nimodipine	F-11 (neuroblastoma N18TG-2 ×	3 nM (I)	g
	dorsal root ganglion)	2 μM (R)	g
Opioids			
Leu-enkephalin	NG108-15 (neuroblastoma × glioma)	8.8 nM	h
Peptide toxins			
ω-Conotoxin GVIA	Hippocampal neuron		i
	Dorsal root ganglion		i
	Sympathetic neuron		i
Benzodiazepines			
Chlordiazepoxide	N1E-115 (neuroblastoma)	420 μM	j
Barbiturates			
Phenobarbital	N1E-115	~400 μM	k
Chlorpromazine	N1E-115	0.6 μM (I)	l
		15 μM (R)	l
Alcohols			
Ethanol	N1E-115	~400 mM	m
	NG108-15		
Octanol	N1E-115	~200 μM	n
Inorganic cations			
Co^{2+}	N1E-115	560 μM	o
Ni^{2+}	N1E-115	280 μM	o
Cd^{2+}	N1E-115	7 μM	o
La^{3+}	N1E-115	0.9 μM	o
Pb^{2+}	SH-SY-5Y (human neuroblastoma)	1 μM	p

[a] B. P. Bean, *Proc. Natl. Acad. Sci. U.S.A.* **81,** 6388 (1984); I, inactivated channel; R, resting channel.

[b] C. J. Cohen, S. Spires, and D. Van Skiver, *in* "Molecular and Cellular Mechanisms of Antiarrhythmic Agents" (L. Hondeghem and B. G. Katzung, eds.). Futura Press, 1989.

[c] G. Loirand, C. Mironneau, J. Mironneau, and P. Pacand, *J. Physiol. (London)* **412,** 333 (1989).

[d] C. Cognard, G. Romey, J. P. Galizzi, M. Fosset, and M. Lazdunski, *Proc. Natl. Acad. Sci. U.S.A.* **83,** 1518 (1986).

[e] C. J. Cohen and R. T. McCarthy, *J. Physiol. (London)* **387,** 195 (1987).

[f] R. T. McCarthy, *in* "Nimodipine and Central Nervous System Function: New Vistas" (J. Traber and W. Gispen, eds.), p. 35. Schattauer, Stuttgart and New York, 1990.

[g] L. M. Boland and R. Dingledine, *J. Physiol. (London)* **420,** 223 (1990).

some neuronal membranes but not in cardiac, skeletal, or smooth muscle membranes.

N-Type Calcium Channels. The pharmacology of N-type calcium channels is characterized by a high sensitivity to ω-conotoxin GVIA,[171] a 27-amino acid peptide isolated from the venom of the fish-hunting marine snail *Conus geographus.*[196,197] Although ω-conotoxin blocks some of the L-type calcium channels in neurons, N-type calcium channels of all neurons so far examined are sensitive to the toxin. These N-type channels include dorsal root ganglion and sympathetic neurons[198]; dorsal root ganglion × neuroblastoma hybrid (F-11) cells[199]; sensory neurons[200]; and human neuroblastoma (IMR32) cells.[201] ω-Conotoxin is the only chemical agent that has been clearly demonstrated to block N-type calcium channels. There are a few other toxins suspected to block calcium channels, but the identity of the calcium channel type that is blocked is not completely clear. These include β-leptinotarsin D and H, polypeptide neurotoxins isolated from the hemolymph of Colorado potato beetles,[202-204] a polypeptide toxin from the coral *Goniopora,* [205] and atrotoxin from a snake.[206]

[196] B. M. Olivera, J. M. McIntosh, L. J. Cruz, F. A. Luque, and W. R. Gray, *Biochemistry* **23**, 5087 (1984).
[197] B. M. Olivera, W. R. Gray, R. Zeikus, J. M. McIntosh, J. Varga, J. Rivier, V. de Santos, and L. Cruz, *Science* **230**, 1338 (1985).
[198] E. W. McCleskey, A. P. Fox, D. H. Feldman, L. J. Cruz, B. M. Olivera, R. W. Tsien, and D. Yoshikami, *Proc. Natl. Acad. Sci. U.S.A.* **84**, 4327 (1987).
[199] L. M. Boland and R. Dingledine, *J. Physiol. (London)* **420**, 223 (1990).
[200] H. Kasai, T. Adasaki, and J. Fukuda, *Neurosci. Res.* **4**, 228 (1987).
[201] E. Carbone, E. Sher, and F. Clementi, *Pfluegers Arch.* **416**, 170 (1990).
[202] R. D. Crosland, T. H. Hsaio, and W. O. McClure, *Biochemistry* **23**, 734 (1984).
[203] L. Madedder, T. Pozzan, M. Robello, R. Rolandi, T. H. Hsaio, and J. Meldolesi, *J. Neurochem.* **45**, 1719 (1985).
[204] W. O. McClure, B. C. Abbot, D. E. Baxter, T. H. Hsaio, L. S. Satin, S. Siger, and J. Yoshino, *Proc. Natl. Acad. Sci. U.S.A.* **77**, 1219 (1985).
[205] J. Qar, J.-P. Glizzi, M. Fossert, and M. Lazdunski, *Eur. J. Pharmacol.* **141**, 261 (1987).
[206] S. L. Hamilton, M. J. Hawkes, K. Redding, and A. M. Brown, *Science* **229**, 182 (1985).

[h] A. Tsunoo, M. Yoshii, and T. Narahashi, *Proc. Natl. Acad. Sci. U.S.A.* **83**, 9832 (1986).
[i] E. W. McCleskey, A. P. Fox, D. H. Feldman, L. J. Cruz, B. M. Olivera, R. W. Tsien, and D. Yoshikami, *Proc. Natl. Acad. Sci. U.S.A.* **84**, 4327 (1987).
[j] E. Reuveny and T. Narahashi, *Soc. Neurosci. Abstr.* **15**, 355 (1989).
[k] D. A. Twombly, M. D. Herman, and T. Narahashi, *Soc. Neurosci. Abstr.* **13**, 102 (1987).
[l] N. Ogata and T. Narahashi, *J. Pharmacol. Exp. Ther.* **252**, 1142 (1990).
[m] D. A. Twombly, M. D. Herman, C. H. Kye, and T. Narahashi, *J. Pharmacol. Exp. Ther.* **254**, 1029 (1990).
[n] D. A. Twombly and T. Narahashi, *Soc. Neurosci. Abstr.* **15**, 355 (1989).
[o] T. Narahashi, A. Tsunoo, and M. Yoshii, *J. Physiol. (London)* **383**, 231 (1987).
[p] E. Reuveny and T. Narahashi, *Brain Res.* **545**, 312 (1991).

Dynorphin A, a κ-opioid receptor agonist, blocks N-type calcium channels in dorsal root ganglion neurons through interaction with κ-receptors.[207]

T-Type Calcium Channels. Dihydropyridines have been found to block both T- and L-type calcium channels in neurons and in cardiac and smooth muscles. However, the potency is not high on T-type channels, with an ED_{50} value of the order of $0.1-1.0$ μM. Examples include dihydropyridines on hippocampal neurons,[208] dorsal root ganglion $\times$ neuroblastoma hybrids (F-11),[199] hypothalamic neurons,[209] vascular smooth muscle,[178] and cardiac muscle.[177,210]

The pyrethroid insecticide tetramethrin blocks T-type calcium channels of the neuroblastoma N1E-115 line[211] and those of the sinoatrial node.[212] Octanol was reported to block T-type calcium channels selectively in inferior olivary neurons,[213] but in the neuroblastoma N1E-115 line it blocks both T- and L-type calcium channels.[214] Ethanol also blocks T-type as well as L-type calcium channels in the neuroblastoma N1E-115 line, by 40% ant 300 nM.[193] Phenytoin blocks T-type channels selectively without affecting L-type channels but requires about 30 μM to cause a 50% block.[215] Like L-type calcium channels, T-type calcium channels are blocked by various polyvalent cations such as La^{3+}, Pb^{2+}, Cd^{2+}, Ni^{2+}, and Co^{2+}.[173,194] However, the profile of blocking potency is largely similar between the two types of calcium channels with the exception of Cd^{2+}, which exhibits a much higher blocking potency for L-type channels (K_d 7.0 μM) than T-type channels (160 μM).[173] Thus, there is at present no specific and potent T-type channel blocker. This certainly hinders further study of T-type channels.

P-Type Calcium Channels. P-Type calcium channels are characterized by their sensitivity to a low molecular weight toxin (FTX) from funnel web spiders.[185,216] P-type channels are insensitive to ω-conotoxin and dihydropyridines.

[207] R. A. Gross and R. L. Macdonald, *Proc. Natl. Acad. Sci. U.S.A.* **84,** 5469 (1987).
[208] M. Takahashi and Y. Fujimoto, *Biochem. Biophys. Res. Commun.* **163,** 1182 (1989).
[209] N. Akaike, P. G. Kostyuk, and Y. V. Osipchuk, *J. Physiol. (London)* **412,** 181 (1989).
[210] J. Tytgat, J. Vereecke, and E. Carmeliet, *Naunyn-Schmiedebergs Arch. Pharmacol.* **337,** 690 (1988).
[211] M. Yoshii, A. Tsunoo, and T. Narahashi, *Soc. Neurosci. Abstr.* **11,** 518 (1985).
[212] N. Hagiwara, H. Irisawa, and M. Kameyama, *J. Physiol. (London)* **359,** 233 (1988).
[213] R. Llinás and Y. Yarom, *Soc. Neurosci. Abstr.* **12,** 174 (1986).
[214] D. A. Twombly and T. Narahashi, *Soc. Neurosci. Abstr.* **15,** 355 (1989).
[215] D. A. Twombly, M. Yoshii, and T. Narahashi, *J. Pharmacol. Exp. Ther.* **246,** 189 (1988).
[216] H. Jackson and T. N. Parks, *Annu. Rev. Neurosci.* **12,** 405 (1989).

Calcium Channel Activators

There are a few compounds which are unique in the sense that they activate L-type calcium channels selectively.[166,171,175,198,201,217] These chemicals include Bay K8644 and CGP 28392.

The activating effect of Bay K8644 on L-type calcium channels of chick dorsal root ganglion neurons was analyzed in detail using the whole-cell and single-channel patch clamp techniques.[217] The whole-cell L-type current was greatly increased in amplitude by Bay K8644. Although single-channel conductance remained unchanged at 25 pS, the activation voltage was shifted in the direction of hyperpolarization by 14 mV. Furthermore, in the presence of Bay K8644, the type 2 mode of openings characterized by a prolonged open time became predominant. Because of the action to stimulate or increase L-type calcium channel currents, Bay K8644 is now often used to observe the channel currents more easily.

Acknowledgments

The authors wish to thank Vicky James-Houff for unfailing secretarial assistance. The author's studies cited in this chapter were supported by National Institutes of Health Grants NS14143 and NS14144, and Alcohol, Drug Abuse, and Mental Health Administration Grant AA07836.

[217] A. P. Fox, M. C. Nowycky, and R. W. Tsien, *J. Physiol. (London)* **394,** 173 (1987).

[44] Overview of Toxins and Drugs as Tools to Study Excitable Membrane Ion Channels: II. Transmitter-Activated Channels

By TOSHIO NARAHASHI

Acetylcholine-Activated Channels

Among the variety of ligand-activated ion channels, acetylcholine (ACh)-activated channels or ACh receptor (AChR) channels have been studied most extensively and for the longest period of time since the first application of voltage clamp techniques to the muscle end-plate membrane by Takeuchi and Takeuchi.[1,2] This was largely due to the facts that voltage clamp techniques were applicable to skeletal neuromuscular preparations

[1] A. Takeuchi and N. Takeuchi, *J. Neurophysiol.* **22,** 395 (1959).
[2] A. Takeuchi and N. Takeuchi, *J. Physiol (London)* **154,** 52 (1960).

with relative ease and that ACh was well established as the neurotransmitter at the neuromuscular junction. Therefore, most studies of ACh-activated channels have been performed with the nicotinic AChR channels. Voltage clamp experiments for the muscarinic AChR channels have been performed more recently in the early 1980s.[3]

Owing to the long history of studies, a large body of information has been accumulated regarding drugs and chemicals that act on the nicotinic AChR channels. A variety of chemicals which are known to act on other receptors or channels have also been found to affect the AChR channels. For certain chemicals, the AChR channels represent an important target site to exhibit therapeutic or toxic effects. For others, the effects on the AChR channels may be regarded as incidental. Thus, only a limited number of chemicals that have been studied are deemed useful as tools because of their potent and specific actions on the AChR channels. Numerous reviews have been published dealing with the actions of chemicals on AChR channels or neuromuscular junctions in general.[4-13]

Acetylcholine Receptor Channel Blockers

The chemical agents that block the muscle end-plate were in the past divided into two classes, namely, competitive blockers and noncompetitive blockers. The latter were often called depolarizing blockers. Competitive blockers include those agents that compete with ACh for the receptor site; d-tubocurarine and pancuronium are among the examples. Noncompeti-

[3] E. Kato, R. Anwyl, F. N. Quandt, and T. Narahashi, *Neuroscience* **8**, 643 (1983).
[4] E. X. Albuquerque, J. W. Daly, and J. E. Warnick, *in* "Ion Channels" (T. Narahashi, ed.), Vol. 1, p. 95. Plenum, New York, 1988.
[5] D. Colquhoun, *in* "Receptors for Neurotransmitters and Peptide Hormones" (G. Pepeu, M. J. Kuhur, and S. J. Enna, eds.), p. 67. Raven, New York, 1980.
[6] Y. Aracava, K. L. Swanson, R. Rozental, and E. X. Albuquerque, *in* "Neurotox '88: Molecular Basis of Drug and Pesticide Action" (G. G. Lunt, ed.), p. 157. Elsevier, Amsterdam, 1988.
[7] C. E. Spivak and E. X. Albuquerque, *in* "Progress in Cholinergic Biology: Model Cholinergic Synapses" (I. Hanin and A. M. Goldberg, eds.), p. 323. Raven, New York, 1982.
[8] J. H. Steinbach and C. F. Stevens, *in* "Frog Neurobiology" (R. Llinás and W. Precht, eds.), p. 33. Springer-Verlag, Berlin, 1976.
[9] A. Mathie, S. G. Cull-Candy, and D. Colquhoun, *in* "Neurotox '88: Molecular Basis of Drug and Pesticide Action" (G. G. Lunt, ed.), p. 393. Elsevier, Amsterdam, 1988.
[10] K. Peper, R. J. Bradley, and F. Dreyer, *Physiol. Rev.* **62**, 1271 (1982).
[11] J. J. Lambert, N. N. Durant, and E. G. Henderson, *Annu. Rev. Pharmacol. Toxicol.* **23**, 505 (1983).
[12] V. P. Whittaker, *Trends Pharmacol. Sci.* **11**, 8 (1990).
[13] T. Narahashi, *in* "Neurobiology of Acetylcholine" (N. J. Dun and R. L. Perlman, eds.), p. 339. Plenum, New York, 1987.

tive or depolarizing blockers depolarize the end-plate membrane and include succinylcholine and decamethonium. However, with the advancement of our knowledge about how chemicals block the end-plate, the above classification is now deemed inaccurate. It would be more appropriate and precise to divide the blockers into "receptor blockers" and "channel blockers" because we now know which component is the target site for most of the blockers. Exhaustive lists of receptor and channel blockers so far examined are found in review articles.[4,7,10,11]

Acetylcholine Receptor Blockers. d-Tubocurarine is a classic example of an AChR blocker. It competes with ACh for binding to AChR. However, more recently it has been found that d-tubocurarine also blocks the AChR channels, as is described later. Pancuronium blocks the AChR in a manner similar to d-tubocurarine.

The discovery of α-bungarotoxin (α-BuTX) as a highly potent and specific AChR blocker at the end-plate opened an avenue for the study of the molecular structure of AChR.[14] α-Bungarotoxin is a polypeptide toxic component contained in the venom of a snake, *Bungarus multicinctus*. It blocks the end-plate potential without causing membrane depolarization and without impairing nerve conduction by binding to the AChR irreversibly. Thus Professor C. Y. Lee thought that α-BuTX would be a very useful tool for the study of AChR.[14] Despite this important discovery, it was not until the early 1970s that two groups successfully utilized α-BuTX in an attempt to characterize AChR.[15,16] Since then α-BuTX has been used extensively as a tool.

Nicotine was known to activate the nicotinic AChR, thereby causing a depolarization of postsynaptic membranes which evokes repetitive firing leading to a block in both muscle end-plates and ganglion neurons.[17,18] However, at concentrations lower than those that cause end-plate depolarization, nicotine blocks the end-plate current (EPC).[19] The apparent K_d values are estimated to be 60 μM for end-plate depolarization and 5 μM for EPC block. Nicotine appears to block the AChR channels in their open configuration, as flickerings are induced at low concentrations.[6]

Acetylcholine Receptor Channel Blockers. The first demonstration of AChR channel block was performed by Steinbach[20,21] using lidocaine

---

[14] C. C. Chang and C. Y. Lee, *Arch. Int. Pharmacodyn. Ther.* **144,** 241 (1963).

[15] J. P. Changeux, M. Kasai, and C. Y. Lee, *Proc. Natl. Acad. Sci. U.S.A.* **67,** 1241 (1970).

[16] R. Miledi, P. Molinoff, and L. T. Potter, *Nature (London)* **229,** 554 (1971).

[17] W. D. M. Paton and W. L. M. Perry, *J. Physiol (London)* **119,** 43 (1953).

[18] S. Thesleff, *Acta Physiol. Scand.* **34,** 218 (1955).

[19] C. M. Wang and T. Narahashi, *J. Pharmacol. Exp. Ther.* **182,** 427 (1972).

[20] A. B. Steinbach, *J. Gen. Physiol.* **52,** 144 (1968).

[21] A. B. Steinbach, *J. Gen. Physiol.* **52,** 162 (1968).

(Xylocaine)-treated end-plates. It was proposed that lidocaine blocked the AChR channels in their open configuration. In 1973, histrionicotoxin isolated from a dendrobatid frog, *Debdrobates histrionicus*, was found to block AChR channels.[22] A large number of toxins, chemicals, and therapeutic drugs have since been shown to cause the block of AChR channels. These include local anesthetics (procaine, QX-314, QX-222), competitive or nondepolarizing neuromuscular blocking agents (*d*-tubocurarine, gallamine), antimuscarinic agents (atropine, scopolamine, quinuclidinyl benzilate), anesthetics (pentobarbital, thiopental, ketamine), other central nervous system (CNS) agents (phenytoin, diazepam, phencyclidine), antibiotics (polymyxin B, clindamycin), antiviral agents (amantadine), anticholinesterases (neostigmine, physostigmine), ganglionic blocking agents (trimethaphan, hexamethonium, tetraethylammonium), depolarizing neuromuscular blocking agents (decamethonium, nicotine), natural toxins (perhydrohistrionicotoxin, gephyrotoxin, anatoxin *a*), antiplasmodiums (quinacrine), and guanidine derivatives.[4,7,10,11,13]

One of the important questions about the channel block is the channel configuration for which the blocking agent has the highest affinity. Some chemicals block the channel when it is at the resting state, whereas others block the channel only when it is open. This "state dependence" of channel block is important as it provides us with a clue to the molecular mechanism of channel block, and it has been studied extensively.

Measurement of the time course of decay of the end-plate current can distinguish between open and closed channel block. If a chemical blocks the AChR channels only when they are open, the time course of EPC decay is expected to be shortened, because as channels open during the rising phase of EPC the blocking molecules bind to a site inside the channel, thereby curtailing the peak and accelerating the decay phase of EPC. This group of chemicals may be divided into one class that causes a biphasic EPC decay and another that causes a simple shortening of EPC decay. The chemicals that change the EPC decay from a single-exponential function to a dual-exponential function are represented by QX-222, procaine, lidocaine, amylguanidine, gallamine, scopolamine, pentobarbital, and decamethonium.[11,23] In single-channel recording experiments, QX-222 was found to cause bursts of openings and closings, indicating that both blocking and unblocking of the channel occur at comparable rates which are slow enough to be recorded without much distortion.[24] The chemicals that

[22] E. X. Albuquerque, E. A. Barnard, T. H. Chiu, A. J. Lapa, J. O. Dolly, S.-E. Jansson, J. Daly, and B. Witkop, *Proc. Natl. Acad. Sci. U.S.A.* **70**, 949 (1973).

[23] J. M. Farley, J. Z. Yeh, S. Watanabe, and T. Narahashi, *J. Gen. Physiol.* **77**, 273 (1981).

[24] E. Neher and J. H. Steinbach, *J. Physiol. (London)* **277**, 153 (1978).

simply shorten the single-exponential EPC decay include QX-314, d-tubo-curarine, methylguanidine, octylguanidine, atropine, thiopental, ketamine, amantadine, tetraethylammonium, and histrionicotoxin.[11,23] Contrary to QX-222, which causes burts of single-channel openings and closings, QX-314 does not cause bursts.[24] This is interpreted as arising from the fact that the blocking rate is much faster than the unblocking rate.

Closed channel block has also been found with some chemical agents, including procaine, histrionicotoxin, phencyclidine, amantadine, d-tubo-curarine, tetraethylammonium, physostigmine, decamethonium, and oc-tylguanidine.[4,23,25] It should be noted that many of these agents also cause open channel block as described above. The closed channel block is characterized, with some exceptions, by the absence of change in EPC decay time course and in single-channel lifetime. It is sometimes associated with a change in the peak current–voltage $(I-V)$ relationship; the $I-V$ relation is normally almost linear, but in the presence of some closed channel blockers (e.g., histrionicotoxin) rectification of the $I-V$ curve is observed, with the degree of block increasing with hyperpolarization.[11] It is not clear whether the rectification is related to closed channel block.

Another important aspect of end-plate channel block is voltage and current dependence. Many chemicals were found to change the EPC–voltage relationship from the normal straight line to a nonlinear curve, block being intensified with hyperpolarization, for example, $H_{12}HTX$ as observed by Spivak et al.[26] Such a nonlinear $I-V$ curve in the presence of a blocking agent was traditionally interpreted as being due to voltage dependence of the block. However, some such cases have been found to be due to current dependence of the block. Careful analyses of the EPC block unequivocally indicated that methylguanidine and ethylguanidine block the EPC in a manner dependent on the direction of current flow, whereas longer chain guanidine derivatives such as octylguanidine block the EPC in a voltage-dependent manner.[13,27]

Neuronal Acetylcholine Receptor Blockers. Unlike the skeletal muscle AChR, neuronal AChR is not very sensitive to α-BuTX.[28,29] However, κ-bungarotoxin (κ-BuTX), another component contained in the venom of *Bungarus multicinctus* together with α-BuTX, experts a potent blocking action on AChR of some neurons, including avian ciliary ganglion neu-

[25] P. R. Adams, *J. Physiol. (London)* **268,** 291 (1977).
[26] C. E. Spivak, M. A. Maleque, A. C. Oliveira, L. M. Masukawa, T. Tokuyama, J. W. Daly, and E. X. Albuquerque, *Mol. Pharmacol.* **21,** 351 (1982).
[27] S. M. Vogel, S. Watanabe, J. Z. Yeh, J. M. Farley, and T. Narahashi, *J. Gen. Physiol.* **83,** 901 (1984).
[28] D. A. Brown and L. Fumagalli, *Brain Res.* **129,** 165 (1977).
[29] E. Kato and T. Narahashi, *Brain Res.* **245,** 159 (1982).

rons[30] and rat sympathetic ganglion neurons.[9] However, κ-BuTX has only a weak blocking action on bovine adrenal chromaffin cells.[31]

Acetylcholine Receptor Channel Agonists

There are few compounds that have been shown to stimulate the AChR channel system other than the classic agonists such as ACh, nicotine, and carbachol. Anatoxins, isolated from the blue green algae *Anabaena flosaquae*, have been shown to open the AChR channels. (+)-Anatoxin *a* causes openings of the channels at a concentration of 20 nM, whereas *(S)N*-methylanatoxinol does so only at higher concentrations (1 – 20 μM).[6] At a higher concentration 3.2 μM), (+)-anatoxin *a* causes desensitization and block of the AChR. Channels open in clusters separated by long, silent periods. The block appears to occur at the open state of the channels.

Inhibitory Amino Acid Receptor – Channel Complexes

Inhibitory amino acid receptor – channel complexes may be divided into three broad categories, γ-aminobutyric acid$_A$ (GABA$_A$), GABA$_B$, and glycine systems. In mammals, these inhibitory systems are located in the central nervous system; glycine is the major inhibitory neurotransmitter in the spinal cord and brain stem, and GABA is an inhibitory neurotransmitter in the spinal cord, cortex, midbrain and cerebellum. In arthropods, GABA is also an inhibitory neurotransmitter at neuromuscular junctions.

Two types of GABA systems, GABA$_A$ and GABA$_B$, are distinguished on the basis of differences in the actions of agonists and antagonists, ion channels involved, and coupling mechanisms. The GABA$_A$ system is present primarily on the postsynaptic side, whereas the GABA$_B$ system is found in both presynaptic and postsynaptic elements. Several subtypes of GABA$_A$ receptors are known, and all are associated with GABA-activated chloride channels. By contrast, GABA$_B$ receptors present on postsynaptic membranes activate potassium channels via G proteins. Presynaptic GABA$_B$ receptors appear to close calcium channels, thereby causing synaptic inhibition.

γ-Aminobutyric Acid Receptor – Channel Complexes

A variety of chemicals and toxins are known to interact with the GABA$_A$ receptor – channel complex. This complex comprises four sites: the GABA site, benzodiazepine site, barbiturate site, and chloride chan-

[30] V. A. Chiappinelli, *Brain Res.* **277**, 9 (1983).
[31] L. S. Higgins and D. K. Berg, *J. Neurosci.* **7**, 1792 (1987).

TABLE I

CHEMICALS ACTING ON VARIOUS SITES OF γ-AMINOBUTYRIC ACID$_A$
RECEPTOR–CHANNEL COMPLEXES[a]

Site	Agonist	Antagonist	Allosteric inhibitor
GABA	GABA Muscimol	Bicuculline Pitrazepin RU 5135 SR 95531	
Benzodiazepine	Diazepam Flunitrazepam Flurazepam	Ro15-1788 Ro15-4513	β-Carbolines GABA-modulin DBI[b]
Barbiturate	Pentobarbital Etazolate Alphaxolone		
Chloride channel		Picrotoxin Pentylenetetrazol Lindane Dieldrin TBPS[c]	
Unknown	Ethanol Higher alcohols Halothane Enflurane Isoflurane Hg^{2+}		

[a] Data from R. W. Olsen, M. Bureau, R. W. Ransom, L. Deng, A. Dilber, G. Smith, M. Krestchatisky, and A. J. Tobin, *in* "Neuroreceptors and Signal Transduction" (S. Kito, T. Segawa, K. Kuriyama, M. Tohyama, and R. W. Olsen, eds.), p. 1, Plenum, New York, 1988; and A. T. Eldefrawi and M. E. Eldefrawi, *FASEB J.* **1,** 262 (1987).
[b] Diazepam binding inhibitor.
[c] *tert*-Butylbicyclophosphorothionate.

nel.[32] Some of the chemicals known to act on each site are given in Table I.[32,33]

γ-Aminobutyric acid and muscimol are agonists that bind to the GABA site of the GABA$_A$ receptor–channel complex, thereby opening the chloride channels. Bicuculline binds to this site and acts as a competitive antagonist. Benzodiazepines such as diazepam and flunitrazepam bind to the benzodiazepine site and potentiate the action of GABA. Ro15-1788 is

[32] R. W. Olsen, M. Bureau, R. W. Ransom, L. Deng, A. Dilber, G. Smith, M. Krestchatisky, and A. J. Tobin, *in* "Neuroreceptors and Signal Transduction" (S. Kito, T. Segawa, K. Kuriyama, M. Tohyama, and R. W. Olsen, eds.), p. 1. Plenum, New York, 1988.
[33] A. T. Eldefrawi and M. E. Eldefrawi, *FASEB J.* **1,** 262 (1987).

an antagonist that binds to the benzodiazepine site. β-Carboline 3-carboxylate esters are allosteric inhibitors or inverse agonists of the benzodiazepine site and act as convulsants or proconvulsants. Barbiturates such as pentobarbital bind to the barbiturate site, thereby potentiating the action of GABA. They also open the chloride channels, causing Cl^- ions to flow. There are several other chemicals which are known to potentiate the action of GABA and to cause a sizable increase in GABA-evoked chloride current or in GABA-induced Cl^- flux; these include ethanol,[34-38] longer chain alcohols,[39] and the general anesthetics halothane, enflurane, and isoflurane.[40] The exact site of binding of these chemicals is unknown, but it is likely that they interact with the barbiturate site. The chloride channel is directly blocked by picrotoxin, pentylenetetrazol, lindane, cyclodienes such as toxaphene, dieldrin, and endrin, and *tert*-butylbicyclophosphorothionate (TBPS). The anthelminthic and insecticidal avermectin B_{1a} is a potent partial agonist acting on the $GABA_A$ system. It activates the receptor–channel complex, inhibits binding of [^{3}H]muscimol and [^{35}S]TBPS, and potentiates binding of [^{3}H]flunitrazepam.[41] The exact site of avermectin B_{1a} in the $GABA_A$ receptor–channel complex remains to be seen.

The $GABA_A$ receptor–channel complex is now known to be composed of at least three subunits, α, β, and γ. The first study of isolation and sequencing of cDNAs reported one α and one β subunit,[42] but at least four α subunits, three β subunits, and two γ subunits were identified later.[43,44] The β_1 subunit has been reported to be sufficient for the activity of the chloride channel.[45] In neurons coexpressing α_1 and β_1, the GABA-induced current is inhibited by bicuculline and picrotoxin but not by β-carboline and 4′-chlordiazepam, whereas in neurons coexpressing α_1, β_1, and γ_2, the

[34] J. P. Huidobro-Toro, V. Bleck, A. M. Allan, and R. A. Harris, *J. Pharmacol. Exp. Ther.* **242**, 963 (1987).

[35] A. K. Mehta and M. K. Ticku, *J. Pharmacol. Exp. Ther.* **246**, 558 (1988).

[36] P. D. Suzdak, R. D. Schwartz, P. Skolnick, and S. M. Paul, *Proc. Natl. Acad. Sci. U.S.A.* **83**, 4071 (1986).

[37] M. K. Ticku, P. Lowrimore, and P. Lehoullier, *Brain Res. Bull.* **17**, 123 (1986).

[38] M. Nishio and T. Narahashi, *Brain Res.* **518**, 283 (1990).

[39] M. Nakahiro, O. Arakawa, and T. Narahashi, *J. Pharmacol. Exp. Ther.* **259**, 235 (1991).

[40] M. Nakahiro, J. Z. Yeh, E. Brunner, and T. Narahashi, *FASEB J.* **3**, 1850 (1989).

[41] I. M. Abalis, A. T. Eldefrawi, and M. E. Eldefrawi, *J. Biochem. Toxicol.* **1**, 69 (1986).

[42] P. R. Schofield, M. G. Darlison, N. Fujita, D. R. Burt, F. A. Stephenson, H. Rodriguez, L. M. Rhee, J. Ramachandran, V. Reale, T. A. Glencorse, P. H. Seeburg, and E. Barnard, *Nature (London)* **328**, 221 (1987).

[43] P. R. Schofield, *Trends Pharmacol. Sci.* **10**, 476 (1989).

[44] R. W. Olsen and A. J. Tobin, *FASEB J.* **4**, 1469 (1990).

[45] E. Sigel, R. Baur, G. Trube, H. Mohler, and P. Malherbe, *Neuron* **5**, 703 (1990).

current is inhibited by bicuculline, picrotoxin, β-carboline, and 4′-chlordiazepam.[46] More data are now being accumulated regarding which subunit is necessary for which function, such as ligand binding and channel function, but the final conclusion awaits further experimentation.

Earlier electrophysiological studies performed with CNS neurons using fluctuation analysis have indicated that GABA and 17 GABA-mimetics open the chloride channel with a uniform conductance at 16.5 pS but with a wide variation of the mean open time. Only β-alanine and glycine open the channel with a much larger conductance at 30 pS.[47] Pentobarbital and diazepam open the chloride channel with a single-channel conductance similar to that activated by GABA, but pentobarbital keeps the channel open longer than GABA.[47] In the rat spinal cord neurons, the steroid anesthetic alphaxalone increases the GABA-induced current by prolonging the mean open time of single channels.[48]

Patch clamp experiments performed more recently have disclosed some features of interactions of chemicals and ligands with the GABA-activated chloride channel. The $GABA_A$-activated chloride channels of mouse spinal cord neurons have been analyzed in detail.[49] The channels are selective for anions, with the permeability ratio of K^+ to Cl^- at 0.05. Measurements of the permeability sequence for various large anions such as formate, bicarbonate, and propionate have led to an effective pore diameter of 5.6 Å. Open channels show multiple conductance states ranging from 10 to 44 pS. Glycine-activated channels in the same neurons exhibit characteristics similar to those of GABA-activated channels. Pentobarbital increases the GABA-evoked whole-cell current in rat cortical neurons by prolonging the mean open time but without changing the single-channel conductance or the frequency of channel openings.[50] Detailed kinetic analyses with mouse spinal cord neurons have shown that there are three states of GABA-activated channels[51] and that pentobarbital and phenobarbital increase the relative frequency of occurrence of the longest burst state, thereby increasing the whole-cell current.[52] Picrotoxin

[46] G. Puia, M. R. Santi, S. Vicini, D. B. Pritchett, P. H. Seeburg, and E. Costa, *Proc. Natl. Acad. Sci. U.S.A.* **86**, 7275 (1989).

[47] J. L. Barker and D. G. Owen, *in* "Benzodiazepine/GABA Receptors and Chloride Channels: Structural and Functional Properties" (R. W. Olsen and J. C. Venter, eds.), p. 135. Alan R. Liss, New York, 1986.

[48] J. L. Barker, N. L. Harrison, G. D. Lange, and D. G. Owen, *J. Physiol. (London)* **386**, 485 (1987).

[49] J. Bormann, O. P. Hamill, and B. Sakmann, *J. Physiol. (London)* **385**, 243 (1987).

[50] J. M. Mienville and S. Vicini, *Brain Res.* **489**, 190 (1989).

[51] R. L. Macdonald, C. J. Rogers, and R. E. Twyman, *J. Physiol. (London)* **410**, 479 (1989).

[52] R. L. Macdonald, C. J. Rogers, and R. E. Twyman, *J. Physiol. (London)* **417**, 483 (1989).

and pregnenolone sulfate block the GABA-induced chloride current by decreasing the frequency of channel openings.[50] Diazepam-induced potentiation of GABA-evoked current is due to a prolongation of the mean open time; the frequency of openings is not altered.[53]

γ-Aminobutyric Acid$_B$ Receptor–Channel Complex

Whereas the inhibitory action of GABA was believed to involve an increase in chloride permeability, several studies of baclofen, a β-(p-chlorophenyl) analog of GABA, indicated that there was another type of presynaptic inhibition that is mediated by a GABA receptor–channel complex insensitive to bicuculline.[54] Studies of GABA and baclofen have led to the conclusion that in addition to the GABA$_A$ chloride channel system there is a GABA$_B$ receptor which activates a potassium channel and closes a calcium channel. The GABA$_B$ receptor system is involved in a wide variety of physiological functions including inhibition of neurotransmitter release, mediation of slow inhibitory postsynaptic potentials (IPSPs), inhibition of release of some hormones, increase in gastric motility, suppression of hippocampal epileptiform activity, and relaxation of bronchi and urinary bladder, to mention a few.[55]

A variety of compounds have been found to act on the GABA$_B$ receptor–channel complex as either agonists or antagonists. Some of them are given in Table II.[55] Potency and selectivity may vary greatly among them. For instance, $(-)$-β-hydroxy-GABA is an agonist acting on both GABA$_A$ and GABA$_B$ systems with similar affinity; kojic amine acts on the GABA$_B$ system as an agonist but antagonizes the GABA$_A$ system; and δ-aminovaleric acid is an antagonist on the GABA$_B$ system and an agonist on the GABA$_A$ system. Among the most notable and/or most commonly used are the following (with IC$_{50}$ for displacement of GABA$_B$ binding in parentheses): $(-)$-baclofen ($0.08-0.13$ μM), selective agonist on GABA$_B$; muscimol ($2.4-5.3$ μM), more potent as GABA$_A$ agonist by a factor of 100; 3-aminopropylphosphinic acid (0.001 μM), selective and the most potent agonist; phaclofen (118 μM), selective but weak antagonist on GABA$_B$; 2-hydroxysaclofen (5.1 μM), selective antagonist on GABA$_B$.

The GABA$_B$ receptor–channel complex is known to be present in both presynaptic and postsynaptic sites and to be linked with potassium channels and/or calcium channels via G proteins.[55–57] However, it remains to be seen how these two sites of GABA$_B$ receptors are associated with the two

[53] R. E. Twyman, C. J. Rogers, and R. L. Macdonald, *Ann. Neurol.* **25**, 213 (1989).
[54] N. G. Bowery, *Trends Pharmacol. Sci.* **3**, 400 (1982).
[55] N. Bowery, *Trends Pharmacol. Sci.* **10**, 401 (1989).
[56] P. Dutar and R. A. Nicoll, *Neuron* **1**, 585 (1988).
[57] R. A. Nicoll and P. Dutar, in "Allosteric Modulation of Amino Acid Receptors: Therapeutic Implications" (E. A. Barnard and E. Costa, eds.), p. 195. Raven, New York, 1989.

TABLE II
CHEMICALS ACTING ON γ-AMINOBUTYRIC ACID$_B$ RECEPTOR–CHANNEL COMPLEXES[a]

Compound	GABA$_B$		GABA$_A$	
	Agonist	Antagonist	Agonist	Antagonist
3-Aminopropylphosphinic acid	++++[b]			
(−)-Baclofen	+++			
GABA	+++		+++	
(−)-β-Hydroxy-GABA	++		++	
Muscimol	+		+++	
Saclofen		++		
2-Hydroxysaclofen		++		
Phaclofen		+		

[a] Data from N. Bowery, *Trends Pharmacol. Sci.* **10**, 401 (1989).
[b] Potency increases with the number of + signs.

kinds of channels. It is of interest to note that the GABA$_B$ receptor and the serotonin (5-HT$_{1a}$) receptor share the same potassium channel via G proteins.

Excitatory Amino Acid Receptor–Channel Complexes

The era of excitatory amino acids (EAAs) began in the early 1950s when Hayashi discovered the direct stimulating action of L-glutamate and L-aspartate on the cortex.[58,59] Stimulation of single brain neurons by L-glutamate and other amino acids was later demonstrated by Curtis *et al.*[60] However, it was not until the 1980s that studies of excitatory amino acids became a very hot subject of investigation. This rapid progress was due largely to the development of many agonists and antagonists that acted on the EAA receptors and to the introduction of patch clamp techniques which allowed the activity of various ion channels to be recorded from neurons. Several excellent review articles have been published.[61–65]

[58] T. Hayashi, *Jpn. J. Physiol.* **3**, 46 (1952).
[59] T. Hayashi, *Keio J. Med.* **3**, 183 (1954).
[60] D. R. Curtis, J. W. Phillis, and J. C. Watkins, *Nature (London)* **183**, 611 (1959).
[61] G. L. Collingridge and R. A. J. Lester, *Pharmacol. Rev.* **40**, 143 (1989).
[62] M. L. Mayer and G. L. Westbrook, *Progr. Neurobiol.* **28**, 197 (1987).
[63] D. T. Monaghan, R. J. Bridges, and C. W. Cotman, *Annu. Rev. Pharmacol. Toxicol.* **29**, 365 (1989).
[64] T. W. Stone and N. R. Burton, *Progr. Neurobiol.* **30**, 333 (1988).
[65] J. C. Watkins, *in* "Excitatory Amino Acids" (P. J. Roberts, J. Storm-Mathisen, and H. F. Bradford, eds.), p. 1. MacMillan, Chichester, 1986.

Excitatory amino acid receptors may be classified into three major categories based primarily on the affinity for agonists, not on the type of responses to each ligand. All EAA receptors can be activated by L-glutamate; NMDA receptors are activated by N-methyl-D-aspartate (NMDA) and have been studied most extensively for their physiological and pharmacological characteristics. Kainate receptors are activated by kainate (KA), and quisqualate or AMPA receptors are activated by quisqualate (QA) and α-amino-3-hydroxy-5-methyl-4-isoxazole propionate (AMPA). A fourth receptor, the AP4 receptor, activated by 2-amino-4-phosphonobutanoate (AP4), is not well characterized yet.

N-Methyl-D-Aspartate Receptors

Receptors for NMDA can be activated by NMDA, L-glutamate, L-aspartate, and ibotenate, and they are unique in that NMDA activation is greatly potentiated by glycine and suppressed by Mg^{2+} ions. At least five distinct sites are recognized in the NMDA receptors, namely, the NMDA site, glycine site, phencyclidine site, Mg^{2+} site, and Zn^{2+} site. The four agonists mentioned above bind to the NMDA site.

Glycine is an inhibitory transmitter that binds to a strychnine-sensitive site. However, glycine has recently been found to play an important role in the modulation of NMDA receptors.[66] It potentiates the action of NMDA on the receptor, the effect being noticeable even at 10 nM and reaching a maximum at 1 μM. This glycine site in the NMDA receptor is insensitive to strychnine, so it is different from the inhibitory glycine site. It appears that glycine binds to an NMDA site as a coagonist while NMDA binds to another site. The physiological role of glycine is unclear as its normal level exceeds 1 μM, the concentration at which glycine causes a maximal potentiating effect. Antagonists that act on the glycine site include kynurenate, 7-chlorokynurenate, 6-cyano-7-nitroquinoxaline-2,3-dione (CNQX), 6,7-dinitroquinoxaline-2,3-dione (DNQX), and 3-amino-1-hydroxy-2-pyrrolidone (HA-966).

Glycine has been found to reduce NMDA desensitization in mouse hippocampal neurons.[67] This is interpreted by a cyclic model in which the glycine-bound receptors are regarded as the "nondesensitized, low-affinity" receptors and the glycine-free receptors as the "desensitized, high-affinity" receptors.[68,69] However, controversial results have been obtained

[66] J. W. Johnson and P. Ascher, *Nature (London)* **325**, 529 (1987).
[67] M. L. Mayer, L. Vyklicky, Jr., and J. Clements, *Nature (London)* **338**, 425 (1989).
[68] L. Vyklicky, Jr., M. Benveniste, and M. L. Mayer, *J. Physiol. (London)* **428**, 313 (1990).
[69] M. Benveniste, J. Clements, L. Vyklicky, Jr., and M. L. Mayer, *J. Physiol. (London)* **428**, 333 (1990).

[45] Biosynthesis of Ion Channels in Cell-Free and Metabolically Labeled Cell Systems

By William B. Thornhill and S. Rock Levinson

Introduction

Sodium channels have been purified from a number of tissues and have been found to have significant nonprotein domains (e.g., see Levinson *et al.*[1]). In our studies, sodium channels isolated from the electric organ of the eel have been shown to be large monomeric proteins of about M_r 260,000–290,000 that are 30% carbohydrate, 40% of which is sialic acid,[2] and about 6% lipid by weight.[1] The rather extensive nature of these nonprotein domains has prompted studies of their possible roles in channel biology. Attachment of these domains occurs posttranslationally in the endoplasmic reticulum and Golgi apparatus, and the generally considered view is that they play a role in protein folding and targeting to the appropriate area of the plasma membrane.[3] On the other hand, some evidence supports the notion that these domains may be specific parts of the molecular mechanism of the channel as well.[1]

Below we describe techniques that we have used to deduce the nature of nonprotein domains and investigate their biosynthesis. Because the sodium channel is so heavily posttranslationally modified, these methods focus on the biosynthetic events occurring in the rough endoplasmic reticulum (RER) and Golgi apparatus as the channel molecule is processed from an initial immature core polypeptide to a mature lipoglycoprotein. Insofar as other ion channel types have also been found to be posttranslationally modified by carbohydrate and fatty acylation, the methods described here should be of more general interest and can be readily adapted to other ion channel systems.

Biosynthesis of Ion Channels in Cell-Free Systems

Evidence of posttranslational modification may be obtained by comparing the physicochemical characteristics of the mature protein as isolated from the plasma membrane with the unmodified protein core. An effective way of doing this is to use mRNA to direct the synthesis of the core protein

---

[1] S. R. Levinson, W. B. Thornhill, D. S. Duch, E. Recio-Pinto, and B. W. Urban, *in* "Ion Channels" (T. Narahashi, ed.), Vol. 2, p. 33. Plenum, New York, 1990.

[2] J. A. Miller, W. S. Angew, and S. R. Levinson, *Biochemistry* **22,** 462 (1983).

[3] S. Fleischer and B. Fleischer, eds., this series, Vol. 96.

in a cell-free system lacking the posttranslational machinery. This is followed by electrophoretic analysis of both the core and mature protein using sodium dodecyl sulfate–polyacrylamide gel electrophoresis (SDS-PAGE).

The most commonly used cell-free systems to study the biosynthesis of proteins *in vitro* are the wheat germ lysate system (WGLS) and the rabbit reticulocyte lysate system (RRLS). The preparation of both of these systems has been described in this series.[4,5] The RRLS may be used for the translation of most mRNAs, whereas the WGLS should be limited to translation of messages coding for proteins less than 100K since full-length translation products may not be produced. Both the RRLS and WGLS are commercially available from numerous vendors including Amersham (Arlington Heights, IL), New England Nuclear (Boston, MA), Bethesda Research Laboratories (Gaithersburg, MD), and Promega (Madison, WI).

Because the eel electroplax sodium channel is a protein of M_r 260K–290K, we have used the nuclease-treated RRLS programmed with poly(A) mRNA or total RNA derived from eel electric organ to synthesize the sodium channel core polypeptide (NaChCP). Total RNA and poly(A)RNA were isolated as previously described.[6] The NaChCP may then be immunoaffinity purified from the translation mixture, using rabbit polyclonal antisera to the highly purified sodium channel, and analyzed on SDS gels.[6] Alternatively, if one has homogeneous mRNA (e.g., message derived from *in vitro* transcription of a full-length clone), immunoprecipitation is unnecessary when using the nuclease-treated RRLS, and an aliquot of the translation mixture may be directly analyzed by SDS-PAGE.

The RRLS system should be titrated to determine the optimal concentrations of RNA, potassium acetate, and magnesium acetate to be used; this is done by monitoring the incorporation of trichloroacetic acid (TCA)-insoluble radioactivity into proteins.[5] We have also found that addition of tRNA, RNase inhibitor, and a cocktail of protease inhibitors all combine to give the most optimal production of NaChCPs. Thus for the synthesis of electroplax NaChCP we use a translation volume of 100 μl with the following components: potassium acetate (100–150 mM), magnesium acetate (1.5–2.0 mM), tRNA (1–3 A_{260}/ml, calf liver tRNA from Boehringer-Mannheim, Indianapolis, IN), RNase inhibitor (16 units/ml), protease inhibitors (pepstatin A, chymostatin, antipain, and leupeptin at 0.1 μg/ml), [^{35}S]methionine (2.0 mCi/ml), and electroplax total RNA (4–8 mg/ml) or poly(A) mRNA (1–4 mg/ml). Translation is carried out for

[4] A. H. Erickson and G. Blobel, this series, Vol. 96, p. 38.
[5] R. J. Jackson and T. Hunt, this series, Vol. 96, p. 50.
[6] W. B. Thornhill and S. R. Levinson, *Biochemistry* **26**, 4381 (1987).

2–3 hr at 31° and terminated by adding SDS (stock solution of 25%) to a final concentration of 4% and heating to 100° for 2 min.

To immunoprecipitate NaChCP, an equal volume of sterile water is added, followed by 4 volumes of immunoprecipitation buffer [190 mM NaCl, 6 mM EDTA, 50 mM Tris-HCl (pH 7.5), 2.5% Triton X-100, and 100 units of aprotinin/ml]. To minimize nonspecific background interaction of labeled protein with immunoprecipitation components, the translation mixture should first be precleared with preimmune sera (5 μl) and protein A-Sepharose (30 μl of a 1 : 1 slurry) for 3 hr with shaking at 4°. This is followed by incubation with sufficient specific sodium channel antibody to clear quantitatively all sodium channel polypeptides from the solution. In our system this is usually 1–5 μl antisera for at least 16 hr at 4°. Protein A-Sepharose (30 μl of a 1 : 1 slurry) is then added for 3 hr with shaking; the beads are pelleted in a microcentrifuge for 5–10 sec and washed 5 times (1 ml per wash) with immunowash buffer [150 mM NaCl, 6 mM EDTA, 50 mM Tris-HCl (pH 7.5), 0.1% Triton X-100, 0.02% SDS, and 100 units of aprotinin/ml]. Radiolabeled sodium channels are eluted from the beads with SDS sample buffer, run on an SDS–polyacrylamide gel, and finally visualized by fluorography.[6]

Representative results of these procedures may be seen in Fig. 1. Total proteins synthesized from the addition of electroplax RNA are shown in lane 2 (lane 5, no addition of RNA). Antisera raised to the mature sodium channel precipitated a protein of about 230K from the translation mix (lane 3). The specificity of this precipitation was demonstrated by the lack of a similar band in the lane using channel antisera that was preblocked with biochemically purified sodium channel (Fig. 1, lane 6). Such controls are essential in immunoprecipitation procedures, since spurious bands may arise from several causes (e.g., inadequate washes of protein A-Sepharose to remove trapped protein). Rigorous controls may also provide more detailed information than a poorly controlled experiment. For example, with longer exposure times of the film to the gel a ladder of polypeptides was observed below the 230K protein in lane 3 of Fig. 1. Because this ladder appearance was not found in the preblocked antibody control, it may be suggested that these polypeptides are incomplete translation products of the NaChCP and result from ribosomes prematurely terminating toward the 3′ end of the large sodium channel mRNA in the cell-free system.

In our hands the addition of electroplax RNA to the translation mixture stimulated protein synthesis by 20 to 30-fold over control values, and about 200,000 counts/min (cpm) per microliter of translation mix was incorporated into TCA-precipitated proteins. Of the total radiolabel incorporated into proteins, about 0.1% was associated with NaChCP. This was

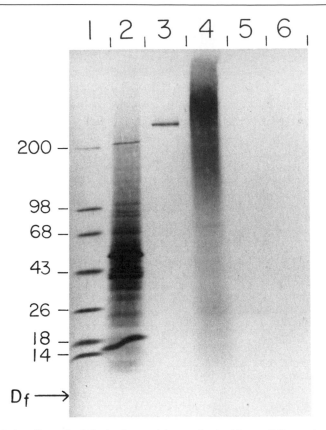

FIG. 1. Autoradiograph of electroplax proteins synthesized in a cell-free system. Lane 2, Total proteins; lane 3, core polypeptide precipitated by sodium channel antisera. Controls: lane 6, antisera preblocked with purified sodium channel; lane 5, proteins synthesized in the absence of electroplax RNA; lane 4, ^{125}I-labeled mature sodium channel; lane 1, ^{14}C-labeled protein standards. Samples were run on a 5–30% polyacrylamide gradient, SDS-containing gel with low cross-linking.

calculated by comparing the precipitated TCA counts with the antisera-precipitated activity (corrected for nonspecific precipitation as seen in the preblocked control).

One may use cell-free synthesis to gain insight into the nature of posttranslational modifications of a purified ion channel. For example, comparison of the electrophoretic properties of the purified mature protein (Fig. 1, lane 4) with the core molecule (lane 3) suggests that the highly disperse, microheterogeneous nature of the mature molecule is posttranslationally acquired, consistent with a high degree of glycosylation. In addi-

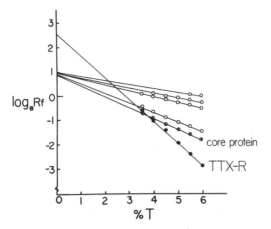

FIG. 2. The sodium channel core polypeptide has normal electrophoretic free mobility. This is a Ferguson analysis where the natural logarithm of the relative electrophoretic mobility (R_f) is plotted as a function of the percent acrylamide concentration used to cast the separating SDS gel. The open circles are the ^{14}C-labeled protein standards, and lines for the core protein synthesized in the cell-free system and the mature ^{125}I-labeled sodium channel (TTX-R) are labeled.

tion, the mature electroplax sodium channel exhibits an unusually high electrophoretic *free* mobility. (EFM) on SDS–polyacrylamide gels[2] owing to the binding of high amounts of SDS to the mature channel.[6] We synthesized the NaChCP to test whether the *in vitro* synthesized channel also displayed a high EFM, indicating that it, too, bound high amounts of SDS. As shown in Figure 2 the NaChCP displayed near normal EFM when analyzed by a Ferguson plot versus the mature sodium channel; thus, it may be concluded that posttranslational modifications are required before the sodium channel displays this unusual characteristic.[6] It was this result that motivated us to analyze purified, mature sodium channel directly for the presence of posttranslationally attached lipid moieties to account for this inferred hydrophobicity.

Biosynthesis of Ion Channels in Cell Systems

Often more detailed information is required about the steps involved in processing the translation product to a mature channel. For example, such detail might yield clues about proteolytic cleavage, posttranslational regulation of channel expression, or subunit assembly. Although some insight might be gained by adding rough intracellular microsomes to the cell-free system, use of this approach may be complicated by the fact that voltage-dependent channels lack a hydrophobic leader sequence as is apparently required for translocation into the RER via a signal recognition particle (SRP)-mediated system. Instead, one may use metabolic labeling techniques to study the time-dependent synthesis of channels in cultured cells.

We investigated the biosynthesis of electroplax sodium channels both in isolated cells from the electric organ of the eel and in *Xenopus* oocytes injected with electroplax mRNA. In either case the goal is to define the biosynthetic history of the sodium channel as it is processed through the RER and Golgi apparatus. The general approach is to incubate cells in [^{35}S]methionine for a short time (pulse) followed by unlabeled medium for different periods of time (chase). The cells are then solubilized at various times and the radiolabeled sodium channels immunopurified and biochemically characterized to determine the extent of posttranslational modifications. For example, glycosylated channel intermediates of the RER would be expected to bind to lentil lectin and to be sensitive to enzymatic treatment by endoglycosidase H (Endo H), whereas Golgi intermediates would bind wheat germ agglutinin (WGA) and be less sensitive to Endo H. Pulse–chase protocols would help to define the sequence of processing events that take place as the protein is transported through the RER, Golgi, and finally to the cell surface. Ferguson analysis may also be used to determine when the sodium channel acquires the high EFM characteristic of the mature molecule.

Electrocytes

The organ of Sachs from an eel 2–3 feet long contains large electrocytes (15–30 by 3 mm) that are straightforward to isolate.[7] About 30–40 electrocytes may be isolated in 1–2 hr and used in primary culture. An example of a pulse–chase experiment using isolated electrocytes is as follows. Electrocytes are incubated in electrocyte buffer [160 mM NaCl, 5 mM KCl, 3 mM CaCl$_2$, 1.5 mM MgCl$_2$, 1.5 mM NaHPO$_2$, 15 mM HEPES (pH 7.4), 10 mM glucose, 1 mM pyruvate, 200 μg/ml gentamicin] plus 1 mCi/ml [^{35}S]methionine at 25°–27° for 45 min and then chased with unlabeled methionine (10 mM) for 1, 3, 8, and 24 hr. At these different times 3 electrocytes are removed from the culture dish, rinsed in unlabeled medium, homogenized in a ground glass homogenizer in 0.5 ml solubilization buffer [4% SDS, 10 mM EDTA, 1 mM phenylmethylsulfonyl fluoride, and 50 mM Tris-HCl (pH 7.5), and heated at 100°. An equal volume of sterile water is added plus 4 volumes of immunoprecipitation buffer as outlined above. Insoluble material may be removed by centrifugation, and

[7] E. Schoffeniels and E. Nachmansohn, *Biochim. Biophys. Acta* **26**, (1957).

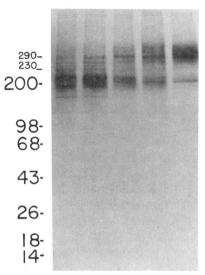

FIG. 3. Sodium channel biosynthesis in eel electrocytes. Electrocytes were incubated in [³⁵S]methionine for 45 min (lane 1) and then chased with unlabeled methionine for 1, 3, 8, and 24 hr, as shown in lanes 2, 3, 4, and 5, respectively. The labeled molecules were precipitated with specific antisera.

the solution may then be precleared with preimmune sera and protein-A Sepharose as outlined above. Radiolabeled sodium channels are then immunopurified, run on SDS–polyacrylamide gels and visualized by fluorography as outlined above (Fig. 3). After labeling for 45 min (lane 1, Fig. 3) the sodium channel antibody precipitated a complex array of both tightly and diffusely banded proteins ranging in M_r from 180K to 350K, although most of the radiolabel was associated with a diffuse band at M_r 200K. The bands of greater and lesser M_r than the core polypeptide (230K) are thus inferred to be posttranslationally processed polypeptides. At the 24-hr time point (lane 5) most of the label had chased to produce a diffuse band at about 290K characteristic of the mature native channel, indicating that the intermediates seen in early time points had been processed to mature molecules.[6]

Xenopus Oocytes Injected with Electroplax mRNA

Frog oocytes may also be used to study the biosynthesis of ion channels using the same general approaches outlined above. Stage V and VI oocytes are isolated, injected with RNA, and incubated in modified OR2 medium as described elsewhere in this volume.[8]

Generally, denuded oocytes (follicular layer removed) are injected with electroplax poly(A) mRNA (1–5 mg/ml) or total RNA (5–8 mg/ml) and incubated in OR2 for at least 2 hr (or overnight) before labeling with [35S]methionine. This preincubation allows time for the injected mRNA to bind to ribosomes as well as time for any damage to the oocyte from the injection procedure to be more readily apparent. The general protocol that we use is to immunopurify radiolabeled sodium channels from five injected oocytes for detection on SDS–polyacrylamide gels by fluorography. To label the proteins synthesized in oocytes to a high specific activity they should be incubated in a small volume, generally 10–40 μl per oocyte. Thus, if 20 oocytes are to be used they should be transferred to a small culture dish (e.g., 48-well dish) containing 200–800 μl of medium. We routinely label oocytes in 2–5 mCi/ml [35S]methionine (> 1100 Ci/mmol, at 50 mCi/ml). At this high specific activity the radiolabel is depleted from the medium in about 5–8 hr, so additional label should be added for longer incubations if continuous labeling of proteins is required. For a 15-hr incubation of oocytes in high specific activity [35S]methionine, each oocyte will incorporate about $5–40 \times 10^6$ cpm into TCA-precipitated protein. However, this number will vary with the overall quality of the ovary and health of individual oocytes.

After the incubation period, oocytes are transferred to a large culture dish with unlabeled medium. For each five oocytes, transfer the oocytes to a 1.5-ml microfuge tube and remove any excess medium. Add 120 μl of solubilization buffer and homogenize the oocytes by pipetting up and down using a 200 μl pipette tip (a microhomogenizer may also be used). The sample is then heated to 100° for 2 min. Add an equal volume of sterile water and 4 volumes of immunoprecipitation buffer as outlined above. The particulate matter is cleared from the solution either by filtering through a sterile 0.45-μm nylon filter into another 1.5-ml microfuge tube or by centrifugation. Preclear the solution with preimmune sera and protein A-Sepharose as described above. Sodium channel antisera (5 μl for five oocytes) is then added, and radiolabeled molecules are precipitated and analyzed on SDS-containing gels. Noninjected or sham-injected oocytes should also be used as controls to determine the extent of any cross-reactivity of the antisera to endogenous oocyte proteins.

<hr>

[8] H. Soreq and S. Seidman, this volume [14].

Example of Labeling Oocytes

Oocytes were injected with electroplax RNA and incubated in [^{35}S]methionine for up to 20 hr. Because [^{35}S]methionine was only added at the start of the incubation period, the radiolabel is expected to be depleted from the medium by 10 hr. Oocytes were removed from the incubation medium and processed at 10 and 20 hr. The sodium channel-related polypeptides synthesized by the oocyte at 10 or 20 hr are shown in Fig. 4, lanes 1 and 5, respectively. The pattern of sodium channel polypeptides synthesized by the oocyte is similar to that seen as early processing events detected in eel electrocytes (Fig. 3, lanes 1 and 2). Again polypeptides of greater and lesser M_r than the core polypeptide were detected, suggesting that these were posttranslationally processed polypeptides. The differences in the polypeptide banding pattern at 10 versus 20 hr is presumably the result of the depletion of radiolabeled methionine from the medium and probably represents a partial chase. For example, after a 10-hr incubation

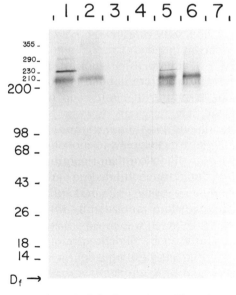

FIG. 4. Sodium channel biosynthesis in frog oocytes. The polypeptides precipitated with antisera after a 10- or 20-hr incubation are shown in lanes 1 and 5, respectively. Sodium channel glycoproteins that bound to WGA beads after 10 or 20 hr are shown in lanes 2 and 6, respectively. Controls: lane 3, antisera preblocked with sodium channel; lanes 4 and 7, WGA beads blocked with 0.1 *M N*-acetylglucosamine.

there are prominent bands at M_r 230K and 290K (Fig. 4, lane 1), whereas after 20 hr these bands are present in much smaller amounts, with the diffuse band at 200K being more prominent (lane 5).

To assist in ascertaining which of the polypeptides are glycosylated, reaffinity purification with lectins may be used. Sodium channels that have been immunopurified are eluted from the protein A-Sepharose beads with 3 successive applications of 50 μl of 2% SDS. This procedure efficiently elutes the antigen but relatively little of the antibody.[9] The eluted material is heated to 100°, and 1.3 ml of lectin buffer [250 mM KCl, 10 mM CaCl$_2$, 2.5% Triton X-100, and 20 mM HEPES (pH 7.4)] is added followed by 100 μl of wheat germ agglutinin (WGA)-Sepharose beads (2 mg/ml) in the presence or absence of 0.1 M N-acetylglucosamine. The beads are incubated, with shaking, for 5 hr and pelleted out of solution in a microfuge for 5 sec. This is followed by washing the beads 5 times in lectin wash buffer (same as lectin buffer, except 150 mM KCl and 0.2% Triton X-100) in the presence or absence of 0.1 M N-acetylglucosamine. Bound glycoproteins are then eluted in SDS sample buffer. Alternatively, total solubilized proteins from oocytes may be incubated with WGA beads and WGA-binding proteins eluted with N-acetylglucosamine and prepared for immunopurification.

The immunopurified sodium channel polypeptides that specifically bound to WGA are shown in Fig. 4 (lanes 2 and 6). The diffuse band at M_r 200K bound to WGA whereas the sharply banded protein of M_r 230K did not, supporting its assignment as the unprocessed core polypeptide seen in cell-free systems. The band at 290K also did not bind to WGA (Fig.4, lane 2) although it did bind lentil lectin (data not shown). For enzymatic deglycosylation, immunopurified channel intermediates may be eluted from the protein A-Sepharose beads and treated with different glycosidases such as Endo H.[10] Finally, the major intermediate sodium channel bands at 290K, 230K, and 200K were also submitted to Ferguson analysis to determine whether they exhibited the high EFM of the native mature channel. It was found that only the 200K band displayed a higher EFM than standard proteins, although not so great as the mature channel (data not shown).[6]

In summary, the 290K band did bind to lentil lectin but not to WGA, while displaying a normal EFM on SDS–polyacrylamide gels; from this one can infer that this component is a high mannose RER intermediate. On the other hand, the 200K band did bind to WGA and exhibited a high EFM suggestive of partial lipidation; both characteristics are consistent

[9] D. J. Anderson and G. Blobel, this series, Vol. 96, p. 111.
[10] D. J. Anderson and G. Blobel, this series, Vol. 96, p. 367.

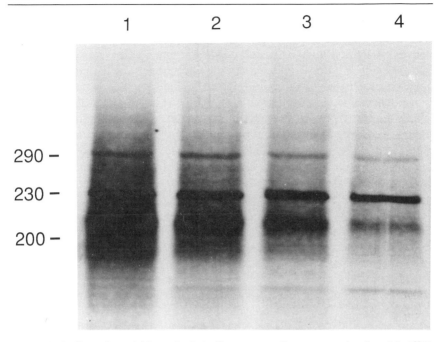

FIG. 5. Sodium channel biosynthesis in frog oocytes. Oocytes were incubated in [³⁵S]-methionine for 16 hr (lane 1) and then chased with unlabeled methionine for 20, 50, and 75 hr, as shown in lanes 2, 3, and 4, respectively. The labeled molecules were precipitated with specific antisera.

with an assignment to the medial Golgi for this component.[6,11] Additionally, comparison of the electrophoretic properties of this band with those of the mature channel allows one to conclude that the sodium channel must be further processed in the Golgi, since extensive carbohydrate and hydrophobic domains must still be added to the 200K band for it to exhibit the characteristic physicochemical appearance of the mature channel on SDS–polyacrylamide gels.

Using the above methods, we have been unable to detect the synthesis of fully processed electroplax sodium channels in frog oocytes injected with electroplax RNA in long pulse–chase protocols, despite the fact that oocytes very efficiently process channels to apparent Golgi intermediates. An example of this is shown in Fig. 5. Oocytes injected with electroplax RNA were incubated in [³⁵S]methionine for 16 hr and chased with unlabeled methionine for 20, 50, and 75 hr (lanes 2, 3, and 4, respectively). The radiolabel associated with the diffuse band at 200K at 16 hr did not chase

[11] W. G. Dunphy, R. Brands, and J. E. Rothman, *Cell (Cambridge, Mass.)* **40,** 463 (1985).

to a diffuse band at 290K (as seen in the electrocyte) but rather disappeared with time, suggesting that channel protein was degraded. Thus, it appears that the majority of eel electroplax sodium channels are not processed to mature channels; this is presumably the explanation for the lack of detectable sodium currents in oocytes injected with electroplax mRNA.[12,13]

In summary, this chapter has described the basic methods for investigating the biosynthesis of sodium channels, as well as outlining the use of these procedures to deduce certain characteristics of these complex molecules. Providing that one has the appropriate means of detection, these methods should be readily adaptable for similar studies on other ion channels.

Acknowledgments

This work was supported in part by the National Institutes of Health, Grants NS-15879 (S.R.L.) and NS-23509 (W.B.T.).

[12] W. S. Agnew, *Nature (London)* **322,** 770 (1986).
[13] M. Noda, T. Ikeda, H. Suzuki, H. Takashima, H. Takahashi, M. Kuna, and S. Numa, *Nature (London)* **322,** 826 (1986).

Section V

Recording of Ion Channels of Cellular Organelles and Microorganisms

[46] Patch Clamp Techniques to Study Ion Channels from Organelles

By BERNHARD U. KELLER and RAINER HEDRICH

Introduction

Ionic transport across intracellular membranes is thought to be involved in a variety of cellular responses, such as synaptic transmission, stimulus–secretion coupling, or muscle contraction. To study ionic processes involved in the various aspects of intracellular signal transduction, it is desirable to utilize the molecular resolution provided by the patch clamp technique. Until recently, however, it has been a considerable problem to investigate ionic signals across intracellular membranes on the basis of single channels, mainly because organelles are too small and difficult to access within the living cell. Here, several recently developed strategies are presented to isolate intracellular membranes and yield organelles large enough for patch clamp experiments. In this chapter, we focus on the methods utilized for patch clamping membranes of mitochondria, endoplasmic reticulum, and plant vacuoles. In addition, the potential application of these techniques for other intracellular membranes is discussed.

Osmotic Swelling Techniques: Mitochondria and Chloroplasts

Patch Clamp of Mitochondrial Membranes

The most important function of mitochondria is to supply the cell with adenosine triphosphate (ATP). ATP synthesis is driven by the protonmotive force, which is maintained across the inner mitochondrial membrane by the activity of the respiratory chain. It was thought to be unlikely that the inner mitochondrial membrane would contain ion channels like those present in the plasma membrane, because the high rates of ion transport characteristic of open channels would dissipate the protonmotive force. By patch clamping the inner mitochondrial membrane, it is possible to test this concept directly and, moreover, investigate in more general terms the molecular mechanisms of ion transport across the inner and outer mitochondrial membranes.

To patch clamp the mitochondrial membranes, two problems need to be solved. First, the mitochondria have to be enlarged to reach a suitable size, and second, the outer and inner mitochondrial membranes have to be separated. Both problems can be solved by osmotic swelling of mitochon-

METHODS IN ENZYMOLOGY, VOL. 207

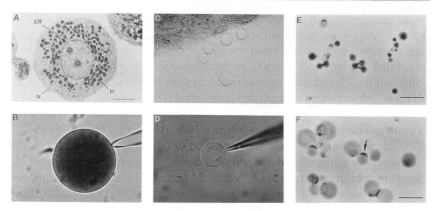

FIG. 1. Micrographs of intracellular membranes utilized for patch clamp measurements. (A) Electron micrograph of a liver cell showing the different intracellular membranes (ER, endoplasmic reticulum; N, nucleus; M, mitochondria). (B) Patch clamped vacuole isolated from sugar beet as described by R. Hedrich and E. Neher, *Nature (London)* **329,** 833 (1987). (C) Giant vesicles obtained by a dehydration/rehydration cycle, forming at the edge of a lipid film. (D) Giant vesicle, obtained with the hydration technique, of approximately 30 μm diameter in the attached-patch configuration. (E) Parent mitochondria as obtained after isolation from liver cells. Bar, 10 μm. (F) Swollen mitoplasts obtained after osmotic swelling of parent mitochondria. Note the residual fraction of outer mitochondrial membrane at the "cap" region of a mitoplast (arrow). Bar, 10 μm.

drial membranes. The first experiments to patch clamp the inner membrane of mitochondria were performed on mitochondria obtained from liver cells of mice fed with cuprizone, a compound which is known to induce the formation of larger liver mitochondria. In a later series of experiments, however, cuprizone was found to be unnecessary for successful experiments, mainly because osmotic swelling itself yielded mitochondria which were large enough for patch clamp measurements.[1-3]

In a typical experiment, mitochondria were isolated from liver cells by standard centrifugation procedures as described by Sorgato *et al.*[1] and references therein. Before swelling, isolated mitochondria (Fig. 1E) were loaded with a solution containing 150 mM KCl, 20 mM HEPES–KOH at pH 7.2. Osmotic shock was performed by exposing mitochondria to a 5- to 10-fold KCl gradient for 2–5 min. After that, inner mitochondrial mem-

[1] M. C. Sorgato, B. U. Keller, and W. Stühmer, *Nature (London)* **330,** 498 (1987).
[2] M. C. Sorgato, O. Moran, V. De Pinto, B. U. Keller, and W. Stühmer, *J. Bioenerg. Biomembr.* **21,** 485 (1989).
[3] W. Stühmer, B. U. Keller, G. Lippe, and M. C. Sorgato, *in* "Hormones and Cell Regulation" (J. Nunes, J. E. Dumont, and E. Carafoli, eds.), Vol. 165, p. 89. Libbey Eurotext, Paris, 1988.

branes had usually unfolded, yielding large vesicles (mitoplasts) of $3-6\ \mu m$ in diameter. Moreover, fractions of outer mitochondrial membranes were visible as dark "cap" regions on top of the swollen mitoplasts (Fig. 1F, arrow).

For patch clamp experiments, aliquots containing $5-10\ \mu l$ of mitochondria or mitoplasts were layered on the glass bottom of the recording chamber and then diluted with $200-300\ \mu l$ of bath solution, which usually contained 150 mM KCl, 0.1 mM CaCl$_2$, 20 mM HEPES–KOH at pH 7.2. After allowing the mitoplasts to adhere to the bottom of the chamber for a minimum of 15 min, the chamber was mounted on the patch clamp setup and extensively perfused with the bath solution. The pipette was usually filled with the same solution as in the bath. Usually, pipettes with resistances of $5-18$ MΩ were used for patch clamp experiments. Seals with resistances greater than 10 GΩ formed relatively easily, provided the chamber had been extensively perfused.

In such recordings, voltage-dependent ion channels could be identified in the inner mitochondrial membrane (IMM). At membrane voltages negative to -20 mV (mitoplast inside negative), IMM channels were completely closed (Fig. 2A). For depolarizations positive to 0 mV, continuous openings and closings of single channel molecules could be observed (Fig. 2B). From reversal potential measurements, IMM channels were found to be anion selective, with a conductance of 107 pS in 150 mM KCl. Since the first findings of these channels in the inner mitochondrial membrane,[1] several other electrophysiological studies have confirmed the existence of such entities.[4,5] Although the physiological role of this channel *in vivo* is not completely understood, it may represent a suitable pathway for the transport of small molecules or metabolites into the mitochondria.[6]

Patch Clamp of Photosynthetic Membranes of Plants

Bulychev *et al.*[7] reported small membrane potentials ($10-15$ mV relative to the cytoplasm) across the chloroplast membranes when giant chloroplasts from leaves of *Peperomia metallica* were impaled by microelectrodes. Because the magnitude of these potentials depended on light, the tip of the electrode was likely inside the thylakoid. Measurements of pH demonstrated that photosynthetic electron transport of the thylakoid generated a proton gradient which in turn drove the ATP synthase. Surpris-

[4] M. Thieffry, J. Chich, D. Goldschmidt, and J. Henry, *EMBO J.* **7**, 1449 (1988).
[5] K. W. Kinnally, M. L. Campo, and H. Tedeschi, *J. Bioenerg. Biomembr.* **21**, 497 (1989).
[6] S. J. Singer, P. A. Maher, and M. P. Yaffe, *Proc. Natl. Acad. Sci. U.S.A.* **84**, 1015 (1987).
[7] A. A. Bulychev, V. K. Andrianov, G. A. Kurella, and F. F. Litvin, *Biochim. Biophys. Acta* **420**, 336 (1976).

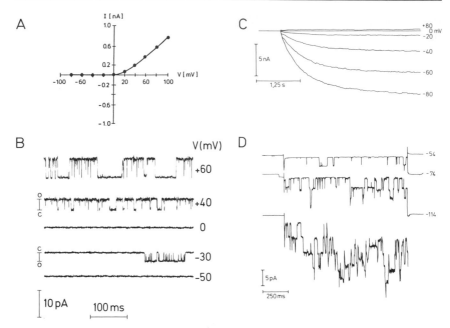

FIG. 2. (A) Whole-mitoplast current I flowing through the mitoplast membrane at different applied voltages V. Note the increase in membrane current for applied voltages more positive than 0 mV. (B) Openings and closings of a single ion channel in the inner mitochondrial membrane in the attached-patch configuration. The applied voltage is indicated in millivolts at right. Depolarizing voltages evoked voltage-dependent ion channels with a conductance of 107 pS. Note the close correlation between single channel openings and the increase in whole-mitoplast currents shown in (A). (C) Whole-vacuole currents in response to a series of voltage steps from a holding potential with the vacuole exposed to 10^{-4} M Ca^{2+} on the cytoplasmic side of the membrane. (D) Single channel openings and closings recorded from an outside-out patch excised form the vacuole shown in (C) using the same voltage protocol. Hyperpolarizing voltages evoked voltage-dependent ion channels with a conductance of 60 pS. The slow activation and voltage dependence of the 60 pS channel reflect the macroscopic currents shown in (C).

ingly, the membrane potential only transiently exceeded 20 mV. Thus, it was proposed that fluxes of counterion across the thylakoid such as chloride and magnesium may have short-circuited photosynthetic H$^+$ transport.

To patch clamp the photosynthetic membrane, leaf slices of *Peperomia metallica* were incubated in 2% cellulase Onozuka R-10 and 1% Mazerozyme R-10 (Yakalt Honsha, Tokyo, Japan), 0.5% bovine serum albumin (BSA), 0.27 M sorbitol, 1 mM CaCl$_2$ for about 1 hr. Resulting protoplasts were washed with 0.35 M sorbitol, 1 mM CaCl$_2$, and separated from debris by filtration through 200- and 25-μm nylon nets followed by centrifugation (7 min at 100 g). The protoplasts were suspended in wash medium, stored on ice, and used within 4 hr. Protoplast suspensions (5 – 10 μl) were rapidly mixed with a 100-fold volume of bath solution {20 mM KCl, 5 mM MgCl$_2$, 2 mM MOPS [3-(N-morpholino)propanesulfonic acid)] – KOH, pH 6.9} in the recording chamber. After 15 min of osmotic swelling in the dark, plasma membranes, vacuoles, and thylakoid envelopes had ruptured and thylakoids formed large blebs. The chamber was perfused with bath solution, and blebs adhering to the bottom of the chamber were used for patch-clamp studies. Experiments were performed with 20 mM KCl, 5 mM MgCl$_2$, 2 mM MOPS – KOH, pH 6.9, inside the pipette.

By applying this protocol to *Peperomia metallica* chloroplasts, Schönknecht *et al.*[8] gained access to the thylakoid membrane with patch pipettes. In such experiments, voltage-dependent chloride-selective channels with a conductance of 80 – 100 pS in 100 mM KCl were identified. As has been discussed in the literature, these anion channels may be essential to balance the transthylakoid potential and to establish a pH gradient across the thylakoid membrane.[7]

Hydration Technique: Intracellular Membranes

Intracellular membranes like the endoplasmic reticulum (ER) are thought to be the central store for intracellular Ca^{2+}, which is released in response to intracellular second messengers. Also, the modulation of calcium release by monovalent ion concentrations has suggested the existence of transport pathways for monovalent cations and anions in ER membranes.[9] These transport pathways can be investigated by using a fusion strategy based on a dehydration/rehydration cycle[10] of isolated membrane preparations.

In a typical experiment, a lysed membrane preparation was isolated by standard centrifugation techniques for endoplasmic or sarcoplasmic reticulum membranes.[11] Isolated membranes were suspended in 5 μl fusion buffer containing 10 mM MOPS adjusted to pH 7.4. To protect the protein maximally against damage by water loss, an additional 10 μl of fusion

[8] G. Schönknecht, R. Hedrich, W. Junge, and K. Raschke, *Nature (London)* **336,** 589 (1988).
[9] S. K. Joseph and J. R. Williamson, *J. Biol. Chem.* **261,** 14658 (1986).
[10] B. U. Keller, R. Hedrich, W. L. C. Vaz, and M. Criado, *Pfluegers Arch* **411,** 94 (1988).
[11] M. Criado and B. U. Keller, *FEBS Lett.* **224,** 172 (1987).

buffer containing 10% ethylene glycol v/v was added to the suspension. Subsequently, membranes were deposited on a glass slide to form a circle about 8 mm in diameter. The membrane-containing drop was dehydrated at 4° in a desiccator containing $CaCl_2$ granules. After 3 hr, a partially dehydrated lipid film could be observed, which was subsequently hydrated by covering it with 20 μl of 100 mM KCl. Complete rehydration of the lipid film was performed by incubating it for several hours at 4° in a closed petri dish. A wet filter paper pad was placed under the glass slide to ensure full rehydration. Usually, cell size vesicles could be observed after 2–3 hr at the edge of the lipid film (Fig. 1C).

To either reduce the density of ion channels or facilitate the fusion process, it was often desirable to dilute the isolated membranes by exogenous lipids. In this case, crude L-α-lecithin from soybean was obtained from Sigma (St. Louis, MO; type II-S) and suspended in water using a Branson sonifier at 40 W for 5 min. The stock solution was prepared with 100 mg/ml lecithin. For the formation of lipid vesicles, 10 mg/ml lecithin was dissolved in 1% CHAPS {3-[(3-cholamidopropyl)dimethylammonio]-1-propanesulfonic acid}, 100 mM NaCl, and 20 mM MOPS adjusted to pH 7.4. Subsequently, dialysis was performed for 24 and 48 hr against 500 volumes of dialysis buffer containing 100 mM NaCl and 20 mM Tris-Cl adjusted to pH 7.4. The resulting small lipid vesicles were stored at $-80°$ until use. In a typical experiment, isolated membranes were diluted with lipid vesicles at the desired concentration and centrifuged for 40 min at 15,000 rpm. After adding the fusion buffer, giant vesicles of diluted membranes were performed as previously described.

Figure 1C,D display giant vesicles formed by fusing ER membranes using the hydration technique. Vesicles up to 100 μm in diameter could be readily observed at the edge of the dehydrated film. For patch clamp experiments, a few microliters of fused vesicles were removed with a pipette and diluted in 300 μl filtered buffer solution. Usually, the buffer contained 50 mM KCl, 0.1 mM $CaCl_2$, and 5 mM HEPES–KOH adjusted to pH 7.2. Large vesicles 10–20 μm in diameter were preferably used for patch clamp experiments. Before starting electrophysiological experiments, vesicles were allowed to settle firmly on the glass surface of the recording chamber. Single-channel recordings were carried out using standard patch clamp equipment.[12] Channel-free membrane patches usually displayed ohmic resistances in the range $R_P > 10$ GΩ. In a typical ER experiment, more than 40% of all patches showed steplike current activities of one or several ion channels.

[12] O. P. Hamill, A. Marty, E. Neher, B. Sakmann, and F. J. Sigworth, *Pfluegers Arch.* **391**, 85 (1981).

Mechanical Isolation of Large Organelles: Plant Vacuoles

Patch Clamp of Vacuolar Membrane

Mature plant cells are characterized by the presence of large central vacuoles. The storage of solutes in vacuoles and their subsequent release are important in cell metabolism and play a fundamental role in the balance of the osmotic pressure and the control of the electrical potential difference across the vacuolar membrane. The intracellular location of this organelle has, until recently, complicated the study of the electrical properties of the vacuolar membrane from higher plant cells by standard electrophysiological techniques. Improvement of methods for the isolation of stable vacuoles and the application of the patch clamp technique made possible much more incisive studies on the electrophysiology of the vacuolar membrane.

The fastest method for the isolation of small numbers of intact vacuoles directly from intact tissue is the one described by Coyaud et al.[13]. In short, the surface of a freshly cut tissue slice is rinsed with buffer solution[14] to wash the liberated vacuoles directly into the recording chamber. Thus, fresh vacuoles can be isolated for each experiment and vacuole isolation and seal formation can be performed within 2–5 min. The whole-cell configuration of the patch clamp technique was established after patch pipettes were sealed against an isolated vacuole, and the underlying membrane was broken by alternate ± 0.6 V pulses 1–3 msec in duration. After access to the lumen of the vacuoles was gained by patch pipettes, the pipette solution equilibrated with the vacuolar sap. Using solutions with symmetric ion compositions on both sides of the membrane, steady-state conditions were indicated by a resting potential of 0 mV (which was reached within 1–5 min for a 20-pF vacuole.

Vacuoles were exposed to solutions containing either 200 mM KCl or KNO$_3$. Both bathing media included 5 mM MgCl$_2$, 0 or 0.1 mM CaCl$_2$, and 5 mM Tris–MES (4-morpholinoethanesulfonic acid) or citrate–KOH buffered to pH 7.5. The vacuole was equilibrated with either 200 mM KCl including 5 mM MgCl$_2$, 1 mM CaCl$_2$, and 5 mM MES–Tris, pH 7.5 and 5.5, or citrate–KOH, pH 3.5 and 4.5. A patch clamp survey of the electrical properties of the vacuolar membrane from a large variety of plant materials has demonstrated the presence of voltage-dependent ion channels and electrogenic pumps as general features of ion transport in higher plant vacuoles.

---

[13] L. Coyaud, A. Kurkdjian, R. Kado, and R. Hedrich, *Biochim. Biophys. Acta* **902**, 263 (1987).
[14] R. Hedrich and E. Neher, *Nature (London)*, **329**, 833 (1987).

At high cytoplasmic Ca^{2+} ($>0.3\ \mu M$), the ionic conductance of the vacuolar membrane was found to be entirely accounted for by currents directed into the vacuole. These currents are activated at negative voltages (negative inside the vacuole), as well as at slightly positive potentials (<20 mV, Fig. 2C). The kinetics of activation of these currents are slow (τ 100–200 msec) and were therefore termed "SV" (slow vacuolar) type currents.[14] Using excised patches, SV-type channels could be resolved at the single-channel level (Fig. 2D). The reported single-channel conductance obtained under different patch clamp configurations averaged 60–80 pS in 50–100 mM salt solutions. The voltage dependence and time course of activation and deactivation, as well as the permeability sequence of the single channels, were in agreement with measurements of the whole-vacuolar currents, indicating that SV currents are carried by unitary SV-type channels.[13]

Discussion

One of the objectives of our approach has been to understand ionic transport across intracellular membranes in an equally detailed way as the patch clamp has provided for the study of ionic transport across the plasma membrane. Three principally different strategies to patch clamp intracellular ion channels are proposed in this chapter.

One method, which can be preferably used for highly folded membranes, is the osmotic swelling technique, and it has been successfully applied to membranes from mitochondria and thylakoids. As this method does not employ isolation procedures for membrane proteins, it is appropriate if the membranes allow an osmotic swelling procedure. Although for rat liver mitochondria the copper chelator cuprizone was originally used to additionally enlarge these membranes, the osmotic swelling technique can be employed as well for untreated membranes isolated from cells with standard isolation procedures.

The dehydration/rehydration cycle composing the hydration technique provides a general approach to investigate intracellular ion channels. So far, quite a large number of ion channels have been investigated with the technique. The broad variety of preparations to which this technique has been applied is represented by studies of ion channels in ER,[15] synaptosomes,[16] mitochondria,[2] and *Escherichia coli* membranes.[17] A major ad-

[15] A. Schmid, M. Dehlinger-Kremer, I. Schulz, and H. Gögelein, *Nature (London)* **346**, 374 (1990).
[16] E. Tarelius, W. Hanke, and H. Breer, *Eur. Biophys. J.* **19**, 79 (1990).
[17] B. Martinac, M. Buechner, A. H. Delcour, J. Adler, and C. Kung, *Proc. Natl. Acad. Sci. U.S.A.* **84**, 2297 (1987).

vantage of the hydration technique is demonstrated by the fact that the preparation of interest can be biochemically modified. With this additional tool, channel analysis may range from the study of ion channels in a purely synthetic environment to channels in an environment which largely resembles the physiological situation. The hydration technique may, therefore, provide a basis to investigate ion channel structure and function in its original environment.

In some cases, when the organelles are large enough, they can be patch clamped directly without any manipulation to enlarge their size. In the case of plant vacuoles mechanically isolated from plant tissue, they can be patch clamped in an osmotically adapted bathing solution. Owing to the large size of vacuoles (10–15 μm), the classic patch clamp configurations can be employed directly in that system. In a similar approach, Mazzanti et al.[18] studied ionic channels in the nuclear membrane. In their experiments, standard microelectrodes were inserted into the nuclcus so that the nucleus adhered to the micropipette when it was withdrawn from the cell. After removal of the nucleus, the nuclear membranes could be directly accessed by patch pipettes.

Taken together, these experiments demonstrate that there is no general strategy for patch clamping different organelles. Instead, the specific structure of each organelle has to be considered before choosing an appropriate patch clamp strategy. However, the methods presented here promise to serve as versatile tools for a better understanding of intracellular ion transport and signal transduction.

[18] M. Mazzanti, L. J. De Felice, J. Cohen, and H. Malter, *Nature (London)* **343**, 764 (1990).

[47] Patch Clamp Studies of Microbial Ion Channels

By Yoshiro Saimi, Boris Martinac, Anne H. Delcour,
Peter V. Minorsky, Michael C. Gustin,
Michael R. Culbertson, Julius Adler, and Ching Kung

Introduction

Interest in microbial ion channels stems from three sources: (1) microbial biology, (2) channel evolution, and (3) the use of microbes to study the principles of channel structures, functions, and regulations. We have pioneered patch clamp studies of *Paramecium* (a ciliated protozoan), budding yeast (a fungus), and *Escherichia coli* (a gram-negative bacterium). Meth-

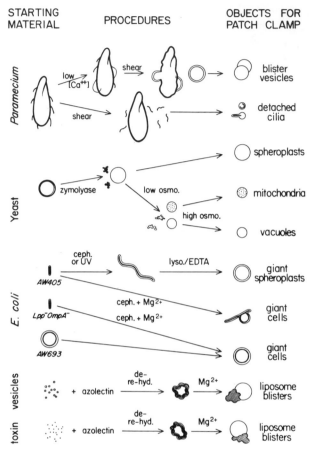

STARTING MATERIAL   PROCEDURES   OBJECTS FOR PATCH CLAMP

FIG. 1. Outline of procedures. The starting material (left) *Paramecium,* yeast, *E. coli,* membrane vesicles, and yeast killer toxin have dimensions of the order of 100, 10, 1, 0.1 and 0.01 μm, respectively. The procedures (center), detailed in the text, convert them to objects used successfully in patch clamp experiments. These objects (right) are all about 5 to 15 μm in diameter except detached cilia from *Paramecium* (~ 2 μm) and yeast mitochondria and vacuoles (~ 3 μm). AW405 is a wild-type *E. coli* strain. Lpp$^-$ OmpA$^-$ and AW693 are mutants. Osmo., osmolarity; ceph., cephalexin; lyso./EDTA, lysozyme and EDTA; de- re-hyd., dehydration/rehydration. Objects not drawn to scale.

ods we developed are diagrammed in Fig. 1 and described below. Activities of channel types we found have been reviewed frequently[1-8] and are summarized in Table I.[9-20]

[1] Y. Saimi, B. Martinac, M. C. Gustin, M. R. Culbertson, J. Adler, and C. Kung, *Trends Biochem. Sci.* **13,** 304 (1988).

[2] B. Martinac, Y. Saimi, M. C. Gustin, and C. Kung, *in* "Calcium and Ion Channel Modulation" (A. D. Grinnell, D. Armstrong, and M. B. Jackson, eds.), p. 415. Plenum, New York, 1988.

TABLE I
MICROBIAL CHANNELS RECORDED USING PATCH CLAMP METHODS

Organism	Channel	Conductance (approx.) (pS)	Ion specificity	Gating principle	Refs.
Paramecium	K$^+$ channel	150	K $\gg$ Na	Ca^{2+}, depolarization	3,4
	K$^+$ channel	70	K $\gg$ Na	Ca^{2+}, hyperpolarization	9
	Na$^+$ channel	19	Na $\gg$ K	Ca^{2+}	10
	Cl$^-$ channel	<430	Cl $\gg$ K	?	3,4
	Cation channel	150	K $\approx$ Na $\gg$ Cl	ATP, depolarization	3,4
Yeast	K$^+$ channel	20	K $\gg$ Na	Depolarization	4,11
	MS channel[a]	40	K $\approx$ Na $\gg$ Cl	Tension[g]	4,12
	Mitochondrial VDAC[b]	400	Cl $\gg$. K	Voltage	13
	Vacuolar channel 1[c]	465	K > Ca $\gg$ Cl	Ca^{2+}, voltage	13,14
	Vacuolar channel 2[d]	160	K $\approx$ Na $\gg$ Cl	Ca^{2+}, voltage	13,14
	Killer toxin	120	K $\approx$ Na $\gg$ Cl	None	15
E. coli	MS channel[a]	900	Cl > K	Tension,[g] depolarization	4,7,16–18
	Cation channel 1[e]	90	K > Cl	Depolarization	4,17–20
	Cation channel 2[f]	30	K > Cl	Voltage	18
B. subtilis	MS channel[a]	100	?	Tension[g]	7

[a] MS channel, mechanosensitive channel.
[b] Mitochondrial VDAC, mitochondrial voltage-dependent anion channel.
[c] Vacuolar channel 1, vacuolar Ca^{2+}-release channel, which has a 465-pS conductance for K$^+$, 100 pS for Ba^{2+} and Ca^{2+}, and is activated by millimolar vacuolar Ca^{2+}, micromolar cytoplasmic Ca^{2+}, and cytoplasmic depolarization.
[d] Vacuolar channel 2, vacuolar cation channel.
[e] Cation channel 1, outer membrane channel, most likely a porin, which shows cooperative closing and reopening on depolarization.
[f] Cation channel 2, location unknown.
[g] Tension, membrane tension.

[3] B. Martinac, Y. Saimi, M. C. Gustin, M. R. Culbertson, J. Adler, and C. Kung, *Period. Biol.* **90,** 375 (1988).
[4] Y. Saimi, B. Martinac, M. C. Gustin, M. R. Culbertson, J. Adler, and C. Kung, *Cold Spring Harbor Symp. Quant. Biol.* **53,** 667 (1988).
[5] C. Kung, Y. Saimi, and B. Martinac, *Curr. Top. Membr. Transp.* **26,** 145 (1990).
[6] C. Kung, *in* "The Evolution of the First Nervous Systems" (P. A. V. Anderson, ed.), p. 203. Plenum, New York, 1990.
[7] B. Martinac, A. H. Delcour, M. Buechner, J. Adler, and C. Kung, *in* "Comparative Aspects of Mechanoreceptor Systems" (F. Ito, ed.), Springer-Verlag, in press.
[8] R. R. Preston, J. A. Kink, R. D. Hinrichsen, Y. Saimi, and C. Kung, *Annu. Rev. Physiol.* **53,** 309 (1991).
[9] Y. Saimi and B. Martinac, *J. Membr. Biol.* **112,** 79 (1989).
[10] Y. Saimi and K.-Y. Ling, *Science* **249,** 1441 (1990).
[11] M. C. Gustin, B. Martinac, Y. Saimi, M. R. Culbertson, and C. Kung, *Science* **233,** 1195 (1986).
[12] M. C. Gustin, X.-L. Zhou, B. Martinac, and C. Kung, *Science* **242,** 762 (1988).
[13] P. V. Minorsky, X.-L. Zhou, M. R. Culbertson, and C. Kung, unpublished observations (1989).
[14] P. V. Minorsky, X.-L. Zhou, M. R. Culbertson, and C. Kung, *Plant Physiol.* **89,** 148 (Abstr.) (1989).
[15] B. Martinac, H. Zhu, A. Kubalski, X.-L. Zhou, M. Culbertson, H. Bussey, and C. Kung, *Proc. Natl. Acad. Sci. U.S.A.* **87,** 6228 (1990).

Paramecium Blister Membranes

Paramecium tetraurelia is a large cell, over 100 μm long and 50 μm in diameter. Macroscopic currents can be measured readily with a conventional two-electrode voltage clamp.[21] Attempts to patch clamp live paramecia, ciliated or deciliated, have failed. Isolated membrane vesicles in the form of detached surface blisters have been patch-clamped successfully.[9,22] *Paramecium* can be grown in several types of medium, such as 0.1 g/liter glucose, 0.1 g/liter casamino acid, 0.5 mM K$_2$HPO$_4$, 0.2 mM MgSO$_4$, 0.2 mM CaCl$_2$, 5 mg/liter stigmasterol, 7.5 mg/liter phenol red, 5 mM HEPES (pH adjusted to 7.0–7.5 with NaOH). The medium (5 ml in a test tube) is bacterized with a tiny scoop (~ 10 μl) of *Aerobacter aerogenes* (food bacterium) from a plate with a platinum loop 1 day before the *Paramecium* inoculation and 2 days before experiments. Add as aseptically as possible 0.5–1 ml of *Paramecium* culture or 500–1000 cells to the bacterized medium. Higher growth temperatures (around 31°) and stigmasterol supplement (3–10 mg/liter) in the medium appear crucial. These factors may affect the rigidity of the membrane and therefore the stability of the patch.

Tens of cells are transferred to a chamber where blisters are induced in 100–150 mM sodium glutamate or NaCl, 10^{-5} M free Ca^{2+} (buffered with EGTA), 5 mM HEPES (pH $\sim$7.0). The blistering process, which takes place in 5 min, can be monitored under a phase-contrast microscope with a magnification of $\times$300. If the free Ca^{2+} level is below 10^{-6} M or above 10^{-4} M, the cells do not blister consistently. Also avoid cells in the stationary growth phase, which is marked by the absence of dividing cells in the culture. Cells in the late logarithmic to early stationary phase, where unperturbed cells start to descend and spread in the culture tube, are preferred for blistering. The blisters can be detached from the actively swimming cell by passing the cell in and out of a glass suction pipette whose bore is slightly smaller than a paramecium.

Several detached vesicles 5–20 μm in diameter are then picked up with the same suction pipette and laid down on the bottom of a separate

[16] B. Martinac, M. Buechner, A. H. Delcour, J. Adler, and C. Kung, *Proc. Natl. Acad. Sci. U.S.A.* **84**, 2297 (1987).
[17] M. Buechner, A. H. Delcour, B. Martinac, J. Adler, and C. Kung, *Biochim. Biophys. Acta* **1024**, 111 (1990).
[18] A. H. Delcour, B. Martinac, J. Adler, and C. Kung, *Biophys. J.* **56**, 631 (1989).
[19] A. H. Delcour, B. Martinac, J. Adler, and C. Kung, *J. Membr. Biol.* **112**, 267 (1989).
[20] A. H. Delcour, J. Adler, and C. Kung, *J. Membr. Biol.* **119**, 267 (1991).
[21] D. O. Oertel, S. J. Schein, and C. Kung, *Nature (London)* **268**, 120 (1977).
[22] A. Kubalski, B. Martinac, and Y. Saimi, *J. Membr. Biol.* **112**, 91 (1989).

chamber containing the seal-forming solution, which may have (in mM) 100 K$^+$ or Na$^+$, 10 MgCl$_2$, 5 HEPES (pH $\sim$ 7.0), an appropriate amount of anion, and 10^{-8} to 10^{-4} M free Ca^{2+}, depending on the type of experiments. The patch pipette often contains a similar solution. The blisters are double layered. Depending on the osmolarity difference between the blistering and gigohm seal-forming solutions, the blisters may stay round or start to peel off to give rise to a vesicle (with the inner, probably the alveolar membrane) capped with hemispherical membrane (outer, the plasma membrane). Gigohm seals can be formed on either membrane. The inner membrane appears electrically silent in our survey; the outer membrane contains channels (Table I). After gigohm seal formation, the membrane patches can be excised by air exposure or shaking the vesicle off by tapping the pipette holder. If the free Ca^{2+} level in the gigaseal forming solution is higher than 10^{-6} M, the excised membrane may be closed into vesicles. Fire-polished, coated pipettes of Boralex glass (Dynalabs, Rochester, NY, Cat. No. 5068) are favored although other types of pipettes have been used successfully.

Paramecium Cilia

Each *Paramecium* is covered with some 5000 cilia, whose membranes are continuous with the body membrane. Forcing live paramecia through a narrow-bore pipette as above can shear off cilia. Each detached cilium retains its microtubular axoneme inside, and its membrane apparently reseals, since it appears osmotically active. These ciliary vesicles can be transferred to the recording chamber, form seals with recording pipettes, and show activities of several channel types.[2]

Yeast Spheroplasts

Cells of budding yeast (*Saccharomyces cerevisiae)* are about 7 μm in diameter and have external cell walls. For preparation of yeast spheroplasts,[11,12] haploid or diploid yeast is cultured in liquid YEPD media [1% yeast extract, 2% Bacto-peptone (Difco, Detroit, MI), 2% dextrose] at 30° overnight. A second culture is prepared the next morning by a 1:20 dilution into fresh YEPD. After incubation at 30° for 90 to 120 min, the cells are washed twice with distilled water by centrifugation (2000 rpm, for 5 min in a tabletop centrifuge), and the final pellet is resuspended in water at an A_{600} of 1.5. This suspension is diluted 1:1 (final volume 0.75 ml) with a solution containing 0.8 M sorbitol plus 0.133 mg/ml Zymolyase (100T, ICN, Costa Mesa, CA) to remove the cell wall. After incubation at 30° for 18 min, 3 ml of solution A (in mM, 120 KCl, 50 MgCl$_2$, 5 HEPES, pH 7.2) is added. The spheroplasts are then spun down and resuspended in 0.5 ml of solution A. A small aliquot of this suspension is diluted into a chamber containing the bath solution (solution A plus 2 mM CaCl$_2$).

Spheroplasts are used only on the day of preparation. Various recording configurations (on-cell, excised inside-out and outside-out, and whole-cell modes) can be and have been used. Seals are usually obtained on free-floating spheroplasts; a spheroplast is moved off the bottom by slight positive pressure through the recording pipette and then captured in the mouth of the pipette by suction. High gigohm seals (> 10 GΩ) will form but take longer and probably require more suction than with animal cells. A 3-ml syringe with a two-way valve connected by tubing to the port on the pipette holder is used to apply the strong and prolonged suction necessary. A pressure transducer (PX143, Omega Engineering, Stamford, CT) in line is used for monitoring the applied pressure. At least 5–20 cm Hg appears to be necessary for seal formation. Often during maintained application of suction the resistance will abruptly decrease (probably owing to the pinching off of surface membrane blebs), only to start slowly increasing again. Gigohm seals will still usually develop on such spheroplasts. Recording at seal resistances less than 10 GΩ is not recommended.

Patch pipettes are made from Boralex glass, pulled to a bubble number[23] around 3.5 (8–10 MΩ). Two sets of solutions have been used most often: pipette hypotonic [bath, solution A; pipette, (in mM) 150 KCl, 5 MgCl$_2$, 0.1 EGTA, 5 HEPES, pH 7.2] and pipette hypertonic[24] (bath, 10 potassium glutamate, 390 mannitol, 2 MgCl$_2$, 1 CaCl$_2$, 10 MES, pH 5.5; pipette, 100 potassium glutamate, 2 EGTA, 2 ATP, 4 MgCl$_2$, 250 mannitol, 10 HEPES, pH 7.2). Seals form more easily with the former, but whole-cell recordings are more stable with the latter.

Yeast Mitochondria and Vacuoles

Each yeast cell has a giant mitochondrion and a large vacuole. When yeast spheroplasts are lysed under the microscope in a hypoosmotic solution of (in mM) 50 sorbitol, 5 MgCl$_2$, 50 KCl, 5 HEPES, 0.1 EGTA, pH 7.2, two types of vacuolelike structures are evident. The less abundant type appears colorless in plain light optics, has a more rigid membrane, and encloses many particles in Brownian motion. When reshrunk in the seal-forming solution (see below) and patch clamped, these organelles are found[14] to possess an ion channel with a conductance (~390 pS in sym-

[23] D. P. Corey and C. F. Stevens, in "Single Channel Recording" (B. Sakmann and E. Neher, eds.), p. 53. Plenum, New York, 1983.
[24] J. I. Schroeder, J. Gen. Physiol. 92, 667 (1988).

metrical 100 mM KCl) and voltage dependency virtually identical to those reported for the voltage-dependent anion channel (VDAC) found in the outer membrane of all eukaryotic mitochondria so far examined.[25] The more abundant organelle released following lysis is extremely elastic, contains few or no particles, appears lavender, and has a complement of channel activities entirely different from those of the mitochondrial or plasma membrane. The kinds of channels encountered indicate that these organelles are most likely vacuoles.

Following shrinkage of the isolated vacuoles by perfusion of the gigohm-forming solution (in mM; 300 potassium glutamate, 5 MgCl$_2$, 5 HEPES, adjusted to pH 7.2 with KOH), excised, vacuolar-side-out patches can be attained simply by touching the membrane with a fire-polished Boralex pipette filled with (in mM) 300 potassium glutamate, 0.1 MgCl$_2$, 5 EGTA, variable CaCl$_2$, 5 HEPES, adjusted to pH 7.2 with KOH. We have succeeded in establishing reliably the on-vacuole or vacuole-side-out excised recording configuration but not other modes. Our studies indicate that the vacuolar membrane possesses several distinct channel types, but we have only examined two conductances in detail.[13,14]

Escherichia coli Giant Cells or Giant Spheroplasts

Escherichia coli is rod shaped, 0.5 μm in diameter and 2 μm long. Patch clamping such cells directly was unsuccessful. We have developed five methods[16,17] to generate giant cells or giant spheroplasts 5 to 10 μm in diameter for patch clamp experiments.

Giant Spheroplasts from Cephalexin-Treated Cells

Wild-type *E. coli,* such as strain AW405, grown in modified Luria–Bertani medium (MLB) [1% Bacto-tryptone (Difco), 0.5% yeast extract, 0.5% NaCl] at 35° (up to an A$_{590}$ of 0.5–0.6) is used to inoculate a culture (1 : 10 dilution of bacteria) in MLB plus 60 μg/ml of cephalexin at 42° until unseptated filaments of 50–150 μm observed under a microscope ($\times$400) are formed (in 2.5–3 hr). The harvested filaments are resuspended in 0.8 M sucrose and digested with lysozyme (200 μg/ml) in the presence of (in mM) 50 Tris buffer and 6 NaEDTA, pH 7.2, to hydrolyze the peptidoglycan layer (cell wall) at room temperature for 7–10 min. The progress of spheroplast formation is followed under a microscope ($\times$400).[26] This treatment apparently does not digest the peptidoglycan completely, but

[25] M. Colombini, *J. Membr. Biol.* **111**, 103 (1989).
[26] H.-J. Ruthe and J. Adler, *Biochim. Biophys. Acta* **819**, 105 (1985).

clips the polymers and weakens the cell wall, thereby allowing the swelling of *E. coli* spheroplasts.

A similar lysozyme treatment followed by overnight growth and activation of autolysin can also generate giant spheroplasts of *Bacillus subtilis,* a Gram-positive bacterium.[7] *Bacillus subtilis* is grown overnight in 10 ml of 2% Bacto-peptone (Difco) in the presence (crucial) of 0.5% (w/v) NaCl and 25 mM KCl. A 1:100 dilution of the culture is made into 80 ml of the same fresh medium as above, and the bacteria are grown at 35° to an A_{590} of 0.8–0.9. Cells are harvested by centrifugation in a Sorvall SS-34 rotor at 5000 rpm for 15 min. The cell pellet is resuspended in 70 ml of 0.5 M sucrose, 16 mM MgSO$_4$, and 0.5 M potassium phosphate buffer (pH 7.0) in a 1-liter flask. The cell suspension is incubated at 35° for 1 hr in the presence of 2 mg/ml lysozyme. Spheroplasts are harvested by centrifugation in a Sorvall SS-34 rotor at 5000 rpm for 15 min. The spheroplasts are resuspended in 1.6 ml of the same solution as above. The spheroplasts must be grown further to larger sizes in order to be patch-clamped. A 1:25 dilution (4–10 ml total) of the spheroplast suspension is made with the succinate medium (0.5 M sodium succinate, 0.1 M KCl, 1 mM MgSO$_4$, 10 mg/ml casamino acid, 0.6 mg/ml fructose bisphosphate) and then added (per ml) with 40 μl of 5% casamino acid and 10 μl of 10% yeast extract. The culture is shaken slowly (50–75 rpm) at 33° for 24 hr before the giant spheroplasts are harvested by centrifugation in a Sorvall SS-34 rotor at 5000 rpm for 10 min. The spheroplast pellet is resuspended gently into 10 ml of the activating solution (1 M sucrose, 6 mM KCl, 20 mM NaHCO$_3$, and 10 mM MgSO$_4$, pH 7.5), and incubated at 35° for 90 min. This treatment activates the autolysin. RNase and DNase (0.8 mg/ml each) can be added to the suspension during this time to reduce viscosity, but this is optional. Spheroplasts are diluted directly into the chamber, where gigohm seals are formed on them with patch pipettes.

Giant Spheroplasts from Cells Treated with Ultraviolet Light

Escherichia coli cells irradiated with UV light also form unseptate filaments. Bacterial cultures (A_{590} 0.1–0.2, 5 mm in depth) in plastic petri dishes are shaken at 40 rpm under UV light (254 nm, 124 erg mm^{-2} sec^{-1}) for 3 min, pooled into 9 volumes of a growth medium (MLB), and cultured at 42° until filaments 80–100 μm long developed (2.5–3 hr). Spheroplasts are prepared with lysozyme as above.

Magnesium-Induced Giant Cell

Cells are grown in LB (1% Bacto-tryptone, 0.5% yeast extract, 1% NaCl) plus 50 mM MgCl$_2$ at 35° to an A_{590} of 0.5, then diluted 1:10 into same medium with the addition of 60 μg/ml cephalexin, and grown for 2 hr by shaking at 42°. This culture yields a mixture of giant round cells (5 – 10 μm in diameter) and long filaments, some with bulges on the sides. These bulges and giant cells can be used directly for patch clamp studies.

lpp⁻ ompA⁻ Giant Cells

Cells of a mutant *E. coli* strain which lacks both Lpp and OmpA proteins, two major components of the outer membrane, round up in the presence of 30 mM MgCl$_2$. Mutant cells are cultured in LB plus 30 mM MgCl$_2$ at 35° to an A_{590} of 0.5, then diluted 1 : 10 in LB plus 30 mM MgCl$_2$ containing 60 μg/ml cephalexin, and grown for 4 hr to form giant round cells 5 – 10 μm in diameter. They are usually washed twice with 0.8 M NaCl by centrifugation in a Sorvall SS-34 rotor at 5000 rpm for 3 min. We found an increase in seal resistance after this salt wash.

Giant Cells of an Osmotic-Sensitive Mutant

Escherichia coli AW693 is selected for its failure to grow at high osmolarity.[13] AW693 cells are grown in LB plus 400 mM KCl to an A_{590} of about 0.6 and incubated overnight at 4° without shaking. A few of the cells are large, some large enough (4 – 8 μm in diameter) to be patch clamped directly. The number of these large cells increases to about 10% of the total by the inclusion of 10 mM MgCl$_2$ in the growth medium.

Most recordings are conducted in the on-cell or excised inside-out mode. As with yeast spheroplasts, seals form more slowly with these bacterial preparations than with most animal cells. Seals of several gigohms can nonetheless be formed routinely. We found that giant spheroplasts from cephalexin- or UV-treated cells and from *lpp⁻ ompA⁻* cells gave higher seal resistances. A mechanosensitive channel is most commonly encountered in all five preparations. A voltage-gated channel, which tends to close cooperatively, is encountered less frequently. Note that *E. coli* has an outer and an inner membrane. We have provided evidence that the mechanosensitive and voltage-gated channels are located in the outer membrane.[17] We also found a mechanosensitive channel in *B. subtilis*.[7]

Reconstitution of Microbial Channels in Liposome Blisters

We reconstitute channels in liposomes by a fusion procedure[18] similar to that of Criado and Keller[27] except that azolectin is prepared without the use of detergent. High gigohm seals are formed more reproducibly with this preparation than when azolectin is prepared with a detergent.

---

[27] M. Criado and U. B. Keller, *FEBS Lett.* **224**, 172 (1987).

Azolectin (Sigma) of 10 mg/ml is sonicated to clarity by a Branson probe sonicator for 5 min in the presence of 5 mM Tris-HCl (pH ~7.2). Aliquots of 1 ml are twice frozen in dry ice–acetone (5 min each) and thawed (15 min each). This process yields multilamellar liposomes. Aliquots of bacterial native membrane vesicles[18-20] are mixed with 0.5 ml of freeze–thawed azolectin at a desired protein-to-lipid ratio (w/w; 1:600 for the outer membrane protein, 1:6 for the inner membrane protein), and the mixture is pelleted for 1 hr at 95000 g at 4° (add 5 mM Tris-HCl, pH 7.2, as necessary). Large membrane vesicles, such as those obtained by a freeze–thaw procedure, pellet better and are needed to obtain a reproducible recovery of lipid.

To form giant liposomes, the pelleted membranes are resuspended into 25 μl of buffer containing 10 mM MOPS and 5% (w/v) ethylene glycol (pH 7.2). Aliquots of the suspension are placed onto a clean glass slide and subjected to a 4-hr dehydration in a desiccator at 10°C. The dehydrated lipid film on the slide glass is then rehydrated at 4° overnight with a solution of 150 mM KCl, 0.1 mM EDTA, 10^{-5} M CaCl$_2$, and 5 mM HEPES (pH 7.2). We found that a lipid concentration of at least 90 mg/ml is needed during the rehydration step to produce giant liposomes.

The liposomes are not suited for gigohm seal formation, but we have developed a protocol whereby unilamellar blisters are induced to grow out of these liposomes. The blisters form high-resistance seals readily and very reproducibly with patch pipettes. Blisters are formed by placing a few microliters of the rehydrated suspension in the patch clamp chamber, which contains the experimental buffer plus 20 mM MgCl$_2$. The presence of MgCl$_2$ causes the liposomes to collapse. Within a few minutes, faint and most likely unilamellar blisters emerge from the sides of the collapsed liposomes. Blisters form rapidly (3–10 min) at low protein-to-lipid ratios, but slowly (30–60 min) at high protein-to-lipid ratios. Once formed, the blisters are stable for hours, as long as MgCl$_2$ is in the buffer. We found that blisters made form pure synthetic lipids alone tended to yield artifactual channellike currents. However, experiments done with azolectin (a mixture of lipids extracted from soybeans), in the absence of fused native membranes or toxins, yield high-resistance seals and quiet background. Care must be taken to use very clean, sterile solutions, especially during the rehydration procedure, to avoid contamination by exogenous microbial membrane fragments.

Bacterial membrane vesicles can be prepared mostly inside-out from French-pressed cells or mostly outside-out from sonicated spheroplasts. Inner and outer bacterial membranes fractions can be separated via sucrose gradient centrifugation. All membrane fractions can be stored at −80° for 2 to 3 months without loss of channel activity. Using this method, detailed elsewhere,[18,19] we have reconstituted three types of *E. coli* channels: a mechanosensitive channel, a voltage-sensitive channel, and a small cation-selective channel. The first two have also been observed during recordings of outer membrane activity using live cells and spheroplasts, and they have retained their native properties. The third type has been seen only in reconstituted membranes so far and is, most likely, an inner membrane channel.

This method can be used for reconstitution of channels from other sources. Killer strains of yeast harbor a double-stranded RNA virus, which produces a dimeric toxin capable of killing the virus-free strains. This toxin forms channels. We showed this, in part, by incorporating it into liposome blisters.[15] Partially purified toxin or a toxin-containing concentrated filtrate of a killer-yeast culture is added to azolectin (toxin:lipid, 1:1000, molar ratio) in 100 mM KCl plus 1× Halvorson salts.[28] The rest of the procedure is as above.

Conclusions

It appears that all plasma membranes and organelle membranes, including those of microbes, are equipped with ion channels. Thus, ion channels apparently evolved early. Patch clamp experiments have revealed many types of ion channels in different microbial preparations. These channels appear to underlie large sectors of interesting biology yet to be explored. The methods described here have been developed for this exploration.

Acknowledgments

We thank M. Buechner, C. Hirscher, A. Kubalski, X.-L. Zhou, and H. Zhu for their technical or other assistance in developing the methods described here. The work in our laboratories described here was supported by National Institutes of Health Grants GM22714, GM32386, GM37925, DK93121, and a grant from the Lucille P. Markey Trust.

[28] Composition of Halvorson salts [H. O. Halvorson, *Biochim. Biophys. Acta* **27**, 267 (1958)]: in mM, 30 $(NH_4)_2SO_4$, 50 K_2HPO_4, 50 succinic acid, 2.71 $CaCl_2$, 4.14 $MgSO_4$, plus trace metals [in μM, 7.7 $Fe_2(SO_4)_3$, 16.6 $MnSO_4$, 15.5 $ZnSO_4$, 15.7 $CuSO_4$].

[48] Studies on Intact Sacroplasmic Reticulum: Patch Clamp Recording and Tension Measurement in Lobster Split Muscle Fibers

By J. M. TANG, J. WANG, and R. S. EISENBERG

Introduction

The sarcoplasmic reticulum (SR) of muscle fibers is of critical importance to contraction: it regulates the intracellular movement of calcium ions that links the signals sent by the nervous system to the chemical reactions that fuel contraction. The SR membranes thus need to be understood in detail even though they are inside a muscle fiber, inaccessible to typical experimentation.

Sarcoplasmic reticulum membranes are accessible in skinned muscle fibers. Skinned muscle fibers are broadly defined as preparations in which removal or disruption of the sarcolemma exposes the highly ordered intracellular space to extracellular solutions. Skinned fibers bridge the gap between intact and isolated systems. Biochemical or physical properties characterized in isolated systems under well-controlled conditions have additional constraints *in situ,* imposed by geometry, diffusion, and interaction with other pathways, whereas many parameters in the intact fiber are unknown or uncontrolled. Skinned fibers can bridge the gap, maintaining a topologically realistic matrix with many of the normal constraints on specific reactions and the relations between them.

One way to examine the influence of Ca^{2+} ions on the properties of the contractile machinery is to employ tension measurements. This method bypasses the T-membrane depolarization step of excitation–contraction coupling; it activates isolated bundles of myofibrils directly in Ca^{2+}-buffered solutions. Activation by this procedure has the advantage that one can readily control the ionic environment around the contractile elements. In tension experiments, lobster skinned fibers respond to caffeine in a dose-dependent manner. They sustain many cycles of contractures and reloading.

The patch clamp is a powerful tool for studying ion channels with molecular, even atomic resolution. The technique has not often been used to study channels from internal membranes (such as SR and endoplasmic reticulum) because they are inaccessible in intact fibers, hidden behind the plasma membrane. If the plasma membrane of vertebrate muscle fibers is removed, giving the patch pipette access to the SR, gigaseals are hard to form, presumably because of mechanical interference from the myofibrils (however, see Stein and Palade[1]) which fill over 90% of most muscle fibers.[2]

Gigaseals might form more easily in muscles with fewer myofibrils and more SR, so we investigated muscles evolved to produce sound. They have few myofibrils and profuse SR,[3-5] probably because they are synchronous and fast, contracting at more than 100 Hz.[6,7] The remotor muscle of the lobster second antenna was chosen because (1) it has the highest reported content of SR ($\sim 70\%$, v/v),[3,5] compared to about 34% in synchronous insect muscle[8] and probably a similar figure in the brain heater muscle of billfish.[9,10] (2) Excitation–contraction coupling in crustacean muscle is quite similar to that in vertebrate skeletal muscle.[11-16] We have split such fibers, exposing the SR, and used the patch clamp technique to examine channels in their native state. We use the words split and skinned in this chapter to imply the mechanical removal of the sarcolemma by dissection. The remotor muscle is a practical preparation: it is large enough to handle, and it is easy to obtain because lobsters are widely distributed commercially. Fibers were prepared by microdissection and split in relaxing saline.[5,17-19] Pipettes readily formed gigaseals to this preparation, allowing the study of the behavior of single channels from the SR membrane.

In both tension measurement and patch clamp recording, the skinned lobster remotor muscle fibers seem as viable as most skinned preparations. Some of these results have already been previously presented,[20-23] and

[1] P. Stine and P. Palade, *Biophys. J.* **54**, 357 (1988).
[2] B. R. Eisenberg, *in* "Handbook of Physiology, Section 10: Skeletal Muscle" (L. D. Peachey and R. H. Adrian, eds.), p. 73. Williams & Wilkins, Baltimore, Maryland, 1983.
[3] J. Rosenbluth, *J. Cell Biol.* **42**, 534 (1969).
[4] D. J. Scales, P. Kidd, T. Yasumura, and G. Inesi, *Tissue Cell* **14**, 163 (1982).
[5] M. Villaz, M. Ronjat, M. Garrigos, and Y. Dupont, *Tissue Cell* **19**, 135 (1987).
[6] M. Mendelson, *J. Cell Biol.* **42**, 548 (1969).
[7] D. Young and R. K. Josephson, *Nature (London)* **309**, 286 (1984).
[8] R. K. Josephson and D. Young, *J. Exp. Biol.* **118**, 185 (1985).
[9] B. Block, *News Physiol. Sci.* **2**, 208 (1987).
[10] B. Block, G. Meissner, and C. Franzini-Armstrong, *Fed. Proc.* **46**, 812 (1987).
[11] C. C. Ashley and E. B. Ridgway, *J. Physiol. (London)* **209**, 105 (1970).
[12] J. P. Reuben, P. W. Brandt, M. Berman, and H. Grundfest, *J. Gen. Physiol.* **57**, 385 (1971).
[13] P. W. Brandt, J. P. Reuben, and H. Grundfest, *J. Gen. Physiol.* **59**, 305 (1972).
[14] T. J. Lea, *Pfluegers Arch.* **406**, 315 (1986).
[15] T. J. Lea and C. C. Ashley, *Membr. Biol.* **61**, 115 (1981).
[16] M. P. Timmerman and C. C. Ashley, *FEBS Lett.* **209**, 1 (1986).
[17] M. Endo, M. Tanaka, and Y. Ogawa, *Nature (London)* **228**, 34 (1970).
[18] M. Endo and Y. Nakajima, *Nature (London) New Biol.* **246**, 216 (1973).
[19] M. Endo, *Physiol. Rev.* **57**, 71 (1977).
[20] J. M. Tang, J. Wang, and R. S. Eisenberg, *Biophys. J.* **51**, 48a (1987).
[21] J. M. Tang, J. Wang, and R. S. Eisenberg, *J. Gen. Physiol.* **94**, 261 (1989).
[22] J. M. Tang, J. Wang, and R. S. Eisenberg, *Biophys J.* **57**, 171a (1990).

parts of this chapter closely follow Tang et al.[21] This chapter provides references to only some of the literature on experimentation with lobster SR. Appropriate articles are cited in each section.

Materials and Methods

American lobsters, *Homarus americanus,* are obtained from a commercial fish dealer and maintained in refrigerated (at 12°), recirculating artificial seawater in an aquarium until used, usually within a week. The animals are sacrificed by decapitation, opened by removal of the dorsal part of the carapace, and are cleaned of viscera to expose the remotor muscle of the coxa of the antenna. The muscle of the second antenna is a large, prominent muscle (mass ~550 mg each) which has its origin posterolateral to the base of the antenna.[6] The remotor from both sides of the head are removed intact with the overlying exoskeleton. The muscles are cleaned of blood vessels, connective tissues, and nervous tissues and put in lobster saline (composition given in Table I). Fibers sometimes give spontaneous long contractures or spasms of rapid twitches. Muscle from one side is used immediately, and the other is stored in lobster saline at 4° for up to about 4 hr before use. The rest of the musculature is frozen for later, more conventional use.

Preparation for Patch Clamp Experiments

The preparation is soaked in 460 mM potassium glutamate, so-called relaxing saline (Table I, calculated 100 nM free Ca^{2+}), for several minutes. After the initial K contracture, the fibers show no mechanical activity in relaxing saline. A short section of the fiber bundle (5–10 fibers ~10 mm in length) is cut from the whole muscle. Single fibers (~400 μm in diameter) are isolated with a 27-gauge hypodermic needle and are teased out from the muscle. To obtain a split preparation, one end of a single fiber is split along its longitudinal axis into two pieces. With the pieces held in a pair of fine forceps, the fiber is torn into two strips. The splitting procedure is repeated until a preparation about 50 μm in diameter is left, measured using a Nikon microscope at a total magnification of ×250.

Single skinned fibers are mounted between two balls of grease (Leitz #465) on a Sylgard disk. The disk is pinned down for patch clamp recording with stainless steel insect pins (#00), into a Sylgard-lined acrylic plastic chamber filled with 460 mM potassium glutamate relaxing saline, on a microscopic stage. The split muscle fiber preparations are observed during

[23] J. Wang, J. M. Tang, and R. S. Eisenberg, *Biophys. J.* **55,** 207a (1989).

TABLE I
COMPOSITION OF SOLUTIONS FOR LOBSTER SKINNED FIBER PATCH CLAMP EXPERIMENTS

Solution[a]	Concentration (mM)									
	Potassium glutamate	KCl	Sodium glutamate	NaCl	MgATP	CaCl$_2$	MgCl$_2$	K$_2$EGTA	HEPES	pCa
Lobster saline	—	10	—	450	—	16	7	—	25	—
Potassium glutamate	460	—	—	—	1	1.2	0.9	5	25	7.0
Sodium glutamate	—	—	460	—	1	1.2	0.9	5	25	7.0

[a] pH 7.4 for lobster saline and pH 7.0 for glutamate solutions.

TABLE II
COMPOSITION OF SOLUTIONS FOR LOBSTER SKINNED FIBER TENSION EXPERIMENTS

Solution[a]	Concentration (mM)									
	Potassium glutamate	KCl	Sodium glutamate	Na$_2$CP	Na$_2$ATP	CaCl$_2$	MgCl$_2$	K$_2$EGTA	HEPES	pCa
Basic	68	87	74	10	5	0	1	0.1	20	—
Relaxing	68	87	74	10	5	0.059	1	0.1	20	7.3
Loading	68	87	74	10	5	0.011	1	0.1	20	6.3

[a] pH 7.0 for all solutions.

single-channel experiments using a modified fold-back Nikon (Labophot) Hoffman modulation microscope at a total magnification of ×250.

Single-Channel Recording

Patch pipettes are made from Corning 7052 glass (outside diameter 1.65 mm, inside diameter 1.15 mm, purchased from Garner Glass, Claremont, CA) in a two-stage pulling process, using a patch pipette puller. Immediately before use, the pipettes are coated with Sylgard 184 (Dow Corning, Midland, MI) and heat polished to a nipple shape with a final inside tip diameter of approximately 0.5 μm. The Sylgard coating is thought to reduce pipette capacitance to the bath, and to prevent creeping of fluid up the shank of the patch pipette. The pipettes, typically filled with 460 mM potassium glutamate relaxing saline, have resistances in the range of 15 to 20 MΩ. Pipette tips are filled by strong backward suction and the shanks backfilled with a fine hypodermic syringe needle. Gigohm seals are obtained using very light suction from a syringe, with seal resistance between 10 and 20 GΩ. In some "better experiments" (about one-fifth), a gigaseal forms without any suction.

A patch clamp amplifier is used for measuring current. The voltage signals are displayed on a digital oscilloscope and stored on magnetic tapes with the bandwidth dc to 5 kHz for further analysis and graphical display. Data are digitized every 100 μsec after passing through a low-pass 8-pole Bessel filter, −3 dB at 1 kHz. Input resistance and resting potential are measured in some experiments: a voltage pulse of −400 to −500 mV is applied to the pipette to break down the membrane, that is, to remove the impedance of the membrane patch. The input resistance of the SR is measured by applying a 20-mV voltage pulse. For the resting potential measurement, the voltage control circuitry is turned off, the current through the pipette is set to zero, and the resulting "open circuit" voltage is measured. This resting potential is stable for at least 15 min, there being a drift of 2 to 3 mV in that time. Liquid junction potentials and offset currents through the gigaseal undoubtedly limit the precision of our estimates. All experiments are carried out at room temperature of about 20°.

We form seals on the cytoplasmic side of the SR membrane of the split muscle fiber (an on-SR patch), probably the equivalent of the cis side of the reconstituted SR preparations as studied in the laboratories of Miller and Williams.[24-28] The other side of the on-SR patch is the SR lumenal side,

[24] C. Miller, in "Current Methods in Cellular Neurobiology" (J. L. Barker and J. F. McKelvy, eds.), Vol. 3, p. 1. Wiley, New York, 1983.

[25] C. Miller, J. E. Bell, and A. M. Garcia, in "Current Topics in Membranes and Transport" (F. Bronner and W. D. Stein, eds.), Vol. 21, p. 99. Academic Press, New York, 1984.

probably equivalent to the trans side in experiments on reconstituted systems. Excised patches are formed by pulling the electrode tip away from the SR membrane after the gigaseal is formed. Such "inside-out patches" [29] have the SR lumenal side exposed to the bath. Our voltage convention places ground (zero potential) on the bath side, and the pipette side could be clamped at a range of voltages relative to virtual ground. Thus, depolarization of the SR membrane, which makes the sarcoplasm more positive, can be produced by (negative) Ca^{2+} current flowing into the sarcoplasm down its concentration gradient across the SR membrane, just as a depolarizing action potential can be produced by (negative) Ca^{2+} current flowing into the sarcoplasm down its concentration gradient across the fiber membrane.

Preparation for Tension Measurement

A small bundle of fibers, several millimeters in length, is cut from the intact muscle, and excess fluid is removed by blotting lightly with Kimwipes tissue. The small bundle of fibers is placed in light mineral oil. A single fiber is teased out of the bundle and skinned by mechanical removal of the sarcolemma as in patch clamp preparation. The skinned preparation used for tension measurement is about 200 μm in diameter. Skinned fiber less than 100 μm in diameter usually gives little or no tension, perhaps because an outer annulus of 100 μm is damaged in the skinning procedure.

Measurement and Recording of Tension

The preparation is switched to relaxing saline for mounting. Care is taken to keep the preparation under solution during the mounting process. A single skinned fiber is mounted on a force transducer (Cambridge Technology, Watertown, MA, Model 400). One cut end of the skinned fiber is held stationary with a pair of forceps, controlled by a screw clamp. The other cut end of the preparation is held by cyanoacrylate glue (gel form) to the stainless steel rod, which is attached directly to the transducer. The forceps and the transducer are attached to separate manipulators which are positioned beforehand so that the opposing edges of the forceps and the rod are parallel to each other and perpendicular to the fiber axis. The fiber is stretched to about 115% of the slack length ($\sim$ 1 mm).

[26] B. Tomlins and A. J. Williams, *Pfluegers Arch.* **407**, 341 (1986).
[27] B. Tomlins, A. J. Williams, and R. A. P. Montgomery, *J. Membr. Biol.* **80**, 191 (1984).
[28] M. A. Gray, R. A. P. Montgomery, and A. J. Williams, *J. Membr. Biol.* **88**, 85 (1985).
[29] O. P. Hamill, A. Marty, E. Neher, B. Sakmann, and F. J. Sigworth, *Pfluegers Arch.* **391**, 85 (1981).

The method employed for changing solutions is similar to that of Ashley and Moisescu.[30] The bathing salines are contained in 1.5-ml vials (Nalgene cryovial, Halge Company, Rochester, NY) that fit into a series of cylindrical wells drilled out of an aluminum block. This block is assembled so that it could be rotated horizontally about its axis and is attached to a large, adjustable Palmer stand (Harvard Apparatus, South Natick, MA) so that it is easily raised or lowered. All solutions are cooled to 12° and are changed by switching the wells on the mechanical stage. The output from the transducer goes to a chart recorder and to a magnetic tape recorder. Standard protocol starts with a wash in relaxing solution (pCa 7.3) for 30 sec followed by contracture in caffeine solution (20 mM), a wash in relaxing solution for 30 sec, and loading in loading solution (pCa 6.3) for 10 min.

Solutions

The dissected intact lobster muscle is kept in lobster saline,[31] containing 450 mM NaCl, 10 mM KCl, 16 mM $CaCl_2$, 7 mM $MgCl_2$, and 25 mM N-2-hydroxyethylpiperazine-N'-2-ethanesulfonic acid (HEPES), 942 mOsm/(kg H_2O), adjusted to pH 7.4 by adding NaOH, typically 11 mM.

In patch clamp experiments, potassium glutamate relaxing saline (Table I) contains 460 mM potassium glutamate, 5 mM ethylene glycol bis (β-aminoethyl ether)-N,N,N',N'-tetraacetic acid (K_2EGTA), 1.2 mM $CaCl_2$, 1 mM MgATP, 0.9 mM $MgCl_2$, and 25 mM HEPES, with 100 nM free Ca^{2+} and 1 mM free Mg^{2+}, calculated from the apparent dissociation constants,[32-35] with osmolality 922 mOsm/(kg H_2O), adjusted to pH 7.0 by adding KOH, typically 7.5 mM, with a total K^+ concentration ([K^+]) of approximately 480 mM. Sodium glutamate relaxing saline contains 460 mM sodium glutamate, 5 mM Na_2EGTA, 1.2 mM $CaCl_2$, 1 mM MgATP, 0.9 mM $MgCl_2$, and 25 mM HEPES, with osmolality 907 mOsm/(kg H_2O), adjusted to pH 7.0 by adding NaOH, typically 10 mM. The osmolality of the solution is routinely monitored with a high precision osmometer (Model 3 MO, Advanced Instruments, Needham, MA). The relaxing solutions are kept hypoosmotic, presumably swelling the SR lumen and making gigaseals easier to form.[29] Gigaseals are stable and well behaved: no irregular bursts of fast current transients are observed.

Three different solutions are used in tension experiments with skinned muscle fibers: relaxing, loading, and releasing. All solutions (Table II) are

[30] C. C. Ashley and D. G. Moisescu, *J. Physiol. (London)* **270,** 627 (1977).

[31] R. A. DeRosa and C. K. Govind, *Nature (London)* **273,** 676 (1978).

[32] A. E. Martell and R. M. Smith, "Critical Stability Constants." Plenum, New York, 1977.

[33] A. Fabiato and F. Fabiato, *J. Physiol. (Paris)* **75,** 463 (1979).

[34] R. Y. Tsien and T. J. Rink, *Biochim. Biophys. Acta* **599,** 623 (1980).

[35] A. Fabiato, this series, Vol. 157 [31].

designed to mimic intracellular conditions and have composition and pH close to lobster muscle.[36,37] The relaxing solution is designed to mimic resting intracellular conditions. The loading solution is designed to facilitate accumulation of calcium in the SR. The release solution is designed to induce calcium release from the SR and is made by adding caffeine as the dry powder to a volume of basic solution. The actual salts and their concentrations are shown in Table II.

Conclusion

The utility of this preparation and these procedures is documented by the results obtained.[20-23,38] The skinned lobster remotor muscle seems as viable as most skinned preparations as judged by usual criteria, namely, the tension generated by caffeine: responses are vigorous, and the SR can be reloaded many times. In addition, single K^+ and Ca^{2+} channels can be studied in their native membrane by the patch clamp technique. The skinned lobster remotor preparation can be studied with an unusually powerful combination of techniques and so perhaps can yield some unusual information.

[36] P. B. Dunham and H. Gainer, *Biochim. Biophys. Acta* **150,** 488 (1968).
[37] J. D. Robertson, *J. Exp. Biol.* **38,** 707 (1961).
[38] J. M. Tang, Ph.D thesis (1991).

[49] Planar Bilayer Recording of Ryanodine Receptors of Sarcoplasmic Reticulum

By ROBERTO CORONADO, SEIKO KAWANO, CHEOL J. LEE, CARMEN VALDIVIA, and HECTOR H. VALDIVIA

Introduction

Three situations in ion channel analysis require a cell-free recording or planar bilayer technique: (1) in the case of channels confined to regions of the cell which are largely inaccessible to patch electrodes such as narrow tubules and intracellular organelles[1,2]; (2) to control solutions on both faces of a channel with analytical precision[3]; and (3) to test the ionophoric

[1] H. Valdivia and R. Coronado, *J. Gen. Physiol.* **95,** 1 (1990).
[2] C. Valdivia, H. Valdivia, B. V. L. Potter, and R. Coronado, *Biophys. J.* **57,** 1233 (1990).
[3] J. S. Smith, T. Imagawa, J. Ma, M. Fill, K. P. Campbell, and R. Coronado, *J. Gen. Physiol.* **92,** 1 (1988).

properties of purified proteins.[4] In this chapter we describe methods to incorporate sarcoplasmic reticulum (SR) Ca^{2+} channels, also called ryanodine receptors, into planar bilayers. Decane-containing bilayers are described exclusively since, in our experience, they are the most reproducible for this application. The technique is illustrated with recordings of ryanodine receptors from rabbit skeletal muscle and bovine heart.[5]

Bilayer Hardware

Planar bilayers useful for recording SR channels[6] are formed spontaneously after depositing a small volume of phospholipids dissolved in decane in a pinhole or aperture under water. The thermodynamic forces that govern thinning of the originally thick film into a bimolecular film have been described in detail by White.[7] Lipid films formed in a mechanical support require a transition zone or annulus between the macroscopic-sized support and the bilayer film. Critical to the stability of planar bilayers is the thickness and composition of the aperture in contact with the annulus. A lipid-wetting surface such as that of TPFE fluorocarbon (Teflon) is not adequate since this material is too soft for drilling clean holes. A hardened version of PTFE called Delrin AF blend (VT Central Plastics, Inc., Chicago, IL) has excellent machinability and provides an acceptable surface for bilayer assembly. Compared side by side, bilayers formed in Delrin holes last 2–5 times longer than those assembled in Teflon holes. A cup serving as inner chamber[8] is machined from a $\frac{7}{8}$ inch diameter Delrin rod into a cylinder with approximate dimensions 2.2 cm outer diameter, 1.9 cm inner diameter, and 1.9 cm height. One side of the cup wall is shaved down to about 25 μm in its thinnest section. A 0.0135 inch diameter drill is used to perforate a hole at the center of the thinnest section. The hole should be free from plastic burrs or residue on inspection under low power magnification and should have a diameter no less than 300 μm and no more than 400 μm. For convenience, the hole should lie approximately 0.7 cm from the bottom of the cup. A block machined from PVC or from any other dark PV-type material such as Acetron Acetal (VT Central

[4] T. Imagawa, J. S. Smith, R. Coronado, and K. P. Campbell, *J. Biol. Chem.* **262**, 16636 (1987).
[5] M. Fill and R. Coronado, *Trends Neurosci.* **11**, 453 (1988).
[6] C. Miller, *J. Membr. Biol.* **40**, 1 (1978).
[7] S. H. White, *in* "Ion Channel Reconstitution" (C. Miller, ed.), p. 3. Plenum, New York, 1986.
[8] J. S. Smith, R. Coronado, and G. Meissner, this series, Vol. 157, p. 480.

Plastics, Inc.) serves as outer chamber and should provide a snug fit for the Delrin cup. The PVC block fits on a brass block close to the electrode inputs of a head stage amplifier.

A metal box with a lid serves as grounding shield for the head stage amplifier, electrodes, and solutions. To avoid condensation of water in the vicinity of the head-stage input that causes drift, this box should be no smaller than 30 cm wide by 30 cm deep by 15 cm high. The box is machined from 1.2 mm thick stainless steel and is bolted to a 2 cm thick lead base that provides mechanical stability. A 12 V dc, 125 rpm motor under the box moves a magnet, in turn used to spin a small magnetic stir bar (3 mm diameter, 12.7 mm length) placed inside the bilayer cup. Magnetic bars can be purchased from Fisher Scientific (Pittsburgh, PA). The dc motor can be purchased from Edmund Scientific (Barrington, NJ). The box rests in an air-lifted table to insulate it from mechanical vibration. Homemade circuits for monitoring planar bilayer currents at high gain, built from standard electronic components, have been described in detail elsewhere.[8] However, a regular patch clamp amplifier is also adequate. Software packages for single-channel analysis such as PClamp 5.5 distributed by Axon Instruments (Fullerton, CA) is highly desirable since it serves for data acquisition and for delivering voltage pulses.

Electrodes and Noise

Macro Ag/AgCl electrodes are encased in agar-filled polyethylene tubing to minimize liquid junction potentials and to avoid contamination of solutions with Ag^+ ions. Electrodes are made from 0.25 mm diameter silver wire (~ 5 cm) soldered to a miniature gold-plated pin for hookup to amplifier (a diagram appears in Ref. 8). The silver metal is coated with silver chloride in an electrolytic cell filled with 0.1 N HCl. A coated electrode is inserted into 1 mm diameter polyethylene tubing which has been filled with 0.2 M KCl in 2% agar. Carboxylate cement is used to glue the tubing to the pin connector. This assembly is durable and may last for several weeks if the electrode tip is always kept in a 0.2 M KCl solution.

Peak-to-peak noise of an assembled bilayer should be no more than 10 pA at 2 kHz or 0.2 pA at 400 Hz low-pass filtering for the recommended aperture size. Appropriate grounding of the bilayer box is sufficient to eliminate the 60 cycles noise common to most electrophysiological setups. Other sources typical of planar bilayers are mechanical and microphonic noise. Both are the result of small changes in area (capacitance) produced when the solutions vibrate and the bilayer film wobbles. Mechanical noise is usually of low frequency and many times erratic. The bilayer setup should be isolated from other fixtures in the room. Ideally, the setup should rest on an air-lifted table (see list of equipment). Electrodes should be fitted tightly to the head stage. The head stage should be glued or screwed to the bilayer box. Microphonic noise increases as the bilayer capacitance increases. Avoid placing a setup next to an elevator shaft, equipment room, or cold room.

Bilayer Stability

The quality of the pinhole is the number one source of day-to-day technical difficulty in assembling long-lasting stable bilayers. In an acceptable cup, a thin bilayer should last no less than 30 min at 0 mV applied potential in the absence of added protein. Cups should be inspected regularly to ensure that the hole is free from dirt and that cracks have not formed along the edges of the hole. Cups are rinsed only with high quality methanol. Avoid the use of detergents or chloroform. Users should have a set of no less than 20 machined cups at their disposal at any given time. Cups showing signs of wear should be discarded.

A good commercial source of phospholipids is Avanti Polar Lipids (Birmingham, AL). Lipids should be purchased dissolved in chloroform and shipped in N_2-sealed vials in aliquots of 1 ml of 5 mg/ml phospholipid. Phosphatidylethanolamine (PE) and phosphatidylserine (PS) purified from brain are ideal for bilayer stability and single-channel recording. Lipids are mixed at a 1:1 weight ratio, dried under a stream of N_2, and resuspended in ultrapure n-decane (Aldrich Chemical Co., Milwaukee, WI) at a concentration of 20 mg/ml phospholipid. A 100 μl lipid solution in decane should be prepared daily. Fresh vials of PE and PS should be opened every 2 weeks and old lipids discarded.

Lipid solution is applied to the aperture with a 0.2 cm diameter Delrin stick cut diagonally at the tip. Bilayer thinning, from a thick to a thin film, is monitored by capacitance measurement. Capacitance rises from approximately 100 to 500 pF in the apertures of the size recommended. Ultrapure grade chloride salts of monovalents or divalents are recommended for planar bilayer recording. A good source is Johnson Matthey Chemicals Ltd. (Hereforshire, England) distributed in the United States by Alpha Products (Morthon Thiokol, Inc., Danvers, MA). The recommended concentration of EDTA (ethylenediaminetetraacetic acid) for Ca^{2+} buffering is 1 mM. pH buffers such as HEPES (N-2-hydroxyethylpiperazine-N'-2-ethanesulfonic acid), MES (2-N-morpholinoethanesulfonic acid), or Tris [Tris(hydroxymethyl)aminomethane] should not be used in excess of 10 mM. Water should be distilled in the laboratory in an all-glass still.

Fractionation of Sarcoplasmic Reticulum

Light and heavy SR contains K^+ and Cl^- channels,[5] but ryanodine receptors are only found in heavy SR.[9] Procedures that use Ca^{2+}-precipitating agents such as oxalate or phosphate for loading Ca^{2+} into the SR (transport-specific fractionation) should be avoided since intravesicular Ca^{2+} impairs SR incorporation into bilayers. A procedure to fractionate light and heavy SR of skeletal or cardiac muscle. which yields active ryanodine receptor channels in planar bilayers, is as follows.

Tissue taken from the animal is immersed in ice-cooled saline and taken to a cold room where the homogenization step is carried out. White muscle from the back and hind legs of one adult New Zealand rabbit or one fresh bovine heart collected no more than 20 min after sacrifice is minced and ground in a food processor. Each of 4–5 portions of 40 g are homogenized for 60 sec in 300 ml of 0.3 M sucrose, 0.5 mM EGTA, 20 mM $Na_4P_2O_7$, 20 mM NaH_2PO_4, 1.0 mM $MgCl_2$, pH 7.1, using a Waring blender at maximum speed. The following protease inhibitors are added during homogenization: pepstatin A (1 μM), iodoacetamide (1 mM), phenylmethylsulfonyl fluoride (PMSF) (0.1 mM), leupeptin (1 μM), and benzamidine (1 mM).

The homogenate is spun for 15 min at 9000 rpm in a Sorvall GSA rotor. The supernatant is filtered through four layers of gauze and centrifuged at 14,000 rpm for 30 min in the same rotor. Pellets are resuspended to a final volume of 60 ml in the same medium used for homogenization and layered on top of a step sucrose gradient composed of 7 ml of 27% (w/v) sucrose, 7 ml of 32% sucrose, 14 ml of 38% sucrose, in 20 mM $Na_4P_2O_7$, 20 mM NaH_2PO_4, 1 mM $MgCl_2$, pH 7.1. Gradients are centrifuged for 16 hr at 20,000 rpm in a Beckman SW 28 rotor at 4°. The heavy microsome fraction is collected from the 32–38% sucrose interface. After gentle dilution with 3 volumes of homogenization medium without sucrose, a pellet is obtained by centrifugation at 30,000 rpm for 40 min using a Beckman 35 rotor. Microsomes are resuspended in 0.3 M sucrose, 5 mM HEPES–KOH, pH 7.0, aliquoted, frozen in liquid nitrogen, and stored at −80° for up to 3 months. When prepared from rabbit skeletal muscle,[2,10] this preparation has a [^{3}H]PN200-110 binding capacity of 12 pmol/mg protein, a [^{3}H]ryanodine binding capacity of 9 pmol/mg protein, and an ATP-dependent Ca^{2+} uptake capacity of 15 nmol/mg protein.

[9] G. Meissner, *J. Biol. Chem.* **259**, 2365 (1984).
[10] H. Valdivia, C. Valdivia, J. Ma, and R. Coronado, *Biophys. J.* in press (1990).

Solutions and Incorporation of Channels

Recordings of ryanodine receptors are best made using CsCl as current carrier[1-3,10,11] instead of the Ca–HEPES and Tris–HEPES solutions suggested earlier.[8,12] Insights into the mechanism of SR vesicle incorporation into bilayers may be found in Ref. 13. The previous recording technique assumed that only Ca^{2+} was permeable through ryanodine receptors, and therefore large concentrations of divalents (Ca^{2+} or Ba^{2+}) were used to increase current amplitude. This assumption turned out to be incorrect: we now know that the channel is highly nonselective.[3] Divalent cations can be thus kept at physiological levels without loss of conductance if a monovalent cation is present in both chambers (Fig. 1[14]).

Other complications in the previous technique arose from the use of large cations and anions like $Tris^+$ and $HEPES^-$ to block SR K^+ and Cl^- channels. Because SR incorporation into bilayers is rare in the absence of Cl^-, bilayer chambers had to be perfused with Cl^--free solutions after SR incorporation in Cl^- media. The perfusion maneuver tended to break bilayers even when a circuit was developed to bypass the electronics.[8] This made the procedure tedious, and the number of successful recordings was limited. The use of CsCl instead of Ca–HEPES and Tris–HEPES eliminated the need for perfusion and eliminated the need for large and unphysiological gradients of Ca^{2+}, which severely inactivate the channel (in high trans Ca^{2+}, the open probability is around 10-fold lower than when the trans free Ca^{2+} is kept at 10 μM).

Finally, Cs^+ blocks SR K^+ channels, and because it has a higher conductance than Ca^{2+} or Na^+ through ryanodine receptors ($g_{Cs}/g_{Ca} = 2$),[3] it improved the signal-to-noise ratio. Sarcoplasmic reticulum Cl^- channels can be separated from ryanodine receptors on the basis of reversal potential. In the solutions recommended, E_{Cl} is $+59$ mV and E_{Cs} is -59 mV. The reversal potential for SR Cl^- channels is approximately $+30$ mV. Recordings in the presence of Cl^- channels can thus be made by holding the membrane potential close to the Cl^- equilibrium potential (more positive than $+20$ mV).

Sarcoplasmic reticulum preparations are thawed by hand and kept on ice. Approximately 100 μg of skeletal SR or 500 μg of cardiac SR protein are added to the cup-side solution, designated the cis side. Cis solution is

[11] M. Fill, R. Coronado, J. R. Mickelson, J. Vilven, J. Ma, B. A. Jacobson, and C. F. Louis, *Biophys. J.* **50,** 471 (1987).

[12] J. S. Smith, R. Coronado, and G. Meissner, *Nature (London)* **316,** 446 (1985).

[13] F. S. Cohen, M. H. Akabes, J. Zimmerberg, and A. Filkenstein, *J. Cell Biol.* **98,** 1054 (1984).

[14] S. Kawano and R. Coronado, *J. Physiol.* (1992) In press.

A   B

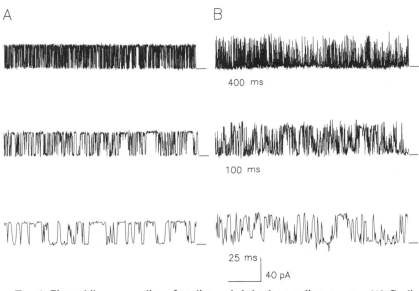

400 ms

100 ms

25 ms

40 pA

FIG. 1. Planar bilayer recording of cardiac and skeletal ryanodine receptor. (A) Cardiac ryanodine receptor from bovine heart at +20 mV holding potential in cis 0.5 M CsCl, trans 50 mM CsCl. Cis and trans Ca^{2+} was 3.5 μM. (B) Skeletal ryanodine receptor from rabbit skeletal muscle under same conditions. (From Ref. 14 with permission.)

typically composed of 3 ml of 0.5 M CsCl, 10 to 100 μM CaCl$_2$, and 10 mM HEPES – Tris, pH 7.5. The bath-side solution, designated trans side, is the same except the CsCl is 50 mM. In the recordings of Fig. 1, a List L/M EPC 7 amplifier (List Electronics, Eberstadt, Germany) was connected to the interior of the cup; the bath side was held at ground. Recordings were filtered through a low-pass Bessel filter (Frequency Devices, Haverhill, MA) at a front panel setting of 1 kHz and digitized at 100 to 250 μsec per point.

The protocol to incorporate SR channels is according to points (1) through (4) as follows. (1) Bilayers are first formed in symmetrical 50 mM CsCl buffer (the trans solution). Wetting the dry hole with a drop of lipid, before pouring a solution into the cup, considerably helps the formation of the first bilayer. Bilayer capacitance is continuously monitored to ensure an appropriate degree of thinning. The increase in bilayer capacitance or "thinning of the bilayer" is favored by large positive or negative potentials (±100 mV). After thinning is completed, the CsCl in the cis solution is increased to the final value. (2) Sarcoplasmic reticulum protein is the last component added after adjustment of free Ca^{2+} with CaEGTA buffers. The cis chamber is stirred for 30 to 60 sec. Membrane potential is kept constant at 0 mV for cardiac SR or between 0 mV and $+20$ mV for skeletal SR. Cardiac SR makes bilayers unstable more often than skeletal SR does. Less applied voltage is recommended when using cardiac SR protein. When a bilayer breaks, or there is a shift in baseline indicative of a leak, or when the capacitance decreases spontaneously, a new membrane should be made in the same solutions. If this maneuver is repeated too many times, however, there will be a substantial mixing of the cis and trans solutions. (3) In an typical case, Cl^- channels incorporate within the first 10 min and ryanodine receptors incorporate afterward within 30 min. Recordings of Cl^- channels should be continued until ryanodine receptors appear. The frequency of presentation of ryanodine receptors without Cl^- channels is 1 out of 5 recordings. (4) When channels do not incorporate, the bilayer is broken and a new one is reformed in the same solutions. Breakage and reformation is repeated up to four times. Afterward, the cup is removed and rinsed with methanol before repeating the protocol.

The polarity of channels incorporated into the bilayer is constant, in the majority of cases. The myoplasmic end of the receptor faces into the cis solution, and the intravesicular end faces into the trans solution. Polarity can be easily confirmed by the cis-activation of channels by ATP and micromolar Ca^{2+}, which are myoplasmic activators of ryanodine receptors.[9] Four useful characteristics that serve to identify ryanodine receptors are the following: (1) a linear current–voltage relationship with a slope conductance of 600 pS for cardiac SR and 700 pS for skeletal SR and reversal at potentials more negative than -50 mV in the solutions recommended[14]; (2) an activatory effect of adenine nucleotides and Ca^{2+} and the inhibition by Mg^{2+}.[11,14] Cis 5 mM ATP should increase the open probability about 2- to 5-fold; cis Ca^{2+} should increase open probability from virtually null at pCa 9 to typical values of 0.03 for skeletal and 0.2 for cardiac receptors at pCa 6; cis 1 mM free Mg^{2+} should decreases activity around 10-fold; (3) an inhibition by cis 1 μM ruthenium red which decreases activity 100-fold[11,12]; (4) an activation by ryanodine, which at a concentration of 100 nM should produce an irreversible decrease in conductance and an increase in mean open time.[3]

List of Equipment

A minimal list of equipment for planar bilayer recording and suggested sources are given below. Preparative biochemical instruments for microsome purification are not included.

Bilayer cups, PVC block, brass block, and steel box with lead base, to be machined at a local shop

Vibration-isolation table (Technical Manufacturing Corporation, Peabody, MA, Model Micro-G)

Patch clamp amplifier (Axon Instruments, Burlingame, CA, Model Axopatch 1D with head stage for planar bilayers)

VCR-based instrumentation recorder (Medical Systems Corporation, Greenvale, NY, Model PCM-2 Recorder adapter, VCR not included)

286-based PC with MS DOS, full memory, coprocessor, 40 Mbyte hard disk drive and EGA graphics (Standard Brand Products, Austin, TX, model Standard 286)

Acquisition software (Axon Instruments, Model TL-125 A/D and D/A interface, PClamp software version 5.5)

Laser printer (Hewlett Packard, Sunnyvale, CA, Model Laserjet series II)

Section VI

Data Storage and Analysis

[50] Software for Acquisition and Analysis of Ion Channel Data: Choices, Tasks, and Strategies

By ROBERT J. FRENCH and WILLIAM F. WONDERLIN

Introduction

A laboratory computer, running suitable software, frees an investigator from the repetitive tedium of the details of controlling acquisition of data, and of directing its routine mathematical transformation and collation. This allows attention to be focused on the biological implications of the macroscopic waveforms, or unitary fluctuations, that appear in response to changes in transmembrane voltage or applications of agonists. However, the importance of an efficient acquisition system extends beyond relieving tedium. Often, it actually makes possible the collection of well-controlled data sets within the limited lifetime of an evanescent experimental preparation, be it a bilayer with a reconstituted channel, a native membrane patch, or a whole cell, stressed by unnatural, but experimentally essential, surroundings. Without the computer, the experiment could not be done. Similarly, efficient data processing enables systematic fitting of data and testing of interpretive models using well-defined statistical criteria where, without such computational means, only subjective, qualitative analyses would have been possible. Certainly, blind acceptance of results of computer analyses, without testing against the standards of qualitative and intuitive appraisal, may be as misleading as neglecting to apply quantitative methods. Nonetheless, there is no doubt that a computer is an essential and integral part of a contemporary electrophysiological setup being used to explore details of ion channel function.

A review of details of software would be doomed to obsolescence even before publication. Thus, rather than attempting to provide a comprehensive description of the rapidly increasing number of available software packages, we discuss some criteria that might usefully be considered in selecting software. The overriding goal is to obtain a set of programs which will enable efficient accomplishment of routine tasks, with a minimum of involvement of the investigator, but will still offer enough flexibility to allow a variety of experimental protocols or analyses to be explored should the need arise. In the following sections we outline several commonly used modes of data acquisition and a variety of general analytical procedures to indicate some of the key tasks that should be performed easily and quickly by a comprehensive package of acquisition and analysis software. More detailed discussions of analyses are available in other chapters in this

METHODS IN ENZYMOLOGY, VOL. 207

section and in a recent review by Wonderlin *et al.*[1] Finally, we list some of the commercial options and other approaches that are available to meet these needs.

Acquisition

Analog signals representing the time course of changes in transmembrane current and voltage must be converted to digital form before they can be stored and manipulated by a computer. This process of digitization is usually carried out on-line during an experiment by the data acquisition system.

We divide data acquisition tasks into three classes. When data are collected in separate episodes, each synchronized with the imposition of a physiological stimulus, acquisition can be termed stimulus-driven or evoked. A stimulus may take the form of a change in voltage, a light flash, a mechanical disturbance, or the application of a chemical agonist or modulator. When precisely synchronized control of several variables is not needed, or when the time of occurrence of events of interest cannot be predicted, data acquisition may be either continuous or event-driven. Continuous acquisition may be used throughout an experiment, with subsequent subdivision of the data into logical segments for analysis. Event-driven acquisition implies that a detector recognizes the occurrence of an event of interest and then signals the acquisition system to collect data for a predetermined period.

For each mode of acquisition, one must decide on the necessary rate of sampling, the number of data channels, and the time over which acquisition must be continued. These factors determine the number of data points and hence the data storage requirements. Ultimately, these capabilities are limited by the hardware used for data acquisition and storage.

Multiple input channels are provided most economically by multiplexing input to a single analog to digital (A–D) converter, so that n channels are sampled in a sequence that is repeated cyclically for the duration of the acquisition. The maximum rate of acquisition per channel is then the maximum rate of conversion for the A–D converter multiplied by $1/n$, where n is the number of channels. If absolutely synchronous acquisition of data on more than one channel is required, then a more expensive option, in which there is a separate A–D converter for each channel, must be employed. However, for most purposes, the offset of, for example, current and voltage data points by one sample interval at the maximum

---

[1] W. F. Wonderlin, R. J. French, and N. J. Arispe, *in* "Neuromethods" (A. A. Boulton and G. B. Baker, series eds.; C. H. Vanderwolf, volume ed.), Vol. 14, p. 35, 1990.

rate of the A–D converter presents no problem. Thus, multiplexing of the A–D input is the most common means of implementing multichannel acquisition.

Stimulus-Driven Acquisition

The most common mode of stimulus-driven acquisition is the recording of currents flowing during a square voltage clamp command pulse. Command waveforms other than square pulses are, however, also widely used. Relatively slow voltage ramps allow direct recording of current–voltage relations, whereas brief, faster ramps offer a way to measure directly the capacitance of a preparation as the scaled amplitude of the square current pulse driven through the membrane capacitance. This allows rapid, routine monitoring of the thinning of a bilayer or of the entry into whole-cell recording mode after forming a seal on a cell-attached patch. Other command waveforms may also be required such as recorded or simulated action potentials, or pseudorandom binary sequences (PRBS) used to obtain frequency domain information.[2] Flexibility to generate arbitrarily shaped stimuli is offered by software systems that drive D–A output from a memory buffer that can be loaded with a numerical array representing any chosen waveform. Rapid and easy use is enabled by experimental programs which should be set up to allow generation of standard stimuli (e.g., families of step sequences, or of ramps, together with associated logic outputs) with a minimum of input parameters. In addition, it should be possible to store and recall protocols rapidly once they have been defined. The ingenuity in designing such a system lies in making routine tasks as simple and rapid as possible, while not sacrificing flexibility to meet unusual, perhaps unanticipated demands.

Frequently, different modes of stimulation will be combined, and two or more will be used under the simultaneous control of the data acquisition system. For example, valves allowing rapid switching of solutions flowing over the preparation may be activated by a logic pulse while a specified voltage command is applied. Similarly, agonists or other drugs may be applied by precisely timed electrophoretic or pressure pulses. Responses of ion channels are directly measured as current flow, but many investigations demand simultaneous recording of a mechanical response, an optical signal, or some other parameter. Thus, a general purpose data acquisition system will offer several logic outputs to trigger external devices, one or more channels of analog output to allow incremental control of either transmembrane voltage or iontophoretic drug application, and one or more analog input channels to accept data.

[2] J. M. Fernandez, F. Bezanilla, and R. E. Taylor, *J. Gen. Physiol.* **79,** 41 (1982).

Continuous Acquisition to Tape or Disk

Much single-channel data can simply be recorded continuously during the experiment. Generally, this mode of recording is used for relatively long periods under steady-state conditions, rather than for episodically generated data. At present, the most economical medium for continuous acquisition is video cassette tape, with the signal processed by a pulse code modulator (PCM), then input to a commercially available video cassette recorder (VCR). The PCM must be modified to handle signals from dc upward, as opposed to the normal lower limit of about 10–20 Hz used for audio signals.[3,4] The standard sampling frequency has been 44 kHz, making the usable signal bandwidth 22 kHz. However, oversampling at rates that are 2-fold or more higher has been recently introduced in some recorders, and this provides added flexibility for biological studies. Digital audio tape recorders, which are now being adapted by some companies, may also fill this role in data acquisition.

Once data are stored on VCR tape, they must then be transferred to computer disk for processing. At this stage, it is convenient to divide the continuous data stream from the tape into multiple separate files for analysis — one for each of the experimental conditions used. Transcription to disk can be accomplished either by directly reading in the digitized data via a dedicated interfacing card or by playing the analog output from the VCR/PCM system into the input of the A–D converter of the computer. Because one can use, for this purpose, the same A–D system that allows stimulus-driven direct acquisition, this saves the cost of a dedicated board and frees a slot on the computer bus. It also allows one to choose, retrospectively, the filter band-width and sampling rate for the transcribed data. Although there is, in principle, some loss of fidelity in the repeated D–A and A–D conversion, the resolution in these systems is so good that this is not a concern for most purposes. In general, the signal-to-noise ratio is limited by the characteristics of the preparation and the patch or voltage clamp head stage (e.g., see discussion of noise in single-channel recording systems by Wonderlin et al.[5]). Once transcribed onto the computer disk, the data can be conveniently edited and analyzed at will. Transcription usually embodies the first stage of editing and saves some time during the later analysis, because only those portions of the record that contain data relevant to the subject being analyzed need be transcribed. Also, it may not be necessary to transcribe all of the data at the maximum sampling rate

[3] F. Bezanilla, *Biophys. J.* **47**, 437 (1985).
[4] T. D. Lamb, *J. Neurosci. Methods* **15**, 1 (1985).
[5] W. F. Wonderlin, A. Finkel, and R. J. French, *Biophys. J.* **58**, 289 (1990).

and bandwidth, thus saving disk space and later processing time. For example, channel block by high affinity toxins can be analyzed at a fraction of the bandwidth required for the rapid intrinsic gating processes of many channels.

Direct to disk recording will undoubtedly become a more widely used option, even for continuous acquisition, as high capacity, fast storage media become more economical. Retrieval and editing of data, with a program such as Axotape, are much quicker when the data are acquired directly to either a magnetic or optical computer disk. However, it is necessary that the primary archival storage contain the data sampled at the maximum rate needed for any of the possible analyses, and thus the demands on storage are considerable (about 7 Mbytes/hr at 1 kHz) when continuous data acquisition is necessary.

Event-Driven Acquisition

Event-driven acquisition can be an economical way of collecting what is essentially a continuous record. Event-driven acquisition is most useful for widely spaced, randomly occurring events such as quantal responses at synapses and in photoreceptors and for recording occasional bursts of activity by single channels that show a low overall open probability. Event-driven acquisition[6] provides a pseudocontinuous record if the time intervals between periods of acquisition are recorded. This can drastically reduce storage requirements for activity occurring in infrequent bursts, but it entrusts the recognition of events to an automated event detector, with preselected criteria (generally a combination of minimum amplitude and duration) for event acceptance. Event-driven acquisition is most effectively implemented in conjunction with a circular memory buffer that enables the acquisition system to "look back" in time and record the interval immediately preceding, as well as that following, the detected event.

Analysis

Although there is some variability among the ways that different packages are divided into functional modules, we shall arbitrarily divide analysis into two groups of operations: (1) A "preprocessing" stage in which the data are viewed, filtered, and sorted and (2) the quantitative analysis of identified components of the ionic current.

[6] F. J. Sigworth, in "Single-Channel Recording" (B. Sakmann and E. Neher, eds.), p. 301. Plenum, New York and London, 1983.

Overview: Some General Requirements of Analysis Software

One essential in programs for analysis of either macroscopic or single-channel data is the facility for fast and flexible visual examination of the data. The program(s) should be able to display single or multiple records with easy control over the amplitude scaling and time base. It should be possible to sort records into groups (e.g., separating blanks from active single-channel sweeps) and to exclude from the analysis any that might be excessively noisy, or otherwise unsuitable. Careful prescreening of the data can save time in later, more computationally intensive and time-consuming sections of the analysis by allowing them to be automated more than might otherwise be safe.

Other general preprocessing operations common to analysis of macroscopic and single-channel records might include baseline subtraction, digital filtering, and making cursor-controlled measurements on individual records. Such measurements may be used to check specific points or to provide setup information for later parts of the analysis.

Fundamentally, processing of records requires efficient arithmetic operations on arrays. The arrays may consist of either complete records or sections of records definable by timing, cursor placement, or occurrence of marker events such as trigger pulses. Necessary operations range from simple point by point addition or subtraction of arrays stored in a memory buffer to multiplication by a constant or more complex operations such as Fourier transformation to allow generation of power spectra, or digital filtering to reduce noise. Digital filtering offers a powerful way to optimize signal processing and simplifies data collection. Data may be collected uniformly at the highest bandwidth likely to be used, and the digital filter can than be set optimally for different sections of the experiment depending on the amplitude and kinetics of the signal, both of which may vary significantly with voltage or environmental conditions. More detailed discussions of the strategies for optimally filtering records, and of the effects of filtering on single-channel event detection may be found elsewhere.[1,7,8]

Simple statistical and sorting functions to determine such quantities as means, standard deviations, and minima and maxima of a record segment should be readily available. Records should be locatable by their position in a protocol, or accessible for selection based on inspection; these two options offer both efficiency and flexibility in analysis.

After selecting appropriately filtered records that reflect a chosen aspect of channel behavior, details of procedures for macroscopic and single-

[7] D. Colquhoun and F. J. Sigworth, *in* "Single-Channel Recording" (B. Sakmann and E. Neher, eds.), p. 191. Plenum, New York and London, 1983.
[8] O. B. McManus, A. L. Blatz, and K. L. Magleby, *Pfluegers Arch.* **410,** 530 (1987).

channel records differ. However, there are conceptual parallels between the isolation, from a macroscopic record, of a component due to a particular channel type, and the step of event recognition in single-channel analysis. Parallel roles are also played by the conductance determinations from amplitude measurements on either macroscopic or single-cell channel data, and by kinetic measurements based on fits to either macroscopic transients or single-channel open- and closed-time distributions. We elaborate further on the handling of macroscopic and single-channel data in the following paragraphs.

Analysis, especially of single-channel records, is time consuming, and many procedures require careful visual monitoring. Nonetheless, many parameters required to set up the details of an analysis protocol can be usefully saved in files to be recalled by a keystroke or the touch of a mouse button and reused with only minor editing for analysis of similar data sets. This tactic, similar to the saving of parameter files which define stimulus and acquisition protocols, can save much time and allow the investigator to focus attention on less routine decisions.

Analysis of Ion Channel Currents

The first step in analysis of either macroscopic or single-channel records is to recognize and isolate the ionic current of interest. This usually requires digital subtraction of other components following one or another of the tactics that are outlined below.

Macroscopic and Gating Currents. To isolate current associated with a specific type of channel, the following components must be removed from the record: transient capacitive currents that charge or discharge the membrane when the command voltage is changed; currents through nonspecific leakage conductances; and currents through other well-defined ion channels. Leakage and capacitive current components are usually assumed to depend linearly on voltage and thus can be removed by addition/subtraction of the current elicited by a voltage step of similar size, applied in such a range that no channels are activated. Linear capacitive and leakage currents may also be removed by a divided pulse procedure.[9] Alternatively, the experimental voltage protocol may be repeated after blocking all the known ion channels and the resulting records subtracted from corresponding records in which there was no block. Specific channel currents may be blocked pharmacologically, or in some cases may be removed by a voltage protocol that selectively inactivates one channel type. Isolation of a specific current component, then, requires software that performs array arithmetic on selected records.

[9] F. Bezanilla and C. M. Armstrong, *J. Gen. Physiol.* **70**, 549 (1977).

Current–voltage data are usually extracted from macroscopic voltage step–response records in one of two ways: (1) by extracting maxima or minima (to give a peak I–E relation) or (2) by measuring amplitudes at a fixed time (isochronally), usually either as soon as possible after a perturbation such as a voltage step (to yield an instantaneous I–E) or at very long times (to yield a steady-state I–E). In some cases, it may be possible to record complete I–E relations directly as the response to a voltage ramp command (cf. single-channel section below), but this is not generally useful for voltage- and time-dependent channels.

Kinetic parameters generally are determined from macroscopic records by fitting the time courses to a theoretical curve consisting of a sum of exponential terms and a constant. Because this process will be repeated many times for each experiment, the processes of selecting the fitting function and the record segment to be fitted, and the setting of starting parameters should be made as efficient and convenient as possible.

Single-Channel Records. Single-channel events are usually recognized by their specific and reproducible size. Thus, in steady-state continuous recordings, the only current separation required is to subtract the current level corresponding to the closed state from the record. In stimulus-evoked responses, this background leakage current and the capacitive current are usually subtracted together in a single operation.

(a) Continuous (or Event-Driven) Data. Analysis is usually divided into, first, direct operations on the data records and, second, processing of "events lists" consisting of tabulated parameters describing the single-channel events identified in the record.

(i) Measurement of Amplitudes and Open/Closed Probabilities, and Event Detection. Two common calculations are performed on raw records. All data points may be binned by current level to form a histogram whose identifiable peaks indicate discrete open or closed levels. Fitting these distributions to a sum of Gaussian components yields these levels, the standard deviations, and the number of points falling in each component. These last mentioned parameters allow calculation of the fraction of time spent in each level. The other general operation performed on the current records is to detect the transitions between the discrete levels, usually by monitoring crossings of a threshold set at 50% of the distance between adjacent levels. Amplitudes, noise standard deviations, and durations of sojourns in each successive level are then stored in an events list for later processing. Event detection may be automated or may be performed under continual visual inspection with the option to override the decisions of the program. Colquhoun[10] makes a graphic argument for the latter approach;

[10] D. Colquhoun, *Trends Pharmacol. Sci.* **9,** 157 (1988).

Sachs[11] gives an overview of some issues involved in automating single-channel analysis. Our preference is to have both options available, and to be able to switch instantly between automatic detection and visual accept/reject operation during the course of analysis. This flexibility makes it much easier to avoid such errors as having the audio pickup of someone coughing being unwittingly interpreted by the program as an agonist-induced burst.

A useful way to monitor the stationarity of behavior is to calculate the mean current on an epoch-by-epoch basis. When normalized by dividing by the single-channel amplitude, i, this gives the product nP_o, where n is the number of channels present and P_o is the mean open probability. Plots of nP_o provide a convenient way of monitoring slow changes in gating (e.g., the "modal" gating of Ca^{2+} channels studied by Hess et al.[12]). Verification of stationarity, using plots of nP_o versus time, also gives an indication of whether it is valid to lump different segments of record together for analysis.

(ii) Analyzing the Events Lists. A variety of manipulations of the events lists are possible. With no further editing, distributions of event amplitudes (as distinct from the all-points current level histogram), or of dwell times in each conducting state, may be constructed. For a single, archetypal two-state (open or closed) channel, three basic kinds of information are provided, namely, open times, closed times, and single-channel amplitude. If kinetics are more complicated, additional processing might involve sorting the events into bursts separated by long closures, and perhaps even groups of bursts into longer epochs. This depends on determining criteria for distinguishing between intraburst events and interburst intervals, and these decisions are best made on the basis of examination of the complete dwell time histograms, preferably on a logarithmic time scale to allow event durations over a very long time scale to be grouped into separate exponential components.[13] The distributions of event durations, burst durations, or interburst intervals are generally fitted with theoretical distributions, composed of sums of exponential terms, in order to reconcile the channel gating fluctuations to a Markovian model.

(b) Evoked (Stimulus-Driven) Data. Responses evoked by steps in voltage, from defined initial conditions, allow study of channels that are activated only transiently in response to a stimulus, yielding unique information on the kinetics of channel gating and open channel properties.

[11] F. Sachs, in "Single-Channel Recording" (B. Sakmann and E. Neher, eds.), p. 265. Plenum, New York and London, 1983.
[12] P. Hess, J. B. Lansman, and R. W. Tsien, Nature (London) **311**, 538 (1984).
[13] F. J. Sigworth and S. M. Sine, Biophys. J. **52**, 1047 (1987).

Application of a voltage ramp command offers a quick way of collecting $i-E$ relations that reflect open channel properties.

(i) Step Responses. Separation of single-channel currents from capacitive and leakage components in voltage step-evoked records may often be accomplished by subtraction of blanks — records in which, by chance, no openings occurred. Blanks are located, averaged, and then the average is subtracted from individual records that show openings. An alternative procedure is to fit the average blank with a smooth curve and then subtract this noiseless curve to remove the idealized leakage and capacity components. As addition or subtraction of records increases the noise level, the latter procedure has the advantage that it yields a record of the isolated ionic current without additional noise.

Events lists may be constructed and processed from step response data as for continuous steady-state records, but one specific kind of analysis has received particular attention in the study of inactivating voltage-activated channels. This involves analysis of the first latency distributions, that is, the distributions of times to the first channel opening after the voltage step. It is this distribution that gives the most direct information about the rates of transition from the closed states to the open state(s), uncontaminated by transitions from inactivated states which the channel might enter later in the pulse.

(ii) Ramp Responses. A relatively slow voltage command ramp elicits a changing current output which is the sum of resistive open channel and leakage currents plus the capacitive current that is required to change the membrane potential. If a channel opens or closes during the course of the voltage ramp, the current amplitude will jump suddenly and the slope will change. The open channel $i-E$ is obtained directly by subtraction of the ramp response when the channel is absent or closed. In practice, during any ramp sweep, there are often several open–closed transitions. Thus, to extract the open channel $i-E$ requires the editing of the ramped sweeps into open and closed segments. This is usually done under cursor control, but for good quality records the procedure could be automated, recognizing transitions as a discontinuity in the derivative of the current ramp. The separated open and closed current levels are then averaged at each voltage point for an arbitrary number of sweeps, and the open channel $i-E$ relation obtained by subtracting the closed level from the open.

Manipulating Databases

The above analyses generate a variety of data to be plotted, fitted, and perhaps processed further. Many general purpose spreadsheet, graphics, curve-fitting, and statistics programs are available, and one of these may provide the only readily available option when some unanticipated need arises. However, fitting and plotting of such things as amplitude and dwell time distributions is required so often that these functions should be highly efficient and accessible with the minimum of input from the user. Routines for these purposes are essential to any single-channel analysis package.

Meeting Software Needs

In setting up a laboratory acquisition and analysis system one can choose, at one extreme, to code from scratch a polyglot of programs to handle various needs as they arise, or one can buy a commercial package and hope that it will handle all of the computing chores without the need for in-house programming. There are a variety of options in between. Choosing involves making compromises and balancing competing demands to get established quickly, to collect data and analyze it efficiently, and from time to time to export data to other programs for further analysis. Some of the options in a search for efficiency, flexibility, and compatibility are mentioned below. Different software packages are listed in Tables

TABLE I
APPLICATIONS PACKAGES OFFERING BOTH ACQUISITION AND ANALYSIS

Package	Company[a]	Notes
Bio-Patch Bio-Clamp	Bio-Logic	Acquisition and analysis of patch clamp or voltage clamp data
Strathclyde	Dempster/Dagan	Five modules for evoked, continuous, and event-triggered acquisition and analysis, including spectral analysis
M2 LAB	Instrutech	For the Atari system. A variety of programs, including versions customized to work conveniently with several different commercial patch clamps
Patch	Med Systems	Two programs, one for recording and idealizing records, the second for subsequent analysis
Patch & Voltage Clamp	CED	Integrated system operating from a single menu for evoked and continuous acquisition and analysis
pCLAMP	Axon	A suite of six programs. Acquisition, analysis, and fitting for continuous or evoked data
Satori, Active	Intracel	Acquisition and analysis of patch clamp or voltage clamp data

[a] See Table IV for locations.

TABLE II
SPECIAL PURPOSE APPLICATIONS PROGRAMS

Program	Company[a]	Notes
AXOTAPE	Axon	Data files are in pCLAMP (FETCHEX) format
IPROC	Axon	Processes pCLAMP (FETCHEX) or binary data-only files. Output processible by pStat
LPROC	Axon	Operates on pCLAMP (FETCHAN) or IPROC events lists. Programmable analysis of bursting kinetics
CSIM	Axon	Markov simulation of single-channel fluctuations. Data files may be generated to test analysis routines

[a] See Table IV for locations.

I–III, and addresses of companies whose products are mentioned are given in Table IV.

Packaged Systems

Since personal computers and programmers began to proliferate, an investigator assembling a laboratory to study ion channels has had available progressively more options from which to satisfy software needs for data acquisition and analysis. For those not wishing to reinvent the analysis program, several software packages are available. These range from tape

TABLE III
DEVELOPMENT SYSTEMS

Package	Company[a]	Notes
ASYST	Asyst	Optionally program in RPN or in algebraic notation. Array operations speed executions and simplify programming. Different modules are available separately. Accesses LIM expanded memory
AxoBASIC	Axon	QuickBASIC environment, enhanced with array operations and graphics and acquisition functions. Commands parallel BASIC-23 as far as possible in the DOS/PC environment
BASIC-FASTLAB	Indec	QuickBASIC environment, similar philosophy to BASIC-23
C-LAB II	Indec	C language-based system, with added high level commands for acquisition and analysis
KHF BASIC	Instrutech	BASIC-23 emulator for the Atari

[a] See Table IV for locations.

TABLE IV
SOURCES OF SOFTWARE

Abbreviated name	Company name and address
Asyst	Asyst Software Technologies, Inc., 100 Corporate Woods, Rochester, NY 14623
Axon	Axon Instruments, Inc., 1101 Chess Drive, Foster City, CA 94404
BioLogic	Bio-Logic, 4 rue Docteur Pascal, Z.A. du Rondeau, 38130 Echirolles, FRANCE (in the United States contact Molecular Kinetics, Inc., P.O. Box 2475 C.S., Pullman, WA 99165-2475
Strathclyde	Dr. John Dempster, University of Strathclyde, Glasgow G1 1XW, UK (in the United States contact Dagan Corporation, 2855 Park Avenue, Minneapolis, MN 55407)
IMSL	IMSL, 2500 Permian Tower, 2500 City West Boulevard, Houston, TX 77042-3020
Indec	Indec Systems, Inc., 1283A Mt. View/Alviso Road, Sunnyvale, CA 94089
Instrutech	Instrutech Corp., 425 Meacham Avenue, Elmont, NY 11003
Intracel	INTRACEL Limited, Unit 4, Station Road, Shepreth, Royston Herts SG8 6PZ, United Kingdom
Jandel	Jandel Scietific, 65 Koch Road, Corte Madera, CA 94925
MathSoft	MathSoft Inc., 201 Broadway, Cambridge, MA 02139
Med Systems	Medical Systems Corp., One Plaza Road, Greenvale, NY 11458
CED	Cambridge Electronic Design, Science Park, Milton Road, Cambridge CB4 4FE, UK

emulators with powerful data review and plotting capabilities to comprehensive acquisition and analysis packages designed specifically with the electrophysiologist in mind. For those who demand flexibility but would generally prefer to avoid dealing with devices on a bit-by-bit basis, several development systems using a selection of programming languages are available. Most of these include application programs that can be used as is or as examples of solutions to programming problems that routinely arise.

We deliberately choose not to give a detailed list of features of the programs listed. Given the pace of continued development, such a list would be far out of date by the time of printing. Rather, we hope to give an idea of the range of possibilities and of some useful sources from which further inquiries can be made. With the exception of M2 LAB (including KHF BASIC, a BASIC-23 emulator) which runs on the Atari ST (based on a Motorola 68000 series microprocessor), these packages generally require an IBM PC/AT compatible computer (based on Intel 80286, 80386, or 80486 microprocessors) with at least an EGA display screen and a suitable hardware interface. More options are also becoming available for the Apple Macintosh family of computers, but these have been less widely used in laboratories to date. Individual companies should be contacted regarding the specific interfaces that are supported and their current specifications.

The Atari system offers an unique opportunity in that it is the computer used to control the revolutionary EPC-9 patch clamp, making a "no knobs," completely software-controlled patch clamp and acquisition system. At the time of writing the software is still being refined, but this integrated system, developed with enormous technical resources, will offer a versatile option for the various modes of patch clamping in the future.

Turnkey Programs (Axotape, pClamp, Applications Programs Provided with Development Systems). Although turnkey programs offer the quickest way to get experiments started, the term should not be taken too literally. Not only must one budget some time to learn the options available and how best to use them, but tests of specific calculations that will be used should always be run with known data files when beginning to use such a complex system. Ready-to-run programs are available for prices in the range of approximately $U.S.1500–2000. Special purpose modules generally cost from about $200 to a few hundred dollars. General features are summarized in Tables I, II, and III.

Development Systems (AxoBASIC, ASYST, BASIC-FASTLAB, C-LAB). Development systems to suit a variety of linguistic tastes are available (see Table III). Several come with application programs specifically designed for electrophysiological studies, and another (ASYST) has been aimed at a broader market. AxoBASIC and KHF BASIC emulate the highly successful BASIC-23 system written to run under the Digital Equipment Corporation RT-11 operating system. All of these systems have in common the philosophy of providing array-oriented mathematical operations and high-level commands to control data acquisition devices to facilitate program development. The code for these operations is generally written in assembly language to speed execution. Also generally true is that both interpretive (for quick, interactive testing and development) and compiled operation (for more rapid execution of completed programs) are available. However, as compilers and debuggers become more user friendly and efficient, and the complexity of many programs demands compilation for sufficiently rapid execution, the need for a truly interpretative mode of operation seems to us less pressing than it did a few years ago. Several companies now offer integrated programming environments which include editing, compiling, and debugging capabilities that allow optimized, compiled code to be generated almost as easily and directly as a program in an interpretive system. Interpretive systems, however, still offer an unmatched flexibility to respond, in mid experiment, to circumstances unforeseen when the program was written.

Building Blocks

For those who wish to design their own data-handling software, the development process can be made vastly more efficient and less tedious than it might have been a decade ago. The number and power of useful options in languages and libraries has burgeoned, greatly facilitating the task of creating special purpose software. Although the most direct approach to "writing your own" would be to adopt one of the development systems already mentioned, those systems can provide useful guidelines even if one wishes to begin closer to the base of the design process. Useful ideas are to be found in the choices others have made on the following issues. One must decide what language processor to use. Can the best combination of ease of development and testing and powerful and efficient operation be obtained with an interpreter–compiler system, or with a new generation compiled language that offers rapid and easy compiling, running, and debugging of programs? What libraries are available to speed up the assembly of a larger program? At some point, if experiments are to be done, the decision will be made to rely, at least in part, on someone else's code. Libraries of routines are available for operations ranging from array-oriented mathematics to graphics, creation of menus, cursors and icons, and the internal "housekeeping" required in TSRs (terminate-and-stay-resident routines[14]). These resources range from the widely used and continually updated International Mathematical and Statistical Library (IMSL) to the inexpensive, unsupported, but nonetheless very useful *Numerical Recipes* series.[15] Although there may be satisfaction in programming one's own digital filter, interpolation, or curve-fitting algorithms, it is not necessary to recreate such items, nor, at present, does it have to be expensive to save oneself this labor (see also Colquhoun and Sigworth[7]).

Simulation

Programs that generate simulated single-channel records serve two purposes. They allow exploration of the behavior of well-defined schemes that might represent the kinetic basis of experimental observations. In addition,

[14] M. Davidson, *Comput. Language,* 7, 127 (1990).
[15] W. H. Press, B. P. Flannery, S. A. Teukolsky, and W. T. Vetterling, "Numerical Recipes in C: The Art of Scientific Computing." Cambridge Univ. Press, Cambridge and New York, 1988.

they are a source of data based on known parameters that can be used to test the effectiveness of analysis software.

Exploring Models

CSIM is a powerful Markov simulator program that allows fluctuations to be calculated for Markov kinetic schemes with up to 12 kinetic states, each having its own conductance and reversal potential. Step responses may be simulated by defining matrices of rate constants to be applied before and after a step in, for example, voltage or agonist concentration. Realistic noise simulation can be obtained by defining multiple conducting states, or by adding white noise to the simulated record. The complications of recording from multiple channels of one or several channel types may be explored. Whole-cell records can be simulated by defining up to 12 models with as many as 100 channels of each type simultaneously active. Output is generated optionally as a data-only binary file or in FETCHEX format for ease of analysis. ASCII histogram files of the dwell times are also generated. Such a simulation program offers a convenient and powerful means of testing the possibility that specific models might account for particular experimental observations.

Testing Analysis Programs

No analysis program, no matter who the programmer, should be used without thorough testing on data of the type with which it will be challenged. Such testing is most easily done with simulated data which can be noise free and generated from known parameters. The ability of the analysis routines to extract the original rate constants and conductance levels from the data can be precisely tested for records of arbitrary length. Complications of adding noise can be directly examined.

Compatibility and Portability: General Purpose Software (CAD Systems, Graphics, Curve-Fitting, Spreadsheets)

Although customized routines are essential for computationally intensive tasks that are performed repetitively, research problems inevitably lead us to look for new relationships that suggest themselves as we sift through megabyte on megabyte of data. To handle the unforeseen or simply occasional tasks that do not justify the demands on programming time or the memory space that would be needed to include them in a specialized analysis package, many general purpose commercial software packages have proven invaluable. These include programs for graphics and curve-fitting, notebook format calculations, and spreadsheet and database systems. Easy interchange of files, in compatible formats, should be a central criterion when considering such programs. Critical reviews of these popular programs appear regularly in magazines devoted to personal computing, such as *BYTE, PC World,* and *Computer Language.*

Among the programs that we have found extremely useful in these categories are Sigmaplot (graphics and curve-fitting) and MathCAD (notebook format computations and easy data entry and extensive mathematical functions, including equation solving and array operations). Also a variety of mathematical programs that perform complicated symbolic operations as well as numerical calculations are now available. In some research projects, there will be a daily need for spreadsheet and data base programs. And of course, beyond writing manuscripts, a word processor or editor with windowing capability allows easy examination, and cutting and pasting of data files that are in ASCII format. For example, editing header information may be required to transfer data between programs.

Overview: Choosing a System

We present below a brief checklist of questions intended to elucidate the requirements for an acquisition and analysis system, including both hardware and software. We hope that an adequate framework for finding individual answers to these questions will be found in the body of this chapter, within its attendant references, and in the other chapters in this section.

Acquisition: What is the duration of data collection, and what is the range of sampling rates needed? How many channels of data input are needed? Must acquisition on different channels be synchronous, or may the channels be multiplexed? What resolution in the digitization of data and stimuli is needed? (Twelve-bit digitization is most common, lower resolution is generally not adequate, and up to 16-bit resolution is available and may be useful if a wide dynamic range is needed.) Are data to be stimulus evoked or collected continuously? What form of stimulus signals are required? Is more than one analog stimulus channel required? For example, do any ancillary instruments require a command signal with a continuously variable amplitude (e.g., electrophoretic drug delivery), or is a binary logic signal (e.g., to control the off–on positioning of a valve) all that is required? How many independently timed logic signals are required to synchronize various instruments with the acquisition system?

The answers to these questions will ultimately set limits on the kinds of hardware that will be satisfactory, but, having chosen suitable hardware, one must also ensure that the available software provides control over all features that are to be used. This is not an assumption that can be taken for granted, even when purchasing an "integrated" hardware/software package from a single company.

Analysis: The following questions should help to delimit the requirements for analysis software. Are macroscopic records, single-channel data, or both to be analyzed? Must records with more than two conductance levels be analyzed? (Some programs handle only records that represent a single open–closed channel, i.e., one exhibiting only two states based on conductance, closed and open. Programs handling more than two conductance levels are required in order to analyze records from several two-state channels, or from single channels with several conductance substates.) Is noise analysis, either of power spectra by Fourier transform methods or of evoked records by nonstationary fluctuation analysis, needed? Are single-channel data to be collected continuously in the steady state, or episodically evoked by stimuli? (In the latter case, nonstationary methods, such as first latency analysis, will probably be required.) What level of interactive control is required at different stages of analysis and acquisition? Are analysis procedures well defined and fixed, or is the flexibility to explore many possibilities required? (In the latter case, the use of a development system for generating programs from more basic tools may be the approach of choice.) Are there special hardware requirements for data input/output or display? If so, does the software generate output in a device-independent, easily utilized form, or communicate directly to the selected devices? Will channel analysis output be used to assemble large databases for further statistical analysis, requiring easy interchange of data among different programs?

Acknowledgments

The authors are members of the Department of Medical Physiology and the Neuroscience Research Group, Faculty of Medicine, University of Calgary, Alberta, Canada. The research was supported by grants from the Medical Research Council of Canada (Grant MA-10053) and from the Heart and Stroke Foundation of Alberta. The Alberta Heritage Foundation for Medical Research provided support in the form of a Scholarship (R.J.F.) and a Fellowship (W.F.W.).

[51] Stationary Single-Channel Analysis

By MEYER B. JACKSON

Introduction

The opening and closing of ion channels in artificial and cell membranes generates a signal of membrane current that varies in a stepwise manner with time. The stochastic manner in which this signal varies depends on the underlying channel-gating mechanisms. Thus, analysis of records of single-channel current can provide much insight into the mechanisms by which channels open and close. Because channel gating is a stochastic process, the methods of analysis are statistical.

This chapter describes a number of useful methods of kinetic analysis of stationary single-channel data. In a stationary system, the properties of the ensemble do not change with time; the mean probabilities of occupancy of specific configurations remain constant. Such a system must satisfy a condition of detailed balance, in which the frequency of transitions toward a particular configuration is equal to the frequency of transitions out. In nonstationary systems, detailed balance is not satisfied, and the mean probabilities of occupancy of a particular configuration change with time. The practical implication is that if the data are stationary, one simply can collect the data continuously for analysis. Nonstationary conditions are often achieved by perturbing the system, most commonly with changes in voltage. Methods for the analysis of nonstationary single-channel data are presented elsewhere (see [52], this volume).

Single-channel kinetic analysis can be divided into three distinct steps. The first step involves the theoretical derivation of predictions of kinetic models. Examples of such theoretical predictions include the probability distribution function for open times and the joint probability density function for the lifetimes of two consecutive events. The second step of single-channel kinetic analysis is to extract something from real data that can be compared to theoretical predictions. These operations generally start with raw single-channel data and generate open time and closed time histograms, or more complicated distillates of channel behavior. One thus obtains an experimental result in a form that can be directly compared with a theoretical prediction. This comparison, which most often involves some form of curve fit, is the third and final step of single-channel kinetic analysis. The insights into channel gating processes are obtained through a comparison of distillates of single-channel data with the theoretical predictions of models.

METHODS IN ENZYMOLOGY, VOL. 207

Single-channel analysis is very software intensive, and a large number of computer programs have been developed in many laboratories and by many companies. Many software packages are commercially available. The number of programs is so great, and the purposes so varied, that it is not possible to evaluate and compare them. The emphasis is rather on the specific tasks for which the programs are written. It is hoped that after reading this chapter, the reader will know what to expect from the software, how to use it intelligently, and what are the possibilities for developing programs for more specialized needs.

Summary of Kinetic Theory

The most widely used theories of single-channel kinetics treat channel gating as a finite Markov process. The complex trajectory of a channel through its protein conformational space is modeled as a series of jumps or transitions between states within which the conductance is relatively constant. Such models for channel gating then consist of a finite number of states that are either open, closed, or have an intermediate conductance. Each state can then interconvert to another with a finite rate. This is represented in a kinetic scheme by an arrow between the two states that can interconvert. When a rate is zero in both directions, then the two states can be considered unconnected. Within this framework, the relevant questions about a channel-gating mechanism are the following: (1) How many states are there with a given conductance? (2) How are they connected? (3) What are the rate constants for each step?

This description defines the term "state," which is used extensively in the ensuing discussion. Relative to the time scales accessible in a single-channel experiment (> 20 μsec), the transitions between states are instantaneous, whereas the states themselves are stable enough to observe. According to the Markov assumption, the probability of a particular transition depends only on which state the system is in. This model for channel gating corresponds to a specific physical picture which at the present time appears to be consistent with essentially all experiments on channels.

One of the most useful results of the kinetic theory of channel gating provides a method of estimating the number of states with a particular conductance. Consider a membrane system with a single channel having n open states and m closed states. The open time and closed time distributions are represented by sums of n and m exponentials, respectively:

$$P_o(t) = \sum_{i}^{n} x_i \, e^{\mu_i t} \tag{1a}$$

$$P_c(t) = \sum_{}^{m} y_j \, e^{v_j t} \tag{1b}$$

where $P_o(t)$ and $P_c(t)$ represent the probabilities of open times or closed times lasting a time t or longer.[1,2] The uppercase symbols are used for the probability distribution, whereas the probability density function, obtained by differentiating the distribution, is denoted by lowercase letters. The parameters μ_i, v_j, x_i, and y_j depend on the rate constants and on the details of how the open and closed states are connected in the gating scheme. However, the form of Eqs. (1a) and (1b) does not depend on these details and is independent of the parameters and connectivity of the gating scheme. In effect, when one determines the number of exponentials that are needed to fit an observed distribution, one has "counted" the number of states with a particular conductance. More precisely, this method provides a lower bound to the number of such states, since one is always faced with the possibility that some of the terms of Eqs. (1a) and (1b) are too small, too rapid, or too slow to detect. It is also possible that the system has two time constants that are too similar to be resolved.

For even three closed or open states the values of the time constants and coefficients in Eqs. (1a) and (1b) can be very complicated functions of the rate constants for transitions between states. Thus, for complex gating mechanisms it is often very difficult to go beyond state counting to a quantitative determination of specific rate constants. With complex gating behavior, it is often so arduous to determine the rate constants that, depending on the questions at hand, it may be prudent to focus on simpler quantities such as mean open time, mean closed time, or mean burst duration.

Equations (1a) and (1b) are examples of one-dimensional probability distributions; they are functions of a single experimental variable. One can also make use of higher dimensional probability distributions and density functions. For a closed time probability density function $p_c(t_c)$ and an open time probability density function $p_o(t_o)$, where t_c and t_o are independent (no correlations), the joint, two-dimensional probability density function for observing both t_o and t_c is the product $p_o(t_o)p_c(t_c)$. If the values of t_o and t_c are correlated, then the joint probability density function for consecutive open and closed times is no longer simply the product of the two one-dimensional probability density functions. This is a qualitative difference, which can be used to estimate the number of distinct gating pathways and to make distinctions between these pathways. In general, correlation anal-

[1] D. Colquhoun and A. G. Hawkes, *Proc. R. Soc. London, Ser. B* **199**, 231 (1981).
[2] D. Colquhoun and A. G. Hawkes, *in* "Single-Channel Recording" (B. Sakmann and E. Neher, eds.), p. 135. Plenum, New York, 1983.

ysis is used more to answer questions about the basic topology or connectivity of a channel-gating mechanism, rather than to estimate the values of specific rate constants.

One important example of correlation analysis is highly analogous with state counting described above, in that it provides an estimate of the minimum number of gating pathways of a channel. This arises from the theoretical result that the number of terms in the covariance function of two event lifetimes as a function of the number of intervening time intervals is equal to one less than the number of distinct pathways between the open and closed states. The number of pathways is defined as the minimum of the number of closed states that can open and open states that can close.[3,4] Thus, when one determines the number of exponentials that are needed to fit an observed covariance function, one has "counted" (or, more precisely, estimated a lower bound to) the number of gating pathways.

Correlation and covariance functions can be derived with higher dimensions (i.e., with more time intervals). However, there is no additional information about the underlying channel-gating mechanism in these higher dimensional functions.[3] Thus, they have never been used in single-channel analysis. As the number of states of a model increases, the number of kinetic parameters and the number of topological possibilities for connecting the states increase. Single-channel analysis is limited in the number of parameters that can be determined, and in the power to discriminate models.[3,4] Nevertheless, a great deal of information can be obtained purely from statistical analysis of single-channel records. To go beyond the limits circumscribed by single-channel kinetic theory, it is necessary to combine the stationary kinetic analysis with other approaches, such as nonstationary kinetic analysis and dose–response behavior.

Idealization of Single-Channel Data

The first step in single-channel kinetic analysis is the idealization of the data. Raw single-channel data represent the superposition of background noise and stepwise changes in channel current. When the data have been idealized, the background noise has been eliminated and only the stepwise changes caused by channel opening and closing remain. With an ideal channel record, no information is lost by replacing the current versus time signal with a two-column list consisting of a time and a current for each

[3] D. R. Fredkin, M. Montal, and J. A. Rice, in "Proceedings of the Berkeley Conference in Honor of Jerzy Neyman and Jack Kiefer" (L. M. Lecam and R. A. Olshen eds.), Vol. 1, p. 269. Wadsworth, Belmont, California, 1985.
[4] D. Colquhoun and A. G. Hawkes, Proc. R. Soc. London B **230**, 15 (1987).

interval at a particular current level. Idealization is then the reduction of a single-channel data record to such a list.

The ease with which single-channel data are idealized is critically dependent on the ratio of the single-channel current amplitude to the background noise. For noisy data, no ideal idealization method exists. Idealization can, and often is, done by hand, or with a computer in a semiautomatic, event-by-event mode. In this mode the computer will search the data for threshold crossings and then display the event, indicating guesses for the precise time of the transition and the levels. A person running the program is then given the opportunity to accept, reject, or modify the event and the initial guesses. Examination of channel current data event by event is essential with data from a new type of experiment where the basic properties are unknown. When one attempts an event-by-event analysis, even of fairly good data, one encounters ambiguous events. There is always a chance that a random background fluctuation will occur that is comparable in size to the single-channel current. For poor data where the channel current is only 3 or 4 times the peak-to-peak background noise, such fluctuations will occur frequently enough to make idealization very difficult. Ambiguities arise more often for brief intervals.[5]

Analyzing records with thousands of events in an event-by-event mode is simply too time consuming to be useful if a systematic study is undertaken. A fully automated computer program for idealization is then required. Such programs are available from Axon Instruments (Foster City, CA) and Instrutech (Elmont, NY), as well as from other companies. Most programs allow the user to toggle between two modes, one of which is a semiautomated, event-by-event analysis by the operator. The other mode is fully automated. This toggling is an essential feature, because it is necessary to check the performance of these programs. The lists of numbers produced by a fully automated idealization program can be very deceptive. Glitches in data, such as those caused by power surges, seal breakdown, or bubbles in the perfusion system, will produce crossings of the threshold used to signify channel-gating transitions. Such events are easily recognized as spurious by the human eye, owing to their nonstepwise character. In an event-by-event analysis, an investigator can almost always avoid such mistakes, but to most computer programs, all threshold crossings look the same.

Another source of false transitions has been described by Colquhoun.[6] If the records contain subconductance states (or for that matter infrequent appearances of another smaller channel) that are near the threshold for

[5] D. Colquhoun and F. J. Sigworth, in "Single-Channel Recording" (B. Sakmann and E. Neher, eds.), p. 191. Plenum, New York, 1983.
[6] D. Colquhoun, Trends Pharmacol. Sci. 9, 157, (1988).

event detection, a single long dwell time at this level will produce a large number of very brief events in the output list. To avoid these and other errors, the raw data should be spot checked, and the output lists should be examined for anomalous behavior. The performance of idealization programs should be compared for different values of the event detection parameters. The amplitude and lifetime distributions should be checked visually with every data set for unusual peaks or components. An anomalous excess of points somewhere in the distribution can almost always be traced to a problem in the data, and editing these events will improve the quality of analysis. As Colquhoun succinctly put it, "single-channel analysis costs time,"[6] and the temptation to turn a software package loose, and publish the results without careful inspection and monitoring along the way, should be resisted.

Because idealization is the stage of analysis where the noise level will have its primary impact, the bandwidth and sampling frequency should be selected to optimize the quality of idealization. The Nyquist criterion of digitization at twice the corner frequency of the low-pass filter is inappropriate for single-channel analysis.[5] Most investigators have determined empirically that one should digitize at 8 to 10 times the corner frequency of the filter. At an early stage in a study, a data set should be analyzed with different bandwidths and digitization frequencies to see where problems develop. When a channel exhibits gating activity on a rapid time scale, the bandwidth must be widened to observe these processes, but if the noise increases enough to impede the idealization, then better data form the best solution.

Curve-Fitting Lifetime Distributions

Once data have been idealized it is generally straightforward to produce a list of all the open times or all the closed times. At this point one is ready to find a sum of exponentials that best describes the relative frequencies of different event lifetimes. One can either fit to a histogram or work with the list of lifetimes directly using likelihood maximization. The first goal of this procedure is to estimate the number of exponential terms needed to describe the open time and closed time data, and thus obtain a count of the minimum number of states. When the number of open or closed states is one or two, it may also be useful to determine the parameters to relate them to specific rate constants of the channel-gating process.

Fitting to Histograms

From a list of state durations, one can construct a histogram, $N(t_i, t_{i+1})$, where t_i and t_{i+1} are times that mark the boundaries of the bins. When the bins are narrow, then $p(t)(t_{i+1} - t_i)$ becomes a viable theoretical estimate of

$N(t_i, t_{i+1})$. The sum of squares difference between these two quantities can can then be used as an index of the quality of the fit. Fitting then proceeds by minimizing the sum of squares with respect to the parameters of $p(t)$. A function minimization program can then be used to carry out this optimization. In most programs, the sum of the coefficients in the probability density function [x_i or y_j of Eqs. (1a) and (1b)] is constrained to be one, to reduce the number of free parameters. Minimizing χ^2 has also been suggested for curve fitting.[5] A likelihood maximization method can also be used for fitting to histograms (see below). A χ^2 goodness-of-fit test can be used to evaluate the fit and the improvement in the fit resulting from additional exponential terms. The number of exponential components in a distribution is estimated by comparing the goodness-of-fit for different numbers of terms.

With a small data set, one has fewer events per bin, and the statistical fluctuations in N become large. The histogram will then appear noisy, the curve fitting will not yield good parameter estimates, and low amplitude exponential terms will be difficult to detect. If an attempt is made to widen bins to remedy this problem, then $p(t)(t_{i+1} - t_i)$ will be a poor approximation for the predicted value of N, and this expression should be replaced by the integral of $p(t)$ over each bin, or $P(t_i) - P(t_{i+1})$. When bins are too wide, there will be too few data points to determine the fitting parameters accurately. Thus, fitting to histograms is problematic with small data sets.

Likelihood Maximization

For n event lifetimes, t_i, and a probability density function $p(t)$, the likelihood is

$$L = \prod_{i}^{n} p(t_i) \qquad (2)$$

A curve fit proceeds by variation of the parameters of $p(t)$ to maximize L.[5,7] In practice, the method of likelihood maximization is carried out with the logarithm of the likelihood, because L itself exhibits such extreme values that its computation can be difficult. The likelihood ratio test[8,9] can then be used to estimate the significance of the improvement of a fit by an additional exponential term.

This method does not involve binning data, as does the sum of squares method with histograms. It is thus advantageous with small data sets. The time required to compute L increases with the size of the data set, and the scatter of N in histograms becomes smaller, so that with large data sets it is

[7] R. Horn and K. Lange, *Biophys. J.* **43**, 207 (1983).
[8] C. R. Rao, "Linear Statistical Inference and Its Applications." Wiley, New York, 1973.
[9] M. B. Jackson, *Biophys. J.* **49**, 663 (1986).

better to use histograms. However, the method of likelihood maximization can still be used with histograms. The above likelihood function [Eq. (2)] then becomes[10]

$$L = \prod^{i} p(t_i)^{n_i} \tag{3}$$

where n_i is the number of events in bin i. This can be computed much more rapidly for a very large event lifetime list than it can using Eq. (2) and is thus a convenient function to use in the fitting of binned histogram data. With large data sets maximum likelihood and least-squares techniques become equivalent, and there are no compelling reasons for preferring one method over the other.

Out-of-Range Events

There will always be some events which are too brief to observe. There will also be events with lifetimes right at the limit of detection, such that the frequency with which such events occur is not accurately determined. In addition, it is often important to exclude from analysis the longest times, which can have anomalous effects on curve fitting. When using the sum of squares method, the correction for out-of-range events is trivial. One simply does not include the N values for bins that are out of range. For likelihood maximization, this procedure is not correct; simply ignoring such events is equivalent to assigning zero values to the experimentally observed frequencies. To correct for out-of-range events, one must fit to a probability density function which includes the condition that the interval is in the specified range.[5] Thus the likelihood for a single event of duration t becomes

$$L = p(t) \bigg/ \int_{t_s}^{t_1} p(t)\,dt = p(t)/[P(t_s) - P(t_1)] \tag{4}$$

where t_s and t_1 are the short- and long-time cutoffs, respectively. When the adaptation of likelihood maximization to binned data [Eq. (3)] is corrected in this fashion, one obtains the following index for the quality of fit.[10,11]:

$$\ln(L) = \sum^{i} n_i \ln \left\{ \frac{P(t_i) - P(t_{i+1})}{P(t_s) - P(t_1)} \right\} \tag{5}$$

Many investigators increase the bin width exponentially to compensate for fewer events at longer durations ($t_{i+1} = \delta t_i$, where δ is approximately 1.25). This provides constant binwidth on a logarithmic time axis. For event lifetime lists of more than a few thousand, this implementation of

[10] F. J. Sigworth and S. M. Sine, *Biophys. J.* **52**, 1047 (1987).
[11] O. B. McManus, A. L. Blatz, and K. L. Magleby, *Pfluegers Arch.* **410**, 530 (1987).

likelihood maximization [Eq. (5)] is at present the method of choice in fitting to a sum of exponentials.

Careful attention must be given to the precise value of the short-time cutoff in these expressions [Eqs. (4) and (5)]. When lifetimes are determined as multiples of a sampling frequency, then the cutoff must be $n + 0.5$ times the sampling frequency,[9] where n is usually an integer between 3 and 6. This form of cutoff is necessary because the true lifetimes are effectively rounded up or down by carrying out the idealization procedure on digitized data. The true cutoff is then at the rounding point. When the idealization procedure employs an interpolation procedure for estimating the lifetimes more accurately, then there is no longer such a restriction on the cutoff time. However, it is then important to evaluate the interpolation procedure carefully to ascertain that it interpolates uniformly. Digitization and binning can introduce another error called the sampling promotion error when the fastest time constant approaches the bin width or digitization time interval.[11,12]

Additional Corrections for Brief Events

The above discussion presented a correction for the primary error produced by out-of-range events in the curve fitting of probability density functions. However, intervals that are too brief to detect can lead to additional errors, and there has been much analysis of how missed intervals influence single-channel analysis. The most commonly encountered problem is that if a brief open (closed) time is missed, then the two adjacent closed (open) times appear as one closed (open) event with a lifetime equal to the sum of the three event lifetimes. A simple and effective correction can be used for missed closures when there is a single exponential distribution of open times.[13,14] The open times will remain exponentially distributed despite the missed brief closures, but with a lengthened apparent time constant. The first step in the correction is to estimate the number of missed brief closures from the zero-time intercept of the best-fitting closed time distribution. The corrected time constant, τ_{cor}, is then related to the observed value, τ_{obs}, by the expression

$$\tau_{cor} = \tau_{obs} N_{obs}/(N_{obs} + N_{missed}) \qquad (6)$$

where N_{obs} is the number of observed open times and N_{missed} is the number of missed brief closures. Treatment of the analogous case for closed times interrupted by brief openings is essentially the same.

[12] S. Sine and J. H. Steinbach, *J. Physiol. (London)* **373**, 129 (1986).
[13] E. Neher, *J. Physiol. (London)* **339**, 663 (1983).
[14] F. Sachs, J. Neil, and N. Barkakati, *Pfluegers Arch.* **395**, 331 (1982).

For multiple exponential distributions the problem becomes more complicated. Equation (6) is still correct for the mean open time, but it is difficult to extend this correction to each time constant of the open time distribution. If one can demonstrate by correlation analysis (discussed below) the extent to which the brief closures are associated with the different open states, then Eq. (6) can be implemented by replacing N_{obs} with the number of observed openings corresponding to a particular component, τ_{obs} with the time constant of that component, and N_{missed} with the number of missed closures estimated to be associated with that component.

The general problem of missed events has been treated systematically by Roux and Sauve.[15] Blatz and Magleby[16] have also presented some useful results.

Plotting Exponential Distributions

In a simple linear plot of a lifetime distribution, it is often difficult to recognize different exponential components (Fig. 1A). For this reason transforms have been developed to facilitate visual inspection. Semilogarithmic plotting provides some improvement by transforming an exponential to a line. However, when the time constants differ by more than a factor of 100 it is difficult to use a linear time axis. A particularly useful means of displaying such data is in a modified log plot (Fig. 1C), in which bin size increases exponentially with time, and n_i is *not* scaled to the size of the bin.[10,11] Further transforming the y axis by taking the square root[10] produces a variance-stabilized plot, that is, the deviation or scatter in the data becomes independent of the y value (Fig. 1D).

Fitting to Data from System with Many Channels

The theory employed above is strictly valid only for a system with one channel.[1,2] When a system has many channels this theory is no longer valid. When channel openings are well spaced from one another, and simultaneous openings are very infrequent, it has been assumed,[17] and can be shown,[18] that the one-channel kinetic theory provides a reasonable approximation to certain aspects of the gating process.

The problems arising in the analysis of a many channel system can be illustrated with a few examples. When examining a closed time interval in a system known to have more than one channel, there is no way of knowing whether the channel that closed to begin the interval is the same

[15] B. Roux and R. Sauve, *Biophys. J.* **48,** 149 (1985).
[16] A. L. Blatz and K. L. Magleby, *Biophys, J.* **49,** 967 (1986).
[17] D. Colquhoun and B. Sakmann, *J. Physiol. (London)* **369,** 501 (1985).
[18] M. B. Jackson, *Biophys. J.* **47,** 129 (1985).

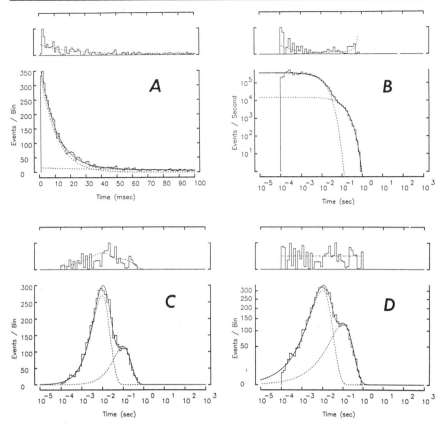

FIG. 1. Four representations of a dwell time distribution with two exponential components (from Ref. 10). A total of 5120 random numbers were generated according to a distribution with time constants of 10 msec (70% of the events) and 100 msec (30%) and binned for display as histograms in the lower part of each figure. The theoretical probability density functions for each component (dashed curves) and their sum (continuous curve) are superimposed with each curve. In each part of the figure the upper panel shows the absolute values of the deviation of the height of each bin from the theoretical curve, with dashed curves showing the expectation value of the standard deviation for each bin. The upper panels were plotted with expansion factors of 2.1, 5.4, 3.1, and 4.9, respectively. (A) Linear histogram. Events were collected into bins of 1 msec width and plotted on a linear scale. The 100-msec component has a very small amplitude in this plot. (B) Log–log display with variable width (logarithmic) binning. The number of entries in each bin was divided by the bin width to obtain a probability density in events/second, which is plotted on the ordinate. (C) Direct display of logarithmic histogram. Events were collected into bins of width $\delta x = 0.2$ and superimposed on the histogram as the sum of two functions, as in Eq. (11) of Ref. 10. (D) Square-root ordinate display of a logarithmic histogram, as in (C). Note that the scatter about the theoretical curve is constant throughout the display.

channel that opened to end the interval. This produces a systematic error in closed time distributions. When two channel openings overlap, there is no way of assigning the open time. Neglecting such events produces a systematic error in open time distributions.

For a system with N identical and independent channels, with a single-channel closed time probability density function of the form $\Sigma^m v_j y_j e^{-v_j t}$ [the derivative of Eq. (1b)], the closed time probability density function is[18]

$$P_{Nc}(t) =$$

$$\frac{(\Sigma\, v_j y_j\, e^{-v_j t})[\Sigma(y_j/v_j)\, e^{-v_j t}]^{N-1} + (N-1)(\Sigma\, y_j\, e^{-v_j t})^2[\Sigma(y_j/v_j)\, e^{-v_j t}]^{N-2}}{[\Sigma(y_j/v_j)]^{N-1}} \quad (7)$$

When the number of channels is known, Eq. (7) can be fitted to data by likelihood maximization, as has been demonstrated for a stretch-activated channel.[19] With a single-channel open time probability density function of $\Sigma\, \mu_i x_i\, e^{-\mu_i t}$ [the derivative of Eq. (1a)], the probability density function for isolated channel openings is[18]

$$p'(t) = \Sigma\, x_i \mu_i\, e^{-\mu_i t}\left[\frac{\Sigma\,(y_j/v_j)\, e^{-v_j t}}{\Sigma\,(y_j/v_j)}\right]^{N-1} \quad (8)$$

where a prime is used to distinguish the probability for isolated openings from the true single-channel open-time probability density function.

A fit with Eq. (8) requires input of parameters obtained from a fit to closed times, and knowledge of N. When N is difficult to determine, as is generally the case when N is large, then one can approximate $N - 1$ as N, which leads to

$$p'(t) \cong (\Sigma\, x_i \mu_i\, e^{-\mu_i t})[\Sigma(z_j/\omega_j)\, e^{-\omega_j t}]/\Sigma(z_j/\omega_j) \quad (9)$$

where the parameters ω_j and z_j are taken from a fit of $\Sigma\, \omega_j z_j\, e^{-\omega_j t}$ to the closed times from the same data set.[20] The results of such a fit are shown in Fig. 2. When the frequency of channel opening is low, there is little deviation of the best fitting one-channel open time distribution from the observed histogram of isolated open times. With data from a patch where the channel opening frequency is high, the observed histogram of isolated open times departs substantially from the best-fitting distribution, indicating that the distribution of isolated openings is a poor representation of the kinetics of closure of a single channel.[20] The similarity between the two functions (the solid curves of Fig. 2A,B) obtained by fitting Eq. (9) indi-

[19] F. Guhary and F. Sachs, *J. Physiol. (London)* **363**, 119 (1985).
[20] M. B. Jackson, *J. Physiol. (London)* **397**, 555 (1988).

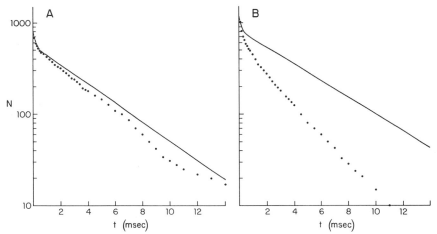

FIG. 2. Comparison of cumulative distributions of lifetimes of isolated openings (dotted curve) with the best-fitting single-channel open time probability density function (smooth curve), obtained by curve fitting with Eq. (8) (from Ref. 20). In (A) the level of channel activity was low, and there were almost never simultaneous channel openings. In (B) the level of channel activity was high; most channel openings were interrupted by other channel openings, and isolated openings were relatively rare. The data were obtained from cultured mouse muscle using 2 μM carbachol. The difference in channel activity between the two patches reflects primarily differences in receptor density. Note that whereas the isolated open time distributions are very similar. The parameters for the best fits in (A) were $\mu_1 = 6.0$ msec^{-1}, $\mu_2 = 0.24$ msec^{-1}, $x_1 = 0.26$ for open times, with a frequency of channel opening of 0.029 msec^{-1}. In (B) the parameters were $\mu_1 = 5.5$ msec^{-1}, $\mu_2 = 0.021$ msec^{-1}, and $x_1 = 0.30$, with a frequency of channel opening of 0.21 msec^{-1}.

cates that with both sparse and crowded channel data, the same kinetics operates at the single-channel level.

When fitting to these equations using likelihood maximization, it is important to incorporate the out-of-range event correction in an analogous manner to that described for Eq. (4). The integrations are lengthy but straightforward. Equation (9) can be used to analyze open times in very crowded data sets, provided that at least several hundred isolated openings are observed. When the frequency of channel opening is so high that the slowest time constant of the observed closed time distribution is shorter than the slowest time constant of the single-channel open time distribution, then the parameters determined by fitting Eq. (9) will have large standard errors.[20] It will then be necessary to carry out such fits on many data sets to obtain accurate estimates of parameters.

Burst Analysis

Many channels exhibit "bursting behavior," in which openings occur in clusters. The openings within such a burst are separated by short closed time intervals. A great deal of jargon has arisen around this phenomenon. Terms such as "flickers," "gaps," and "nachschlags" refer to various types of events in a "bursty" channel record. In the case of chemically activated channels, the bursts of openings are generally thought to be repeated openings of the same liganded receptor–channel complex. This mechanistic interpretation has provided a framework for a useful simplified approach to the kinetic analysis. Treating ligand binding and channel gating as separate steps of a kinetic scheme gives the following model:

$$\text{R} + n\text{A} \underset{k_{-1}}{\rightleftharpoons} \text{RA}_n \underset{\alpha}{\overset{\beta}{\rightleftharpoons}} \text{R*A}_n \tag{10}$$

where A represents the ligand, R the receptor, and R* indicates that the channel is open. A gap or flicker is a sojourn in state RA_n between two openings. The mean burst duration, t_b, mean gap duration, t_g, and mean number of gaps per burst, n_g, can be derived for this model and solved for the kinetic parameters to give[21]

$$\begin{aligned} \alpha &= (n_g + 1)/(t_b - n_g t_g) \\ k_{-1} &= 1/[2t_g(n_g + 1)] \\ \beta &= 2k_{-1}n_g \end{aligned} \tag{11}$$

It then remains to estimate t_b, n_g, and t_g from the data. The mean gap duration is best determined as the time constant of the fast component of the best-fitting closed time distribution. By inspecting a closed time distribution, one can define an operational cutoff time and assign all shorter closed intervals as gaps. One can then reconstruct bursts and determine burst duration and number of gaps. Although the assignment of all closures shorter than a cutoff as gaps may seem arbitrary, when there are two time constants of a closed time distribution differing by more than a factor of 10, then the number of incorrect assignments becomes insignificant. This type of burst analysis has been used in the study of both acetylcholine[17] and γ-aminobutyric acid (GABA)[22] receptor channel data. Burst analysis is particularly useful with data showing frequently occurring subconductance states, or with which idealization is difficult owing to a low signal-to-noise ratio. By including subconductance states as part of a burst, one can avoid the problem of how to idealize these events, which would require treating the subconductances as separate states.

[21] D. Colquhoun and A. G. Hawkes, *Philos. Trans. R. Soc. London, Ser. B* **300,** 1 (1982).
[22] J. Bormann and D. E. Clapham, *Proc. Natl. Acad. Sci. U.S.A.* **82,** 2168 (1985).

Burst analysis has the advantage of being simple and the disadvantage of being too simple. Kinetic models very different from Eq. (10) can generate bursting behavior. Equation (11) may not be valid for these models. If there are other pathways leading away from state RA_n, such as a desensitization step,[23] then what is determined as k_{-1} will actually be the sum of k_{-1} and the rates of the other steps leading away from RA_n. If there are multiple open states, but only one state that can gate, then burst analysis will still work. If there are three or more components in the closed time distribution, then it is very difficult to decide where to place the cutoff in closed times to classify gaps. If there is an open channel blocking process, then t_g will no longer be a function solely of k_{-1} and α. If experiments have established the validity of the model embodied in Eq. (10), then burst analysis can be carried out on the system under a wide variety of conditions to see how the rate constants α, β, and k_{-1} are affected by experimental manipulations.

Correlation Analysis

Correlations are evident in much single-channel data. Bursts of long openings may be separated by very brief closures, whereas brief openings are often separated by long closed times. To detect these correlations in event lifetimes, one employs the linear correlation coefficient or covariance function, which are, respectively $(\overline{t_a t_b} - \overline{t_a} \overline{t_b})/\sigma_a \sigma_b$ and $\overline{t_a t_b} - \overline{t_a} \overline{t_b}$. In general, it is easy to write a program that will accept as input an idealized data set and compute these functions for pairs of events related in any way. The estimation of the statistical significance of a linear correlation coefficient is a standard part of linear regression analysis. Alternatively, events can be classified as long and short with reference to a cutoff time, and a two-by-two contingency test can be used to evaluate the significance of the correlation.[24]

The simple demonstration of statistically significant correlations allows one to make an important statement about the gating mechanism: there is more than one gating pathway. This can most easily be established by determining the correlation coefficient of adjacent open and closed times, or for successive openings separated by a single closed time.[24] For short closed times, there is a high probability that the two successive open times are of the same channel. Thus, focusing on the open intervals separated by relatively brief closures improves the detection of significant correlations.

[23] J. Dudel, C. Franke, and H. Hatt, *Biophys. J.* **57**, 533 (1990).
[24] M. B. Jackson, B. S. Wong, C. E. Morris, H. Lecar, and C. N. Christian, *Biophys. J.* **42**, 109 (1983).

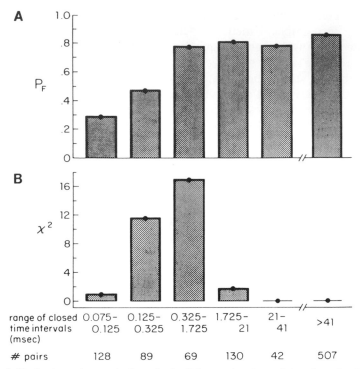

range of closed time intervals (msec)	0.075–0.125	0.125–0.325	0.325–1.725	1.725–21	21–41	>41
# pairs	128	89	69	130	42	507

FIG. 3. Single-channel currents through nicotinic receptor channels in cultured embryonic mouse muscle were activated by the agonist suberyldicholine and analyzed for correlations in successive open times (from Ref. 25). Pairs of openings were separated on the basis of the intervening closure lifetime (abscissa). The probability that a particular opening arose from the rapid or slow exponential component was calculated for a sum of two exponentials fitted to all open times. This probability, denoted as P_F, is plotted in (A) versus the closed time between pairs of openings. Openings adjacent to long closed times have higher values of P_F, meaning they are more likely to be brief. Openings adjacent to brief closed times have lower values of P_F, meaning they are more likely to be long. This indicates an inverse correlation between open times and the adjacent closed time. (B) Correlation between successive open times was tested with a two-by-two contingency test. χ^2 reflects the deviation from random mixing of long duration and short duration consecutive open times. A high χ^2 means that pairs of consecutive openings are more often similar in open time; members of each exponential component are segregated. A low χ^2 suggests that long and short open times are mixed randomly. Pairs of openings separated by closed times between 0.125 and 1.725 msec are highly correlated in open time. This indicates that there are at least two distinct gating pathways. Pairs of openings separated by long open times are uncorrelated, as expected for the gating of independent channels. Pairs of openings separated by very brief closures are uncorrelated, suggesting that most of these brief closures gate only to a long duration open state. This is consistent with the analysis shown in (A).

A measure of the correlation for successive openings of acetylcholine receptor channels is displayed for different values of the intervening closed time[24] (Fig. 3[25]). Significant correlations for intermediate closed times (Fig. 3B) show that there is more than one gating pathway. The same result can be inferred from Fig. 3A, which shows that long duration openings are more likely to be adjacent to short duration closures. Figure 3A thus shows that one of the gating pathways is between a short lifetime closed state and a long lifetime open state. Similar results have been obtained for a Ca^{2+}-activated K^+ channel.[26] Figure 3B suggests that closed states with very short durations have access to only one gating pathway but that closed states with intermediate durations can gate by two pathways to either long duration or short duration open states. With long duration closed times there are no significant correlations between the durations of the two openings, as expected, since such openings are likely to be openings of different channels.

When large numbers of events are available (typically more than 10,000), the condition on the lifetime of a neighboring event can be exploited much more effectively. The quantitative dependence of the lifetime distribution can be examined. This method has been termed adjacent state analysis, and it has been used to discriminate between very complex models in the gating of a Cl^- channel from rat skeletal muscle.[27]

As noted above, the covariance as a function of n is a sum of exponentials, where n is the number of events separating two events for which the covariance is being computed. The number of exponential terms in the covariance function is one less than the minimum number of gating pathways.[3,4] Although specifying the number of closed states, open states, and gating pathways is often deemed substantial progress toward understanding the gating mechanism, it generally does not identify a unique model. Correlation analysis can be pushed further to discriminate between models. The most general way to obtain additional information starts with a derivation of the two-dimensional probability density function for adjacent open and closed times for each candidate model with the requisite number of open states, closed states, and gating pathways.[3] For some of the candidate models, there may be no difference in the form of the two-dimensional probability density function, and discrimination between these models cannot be achieved solely by the analysis of stationary single-channel data.[3,4] The two-dimensional probability density functions can be curve fitted, using likelihood maximization, to the relevant list of pairs of dwell

[25] M. B. Jackson, *Adv. Neurol.* **44,** 171 (1986).
[26] O. B. McManus, A. L. Blatz, and K. L. Magleby, *Nature (London)* **317,** 625 (1985).
[27] A. L. Blatz and K. L. Magleby, *J. Physiol. (London)* **410,** 561 (1989).

times, prepared from the idealized data record.[3,28] The fits provided by the different models can then be compared and ranked according to a goodness-of-fit criterion.[28] These methods have been developed in detail and applied to the glutamate receptor channel of locust muscle.[29]

Acknowledgments

I thank Larry Trussell for critical comments on the manuscript.

[28] F. G. Ball and M. S. P. Sansom, *Proc. R. Soc. London B* **236,** 385 (1989).
[29] S. E. Bates, M. S. P. Sansom, F. G. Ball, R. L. Ramsey, and P. N. R. Usherwood, *Biophys. J.* **58,** 219 (1990).

[52] Analysis of Nonstationary Single-Channel Currents

By F. J. Sigworth and J. Zhou

Introduction

A nonstationary process is a random process whose behavior changes with time. In this chapter we consider the analysis of channel activity as a particular kind of nonstationary process. A channel is prepared in a certain state or configuration of states, for example, by holding the membrane at a negative potential. Then a sudden perturbation is applied, for instance, a step depolarization, causing the channel to relax from this configuration, passing through various states on its way to a new equilibrium configuration; the time course of this channel activity is what is analyzed. Some kinds of channels must be studied in pulsed experiments of this kind. Sodium channels and A-current potassium channels inactivate very strongly, so that one has little hope of observing much channel activity unless one gives pulsed depolarizations. Similarly, some neurotransmitter receptors desensitize rapidly, making pulsed application of agonist necessary. Analyzing the results of such pulsed experiments involves extra difficulty compared to continuous recordings: there are artifacts from the stimulus to deal with, and special consideration has to be given to dwell times immediately after the stimulus. However a great deal of extra information can be gained from such data. For example, if the channel starts in a particular closed state s_0, the time before the channel first opens (the first latency) can be analyzed to give a description of the kinetic steps between s_0 and the open states of the channel.

In many respects the analysis of nonstationary single-channel recordings is the same as the analysis of stationary records. Single-channel current amplitudes and open and closed dwell-time distributions can be obtained and interpreted as explained in [51] and [53] in this volume. This chapter is concerned with procedures in the acquisition and analysis of single-channel activity that are specific to the nonstationary case.

Issues in Acquisition of Data

Removing Artifacts

When a step depolarization is applied with a patch clamp, a very large capacitive current flows. The total capacitance of the membrane patch and pipette might be 1 pF, and if a 100-mV depolarization is to be applied within 10 μsec a current of approximately 10 nA is required. This current is very much larger than the picoampere-size single-channel currents! The current is large because the capacitance of the pipette is much larger than the capacitance of the patch membrane itself (only about 0.01 pF/μm^2). For this reason all commercial patch clamp amplifiers contain an adjustment that allows one to cancel most of this capacitive transient. Even after setting this knob correctly, however, one is usually left with a residual capacitive transient that decays on a slower time scale of around 1 msec (Fig. 1). This transient results mainly from dielectric relaxation in the glass wall of the pipette; it is smaller in types of glass that have low dielectric loss (e.g., Corning 7052; see [3] in this volume) and is smaller when the immersed part of the pipette is well covered with coating such as Sylgard (Dow Corning, Midland, MI). In general, measures that reduce the size of this transient will reduce the background noise of the recording as well.

To be rid of the residual capacitive artifacts, as well as the "pedestal" in the trace that arises from the leakage conductance, some sort of leak subtraction may be used. The most elegant subtraction method is to accumulate the average of sweeps in which no channel activity occurs and then subtract this average from each sweep to be analyzed. Because the average is accumulated from a finite number of sweeps, it will contain some noise which will contribute to the total noise in the subtracted traces. If n sweeps are used in the average, the standard deviation of the noise [i.e., the root mean square (rms) noise] will be increased by the factor $(1 + 1/n)^{1/2}$; this noise increase is negligible for n equal to about 10 or greater.

A useful alternative to averaging null sweeps is to use $P/4$ subtraction. This technique was introduced by Bezanilla and Armstrong[1] to remove

[1] F. Bezanilla and C. M. Armstrong, *J. Gen. Physiol.* **70,** 549 (1977).

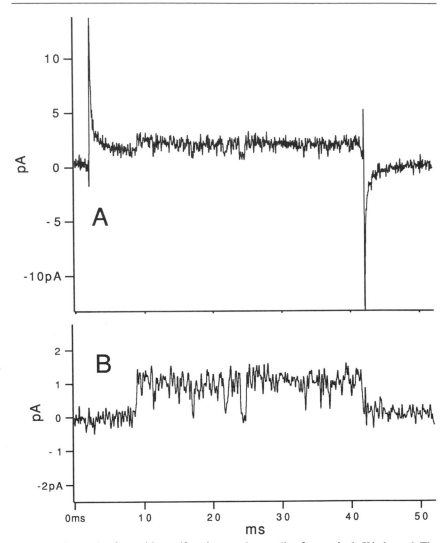

FIG. 1. Example of capacitive artifacts in a patch recording from a single K$^+$ channel. The membrane potential was stepped from -80 to $+20$ mV for 40 msec and then returned to -80 mV. (A) Pipette current as recorded by the patch clamp. The fast capacitance transient was mostly canceled by the C-Fast adjustment, but large decaying artifacts remain at the times when the potential was changed. (B) The same trace after removing the artifacts, in this case by subtracting the average of null records. The original recording bandwidth was 5 kHz; the trace in (B) was digitally filtered to 2 kHz bandwidth. The pipette was fabricated from Kimax glass (Corning).

artifacts from their recordings of the small gating currents in squid axons. The idea is to give a number of small, scaled-down copies of the test pulse and to use the scaled average response to these control pulses for subtraction. Traditionally four pulses are given with a scaling of one-fourth, hence the name $P/4$. The control pulses are given in a potential range too negative to cause the activation of voltage-dependent conductances. For example, in an experiment in which the test pulse is a depolarization from -80 to 0 mV, the $P/4$ pulses might start at -120 mV and step to -100 mV. For subtracting artifacts in single-channel recordings one needs to accumulate the average of a large number of $P/4$ pulses. Because the average must be scaled up by a factor of 4 (which scales the noise variance by 16) the root mean square noise, after subtracting the scaled average of $nP/4$ responses, is increased by the factor $(1 + 16/n)^{1/2}$. It is a good idea to average 20–40 groups of four responses to obtain the subtraction trace.

The $P/4$ subtraction relies explicitly on the responses being a linear function of the voltages applied. This requirement is definitely *not* met when some element of the recording system goes into saturation, that is, receives such a large signal it cannot amplify or filter it linearly. Some patch clamps have a "clipping" light which shows when an internal amplifier goes into saturation; this is valuable because sometimes the patch clamp amplifier saturates on very large, fast current transients that are not visible once the signal is filtered. It is therefore important that all transient signals stay within the linear range (typically ± 10 V) of all the electronic devices in the recording system.

When Does the Pulse Start?

Judging when a pulse starts is important when one wants to characterize the time between the start of the pulse and the beginning of channel openings (the first latency). It is a nontrivial issue because of the many steps between the generation of the stimulus and the measurement of the response. The stimulus pulse is typically generated by a digital-to-analog (D–A) converter in a computer; it commands the patch clamp to change the potential at the pipette which in turn changes the patch membrane potential. The recorded current is delayed by the finite response time of the patch clamp and any filters being used, and it is finally sampled and digitized by an A–D converter. What concerns us is the relative time between the output of the stimulus and the corresponding value sampled by the A–D converter. Let us first consider simple ways to measure the delay; then we consider what is the expected magnitude of the various delays in the recording system; finally we consider the dangers of slowly rising stimulus pulses.

A good way to characterize the overall delay in the recording system is to record the "instantaneous" current change in a channel (or ensemble of channels) that is open at the time of an applied potential step. An example of such a measurement is shown in the leak-subtracted traces of Fig. 2A,B. A patch containing many K^+ channels was depolarized from -80 to $+120$ mV, and then after 10 msec the potential was switched back to 0 mV. Under these conditions the conductance change from the step back to 0 mV is quite slow (~ 1 msec time constant; see the slow "tail" of the current), and we can identify the rapid current change with a change in the current through open channels. The midpoint of the current change occurred about 53 μsec after the command in this case; we take this to be an estimate of the delay in the recording system. A similar kind of measurement can be made from recordings of single-channel sweeps: select those sweeps in which a channel is open during a potential jump, average those sweeps, and then find the time at the midpoint of the rapid current change.

Another way to characterize the delay is to look at fast capacitive transients. If you misadjust the fast capacitance cancellation on the patch clamp, a brief pulse will result. An estimate of the delay can be obtained as the time to the peak of this pulse. The unsubtracted traces of Fig. 2C,D were obtained with a slight misadjustment of C-Fast; the peak of the transient is seen to occur about 55 μsec after the time of the depolarization.

What are the origins of the delay? One aspect is the relative timing between the stimulus output by the D–A converter of the computer and the time of sampling by the A–D converter. For example, the ITC-16 interface (Instrutech Corp, Elmont, NY) that we use takes its A–D sample exactly 1 μsec *before* the corresponding D–A sample is sent out. Depending on how the programs are written that use the interface, the relative timing might also differ by one or more sample intervals. For example, a program might discard the first A–D sample in a sweep, knowing that it is taken before the first D–A value is sent out. This would mean that the A–D sample is taken one sample interval, less 1 μsec, after the D–A value. Hence, the apparent delay might depend on the sampling interval chosen.

In patch-clamp recordings the delay in the application of a voltage stimulus to the membrane is usually negligible. In some patch clamps the stimulus input is passed through a filter to reduce the maximum rate of rise. The EPC-7 and EPC-9 patch clamps have such a filter that can be set to give a rise time of either 2 or 20 μsec, yielding a delay of either 1 or 10 μsec, respectively, between the stimulus input and the potential at the pipette. The patch membrane, in turn, should be very rapidly charged (with a time constant under 1 μsec) once the pipette potential is established. An exception to this is when there is a substantial series resistance in series with the membrane, such as a second membrane formed when an

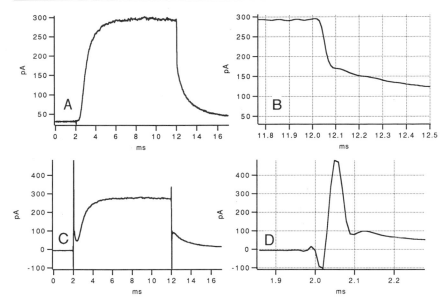

FIG. 2. Measuring the stimulus–response delay. (A) Current in a "macropatch" from a *Xenopus* oocyte expressing a high density of potassium channels. A depolarization from −80 to +120 mV was given 2.0 msec into the sweep, and a step back to 0 mV was given at 12.0 msec. *P*/4 leak subtraction was used. (B) An expanded plot of the same trace shows that the midpoint of the "instantaneous" current change occurs at 12.053 msec, suggesting that the total delay is 53 μsec. (C) The same recording but without leak subtraction has an incompletely canceled fast transient. (D) The peak of the transient is seen to occur about 55 μsec after the depolarizing stimulus was commanded.

inside-out patch is excised. If this slows the membrane charging by a large amount, it may introduce a delay large enough to be visible in the "instantaneous" current time course described above. Under normal conditions one expects that when the stimulus filter is set to the fast position, the delay in the stimulus pathway will be typically under 2 μsec.

It might be added that it is sometimes very important that "step" changes in the applied membrane potential occur very quickly; otherwise there can be serious distortions in the time course of gating. In some channels the time constants and state occupancies can change substantially with a few millivolts change in membrane potential. If one applies a 100-mV step through a patch clamp (or voltage clamp) system with a time constant of 10 μsec, it will be some 40 μsec before the potential has settled to within a few millivolts of the final value; for steeply activating channels with fast kinetics this causes a delay in activation of about this length of time, and a substantial distortion of the activation time course.

The delay in the current-measuring pathway, meanwhile, is essentially all due to filtering. Patch clamp amplifiers have an intrinsic delay that is quite small, perhaps 2 μsec. The EPC-7 has an internal 10-kHz Bessel filter which, if switched in, adds a delay of about 20 μsec. The delay due to a filter can be substantial. For example, an 8-pole Bessel filter introduces a delay of $0.51/f_c$ sec when the -3-dB frequency f_c is expressed in Hertz. Thus a 1-kHz filter of this type introduces 510 μsec of delay. In the recordings of Fig. 2 the delay is well accounted for by the approximately 51-μsec delay of the 10-kHz filter used.

First-Latency Distribution

Relationship to Channel Kinetics

The first latency is the time between the application of the stimulus and the first opening of a channel. If a channel has only one open state, the first latency has special significance. It represents the time taken by the various steps that the channel goes through to get to the open state from the state (or configuration of states) that the channel started in. The first latency is therefore a special kind of closed time. All subsequent closed times are different because they are "reopening times," that is, times between the channel leaving the open state and reentering it. These reopening times are the only kind of closed times that are measured in a stationary experiment, where no pulsed stimulus is applied and one simply waits for repeated openings of the channel to occur. Thus, in the case of only one open state the first latency contains all of the special information to be gained from nonstationary analysis.

For example, consider the following kinetic Scheme (I) for a channel.

$$C_1 \rightleftharpoons C_2 \rightleftharpoons C_3 \underset{\beta}{\overset{\alpha}{\rightleftharpoons}} O_4 \rightleftharpoons C_5$$

SCHEME I

Let us assume that the channel is in the leftmost closed state C_1 when a depolarization starts at $t = 0$. During the depolarization the rate constants are such that the channel is brought to the open state O_4 eventually. The probability of arriving in the open state for the first time in the time interval $(t, t + dt)$ is

$$f_1(t) = \alpha p_3'(t)$$

where p_3' is the probability of being in state C_3, under the condition that the reverse rate β is set to zero (this artifice keeps us from counting reopenings). The function f_1 is called the probability density function for the first latency. It is independent of steps involving states (like C_5) that are not visited before the first opening. Because it is proportional to the occupancy of C_3, it allows us to "see" one step before the open state in the gating process.

The probability distribution function for the first latency is

$$F_1(t) = \int_0^t f_1(\tau) \, d\tau$$

which would be equal to the probability of being in the open state O_4 in the reduced Scheme II. Measuring the distribution function F_1 is therefore like

$$C_1 \rightleftharpoons C_2 \rightleftharpoons C_3 \xrightarrow{\alpha} O_4$$

SCHEME II

measuring the activation kinetics of a simplified channel in which all transitions have been eliminated that lead away from the open state.

Error Bounds for Cumulative Histograms

It is difficult to accumulate many events to put into a histogram of first latencies, because one obtains only one measurement per pulse. Traditionally the kinetic work on inactivating Na^+ and K^+ channels[2,3] has been presented using cumulative "histograms." Instead of plotting the number of measured times that fall into narrow "bins," for example, from t_0 to $t_0 + 0.1$ msec, these cumulative distributions show the total number of events shorter (or longer) than t_0, plotted as a function of t_0. Plotting sparse data in this way is popular because the curves obtained with a limited number of points appear smoother than the equivalent conventional histograms. The smoothness of such histograms can, however, conceal large errors. We consider here the nature of statistical scatter in such histograms.

Figure 3A shows a conventional histogram of 100 first latencies from a simulation of Scheme I, with the appropriately scaled probability density function $f_1(t)$ shown superimposed. The histogram appears very ragged because of the relatively small number of events it contains. The number of events in each bin of such a histogram is essentially Poisson distributed, which means that if the expectation value of the number of events in a bin is n, the standard deviation of the number in that bin will be $n^{1/2}$. The errors in the different bins are independent, so that even in the absence of the theoretical curve we can gain an impression of the errors by looking at the differences from one bin to the next.

[2] R. W. Aldrich, D. P. Corey, and C. F. Stevens, *Nature (London)* **306**, 436 (1983).
[3] W. N. Zagotta and R. W. Aldrich, *J. Gen. Physiol.* **95**, 29 (1990).

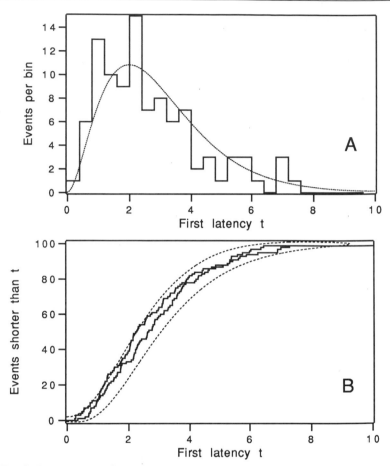

FIG. 3. Two representations of simulated first-latency data from Scheme I. All forward rates were set to 1.0, and all reverse rates were zero in the simulation. (A) Histogram of 100 first latencies, with the theoretical result superimposed. (B) Cumulative distributions from the same 100 first latencies (top curve) and a second, independent simulation. Dotted curves indicate error bounds (± 2 S.E.) about theoretical distribution function F_1.

The corresponding cumulative histograms (Fig. 3B) look much smoother. Here we have plotted the the number of measured latencies that are smaller than a reference time t, as a function of t. The cumulative histogram is the integral of the conventional histogram. How can we compute error bars for such a curve? For a total of N measurements, the expectation value for the number of first latencies shorter than t_0 is $NF_1(t)$, and the variance is

$$\sigma^2(t) = NF_1(t)[1 - F_1(t)] \tag{1}$$

The standard deviation is the square root of the variance. Notice that the variance is zero when F_1 is zero (i.e., at $t = 0$) and also when F_1 approaches 1 with large t; the variance is maximum in the middle of the curve when F_1 equals $\frac{1}{2}$.

Also shown plotted in Fig. 3B are dotted curves representing bounds of ± 2 standard deviations. One of the two cumulative distributions plotted in Fig. 3B is seen to lie about two standard deviations above the theoretical values for almost half of its length. This sort of situation can arise because the points plotted are not statistically independent, but instead are highly correlated: if one point in the curve is too high (because of a few excess events), all succeeding points will tend to deviate in the same direction. Thus, the smoothness of the cumulative histogram gives an impression of precision that can be misleading. So which kind of histogram is better? Cumulative histograms are related to the probability of occupancy of the open state, and therefore can be compared directly with some measurements of the ensemble mean current. However, the conventional histograms (Fig. 3A) have better statistical properties for fitting and for evaluating the quality of data.

The best way to fit experimental first-latency data is to use maximum-likelihood fitting. This does not involve making histograms at all, but instead involves computing the likelihood of a particular form of the probability density f based on the data. The log likelihood L is computed from N measurements t_i of the first latency according to

$$L = \sum_{i=1}^{N} \ln f_1(t_i) \tag{2}$$

where f_1 is the theoretical probability density of the first latency. To find the best values of parameters (e.g., rate constants) that affect f, values of L are computed as the parameters are varied, and the parameters which maximize L are taken to be the most likely values. See Colquhoun and Sigworth[4] for an introduction to this method.

Estimating Closed and Open Times

Apart from the special considerations of first-latency measurements, the estimation of closed and open times in nonstationary channel recordings can be done in the same ways as with stationary data, as discussed in [51] and [53] in this volume. In the interpretation of dwell-time distribu-

---

[4] D. Colquhoun and F. J. Sigworth, *in* "Single-Channel Recording." (B. Sakmann and E. Neher, eds.), p. 191. Plenum, New York, 1983.

tions two things should, however, be kept in mind. First, as Magleby ([53], this volume) points out, there may be information contained in the particular sequence of open and closed intervals. If a channel has multiple, kinetically distinct open states, there may be correlations in open and closed times such that, for example, shorter closed times might tend to follow longer open times. In the nonstationary case there is the additional possibility that open or closed times might show a trend during each pulse. For example, it might be that more brief openings occur early in a trace and longer openings occur later. This will not occur if there is only one open state or only one closed state of the channel, however.

A second issue arises from the limited duration of the stimulus pulse. How are we to evaluate a channel open interval that is cut short by the end of the pulse? Fukushima[5] first addressed this problem in an analysis of inwardly rectifying potassium channels. An elegant approach which can be used in maximum-likelihood estimation of the true dwell-time distribution is the variable-censor method.[6] Here the individual measurements of dwell times t_i are sorted into sets according to whether they were unaffected (**u**) or truncated (**t**) by the end of the pulse. The log likelihood L is then calculated by summing over the two sets of intervals as

$$L = \sum_{i \in \mathbf{u}} \ln f(t_i) + \sum_{i \in \mathbf{t}} \ln[1 - F(t_i)] \tag{3}$$

where $f(t)$ is the probability density function and $1 - F(t)$ is the "survivor function," the probability that a dwell time is longer than the value t. The two functions are related by

$$f(t) = dF/dt$$

The most likely underlying distribution of dwell times is found by maximizing L by varying the parameters of the functions f and F.

Problem of Multiple Channels

When more than one channel is present in a patch, the interpretation of open and closed times becomes quite complicated. In principle, much of the kinetic behavior of a single channel can be deduced from the activity in a patch containing multiple channels, provided the number n of channels is known. Estimating the number of channels is not always straightforward. In a recent investigation of statistical estimators, Horn[7] found that

[5] Y. Fukushima, *Nature (London)* **294**, 368 (1981).
[6] T. Hoshi and R. W. Aldrich, *J. Gen. Physiol.* **91**, 107 (1988).
[7] R. Horn, *Biophys. J.* **60**, 433–439 (1991).

simply assigning n to be the maximum number k_{max} observed of simultaneously open channels is one of the best estimators, especially for $n \leq 4$. Clearly this method will fail when the open probability of an individual channel is very low. To be confident that the estimate is likely to be correct, Horn introduces the f_{max} criterion, which considers the fraction of time the maximum overlap is observed. When f_{max} is 0.1 or larger, k_{max} is quite likely to be equal to n.

Given an estimate for n, the first latency distribution F_1 for one of n independent, identical channels is readily computed[2] according to

$$1 - F_1(t) = [1 - F_{obs}(t)]^{1/n} \tag{4}$$

For other closed and open times, estimating the underlying single-channel distributions is not so simple. Jackson[8] has shown how to obtain the distributions in the stationary case, and presumably this theory could be applied to the nonstationary case. Another approach altogether is presented by Horn and Lange,[9] where a system of n independent channels is described by a multistate kinetic scheme having $n + 1$ conductance levels. If each channel has m kinetic states, such a composite scheme has of the order of m^n states. If n is known and is not too large, this sort of method can be used to fit models to dwell-time data.

Analyzing Modal Behavior

Many types of channels show changes in their gating behavior that occur on time scales of seconds to minutes. Ligand-gated channels such as acetylcholine receptors show long silent periods owing to desensitization,[10] and switching among gating modes has been particularly well characterized in long recordings from a voltage-gated Cl^- channel.[11] In pulsed recordings the identification of such slow processes is more difficult because the behavior of a channel is not observed continuously, but only during pulses that are applied at regular intervals. For example, recordings of Na^+ channels may show consistent channel activity during several successive pulses but then no activity during several subsequent pulses. How can we know whether such a pattern is due to a slow process (e.g., a channel goes into "hibernation" for some time) rather than being a pattern that arises by chance from channels that simply do not open in every pulse? Further, given that there is a slow process, can we obtain information about it?

[8] M. B. Jackson, *Biophys. J.* **47**, 129 (1985).
[9] R. Horn and K. Lange, *Biophys. J.* **43**, 207 (1983).
[10] B. Sakmann, J. Patlak and E. Neher, *Nature* **286**, 71–73 (1980).
[11] A. L. Blatz and K. L. Magleby, *J. Physiol. (London)* **378**, 141 (1986).

Runs Analysis

One approach that has been applied to voltage-gated Na$^+$ channels[12] is runs analysis. A run is a sequence of like elements; we would identify one kind of element with null traces (having no channel activity) and another element with traces showing activity. When two groups of different elements are arranged randomly, the probability of having R runs can be calculated analytically. If the sample number is more than about 40 the distribution of R can be approximated by an asymptotic distribution, forming a normalized random variable Z having a mean of zero and variance of one. Z can be calculated as

$$Z = \frac{R - 2Np(1 - p)}{2N^{1/2}p(1 - p)} \tag{5}$$

where R is the number of runs, N is the total number of trials, and p is the probability of finding one of the two kinds of element. The expected value of R is $2Np(1 - p)$, giving Z a value of zero. Positive values of Z indicate the clustering of like elements, whereas negative Z values indicate that the two kinds of elements tend to alternate with each other. When Z is greater than 1.6 there is statistically significant clustering (at the level of $p < 0.02$). In the study of Horn *et al.,*[12] the number of runs observed experimentally was significantly smaller than the expected value from random arrangements without any slow process, implying that a slow process was causing clusters of sodium channel activity in successive traces.

The runs analysis can also be applied to characterizing modal changes in gating behavior. Nillius[13] and Plummer and Hess[14] scored traces of channel currents according to whether they contained long or short channel openings, and they applied runs analysis to the occurrence of these opening types. A limitation of runs analysis is that although it can show whether there are clusters of activity, it gives little information about the underlying process that gives rise to the clusters.

Autocovariance of Scoring Function

Another approach is to classify pairs of traces according to the kinds of activity they show. Suppose that we find that the probability of a null trace is some value p_0 and the probability of a trace having one or more channel openings is p_1. Then if there is no correlation among successive traces then the probability of seeing two successive nulls will be p_0^2 and the probability of two active traces will be p_1^2. The probablity of an active trace followed by a null will be p_0p_1, which also equals the probability of observing a null

[12] R. Horn, C. A. Vandenberg, and K. Lange, *Biophys. J.* **45,** 323 (1984).
[13] B. Nilius *Biophys. J.* **53,** 857 (1988).
[14] M. R. Plummer and P. Hess, *Nature (London)* **351,** 657 (1991).

followed by an active trace. Now suppose instead that there is a tendency for active traces to be grouped together: then the probability of an active trace being followed by a null trace will be lower than that given by the product of the individual probabilities. A statistical test based on this idea was used as a test for correlations in calcium channel activity by Plummer and Hess.[14]

One could also look for deviations from independence for the occurrence of active and null traces that are not successive, but separated by several intervening traces. This could give further information about the time scale of changes in the channel behavior. The information is contained in an autocovariance function, the theory for which we present here. Let us define a scoring function $s(i)$, where i denotes the trace number, $i = 1 \ldots N$. We let s depend on the channel activity such that

$$s(i) = \begin{cases} 1 & \text{if any channel opened} \\ 0 & \text{if no activity} \end{cases}$$

The autocovariance of s can then be calculated as

$$A_s(j) = \frac{1}{N-1} \sum_{i=1}^{N} s(i)\, s(i+j) - \bar{s}^2 \tag{6}$$

where $\bar{s}$ is the average value of s, averaged over all N traces.

Let us first consider the case in which there are no correlations between traces. In each trace let us assume that there is activity with probability p. Then the expectation values (denoted by angular brackets) for average and autocovariance are

$$\langle \bar{s} \rangle = p$$

$$\langle A_s(j) \rangle = \begin{cases} p(1-p), & j = 0 \\ 0, & j \neq 0 \end{cases} \tag{7}$$

In other words, the autocovariance should have a nonzero value for the lag variable $j = 0$ but should otherwise be zero. Owing to finite N the values of A_s will not be exactly zero; the standard deviation, computed as the square root of the variance, can be used to put error bars on a plot of A_s. The variance of A_s is approximately

$$\text{Var}\{A_s(j)\} \approx \begin{cases} \dfrac{1}{N-1} p(1-p)(1-2p)^2, & j = 0 \\[2mm] \dfrac{1}{N-1} p^2(1-p)^2, & j \neq 0 \end{cases} \tag{8}$$

where terms of order $(1/N)^2$ and smaller have been ignored.

Now let us consider the case where there are correlations among traces. Specifically, we assume there is a single channel in the patch and that it makes slow transitions between available and unavailable states according to Scheme III, where we say that an available channel is not always open,

$$U \underset{\beta}{\overset{\alpha}{\rightleftharpoons}} A$$

SCHEME III

but will be observed to give a trace with activity (i.e., $s = 1$) with probability p. In this case the expectation values are

$$\langle \bar{s} \rangle = p \frac{\alpha}{\alpha + \beta}$$

$$\langle A_s(j) \rangle = \begin{cases} p \dfrac{\alpha}{\alpha + \beta} \left(1 - p \dfrac{\alpha}{\alpha + \beta} \right), & j = 0 \\[2ex] \dfrac{\alpha \beta p^2}{(\alpha + \beta)^2} e^{-(\alpha + \beta)jT}, & j \neq 0 \end{cases} \tag{9}$$

where T is the time interval between traces. Thus when there is a slow process that causes a single channel to change state, this will be reflected as an exponential decay in the autocovariance function. From the value of the autocovariance at $j = 0$ and the amplitude and rate of the decay, all three parameters p, α, and β can be determined.

Autocovariance of Integrated Openings

If there is more than one channel in a patch, the scoring function s does not provide a good reflection of the activity of a single channel, and the interpretation of the autocovariance is not straightforward. An alternative approach is to use a measure of activity that is proportional to the number of channels active in any given trace. A simple measure of this sort is the integral of the channel activity in a trace. This could be, for example, the integral t of the current in the jth trace divided by the single-channel current, yielding the number m of active channels times the average total open time t_o for each active channel,

$$t = mt_o \tag{10}$$

By the same theory that underlies standard fluctuation analysis it can be shown that the autocovariance of t for an n-channel patch has the same form as the autocovariance of a one-channel patch, only is n times larger. We can compute the autocovariance from a set of N sweeps according to

$$A_t(j) = \frac{1}{N-1} \sum_{i=1}^{N} t(i)t(i+j) - \bar{t}^2 \tag{11}$$

as in Eq. (6) above.

Let p be the probability that a channel is active and τ be the mean total open time during a trace. Then if channel activity is not correlated from trace to trace,

$$\langle \bar{t} \rangle = np\tau$$

$$\langle A_t(j) \rangle = \begin{cases} np(2-p)\tau^2, & j=0 \\ 0, & j \neq 0 \end{cases} \tag{12}$$

The variance of the A_t values is useful in order to place error bounds on the computed A_t values. To estimate the variance the probability distribution of the total open times t_o must be known. The broadest distribution results when a single opening or a single burst of openings occurs in each trace; in this case t_o is exponentially distributed and gives the maximal variance values

$$\mathrm{Var}\{A_t(j)\} \approx \begin{cases} \dfrac{n}{N-1} 4p(2-p)(3-2p+p^2)\tau^4, & j=0 \\ \dfrac{n}{N-1} p(2-p)\tau^2, & j \neq 0 \end{cases} \tag{13}$$

If, on the other hand, channel activity is correlated because of a slow process as in Scheme III above, the result analogous to Eq. (9) is

$$\langle \bar{t} \rangle = np\tau\frac{\alpha}{\alpha+\beta}$$

$$\langle A_t(j) \rangle = \begin{cases} p\tau^2\dfrac{\alpha}{\alpha+\beta}\left(2 - p\dfrac{\alpha}{\alpha+\beta}\right), & j=0 \\ \dfrac{\alpha\beta p^2\tau^2}{(\alpha+\beta)^2} e^{-(\alpha+\beta)jT}, & j \neq 0 \end{cases} \tag{14}$$

Again, a relaxation in the autocovariance results if an underlying slow process is present.

Example

Figure 4 shows simulations of channel activity and the resulting auto-covariance functions. In Fig. 4A the channel had no long-term correlations but was active only 40% of the time. The resulting autocovariance $A_t(j)$ (Fig. 4C) is zero except for the point at $j=0$, as expected. The value of Z, computed from the same set of 1000 simulated traces, was 0.26. In Fig. 4B a "hibernating" channel gave rise to a decaying autocovariance whose time constant (5 sec) was expected for the values of α and β used. For these traces the value of Z (17.1) was very large, so the clustering of activity was highly significant by this test. In summary, it appears that the runs analysis

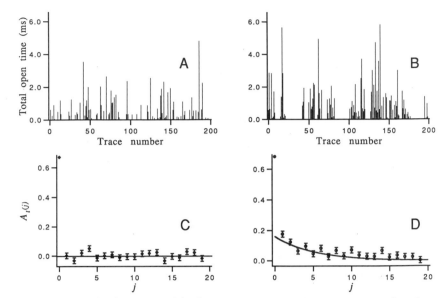

FIG. 4. Analysis of channel activity from sweep to sweep with autocorrelation functions. (A) Simulated total open times for each of 200 simulated traces. When active, the single channel opened once with an average open time of 1 msec; the probablity of being active in each trace was 0.4. (B) Corresponding plot for an identical channel except that the channel switched between unavailable and available states with rates $\alpha = \beta = 0.1$ sec^{-1}, and the probability of the available channel being active was $p = 0.8$, yielding the same average level of activity as in (A). Traces were assumed to be elicited once per second. (C) Autocovariance of integrated open times computed from $N = 1000$ traces for the channel in (A). All points are near zero except the one at $j = 0$. (D) Autocovariance from 100 traces as in (B). The exponential decay predicted by Eq. (14) is shown superimposed. Error bars in (C) and (D) are the standard deviations computed using Eq. (13).

is much more sensitive than the autocorrelation analysis for detecting modal behavior, but when there are sufficient data the autocorrelation analysis promises to be useful for characterizing the kinetics of mode switching.

Acknowledgments

We thank N. Schoppa for the data shown in Figs. 1 and 2, and K. McCormack for helpful comments. This work was supported by National Institutes of Health Grants NS21501 and HL38156.

[53] Preventing Artifacts and Reducing Errors in Single-Channel Analysis

By K. L. MAGLEBY

Introduction

Ion channels are integral membrane proteins that provide an aqueous pore through which selected ions can move from one side of the membrane to the other.[1] Channels control the flux of ions by gating their pores in response to various factors, such as agonists or membrane potential.[1] The gating is thought to arise from a series of conformational changes, perhaps analogous to some of the conformational changes associated with the actions of enzymes.[1-3]

One important goal in the analysis of ion channels is to obtain a kinetic model for the gating. Such kinetic models provide a working hypothesis for the various steps and conformational changes associated with the opening and closing of the channel, and thus provide a starting point for studies relating structure to function.[4-6] Kinetic models for channel gating are also useful for a number of other reasons: the quantitative description provided by a model that describes the channel kinetics allows the activity of the channel to be predicted (on average) so that the contribution of the channel to the function of the cell can be calculated. Kinetic models also provide an important tool with which to study and interpret the actions of the many factors that modulate channel activity.

Because kinetic models form a cornerstone for the study of ion channels, it is important that the models give the best possible representation of the underlying gating mechanism. The purpose of this chapter is to consider the various methods used to develop such models, list the potential artifacts and errors associated with these methods, and indicate how the artifacts and errors can be reduced or prevented. The objective is not to present all the specific details, but to outline the problems and refer interested readers to appropriate sources.

[1] B. Hille, "Ionic Channels of Excitable Membranes." Sinauer, Sunderland, Massachusetts, 1992.

[2] K. L. Magleby and C. F. Stevens, *J. Physiol. (London)* **223,** 173 (1972).

[3] R. S. Eisenberg, *J. Membr. Biol.* **115,** 1 (1990).

[4] B. Sakmann, C. Methfessel, M. Mishina, T. Takahashi, T. Takai, M. Kurasaki, K. Fukuda, and S. Numa, *Nature (London)* **318,** 538.

[5] K. Yoshii, L. Yu, K. M. Mayne, N. Davidson, and H. A. Lester, *J. Gen. Physiol.* **90,** 553, (1987).

[6] W. N. Zagotta and R. W. Aldrich, *J. Neurosci.* **10,** 1799 (1990).

Defining Markov Models for Channel Gating

One goal of single-channel analysis is to find the most likely model that describes the kinetic behavior of the channel (the activity of the channel in time). Such kinetic models are typically based on the assumption that the gating of the channel arises from transitions among a discrete number of states, with the transition rates among the states remaining constant in time for constant experimental conditions.[1,7,8] Such Markov models have proved useful for the study of channel kinetics.[1,2,6,9-13] (The relationship between the kinetic states in kinetic models and the conformational and chemical states of the channel protein has been discussed elsewhere.[14,15]

Scheme I, which has two open (O) and three closed (C) states, presents an example of a kinetic model which can approximate some of the kinetic features[16] of the large conductance Ca^{2+}-activated K^+ channel (BK channel) at a single concentration of Ca^{2+}. The development of such a kinetic

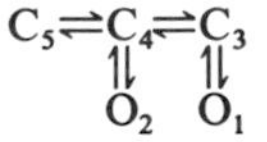

$$C_5 \rightleftharpoons C_4 \rightleftharpoons C_3$$
$$\updownarrow \quad \updownarrow$$
$$O_2 \quad O_1$$

SCHEME I

model requires determining four types of information: (1) the number of open and shut kinetic states; (2) the transition pathways among the various states, (3) the rate constants for the indicated transitions, and (4) the effects of various factors, such as membrane potential, agonists, and antagonists, on the various rate constants. The first two types of information, the numbers of states and the connections between them, are referred to as model identification. The second two types of information, the values of the rate constants and their voltage and/or agonist dependence (if any), are referred to as parameter estimation.[16a] Whereas parameters cannot be estimated without a model, the initial steps of model identification do not necessarily require parameter estimation. In practice, model identification and parameter estimation are often carried out simultaneously.[10,13]

The analysis methods emphasized in this chapter are typically applied

[7] E. Neher and C. F. Stevens, *Annu. Rev. Biophys. Bioeng.* **6**, 345 (1977).
[8] D. Colquhoun and A. G. Hawkes, *Proc. R. Soc. London B* **300**, 1 (1982).
[9] A. L. Hodgkin and A. F. Huxley, *J. Physiol. (London)* **117**, 500 (1952).
[10] R. Horn and C. A. Vandenberg, *J. Gen. Physiol.* **84**, 505 (1984).
[11] F. Bezanilla, *J. Membr. Biol.* **88**, 97 (1985).
[12] D. Colquhoun and B. Sakmann, *J. Physiol. (London)* **369**, 501 (1985).
[13] A. L. Blatz and K. L. Magleby, *J. Physiol. (London)* **378**, 141 (1986).
[14] O. B. McManus and K. L. Magleby, *J. Physiol. (London)* **402**, 79 (1988).
[15] O. B. McManus and K. L. Magleby, *J. Gen. Physiol.* **94**, 1037 (1989).
[16] K. L. Magleby and B. S. Pallotta, *J. Physiol. (London)* **344**, 585 (1983).
[16a] F. G. Ball and M. S. P. Sansom, *Proc. R. Soc. London B* **236**, 385 (1989).

to channels from which it is possible to obtain long records of activity from a single channel, such as for the BK channels that give rise to Scheme I. However, most of the artifacts and errors discussed also apply to other forms of analysis, such as those applied to inactivating channels or to selected data, such as to bursts of channel activity from multichannel patches.

Extracting Kinetic Information in Single-Channel Current Record

The now familiar step changes in current recorded from single channels are thought to indicate the opening and closing of the pore of the channel. Thus, the single-channel current provides a direct window to indicate the times when the channel opens and closes, and the durations (intervals) between the openings and closings reflect time spent in the various steps which precede the opening and closing. Model identification and parameter estimation are based on this kinetic information. Consequently, a first step in single-channel analysis is to obtain a sequential file which contains the successive durations of all open and closed intervals, together with the conductance level of each opening. Such a file would contain the useful kinetic information in the single-channel current record. The first part of this chapter considers potential errors and artifacts associated with obtaining such a file. Many of the errors summarized here (and some that will not be considered) have been described in depth by Colquhoun and Sigworth[17] in their comprehensive chapter on single-channel analysis. Their chapter is necessary reading for anyone interested in single-channel analysis.

Dead Time and Errors in Estimating Level of Filtering

Without filtering to reduce high frequency noise, or, alternatively, without a method to extract the useful information from noisy data, it would be difficult, if not impossible, to distinguish the channel openings and closings from the noise in the single-channel current record. Unfortunately, the filtering used to reduce the noise also distorts the current record[17-22] (to be discussed later). Model identification and parameter

[17] D. Colquhoun and F. Sigworth, *in* "Single-Channel Recording" (B. Sakmann and E. Neher, eds.), p. 191. Plenum, New York and London, 1983.
[18] D. L. Wilson and A. M. Brown, *IEEE Trans. Biomed. Eng.* **BME-32**, 786 (1985).
[19] B. Roux and R. Sauve, *Biophys. J.* **48**, 149 (1985).
[20] A. L. Blatz and K. L. Magleby, *Biophys. J.* **49**, 967 (1986).
[21] K. L. Magleby and D. S. Weiss, *Biophys. J.* **58**, 1411 (1990).
[22] O. B. McManus, A. L. Blatz, and K. L. Magleby, *Pfluegers Arch. Eur. J. Physiol.* **410**, 530 (1987).

estimation require that corrections be made for the distortion due to the filtering. These corrections can only be made if the level of filtering is known.

The level of filtering is most effectively expressed in terms of dead time. The dead time for a given level of filtering can be determined empirically by reducing the duration of a square pulse until the filtered amplitude of the pulse falls to 50% of the unfiltered amplitude.[17] The dead time of the filter is then the duration of the pulse that gives the 50% response. The level of filtering and the resulting dead time must be measured for the entire recording and analysis system, including patch clamp, tape recorders, amplifiers, and any additional filters. The dead time of the system can be determined by finding the duration of an interval which just reaches 50% threshold, by using rise times, or by considering the cascade of filtering introduced by each component in the system.[17] An error that is sometimes made when stating the level of filtering is to assume that the last filter used for analysis sets the level of filtering. Such an assumption ignores the filtering of all the other components and is readily apparent when the rise times of the channel openings and closings in the figures of a paper are appreciably slower than would be expected for the stated levels of filtering.

Errors Associated with Sampling

For channels with a negligible number of sojourns to subconductance states, the open and shut interval durations are conveniently measured by computer. The computer samples the current amplitude at discrete points in time. When the current amplitude is greater than the 50% current level for opening, the channel is assumed to be open, and when the current level is less than the 50% threshold level, the channel is assumed to be closed.[17] Errors are introduced in such half-amplitude (50% threshold) analysis when the sampling rate is too slow. Brief events that cross the 50% level can go undetected if they do not occur when a sample is taken, giving a sampling detection error,[22] and the durations of detected events can be under- or overestimated by almost a sample period, leading to a sampling promotion error.[22,23]

Sampling at the Nyquist rate (usually erroneously taken to be twice the cutoff frequency of the filter rather than twice the highest frequency component in the record, f_o) will not prevent sampling errors.[17] The Nyquist criterion requires that there be no frequencies higher than the selected frequency f_o and that the signal be reconstructed from the sampled data before analysis.[17] These two criteria are seldom met in single-channel analysis. In general, the sampling rates should be many times faster than

[23] S. M. Sine and J. H. Steinbach, *J. Physiol. (London)* **373,** 129 (1986).

the -3 dB frequency of the filter to prevent sampling errors. When a cubic spline function is used to interpolate (reconstruct) between points to determine the time of threshold crossing, the sampling rate should be at least 5 times the -3 dB frequency of the filter to prevent sampling error.[17]

When no interpolation is used, the sampling period should be less than 10–20% of either the dead time or the fastest time constant in the distributions of interval durations, whichever gives the fastest sampling rate.[22] Sampling at 20–40 times the -3 dB frequency of the filter is usually sufficient to meet this criterion. One method of obtaining fast sampling rates with standard computers is to record the single-channel current record and then analyze the record at greatly reduced tape speed.[13] For example, playing the tape 16 times slower gives an effective sampling period of 5 μsec for an analog-to-digital converter that is running at 80 μsec. Playing the data at reduced speed into the computer also allows additional time to inspect the current record visually for artifacts during the analysis.

Commercial programs do not necessarily sample fast enough to prevent sampling errors, and consequently, investigators using such programs should make sure that the sampling rate is sufficiently fast to prevent sampling errors.

Errors Resulting from Other than 50% Threshold Detection Level

Half-amplitude threshold sampling has the advantage over the use of other threshold levels, in that the time distortion of the filtered response at the 50% level during the rising phase is equal and opposite to the time distortion of the response during the falling phase, so that the time distortion by filtering cancels out for intervals with durations greater than four dead times.[17,22,24] With thresholds other than 50%, the time distortion of the rising and falling phases is unequal, leading to such errors as overestimation or underestimation of the duration of open intervals with durations greater than four dead times by about one dead time for 25 and 75% threshold levels, respectively.

Even with 50% threshold detection, intervals with true durations less than the dead time are missed, and the measured durations of intervals with true durations less than four dead times are underestimated. Consequently, a correction must be applied for the missed intervals (to be discussed later), and the detected intervals with durations less than four dead times are typically excluded from the fitting. Numerical corrections for the distortions of the durations of detected intervals can be made,[17] but they are only approximate owing to the effects of noise.[22]

[24] F. Sachs, J. Neil, and N. Barkakati, *Pfluegers Arch. Eur. J. Physiol.* **395**, 331, (1982).

*Necessity of Visually Inspecting Current Record and Testing Current
Measuring Programs*

The computer guideline of "garbage in/garbage out" is especially applicable to computer analysis of single-channel current records. Even the most sophisticated automated programs cannot identify and exclude all artifacts. Consequently, it is always necessary to inspect the single-channel current record during the analysis. Errors that can result from automated analysis[25] and methods to avoid them have been discussed previously.[15] Independent of any precautions taken when analyzing the current record, it is always necessary to test the program with an artificial current record generated from a known gating mechanism and with noise and filtering characteristics similar to the current record being analyzed.[21] The most important aspect of single-channel analysis is to test and define the limitations of the analysis programs being used. Spending 20–50% of the total analysis time on testing and defining the limits of the programs and methods is not unreasonable. This is especially true if commercial programs are used, since their limitations are usually unknown.

Setting Filtering Level to Reduce Noise

Each noise peak that crosses the 50% threshold introduces an erroneous event. The optimal level of filtering to reduce noise depends on the questions being asked. For example, if the goal of the experiment is to determine whether very brief open intervals occur during long shut intervals, then it would be advantageous to set the filtering so that essentially no noise peaks exceed the 50% threshold level. The appropriate level of filtering for this could be determined by setting the threshold level on the opposite side of the current record for channel opening at an absolute distance equal to 50% of the single-channel current amplitude. The data would then be analyzed and the filtering adjusted until no noise events are detected. This level of filtering would then be used to analyze the data with the threshold now set above the baseline. For symmetrical noise, a level of filtering in which no noise events exceed the 50% level below the baseline should also result in no noise events exceeding the 50% level above the baseline. Colquhoun and Sigworth[17] can be consulted for additional methods of selecting the optimum filtering level.

Time Course Fitting

Computer-assisted fitting of the time course of each interval[17] provides a means to overcome many of the artifacts that can be associated with the

[25] D. Colquhoun, *Trends Pharmacol. Sci.* **9**, 157 (1988).

initial analysis of the data, and such visual analysis can be especially useful for channels that frequently enter subconductance states, or for current records that contain numerous obvious artifacts. One difficulty with time course fitting, however, is that so much of the investigator's time is required to perform the analysis that it is not realistically possible to obtain more than a limited number of intervals. Yet large numbers of intervals are often necessary to define models and distinguish among gating mechanisms because of the stochastic variation in single-channel data.[14] Time course fitting also introduces operator bias. For example, decisions may be made as to whether an event is noise or data, or whether an unusual looking event is an artifact or the combination of several brief events. Because of the limited numbers of events and the potential for operator bias, time course fitting is not necessarily better than automated analysis procedures and, for many channels, will be worse. The decision to use time course fitting or automated analysis will depend on whether the single-channel current record is suitable for automated analysis.

Identification of Single-Channel Patches

The analysis of data from single channels is greatly simplified if the patch contains a single active channel. Identification of single-channel patches is performed by observing current steps to a single-current level under experimental conditions in which the channel is open a majority of the time. Quantitative criteria for identification of single-channel patches are presented by Colquhoun and Hawkes.[26]

Combining Single-Channel Data from Two or More Channels May Lead to Artifacts

It is becoming increasing apparent that individual channels of (presumably) the same kind can show different kinetics.[27-29] Thus, combining data from more than one channel can lead to artifacts in the analysis if the properties of the channels are sufficiently different. For example, if intervals obtained from two different channels are analyzed together, then the number of kinetic states may be overestimated if the kinetic properties of the two channels are sufficiently different that the two channels generate exponential components with different time constants.

[26] D. Colquhoun and A. G. Hawkes, *Proc. R. Soc. London B* **240**, 453 (1990).
[27] R. P. Hartshorne, B. U. Keller, J. A. Talvenheimo, W. A. Catterall, and M. Montal, *Proc. Natl. Acad. Sci. U.S.A.* **82**, 240 (1985).
[28] J. B. Patlak, M. Ortiz, and R. Horn, *Biophys. J.* **49**, 773 (1986).
[29] D. S. Weiss and K. L. Magleby, *J. Physiol. (London)* **426**, 145 (1990).

Identifying Moding and Unstable Channels

At this stage in the analysis, the single-channel current record has been measured and transformed into a sequential file containing the durations of the open and shut intervals, including the open channel conductance for each opening. For the next few sections it will be assumed that the conductance levels of all openings are the same. The next step in single-channel analysis is to test for stability and moding[14] of the single channel with stability plots. The mean durations of the open and shut intervals are plotted against interval numbers for means of 10–500 sequential open or shut intervals. Abrupt changes in the mean durations (which return just as abruptly to the normal level) indicate moding.[13,14]

Drift or instability in channel kinetics can be identified by slow drift in the means in the stability plots. BK and fast Cl⁻ channels from cultured rat skeletal muscle typically display relatively stable kinetics in between the infrequent sojourns to the different modes.[13,14,29] For an occasional membrane patch, however, there can be considerable slow drift in the mean intervals for unknown reasons. Excluding such membrane patches from analysis prevents artifacts that would arise from analyzing unstable data.

By using stability plots to identify channel activity in the different modes, it is possible to analyze the kinetics of normal channel activity and the activity in the different modes separately.[13,14] This improves the resolution of the analysis and can prevent artifactual models which might arise if the moding activity were not detected. Stability plots can also be used to identify sojourns to some of the subconductance states. The noise in the current record can lead to multiple threshold crossings for subconductance states[25] that are indicated in the stability plots as pronounced decreases in the means of both the open and shut intervals.

Preventing Artifacts When Binning Interval Durations for Dwell-Time Distributions

Once stable data have been identified and separated into the various modes, the next step is to display the open and shut interval durations as separate dwell-time distributions. The dwell-time distributions are then fitted with sums of exponentials in order to estimate the number of kinetic states. For Markov models, the number of detected exponential components in the dwell-time distributions gives a minimal estimate of the number of kinetic states.[8,14] Display of the interval durations as dwell-time distributions[22] (or transformed dwell-time distributions[30]) requires that

---

[30] F. Sigworth and S. M. Sine, *Biophys. J.* **52**, 1047 (1987).

intervals of similar durations be combined (binned) into frequency histograms. It is not necessary to combine intervals into bins for the fitting of sums of exponentials with the method of maximum likelihood,[17] but the combining of intervals into bins can save enormous amounts of computer time when large numbers of intervals are being fitted.[22]

Two types of artifacts that can arise when intervals are combined into bins will be considered. The first type of error is referred to as a binning promotion error,[22] since its cause and effect are similar to the sampling promotion error.[22,23] When intervals are binned, each of the bins contains intervals with a range of durations, with the amplitude of each bin determined by the number of intervals in that bin. Because the durations of the binned intervals are drawn from a continuous distribution that decays progressively slower as interval duration increases, the tops of the bins for the binned data are higher than the continuous distribution at the midpoint of the bin.[22] Hence, binning promotes the amplitude of the bin over the actual amplitude of the continuous function. Thus, if the tops of the bins are fitted or plotted, errors can result. Methods to correct for the binning promotion error are presented in McManus et al.[22] Alternatively, if the bin width is less than 10–20% of the fastest exponential time constant in the distribution of intervals, then the binning promotion error will be negligible, and no correction is necessary.[22] This criterion for prevention of binning promotion error also requires that the necessary criteria to prevent sampling promotion error, as discussed in a previous section, have been met.

The second type of artifact associated with binning can arise when the interval durations are measured by sampling without interpolating between the sample points to determine the time of threshold crossing. This error is a form of aliasing error. This binning aliasing error, which is sometimes mistakenly referred to as a sampling promotion error, arises when the bin width is not an integral multiple of the sampling interval. For example, assume that the bin width is 1 msec to give bins which range from 1–2, 2–3, 3–4, 4–5 msec, etc. Also assume that the sampling interval is 1.4 msec so that possible measured interval durations would be 1.4, 2.8, 4.2, 5.6 msec, etc. Now, notice that the bin from 3 to 4 msec will never contain any events, giving the false impression that intervals with this range of durations do not exist. They do exist, but, owing to the sampling interval of 1.4 msec, they are placed in the bins adjacent to the 3–4 msec bin. Thus, the 2–3 msec bin and the 4–5 msec bin will contain about 50% more intervals than they should, and the 3–4 msec bin will be empty.

For bin widths and sampling intervals different from those in the example, it is possible that some bins can contain data from two sampling intervals and other bins can contain data from only one (there are also other possibilities). The bins in which two sampling intervals fall will contain about twice as many events as bins in which one sampling interval falls, giving an artifactual alteration in the frequency of occurence of different interval durations. This binning aliasing error is easily prevented by making the bin width an integral multiple of the sampling period, or by interpolating between the sampling points with continuous time resolution. With log binning (see below) it is not possible to have the bin width be an integral number of sampling periods. A method[22] to prevent binning aliasing error with log binning is discussed in a later section.

Maximum Likelihood Fitting of Sums of Exponentials to Dwell-Time Distributions: Least-Squares Fitting Can Introduce Unnecessary Errors

Colquhoun and Sigworth[17] have detailed the reasons and methods for maximum likelihood fitting. Parameters estimated by maximum likelihood have the greatest likelihood of being correct. Least-squares fitting can be adequate in some cases but has a number of disadvantages. First, it can easily be shown, by independently fitting experimental data using the method of least squares and the method of maximum likelihood, that the estimated parameters are different for the two methods. In some cases the differences are small; in others they can be appreciable. Because maximum likelihood fitting gives the most likely estimates of the parameters, it follows that the estimates typically obtained by least squares are less likely. Second, least-squares fitting requires that a minimum number of events be in each bin. This requirement places unnecessary restrictions on the analysis and interferes with the ability to detect exponential components defined by a few intervals of long duration. Hence, least-squares fitting can introduce artifacts by not detecting components of long durations and small area. Furthermore, changing the binning to combine different numbers of intervals for least-squares fitting often changes the estimated parameters. Maximum likelihood fitting does not have these limitations since each interval is considered separately in the fitting; empty bins have no effect, and it is not necessary to bin the data for fitting.

In determining the number of exponential components with maximum likelihood fitting, the distributions are fitted with increasing numbers of components until an additional component does not significantly improve the fit. A penalty is applied for each additional free parameter. Examples of such fitting can be found in McManus and Magleby,[14] where cumulative error plots are also presented to show the improvement in the fit, if any, with each additional exponential component. Penalties for each additional exponential component can be assessed according to the Akaike information criterion[10,31] or the Schwarz criterion.[16a,32] The Schwarz criterion has a greater penalty factor for each additional exponential component than the Akaike criterion, and it favors the detection of fewer kinetic states. The number of significant exponential components required to describe the dwell-time distributions then gives an estimate of the minimum number of kinetic states.[8]

Log-Binning Method That Prevents Artifacts Associated with Binning, Plotting, and Fitting Dwell-Time Distributions

Frequency histograms, which plot numbers of intervals versus their durations (dwell-time distributions), are often used to display single-channel data. Because both the durations and numbers of intervals recorded from single channels can range over many orders of magnitude,[14] linear binning of intervals for the frequency histograms is usually inappropriate, since the bins at the longer times can contain few or no intervals. Plots made from such linear histograms thus become very noisy at long times. The other difficulty with linear binning is that hundreds of thousands of bins would be required to retain sufficient resolution at brief times while capturing the intervals at long times. Making the bins wider can reduce the fluctuation between successive bins and reduce the number of required bins, but it can lead to binning errors and loss of resolution.

McManus *et al.*[22] have developed a log-binning method that resolves these problems. Intervals are binned according to the logarithm of their durations. Such binning leads to an increase in bin width as interval durations increase, while retaining a relatively constant relationship between bin width and bin midtime. If bin width is selected to be less than about 20% of bin midtime, then there are essentially no binning errors, as each exponential component decays to negligible levels before bin width exceeds 20% of the time constant of that component. A decoding file that keeps track of the midtime of each log bin for the sampled data in that bin and also keeps track of the number of sampling periods in each log bin is used to correct the data during plotting to prevent binning aliasing errors. Maximum likelihood fitting can be applied directly to the log-binned data with negligible error, provided that there are more than about 15–25 bins per log unit.[22] Fitting such log-binned data can reduce the fitting time by many orders of magnitude for large data sets, when compared to the time required to fit the intervals individually.

Sigworth and Sine[30] have also developed a log-binning and plotting

[31] H. Akaike, *IEEE Trans. Autom. Control* **AC-19**, 716 (1974).
[32] G. Schwarz, *Ann. Stat.* **6**, 461 (1978).

method. Their method differs from the method of McManus et al.[22] by plotting the numbers of events in each bin directly without correcting for the increase in bin size with log binning. This ingenious transform gives a peak at the time constant of each plotted exponential component, with the amplitudes of the peaks being proportional to the areas of the components. The log-binning methods of McManus et al.[22] and Sigworth and Sine[30] both use maximum likelihood fitting and should give the same estimates of the parameters if used correctly.

The Sigworth and Sine method,[30] as presented, is applicable to intervals measured with continuous time resolution, that is, for interval durations measured with interpolation or exceedingly fast sampling rates. Although the method does not have corrections for use with data sampled without interpolation, it is often used by other investigators for such data. Consequently, an aliasing error is often present in the plotted figures, as indicated by pronounced fluctuation in the amplitudes of adjacent bins at the beginnings of the plots. This error could be eliminated by interpolating when measuring interval durations with sampling or by using a decoding file of the type described by McManus et al.[22] to correct for the fact that the number of sampling periods falling in each log bin is not always directly proportional to bin width.

Increasing Number of Analyzed Events Reduces Errors Arising from Stochastic Variation

The ability to detect exponential components in dwell-time distributions properly, identify models, and estimate parameters increases with the number of analyzed intervals. The more events the better, provided that the data are stable. However, there does become a point of diminishing returns because the errors tend to go as the square root of the numbers of events. Hence, obtaining 10 times as much data may only reduce the statistical variation by a factor of 3. Yet, the increased time required to obtain the additional data may add error from other sources, such as from drift in the properties of the channel or experimental conditions.

Nevertheless, to identify infrequently occurring events or processes, there is little alternative to obtaining large amounts of stable data. For example, if an exponential component contributes only 0.1% of the area to a dwell-time distribution, then it will be necessary to obtain 100,000 events in order to have about 100 events to define the component. As another example, if a sojourn to a particular mode occurs about once every 50,000 events,[14] then the analysis of 5000 events is unlikely to reveal the mode, whereas the analysis of 500,000 events is much more likely. Thus, too little experimental data can result in undetected components, overly simplified models, and large stochastic errors in the estimated parameters. Meaningful comparisons of channel kinetics requires that the numbers of analyzed events be similar. This is readily apparent from the plots in McManus and Magleby,[14] which show that the numbers of detected components in the dwell-time distributions can increase with the number of analyzed intervals.

Testing Whether Gating Is Consistent with Markov Assumption That Rate Constants Are Independent of Previous Channel Activity

As indicated in a previous section, most models for the gating of ion channels have been based on Markov models in which the rate constants (and, consequently, the lifetimes of the various states) are assumed to be independent of previous channel activity.[1,7-16a] If the Markov assumption is incorrect, then the models and estimates of the rate constants based on this assumption would be in error. To avoid this possibility, it is necessary to test the Markov assumption. One experimental approach is to analyze the durations of adjacent open and shut intervals in current records from a single channel. If the gating of the channel is described by discrete states with rate constants independent of previous channel activity, then the time constants of the exponential components in the conditional distributions of interval durations should be independent of the durations of the adjacent intervals used to select the intervals in the conditional distributions.[15,33,34] This has been found to be the case for three different channels.[15,34-36] (In applying such a test for Markov gating, it is important to test each considered non-Markov model to determine whether the method would detect the deviation from Markov gating.) Other tests consistent with Markov gating include the observation that there is symmetry of openings within bursts,[12] and that time constants are independent of previous holding potentials.[37,38]

Incidentally, it is often stated that Markov models are memoryless, and this is indeed the case for the rate constants and the lifetimes of the various states, both of which are independent of previous activity. However, Mar-

---

[33] D. R. Fredkin, M. Montal, and J. A. Rice, *in* "Proceedings of the Berkeley Conference in Honor of Jerzy Neyman and Jack Kiefer" (L. M. LeCam and R. A. Olshen, eds.), p. 269. Wadsworth, Belmont, California, 1985.

[34] B. S. Pallotta, *J. Physiol. (London)* **363**, 501 (1985).

[35] A. L. Blatz and K. L. Magleby, *J. Physiol. (London)* **410**, 561 (1989).

[36] D. S. Weiss and K. L. Magleby, *J. Neurosci.* **9**, 1314 (1989).

[37] K. L. Magleby and C. F. Stevens, *J. Physiol. (London)* **223**, 151 (1972).

[38] R. Hanin, *J. Gen. Physiol.* **92**, 331 (1988).

kov models do have a memory; the state to be entered next depends in a probabilistic manner on the current state. This memory results because there are specific transition pathways between the various states. It is this memory of Markov models that forms the basis for the testing of Markov gating with conditional distributions (described above) and also for testing for the number of gateway states (see below) by looking for correlation among successive intervals.

Testing Whether Gating Is Consistent with Microscopic Reversibility

Models for the gating of ion channels often assume that the channels are at thermodynamic equilibrium, obeying the principle of microscopic reversibility. With microscopic reversibility, all the reaction steps will proceed, on average, at the same rate in each direction.[39] Thus, with a cyclic mechanism and microscopic reversibility, the reaction cannot proceed more rapidly in one direction around a loop than in the opposite direction.[39] (States O_1, O_2, C_3, and C_4 would form a loop in Scheme I if transitions were allowed between O_1 and O_2.) The condition of microscopic reversibility can be imposed on a kinetic scheme by constraining the rate constants so that the product of the rate constants in one direction around each loop equals the product of the rate constants in the opposite direction.[39] Because the gating of some channels does not obey microscopic reversibility,[40,41] it is necessary to test for microscopic reversibility for each channel. An assumption of microscopic reversibility, when microscopic reversibility is not obeyed, could lead to errors in models and rate constants.

Microscopic reversibility requires that the data show time reversibility, such that the kinetics are independent of whether the data are analyzed forward or backward in time.[39,42] Such tests for the glutamate-activated channel,[43] the BK channel,[15,34] and for bursts of activity from the acetylcholine receptor channel[12] are consistent with microscopic reversibility. Caution must be used, however, in interpreting the results. An observation of kinetics dependent on the time direction of analysis would indicate deviation from microscopic reversibility. The opposite observation of ki-

[39] D. Colquhoun and A. G. Hawkes, *in* "Single-Channel Recording" (B. Sakmann and E. Neher, eds.), p. 231. Plenum, New York and London, 1983.

[40] A. Hamill and B. Sakmann, *Nature (London)* **294,** 462 (1981).

[41] E. A. Richard and C. Miller, *Science* **247,** 1208 (1990).

[42] I. Z. Steinberg, *J. Theor. Biol.* **124,** 71 (1987).

[43] C. J. Kerry, R. L. Ramsey, M. S. P. Sansom, and P. N. R. Usherwood, *Biophys. J.* **53,** 39 (1988).

netics independent of the time direction of analysis, while consistent with microscopic reversibility, does not necessarily establish it.[42]

Excluding Some Incorrect Classes of Models

At this stage in the analysis, tests have been made for Markov gating and for microscopic reversibility. If the kinetics appear consistent with Markov gating, then an estimate of the minimum number of open and closed kinetic states can be made from the number of significant exponential components required to describe the open and shut dwell-time distributions. This estimate would be carried out separately for activity in each mode of the channel. To simplify the presentation in the remainder of the chapter, it will be assumed that sojourns into modes other than normal are insignificant.

If two exponential components were required to describe the open dwell-time distribution and three for the closed dwell-time distribution, then these observations would suggest a minimum of two open and three shut kinetic states. The next step in the analysis is to list all the possible models with two open and three closed states. Scheme I is an example of one such model. Scheme II is an example of another.

$$C_5 \rightleftharpoons C_4 \rightleftharpoons C_3 \rightleftharpoons O_2 \rightleftharpoons O_1$$
SCHEME II

The possibility of selecting an incorrect model can be reduced if some simple tests are carried out to test certain classes of models before more complex fitting procedures are applied. Scheme II, with one independent transition pathway between the open and closed states (one gateway state) predicts that the durations of open intervals would be independent of the durations of adjacent shut intervals.[15,33,44,45] This is the case since all open intervals are adjacent to shut intervals originating from the same identical shut states. Scheme I, with two independent transition pathways between open and closed states (two gateway states), predicts a dependent relationship between the durations of adjacent open and shut intervals. This is the case for Scheme I, since each open state is connected to a different shut state.

Schemes with a single gateway state can be distinguished from those with two or more gateway states by determining whether the durations of adjacent intervals are dependent.[15,33,44,45] Such correlation has been ob-

[44] O. B. McManus, A. L. Blatz, and K. L. Magleby, *Nature (London)* **317**, 625 (1985).
[45] F. G. Ball, C. J. Kerry, R. L. Ramsey, M. S. P. Sansom, and P. N. R. Usherwood, *Biophys. J.* **54**, 309 (1988).

served for a number of channels including BK channels,[15,34,44] fast Cl⁻ channels,[35,44] acetylcholine receptor channels (correlation of open intervals),[46] glutamate channels,[43] and α-aminobutyric acid (GABA)-activated channels.[36] Thus, these channels have more than one gateway state, so that schemes such as Scheme II can be excluded for these channels.

After listing all the possible models and excluding some classes of models, the next step in the analysis would be to obtain estimates of the most likely rate constants for each model in light of the experimental data (as described below) and then rank the models in terms of likelihood. (A few of the excluded models might also be included in the analysis for purposes of comparison.) Statistical tests[10,16a,47] would then be applied to determine if the likelihood differences among the different models are significant.

Errors Associated with Filtering and Correcting for Missed Events: Assumption of Idealized Filtering

Identifying the most likely model requires finding the most likely set of rate constants for each model. Estimating the most likely rate constants would be relatively straightforward[7,8,10,16a,33] if it were not for the problems of limited time resolution and noise. The single-channel current record has limited time resolution owing to the finite frequency response of the patch clamp, amplifiers, and recording devices. The time resolution is typically reduced even further when additional filtering is added to reduce the noise to acceptable levels before analysis. With limited frequency response the brief open and shut intervals can be greatly attenuated so that they are not detected. Such missed events lead to an increase in the observed durations of the open and shut intervals.[17-20,48,49]

Analyzing data without corrections for missed events can lead to large errors in estimated rate constants, since the observed durations of the intervals in the filtered data can be many times greater than the durations of the actual opening and closings of the channel. Various methods to correct for missed events will be considered in some of the following sections. Although the considered methods give only approximate corrections, one or more of them should be adequate for most applications. The important thing is to apply an appropriate correction for missed events. If

[46] M. B. Jackson, B. S. Wong, C. E. Morris, and H. Lecar, *Biophys. J.* **42**, 109 (1983).
[47] R. Horn, *Biophys. J.* **51**, 255 (1987).
[48] G. F. Yeo, R. K. Milne, R. O. Edeson, and B. W. Madsen, *Proc. R. Soc. London B* **235**, 63 (1988).
[49] R. K. Milne, G. F. Yeo, R. O. Edeson, and B. W. Madsen, *Proc. R. Soc. London B* **233**, 247 (1988).

such a correction is not applied, then it is likely that erroneous conclusions will be reached. For example, an experimental procedure that changes the fraction of shut events that are detected will lead to a change in the observed mean open times. Without correction for missed events, the erroneous conclusion would be made that the procedure altered the open state rather than the shut.

Many of the methods used to correct for missed events assume idealized filtering, such that all intervals with durations less than the dead time go undetected.[17,19,20,48-50] The assumption of idealized filtering can lead to errors because intervals with durations less than the dead time can be detected; subthreshold events can fall on one another and sum to threshold.[51] Furthermore, methods that assume idealized filtering typically do not take into account the effects of noise in the current record. The effects of noise are detailed in a later section.

One of the first correction methods based on the assumption of idealized filtering was the method of Colquhoun and Sigworth.[17] Their method, which is only applicable to channels with two states (open and closed), is preferred[20,52] over the method of Neher.[53] The method of Colquhoun and Sigworth[17] and the extension of their method by Blatz and Magleby[20] for more complex models can introduce the interesting artifact, when correcting for missed events, of obtaining a false solution for the rate constants which can differ from the correct solution by factors of 3–10 or more.[13,17] A false solution can be obtained because the false solution predicts a dwell-time distribution almost identical to that of the correct solution.[51] Which of the two solutions is the correct one can be found by analyzing the data with several different dead times.[13,17]

The method of Blatz and Magleby[20] has proved useful,[13,35,36] but it is limited in its applicability because of the assumption that multistate transitions do not occur during missed events. Crouzy and Sigworth[50] have overcome this problem by introducing virtual (phantom[20]) states and using the results of Kienker[54] to assign the virtual states in complex schemes. Because the method of Crouzy and Sigworth[50] uses virtual states, it would be expected to prevent the false solutions inherent in methods like that of Blatz and Magleby[20] which use the approach of Colquhoun and Sigworth.[17] The method of Crouzy and Sigworth[50] appears to be one of the most powerful approaches available to correct for missed events with an assumption of idealized filtering and no noise.

[50] S. C. Crouzy and F. J. Slgworth, *Biophys. J.* **58**, 731 (1990).
[51] K. L. Magleby and D. S. Weiss, *Biophys. J.* **58**, 1411 (1990).
[52] R. K. Milne, G. F. Yeo, B. W. Madsen, and R. O. Edeson, *Biophys. J.* **55**, 673 (1989).
[53] E. Neher, *J. Physiol. (London)* **339**, 663 (1983).
[54] P. Kienker, *Proc. R. Soc. London B* **236**, 269 (1989).

Roux and Sauve[19] have presented a general method to correct for missed events, but their method is difficult to apply in practice because of the need to numerically invert a Laplace transform. They also present a first-order solution which ignores the time of the missed events but which should be adequate for some applications.

It is absolutely essential to check for errors when correcting for missed events with an assumption of idealized filtering. Such tests could have prevented some of the incorrect results and conclusions reached in the literature for those conditions in which an assumption of idealized filtering gave large errors. Checking for errors is done by simulating a current record with filtering (and noise, see below) for the estimated model and rate constants,[51] and then analyzing the current record in the same identical manner that was used to analyze the experimental current record. The entire process is then repeated several times. The difference between the rate constants used to simulate the current record and the rate constants derived from the simulated current record gives a measure of the errors and uncertainty in the estimated rate constants. Such simulation testing can be very time consuming. However, meaningful kinetic analysis often requires that 20–50% of the total analysis time be devoted to testing programs and assumptions and estimating errors for the applied conditions.

When such testing reveals that the errors arising from the assumption of idealized filtering and/or no noise (see below) are excessive, then the correction methods considered above can no longer be used. Methods applicable to data with large levels of filtering and noise are considered in later sections.

Errors Associated with Assumption of No Noise

The noise in the current record can directly introduce artifacts as well as have profound effects on the corrections for missed events.[22,51] When the noise peaks in the absence of channel activity exceed threshold for detection of intervals, then the noise peaks directly add false events, as detailed by Colquhoun and Sigworth.[17] Such false events from noise can erroneously reduce the durations of observed open and shut intervals and add erroneous fast components to the dwell-time distributions. Even when the noise peaks in the absence of channel activity are less than the threshold level for event detection, noise still introduces error.[22,51] The reason for this is that noise increases the total number of (true) events that are detected, when compared to detection in the absence of noise. The magnitude of this noise detection error depends on the time constants of the exponential components in the distributions, as shown in McManus et al.[22]

The errors associated with an assumption of no noise can be determined by simulation, as described above, to test for errors associated with an assumption of idealized filtering. To test for errors associated with noise, the simulated current record includes noise equivalent to that in the experimental current record. When the errors associated with an assumption of no noise become excessive, then other methods which take noise into account must be used (see later sections).

Errors Associated with Synthetic Approach to Kinetic Analysis

Horn and Vandenberg[10] have detailed the advantages of statistical over synthetic methods of kinetic analysis. The synthetic approach involves the construction of a model with a minimal number of parameters, with the estimation of the parameters performed in a semiquantitative manner and the evaluation of the model performed by visually comparing predictions of the model with the experimental data. Usually only one model, which is found to be "consistent" with the experimental data, is presented. There is little reason to compare multiple models, because the semiquantitative approach does not necessarily allow an adequate basis for comparison.

In contrast, the statistical approach[10] considers large numbers of models and ranks them according to their likelihood, as determined by statistical (maximum likelihood) fitting methods. The statistical approach has a number of advantages over the synthetic approach. First, although investigators often talk of correct and incorrect models, in reality, all models are likely to be incorrect to some degree. The statistical approach allows models to be ranked in terms of the likelihood that the experimental data were generated by the models, so that there is a rational basis for choosing (ranking) different models. With the statistical approach it also possible to calculate the likelihood for a theoretical best description of the data[13] so that the fits of the considered models can be compared to the theoretical best fit. Such comparisons are useful since models many orders of magnitude less likely than the theoretical best fit can still give descriptions of the data that would be considered to be "consistent" with the experimental observation (O. B. McManus and K. L. Magleby, unpublished observations).

Another advantage of the statistical approach[10] of testing and ranking many different models is that it adds an element of reality to the process of model building. A study using the synthetic approach that presents one model "consistent" with the data may, at first glance, appear to be a greater step forward than a study that cannot identify the "correct" model because several of the top ranked models are found to have similar likelihoods. However, the synthetic approach can be misleading, since it is unlikely that the tested model is the most likely, even if it is "consistent" with the data.

In contrast, the second study with multiple models better defines the level of understanding.

It follows from the above (and following discussions) that it is relatively easy to find a model consistent with the experimental observations; the difficult task to is to identify the most likely model or models. Thus, if a single model is tried, it will probably appear consistent with the experimental data but is unlikely to be the most likely model.

Fitting Data Is Preferred over Fitting Parameters Derived from Data

In some types of single-channel analysis, specific parameters are derived from the single-channel data, and it is these specific parameters that are fitted. Such a selective approach can have some advantages, since it allows the investigator to focus on a particular aspect of the kinetics, such as, for example, adjacent interval durations[35] or burst kinetics.[55] It is usually preferred, however, to fit the data, rather than specific parameters derived from the data, since additional error can arise from fitting the specific parameters. For example, the fitting of a model for gating mechanism to the data in dwell-time distributions could be performed in either of two ways: (1) by fitting the experimental observations in the dwell-time distributions directly or (2) by first fitting exponentials to the dwell-time distributions to estimate specific parameters (the areas and time constants of the components) and then fitting the model to the estimated specific parameters.

Because the most likely estimates of the model parameters are obtained by fitting the experimental observations in the distributions directly with maximum likelihood methods, it follows that the estimates of the model parameters obtained by fitting the specific parameters are not the most likely. The reason for this is that some of the specific parameters may be poorly defined so that they can vary severalfold with little effect on the likelihood. When fitting the experimental observations directly with a model, the poorly defined parameters can change as necessary to obtain the overall best fit. If the specific parameters are being fitted directly, then any deviation from the poorly defined parameters will unnecessarily constrain the fitting.

Another difficulty with fitting specific parameters derived from the experimental observations, rather than fitting the experimental observations directly, is that there is no longer a rational basis for ranking models. Even if maximum likelihood fitting is used to fit the specific parameters,

[55] K. L. Magleby and B. S. Pallotta, *J. Physiol. (London)* **344,** 605 (1983).

the model fits to the specific parameters are not likely to give the same answers as maximum likelihood fits to the data. Hence, one model might be erroneously rejected in favor of another because it give a larger error in some particular derived parameter. Yet, the rejected model may well be the most likely of the two if all the data were considered and if the fitting were evaluated in terms of likelihood of the data.

Errors Associated with Analysis Restricted to One-Dimensional Dwell-Time Distributions

One-dimensional dwell-time distributions plot the numbers of observed intervals against their durations, and they have been a major method of presenting single-channel data. However, one-dimensional distributions do not contain all the kinetic information in the single-channel data, as the correlation information is lacking. Such correlation information can be essential to distinguish among models.[16a,33,35,43,45,46,56,57]

If the kinetic analysis of single channels is restricted to the analysis of one-dimensional dwell-time distributions, then it should be anticipated that several different most likely models might be found, each with a likelihood identical to that of the theoretical best fit for one-dimensional distributions.[13,36] Furthermore, it is possible that none of the most likely models which have been found for the one-dimensional distributions are viable models, as they may all be inconsistent with the correlation information in the experimental data. If only one model is tried with one-dimensional analysis, then the findings may be even more misleading, since it will not become apparent that there are probably other models which can give similar excellent descriptions of the one-dimensional distributions. Methods that use the correlation information to increase the ability to identify models, such as full likelihood approaches and fitting of two-dimensional dwell-time distributions, are considered in later sections.

Errors Associated with Estimating Parameters

Methods used to identify models and estimate parameters can typically find a model and associated parameters that appear consistent with the experimental data. As mentioned above, such models may not be the most likely models. It is also the case that the estimated parameters may have large errors. It was already indicated in a previous section that certain types of corrections for missed events can lead to a false solution where the

[56] D. Colquhoun and A. G. Hawkes, *Proc. R. Soc. London B* **199,** 231 (1987).
[57] I. Z. Steinberg, *Biophys. J.* **52,** 47 (1987).

estimated parameters differ typically 3- to 10-fold from the correct solution.[13,17] This type of error can be prevented by analyzing the data at several different dead times,[13,17] or by taking the true effects of filtering and noise into account.[51] It is also possible to obtain incorrect estimates of the parameters because there is insufficient information in the fitted data to define the parameters.[33,58,59] Some of the parameters in loops can be especially difficult to define, and simulations have shown (O. B. McManus and K. L. Magleby, unpublished observations) that small perturbations of the type that might arise from noise or imperfect corrections for missed events can lead to large changes in the values of the poorly defined parameters. Consequently, it is necessary to assess the uncertainty in the estimated parameters by numerical methods,[17] resampling,[47] or simulation.[29]

Comparison of Errors Associated with Simple versus More Complex Models

A common feature of kinetic analysis is that increases in time resolution and/or in the amount of data analyzed typically reveal new phenomena that were previously unresolved. The current "simple" models are then extended or changed to take the new observations into account, resulting in progressively more "complex" models. Such kinetic expansion has given[14,60] and should continue to give increasingly greater insight into the mechanisms by which ion channels gate their pores. With the kinetic expansion it is found that phenomena that were previously associated with single states or single rate constants actually arose from multiple states or rate constants. Although the time resolution for the kinetic analysis of ion channels is now in the microsecond to millisecond range, it is known that motions in proteins extend orders of magnitude faster.[61] Furthermore, the power of techniques to study single-channel kinetics is rapidly increasing. Hence, the analysis of ion channels is in its infancy, and it might be anticipated that kinetic expansion of gating mechanisms will continue. The consequence of such expansion is that there are typically a series of progressively more complex models that are used to describe the gating of any channel at any given time.

As long as it is realized that the states and rate constants in the simpler models can represent multiple states and rate constants, then the simpler models are still useful: they are easier to understand and to apply, and, most importantly, because they have fewer parameters, the estimated pa-

[58] R. J. Bauer, B. F. Bowman, and J. L. Kenyon, *Biophys. J.* **52**, 961 (1987).
[59] L. Goldman, *Biophys. J.* **60**, 519 (1991).
[60] D. Colquhoun and B. Sakmann, *Nature (London)* **294**, 464 (1981).
[61] R. Elber and M. Karplus, *Science* **235**, 318 (1987).

rameters can be determined with less error. This makes for easier comparisons among channels and findings from different laboratories. The simpler models have the disadvantage that they can give misleading information about gating mechanism because various states and transition pathways are missing. Hence, although simpler models can have well-defined parameters, the meaning of the parameters can be misleading because of the lumping of states and rate constants. Simpler models also, typically, only approximate the experimental data.

More complex models have the advantage over simpler models in that they define the current limits of knowledge about the gating mechanism. They also have the advantage of giving better descriptions of the experimental data. More complex models have the disadvantage that the parameters can be less well defined. This is the case since it becomes increasingly more difficult to estimate parameters as the complexity of the model increases, and this is especially the case for models with loops where some of the parameters may only be definable within a wide range.[33] Another disadvantage of more complex models is that the added complexity may actually arise (unknowingly) from drift or artifact in the data rather than from gating mechanism. Presumably, such errors would be detected with further analysis over wider ranges of conditions.

Both simple and complex models have applicability as long as their advantages and disadvantages are appreciated. An alternative approach to some of the problems associated with simple and complex models is to examine complex models with a limited number of free parameters.[62,63] In this way it is possible to have large numbers of states, but with a limited number of parameters, so that the parameters can be defined. Such models are just beginning to be examined.[64]

Importance of Fitting All the Data: Simultaneous Fitting

In some instances a model will be developed with data collected under one set of experimental conditions and then tested with data collected under different conditions. If the model can describe the data under different conditions, then it is said to have predictive power. Although it can be satisfying to test a model in this manner, this approach does not necessarily yield the most likely model or estimates of parameters. The most rigorous identity of models and estimation of parameters will be made when all the available data, including that obtained under different experimental con-

[62] G. L. Millhauser, E. E. Salpeter, and R. E. Oswald, *Proc. Natl. Acad. Sci. U.S.A.* **85**, 1503 (1988).
[63] P. Lauger, *Biophys. J.* **53**, 877 (1988).
[64] G. L. Millhauser, E. E. Salpeter, and R. E. Oswald, *Biophys. J.* **54**, 1165 (1988).

ditions, are simultaneously fitted.[16a,58,59,65] Such simultaneous fitting (global parameter optimization,[66] and see Goldman[59]) places the greatest constraints on models and parameters, and it can allow model discrimination that is not available by other means. Furthermore, it can be better to have fewer data at a number of conditions, rather than more data at each of a few conditions.[16a]

A danger in simultaneous fitting can be to overestimate the amount of new information available in the data obtained under different experimental conditions, and erroneously assume that there is sufficient independent kinetic information to define the parameters when there is not. For example, at first it might be thought that obtaining dwell-time distributions at five different voltages would increase the amount of kinetic information 5-fold for a voltage-dependent channel. However, the increase in information is likely to be less than this since all the rate constants may not be voltage sensitive and the various voltage-dependent parameters may not be independent.[59] Perhaps the prudent approach in single-channel analysis is to assume that the determined rate constants are not necessarily unique unless proved otherwise.

Full Maximum Likelihood Methods Are Preferred over Partial Maximum Likelihood Methods

Single-channel data consist of a series of open and shut intervals, with the durations of successive intervals containing the correlation information. The best model is the one that has the greatest likelihood of generating the observed sequence of intervals[67] (with appropriate penalty for additional free parameters). Analysis methods that maximize the probability of describing the observed sequence of intervals are full maximum likelihood methods. Analysis methods that use maximum likelihood fitting to describe data drawn from the sequence of intervals, such as the dwell-time distributions, are partial maximum likelihood methods. In general, full likelihood methods are to be preferred over partial methods.[16a,67] This is the case because, whereas partial likelihood methods will determine the most likely models for the data they consider, the models developed by the partial likelihood methods may not be the most likely when all the kinetic information in the experimental data is considered.

In a landmark paper Horn and Lange[67] presented the first full maximum likelihood method of single-channel analysis, and Horn and Van-

[65] O. B. McManus and K. L. Magleby, *Biophys. J.* **49**, 171a (1986).
[66] J. R. Balser, D. M. Roden, and P. B. Bennett, *Biophys. J.* **57**, 433 (1990).
[67] R. Horn and K. Lange, *Biophys. J.* **43**, 207 (1983).

denberg[10] then used this method to study Na^+ channel gating. More recently Ball and Sansom[16a] and Chung et al.[68] have also presented full likelihood methods. The applicability of the methods of Horn and Lange[67] and Ball and Sansom[16a] is limited (as presented) because they do not correct for missed events or account for the effects of noise. R. Horn (personal communication, 1990) has since extended his method to include a correction for missed events based on the approach of Roux and Sauve.[19]

The full likelihood method of Chung et al.[68] has a number of powerful features. It fits the experimental current record directly, eliminating the need for initial analysis, except to test for stability and moding. It also allows transitions to multiple subconductance states. These features make the method of Chung et al.[68] attractive for channels that readily enter subconductance states, and it may eliminate the need for computer-assisted visual analysis of subconductance levels. Another advantage of the method of Chung et al.[68] is that it takes noise into account, and the noise levels can be larger than the signal levels. Consequently, the method should be suited for analysis of channels with small conductances where the noise can dominate. The method should also be suited to the analysis of large channels with reduced levels of filtering and increased frequency response, where the noise can also dominate. The disadvantages of the method of Chung et al.[68] appear to be that it does not correct for missed events and that there are significant errors in the detection of brief events in the presence of noise (see Chung et al.[68]). Thus, model identification and parameter estimation will be compromised. However, as less filtering is required with this method than with some methods, errors resulting from lack of correction for missed events may be acceptable in some cases.

Fitting Two-Dimensional Dwell-Time Distributions May Approximate Full Likelihood Methods

Although full likelihood methods would, in general, be preferred over partial likelihood methods, it might not always be practical to use them. For example, full likelihood methods are less readily applied to large numbers of events because of the dramatic increase in computation times. Yet, large numbers of events may be necessary in order to detect kinetic states with limited areas and identify complex models. What is needed, then, is a partial maximum likelihood method that effectively includes all the information used by full likelihood methods. The theoretical study of Fredkin et al.[33] suggests such an approach. Fredkin et al.[33] found that

[68] S. H. Chung, J. B. Moore, L. Xia, L. S. Premkumar, and P. W. Gage, *Philos. Trans. R. Soc. London, Ser. B* **329**, 265 (1990).

two-dimensional dwell-time distributions contain the useful information in the single-channel data that is contained in higher order distributions. (Two-dimensional dwell-time distributions plot the number of observations against the joint probabilities of observing two intervals of specified durations.) Because the experimental observations can be combined (binned) when generating two-dimensional dwell-time distributions, any number of events can be analyzed with a negligible increase in the fitting time, as was the case for one-dimensional dwell-time distributions. Yet, correlation information that is essential to identify many types of models is retained in the two-dimensional dwell-time distributions.[33]

In contrast to the suggestion that two-dimensional distributions may be sufficient, Ball and Sansom[16a] have observed greater errors with the fitting of two-dimensional distributions than when fitting with a full likelihood method. However, these greater errors may have arisen because no corrections were applied for missed events and because the two-dimensional distributions fitted were only for adjacent open and shut interval pairs. The inclusion of additional two-dimensional distributions of adjacent open intervals and of adjacent shut intervals might be required, in practice, as open–shut interval pairs are not sufficient to contain all the useful correlation information. For example, if the openings generated by two different open states for a scheme, such as Scheme I, had identical mean dwell times, then the two open states would not be detected with two-dimensional distributions of adjacent open and shut times. A two-dimensional distribution of adjacent shut times could, however, reveal the correlation information in the kinetic scheme, since, on average, adjacent shut times might be either brief and correlated, from transitions such as $O_1-C_3-O_1-C_3-O_1$, or long and correlated, from transitions such as $O_2-C_4-O_2-C_4-O_2$.

Simulation Provides Means to Correct for True Effects of Filtering and Noise

The true effects of filtering and noise can be taken into account during the analysis by using simulation to predict the expected distributions for a given model and rate constants.[51] For example, comparisons of amplitude histograms of currents from experimental and simulated data can be used to account for filtering and noise if the durations of open and shut intervals are sufficiently brief to yield flickery data.[69,70] Although such current amplitude methods have proved useful for the analysis of two-state blocking models,[69,70] they would lack resolution for more complicated gating mech-

[69] G. Yellen, *J. Gen. Physiol.* **84,** 157 (1984).
[70] D. Pietrobon, B. Prod'hom, and P. Hess, *J. Gen. Physiol.* **94,** 1 (1989).

anisms or for data in which the intervals are not heavily attenuated. The current amplitude methods also do not directly take the correlation information into account. The following section presents a simulation method which takes these factors into account.

Identifying Kinetic Gating Mechanisms for Ion Channels by Using Two-Dimensional Distributions of Simulated Dwell Times: A Method That Includes Correlation Information and Takes True Effects of Filtering and Noise into Account

Magleby and Weiss[51,71] have presented a simulation method which should be able to prevent many of the errors and artifacts associated with single-channel analysis. The experimental current record is first analyzed with 50% threshold detection and the interval durations binned into one or more two-dimensional dwell-time distributions. As discussed in a previous section, the two-dimensional dwell-time distributions retain the correlation information in the single-channel data that is so important for identifying models.[33] The next step in the method is to simulate a single-channel current for a given kinetic scheme and rate constants. The true effects of filtering and noise are taken into account for the simulation. The simulation of the current is done with the same level of filtering and noise as was used for the experimental current record so that the effects of filtering and noise will be the same for both the simulated and experimental data.

The simulated current record is then analyzed in the exact same manner as was used to analyze the experimental current record to produce a two-dimensional distribution of the dwell times measured from the simulated current record. A bin by bin comparison of the simulated and experimental two-dimensional dwell-time distributions is performed to calculate the likelihood that the experimental two-dimensional distributions were obtained from the model used to generate the simulated two-dimensional dwell-time distributions. A search routine is then used to find the parameters, such as rate constants, which maximize the likelihood for the given kinetic scheme. The process is then repeated for the various models to be tested and the models ranked in order of likelihood.

Because the two-dimensional simulation method treats the experimental and simulated current records identically, any errors from noise, filtering, missed events, sampling, and binning are essentially identical (on average) for experimental and simulated data and cancel out. Thus, the simulation method automatically excludes many of the errors and artifacts often associated with single-channel analysis.

---

[71] K. L. Magleby and D. S. Weiss, *Proc. R. Soc. London B* **241**, 220 (1990).

There are a number of other advantages for the two-dimensional simulation method. The simulation method takes into account the true effects of filtering and noise, so that the false solutions which can arise in other methods from the assumption of idealized filtering[13,17] are automatically excluded. The two-dimensional simulation method has a greatly increased ability to identify models over methods which use one-dimensional distributions,[71] since the two-dimensional simulation method retains the correlation information. The two-dimensional simulation method makes no assumptions as to the form of the dwell-time distributions, so it can be applied to non-Markov as well as Markov gating mechanisms, and thermodynamic equilibrium is not a requirement. The two-dimensional simulation method should be expandable to include membrane patches that contain more than one channel by having multiple threshold levels and generating two-dimensional distributions for each level. The two-dimensional simulation method should also be applicable to nonstationary data, since nonstationary responses to voltage or agonist changes are easily calculated by simulation. Two-dimensional distributions constructed at various times during the experimental and simulated data would then be compared. The two-dimensional simulation methjod is also well suited to both small and large amounts of data. There are no limits on the minimal amount of data required by the method, except for those imposed by stochastic variation, which apply to all methods, and the time required for fitting of large data sets is almost independent of the number of data points, because of the log binning.

The two-dimensional simulation method also has some disadvantages. The method, as presented, is not directly applicable to channels which frequently enter subconductance states. However, it is likely that the method could be expanded to include such channels. Another disadvantage of the two-dimensional simulation method is the extended times required for the calculations because of the simulation. Days to weeks of computer time would be required for solutions using 486 PC computers. This can be reduced to hours to days with add-in processors or modestly priced workstations, and the time should be considerably less than this in a few years owing to the rapid advances in affordable computing power. Although extended computing times are a disadvantage, this may be a small penalty to pay for elimination of many of the errors and artifacts associated with single-channel analysis. Another disadvantage of the two-dimensional simulation method is that it is not a full likelihood method. However, the use of two-dimensional dwell-time distributions should allow the two-dimensional simulation method to make effective use of the correlation information.[33] Finally, the two-dimensional simulation method has stochastic error in the estimates of the rate constants because of the stochastic variation inherent in using simulation to predict the two-dimensional distributions. However, the stochastic uncertainty can be reduced to insignificant levels if a sufficiently large number of simulated events is used.

Summary

The power of single-channel analysis techniques has rapidly expanded during the past few years, giving investigators increased ability to identify models and estimate parameters while reducing error and artifacts. At present, however, there is no single best method, as even the most advanced techniques have various limitations which depend on the experimental data and models being examined. Consequently, for the examined models and experimental data, the most critical part of single-channel analysis is to estimate errors and evaluate the ability of the methods used to discriminate among possible gating mechanisms. The magnitudes of the errors and the ability to identify models and estimate parameters depend on the models being examined as well as the experimental conditions and data. Consequently, the evaluation of the errors associated with each method needs to be repeated when the experimental data and examined models change.

Acknowledgments

The writing of this review was supported in part by grants from the National Institutes of Health (AR 32805) and the Muscular Dystrophy Association.

[54] Analysis of Drug Action at Single-Channel Level

By Edward Moczydlowski

Introduction

A large number of pharmacologically active molecules perturb the function of ion channels. These effects may occur as a direct drug–channel interaction or indirectly, via various regulatory pathways. This chapter is primarily concerned with the analysis of direct ligand interactions with channel proteins. In principle, similar approaches can be used to characterize channel regulation by phosphorylation or G proteins; however, these multicomponent systems are best considered in the context of neuromodulation or receptor-mediated signal transduction.[1] Although any small molecule or ion can be considered as a drug, the focus here is on the effects of organic molecules or peptides, since the behavior of inorganic ions that function as poor substrates (i.e., slowly permeating current carriers) or blockers of channel pores is too vast a topic to cover here and is the subject of other specialized reviews.[2-4]

General Considerations

Initial identification of drugs and toxins that affect ion channel currents is usually made by a wide variety of macroscopic techniques for sampling channel populations that include radioactive flux assays, membrane potential measurements, whole-cell voltage clamping, and noise analysis. Such techniques are suitable for characterization of physiological responses to drugs and often permit the identification of a specific type of ion channel that mediates the drug response. However, these macroscopic techniques present difficulties for detailed molecular analysis of drug mechanisms. Macroscopic currents are a function of three parameters: (1) the number of active channels being sampled, (2) the unitary conductance of individual channels, and (3) the average probability of channel opening. Thus, a macroscopic drug effect may result from a change in any or all of these underlying parameters. Even though macroscopic recording techniques can often effectively isolate an ion-selective current, such as a Na^+ or a Ca^{2+} current, any given cell may express several subtypes or isoforms of a given channel type. Molecular cloning studies of Na^+, Ca^{2+}, and K^+ channels have revealed that multigene families underlie the pharmacological and functional diversity of these ionic currents. Therefore, analysis of macroscopic currents entails the uncertainty of whether kinetic heterogeneity is due to multistate behavior of a homogeneous population or a heterogeneous population that has not been completely "dissected" by the use of special voltage protocols or drug and toxin cocktails designed to eliminate contaminating currents.

Even with the assumption that a clean current dissection can be achieved, it is often difficult to determine from macroscopic measurements whether a drug acts by altering the ion permeation properties or the gating behavior of the channel. The advantage of single-channel recording is

[1] L. K. Kaczmarek and I. B. Levitan, eds., "Neuromodulation: The Biochemical Control of Neuronal Excitability." Oxford Univ. Press, New York, 1987.

[2] G. Yellen, *Annu. Rev. Biophys. Biophys. Chem.* **16,** 227 (1987).

[3] T. Begenisich, *Annu. Rev. Biophys. Biophys. Chem.* **16,** 247 (1987).

[4] R. W. Tsien, P. Hess, E. W. McCleskey, and R. L. Rosenberg, *Annu. Rev. Biophys. Biophys. Chem.* **16,** 265 (1987).

particularly evident in situations where channels display multiple substates or where drugs induce new subconductance states. In some cases, it may be virtually impossible to unambiguously extract such details from a multichannel record.

Whereas macroscopic recordings reveal too few molecular details about the underlying channel fluctuations, microscopic recordings of single channels conversely contain too much information. As records of the conformational history of large proteins detected by current fluctuations, high resolution single-channel recordings often exhibit a bewildering array of phenomenology. Reflecting the complexity of protein conformational dynamics, these phenomena may include brief, unresolved closures or openings, multiexponential distributions of open and closed state events, nonstationary patterns of gating behavior (mode behavior), excess noise associated with open states of the channel, and complex patterns of substate behavior. Addition of a drug may perturb any or all of these phenomena and induce new patterns of activity. An exhaustive quantitative description of several hours of a current record from one such channel at different concentrations and voltages can potentially require months of computer-assisted analysis that includes event tabulation, histogram fitting, correction for missed events, model testing, and simulation. After completing a rigorous analysis of a few such records, there may still be doubts as to whether one unknowingly studied a proteolyzed or otherwise deviant channel molecule, whose behavior may not be fully representative of the majority of channels from the population. This latter worry must ultimately be addressed by comparison with macroscopic behavior or consecutive analysis of many single-channel membranes to ensure consistency of the results from patch to patch or bilayer to bilayer.

With these considerations, an appropriate strategy for single-channel analysis often requires compromises in which some details must be sacrificed to achieve an elementary understanding of the primary basis of drug interaction. Such compromises may include purposeful exclusion of records that exhibit abnormally unstable gating activity, exclusion of very rare types of events such as infrequent substates, and exclusion of dwell time events with low amplitude time constants ($<5\%$ frequency) or events with time constants too brief to be resolved. This selective approach has the advantage that the remaining data may be well described by a simple scheme with far fewer states than would be possible for the unrestricted data set. However, because of the danger that significant clues to channel mechanisms may be lost in this process, it is important to specify the assumptions and consider the limitations of such an analysis.

Despite the complex nature of the data, there have been many elegant studies of drug action at the single-channel level. The opening and closing process of ion channels can be spontaneous (subject to thermodynamic constraints), voltage dependent, pressure dependent, or dependent on a chemical stimulus; many drugs act as direct agonists or modifiers of these gating processes. Some examples of drugs and toxins that modify gating processes of ion channels include the following: acetylcholine,[5] glutamate,[6] glycine,[7] and γ-aminobutyric acid,[7] all of which directly activate distinct classes of neurotransmitter-gated ion channels; batrachotoxin[8] and veratridine,[9] which activate and modify ion permeation properties of voltage-dependent Na^+ channels; dihydropyridine derivatives,[10] which may either activate or inhibit a particular subtype of voltage-dependent Ca^{2+} channels; ATP^{11} and ADP, which inhibit a certain class of nucleotide-sensitive K^+ channels; and cyclic GMP,[12] which activates a cation-selective channel in retinal rod cells of vertebrates. As in studies of enzyme kinetics, the basic approach to modeling such drug effects on channel gating is the derivation of a kinetic scheme that predicts both the gating behavior of the normal channel and the altered pattern of gating observed in the presence of the drug.

Many other drugs and toxins act by perturbing the process of ion permeation. Such an effect has widely been described as a channel block or blockade. Although this term implies the seductive notion that the drug acts by direct plugging of the channel pore, in many cases it is far from clear that this is the actual physical mechanism. Binding of a drug to a blocking site can be a sensitive function of voltage, occupancy states of permeant ions in the channel, open and closed gating conformations, and occupation of other allosterically linked ligand binding sites. Because of these site–site interactions, blockers can serve as sensitive molecular probes of channel function. The exploitation of blocking drugs in studies of ion channels is analogous to the role of inhibitors in studies of enzyme turnover mechanisms. Table I provides an abbreviated list of blocking reactions that have been characterized at the single-channel level. These reactions are classified according to whether the residence time of the blocker can be resolved (slow block) or not (fast block) and also whether

[5] D. Colquhoun and D. C. Ogden, *J. Physiol. (London)* **395,** 131 (1988).

[6] P. Ascher and L. Nowak, *Trends Neurosci.* **10,** 284 (1987).

[7] J. Bormann, O. P. Hamill, and B. Sakmann, *J. Physiol. (London)* **385,** 243 (1987).

[8] L.-Y. M. Huang, N. Moran, and G. Ehrenstein, *Proc. Natl. Acad. Sci. U.S.A.* **79,** 2082 (1982).

[9] S. S. Garber and C. Miller, *J. Gen. Physiol.* **89,** 459 (1987).

[10] H. H. Valdivia and R. Coronado *J. Gen. Physiol.* **95,** 1 (1990).

[11] N. B. Standen, J. M. Quayle, N. W. Davies, J. E. Brayden, Y. Huang, and M. T. Nelson, *Science* **245,** 177 (1989).

[12] A. L. Zimmerman and D. A. Baylor, *Nature (London)* **321,** 70 (1986).

TABLE I
SELECTED EXAMPLES OF DRUG AND TOXIN BLOCKING REACTIONS

Channel	Drug/toxin	Type of block	K_D $(M)^a$	Ref.
Acetylcholine receptor	QX-222 (out)	Slow	1.2×10^{-3}	b, c
	Suberyldicholine (out)	Slow	5.2×10^{-3}	d, e
	d-Tubocurarine (out)	Slow, subconductance	1.4×10^{-4}	f
Na$^+$ channel	9-Aminoacridine (in)	Slow	2.1×10^{-5}	g
	QX-222 (in)	Fast	1.4×10^{-2}	h
	QX-314 (in)	Fast	5.8×10^{-3}	h, i
	TEA (in)	Fast	3.0×10^{-2}	j
	Cocaine (in)	Slow	6×10^{-5}	i
	STX (muscle, out)	Slow	4.4×10^{-9}	k
	STX (heart, out)	Slow	1.0×10^{-7}	k
	μ-Conotoxin GIIIA (muscle, out)	Slow	1.1×10^{-7}	l
	Veratridine	Slow, subconductance	5.1×10^{-7} (-30 mV)	m
Maxi K$^+$(Ca^{2+}) channel	TEA (lin)	Fast	3.4×10^{-2}	n
	TEA (out)	Fast	1.2×10^{-4}	n
	Charybdotoxin (out)	Slow	2.2×10^{-8} ($+20$ mV)	o
	Dendrotoxin I (in)	Slow, subconductance	9.0×10^{-8} (-30 mV)	p
Cl$^-$ channel	DNDS	Slow	2.1×10^{-6} (-30 mV)	q

^{a} K_D values are reported for 0 mV unless otherwise indicated.
^{b} E. Neher and J. H. Steinbach, *J. Physiol. (London)* **277**, 153 (1978).
^{c} P. Charnet, C. Labarca, R. J. Leonard, N. J. Vogelaar, L. Czyzyk, A. Gouin, N. Davidson, and H. A. Lester, *Neuron* **2**, 87 (1990).
^{d} S. M. Sine and J. H. Steinbach, *Biophys. J.* **46**, 277 (1984).
^{e} D. C. Ogden and D. Colquhoun, *Proc. R. Soc. London B* **225**, 329 (1985).
^{f} G. J. Strecker and M. B. Jackson, *Biophys. J.* **56**, 795 (1989).
^{g} D. Yamamoto and J. Z. Yeh, *J. Gen. Physiol.* **84**, 361 (1984).
^{h} E. Moczydlowski, A. Uehara, and S. Hall, in "Ion Channel Reconstitution" (C. Miller, ed.), p. 405. Plenum, New York, 1986.
^{i} G. K. Wang, *J. Gen. Physiol.* **92**, 747 (1988).
^{j} W. N. Green, L. B. Weiss, and O. S. Andersen, *J. Gen. Physiol.* **89**, 841 (1987).
^{k} X. Guo, A. Uehara, A. Ravindran, S. H. Bryant, S. Hall, and E. Moczydlowski, *Biochemistry* **26**, 7546 (1987).
^{l} L. Cruz, W. R. Gray, B. M. Olivera, R. D. Zeikus, L. Kerr, D. Yoshikami, and E. Moczydlowski, *J. Biol. Chem.* **260**, 9280 (1985).
^{m} G. Wang, M. Dugas, B. I. Armah, and P. Honerjager, *Mol. Pharmacol.* **37**, 144 (1990).
^{n} A. Villarroel, O. Alvarez, A. Oberhauser, and R. Latorre, *Pfluegers Arch.* **413**, 118 (1988).
^{o} C. S. Anderson, R. MacKinnon, C. Smith, and C. Miller, *J. Gen. Physiol.* **91**, 317 (1988).
^{p} K. Lucchesi and E. Moczydlowski, *Neuron* **2**, 141 (1990).
^{q} R. J. Bridges, R. T. Worrell, R. A. Frizzell, and D. J. Benos, *Am. J. Physiol.* **256**, C902 (1989).

subconductance behavior is observed. The latter phenomenon refers to the appearance of resolvable conductance levels which are intermediate between the closed (zero current) and fully open current levels. To provide a basic guide to the theory of channel block, the subsequent discussion outlines methods for analyzing a few important types of commonly observed blocking phenomena.

Woodhull Model of Fast Block

Many small molecules have been found to reduce the conductance of ion channels as observed by an apparent decrease of unitary current in a dose-dependent manner. Figure 1A illustrates this phenomenon for the effect of a small organic cation, butylguanidine, on a single batrachotoxin-activated Na^+ channel from rat muscle as studied by planar bilayer recording.[13,14] Other well-documented examples include the effect of tetraethylammonium (TEA) on large conductance Ca^{2+}-activated K^+ channels[15-17] and the effect of various organic cations and local anesthetic derivatives (such as QX-222 and QX-314) on batrachotoxin-activated Na^+ channels.[13,14,18,19] This particular behavior is often described as a fast block, since it is the expected result for a very rapid reaction in which the individual blocked and unblocked dwell times are not resolvable owing to the limited frequency response of the recording system. Fast block is usually modeled as reversible binding of the drug to a site on an open channel that results in a blocked state with near-zero conductance indistinguishable from the closed state current. Although this model has been widely applied, it has not always been adequately tested. For example, behavior resembling that of a fast block can also be mimicked by an electrostatic screening phenomenon that reduces permeant ion concentration at the mouth of the channel, as is proposed to account for part of the effect of divalent cations in reducing the conductance of single Na^+ channels in planar bilayers.[19] Another example of misinterpretation could occur if a drug induced very brief subconductance states, such as those produced in L-type Ca^{2+} channels by lowering extracellular pH.[20]

[13] E. Moczydlowski, A. Uehara, X. Guo, and J. Heiny, *Ann. N.Y. Acad. Sci.* **479,** 269 (1986).
[14] E. Moczydlowski, A. Uehara, and S. Hall, *in* "Ion Channel Reconstitution" (C. Miller, ed.), p. 405. Plenum, New York, 1986.
[15] C. Vergara, E. Moczydlowski, and R. Latorre, *Biophys. J.* **45,** 73 (1984).
[16] G. Yellen, *J. Gen. Physiol.* **84,** 157 (1984).
[17] A. Villarroel, O. Alvarez, A. Oberhauser, and R. Latorre, *Pfluegers Arch.* **413,** 118 (1988).
[18] G. W. Wang, *J. Gen. Physiol.* **92,** 747 (1988).
[19] W. N. Green, L. B. Weiss, and O. S. Andersen, *J. Gen. Physiol.* **89,** 841 (1987).
[20] D. Pietrobon, B. Prod'hom, and P. Hess, *J. Gen. Physiol.* **94,** 1 (1989).

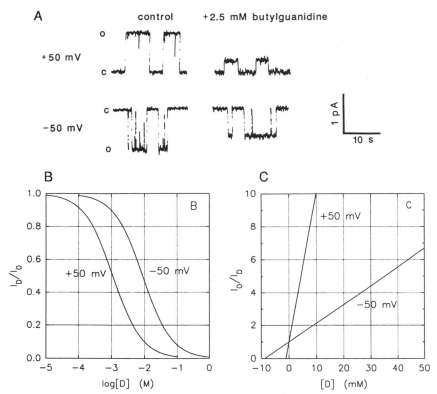

FIG. 1. Characterization of the drug concentration dependence of a fast blocking reaction. (A) Single-channel records of a batrachotoxin-activated Na^+ channel from rat muscle at ± 50 mV in the presence of symmetrical 0.2 M NaCl and after the addition of 2.5 mM butylguanidine to the internal side (c, closed; o, open current level). The slow blocking toxin tetrodotoxin (50 nM) was also present on the external side to facilitate measurement of the unitary current. (Recordings are taken from Ref. 14 with permission.) (B) Simulated behavior of the drug concentration dependence of the butylguanidine fast block according to Eqs. (2) and (3) at ± 50 mV. (C) Simulated behavior of the butylguanidine fast block plotted according to Eqs. (2) and (4) at ± 50 mV. Parameters used in these simulations were $z\delta = -0.53$, $K_D(0) = 3.1$ mM, $K_D(+50$ mV$) = 1.09$ mM, and $K_D(-50$ mV$) = 8.79$ mM.

Several different types of graphical representations have been used to analyze fast block behavior as derived from reaction (1):

$$O + D \xrightleftharpoons{K_D(V)} O \cdot D \tag{1}$$

In reaction (1), $K_D(V)$ is the equilibrium dissociation constant for binding of a drug, D, to a site on an open or unblocked channel, O, which results in a blocked channel, $O \cdot D$. It is widely observed that binding constants of many ligands to ion channels are functions of the transmembrane voltage, V. For example, the observed K_D of a cationic blocker that binds on the intracellular side of a channel often increases with hyperpolarization and decreases with depolarization. Two principal theories have been proposed to explain voltage-dependent binding: (1) a direct effect of voltage on the association and dissociation rates of a charged molecule entering an electric field or (2) an indirect allosteric coupling between the drug binding site outside the field and domains of the protein that do reside in the electric field and directly sense the applied voltage.

In 1973, Woodhull[21] devised a model for voltage-dependent binding according to the first of these theories, which was applied to block of nerve axon Na^+ channels by external protons and Ca^{2+} ions. The basic assumption of this model is that the observed K_D of the blocking ion may be expressed as a Boltzmann function of the applied voltage:

$$K_D(V) = K_D(0) \exp(z\delta FV/RT) \tag{2}$$

where $K_D(0)$ is the value of K_D at 0 mV, z is the charge valence of the blocking ion, δ is the fraction of the electric field which the blocker traverses to reach the site, F is Faraday's constant, R is the gas constant, and T is absolute temperature. At 22°, $RT/F = 25.4$ mV. An expression similar to that of Eq. (2) with a minus sign in the exponent can be used for blockers that bind intracellularly and display enhanced binding affinity at positive voltage. In Woodhull's original description, an Eyring rate model of blocker binding was used with an energy well for the binding site at a relative distance, δ, into the membrane field (Fig. 3B). For some blocking reactions, this physical model for the mechanism of voltage dependence seems to be applicable because divalent ions exhibit about twice the voltage dependence as monovalent blocking ions binding in the same site.[17,22] However, in other cases, there is no demonstrated relationship between the charge distribution of the blocking ion and the observed voltage dependence of the blocking reaction.[23,24] In the absence of such detailed information, many investigators choose not to assume a physical interpretation of Woodhull's model and describe the $z\delta$ parameter as z' or an effective valence of the blocking reaction.

The Woodhull model[21] was originally derived as a one-site – two-barrier

[21] A. M. Woodhull, *J. Gen. Physiol.* **61,** 687 (1973).

[22] C. Miller, *J. Gen. Physiol.* **79,** 869 (1982).

[23] E. Moczydlowski, S. Hall, S. S. Garber, G. R. Strichartz, and C. Miller, *J. Gen. Physiol.* **84,** 687 (1984).

[24] X. Guo, A. Uehara, A. Ravindran, S. H. Bryant, S. Hall, and E. Moczydlowski, *Biochemistry* **26,** 7546 (1987).

Eyring permeation model[25] that allowed for the possibility of H^+ entry and permeation through the Na^+ channel in both directions. Despite this historical feature, the model has been widely applied to voltage-dependent blockers that do not significantly permeate. In this case, Woodhull's paper considered a situation where one of the energy barriers is infinitely high so that entry or exit from that side of the channel is negligible. In this limit, the Woodhull blocking model takes on a very simple form. In terms of receptor site vacancy, the probability, P_U, that the site is unblocked is assumed to be equivalent to i_D/i_0, the ratio of the measured unitary current in the presence of the blocking drug to that in the absence of blocker. Using the mass action expression for this fractional site vacancy in terms of reaction (1), this ratio is equal to

$$P_U = i_D/i_0 = K_D(V)/\{K_D(V) + [D]\} \tag{3}$$

Equation (3) is simply an inverted form of the Langmuir isotherm which states that the fractional occupancy of a one-site system is equal to the term $[D]/\{K_D + [D]\}$. The standard Langmuir relationship may be obtained from Eq. (3) by subtracting both sides of Eq. (3) from 1.

Experimental data testing the applicability of Eq. (3) are commonly displayed in several different graphical representations. Figure 1B illustrates a log-concentration plot of the simulated behavior at $+50$ and -50 mV for block of a batrachotoxin-activated Na^+ channel in planar bilayers by internal butylguanidine, which has been reported[13,14] to exhibit a $K_D(0)$ value of 3.1 mM with $z\delta = -0.53$. A log-concentration plot has the advantage that i_D/i_0 data can be displayed over a wide range of drug concentration. This format provides a stringent test of deviation from a Langmuir isotherm, particularly at high blocker concentration where surface charge effects may become important. A Hill plot of the data according to $\log[(i_0 - i_D)/i_D]$ versus $\log[D]$ can also be used to investigate the possibility of multiple blocker binding sites, which may be indicated by a slope or Hill coefficient different than 1.0. Some authors alternatively prefer to use the reciprocal form of Eq. (3), which produces a linear transformation according to:

$$1/P_U = i_0/i_D = 1 + [D]/K_D(V) \tag{4}$$

A plot of i_0/i_D versus $[D]$ exhibits an ordinate intercept of 1, a slope of $1/K_D$, and an abscissa intercept of $-K_D$ as shown in Fig. 1C.

The voltage dependence of a fast blocking reaction can be analyzed and graphically represented in several ways. Many authors prefer to present

[25] J. W. Woodbury, *in* "Chemical Dynamics: Papers in Honor of Henry Eyring" (J. O. Hirschfelder, ed.), Wiley, New York, 1971.

their data in the form of current–voltage (I–V) measurements at several different blocker concentrations. As shown in Fig. 2A, the presence of an internal cationic blocker such as butylguanidine results in an inward-rectifying I–V behavior, where inward (negative) currents are larger than outward (positive) currents. Conversely, a voltage-dependent cationic blocker on the outside of a channel may result in outward rectification. For many channels, the current–voltage relationship of the open channel in the absence of blockers approximates an ohmic relationship, especially at low

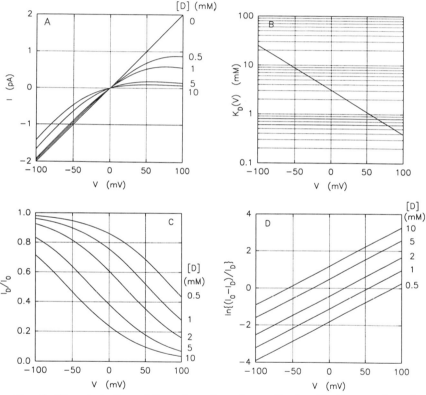

FIG. 2. Characterization of the voltage dependence of a fast blocking reaction. The simulations used the $K_D(0)$ and $z\delta$ parameters for fast block of batrachotoxin-activated Na^+ channels by internal butylguanidine given in Fig. 1. (A) Simulated current–voltage behavior of single batrachotoxin-activated Na^+ channels in the presence of 0.2 M symmetrical NaCl and 0.5, 1, 5, and 10 mM butylguanidine on the internal side. An ohmic conductance of 20 pS is assumed for the control channel in the absence of blocker. Equations (2) and (3) are used to calculate the expected current as a function of voltage and drug concentration. (B) Dependence of the equilibrium dissociation constant on voltage simulated according to Eq. (2). (C) Dependence of the ratio of the unitary current in the presence of blocker to that in the absence of blocker simulated according to Eqs. (2) and (3). (D) Linearized replot according to Eq. (6).

voltage under conditions of symmetrical permeant ion concentrations. Once this ohmic conductance, g, is known, i_0 in Eq. (3) can be computed for any voltage as $i_0 = gV$. Strictly, however, this ohmic relationship is only an approximation. Many theories of ion permeation actually predict non-linear $I-V$ behavior.[25-27] This is commonly observed as a departure from linearity at high voltage but is also exhibited by some channels at low voltages, as in the case of outward-rectifying Cl^- channels.[28] Voltage-dependent block is often directly represented by plotting the apparent dissociation constant on a logarithmic scale versus voltage as in Fig. 2B. Assuming that one-site behavior has been demonstrated according to Eq. (3), the equilibrium dissociation constant at a fixed voltage can be computed from the following rearrangement of Eq. (3):

$$K_D(V) = [D]/\{(i_0 - i_D)/i_D\} \tag{5}$$

Other alternative graphical representations of voltage-dependent fast block are the use of Eq. (3) itself to fit i_D/i_0 versus V at different concentrations of B (Fig. 2C). $K_D(0)$ and $z\delta$ parameters for this latter representation can also be obtained by a linear regression fit to the following transformation of Eqs. (2) and (3) with respect to V, as illustrated in Fig. 2D.

$$\ln\{(i_0 - i_D)/i_D\} = \ln\{[D]/K_D(0)\} - z\delta FV/RT \tag{6}$$

Neher–Steinbach Model of Slow Block

In contrast to the case of fast blockers, the residence times of many drugs and toxin blockers are in the range of 10^{-3} to 10^2 sec, which can be resolved and measured by single-channel recording. An example of this phenomenon is shown in Fig. 3A, where the addition of the guanidinium toxin, saxitoxin (STX), to the external side of a batrachotoxin-activated Na^+ channel from dog heart is observed to induce the appearance of long-lived blocked and unblocked events. Such an effect is often described as a slow or discrete block since the rate of the blocking reaction is slow enough for individual blocker dwell times to be observed. The first example of this phenomenon was reported and analyzed by Neher and Steinbach[29] for block of open acetylcholine receptor channels by the lidocaine derivatives QX-222 and QX-314. Their model and methods of analysis have proved to be generally applicable to many other examples of slow block.

[26] B. Hille and W. Schwarz, *J. Gen. Physiol.* **72**, 409 (1978).
[27] J. A. Dani, *Biophys. J.* **49**, 607 (1986).
[28] D. R. Halm, G. R. Rechkemmer, R. A. Schoumacher, and R. A. Frizzell, *Am. J. Physiol.* **254**, C505 (1988).
[29] E. Neher and J. H. Steinbach, *J. Physiol. (London)* **277**, 153 (1978).

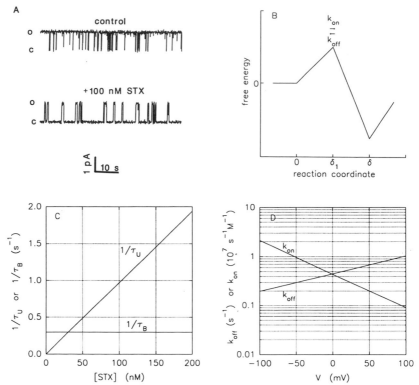

FIG. 3. Characterization of a slow blocking reaction. (A) Effect of 100 nM external saxitoxin (STX) on a single batrachotoxin-activated Na$^+$ channel from dog heart. Conditions: 0.2 M symmetrical NaCl; holding voltage, -50 mV (c, closed; o, open). (Records are unpublished data of L. Schild and E. Moczydlowski, 1991.) (B) Two-state model of a one-site binding reaction. (C) Dependence of reciprocal lifetimes of blocked ($1/\tau_B$) and unblocked ($1/\tau_U$) states on STX concentration. Simulated behavior of STX block of a batrachotoxin-activated Na$^+$ channel from dog heart according to Eqs. (8) and (9) is shown at -50 mV with $k_{off} = 0.30$ s^{-1} and $K_{on} = 9.7 \times 10^6$ s^{-1} M^{-1} as reported by Guo et al.[24] The α rate constant in Eq. (9) is assumed to be insignificant in this example. (D) Voltage dependence of the association and dissociation rate constants for STX block of a batrachotoxin-activated Na$^+$ channel from dog heart simulated according to Eqs. (10) and (11) with $k_{on}(0) = 4.4 \times 10^6$ s^{-1} M^{-1}, $z\delta_1 = 0.40$, $k_{off}(0) = 0.45$ s^{-1}, and $z\delta_2 = 0.21$ as reported by Guo et al.[24]

The basic scheme often proposed for slow block is

$$C \underset{\alpha}{\overset{\beta}{\rightleftharpoons}} O + D \underset{k_{off}(V)}{\overset{k_{on}(V)}{\rightleftharpoons}} O \cdot D \tag{7}$$

The drug-binding step in Eq. (7) is essentially the same as Eq. (1) for a fast blocking reaction with $K_D = k_{off}/k_{on}$. In a slow blocking reaction, the individual association and dissociation rate constants that, respectively, describe the drug blocking (O $\rightarrow$ O $\cdot$ D) and unblocking (O $\leftarrow$ O $\cdot$ D) reactions are experimentally accessible to direct measurement. The drug-independent channel-gating reaction is often abbreviated as a reversible one-step closed–open (C $\leftrightarrow$ O) process. Colquhoun and Hawkes[30] have presented a detailed probabilistic analysis of the predicted behavior of reaction (7) in terms of single-channel events.

Extraction of rate constants from single-channel records involves the compilation and fitting of dwell time histograms of open and shut (closed/blocked) events. Detailed discussions of this topic describing the proper use of cumulative and probability density functions of such even populations have been published.[30,31] Sigworth and Sine[32] have also described the use of logarithmically binned time histograms for analysis and fitting of multiexponential populations, where the various time constants appear as peaks in the distribution. Because there is only a single open state, reaction (7) predicts an exponential distribution of open or unblocked dwell times. The predicted probability distribution for the shut state population is a more complicated sum-of-two exponentials since there are two such states, C (closed) and O $\cdot$ D (blocked). However, because these latter two states are not in direct communication in reaction (7), the two observed shut time constants can be unambiguously identified with the gating or blocking processes. Closed state populations in the absence of drug will exhibit the β time constant associated with channel opening, whereas distributions collected in the presence of the blocker will exhibit a new time constant, τ_B, equal to the mean dwell time of the blocker. The predicted relationships between the measured blocked (τ_B) and unblocked (τ_U) time constants and the underlying rate constants are

$$\tau_B = k_{off}^{-1} \tag{8}$$
$$\tau_U = (\alpha + k_{on}[D])^{-1} \tag{9}$$

Equations (8) and (9) provide a simple test for the applicability of reaction (7) as simulated in Fig. 3C. The reciprocal time constant for the blocked state should be completely independent of drug concentration, and the reciprocal time constant for the unblocked population should be a linear function of [D] with a slope of k_{on} and an ordinate intercept of α.

By analogy to the binding equilibrium constant of a fast blocker, the k_{on} and k_{off} rate constants for binding of a slow blocker are often found to be voltage dependent. This behavior can be described by

$$k_{off}(V) = k_{off}(0) \exp(z\delta_2 FV/RT) \tag{10}$$
$$k_{on}(V) = k_{on}(0) \exp(-z\delta_1 FV/RT) \tag{11}$$

[30] D. Colquhoun and A. G. Hawkes, *in* "Single-Channel Recording" (B. Sakmann and E. Neher, eds.), p. 135. Plenum, New York, 1983.
[31] D. Colquhoun and F. J. Sigworth, *in* "Single-Channel Recording" (B. Sakmann and E. Neher, ed.), p. 191. Plenum, New York, 1983.
[32] F. J. Sigworth and S. M. Sine, *Biophys. J.* **52**, 1047 (1987).

Equations (10) and (11) follow directly from an Eyring rate model of a two-state process with a single energy barrier (Fig. 3B). This barrier represents the transition state for the reversible transition of the blocker from a reference state in solution outside the electric field to a binding site located at a distance δ. Because the transition state is located at a distance δ_1 with respect to the reference state, the fraction of the field traversed by the blocker associating with the site is δ_1 [Eq. (11)] and that for the blocker dissociating from the site is δ_2, which is equal to the difference $\delta - \delta_1$. This model predicts that the k_{off} and k_{on} reactions for a positively charged blocker entering the site from the extracellular side of the channel will exhibit opposite voltage-dependent behavior, namely, k_{off} decreasing with negative voltage and k_{on} increasing with negative voltage. The actual magnitude of the voltage dependence for the on and off reactions is determined by the relative position of the transition state. Figure 3D shows the simulated dependence of k_{on} and k_{off} rate constants for STX block of the heart Na^+ channel which has been previously found[24] to follow Eqs. (10) and (11).

Miller Test for Binding Competition

Once a slow blocking reaction has been characterized with a particular drug or toxin, one may wish to investigate whether other drugs or ions bind competitively at the same site. A simple analysis of this phenomenon is available in cases where the inhibitor to be tested for competition is a fast blocker with lower affinity in comparison to the slow blocking probe molecule. One example of this phenomenon is competition between the fast blocker, QX-314, and the slow blocker, cocaine. Both of these drugs appear to bind to a common local anesthetic binding site accessible from the intracellular side of batrachotoxin-activated Na^+ channels studied by Wang.[18] Another example is competition between the fast blocker TEA and the slow blocker charybdotoxin, which appear to compete for a common extracellular site on maxi Ca^{2+}-activated K^+ channels as described by Miller.[33] The kinetic scheme for such a competitive interaction is

$$O + D \underset{k_{off}}{\overset{k_{on}}{\rightleftharpoons}} O \cdot D \tag{12}$$

$$O + I \overset{K_I}{\rightleftharpoons} O \cdot I$$

In Eq. (12), D is a slow blocker with observable dwell times and I is a fast blocker with brief or unresolvable dwell times. Because binding of the two

---

[33] C. Miller, *Neuron* **1**, 1003 (1988).

ligands cannot take place simultaneously, the mean residence time of the slow blocker ($1/k_{off}$) should be independent of [I]. However, unblocked dwell times in the presence of fixed [D] will increase as a function of [I] because increasing [I] reduces the fraction of time that the channel is in the unoccupied O state. Quantitatively, reaction (12) predicts the following relationships for the observed time constants of exponential distributions of the blocked (τ_B) and unblocked (τ_u) dwell times of the slow blocker, D:

$$\tau_B = k_{off}^{-1} \qquad \text{(independent of [I])} \tag{13}$$

$$\tau_U = (k_{on}[D])^{-1}\{1 + [I]/K_I\} \qquad \text{(directly proportional to [I])} \tag{14}$$

In Miller's analysis[33] of the TEA–charybdotoxin interaction, these predictions were tested by plotting the ratio of the respective blocked or unblocked lifetimes in the presence of I to that in the absence of I; the lifetimes are predicted to have the following behavior for a purely competitive binding interaction:

$$\tau_B(+I)/\tau_B(-I) = 1 \tag{15}$$

$$\tau_U(+I)/\tau_U(-I) = 1 + [I]/K_I \tag{16}$$

Graphical representation of Eqs. (15) and (16) is very similar to that of Eqs. (8) and (9) represented in Fig. 3C. The slope and abscissa intercept of a plot of Eq. (16) are, respectively, equal to $1/K_D$ and $-K_D$. Behavior that strictly follows these relationships over a wide range of [I] is a stringent test for binding competition according to reaction (12). Noncompetitive schemes that permit simultaneous binding of D and I at different sites generally predict that the liftime ratios of Eqs. (15) and (16) will be hyperbolic functions of [I]. Miller also noted that Eq. (16) has the same dependence on [I] as the reciprocal unblocked probability of the fast blocker (I) in the absence of D, according to the previous Eq. (4). If independent analyses of the behavior specified by Eq. (4) and Eq. (16) result in the same equilibrium constant for fast block (K_D) and competition (K_I), this constitutes further evidence that a common site mediates both processes.

Summary

Many drugs interact directly with ion channel proteins to alter gating and permeation functions. Single-channel recording affords resolution of drug-induced functional changes in channel behavior at the molecular level. Drug and toxin molecules that block ion channels are useful probes of channel mechanisms because blocking sites are often coupled to other pharmacologically relevant binding sites. Simple kinetic schemes describing fast block, slow block, and binding competition between two blocking molecules provide useful models of drug-induced blocking processes. From a careful perspective, a single channel is best approached as the analog of a purified enzyme preparation in the hands of an enzymologist. The confidence gained by knowing that one is viewing a single subtype must be weighed against the possibility that the channel could have been altered in the process of patch isolation or bilayer reconstitution. As in all kinetic studies, a curve fit to a two-state scheme is contingent on the possibility that a more complex multistate system can masquerade as the simple cartoon one would like to put forward.

Acknowledgments

This work was supported by grants from the National Institutes of Health (AR38796 and HL38156) and an Established Investigator award from the American Heart Association.

[55] Analysis of Sodium Channel Tail Currents

By GABRIEL COTA and CLAY M. ARMSTRONG

Introduction

Analysis of tail currents has yielded much information about the functional properties of voltage-gated ion channels. Aspects of channel behavior that can be inferred from tail currents include the closing kinetics, the open-channel (or instantaneous) current–voltage relationship, and the voltage dependence of the fraction of open channels. Among other examples, the study of tail currents has been helpful in demonstrations that blocking agents can be trapped in closed channels[1-5] and that a single cell can express distinct types of Na^+ channels,[6] Ca^{2+} channels,[7-11] or K^+ channels.[12]

[1] C. M. Armstrong, *J. Gen. Physiol.* **58,** 413 (1971).
[2] C. M. Armstrong and S. R. Taylor, *Biophys. J.* **30,** 473 (1980).
[3] C. M. Armstrong, R. P. Swenson, and S. R. Taylor, *J. Gen. Physiol.* **80,** 663 (1982).
[4] D. Swandulla and C. M. Armstrong, *Proc. Natl. Acad. Sci. U.S.A.* **86,** 1736 (1989).
[5] R. H. Chow, *J. Gen. Physiol.* **98,** 751 (1991).
[6] W. F. Gilly and C. M. Armstrong, *Nature (London)* **309,** 448 (1984).
[7] C. M. Armstrong and D. R. Matteson, *Science* **227,** 65 (1985).
[8] G. Cota, *J. Gen. Physiol.* **88,** 83 (1986).
[9] M. Hiriart and D. R. Matteson, *J. Gen. Physiol.* **91,** 617 (1988).
[10] D. Swandulla and C. M. Armstrong, *J. Gen. Physiol.* **92,** 197 (1988).
[11] L. Tabares, J. Ureña, and J. Lopez-Barneo, *J. Gen. Physiol.* **93,** 495 (1989).
[12] A. Castellano, J. Lopez-Barneo, and C. M. Armstrong, *Pfluegers Arch.* **413,** 644 (1989).

This chapter focuses on whole-cell patch clamp experiments in clonal pituitary GH_3 cells and illustrates how the analysis of tail currents has helped us to explore some properties of Na^+ channels, including the kinetics of channel inactivation and the interaction of divalent and trivalent cations with the channels. GH_3 cells are derived from a rat pituitary adenoma,[13] and they have proved to be an excellent model for the study of Na^+ channel properties with patch clamp techniques.[14-20]

The specific objectives of our analysis are (1) to determine the voltage dependence of the inactivation step; (2) to decide whether modification of channel gating by external lanthanum ion can be explained by surface charge theory; and (3) to examine the idea that calcium ion serves as a gating cofactor. In most of these experiments, gating of Na^+ channels has been simplified by using intracellular papain to remove inactivation. Details of the methodological procedures that we use to record Na^+ channel currents from GH_3 cells are given in the following section. The last section summarizes the major findings.

Recording of Sodium Channel Currents

Cell Culture

We obtain the GH_3 cells from the American Type Culture Collection (Rockville, MD) and maintain them in polystyrene 25-cm^2 culture flasks (Corning Glass Works, Corning, NY) using Kennetts's HY medium (Cell Center, University of Pennsylvania, Philadelphia, PA, or Hazleton Research Products, Lenexa, KS) supplemented with 5% fetal bovine serum (GIBCO, Grand Island, NY) and 1% glutamine (GIBCO). Cells are cultured in a humidified atmosphere of 5% CO_2–95% air at 37°. The maintenance culture is split every 10 days by using a brief proteolytic digestion in a dilute solution of trypsin (trypsin–EDTA; Flow Laboratories, McLean, VA) to detach the cells, replating the monodispersed cells at 7-fold lower density.

For the electrophysiological experiments, at the time of splitting some cells are plated on slivers (11×2.5 mm) of glass coverslips in 35-mm

[13] A. H. Tashjian, Jr., this series, Vol. 58, p. 527.
[14] J. M. Dubinsky and G. S. Oxford, *J. Gen. Physiol.* **83**, 309 (1984).
[15] J. M. Fernandez, A. P. Fox, and S. Krasne, *J. Physiol. (London)* **356**, 565 (1984).
[16] R. Horn and C. A. Vandenberg, *J. Gen. Physiol.* **84**, 505 (1984).
[17] D. R. Matteson and C. M. Armstrong, *J. Gen. Physiol.* **83**, 371 (1984).
[18] C. A. Vandenberg and R. Horn, *J. Gen. Physiol.* **84**, 535 (1984).
[19] G. Cota and C. M. Armstrong, *J. Gen. Physiol.* **94**, 213 (1989).
[20] C. M. Armstrong and G. Cota, *J. Gen. Physiol.* **96**, 1129 (1990).

plastic petri dishes. We use a subcultivation ratio of 1 : 15 and record from cells cultured for 2–7 days after replating. The culture medium is replenished every day.

Recording Conditions

Coverslips with attached GH₃ cells are transferred from culture dishes to the experimental chamber, which has a relatively small volume (~0.2 ml) and is mounted on the stage of an inverted Diaphot TMD microscope (Nikon Corporation, Tokyo, Japan). The external recording solution (see below) is continually perfused through the chamber by using a gravity-driven flow/suction arrangement. The fluid height in the chamber is controlled by adjusting the flow rate and the position of the suction tube. This tube is a 15-gauge needle with its end beveled at a 45° angle, covered with 1000 mesh gold screen (Ted Pella, Redding, CA). With fast flow, the solution exchange in the chamber is nearly complete in about 30 sec. The temperature of the solution in the chamber is kept at 15° using a controller device connected to a Peltier cooler, with a thermistor in the chamber acting as a temperature sensor.

Cells are visualized at 600× magnification and approached with fire-polished pipettes containing the internal recording solution (see below). We select isolated cells that are almost spherical and 15–25 μm in diameter.

Recording Solutions

The composition of the recording solutions is designed to isolate currents through Na⁺ channels from currents carried by other cation-selective channels. We use K⁺-free solutions to eliminate K⁺ currents, and we sometimes include 0.2 mM CdCl₂ in the external medium in an attempt to suppress current through Ca²⁺ channels. With 2 mM Ca²⁺ in the external solution, the Ca²⁺ current is usually small compared with the Na⁺ current even in the absence of Cd²⁺, and after papain action (see below) channel activity resistant to 1 μM external tetrodotoxin is practically absent.

We use two different internal solutions in the experiments described here. The internal solution A contains (concentrations in mM) 100 NaF, 30 NaCl, 1 CaCl₂, and 10 EGTA–CsOH, and the internal solution B contains 30 NaCl, 9 NaF, 91 CsF, and 10 EGTA–CsOH. These solutions are supplemented with 1 mg/ml papain (see below). The composition of the external solutions is indicated in the figure legends, with concentrations in millimolar units. All solutions also contain 10 mM HEPES acid, which was neutralized to pH 7.30 with CsOH (internal solutions) or NaOH (external solutions).

Whole-Cell Clamping and Data Acquisition

To examine the activity of Na^+ channels, GH_3 cells are subjected to whole-cell patch clamping.[21] Voltage steps are applied to the cell membrane from a holding potential of -80 mV and are practically complete within 50 μsec. This relatively high time resolution for monitoring current is obtained with the combined use of a low (10 Ω) feedback resistance on the head stage amplifier (an OPA-111; Burr Brown Research Corp., Tucson, AZ), low-resistance patch electrodes (see below), and "supercharging," an improved patch clamp circuit.[22]

Supercharging speeds the change of membrane voltage (V_m) by altering the command voltage applied to the positive input of the head stage amplifier. A 15-μsec voltage spike of appropriate size is added to the leading edge of the command step. The spike enhances the speed with which the membrane capacitance is charged by driving current rapidly through the electrode resistance. The spike is terminated when V_m reaches the required level. We adjust the spike amplitude by watching the current transient during the first 200 μsec after a square change in command voltage. After breaking into the cell with the patch electrode, the transient has a very fast component corresponding to charging current for the stray capacitance of the electrode, as well as a slower component with a time constant equal to the product of access resistance times membrane capacitance. The appropriate amplitude of the voltage spike is determined by nulling the slower component. Spike amplitude is readjusted at frequent intervals because the access resistance normally changes during the experiment.

Pulse generation and data acquisition are controlled by an LSI-11/73 computer (Scientific Micro Systems, Mountain View, CA). Membrane current signals are sampled at 10- or 20-μsec intervals, and their linear components are subtracted out using the scaled current response to 50-mV hyperpolarizing steps.

Patch Electrodes

Patch pipettes are fabricated from hard glass capillaries. We have used either aluminosilicate glass (A-M Systems, Everett, WA) or borosilicate glass (Kimax 51; Kimble Div., Owens-Illinois, Inc., Toledo, OH). Pipettes are pulled in two steps using a vertical puller (Kopf Model 700C; David Kopf Instruments, Tujunga, CA), with a coil that is two turns of 1-mm

[21] O. P. Hamill, A. Marty, E. Neher, B. Sakmann, and F. J. Sigworth, *Pfluegers Arch.* **391**, 85 (1981).

[22] C. M. Armstrong and R. H. Chow, *Biophys. J.* **52**, 133 (1987).

nichrome wire (David Kopf Instruments), shaped to 2.5 mm inside diameter and 2.3 mm length. Pipette tips are then carefully fire polished to a bullet shape using a homemade microforge. In most of our experiments, electrode resistance is between 0.4 and 0.7 MΩ. With these electrodes, and after papain action (see below), the access resistance is 0.5–1.0 MΩ, and the estimated series resistance error is usually smaller than 3 mV.

Removal of Inactivation by Papain

To remove the fast inactivation gating of Na$^+$ channels, we apply papain (1 mg/ml; type IV, Sigma Chemical Co., St. Louis, MO) inside the cells. The enzyme is added to the internal solution, contained in the patch pipette. After break-in with the electrode, inactivation is slowly removed over the course of 10 min, making it possible to obtain control traces of Na$^+$ currents before the enzyme acts on Na$^+$ channel gating.[19] After papain action, the cells have large and stable currents for 15–20 min, making them an excellent preparation for many types of experiments.

Inferring Sodium Channel Properties from Tail Currents

It is convenient to start this section by studying Fig. 1, which presents a series of Na$^+$ current traces through Na$^+$ channels with inactivation removed. The channels have activation gates that open in response to membrane depolarization and close on repolarization. The magnitude of the current depends on the number of conducting channels and on the electrochemical force that drives Na$^+$ ions through the channels. In this case the Na$^+$ equilibrium potential had a small positive value ($\sim$4 mV). At -80 mV (the holding potential) all channels are in the gate closed state, and current is zero. In each trace, the channels are opened (activated) by a change in membrane voltage (V_m) from -80 mV to a depolarized level and are closed (deactivated) by the return to -80 mV. In response to depolarization, inward Na$^+$ current activates with a sigmoidal time course as the number of conducting channels increases. On stepping back to -80 mV the current magnitude jumps because of the sudden increase in driving force for Na$^+$ entry. The tail current then decays as the channel gates close, with a time course that can be approximated with a single exponential.

Voltage Independence of the Inactivation Step

With inactivation intact, an open Na$^+$ channel can cease to conduct either because its activation gate closes (deactivation), or because its inactivation gate closes. The rate constant of channel closing is then the sum of the inactivation rate constant (k) and the deactivation rate constant (b).

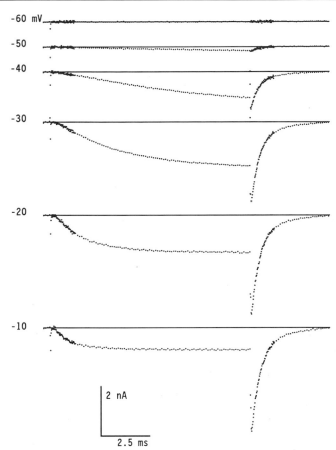

FIG. 1. Opening and closing of sodium channels. The traces are whole-cell Na$^+$ currents recorded from a GH$_3$ cell after removing the inactivation gating of the Na$^+$ channels with internal papain. Channels were opened by 10-msec steps from -80 mV to the indicated voltages and were closed by the return to -80 mV. The initial amplitude of the tail current recorded on repolarization is directly proportional to the number of channels with open gates at the end of the activating pulse. The time course of the tail current marks the closing of the gates. External solution: 150 NaCl, 2 CaCl$_2$; internal solution, internal solution A (see text).

We have determined k by comparing the channel closing kinetics before and after the proteolytic removal of inactivation.[19] We activated the channels with a brief pulse to $+60$ mV and studied the time course of channel closing on changing V_m to a second, usually more negative level. We found that channels no longer close at or positive to -20 mV when the inactivation gate has been removed, which indicates that closing of intact channels at these voltages results exclusively from inactivation. The time course of inactivation at -20 mV can be compared with that at 0, $+40$,

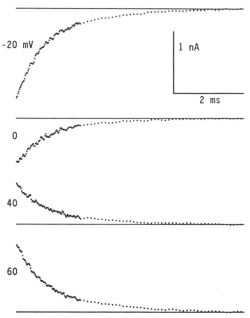

FIG. 2. The inactivation step is not voltage dependent. The traces are Na^+ currents recorded after 0.4-msec pulses to $+60$ mV, before removal of inactivation with internal papain. To analyze the time course of channel closing at different voltages, V_m was first stepped from -80 to $+60$ mV, then maintained at this value (lower trace) or changed to $+40$, 0, or -20 mV. All of the decay in current magnitude at these voltages results from inactivation. This is clear because current magnitude shows no decay at or positive to -20 mV after removing inactivation. There is no significant effect of V_m on the rate of channel inactivation. External solution: 75 NaCl, 75 choline chloride, 2 $CaCl_2$, 0.2 $CdCl_2$; internal solution, internal solution B (see text).

and $+60$ mV in Fig. 2. At every voltage, channel inactivation was well fit by a single exponential. The rate constant (k) was 1.08 $msec^{-1}$ at -20 mV, 1.07 $msec^{-1}$ at 0 mV, 1.03 $msec^{-1}$ at $+40$ mV, and 1.06 $msec^{-1}$ at $+60$ mV. The effect of V_m on k is thus negligible. These results show unequivocally that the rate constant for the transition of an open Na^+ channel into the inactivated state has no significant voltage dependence over a wide range of membrane voltages.

Modification of Channel Gating by Lanthanum

Divalent and trivalent cations have strong effects on Na^+ channel function. They shift the gating behavior of the channel along the voltage axis,[23] and Ca^{2+} (and presumably other ions as well) causes voltage-dependent block of the channels.[24-29] It is usually considered that alteration of gating and blocking are two separate actions, with the effects on gating explained in terms of surface charge theory. In its uniform surface charge version (the version commonly considered), the surface charge theory predicts that all aspects of channel gating should be affected equally. We have tested this prediction in inactivationless Na^+ channels by studying the changes in gating induced by substitution of La^{3+} for Ca^{2+} in the external medium.[20]

To quantify La^{3+} effects, we recorded the Na^+ currents caused by 10-msec activating pulses to various membrane potentials followed by repolarization to -80 mV, as in Fig. 1. We then empirically fitted the late phase of opening of the channels at every voltage with a single exponential, determined the opening (activation) rate (a), and plotted this rate as a function of V_m. We also determined fraction open–V_m curves by plotting the initial amplitude of the tail current at -80 mV as a function of V_m during the activating pulse that preceded the tail measurement. Tail current amplitude is directly proportional to the number of channels that have open gates at the end of the activating pulse. In addition, we determined the voltage dependence of the closing (deactivation) rate (b), from experiments similar to that in Fig. 2.

The three measurable parameters of gating, namely, activation rate, deactivation rate, and the midpoint of the fraction open–voltage relation, were determined in the presence of different external La^{3+} concentrations (from 5 μM to 4 mM), and the shift of each along the voltage axis, relative to their values in 2 mM Ca^{2+}, was quantified. Table I presents results from a representative experiment. It is clear that the three parameters are not equally shifted, as the uniform surface charge theory predicts. This discrepancy may be resolvable by invoking a nonuniform charge distribution. However, at low (5 or 10 μM) La^{3+} concentrations, not only are the shifts of opening and closing kinetics different in size, but they are of opposite sign. This seems impossible to explain by any modification of the surface charge theory. Like Ca^{2+} (see below), La^{3+} also blocks Na^+ channels at negative voltages, an action that lies outside of the surface charge theory.

[23] B. Hille, "Ionic Channels of Excitable Membranes." Sinauer, Sunderland, Massachusetts, 1984.

[24] A. M. Woodhull, *J. Gen. Physiol.* **61**, 687 (1973).

[25] R. E. Taylor, C. M. Armstrong, and F. Bezanilla, *Biophys. J.* **16**, 27a (1976).

[26] D. Yamamoto, J. Z. Yeh, and T. Narahashi, *Biophys. J.* **45**, 337 (1984).

[27] G. N. Mozhayeva, A. P. Naumov, and E. D. Nosyreva, *Gen. Physiol. Biophys.* **4**, 425 (1985).

[28] S. Cukierman, W. C. Zinkand, R. J. French, and B. K. Krueger, *J. Gen. Physiol.* **92**, 431 (1988).

[29] B. Nilius, *J. Physiol. (London)* **399**, 537 (1988).

TABLE I
LANTHANUM-INDUCED SHIFTS OF THREE
PARAMETERS OF SODIUM CHANNEL GATING[a]

$[La^{3+}]$	A	Open	B
5 μM	11.5	-1.0	-7.5
10 μM	16.5	3.8	-4.8
4 mM	53.0	44.5	30.0

[a] Opening rate (A), closing rate (B), and the midpoint of the fraction open–V_m curve (Open) (see text). Voltage shifts are in millivolts. Recording solutions were as in Fig. 1, except that external Ca^{2+} was replaced by the indicated La^{3+} concentration.

Calcium May Be a Gating Cofactor

Results presented above suggest that a new theory of di- and trivalent cation action is needed, and we are attempting to develop one that relates the blocking and gating effects of Ca^{2+} and other multivalent ions on Na^+ channels. Specifically, we propose that both actions, blocking and "shifting," are related to Ca^{2+} entry into the Na^+ channels. Our proposal is based on the finding of a close correlation between Ca^{2+} block and the effects of Ca^{2+} on gating,[30] as described below.

Calcium block was analyzed by determining instantaneous $I-V$ (IIV) curves. The IIV curve is obtained by activating the channels with a large depolarization and then, when most of them are open, changing V_m and measuring the current (i.e., the tail amplitude) at the new voltage before, ideally, the gates of any of the channels have closed. Current is then plotted as a function of V_m in the second step. The method depends on the fact that opening and closing of the activation gates is relatively slow. The blocking reaction, on the other hand, is effectively instantaneous and cannot be time resolved.

The IIV curve in the presence of 2 mM Ca^{2+} was nearly linear between -30 and $+30$ mV, but below -30 mV the curve became sublinear. The curvature of negative voltage is due to the blocking action of Ca^{2+} and was accentuated at higher Ca^{2+} concentration. From the IIV curves it is possible to estimate the fraction of the channels that are blocked at any Ca^{2+} concentration and V_m, as shown in Fig. 3A.

Alterations of Na^+ channel gating by Ca^{2+} were quantified by deter-

[30] C. M. Armstrong and G. Cota, *Proc Natl. Acad. Sci. U.S.A.* **88**, 6528 (1991).

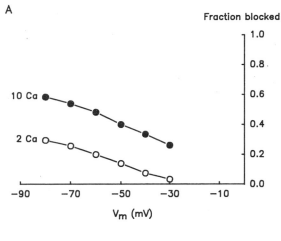

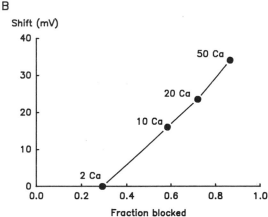

FIG. 3. (A) Voltage dependence of the fraction of sodium channels that are blocked by Ca^{2+}, as derived from IIV curves (see text). (B) Correlation between the effect of Ca^{2+} on Na^+ channel gating and Ca^{2+} block. The Ca^{2+}-induced shift of the midpoint of the fraction open $- V_m$ curve, relative to the value in 2 mM Ca^{2+}, is plotted as a function of the fraction of Na^+ channels that are Ca^{2+} blocked at -80 mV. External solutions with the Ca^{2+} concentration (mM) indicated were appropriate mixtures of the following two solutions: "50 Ca" solution; 50 $CaCl_2$, 80 NaCl; "0 Ca" solution; 80 NaCl and sucrose to raise the osmolarity to 300 mosmol. The internal solution was internal solution A. Raised Ca^{2+} increases the blocked fraction of channels. This action seems to be closely related to a stabilization of the closed conformation of the channel.

mining the midpoint of the fraction open $- V_m$ curve and the deactivation rate constant. The external Ca^{2+} concentration ranged from 2 to 50 mM. At every $[Ca^{2+}]$, normalized activation curves relating the fraction of channels with open gates to voltage were obtained by dividing the Na^+

conductance at the end of 10-msec activating pulses by the corresponding fraction of the channels that were not Ca^{2+} blocked.

Figure 3B shows that the shift of the midpoint of the fraction open $- V_m$ curve is almost directly proportional to the fraction blocked at -80 mV. Curves using block at -50, -60, and -70 mV gave similar correlation lines. There was also a close correlation between the degree of block at a given voltage and the rate at which the channel gates close. Because blocking is obviously related to the rate at which Ca^{2+} ions are entering the channels, these correlations suggest strongly that "shifting" and closing are also so related.

Calcium ion entry into Na^+ channels thus seems to play a part in the opening and closing of these channels. Although details of this role of Ca^{2+} are likely to be complicated, we tentatively suggest that the channel activation gates close stably when the channel is occupied by Ca^{2+} (or a suitable ion substitute), as is the case for potassium channels in squid neurons,[31,32] so that normally the state of the Na^+ channel on opening of the gates is Ca^{2+} blocked, and block is a step in the closing pathway.

[31] C. M. Armstrong and D. R. Matteson, *J. Gen. Physiol.* **87,** 817 (1986).
[32] C. M. Armstrong and J. Lopez-Barneo, *Science* **236,** 712 (1987).

[56] Calculation of Ion Currents from Energy Profiles and Energy Profiles from Ion Currents in Multibarrier, Multisite, Multioccupancy Channel Model

By Osvaldo Alvarez, Alfredo Villarroel, and George Eisenman

Introduction

In the quest for finding the molecular basis of ion transport through membrane channels, the connection between structure and function is the free energy profile of the ion in the pathway.[1,2] This energy profile can be calculated if the structure is known, using the tools of theoretical chemistry.[3-6] On the other hand, the systematic measurements of the

[1] B. Hille, *J. Gen. Physiol.* **66,** 535 (1975).
[2] G. Eisenman and J. A. Dani, *Annu. Rev. Biophys. Biophys. Chem.* **16,** 205 (1987).
[3] S. Sung and P. Jordan, *Biophys. J.* **51,** 661 (1987).
[4] G. Eisenman, A. Oberhauser, and F. Bezanilla, *in* "Transport through Membranes: Carriers, Channels and Pumps" (A. Pullman, J. Jortner, and B. Pullman, eds.), p. 27. Kluwer, Academic Publishers, Dordrecht, The Netherlands, 1988.
[5] S. Furois-Courbin and A. Pullman, *in* "Transport through Membranes: Carriers, Channels

electrical characteristics generate data that contain information on the channel energy profile.[7-9] A key step in the deduction of this energy profile from physiologic data is to be able to calculate currents for a given barrier model. Several treatments of barrier models have been published.[3,9-17] These models involve particular restrictions on the number of barriers, barrier symmetries, and special ion interaction rules. This paper presents a general procedure to state the problem as a computer program, which is applicable to channels with any number of barriers and sites. The model allows for multiple occupancy.[16,18] It includes a simple Gouy–Chapman model of surface charge on the channel edges to stimulate the effect of electrostatic potentials arising from membrane or protein ionizable groups.[19,20] The model accepts one or more ion species, so it can be used to analyze interactions of the ions in the channel and to predict inhibition by nonpermeable ions. Our contribution is to make available a detailed numeric procedure which is general enough, and which can be specialized easily, so it can be used in a wide variety of experimental situations.

The ability to calculate currents from a model under given experimental conditions is important, but searching for the best model that describes a set of results is another step. We describe here AJUSTE, a nonlinear curve-fitting procedure based on the Gauss–Newton method. AJUSTE has evolved starting from a program designed to describe Mössbauer effect spectra.[21] AJUSTE includes ways of fixing or adjusting any of the parameters of the model and allows constraints to be defined on the allowable

and Pumps" (A. Pullman, J. Jortner, and B. Pullman, eds.), p. 337. Kluwer Academic Publishers, Dordrecht, The Netherlands, 1988.

[6] J. Åcquist and A. Warshel, *Biophys. J.* **56**, 171 (1989).

[7] G. Eisenman and J. P. Sandblom, *in* "Physical Chemistry of Transmembrane Ion Motions" (G. Spach, ed.), p. 329. Elsevier, Amsterdam, 1983.

[8] X. Cecchi, R. Latorre, and O. Alvarez, *J. Membr. Biol.* **77**, 277 (1983).

[9] B. W. Urban and S. B. Hladky, *Biochim. Biophys. Acta* **554**, 410 (1979).

[10] P. Läuger, *Biochim. Biophys. Acta* **311**, 423 (1973).

[11] J. Sandblom, G. Eisenman, and E. Neher, *J. Membr. Biol.* **31**, 383 (1977).

[12] J. Sandblom, G. Eisenman, and J. Hägglund, *J. Membr. Biol.* **71**, 61 (1983).

[13] J. Hägglund, G. Eisenman, and J. Sandblom, *Bull. Math. Biol.* **46**, 41 (1984).

[14] T. B. Begenisich and M. D. Cahalan, *J. Physiol. (London)* **327**, 217 (1980).

[15] T. B. Begenisich and M. D. Cahalan, *J. Physiol. (London)* **307**, 243 (1980).

[16] B. Hille and W. Schwarz, *J. Gen. Physiol.* **72**, 409 (1987).

[17] P. Hess and R. W. Tsien, *Nature (London)* **309**, 453 (1984).

[18] G. Eisenman, R. Latorre, and C. Miller, *Biophys. J.* **49**, 509 (1986).

[19] E. Moczydlowski, O. Alvarez, C. Vergara, and R. Latorre, *J. Membr. Biol.* **83**, 273 (1985).

[20] A. Villarroel and G. Eisenman, *Biophys. J.* **51**, 546a (1987).

[21] D. G. Agresti, M. F. Bent, and B. I. Persson, "A Computer Program for Analysis of Mössbauer Spectra." California Institute of Technology, Pasadena, California, 1972.

values of the parameters. It gives information on the errors of the parameter estimations, and also parameter correlations. The program is simple, fast, and has provisions to escape from local minima. We have been using AJUSTE since 1981.[8,18-20,22-27] Here it is described in some detail, and the FORTRAN source code is made available in machine-readable format.

The plan of this chapter is to give first a brief summary of the theory of multibarrier channels, second, to show an example of how to find the parameters of a model channel using our programs, to encourage the less computer-literate researchers, and, finally, to present a detailed description of the programs for those researchers interested in adapting our programs to their specific tasks.

Barrier Models of Channel Permeation

Barrier models describe the ion transport through a channel as a series of ion movements from one binding site of the channel to another binding site.[10] Each of these jumps implies the movement of the ion away from the ligands of one binding site and approach to the ligands of the adjacent binding site. As the ion moves from one site to the next along the reaction coordinate, the system starts from a local energy minimum, passes through a maximum, and reaches another local minimum. The minima, or energy wells, are the binding sites, and the maxima are the energy barriers of the potential energy profile. The rate of ion movement is proportional to the probability of occupancy of a given binding site, and it is exponentially related to the activation energy to leave the site, that is, the energy difference from the bottom of the energy well to the peak of the energy barrier. The channel can have several binding sites, and more than one site can be occupied at any given time.

To calculate the current passing through a channel, the first step is to describe the occupancy states, the allowed state transitions, and the rate constants of the transitions. The next step is to write equations for the rate of change of the probability of occupancy of each state. At steady state these rates of change are zero, that is to say, the rate at which the channel enters any given occupancy state is equal to the rate of leaving it. The rate of transition from state i to state j is the probability of being in state i, p_i,

[22] X. Cecchi, O. Alvarez, and R. Latorre, *J. Gen. Physiol.* **78,** 657 (1981).
[23] X. Cecchi, O. Alvarez, and D. Wolff, *J. Membr. Biol.* **91,** 11 (1986).
[24] X. Cecchi, D. Wolff, O. Alvarez, and R. Latorre, *Biophys. J.* **52,** 707 (1987).
[25] A. Oberhauser, O. Alvarez, and R. Latorre, *J. Gen. Physiol.* **92,** 67 (1988).
[26] A. Villarroel, O. Alvarez, and G. Eisenman, *Biophys. J.* **53,** 259a (1986).
[27] A. Villarroel, O. Alvarez, A. Oberhauser, and R. Latorre. *Pfluegers Arch.* **413,** 118 (1988).

times the rate constant connecting state i with state j, k_{ij}. For a channel with n states of occupancy, there are n equations of the form

$$0 = -\sum_{j=1}^{n} k_{j,i}p_i + \sum_{j=1}^{n} k_{i,j}p_j \tag{1}$$

In addition the sum of the probabilities of occupancy of all the states must add up to 1.0, so the equation for one state can be eliminated. The system of equations, stated in matrix form, can be solved for p_i, using any matrix inversion routine. The current is calculated as the sum of all the rates of transitions in one direction over a chosen barrier, minus the sum of all transitions in the opposite direction over the same barrier.

In a three-barrier, two-site model with single occupancy, there are only three states: (i) empty, (ii) one ion in site 1, and (iii) one ion in site 2. State (i) is connected to states (ii) and (iii), and states (ii) and (iii) are connected. The allowed transitions are among the connected states, and a simple transition diagram can be drawn easily. The number of occupancy states grows rapidly as the number of sites increases; for instance, in a four-site model in the presence of one ion species, there are 16 states. In the presence of two ion species, the number of states is 81, and it is very difficult to keep track of the connections among these 81 states. To solve this problem, we developed the program NCREATE which generates automatically a FORTRAN source code that takes care of all the steps involved in the calculation of currents for a channel with any number of binding sites and ion species.

Example of Usage of Programs

The distribution disk has several programs ready to run on IBM personal computers or compatibles, and the source code is also included to compile the programs for other computers. This section demonstrates how to use the compiled programs, and it is oriented to the researcher who wants to use the programs as distributed, with a minimum of programming.

Hardware. The programs run on any IBM personal computer PC, XT, or AT with 512 kilobytes of RAM, Hercules or EGA graphics adapter, IBM graphics printer or HP laser printer, or compatibles. A mathematical co-processor is not needed but is used if present.

Software. The graphics display program, APLOTS, uses HALO subroutines and device drivers. To run APLOTS, the drivers for the graphics card adapter and for the printer are needed; for recompiling APLOTS, the object file HALODVXX.OBJ and the HALOF.LIB library are needed. HALO is a product of Media Cybernetics, Inc. [8484 Georgia Ave., Ste.

200, Silver Spring, MD 20910, phone (301) 495-3305], from which these files can be bought. The rest of the programs run without any extra support, so they can be used even if HALO files are not available. The programs were compiled using the FORTRAN Optimizing Compiler of Microsoft for the MS-DOS operating system, which is needed if the programs are to be recompiled (Microsoft Corporation, 16011 NE 36th Way, Box 97017, Redmond, WA 98073-9717). A text editor is needed to prepare the input files. The editor must generate ASCII files, with no control characters other than "carriage return" and "line feed." EDLIN, the MS-DOS editor, is suitable.

Program Setup. Create a separate directory on the hard disk to be used as workspace, for example, C:\AJUSTE. Copy from the distribution diskettes to the AJUSTE directory the following files: AJUSTE.EXE, TOLOTUS.EXE, APLOTS.EXE, FIT.BAT, SHAM.DAT, and SHAM.PAR. Copy from HALO diskettes the screen driver HALOIBME.DEV for the EGA adapter or HALOHERC.DEV for the Hercules card, and the printer driver HALOEPSN.PRN for IBM graphics/pro dot matrix printer or HALOLJTP.PRN for HP Laserjet+ or compatible printer.

Demonstration Run. Once these files are copied you may try the demonstration run typing "FIT SHAM" to the DOS prompt. The batch file FIT.BAT copies the input files SHAM.DAT into DATOS.DAT, SHAM.PAR into PAR.DAT, calls AJUSTE which performs curve-fitting and finds the best values for the parameters, calls TOLOTUS to prepare the output files, and finally calls APLOTS, an interactive graphics program useful to evaluate visually the quality of the fit.

Preparing Input Files

DATOS.DAT. The experimental observations, consisting of a series of determinations of current passing through the channel at various voltages and ion concentrations, are used to prepare the data input file; the name of this file must include the extention .DAT, for example, CHANNEL.DAT. This file is copied into DATOS.DAT before AJUSTE is called. SHAM.DAT in the distribution diskettes can be examined to see the file structures, and Table I is an example of a .DAT file that contains data for K^+ currents in the large-conductance Ca^{2+}-activated K^+ channel.[28] It is important to observe the width of the data fields in the file, and the editor used to prepare these files must fill the blank spaces with "spaces" (ASCII 32); use no "tabs" (ASCII 9).

The first two lines control the operation of AJUSTE. The four-letter code on the first line is the name of the function, which is used to select

[28] A. Villarroel, O. Alvarez, and G. Eisenman, *Biophys. J.* submitted (1992).

TABLE I

First 13 Lines of the Experimental Data File (DATOS.DAT)[a]

3B2S Kasion 1 *A*

215	5	2				
−40.0	0.00097	0.00097	0.000	0.000	−2.857	0.350
−30.0	0.00097	0.00097	0.000	0.000	−2.060	0.485
−20.0	0.00097	0.00097	0.000	0.000	−1.281	0.500
−10.0	0.00097	0.00097	0.000	0.000	−0.582	0.500
10.0	0.00097	0.00097	0.000	0.000	0.471	0.500
20.0	0.00097	0.00097	0.000	0.000	0.867	0.500
30.0	0.00097	0.00097	0.000	0.000	1.224	0.500
40.0	0.00097	0.00097	0.000	0.000	1.551	0.500
50.0	0.00097	0.00097	0.000	0.000	1.817	0.500
60.0	0.00097	0.00097	0.000	0.000	1.968	0.500
70.0	0.00097	0.00097	0.000	0.000	1.990	0.500

[a] The first line has a four-letter name of the function used to fit the data, along with the name of the experiment. The second line has the number of data points, the number of independent columns having independent variables, and a printing level code. Only 11 of the 215 lines are printed here. The other lines have the experimental data points. The first 5 columns are the independent variables describing the experimental conditions. In this case the voltage (in mV), the K^+ concentration on the inside (M), the K^+ concentration on the outside, and two unused variables. The sixth column is the measured current (pA), and the last is a weighting factor equal to 1/current but no larger than 0.5. Data are from Villarroel et al.[28]

what expression AJUSTE will use to calculate the currents, a 3B2S model in this case. This function code is needed because AJUSTE has built in several functions such as exponentials, Michalis–Menten functions, and multibarrier channels. The rest of line 1 is the title of the run.

On the second line, the first eight-character field is the number of observations contained in the file, 215 in the case of Table I. The second eight-character field is the number of independent variables, 5 in this case. The third eight-character field is printing code, which controls how verbose the output to the screen will be, 2 in this case.

The succeeding lines have a description of the experimental observations, using a set of five independent variables. The first field is voltage, the second field is the K^+ activity on the inside, and the third field is the K^+ activity on the outside of the channel. Voltage is in millivolts (ground at the outside), and the activities are in molar. Two more independent variables are set to zero and are reserved for the concentration of a second ion species, used for mixed-ion experiments. The sixth field is the experimental current, in picoamperes, and positive current is carried by cations moving from inside to outside. The last field is a weighting factor; in this case it is

TABLE II
INITIAL PARAMETERS FILE (PAR.DAT)[a]

Peak 1	2.0000	0.0000	1.000	10.000
Peak 2	2.0000	0.0000	0.000	10.000
Peak 3	2.0000	0.0000	1.000	10.000
Well 1	−3.0000	0.0010	−10.000	0.000
Well 2	−3.0000	0.0010	−10.000	0.000
D1	0.0000	0.0000	0.000	0.000
D2	0.2500	0.0010	0.050	0.450
D3	0.0000	0.0000	0.000	0.000
D4	0.0000	0.0000	0.000	0.000
D5	0.0000	0.0000	0.000	0.000
Arep	1.0000	0.0000	−10.000	10.000
Rcint	100.000	0.0000	1.000	1000.000
Rcext	10.0000	0.0000	1.000	1000.000

[a] In the first column are the names of the parameters. The second column contains the initial estimates of the parameters. The third column has the ϵ values used to compute the derivatives of the evaluated functioni with respect to that parameter. A zero entry means the parameter is not adjustable. Columns 4 and 5 contain the lower and upper bounds for the parameters.

the reciprocal value of the current, but no greater than 0.5. This weighting factor assures that AJUSTE will minimize the relative deviations. For a flat weighting factor, for example, 1.0, it will minimize the absolute errors. The numbers on all the fields on lines 3 to the end of the file must have an explicit decimal point.

PAR.DAT. AJUSTE needs a set of initial guesses of the parameters to start; these values are read from the input file PAR.DAT, and an example of such a file is in Table II. The first field (first column) is the name of the parameter, used for documentation purposes. The second field is the initial guess of the parameter, to be used by AJUSTE. The third field is an epsilon (ϵ) used by AJUSTE in the numerical calculation of the partial derivative of the function with respect to the parameter. Sometimes it is useful to keep some parameter fixed during an AJUSTE run, especially when the parameters are away from the optimum, in a many-parameter problem. A zero entry for an ϵ means that the parameter is not to be changed by AJUSTE. The last two columns are the lower and upper bounds for the parameter. These bounds are useful to prevent AJUSTE from giving physically meaningless values to the parameters during the search.

Output Files. AJUSTE produces three ASCII output files: LOG.DAT which contains the history of the changes of the parameters and the sum of squares during the run, RESULT.DAT with the final parameters list and the calculated currents, and PAR.NEW which is a file with the same structure as PAR.DAT that is used to restart the next run of AJUSTE. The utility program TOLOTUS makes RESULT.PRN, an special version of RESULT.DAT which is in the proper format to be imported (as numbers) into Lotus 1-2-3 for graphic display.

Finding the Parameters. Before the first run of AJUSTE on a particular problem, all the parameters are unknowns, and AJUSTE needs some starting values to work. If the parameters are too far from describing the data, and all are allowed to change, AJUSTE will converge very slowly or fail completely. A possible strategy to solve the problem is presented in Table III. For the first run reasonable values of the energy parameters are chosen, the barrier and well positions are symmetrical, and a low surface charge is used. Only the wells are allowed to change in this run. The initial sum of squares is more than 10,000; AJUSTE reduccd it to 192.6 and stopped. The result of run 1 and the starting conditions for run 2 are in the column "Run2." Next, peak energy values are also allowed to change, and the sum of squares reached 55.6. The repulsion parameter is now adjusted in run 3, keeping the rest constant. The surface charge is adjusted in run 4,

TABLE III
STRATEGY TO FIND BEST PARAMETERS[a]

	Run1	Run2	Run3	Run4	Run5	Final results	Standard deviation
Peak 1	2.00	2.00*	2.91	2.91*	2.95*	2.7705	0.1172
Peak 2	2.00	2.00*	0.01	0.00*	−5.00*	−5.0000	
Peak 3	2.00	2.00*	3.28	3.28*	3.91*	3.1545	0.2093
Well 1	−3.00*	−4.83*	−7.67	−7.67*	−7.28*	−7.9746	0.2410
Well 2	−3.00*	−3.92*	−7.97	−7.97*	−7.52*	−7.8829	0.2277
D1	0.00	0.00	0.00	0.00	0.00	0.0000	
D2	0.25	0.25	0.25	0.25	0.25*	0.2046	0.0050
D3	0.25	0.25	0.25	0.25	0.25	0.2954	
D4	0.25	0.25	0.25	0.25	0.25	0.2954	
D5	0.25	0.25	0.25	0.25	0.25	0.2046	
Arep	1.00	1.00	1.00*	1.28	1.28*	1.5927	0.0905
Rcint	100.0	100.0	100.0	100.0*	46.94*	66.5272	9.4850
Rcext	10.00	10.00	10.00	10.00*	18.56*	29.0092	2.3276
SUMSQ	>10,000	192.6	55.6	44.8	21.0	12.45	

[a] Because there are so many parameters it is not possible to let all be adjustable at once. On each run, only the parameters marked with an asterisk were adjustable. The result of a run is used as input for the next run. SUMSQ is the weighted sum of squares calculated using the values in the column. Final results are listed with the standard deviations calculated by AJUSTE. Well 2 has no error entry because it hit a bound set at −5 on run 5. Only distance D2 has an errory entry because it was the only adjustable distance; D1 was never adjusted, and the other distances are calculated from D2.

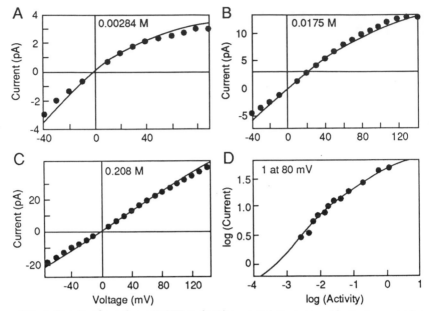

FIG. 1. Results of running AJUSTE to fit K⁺ current data measured in the large-conductance Ca^{2+}-activated K^+ channel (data from Ref. 28). (A–C) Current–voltage curves, (D) log–log plot of the current measured at 80 mV, as a function of K^+ concentration. The experimental observations are plotted as points, and the lines are the currents computed by AJUSTE, using the functions listed in Appendix A and the parameters from Table III. Plots were prepared importing the ASCII file RESULT.PRN into Lotus 1-2-3.

allowing refinement of the well and energy values. The central peak value hit the lower bound in this run. For run 5, the lower bound for this peak was set at -5.0 and all parameters are free to change. The final parameter set gives a sum of squares of 12.45. The central peak was set by AJUSTE to be -5.0, the lower limit, so it is not well known; the other parameters are determined, and the error of the estimation is given in the last column of Table III. Errors for distances 3, 4, and 5 are not stated since these distances are calculated from distance 2. Plots of sample current–voltage curves and current versus concentration for comparison of experiment and theory are shown in Fig. 1.

Description of Programs

The FORTRAN function MBMSMI (and related modules) evaluates the current passing through a multibarrier, multisite, multiion channel. (MBMSMI stands for multibarrier, multisite, multiion.) A specialized version of this program for the case of three barriers, two sites, and one ion is presented in Appendix A. We also describe NCREATE, a program that automatically wrote that part of MBMSMI which is specific for each combination of number of sites and number of ion species. Lastly we describe AJUSTE, a program that uses MBMSMI and experimental data to find the energy profile of a real channel.

Function MBMSMI

MBMSMI is a general function which calculates the current passing through a multibarrier, multisite channel model, with any number of ion species. Part of MBMSMI is totally general, and that part was hand-coded, with the subroutines and functions that depend on the number of sites and the number of ion species being generated automatically by the program NCREATE. For economy of space, only the modules for a three-barrier, two-site, one-ion case are listed in Appendix A. The version distributed in diskettes can evaluate currents for one or two ion species in models with one, two, three, or four binding sites. It executes the proper instructions according to the function name code included in the input DATOS.DAT file. A nonzero concentration entry for a particular ion in DATOS.DAT is used to sense the presence of that ion.

The energy profile is defined by the energy at each point in the ion path at zero applied potential, the effect of applied potential on these energy levels, and by the ion–ion interaction when more than one ion is present in the channel. In addition, surface potential can alter the local concentrations and energy profile. The energies are measured using as a reference state the ion outside the channel at a hypothetical unity mole fraction. EPEAK are the energies at each one of the peaks, and EWELL are the energies at the wells.

The array D contains the differences on "electrical distances" at the various peaks and wells. The electrical distance at each location is the sum of all electrical distances up to the location; for example, the electrical distance at peak 2 is D1 + D2 + D3.

The model allows multiple occupancy, that is, all sites can be occupied at the same time. When one site is occupied it is expected that the energy increases at the other sites and peaks, for example, as a consequence of the electric field around the ion. To describe such an interaction, we assume that the energy shift is proportional to the inverse of the distance measured from the ion to any other point in the channel.[20] (cf. the less physically meaningful Hille–Schwarz rule[16]). Although cooperative interactions would presumably have a different spatial dependence, in the first approximation we assume them to decrement with 1/distance. The interaction parameters therefore need not be strictly repulsive nor species independent. The energy shift is for a distance of 1.0 is the interaction parameter, AINTER, and the "electrical distances" for calculation of the ion interactions inside the channel. The energy shift at any given point is AINTER divided by the electrical distance from the point to the ion. We use one interaction parameter to describe the interaction of each possible ion pair. These parameters are not independent but have the constraint AINTER(i,j) = AINTER(j,i) required for microscopic reversibility.

MBMSMI calculates the current passing through the channels for a given set of channel parameters and a set of independent variables, which are voltage, V, and ion concentrations, CIONS, a two-dimensional array with the molar concentration of each ion species at either side of the channel.

Surface Potential. The program uses the Gouy–Chapman formula for a planar charged surface as an approximation of the surface potential in the mouth of the channel. Surface charge density is defined here as the radius of a circle containing one electron charge.[20] This model is correct only for a plane electrified surface, but it is used for simplicity. As detailed structural details are lacking, no more exact model is used here. Two radii are required to describe the surface charge, RSCIN for the intracellular side and RSCEX for the extracellular side of the channel. The present treatment is restricted to univalent cations, and the surface potential is evaluated using Eq. (2),

$$S.Pot = ARCSINH\{-136/[\pi R^2 (\text{Total Ion Conc})^{1/2}]\} \tag{2}$$

where S.Pot is the surface potential divided by e/kT, π is 3.1416, R is the radius of a circle containing one charge, and Total Ion Conc is the sum of cation concentrations on each side of the channel. Once the surface potential is obtained, the local ion concentrations are calculated using the Boltzmann equation. The concentrations are transformed into mole fractions and are stored in the array XIONS. Also, the applied electrical potential, V, is corrected for the difference in surface potential in both sides of the channel and expressed as a reduced potential, U.

Change in Local Energy as Function of Applied Potential. The program computes first the electrical distances DP and DW for each peak and well. The change in potential energy caused by the applied potential is calculated using the distance parameters and U. These energy shifts are common for both ion species, and the shift is zero for the inside and U for the outside. These shifts are added to the zero-voltage energy levels, EPEAK and EWELL, and stored in the arrays UPEAK and UWELL.

Single-Occupancy Rate Constants. Rate constants for the jump of an ion over an energy barrier can now be calculated as values proportional to the exponential of the activation energy for each ion jump. The activation energy is the difference of energy from the starting well to the peak of the energy barrier. The proportionality factor is taken as unity for the moment, and the correct value is introduced at the end of the calculation when the ion flux is expressed as a current intensity (picoamperes). The entry rate constants are also multiplied by the mole fraction of the ion in the external solution. The rate constants are stored in the three-dimensional array RDIP. The three dimensions are Direction, Ion species, and Peak number. Each element of RDIP has the rate constant for the jumps classified by the direction of the jump, which ion is jumping, and which barrier is jumped over. The forward direction is coded as 1 and backward as 2.

Cases for Particular Combinations of Number of Sites and Ions. MBMSMI is general for any combination of number of sites and ion species, except for the physical limitation of the dimensions of the arrays. In the version of Appendix A, these limits are set to no more than four sites, three ions, and 81 occupancy states. The rest of the program is particular for each combination. In Appendix A the program is set up for the case of two sites and one ion species. NCREATE can write the appropriate modules for other combinations. The distribution diskette has the modules for channels with one, two, three, or four binding sites and 1 or 2 different ion species.

To simplify the notation of the occupancy states, an n-site channel is represented by a string of n characters, one for each site. The characters can be 0, 1, 2, 3, . . . , for empty, or occupied by one ion of species 1, 2, 3, ... For the case of a four-site channel in the presence of only one ion species, the occupancy states are

$$
\begin{array}{cccccccc}
0000 & 1000 & 0100 & 1100 & 0010 & 1010 & 0110 & 1110 \\
0001 & 1001 & 0101 & 1101 & 0011 & 1011 & 0111 & 1111
\end{array}
$$

The states are also numbered, in decimal notation, following this rule

$$\text{State number} = 1 + \text{base}^0\, c_1 + \text{base}^1\, c_2 + \text{base}^2\, c_3 +$$
$$\cdots + \text{base}^{n-1}\, c_n \quad (3)$$

where base is 1 plus the number of ion species, n is the number of sites, and c_i is the numeric value of the character in the ith position. In the example, the state numbers of the first row are 1, 2, 3, 4, 5, 6, 7, and 8. For the case of a five-barrier, four-site, one-ion channel, the allowed transitions from one occupancy state to another are listed in Table IV. The program NCREATE makes these tables, which are stored in the files TRANSITS.DAT and TRANSITN.DAT.

TABLE IV

Allowed Transitions for Four-Site, Multiple-Occupancy Channel Model[a]

FROM	TO . . .	FROM	TO . . .			
0000	1000 0001	1	2	9		
1000	0100 1001 0000	2	3	10	1	
0100	1100 0010 0101 1000	3	4	5	11	2
1100	1010 1101 0100	4	6	12	3	
0010	1010 0001 0011 0100	5	6	9	13	3
1010	0110 1001 1011 1100 0010	6	7	10	14	4 5
0110	1110 0101 0111 1010	7	8	11	15	6
1110	1101 1111 0110	8	12	16	7	
0001	1001 0000 0010	9	10	1	5	
1001	0101 1000 1010 0001	10	11	2	6	9
0101	1101 0011 0100 0110 1001	11	12	13	3	7 10
1101	1011 1100 1110 0101	12	14	4	8 11	
0011	1011 0010 0101	13	14	5	11	
1011	0111 1010 1101 0011	14	15	6	12 13	
0111	1111 0110 1011	15	16	7	14	
1111	1110 0111	16	8	15		

[a] On the left-hand side 0 stands for an empty site and 1 stands for an occupied site. On the right-hand side are the same states in decimal notation. The decimal numbers are calculated using EQ. (3) with base $= 2$ and $n = 4$.

Subroutine R3B2S1

The rate constants $Q(i,j)$, connecting state i and the state j, for a three-barrier, one-ion case are calculated by the subroutine R3B2S1. (RXBYSZ is the generic name for a family of subroutines, X being the number of barriers, Y the number of sites, and Z the number of ion species. These subroutines are made automatically by NCREATE.) The rate constants for the jump over all the barriers are received from MBMSMI in the array RDIP. R5B4S1 deals with the generation of the code for the multioccupied states. The rate constants are of the form exp $[E(\text{peak}) - E(\text{site})]$, and the rate constant in the presence of another ion is equal to the rate constant for the single-occupied channel times exp $[\text{Shift of } E(\text{peak}) - \text{Shift of } E(\text{site})]$, where the Shifts are taken as the excess of potential caused by the second ion at the site or peak. This is calculated from an interaction parameter, AINTER, divided by the distance measured from the peak or site to the second ion location.[20] The interaction parameters are called AINTER(i,j) for the repulsion of one ion of species i caused by an ion of species j. The reciprocal of the distances are called XDPiWj and XDWiWj for the reciprocal values of the distances from peak or site i to site j. Because the ion–ion interaction rule uses the reciprocal of the distance, R5B4S1 calculates all these distances, as can be seen in the program listing.

The next step is to assign the nonzero rate constants $Q(i,j)$. The nonzero rate constants are those called for by the transition table. For instance, for the case of a three-barrier, one-ion case, from state 1 (empty) the channel can go to states 2 or 3 (one ion in well1 or one ion in well2). The transition 1 to 2 is a forward jump of ion 1 over peak 1, and the transition 1 to 3 is a backward jump of ion 1 over peak 3. The rate constant assignments are

$$Q(1,2) = RDIP(1,1,1)$$
$$Q(1,3) = RDIP(2,1,3)$$

The transition 3 to 4 (01 to 11) also involves a forward jump of ion 1 over barrier 1, but with an ion at site 2; thus, the height of barrier 1 is changed by the interaction between ion 1 at peak 1 and ion 1 at site 2, and the rate constant is

$$Q(3,4) = RDIP(1,1,1) * EXP(-AREP(1,1)*XDP1W2)$$

The same logic applies for the rest of the rate constant assignments.

Subroutine M3B2S1

The subroutine M3B2S1 writes Eq. (1) for each of the possible occupancy states of the channel. These equations are written in matrix form:

$$VEC = ELE * Prob \qquad (4)$$

Where the vector VEC contains zeros, the matrix ELE has the coefficients of the probabilities of occupancy and the vector Prob has the (unknown) probabilities of occupancy.

M3B2S1 fills the matrix ELE using Eq. (1), the allowed transitions of the table, and the $Q(i,j)$ values returned by subroutine R3B2S1. It also replaces the last row of Eq. (4) by a row of 1's which includes the condition that the sum of all the occupancy states is 1. M3B2S1 is a member of a family of subroutines named MXBYSZI, where X is the number of barriers, Y the number of sites, and Z the number of ions. These subroutines are generated by NCREATE.

Function C3B2S1

C3B2S1 is a function that calls DMATINV, a matrix inversion subroutine that has to solve the system of equations and put the resulting probability vector on vector VEC. Once this vector is known, the current is calculated by adding all the forward ion jump rates over peak 2 and subtracting all the backward jump rates over the same peak. For the case of two sites and one ion species, the only forward jump over peak 2 is the transition from state 2 to state 3, and the backward jump is from 3 to 2. The rate of jumping is the probability of occupancy of the source state times the rate constant of the jump times the factor 994800.0, which converts the ion jump rate into picoamperes.[1] For example, the rate of transitions from state 2 to state 3 is proportional to $Q(2,3) * VEC(2,1)$. The resulting current is passed back to the calling program in the variable MBMSMI. C3B2S1 is a member of a family of functions named CXBYSZI, where X is the number of barriers, Y the number of sites, and Z the number of ions. These subroutines are generated by NCREATE.

Program NCREATE

The subroutines M3B2S1 and R3B2S1 and the function C3B2S1 were written automatically by NCREATE. These modules are for the case of two sites and one ion species, and the proper modules should be linked to MBMSMI for other combinations of sites and ions. Because NCREATE writes these modules automatically, this package of programs is actually general for multibarrier, multisite, multiion channels, that is, for any number of barriers, sites, and ion species. STATES, RATES, and CURREN are the main logical structures of NCREATE. See Appendix B for full listing of NCREATE.

STATES. In the program NCREATE the occupancy state of an *n*-site channel is represented by a string of *n* characters, one for each site. The states are also given numbers, following the rule stated in Eq. (3).

From each state there can be jumps to one or more states. The numbers of these destination states are stored in the two-dimension array *idest(i,k)*. This array initialized to zeros. To fill this array, the program loops over all the states; the index *i* runs from to 1 to *i*max, the number of states. At the beginning of each turn of the *i* loop, the index *k* is set to 1 and a *n*-character string is created for the *i*th state according to the above definition. Then the program scans the occupancy of the sites and changes the characters according to the transition rules described below. Each time the occupancy state is changed, the state number represented by the new string is saved in an output file, and the original state is restored to find more possible destinations. Once this file is completed, it will contain all the possible transitions arranged in two column entries; the first column is the number of the source state and the second is for the destination state. Each destination state number is also stored in *idest(i,k)*, and the index *k* is incremented.

The transition rules are as follows:

Forward jumps

Over peak 1 from the external solution: If the character in site 1 is '0,' then replace it successively with the character 1, 2, 3, . . . , *maxion.* This generates the occupancy states arising from the movement of an ion of species 1, 2, 3, or maxion from the outside into site 1.

Over peaks j, when j can be 2, 3, . . . , *nsites:* If the character in site j is '0' and the character in site $j-1$ is not '0,' then copy character on site $j-1$ into site j, and put a '0' in site $j-1$. These movements generate the states arising from the forward movement from one site to the next.

Over peak *nsites* + 1 to the external solution: If the character in site *nsites* is not '0' replace it by '0.' This generates the states arising from the exit of an ion from the last site to the solution at the other side of the channel.

Backward jumps

Over peak *nsites* + 1 from the external solution: If the character in site *nsites* + 1 is '0,' then replace it successively with the character 1, 2, 3, . . . , *maxion.* This generates the occupancy states arising from the movement of an ion of species 1, 2, 3, or *maxion* from the outside into site *nsites* + 1.

Over peaks j, when j can be *nsites* - 1, . . . , 3, 2, 1: If the character in site j is '0' and character in site $j+1$ is not '0,' then copy the character on site $j+1$ into site j, and put a '0' in site $j+1$. These movements generate the states arising from the backward movement from one site to the adjacent.

Over peak 1 to the external solution: If the character in site 1 is not '0' replace it by '0.' This generates the states arising from the exit of an ion from the first site to the solution.

STATES creates one output file containing all the possible pairs of source–destination states, to be used by the subroutine RATES, and the forward transitions over peak 1 for two- and three-state channels and over peak 3 for four-site channels, to be used by the subroutine CURRENT. The transition table for the four-site, one-ion case presented in Table IV was written by STATES.

STATES also writes the code to state the set of simultaneous equations of the form of Eq. (1), in matrix format. The output file is the subroutine "M," which takes different names according to the number of sites and ions. For a five-barrier, four-site channel the name of the subroutine is M5B4S1 and the name of the FORTRAN source code is 5B4S1I.FOR.

A two-dimensional array $Q(i,j)$ is used to represent the rate constants. The index i is the number of the source state, and index j is the number of the destination state. The array VEC contains the left-hand side member of Eq. (1), and the two-dimensional array ELE contains the coefficients of p_i in Eq. (1). STATES creates a file with a FORTRAN program. It writes the code to declare the subroutine, dimensions the variables, and writes the code to initialize ELE and VEC to 0.0. To fill the array ELE the program loops over all the occupancy states, the index variable i takes the values from 1 to the number of states. A second loop nested into the i loop uses the index variable j that also goes from 1 to the number of states.

1. If $i = j$ then the program creates a FORTRAN statement that contains the code to define a diagonal element of the matrix.

The string S1 is filled with 'ELE(I,I) = -(', where I is the string representation of the index i. For example, if $i = 2$ the S1 is 'ELE(2,2) = -('.

The string S2 is filled with repetitions of the '$Q(I,D(I,K)) + $' as long as $d(i,k)$ is not zero. In this string, I and $D(I,K)$ are the string representations of the contents of the variables i, the number of the source state, and $d(i,k)$, the number of the destination states.

The FORTRAN statement is created, writing S1 starting from column 7, S2 is catenated, and the last + sign replaced by ')'. If the final line is longer than 72 characters it is split into continuation lines, following FORTRAN syntax. The line(s) is written to the output file.

2. If i is not equal to j then the program scans the array $d(i,k)$ for constant i and over all the values of k; if the condition $d(i,k) = j$ is met, it means that the transition from state j to state i belongs to the allowed transitions, and the program outputs the FORTRAN statement 'ELE(I,J) = Q(I,J)', where I and J are the string representations of the contents of the variables i and j. If the condition is not met, then no output is generated, meaning that the transition from j to i does not belong to the model.

See Appendix A for the listing of 5B4S1I.FOR. In the case of a four-site model with two ion species, the program STATES generates 522 lines of FORTRAN code.

RATES. The program STATES generates the FORTRAN code to solve for the probability of occupancy of each of the states of the model in terms of the rate constants Q. RATES generates the code to calculate some of these rate constants. The rate constants for the single occupancy states are calculated from the energy difference measured from the starting site to the top of the barrier that has to be jumped over. The code to calculate these rate constants is straightforward and it is not generated by RATES. The program deals with the generation of the code for the multioccupied states.

To generate the code for the rate constants $Q(i,j)$, the file created by STATES that contains the pairs of source–destination states is read. For each pair one or more lines of code are created. To determine what type of transition is involved, the state numbers are converted to a n-character string, as in STATES, for the source state From and the destination To state. An equivalence statement makes available the contents of From and To; the characters $F(i)$ and $T(i)$ are the ith characters of the respective state.

> Jump over peak 1: A jump over peak 1 occurs when two conditions are met: $T(1)$ is not equal to $F(1)$ and $T(2)$ equals $F(2)$ (*i.e.*, site 1 changed but the ion did not come from site 2; it must have come from the outside, jumping over peak 1). If these conditions are met, a forward jump occurs when $F(1)$ is equal to '0,' the jump started from an empty site 1, and the ion that jumped is the one found in $T(1)$; otherwise, if $F(i)$ is not '0,' then a backward jump is detected and the ion involved is the one in $F(1)$.
>
> Jump over peak n sites $+$ 1: The logic is equivalent to that used for peak 1.
>
> Jump over other peaks: For a jump over peak i, for $i = 2, 3, \ldots, n$ sites, a transition occurs when site $i - 1$ and i change, because ions can only move from filled to vacant sites. A forward jump occurs for $F(1) = $ '0' and a backward jump for $F(i - 1) = $ '0.' The ions jumping are $T(i)$ and $T(i - 1)$, respectively.

Once the source and destination states are known, as well as the ion and peak involved in the jump, the first line of code can be written. This is the single occupancy part of the statement, which has the form

$$Q(\{\text{FROM-STATE}\},\{\text{TO-STATE}\}) = \text{RDIP}(\{\text{DIRECTION}\},\{\text{ION}\},\{\text{PEAK}\})$$

where the names between braces are replaceable single-digit integer numeric fields. Continuation lines are written if sites other than site 1 are occupied. The program scans these sites and writes out one continuation line for each ion found in the channel. These lines have the general form

$$*\text{EXP}(-(\text{AINTER}(\{\text{IONA}\},\{\text{IONB}\}) * (\text{XDP}\{\text{PEAK}\}\text{W}\{\text{WB}\} - \text{XDW}\{\text{WA}\}\text{W}\{\text{WB}\})))$$

where IONA is the ion involved in the transition and IONB is another ion in the channel. PEAK is the barrier jumped over, and WA and WB are the sites of ions A and B, before the transition. If the transition starts from the outside solution the part of the expression containing WA is not written out.

See Appendix A for the listing of the subroutine generated by RATES for the case of a three-barrier, two-site channel with one ion species. In the case of the four-site, two-ion model RATES generates a subroutine with 1266 lines of code.

CURREN. The function CURREN creates the code to calculate the current when the occupancy of each state has been solved for. CURREN reads a file (C.DAT) created by STATES which contains the state numbers of the initial and final states involved in the forward transitions over the barrier that will be used to calculate the current. Any barrier could be used, but the program uses barrier 2 for two- and three-channels and barrier 3 for four-site channels. See Appendix A for a listing of C3B2S1, which was created entirely by NCREATE.

AJUSTE

MBMSMI can calculate the current passing through a channel for a given set of independent variables and model parameters. AJUSTE is a program that finds the best values for the parameters of a model, given a set of experimental observations of the current. The program is a nonlinear curve-fitting algorithm that minimizes the sum of the squares of the differences between the experimental values of current and the values calculated by MBMSMI. Actually it is more general than that, since MBMSMI is just a subroutine in the program which can be replaced by any other function, such as an exponential, sum of exponentials, or a Michaelis–Menten or Hill function. The code for AJUSTE is extensive, and it is available from the authors in the form of FORTRAN ASCII sources, in personal computer diskette format.

Let us call Y_k a set of experimental observations where k is an index that runs from 1 to N, the number of observations. Let T_k be the corresponding values calculated using MBMSMI, and W_t a statistical weighing factor. The quantity that AJUSTE minimizes is S, the sum of squares:

$$S = \sum_{k=1}^{N} (Y_k - T_k)^2 W_k \tag{5}$$

The set of energy parameters, distances, repulsion factors, and surface charge densities will be called here the P_j. Because T_k is a function of P, then S is also a function of P. To find the best values of the parameters, we must find the minimum of the plane defined by S and the parameters P. This is done using an iterative procedure.

The procedure starts using a set of initial guesses for the parameters; these will be called P_{o_j}, and the sum of squares calculated using these parameters is $S_{(P_o)}$. If each of the parameters is changed by a small amount, a_j, then the new sum of squares is $S_{(P_o+a)}$, given by

$$S_{(P_o+a)} = S_{(P_o)} + \sum_{k=1}^{N_{par}} (\partial S/\partial P_j)_{P_o} a_j \tag{6}$$

where N_{par} is the number of (adjustable) parameters and $(\partial S/\partial P_j)_{P_o}$ are the partial derivatives of S with respect to each one of the N_{par} parameters, evaluated at P_o.[29] This is an approximation based on a Taylor series truncated after the first derivative.

The condition of a minimum in the $S-P$ surface is that all the partial derivatives of S with respect to all P must be zero. The N_{par} partial derivatives have the form

$$(\partial S/\partial P_i)_{P_o+a} = (\partial S/\partial P_i)_{P_o} + \sum_{j=1}^{N_{par}} (\partial^2 S/\partial P_i \partial P_j)_{P_o} a_j \tag{7}$$

Assuming that the condition of a minimum is satisfied for $P_j + a_j$, the partial derivatives are zero and a set of N_{par} "normal equations" can be written. Each one of the N_{par} equations contains N_{par} a_j terms and has the form

$$-(\partial S/\partial P_i)_{P_o} = \sum_{j=1}^{N_{par}} (\partial^2 S/\partial P_i \partial P_j)_{P_o} a_j \tag{8}$$

The partial derivatives of the sum of squares are calculated numerically by AJUSTE from the partial derivatives of T_k, which are calculated by repeatedly calling MBMSMI using small displacements of the parameter values. The relation of the partial derivatives of T and S are

$$(\partial S/\partial P_i)_{P_o} = - \sum_{k=1}^{N_{obs}} (Y_k - T_k) W_k (\partial T_k/\partial P_j)_{P_o} \tag{9}$$

$$(\partial^2 S/\partial P_i \partial P_j)_{P_o} = \sum_{k=1}^{N_{obs}} (\partial T_k/\partial P_i)_{P_o} (\partial T_k/\partial P_j)_{P_o} \tag{10}$$

Once all the partial derivatives are calculated, the N_{par} a_j values are solved for from the set of N_{par} simultaneous normal equations using a matrix inversion routine, multiplying the inverted matrix by the left-hand vector. The square roots of the diagonal elements of the inverted matrix are approximations for the standard error of each parameter. Also, large corre-

[29] E. R. Cohen, R. M. Crowe, and J. W. M. Dummond, "Fundamental Constants of Physics." Wiley (Interscience), New York, 1957.

lations between the parameters appear as large off-diagonal values in the inverted matrix. The new, better set of parameters are obtained from

$$P_j = P_{o_j} + a_j \tag{11}$$

but prior to updating the parameters two checks are taken. First, the final parameter values must be equal to or larger than preset minimum values, and equal to or less than preset maximum values. If this test fails, the parameter takes the value of the closer bound, minimum or maximum. The user defines the ranges for each parameter. The second test is that the sum of squares must be less than the one computed with the initial values of P.

If the Taylor series truncated after the first derivative were exact, these parameters should be the best, that is, give the minimum sum of squares. Because the truncated series is never exact, the new parameters are not the best but are (it is hoped) better than the initial ones. If that is the case, these new sets of parameters are used as initial estimates for a second run. The process is repeated until the changes of the parameters are a small (adjustable) fraction of the standard error of the parameters.

Limitations and Self-Healing Procedure. That is the ideal behavior of the program, but sometimes there are difficulties if there are bad initial estimates or a poor S versus P surface shape. If the initial P values are too bad, the program fails to find a better set and exits with an error code. If the sum of squares increases instead of decreasing, the program tries several parameter changes, as suggested by Chrisman and Tumolillo.[30] For this part of the procedures, instead of using a_j to change the parameters, the program uses STEP times a_j. The values for STEP are -1.0, 0.5, a value taken from the parabola that connects three values of the S plane versus P, 0.01 and 0.001. If some STEP makes the sum of squares decrease, then the best value is used for the next iteration; otherwise the program exits with a warning. These tricks are efficient and can prevent the program from being trapped in local minima. Another useful feature of AJUSTE is that some of the parameters can be held fixed during the initial trials of the problem and relaxed later once some parameters already have good values.

Interfacing MBMSMI to AJUSTE

AJUSTE calls subroutine FUNCT(I) to evaluate $T(I)$, where I defines one experimental observation. This makes AJUSTE a general program because it requires only the proper FUNCT subroutine to fit any function.

---

[30] B. L. Chrisman and T. A. Tumolillo, Preprint, May 1967. Department of Physics, University of Illinois, Urbana, Illinois.

The two-dimensional array $X(j,k)$ contains a list of j independent variables that describe the ith experimental point. For the case of an ion current passing through a three-barrier, two-site, one-ion channel, let us assume that the indices j are 1 for the voltage, 2 for the ion concentration of the solution at the intracellular side of the channel, and 3 for the extracellular solution.

The array PAR contains the parameters whose best values AJUSTE tries to find. In the case of a three-barrier, two-site, one-ion channel these parameters are the zero-voltage single occupancy energies at the three peaks and the two binding sites, five distances to define the positions of the peaks and wells, one ion–ion interaction parameter, and two radii for the surface charge calculations. FUNCT copies the proper elements of the arrays X and PAR to the corresponding variables used by MBMSMI and has MBMSMI evaluate the current passing through the channel, as shown in Appendix A. The version shown in Appendix A forces the peaks and wells to be symmetric around the middle, with the central peak located at 0.5. FUNCT must call subroutines other than R3B2S1, M3B2S1, and C3B2S1 to calculate the currents for other combinations of number of sites and ions.

Acknowledgments

The authors appreciate the effort invested by Andrés Oberhauser, Carmen Alcayaga, and Ricardo Delgado in translating to FORTRAN the early BASIC code of AJUSTE. The authors are indebted to Ximena Cecchi, who patiently debugged the first BASIC version, and to Osvaldo Latorre, who introduced Osvaldo Alvarez to the problem of nonlinear curve fitting and provided the Agresti code that served as starting point in the deployment of AJUSTE. The able technical assistance of Mr. Juan Espinoza is also acknowledged. This work has been supported by Grants USPHS GM 24749, NSF BNS 84-11033, FONDECYT 1112-1989, 1078–1991, and DTI B-2805.

Appendix A

```
C      •••••••••••••••••••••••••••••••••••••••••••••••••••••••••••••••••••
C      ••••••••••••••••••••••••••• FUNCT.FOR •••••••••••••••••••••••••••••
C      •••••••••••••••••••••••••••••••••••••••••••••••••••••••••••••••••••

       SUBROUTINE FUNCT(I)
       INTEGER*2 IABORT
       REAL*8 A(40,42)
       COMMON /DATOS/ X,Y,T,WT
       COMMON NOBS,NVAR,NPAR,IFUNCT,IPRINT,IDIVY,NPTOT,
     2 A,EPSILON,
     3 P,R,E,ISP,SDPAR,PAR,PLOL,PHIL,IABORT
       DIMENSION X(500,6), Y(500), T(500), WT(500)
       DIMENSION EPSILON(40)
       DIMENSION P(40),R(40),E(40),ISP(40),SDPAR(40),PAR(40)
```

```
      DIMENSION PLOL(40),PHIL(40)

      COMMON/FEVAL /FEVAL
      real*8 FEVAL,MBMSMI
      DIMENSION EPEAK(3,5), EWELL(3,5)
      DIMENSION AREP(3,3), D(10), CIONS(3,2)

      CALL AGRANDA

      nsites = 2
      nions = 1
      epeak(1,1)  = par(1)
      epeak(1,2)  = par(2)
      epeak(1,3)  = par(3)
      ewell(1,1)  = par(4)
      ewell(1,2)  = par(5)
      d(1)        = par(6)
      d(2)        = par(7)
      d(3)        = (1./2.) - ( d(2) + d(1) )
      d(4)        = d(3)
      d(5)        = d(2)
      arep(1,1)   = par(11)
      v           = x(1,1)
      rscin       = par(12)
      rscex       = par(13)
      cions(1,1) = x(1,2)
      cions(1,2) = x(1,3)
      feval = MBMSMI
     * (NIONS,NSITES,EPEAK,EWELL,AREP,D,V,RSCIN,RSCEX,CIONS)
      T(I)    = REAL(FEVAL)
      return
      end

C     *****************************************************************
C     ************************** MBMSMI.FOR **************************
C     *****************************************************************

C     REAL*8 FUNCTION MBMSMI
     * (NIONS,NSITES,EPEAK,EWELL,AREP,D,V,RSCIN,RSCEX,CIONS)
C
      REAL*8 Q,ELE,VEC,DETERM,temp
      REAL*8 C3B2S1
      COMMON /ELE   / ELE(81,81), VEC(81,1)
      COMMON /Q     / Q(81,81)

      DIMENSION EPEAK(3,5), EWELL(3,5), UPEAK(3,5), UWELL(3,5)
      DIMENSION AREP(3,3), D(10), DP(5), DW(5), CIONS(3,2), XIONS(3,2)
      DIMENSION RDIP(2,3,5)
      DATA QMOLF,        FI,          EKT  /
     *     0.01801800, 3.14159260, 25.40 /
C
C     Surface Potential Calculations
C     Limit to surface charge density
      IF(RSCIN.LT.1.) RSCIN = 1.
      IF(RSCEX.LT.1.) RSCEX = 1.
```

```
C
C         Sum ion concentrations.
          SUMCIN = 0.
          SUMCEX = 0.
          DO 10 I=1,NIONS
          SUMCIN = SUMCIN + CIONS(I,1)
10        SUMCEX = SUMCEX + CIONS(I,2)
C
C         Compute surface potential per Gouy-Chapman

          PIIN = 2. * RSINH( -136. / (PI*(RSCIN**2) * SQRT(SUMCIN)) )
          PIEX = 2. * RSINH( -136. / (PI*(RSCEX**2) * SQRT(SUMCEX)) )
C
c         Compute local mole fractions for all ions
          DO 12 I=1,NIONS
          XIONS(I,1) = CIONS(I,1) * QMOLF * EXP(-PIIN)
12        XIONS(I,2) = CIONS(I,2) * QMOLF * EXP(-PIEX)
C
C         Compute reduced potential from applied and surface potential
          U     = -( V / EKT  + PIIN - PIEX )
C
c         Calculate the electrical distances for peaks and wells.
          SUM = 0.
          DO 20 I = 0, NSITES-1
            SUM = SUM + D(2*I+1)
            DP(I+1) = SUM
            SUM = SUM + D(2*I+2)
20          DW(I+1) = SUM
          DP(NSITES+1) = SUM+D(2*NSITES+1)
C
C         Calculate the energies from 0-V energies and U
          DO 41, ION =1,NIONS
          DO 40, I=1,NSITES
          UPEAK(ION,I) = EPEAK(ION,I) + DP(I) * U
40        UWELL(ION,I) = EWELL(ION,I) + DW(I) * U
41        UPEAK(ION,NSITES+1) = EPEAK(ION,NSITES+1) + DP(NSITES+1) * U
C
C         Evaluate single-occupancy rate constants
c                   FORWARD
          DO 50 ION = 1,NIONS
          RDIP(1,ION,1) = XIONS(ION,1) * EXP(-UPEAK(ION,1))
          DO 50 IPEAK=2,NSITES+1
          RDIP(1,ION,IPEAK) = EXP(-(UPEAK(ION,IPEAK)-UWELL(ION,IPEAK-1)))
50        CONTINUE
c                   BACKWARD
          DO 60 ION = 1,NIONS
          RDIP(2,ION,NSITES+1)=XIONS(ION,2)*EXP(-(UPEAK(ION,NSITES+1)-U))
          DO 60 IPEAK=1,NSITES
          RDIP(2,ION,IPEAK) = EXP(-(UPEAK(ION,IPEAK)-UWELL(ION,IPEAK)))
60        CONTINUE
C
C         Evaluate current using subroutines R, M and C generated by the
c         program NCREATE, for particular nions-nsites combinations.
          DETERM = 0.
c         .............................................................
```

```
c       This is an example for one ion and two sites
C       Add calls for other nions-nsites combinations
        IF(NIONS .EQ. 1 .AND. NSITES .EQ. 2) THEN
          CALL R3B2S1 (RDIP,DP,DW,AREP)
          CALL M3B2S1
          MBMSMI = C3B2S1(DETERM)
          foo = real(mbmsmi)
        ENDIF

c       Was the calculation ok?
        IF ( DETERM .NE. 0 ) RETURN
        STOP
        END
C
        FUNCTION RSINH(X)
        REAL*8 T
        T = DBLE(X)
        RSINH = REAL(DLOG(T + DSQRT( T * T + 1.0D0 )))
        RETURN
        END
c
c       *****************************************************************
c       ************************** R3B2S1.FOR **************************
c       *****************************************************************

        SUBROUTINE R3B2S1 (RDIP,DP,DW,AREP)
        DIMENSION RDIP(2,3,5)
        DIMENSION DP(5),DW(5)
        DIMENSION AREP(3,3)
        REAL*8 Q
        COMMON /Q        / Q(81,81)
c
c         NIONS       1
c         NSITES      2
c         MAXST       4
C
        XDP1W2 = 1.0 / ABS(DP(1) - DW(2))
        XDP3W1 = 1.0 / ABS(DP(3) - DW(1))
        XDW1W2 = 1.0 / ABS(DW(1) - DW(2))
        XDW2W1 = 1.0 / ABS(DW(2) - DW(1))
c
        Q( 1, 2) = RDIP(1,1,1)
        Q( 1, 3) = RDIP(2,1,3)
c
        Q( 2, 3) = RDIP(1,1,2)
        Q( 2, 4) = RDIP(2,1,3)
     *   *EXP(-AREP(1,1)*XDP3W1)
        Q( 2, 1) = RDIP(2,1,1)
c
        Q( 3, 4) = RDIP(1,1,1)
     *   *EXP(-AREP(1,1)*XDP1W2)
        Q( 3, 1) = RDIP(1,1,3)
        Q( 3, 2) = RDIP(2,1,2)
c
        Q( 4, 2) = RDIP(1,1,3)
```

```
   *    *EXP(-AREP(1,1)*(XDP3W1-XDW2W1))
        Q( 4, 3) = RDIP(2,1,1)
   *    *EXP(-AREP(1,1)*(XDP1W2-XDW1W2))
C
        RETURN
        END
c
c
        *****************************************************************
        ************************* M3B2S1.FOR ****************************
        *****************************************************************

        SUBROUTINE M3B2S1
        REAL*8 ELE(B1,B1),VEC(B1,1),Q(B1,B1)
        COMMON /ELE   / ELE,VEC
        COMMON /Q     / Q
c
c       TRANSITION MATRIX
c
c       NIONS        1
c       NSITES       2
c       MAXST        4
c
c
        DO 10, I=1, 4
        VEC(I,1) = 0.0
        DO 10, J=1, 4
10      ELE(I,J) = 0.0
c
        ELE( 1, 1) = -(Q( 1, 2) + Q( 1, 3) )
        ELE( 2, 1) = Q( 1, 2)
        ELE( 3, 1) = Q( 1, 3)
c
        ELE( 1, 2) = Q( 2, 1)
        ELE( 2, 2) = -(Q( 2, 3) + Q( 2, 4) + Q( 2, 1) )
        ELE( 3, 2) = Q( 2, 3)
        ELE( 4, 2) = Q( 2, 4)
c
        ELE( 1, 3) = Q( 3, 1)
        ELE( 2, 3) = Q( 3, 2)
        ELE( 3, 3) = -(Q( 3, 4) + Q( 3, 1) + Q( 3, 2) )
        ELE( 4, 3) = Q( 3, 4)
c
        ELE( 2, 4) = Q( 4, 2)
        ELE( 3, 4) = Q( 4, 3)
        ELE( 4, 4) = -(Q( 4, 2) + Q( 4, 3) )
c
        DO 20, I=1, 4
        ELE( 4,I) = 1.0
20      CONTINUE
        VEC( 4,1) = 1.0
c
        RETURN
        END
c
```

```
c
c
c          ************************************************************
c          ************************* C3B2B1.FOR **********************
c          ************************************************************

          REAL*8 FUNCTION C3B2S1(DETERM)
          REAL*8 Q,VEC,ELE,TEMP,DETERM
          COMMON /Q      / Q(81,81)
          COMMON /ELE    / ELE(81,81),VEC(81,1)
c
c            NIONS      1
c            NSITES     2
c            MAXST      4
c
          N = 4
          M = 1
          DETERM = 0.
          CALL DMATINV (N,M,DETERM)
          TEMP = 0.
c
          TEMP = TEMP + VEC( 2,1) * Q( 2, 3)
          TEMP = TEMP - VEC( 3,1) * Q( 3, 2)
          C3B2S1 = TEMP * 994800.
          RETURN
          END
c          ************************************************************
c          ************************* END *****************************
c          ************************************************************
```

Appendix B

```
c          ************************************************************
c          ******************** NCREATE.FOR *************************
c          *********************************************** O.A 26-JAN-90 **
c
c          This program creates the subroutines R, M and the function C,
c          which are called by function MBMSMI to calculate ion currents
c          in a multi-barrier multi-site multi-ion channel.
c
          PROGRAM CREATE
          WRITE(*,'(1X,A)') 'MBMSMI CREATE'
          CALL STATES
          CALL RATES
          CALL CURREN
          WRITE(*,'(1X,A)') 'ALL DONE'
          STOP
          END

c          ************************************************************
c          **************** NSTATES.FOR *****************************
c          *********************************************** O.A. 16-JAN-90 ****
c
c          Creates the states for a multi-barrier multi-site model
c          Creates a table of the allowed transitions
```

```
c          Creates a FORTRAN program with the transition matrix
c
C          The state of the channel is described by a 4-character variable.
C          The characters are: 0 for empty, and 1, 2, 3 for occupancy by an
c          ion of species 1, 2, 3 ...
C          The states are also described using a state-number calculated
c          by the formula:
c
C          State Number = 1 + char1 * x**0 + char2* x**1 + char3 * x**2 ...
c
c               x = Number of ions species + 1
c
C          The aray IDEST(I,J) contains the state-numbers of the end-states
c          for transitions starting from state I.
c
           SUBROUTINE STATES
           CHARACTER*4 ESTADO(81), CURST, DESST, NTOS
           CHARACTER*1 SITE(8)
           EQUIVALENCE (CURST,SITE(1))
           EQUIVALENCE (DESST,SITE(5))
           DIMENSION   IDEST(81,10)
           CHARACTER*12 FILNAM
           CHARACTER*12 FNAME(5)
c
c          Make the definitions of the states.
c
           WRITE(*,'(1X,A\)') 'Nions,Nsites? '
           READ(*,*) NIONS,NSITES
           WRITE(*,'(A5,I3,5X,A5,I3)') 'NIONS',NIONS,'NSITES',NSITES

           MAXST  = (NIONS + 1) ** NSITES
           IF(NIONS .LT. 1 .OR. NSITES .LT. 2) THEN
              WRITE(*,'(1X,A)') 'Error: too few ions or sites'
              STOP
           ENDIF
           IF(MAXST .GT. 81 .OR. NIONS .GT. 5 .OR. NSITES .GT. 4) THEN
              WRITE(*,'(1X,A)') 'Error: too many ions or sites'
              STOP
           ENDIF

           OPEN(UNIT=11,FILE='STATEDEF.DAT',STATUS='UNKNOWN')
           WRITE(11,930) 'NIONS',NIONS
           WRITE(11,930) 'NSITES',NSITES
           WRITE(11,930) 'MAXST',MAXST
C
           DO 10, I = 1,MAXST
           CURST = NTOS(I,NIONS,NSITES)
           ESTADO(I) = CURST
           WRITE(11,'(I8,2X,4A1)') I,(SITE(J),J=1,NSITES)
10         CONTINUE
           CLOSE (11)
c
c          CREATES THE TRANSITION TABLE AND WRITES IT INTO "TRANSIT.DAT")
c
           WRITE(*,'(1X,A)') 'CREATING THE TRANSITIONS TABLE'
```

```
C
C       CLEAN THE ARRAYS OF THE DESTINATIONS
        DO 11, I = 1,81
        DO 11, J = 1,10
11      IDEST(I,J) = 0
        FNAME(1) = 'TRANSITS.DAT'
        FNAME(2) = 'TRANSITN.DAT'
        FNAME(3) = 'D.DAT      '
        FNAME(4) = 'C.DAT      '

        DO 12 I=1,4
        IUNIT = I + 9
        OPEN(UNIT=IUNIT,FILE=FNAME(I),STATUS='UNKNOWN')
        WRITE(IUNIT,930) 'NIONS',NIONS
        WRITE(IUNIT,930) 'NSITES',NSITES
        WRITE(IUNIT,930) 'MAXST',MAXST
        IF (I .LT. 3 ) THEN
        WRITE(IUNIT,920) 'C      STATES AND STATE TRANSITION MAP'
        WRITE(IUNIT,920) 'C'
        WRITE(IUNIT,920) 'C      FROM > TO...'
        WRITE(IUNIT,920) 'C'
        ENDIF
12      CONTINUE
C
        DO 40, I = 1,MAXST
        K = 1
        CURST = NTOS(I,NIONS,NSITES)
C       FORWARD TRANSITIONS
        DO 41 J=1,NSITES
        J1 = J + 4
        DESST = CURST
        IF(J .EQ. 1 .AND . SITE(J) .EQ. '0' ) THEN
                DO 42 ION=1,NIONS
                        WRITE(SITE(J1),'(I1)') ION
                        CALL STOD(IDEST,I,K,DESST,NIONS,NSITES)
42              CONTINUE
        ENDIF
        IF (J .GT. 1 ) THEN
                IF(SITE(J) .EQ. '0' .AND. SITE(J-1) .NE. '0' ) THEN
                  WRITE(SITE(J1),910) SITE(J-1)
                  WRITE(SITE(J1-1),'(I1)') 0
                  CALL STOD(IDEST,I,K,DESST,NIONS,NSITES)
                  IF(     (NSITES .LT. 4 .AND. J .EQ. 2)
     *             .OR.   (NSITES .EQ. 4 .AND. J .EQ. 3) )
     *                 WRITE(13,'(2I4)') I,IDEST(I,K-1)
                ENDIF
        ENDIF
        IF (J .EQ. NSITES .AND. SITE(J) .NE. '0') THEN
                WRITE(SITE(J1),'(I1)') 0
                CALL STOD(IDEST,I,K,DESST,NIONS,NSITES)
                IF(NSITES .EQ. 1 ) WRITE(13,'(2I4)') I,IDEST(I,K-1)
        ENDIF
41      CONTINUE
C
C       BACKWARD TRANSITIONS
```

```
        DO 51 J=NSITES,1,-1
        J1 = J + 4
        DESST = CURST
        IF(J .EQ. NSITES .AND. SITE(J) .EQ. 'O' ) THEN
                DO 52 ION=1,NIONS
                        WRITE(SITE(J1),'(I1)') ION
                        CALL STOD(IDEST,I,K,DESST,NIONS,NSITES)
52              CONTINUE
        ENDIF
        IF (J .LT. NSITES) THEN
                IF(SITE(J) .EQ. 'O' .AND. SITE(J+1) .NE. 'O' ) THEN
                WRITE(SITE(J1),910) SITE(J+1)
                WRITE(SITE(J1+1),'(I1)') O
                CALL STOD(IDEST,I,K,DESST,NIONS,NSITES)
                ENDIF
        ENDIF
        IF (J .EQ. 1 .AND. SITE(J) .NE. 'O') THEN
                WRITE(SITE(J1),'(I1)') O
                CALL STOD(IDEST,I,K,DESST,NIONS,NSITES)
        ENDIF
C
51      CONTINUE
c
C       Write transitions out of state I to output files
        CALL OUTSTAT(NSITES,I,K,ESTADO,IDEST)
40      CONTINUE

        CLOSE (10)
        CLOSE (11)
        CLOSE (12)
        CLOSE (13)
C
C       WRITES FORTRAN STATEMENTS TO FILL THE TRANSITION MATRIX
C
        WRITE(*,'(A)') ' CREATING THE MATRIX FORTRAN PROGRAM'
        NPEAK = NSITES + 1
        WRITE(FILNAM,'(3(I1,A1),A)')
            NPEAK,'B',NSITES,'S',NIONS,'I','M.FOR'
        OPEN(UNIT=1,FILE=FILNAM,STATUS='UNKNOWN')
        WRITE(1,'(7X,A,A5)') 'SUBROUTINE M',FILNAM
        WRITE(1,'(7X,A)') 'REAL*8 ELE(B1,B1),VEC(B1,1),Q(B1,B1)'
        WRITE(1,'(7X,A)') 'COMMON /ELE   / ELE,VEC'
        WRITE(1,'(7X,A)') 'COMMON /Q     / Q'
        WRITE(1,920) 'C'
        WRITE(1,920) 'C        TRANSITION MATRIX'
        WRITE(1,920) 'C'
        WRITE(1,'(A,I8)') 'C     NIONS ',NIONS
        WRITE(1,'(A,I8)') 'C     NSITES',NSITES
        WRITE(1,'(A,I8)') 'C     MAXST ',MAXST
        WRITE(1,920) 'C'
        WRITE(1,920) 'C'
        WRITE(1,'(7X,A,I2)') 'DO 10, I=1,',MAXST
        WRITE(1,'(7X,A)') 'VEC(I,1) = 0.0'
        WRITE(1,'(7X,A,I2)') 'DO 10, J=1,',MAXST
        WRITE(1,'(A2,5X,A)') '10','ELE(I,J) = 0.0'
```

```
C
       DO 70, I = 1,MAXST
       WRITE(1,920) 'C'
       DO 70, J = 1,MAXST
       IF( I .EQ. J ) THEN
               CALL DIAG(I,IDEST)
       ELSE
               DO 71, K = 1,10
                       IF(IDEST(I,K) .EQ. J) CALL NODIAG(I,J)
71             CONTINUE
       ENDIF
70     CONTINUE
       WRITE(1,920) 'C'
       WRITE(1,'(7X,A,I2)') 'DO 20, I=1,',MAXST
       WRITE(1,'(7X,A,I2,A)') 'ELE(',MAXST,',I) = 1.0 '
       WRITE(1,'(A2,5X,A)') '20','CONTINUE'
       WRITE(1,'(7X,A,I2,A)') 'VEC(',MAXST,',1) = 1.0   '
       WRITE(1,'(A)') 'C'
       WRITE(1,'(7X,A)') 'RETURN'
       WRITE(1,'(7X,A)') 'END'
       CLOSE (1)
C
       RETURN
C
900    FORMAT(I1)
910    FORMAT(A1)
920    FORMAT(A)
930    FORMAT(A8,I8)
       END
C
C
C      **********************************************************************
C      *********************** NRATES.FOR **********************************
C      ********************************************** 0.A 17-JAN-90 **
C
C      CREATES A FORTRAN PROGRAM WITH THE RATE CONSTANT DEFINITIONS
C      FOR A MBMSMI MODEL
C
       SUBROUTINE RATES
       INTEGER*2       P1,V1,V2
       CHARACTER*7     S1
       CHARACTER*12    FILNAM
       CHARACTER*4     FSTAT,TSTAT,LAST,NTOS
       CHARACTER*1     F(4),T(4)
       EQUIVALENCE     (F(1),FSTAT),(T(1),TSTAT)
       LAST = '////'

       WRITE(*,'(1X,A)') 'CREATING THE RATE CONSTANTS PROGRAM'
       OPEN(UNIT=3,FILE='D.DAT',STATUS='OLD')
       READ(3,'(8X,I8)') NIONS
       READ(3,'(8X,I8)') NSITES
       READ(3,'(8X,I8)') MAXST
       NPEAKS = NSITES + 1
       WRITE(FILNAM,'(3(I1,A1),A)')
     *     NPEAKS,'B',NSITES,'S',NIONS,'I','R.FOR'
       OPEN(UNIT=2,FILE=FILNAM,STATUS='UNKNOWN')
```

```
        WRITE(2,'(7X,A,A5,A)')
    *   'SUBROUTINE R',FILNAM,' (RDIP,DP,DW,AREP)'
        WRITE(2,'(7X,A)') 'DIMENSION RDIP(2,3,5)'
        WRITE(2,'(7X,A)') 'DIMENSION DP(5),DW(5)'
        WRITE(2,'(7X,A)') 'DIMENSION AREP(3,3)'
        WRITE(2,'(7X,A)') 'REAL*8 Q              '
        WRITE(2,'(7X,A)') 'COMMON /Q       / Q(81,81)'
        WRITE(2,'(A)') 'C'
C
        WRITE(2,'(A1,6X,A8,I8)') 'C','NIONS' ,NIONS
        WRITE(2,'(A1,6X,A8,I8)') 'C','NSITES',NSITES
        WRITE(2,'(A1,6X,A8,I8)') 'C','MAXST' ,MAXST
        WRITE(2,'(A)') 'C'
C
C       OUTPUT CODE TO EVALUATE DISTANCES
C
        DO 10,IP =1,NSITES+1
          DO 10,IW =1,NSITES
          IF ( (IW-IP) .GT. 0 .OR. (IP-IW) .GT. 1 ) THEN
            WRITE(2,990) 'XDP',IP,'W',IW,' = 1.0 / ABS(DP(',
    *               IP,') - DW(',IW,'))'
          ENDIF
10      CONTINUE
C
        DO 11,IW1 =1,NSITES
          DO 11,IW2 =1,NSITES
          IF ( IW1 .NE. IW2 ) THEN
            WRITE(2,990) 'XDW',IW1,'W',IW2,' = 1.0 / ABS(DW(',
    *               IW1,') - DW(',IW2,'))'
          ENDIF
11      CONTINUE
990     FORMAT(7X,A,I1,A,I1,A,I1,A,I1,A)
C
C       WRITE(*,'(1X,A)') 'WRITING RATES'
C
100     CONTINUE
C       GET NEW FROM-STATE AND TO-STATE
        READ(3,'(2I4)',END=800) KFROM,KTO
C       SEE IF WE HAVE A TRANSITION (DESTINATION NOT ZERO)
        IF(KTO .NE. 0) THEN
C
C       TURN STATE-NUMBERS INTO STATE-STRINGS
        FSTAT = NTOS(KFROM,NIONS,NSITES)
        TSTAT = NTOS(KTO  ,NIONS,NSITES)
        IF(FSTAT .NE. LAST) WRITE(2,'(A)') 'C'
        LAST = FSTAT
C       ..............................................................
C       JUMPS OVER PEAK ONE:
C       OCCUPANCY OF SITE 1 HAS CHANGED AND OCC. OF SITE 2 NOT CHANGED
        IF(T(1) .NE. F(1) .AND. T(2) .EQ. F(2)) THEN
C
C       FORWARD JUMP:
          IF(F(1) .EQ. '0') THEN
            CALL QLINE(KFROM,KTO,1,T(1),1)
            DO 101 JF=2,NSITES
```

```
              IF(F(JF).NE.'0') CALL REPLIN(T(1),F(JF),1,JF,0)
101           CONTINUE
C       BACKWARD JUMP:
          ELSE
              CALL QLINE (KFROM,KTO,2,F(1),1)
              DO 102 JF=2,NSITES
              IF(F(JF).NE.'0') CALL REPLIN(F(1),F(JF),1,JF,1)
102           CONTINUE
          ENDIF
        ENDIF
C       DONE WITH THE JUMPS OVER PEAK 1
        IF(NSITES .EQ. 1) GOTO 100
C
C       ...............................................................
C
C       JUMPS OVER NON-END PEAKS
C
        DO 200 K = 1,NSITES-1
C       OCCUPANCY OF SITE KT AND KF HAVE CHANGED
        IF ( T(K).NE.F(K) .AND. T(K+1).NE.F(K+1) ) THEN
C
C       FORWARD JUMP:
        IF(F(K+1) .EQ. '0') THEN
            KF = K
            KT = K + 1
            KPEAK = K + 1
            CALL QLINE(KFROM,KTO,1,T(KT),KPEAK)
C       BACKWARD JUMP
        ELSE
            KF = K + 1
            KT = K
            KPEAK = K + 1
            CALL QLINE(KFROM,KTO,2,T(KT),KPEAK)
        ENDIF
        DO 201 JF = 1,NSITES-1
        IF(F(JF).NE.'0' .AND. JF.NE.KF)
     *  CALL REPLIN(T(KT),F(JF),KPEAK,JF,KF)
201     CONTINUE
        ENDIF
200     CONTINUE
C       DONE WITH THE JUMPS OVER NON-END PEAKS
C
C       ...............................................................
C
C       JUMPS OVER PEAK NPEAKS
C       OCCUPANCY OF LAST SITE CHANGE AND
C       OCCUPANCY OF SITE LAST-1 HAVE NOT CHANGED
        IF(    T(NSITES)   .NE. F(NSITES)
     *    .AND. T(NSITES-1) .EQ. F(NSITES-1) ) THEN
C
C       FORWARD JUMP:
        IF(F(NSITES) .NE. '0') THEN
            CALL QLINE(KFROM,KTO,1,F(NSITES),NPEAKS)
            DO 301 JF=1,NSITES-1
            IF(F(JF).NE.'0')
     *      CALL REPLIN(F(NSITES),F(JF),NPEAKS,JF,NSITES)
```

```
301         CONTINUE
C
C       BACKWARD JUMP:
C       ELSE
            CALL QLINE(KFROM,KTO,2,T(NSITES),NPEAKS)
            DO 302 JF = 1,NSITES-1
            IF(F(JF) .NE. '0') CALL REPLIN(T(NSITES),F(JF),NPEAKS,JF,0)
302     CONTINUE

        ENDIF
C       DONE WITH THE JUMPS OVER PEAK NPEAKS
        ENDIF
C       ..............................................................
C
        ENDIF
C       DONE WITH THE MAIN LOOP
        GOTO 100


800     CONTINUE
        WRITE(2,'(A)') 'C'
        WRITE(2,'(7X,A)') 'RETURN'
        WRITE(2,'(7X,A)') 'END'
        CLOSE (1)
        CLOSE (3)
        RETURN

900     FORMAT(A2,I2,A1,I2,A4)
910     FORMAT(7X,A11,A4)
920     FORMAT(A7,A17)
930     FORMAT(A7,A26)
        END
C
C
C       **************************************************************
C       ***************** NCURRENT.FOR *****************************
C       ************************************************ 0.A 25-JAN-90 **
C
C
C       CREATES A FORTRAN PROGRAM TO CALCULATE THE CURRENT
C       ONCE THE ION-JUMP RATES ARE SOLVED
C
        SUBROUTINE CURREN
        CHARACTER*12    FILNAM
        CHARACTER*6     FUNAM

        WRITE(*,'(1X,A)') 'CREATING THE CURRENTS FORTRAN PROGRAM'
        OPEN(UNIT=10,FILE='C.DAT',STATUS='OLD')
        READ(10,'(8X,I8)') NIONS
        READ(10,'(8X,I8)') NSITES
        READ(10,'(8X,I8)') MAXST
        NPEAKS = NSITES + 1
        WRITE(FILNAM,'(3(I1,A1),A)')
     *      NPEAKS,'B',NSITES,'S',NIONS,'I','C.FOR'

        OPEN(UNIT=12,FILE=FILNAM,STATUS='UNKNOWN')
        WRITE(FUNAM,'(A1,A5)') 'C',FILNAM
```

```
        WRITE(12,'(7X,3A)') 'REAL*8 FUNCTION ',FUNAM,'(DETERM)'
        WRITE(12,900) 'REAL*8 Q,VEC,ELE,TEMP,determ'
        WRITE(12,900) 'COMMON /Q      / Q(81,81)'
        WRITE(12,900) 'COMMON /ELE    / ELE(81,81),VEC(81,1)'
        WRITE(12,901) 'C'
C
        WRITE(12,905) 'C','NIONS' ,NIONS
        WRITE(12,905) 'C','NSITES',NSITES
        WRITE(12,905) 'C','MAXST' ,MAXST
        WRITE(12,901) 'C'
C
C       READ THE FORWARD JUMPS OVER SELECTED PEAK
C
        WRITE(12,'(7X,A,I2)') 'N = ',MAXST
        WRITE(12,900) 'M = 1'
        WRITE(12,900) 'DETERM = 0.'
        WRITE(12,900) 'CALL DMATINV (N,M,DETERM)'
        WRITE(12,900) 'TEMP = 0.'

        WRITE(12,901) 'C'
100     CONTINUE
        READ(10,'(2I4)',ERR=110) IFROM,ITO
        WRITE(12,910) IFROM,IFROM,ITO
        WRITE(12,920) ITO,  ITO,  IFROM
        GOTO 100
110     CONTINUE
        WRITE(12,'(7X,2A)')  FUNAM,' = TEMP * 994800.'
        WRITE(12,900) 'RETURN'
        WRITE(12,900) 'END'
        CLOSE (10)
        CLOSE (12)

        RETURN
900     FORMAT (7X,A)
901     FORMAT (A)
903     FORMAT (1X,A8,I8)
905     FORMAT (A1,6X,A8,I8)
910     FORMAT (7X,'TEMP = TEMP + VEC(',I2,',1) * Q(',I2,',',I2,')')
920     FORMAT (7X,'TEMP = TEMP - VEC(',I2,',1) * Q(',I2,',',I2,')')
        END
C
C       ***********************************************************
        SUBROUTINE DIAG (I,IDEST)
C       ***********************************************************
c       Writes diagonal elements of the transition matrix
        INTEGER*2    IDEST(81,10)
        CHARACTER*15 S1
        CHARACTER*70 S2
        CHARACTER*1  S3
        CHARACTER*1  C(70)
        EQUIVALENCE (C(1),S2)
        CHARACTER*150 LINE
C
        KMAX = 0
        LEN = 0
```

```
          DO 10, INDEX = 1,6
          IF( IDEST(I,INDEX) .NE. 0 ) THEN
                  KMAX = KMAX + 1
                  LEN  = LEN + 11
          ENDIF
10        CONTINUE
          LEN = LEN - 2
C
          WRITE(S1,900) 'ELE(',I,',',I,') = -('
          WRITE(S2,910) ('Q(',I,',',IDEST(I,K),') + ', K = 1,KMAX)
          WRITE(S3,920) ')'
          WRITE(LINE,930) S1,(C(L),L=1,LEN),S3
          LEN = 15 + LEN + 1
          CALL PUTFT(LINE,LEN)
          RETURN
900       FORMAT(A4,I2,A1,I2,A6)
910       FORMAT(6(A2,I2,A1,I2,A4))
920       FORMAT(A1)
930       FORMAT(A15,72A1)
          END
C
C         ********************************************************************
          SUBROUTINE NODIAG(I,J)
C         ********************************************************************
c         Writes non-diagonal elements of the transition matrix
          WRITE(1,900) 'ELE(',J,',',I,') = Q(',I,',',J,')'
900       FORMAT(7X,A4,I2,A1,I2,A6,I2,A1,I2,A1)
          RETURN
          END
C
C         ********************************************************************
          FUNCTION NTOS(NSTAT,NIONS,NSITES)
C         ********************************************************************
c         Returns a the 4-character code for a number-code
          CHARACTER*4 NTOS
          CHARACTER*4 CURST,SSTAT
          CHARACTER*1 S(4)
          EQUIVALENCE (SSTAT,S(1))
          DIMENSION IX(4)

          IBASE = NIONS + 1
          N = NSTAT - 1
          DO 10 I = NSITES-1,0,-1
                  IDIV    = IBASE**I
                  IX(I+1) = INT(N/IDIV)
                  N =     N - IX(I+1) * IDIV
10        CONTINUE
          WRITE(SSTAT,'(4A1)') '0','0','0','0'
          IF (IX(4) .NE. 0 ) WRITE(S(4),'(I1)') IX(4)
          IF (IX(3) .NE. 0 ) WRITE(S(3),'(I1)') IX(3)
          IF (IX(2) .NE. 0 ) WRITE(S(2),'(I1)') IX(2)
          IF (IX(1) .NE. 0 ) WRITE(S(1),'(I1)') IX(1)
          NTOS = SSTAT
          RETURN
          END
```

```
c
c        ***************************************************************
         SUBROUTINE PUTFT(LINE,LARGO)
c        ***************************************************************
c        Cuts long lines to 72-character lines and writes on file.
         CHARACTER*150 LINE
         CHARACTER*150 RLINE
         CHARACTER*1   C(150)
         EQUIVALENCE (C(1),RLINE)
c
         RLINE = LINE
         LEN = LARGO + 7
         IF( LEN .GT. 72 ) THEN
                 LEN = LEN - 72
                 WRITE(1,900) (C(I), I = 1,65)
900              FORMAT(7X,65A1)
         ELSE
                 WRITE(1,900) (C(I), I = 1,LEN)
                 LEN = 0
         ENDIF
c
         IF(LEN .NE. 0 ) THEN
                 WRITE(1,910) '*',(C(I+65), I = 1,LEN)
910              FORMAT(5X,A1,1X,65A1)
         ENDIF
         RETURN
         END
c
c        ***************************************************************
         SUBROUTINE QLINE(KFROM,KTO,JDIR,SION,IPEAK)
c        ***************************************************************
c        Writes the single occupancy part of a rate constant
         CHARACTER*1 SION
         WRITE (2,900) 'Q(',KFROM,',',KTO,') = RDIP(',JDIR,SION,IPEAK
         RETURN
900      FORMAT(7X,A,2(I2,A),I1,',',A1,',',I1,')')
         END
c
c        ***************************************************************
         SUBROUTINE REPLIN(IONA,IONB,P1,W1,W2)
c
c        ***************************************************************
c        Writes a continuation line to correct rates for ion repulsion.
         CHARACTER*1 IONA,IONB
         INTEGER*2 P1,W1,W2
         IF(W2 .NE. 0) THEN
         WRITE(2,900) '*EXP(-AREP(',IONA,',',IONB,
      *  ')*(XDP',P1,'W',W1,'-XDW',W2,'W',W1,'))'
         ELSE
         WRITE(2,910) '*EXP(-AREP(',IONA,',',IONB,')*XDP',P1,'W',W1,')'
         ENDIF
         RETURN
900      FORMAT(5X,'*',1X,5A,4(I1,A))
910      FORMAT(5X,'*',1X,5A,2(I1,A))
         END
```

```
c
c
      ************************************************************
      SUBROUTINE STOD(IDEST,I,K,DESST,NIONS,NSITES)
c
      ************************************************************
c     Adds a destination entry on IDEST
      CHARACTER*4 DESST
      INTEGER IDEST(81,10)
      ISTAT = STON(DESST,NIONS,NSITES)
      IDEST(I,K) = ISTAT
      WRITE(12,'(2I4)') I,ISTAT
      K = K + 1
      RETURN
      END
c
c
      ************************************************************
      FUNCTION STON(CURST,NIONS,NSITES)
c
      ************************************************************
c     Returns the number-code for a 4-character string code.
      CHARACTER*4 CURST,SSTAT
      CHARACTER*1 S(4)
      EQUIVALENCE(S(1),SSTAT)
      DIMENSION IX(4)
      SSTAT = CURST
      READ(S(1),'(I1)') IX(1)
      READ(S(2),'(I1)') IX(2)
      READ(S(3),'(I1)') IX(3)
      READ(S(4),'(I1)') IX(4)
      NSTAT = 1
      IBASE = NIONS + 1
      DO 10, I = 0,NSITES-1,1
10    NSTAT = NSTAT + IX(I+1) * IBASE**I
      STON = NSTAT
      RETURN
      END

c
      ************************************************************
      SUBROUTINE OUTSTAT(NSITES,I,K,ESTADO,IDEST)
c
      ************************************************************
c     Writes source and connected destination states on files
c     TRANSITS.DAT sor 4-character codes ans TRANSITN.DAT for
c     number codes
      CHARACTER*4 ESTADO(81)
      DIMENSION IDEST(81,10)
      IF (NSITES .EQ. 2) THEN
      WRITE(10,'(A6,A2,A5,10(1X,A2))') 'C       ',ESTADO(I),' > ',
     *  (ESTADO(IDEST(I,J)), J=1,K-1)
      ENDIF
      IF (NSITES .EQ. 3) THEN
      WRITE(10,'(A6,A3,A5,10(1X,A3))') 'C       ',ESTADO(I),' > ',
     *  (ESTADO(IDEST(I,J)), J=1,K-1)
      ENDIF
      IF (NSITES .EQ. 4) THEN
      WRITE(10,'(A6,A4,A5,10(1X,A4))') 'C       ',ESTADO(I),' > ',
     *  (ESTADO(IDEST(I,J)), J=1,K-1)
      ENDIF
```

```
C
      WRITE(11,'(A4,I2,A4,10(1X,I2))') 'c      ',I,' >  ',
     *  (IDEST(I,J), J= 1,K-1)
      RETURN
      END
C     *********************************************************************
C     ******************** END OF NCREATE ****************************
C     *********************************************************************
```

Author Index

Numbers in parentheses are footnote reference numbers and indicate that an author's work is refered to although the name is not cited in the text.

G

H

Olivera, B. M., 192, 622, 638, 641, 643(198), 795
Olsen, R. W., 649, 650
Olson, A., 203
Ondrias, K., 584
Orear, J., 140
Oritz, M., 150
Orkand, R. K., 174
Orly, J., 226, 228(9), 236(9), 248(9)
Oron, Y., 226, 260, 265, 382, 384, 385, 386, 387(2, 8), 388(6, 12), 389(8)
Ortiz, M., 769
Osipchuk, Y. V., 642
Osterrieder, W., 184
Oswald, R. E., 785
Othman, I. B., 558, 632, 633
Otsu, K., 613
Owen, D. G., 651
Owens, J. L., 393
Oxford, G. S., 185, 623, 630, 807
Ozon, R., 262, 263, 607

P

Pacand, P., 638, 640, 642(178)
Padan, R., 230, 235, 250(34), 251(34)
Padgett, R. A., 317, 377
Palade, P. T., 529, 545
Palade, P., 693
Palla, F., 233, 242(54)
Pallotta, B. S., 195, 199, 200(30), 201, 204(30), 764, 775, 776(34), 782
Pan, Z. Q., 236
Panicali, D., 409
Paoletti, E., 409
Papazian, D. M., 320, 356, 363(15)
Papke, R. L., 393
Pappone, P. A., 95
Paraker, I., 225
Parcey, D., 559
Parcy, D., 557
Parker, B. A., 395
Parker, I., 232, 237, 250, 280, 285, 320, 339, 382, 389
Parks, T. N., 642
Parmentier, M., 229
Parsons, S. M., 496
Parvari, R., 226, 236(6)
Pasquale, E. B., 502

Passow, H., 230, 249
Patinkin, D., 257
Patlak, J. B., 769
Patlak, J., 4, 150, 632
Paton, W. D. M., 645
Patrick, J., 227, 231(15), 239(15), 243, 393
Patton, D. E., 227, 247, 606, 609(10)
Paul, D. L., 232, 249(50), 257(50), 264(50), 376, 378
Paul, D., 247, 376, 609
Paul, S. M., 650
Paulson, H. L., 393, 394, 406(8, 10), 409
Pauron, D., 453, 454(30)
Payne, R., 203
Peganov, E. M., 624
Pelchen-Matthews, A., 632, 633
Pelhate, M., 630
Pelletier, J., 317
Pellicer, A., 394, 395(11), 405(11)
Pelzer, D., 184, 540, 543(32), 545(32), 559
Peng, S., 433, 436(1), 441(1)
Penit-Soria, J., 372
Pennefather, P., 194, 634
Penner, R., 557, 559, 632
Penner, T. L., 473, 480(21)
Penniston, J. T., 571
Peper, K., 644, 645(10), 646(10)
Peralta, E., 574
Perez-Reyes, E., 309, 538
Perkins, N. M., 655
Perozo, E., 356, 363(15)
Perret, J., 229
Perry, R. P., 299
Perry, W. L. M., 645
Pershadsingh, H. A., 571
Pershke, A., 227
Persson, B. I., 817
Petersen, J., 565, 571(6), 573(6)
Petersen, M., 557, 632
Petersen, O. H., 194, 564
Pethig, R., 203
Pfafinger, P. J., 185
Pfaller, R., 433
Pfanner, N., 433
Philipps, M., 558, 631, 634(119)
Phillipps, M., 454, 459(33)
Phillips, D. M., 259
Phillips, M., 497
Phillis, J. W., 653
Picard, A., 264

Subject Index

A

Access resistance problems, with whole-cell patch clamping, 100–122, 142
Acetylcholine receptor. *See also* Nicotinic acetylcholine receptor
channels
  agonist, 648
    to block M-currents, 633–634
  blockers, 644–648, 795
    open versus closed channel block, 646–647
    versus receptor blockers, 644–645
    time course of end-plate current decay and, 646–647
    voltage and current dependence of, 647
  correlations in gating mechanisms of, analysis of, 744–745
Acid guanidinium thiocyanate-phenol-chloroform extraction, 281–284
Aconitine, 625
Action potential
  calcium as second messenger of, 529
  recorded, in data acquisition, 713
  simulated, in data acquisition, 713
A-current, 632–633
Adenosine triphosphate synthesis, 673
Adjacent state analysis, 745
Affinity chromatography
  of calmodulin, 565–567, 571–572
  in IP$_3$ neuronal receptor studies, 584–591
  in Triton X-100, 551–552
Agarose gel system, 299
  preparation of, 289, 600
Air suspension table, in patch clamp setup, 25
AJUSTE (software), 817–824, 834–837
  interfacing to function MBMSMI, 836–837
β-Alanine, 651
Aliasing, 61–63
Alphaxalone, 651

Aluminosilicate, in glass used for patch clamping, 72–75, 78, 81, 88
Amantadine, 646
Amino acid receptor-channel complex
  excitatory, 653–658
    classification of, 654
  inhibitory, 648–653
9-Aminoacridine, 795
γ-Aminobutyric acid, 649, 651
γ-Aminobutyric acid$_A$ receptor
  characteristics of, 648
  expression in *Xenopus* oocytes, 295–296
γ-Aminobutyric acid$_A$ receptor-channel complex, 648–652
  chemicals acting on, 649
  sites of, 648–649
  subunits of, 650–651
γ-Aminobutyric acid$_B$ receptor, characteristics of, 648
γ-Aminobutyric acid$_B$ receptor–channel complex, 652–653
  chemicals acting on, 652–653
γ-Aminobutyric acid$_B$ receptor system, physiological functions of, 652
α-Amino-3-hydroxy-5-methyl-4-isoxazole receptors. *See* Quisqualate receptors
Aminopyridine, 630, 633
Amphotericin B, 152–155
Analog-to-digital (A-D) converter, 749–750
Analysis of variance, versus current plots, 139–141
Anatoxin, 648
Angiotensin II receptor, cell surface receptor binding of [^{125}I]SARILE to, 373–374
Antibiotic, as nicotinic acetylcholine receptor channel blocker, 646
Anticholinesterase, as nicotinic acetylcholine receptor channel blocker, 646
Antimuscarinic agent, as nicotinic acetylcholine receptor channel blocker, 646
Apamin, 634
APLOTS (software), 819
Argiopine, 657

in single-channel analysis
  and dead time, 766
  and delay in current-measuring pathway, 752
  errors associated with, 778–780
  and noise reduction, 768
  stimulus input passed through, 750–751
  true effects of, 788–791
Fire polishing, of patch clamp glass electrode tips, 69–71, 76–77
First latency
  definition of, 749, 752
  importance in nonstationary analysis, 752
Fluorescence quenching, determination of ion permeability by, 501–510
  advantages of, 501–502, 510
  calculation of permeability from fluorescence values, 508
  channel distribution and, 507
  choosing a fluorophore for, 503–505
  equipment for, 503
  experimental design for, 508–510
  experimental method for, 506–507
  factors affecting, 502
  fluorescent probes used in, 503–504
    anionic type, 504–505
      quenchers of, 505
    cationic type, 504–505
      quenchers of, 505
    leakage of, 507
    loading of, 506
  quencher ion used in, 505
    as least permeable species, 509
    as most permeable species, 509–510
    nonpermeant, 505
Fluorescent indicator, used to measure ion fluxes across biological membranes, 502
Fluorophore. See Fluorescence quenching
Focal stimulation, 3
Follicle cells, Xenopus oocyte
  location of, 258–259
  removal of, 272–273, 324, 377
Frog Ringer's solution, normal, 276
  recipe for, 278
Function C3B2S1, 829–830, 842
Function MBMSMI, 824–827, 837–840
  interfacing to AJUSTE (software), 836–837

G

Ganglionic blocking agent, as nicotinic acetylcholine receptor channel blockers, 646
Gap junctional channels, formation of, on Xenopus oocytes, 249, 257–258, 376
Gap junctional proteins, expressed in Xenopus oocyte pairs, 376–380
  in vitro synthesis of messenger RNA for, 376–377
  methods for, 376–378
  potential problems with, 378–380
  preparation of oocytes for, 377–378
  results of, 378
Gapped duplex method, 437
Gating currents
  definition of, 353, 471
  measurable parameters of, 813
  measurement of, oocyte expression system for, 354–356
  recording of, 353–368
    charge immobilization, triple-pulse protocol for, 364–365
    current amplification and, 357–358
    hardware-oriented considerations, 357–358
    high channel density and, 355
    linear correction procedures for, 358–361
      on-line correction, 358–360
    low-noise patch recordings, 357
    macropatch performance in, 355–356
    measurement of gating noise, 366–368
    noise considerations in, 353, 366–368
    nonlinear correction procedures for, 361–365
      P/n at very positive potentials, 363–364
      off-line correction, 360–361
      nonlinear, 362–363
      stimulation protocol for, 360–361
    on-line correction
      alternating leak records and variable P/n, 359–360
      standard P/4 methods and, 358–359
    removal of ionic currents and, 356
    sodium versus potassium channels in, 365–366
Gauss-Newton method, 817

Xenopus oocyte microinjection study of, applications of, 231, 279–280
Ion channel selectivity, 92–100, 432
  determination of, 93–96
    biionic conditions and, 95–96
    with constant but not known internal ions, 96
Ion permeability, and fluorescence quenching. *See* Fluorescence quenching
Ion receptors, tissue RNA as source of, 297–309
IP$_3$
  formation of, 573
  neuronal receptors for. *See* Neuronal receptors for IP$_3$
IPROC (software), 721
[^{125}I]SARILE, 373–374
Isethionate, 94

J

Jandel Scientific, 723
Junction field-effect transistor (JFET), in patch clamp headstages, 28–29, 39–41
Junction potential, liquid. *See* Liquid junction potential

K

Kainate receptors, 657–658
KHF BASIC (software), 722
Kinetic theory
  of channel gating, and conductance states, 730
  summary of, 729–732
König's polyanion, 444

L

LAB++ (software), 722
Laboratory equipment, preparation of, to minimize contamination, 281, 298
Laboratory workers
  guidelines for handling RNA, 281, 298
  guidelines for handling vaccinia virus, 410–411
Langmuir isotherm, 490–491, 799
Lanthanum, modification of channel gating by, 812–814

Lead content, of glass used for patch clamping, 72–75, 77, 82, 89, 92
Leak pulses, 137, 359–360
  signal-to-noise ratio and, 360
Leak subtraction, 137–138, 165–167, 358–361, 747
  P/N protocol for, 166–167, 359–360, 363–364
β-Leptinotarsin, 641
Leu-enkephalin, 639
Ligand-binding assays
  basic principles of, 369–370
  in *Xenopus* oocytes, 368–375
    choice of radioligand and, 370–371
    nonspecific binding and, 371–373
      receptors found in follicular layer and, 371–373
      reduction of, 372–373
        specific method for, 373–374
    procedures for, 370–373
    specific protocols for, 373–375
Lipid bilayers, 447. *See also* Planar lipid bilayers
  assembly from monolayers, technique for, 448–449
  bacterial contamination of, 453–454
  electrostatic potential across membrane of, 483–485
  formation of, 448
    lipid requirements for, 449–450, 462–463
  painted, 448, 464
    presence of solvent in, 448–449, 454–455, 460
  on patch pipettes
    formation of, 464–465
    types of, 464
  storage of, 450
Lipofection, 394, 402–405, 412–414
Liposomes, microbial channels reconstituted into, 689–690
Liquid junction potential, 94
  causes of, 123
  correction for, in patch clamp experiments, 123–131
  measurement of, 129–130
  patch clamp amplifiers and, 125–126
  of standard saline, with respect to exemplary solutions, 130

Lithium chloride/urea method, of isolating RNA, 302–303
L-15 medium, 276
  recipe for, 278
Lobster
  skinned muscle fibers
    patch clamp experiments
      materials and methods, 694–696
      single-channel recording, 696–697
      solutions for, 694–695, 698–699
      preparation of, 694–695
      solutions for, 694–695, 698–699
    tension measurements
      preparation for, 697
      protocol for changing solutions, 698
      solutions for, 694–695, 698–699
      technique, 697–698
    viability of, 692–693, 699
  split muscle fibers, studies on intact sarcoplasmic reticulum from, 692–699
Local anesthetic, as channel blocker, 621–622, 646
Local current density, measurement of, 3
Local perfusion, 126–127
Loose-seal recording. See Patch voltage clamp, loose-seal
Loose seals, versus tight seals, 158–159
Loss factor-dielectric constant of glass, 79–81
Low-resistance seals, measurements using, drawbacks of, 4
LPROC (software), 721

M

Macroscopic currents, 132, 150, 464
  parameters of, 792
  to reveal permeation properties of pores, 92–93
Macroscopic recordings
  current–voltage data, subtraction of, 718
  data analysis, 716–717
    and drug action at single-channel level, 792–793
    isolation of specific current components, 717–718
  kinetic parameters, determination of, 718
Magnesium ions, as NMDA receptor blockers, 655–656
Marker rescue, technique of, 412–415

Markov kinetic models, 726, 730, 770
  assumption that rate constants are independent of previous channel activity, 775–776
  definition of, 764–765
Mast cell degranulating peptide (MCD), 557–558
MathSoft Inc., 723
Maximum likelihood fitting, of single-channel data, 735–736, 755–756
  full versus partial methods, 786–787
M-current, 633–634
Media Cybernetics, Inc., 819–820
Medical Systems Corp., 723
Membrane, biological, origin of fixed charges in, 472, 481
Membrane area, and capacitance measurements, 13
Membrane capacitance, 102–105, 114, 340
  measurement of, 13
Membrane conductance, 3, 483–484
Membrane current, 165, 170, 341
  errors in, not eliminated by signal averaging, 168–169
Membrane current noise, eliminated by signal averaging, 167
Membrane potential
  electrophysiological measurements from oocytes and, 324
  in identification of calcium-activated potassium channel, 197
  in loose-seal voltage clamp, 164–165
  measurement of
    during cell-attached patch recording, 152
    corrections for liquid junction potentials, 123–125, 130
  negative, and calcium buffering, 204
Membrane protein. See also DMB protein
  electrophoresis of, 167
Messenger RNA
  in biosynthesis of proteins, 660
  from brain tissue
    extraction of, 280–281
    large-scale preparation of, 284–291
    small-scale preparation of, 281–284
  electroplax, in Xenopus oocyte microinjection studies of ion channel biosynthesis, 665–666
  injection of. See Xenopus oocytes, microinjection